Complete Solutions Manual

CHEMISTRY

Principles & Practice

SECOND EDITION

REGER • GOODE • MERCER

John DeKorte
Glendale Community College

SAUNDERS GOLDEN SUNBURST SERIES

Saunders College Publishing
Harcourt Brace College Publishers

Fort Worth Philadelphia San Diego New York Orlando Austin
San Antonio Toronto Montreal London Sydney Tokyo

Printed in the United States of America.

DeKorte; Complete Solutions Manual to accompany Chemistry: Principles and Practice, 2E. Reger, Goode & Mercer

ISBN 0-03-019804-6

7 202 98765432

TO THE STUDENT

Chemistry is a subject that cannot be learned by casual reading or memorization. Rather, you will find that chemistry must be learned by learning the thought processes behind the various lines of reasoning. This can only occur when you study the material to the extent that is necessary to clarify your understanding. In addition, you will soon find that chemistry is a science with many principles that lend themselves to quantitative expression. The only way to master the intricacies of logical approaches to quantitative problem solving is to practice looking for logical connections between the given data and the desired result and to practice solving many problems.

The end-of-chapter questions and exercises in Chemistry: Principles and Practice by Reger, Goode and Mercer have been carefully chosen to test your understanding of chemistry and will require you to continually apply your knowledge to new situations. This Student Solutions Manual contains detailed explanations and solutions for all of the questions and exercises. However, it should only be used to assist you after you have first attempted to answer these questions and exercises on your own or to confirm your answers. This will give you the greatest opportunity to refine your approaches to problem solving and develop confidence in your ability to answer questions and exercises successfully.

The exercises have been solved using the same methods that are used in the text. However, we may occasionally obtain answers that differ slightly. This may be the result of "rounding" answers to intermediate steps. The procedure used in this manual is to report answers for intermediate steps to the correct number of significant figures and to use the rounded answers in subsequent steps.

In many cases, you will find that helpful comments have been added to the solutions to enhance your ability to understand the material. In addition, you will find references to pertinent Sections and Examples in the text that you can use to obtain help while working the exercises. We have included these references to point out the usefulness of the text in working the exercises and to encourage you to use the text to work the exercises before consulting this manual. We also have included "strategy roadmaps" in the solutions for many exercises to help you see the logical connections between the given data and the desired result.

We thank the authors, Reger, Goode and Mercer, for their cooperation and appreciation of our efforts and the Consulting Editor for the project, Kent Porter Hamann, for her assistance. We thank Grant Getz for his assistance with MS Word and our daughter Carrie for her assistance with the structures and diagrams in the manual, as well as her excellent work with the word processing for the first edition. In addition, I would like to wholeheartedly thank my wife Carolyn for doing the vast majority of the word processing for this edition. She did an excellent job of working with my handwritten solutions, and I could not have produced this manual without her commitment and sacrifice. Lastly, I would like to thank Alan Gabrielli of Southern College of Technology for his congenial, thorough and timely accuracy review of the solutions. We have tried to keep this manual as error-free as possible but would appreciate it if you would let us know about anything that may have been overlooked by writing or sending an e-mail message to dekorte@gc.maricopa.edu.

We trust you will find this manual to be very helpful in your study of chemistry. We wish you the best as you discover the power of chemistry to interpret the world in which we live.

John M. DeKorte
Chemistry Department
Glendale Community College
6000 W. Olive Avenue
Glendale, AZ 85302

Contents

Chapter 1: Introduction to Chemistry

1.1 *See Section 1.1.*

Scientists: "Hypothesis" is a possible explanation for an event, even if it is an untested assumption. "Theory" is an explanation for a law of nature.

General public: "Theory " is an educated hunch or guess.

1.2 *See Section 1.1.*

The coach of a football team might spend hours looking at the game films of a future opponent to gather data and learn that linebackers of that team seem to be a bit slow when running laterally or diagonally . He could then hypothesize that they would be no match for speedy receivers running crossing patterns. Basketball coaches devise similar strategies in order to take advantage of the skills of their players in certain match-ups. However, there is no way this type of thinking and acting will ever result in a scientific law or theory .

1.3 *See Section 1.1.*

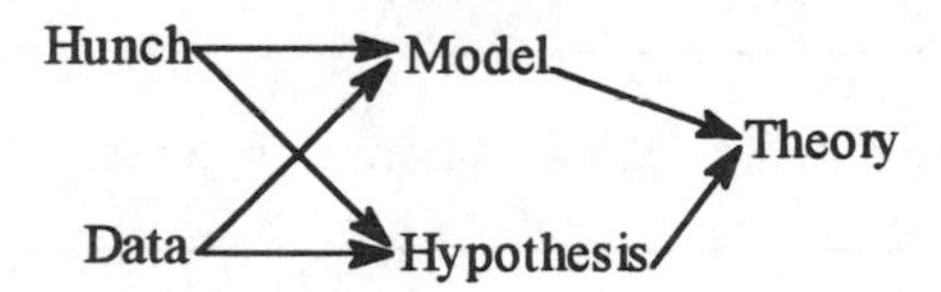

1.4 *See Section 1.1.*

Some predictions that were made by the team of scientists who proposed the hypothesis that the extinction of dinosaurs was caused by a large piece of the solar system striking the earth were that

1. the iridium-clay layer would be found worldwide,
2. the iridium-clay layer would be shown to have been laid down worldwide at the same time,
3. the elemental composition of the clay layer would be shown to be different than that of the clay component of the limestone above and below it,
4. the elemental composition of the clay layer would be shown to be the same worldwide,
5. other animal species became extinct at the same time,
6. some species of plants experienced trauma or extinction at the same time, and
7. the probability of such a collision was very probable on a 65 million-year time scale.

1.5 *See Section 1.1.*

Hypothesis, since it's only a possible explanation for an event and not an explanation for a law of nature.

1.6 *See Section 1.2.*

Intensive properties of milk: color, density, melting point, boiling point, and opaqueness
Extensive properties of milk: mass, volume, cost of producing, cost of shipping, and cost of purchasing

1.7 *See Section 1.2.*

Matter: anything that has mass and occupies space
Mass: measure of the quantity of matter in an object
Weight: the force that results from the gravitational attraction between matter and the earth or any other large object.

1.8 *See Section 1.2.*

The astronauts in space are still composed of matter. They still have the same mass as they do on earth but weigh less because they are further from the earth's gravitational attraction.

1.9 *See Section 1.2.*

Homogeneous mixtures: sugar solution, cough syrup, air
Heterogeneous mixtures: iron filings in sugar, air containing particles

1.10 *See Section 1.2.*

The light given by an electric bulb is an intensive property for a given wattage.

1.11 *See Section 1.2.*

Alloys are homogenous solutions, otherwise different portions of alloys like brass would have differing properties.

1.12 *See Section 1.2 and Figure 1.4.*

A sample of matter that cannot be separated into components by a physical process is called a substance. Compounds and elements are types of substances. Compounds are substances that can be decomposed into their constituent elements by chemical means whereas elements are substances that cannot be further decomposed by chemical means.

1.13 *See Section 1.2 and Figure 1.8.*

Au and Hg should form an alloy like Cu and Zn do, because Au and Hg are next to each other like Cu and Zn are next to each other in the same groups of the periodic table.

1.14 *See Section 1.3.*

group: vertical column in periodic table.
period: horizontal row in periodic table.

Number the columns 1 through 18, starting with Group IA and moving left-to-right across the table to obtain the appropriate Arabic numerals. This gives: Roman = Arabic when Arabic is 7 or less, Roman = VIII when Arabic is 8, 9 or 10 and Roman = (Arabic - 10) when Arabic is 11 or greater.

1.15 *See Section 1.3.*

Cd and Zn are in the same group and behave similarly.

1.16 *See Section 1.4.*

The greater the number of significant figures for a given measurement the greater the precision.

1.17 *See Section 1.4.*

Any example which gives too few or too many figures will suffice. For example, 2.50 x 12.0 gives 25 instead of 25.0, and 9.53/1.3 gives 7.331538462 instead of 7.3.

1.18 *See Section 1.5.*

$$\mathrm{km/hr} \xrightarrow{\text{Step 1}} \mathrm{m/hr} \xrightarrow{\text{Step 2}} \mathrm{m/min} \xrightarrow{\text{Step 3}} \mathrm{m/s}$$

Step 1. Use km to m conversion factor; mutiply by m/km.
Step 2. Use hr to min conversion factor; multiply by hr/min, since hr is in the denominator.
Step 3. Use min to s conversion factor; mutilpy by min/s, since min is in the denominator.

1.19 *See Section 1.4.*

The total length of two mid-size cars is approximately 24 feet.

1.20 *See Section 1.4.*

No, all measured quantities have some uncertainty which depends on the measuring device and/or the person estimating the distance between units.

1.21 *See Section 1.4.*

(a) The diameter of a human hair is about 0.1 mm, so an estimate that is between a μm and a mm is acceptable.

(b) Many people remember that the circumference of the earth is approximately 25,000 miles and that New York and New Zealand are approximately half a world apart. Thus, they are approximately 20,000 km apart when traveling on the earth's surface but only about 8,000 km apart on a line through the earth's center.

(c) Lake Michigan is about 1000 km long, 100 km wide and has an average depth of 100 m. Hence, its volume is about (1000 km x 1000 m/km) x (100 km x 1000 m/km) x 100 m = 10^{13} m^3. Since the density of water is about 1 g/mL = 1 g/cm^3 and 1 g/cm^3 = 10^3 kg/m^3 (Example 1.5), the mass of water in Lake Michigan is about 10^{13} m^3 x 10^3 kg/m^3 = 10^{16} kg.

(d) The volume of a 5-lb bag of table salt is about 1 L. If you realize that it is more than 100 mL and less than 10 L , then you are making reasonable estimates.

(e) The mass of an average house varies from 20,000 kg (small frame house) to well over 200,000 kg. There is no way to estimate this weight, but you might have seen a small house being moved down a highway. The maximum weight on most roads is set at 80,000 lb (about 40,000 kg) by law, so it is reasonable to expect a small house to be below this mass. A large house might be about 10 times larger and 10 times heavier.

1.22 *See Section 1.5.*

The length of a King's foot is easy to comprehend, but the size of a foot would vary from King to King.

1.23 *See Section 1.2.*

(a) intensive, physical (b) intensive, physical (c) intensive, chemical (d) extensive, physical
(e) intensive, physical

1.24 *See Section 1.2.*

(a) intensive, physical (b) intensive, chemical (c) extensive, physical (d) intensive, physical
(e) intensive, chemical

1.25 *See Section 1.2.*

(a) physical (b) chemical (c) chemical (d) chemical

1.26 *See Section 1.2.*

(a) physical (b) chemical (c) physical (d) chemical

1.27 *See Section 1.2.*

(a) chemical (b) physical (c) physical (d) chemical

1.28 *See Section 1.2.*

(a) physical (b) physical (c) chemical (d) physical

1.29 *See Section 1.2.*

(a) chemical (b) physical (c) physical (d) chemical

1.30 *See Section 1.2.*

(a) chemical (b) chemical (c) physical (d) chemical

1.31 *See Section 1.2.*

Chemical: corrosive, reacts with practically all substances, burns some substances giving a bright flame, prevents cavities.

Physical: pale yellow color, melting point of -219.6°C, boiling point of -188.1°C, commonly exists in gaseous physical state.

1.32 *See Section 1.2.*

Chemical: dissolves in alkaline solution with evolution of gas, can be isolated from melt of alumina and cryolite using electric current.

Physical: silver colored solid (metal), density of 2.70 g/cm^3, melts at 660°C, boils at 2467°C, can be vaporized in vacuum and deposited on glass to make mirrors.

1.33 *See Section 1.2.*

Chemical: reacts with water to form sodium hydroxide and hydrogen gas, reacts with air.
Physical: soft, silver-colored, solid (metal) melts at 98°C.

134 *See Section 1.2.*

Chemical: burns in oxygen to form acrid gas that reacts with water to form acid rain.
Physical: yellow nonmetallic element, melted by superheated steam, solidifies when separated from steam.

1.35 *See Section 1.2.*

(a) homogeneous mixture (b) compound (c) homogeneous mixture (d) heterogeneous mixture

1.36 *See Section 1.2.*

(a) compound (b) homogeneous mixture (c) homogeneous mixture (d) element

1.37 *See Section 1.2.*

(a) element (b) heterogeneous mixture (c) homogeneous mixture (d) heterogenous mixture

1.38 *See Section 1.2.*

(a) homogeneous mixture (b) compound (c) heterogeneous mixture (d) homogeneous mixture

1.39 *See Table 1.3 and the list of elements inside the front cover of the text.*

(a) O, oxygen (b) Na, sodium (c) Si, silicon (d) P, phosphorus

1.40 *See Table 1.3 and the list of elements inside the front cover of the text.*

(a) Cl, chlorine (b) Pb, lead (c) K, potassium (d) Ag, silver

1.41 *See Table 1.3 and the list of elements inside the front cover of the text.*

(a) silver, Ag (b) carbon, C (c) sulfur, S (d) iodine, I

1.42 *See Table 1.3 and the list of elements inside the front cover of the text.*

(a) helium, He (b) chlorine, Cl (c) boron, B (d) barium, Ba

1.43 *See Section 1.3 and Figure 1.8.*

(a) H, Li, K, Rb, Cs, or Fr (b) Ru or Os (c) F, Cl, I or At (d) He, Ar, Kr, Xe or Rn

1.44 *See Section 1.3 and Figure 1.8.*

(a) C, Si, Sn, or Pb (b) Be, Mg, Sr, Ba, or Ra (c) Y, La or Ac (d) N, P, Sb or Bi

1.45 ***See Section 1.3 and Figure 1.8.***

(a) Zr or Hf (b) S, Se, Te, or Po (c) Cl, Br, I, or At (d) Be, Mg, Ca or Sr

1.46 ***See Section 1.3 and Figure 1.8.***

(a) He, Ne, Kr, Xe or Rn (b) P, As, Sb or Bi (c) Fe, Ru or Hs (d) Cr, Mo or Sg

1.47 ***See Section 1.3 and Figure 1.8.***

(a) Si, representative (b) Cr, transition (c) Mg, representative (d) Np, inner transition

1.48 ***See Section 1.3 and Figure 1.8.***

(a) Ba, representative (b) Mo, transition (c) F, representative (d) Hf, transition

1.49 ***See Section 1.3 and Figure 1.8.***

(a) Xe, representative (b) Fe, transition (c) K, representative (d) Eu, inner transition

1.50 ***See Section 1.3 and Figure 1.8.***

(a) Br, representative (b) Pt, transition (c) Ru, transition (d) U, inner transition

1.51 ***See Section 1.3 and Figure 1.8.***

(a) Na, sodium (b) Cl, chlorine (c) Ra, radium (d) Ne, neon

1.52 ***See Section 1.3 and Figure 1.8.***

(a) Mg, magnesium (b) Kr, krypton (c) Fr, francium (d) I, iodine

1.53 ***See Section 1.3 and Figure 1.8.***

(a) 6 (b) 5 (c) 14 (d) 32 (e) 3

1.54 ***See Section 1.3 and Figure 1.8.***

(a) 8 (b) 5 (c) 6 (d) 10 (e) 4

1.55 ***See Section 1.3 and Figure 1.8.***

Sr and Ra should have similar properties, since both are members of Group IIA.

1.56 ***See Section 1.3 and Figure 1.8.***

S and Se should have similar properties, since both are members of Group VIA.

1.57 ***See Section 1.3 and Figure 1.8.***

Sn and Pb should have similar properties, since both are metallic members of Group IVA.

1.58 ***See Section 1.3 and Figure 1.8.***

F and I should have similar properties, since both are members of Group VIIA.

1.59 ***See Section 1.4.***

(a) 5 (b) 2; zeros preceding 7 indicate decimal place (c) 2 (d) 3

1.60 ***See Section 1.4.***

(a) 6 (b) 5; zeros preceding 1 indicate decimal place (c) 2 (d) 4

1.61 ***See Section 1.4.***

(a) 1 (b) 5 (c) 6 (d) 2; zeros preceding 4 indicate decimal place

1.62 ***See Section 1.4.***

(a) 5 (b) 3; zeros preceding 3 indicate decimal place (c) 5 (d) 4

1.63 ***See Section 1.4.***

(a) $17.2 \times 12.55 = 2.16 \times 10^2$ Three significant figures in answer, since there are just three significant figures in the quantity 17.2.

(b) $1.4 \times \frac{1.11}{42.33} = 3.7 \times 10^{-2}$ Two significant figures in answer, since there are just two significant figures in the quantity 1.4.

(c) $18.33 \times 0.0122 = 0.224$ Three significant figures in answer, since there are just three significant figures in the quantity 0.0122.

(d) $25.7 - 25.25 = 0.4$ One decimal place in answer, since there is just one decimal place in the quantity 25.7.

1.64 ***See Section 1.4.***

(a) $19.5 + 2.35 + 0.037 = 21.9$ One decimal place in answer, since there is just one decimal place in the quantity 19.5.

(b) $2.00 \times 10^3 - 1.7 \times 10^1 = 1.98 \times 10^3$ Two decimal places in answer, since there is just two decimal places in the quantity 2.00×10^3.

(c) $\frac{15}{25.69} = 0.58$ Two significant figures in answer, since there are just two significant figures in the quantity 15.

(d) $45.2 - 37.25 = 8.0$ One decimal place in answer, since there is just one decimal place in the quantity 45.2.

1.65 ***See Section 1.4.***

(a) $13.51 + 0.00459 = 13.51$ Two decimal places in answer, since there are just two decimal places in the quantity 13.51.

(b) $\frac{16.45}{32.0} + 10 = 11$

No decimal places in answer, since there ar no decimal places in the quantity 10.

(c) $3.14 \times 10^4 - 15.0 = 3.14 \times 10^4$

Two decimal places in answer, since there are just two decimal places in the quantity 3.14×10^4.

(d) $\frac{7.18 \times 10^3}{1.51 \times 10^4} = 0.475$

Three significant figures in answer, since there are three significant figures in each quantity

1.66 ***See Section 1.4.***

(a) $1.88 \times 36.35 = 68.3$

Three significant figures in answer, since there are just three significant figures in the quantity 1.88.

(b) $1.04 \times \frac{3.114}{42} = 7.7 \times 10^{-2}$

Two significant figures in answer, since there are just two significant figures in the quantity 42.

(c) $28.5 + 4.43 + 0.073 = 33.0$

One decimal place in answer, since there is just one decimal place in the quantity 28.5.

(d) $3.10 \times 10^3 - 5.1 \times 10^1 = 3.05 \times 10^3$

Two decimal places in answer, since there are just two decimal places in 3.10×10^3.

1.67 ***See Section 1.4.***

(a) $\frac{(25.12 - 1.75) \times 0.01920}{(24.339 - 23.15)} = 0.377$

Subtraction of quantities in denominator gives 1.19 to two decimal places, and 1.19 leads to just three significant figures for answer to multiplication and division.

(b) $\frac{55.4}{(26.3 - 18.904)} = 7.5$

Subtraction of quantities in denominator gives 7.4 to one decimal place, and 7.4 leads to just two significant figures for answer to division.

(c) $(0.9221 \times 27.977) + (0.0470 \times 28.976) + (0.309 \times 29.974) = 36.42$

The answer for the last two multiplications are restricted to three significant figures and thus two decimal places, since there are just three significant figures in the quantities 0.0470 and 0.309. This limits the answer for the addition to just two decimal places.

1.68 ***See Section 1.4.***

(a) $\frac{(48.35 - 35.18) \times 0.12}{(33.792 - 31.426)} = 0.67$

Two significant figures in answer, since there are just two significant figures in the quantity 0.67.

(b) $\frac{48.33}{(35.2 - 29.0)} = 7.8$

Subtraction of quantities in denominator gives 6.2 to one decimal place, and 6.2 leads to just two significant figures for answer to division

(c) $(0.0742 \times 6.01512) + (0.9258 \times 0.190100) = 0.622$

Three decimal places in answer, since three significant figures in the quantity 0.0742 limits answer for first multiplication to just three significant figures as three decimal places.

1.69 ***See Section 1.4.***

$x = \frac{10^{121}}{10^{-121}} \times 1.01 = 1.01 \times 10^{242}$

Three significant figures in answer, since there are just three significant figures in the quantity 1.01.

1.70 *See Section 1.4.*

$$x = \frac{2.05 \times 10^{-65}}{3.4 \times 10^{51}} + 1.9 \times 10^{-3} = 1.9 \times 10^{-3}$$

One decimal place in answer, since there is just one decimal place in the quantity 1.9×10^{-3}

1.71 *See Sections 1.4, 1.5.*

$$\text{? speed of sound in mi/hr} = \frac{340 \text{ m}}{\text{s}} \times \frac{1 \text{ km}}{10^3 \text{ m}} \times \frac{1 \text{ mi}}{1.609 \text{ km}} \times \frac{60 \text{ s}}{\text{min}} \times \frac{60 \text{ min}}{\text{hr}} = \mathbf{761 \text{ mi/hr}}$$

1.72 *See Sections 1.4, 1.5.*

$$\text{? Na atoms in 28.0 cm} = 28.0 \text{ cm} \times \frac{1 \text{ m}}{10^2 \text{ cm}} \times \frac{1 \text{ Na atom}}{2(1.86 \times 10^{-10}) \text{ m}} = \mathbf{7.53 \times 10^8 \text{ Na atoms}}$$

1.73 *See Sections 1.4, 1.5.*

$$\text{? distance in one light year in mi} = 1 \text{ year} \times \frac{365.25 \text{ days}}{\text{yr}} \times \frac{24 \text{ hr}}{\text{day}} \times \frac{60 \text{ min}}{\text{hr}} \times \frac{60 \text{ s}}{\text{min}} \times \frac{3.00 \times 10^8 \text{ m}}{\text{s}}$$
$$\times \frac{1 \text{ km}}{10^3 \text{ m}} \times \frac{1 \text{ mi}}{1.609 \text{ km}} = \mathbf{5.88 \times 10^{12} \text{ mi}}$$

1.74 *See Sections 1.4, 1.5.*

$$\text{? internuclear distance in } O_2 \text{ in ft} = 1.21 \times 10^{-10} \text{ m} \times \frac{10^2 \text{ cm}}{1 \text{ m}} \times \frac{1 \text{ in}}{2.54 \text{ cm}} \times \frac{1 \text{ ft}}{12 \text{ in}} = \mathbf{3.97 \times 10^{-10} \text{ ft}}$$

1.75 *See Sections 1.4, 1.5.*

Since area = volume/thickness,

$$\text{? area in m}^2 = \frac{4.0 \text{ L}}{8.00 \times 10^{-2} \text{ mm}} \times \frac{10^3 \text{ mL}}{1 \text{ L}} \times \frac{1 \text{ cm}^3}{1 \text{ mL}} \times \left(\frac{1 \text{ m}}{10^2 \text{ cm}}\right)^3 \times \frac{10^3 \text{ mm}}{1 \text{ m}} = \mathbf{50 \text{ m}^2}$$

1.76 *See Sections 1.4, 1.5.*

Since thickness = volume/area and volume = mass/density, thickness = mass/(area x density).

$$\text{? thickness in nm} = \frac{12 \text{ oz}}{75 \text{ ft}^2} \times \frac{1 \text{ lb}}{16 \text{ oz}} \times \frac{453.6 \text{ g}}{1 \text{ lb}} \times \left(\frac{1 \text{ ft}}{12 \text{ in}}\right)^2 \times \left(\frac{1 \text{ in}}{2.54 \text{ cm}}\right)^2$$
$$\times \frac{1 \text{ cm}^3}{2.70 \text{ g}} \times \frac{1 \text{ m}}{10^2 \text{ cm}} \times \frac{10^9 \text{ nm}}{1 \text{ m}} = \mathbf{1.8 \times 10^4 \text{ nm}}$$

1.77 *See Table 1.7.*

(a) mass, kilogram (b) distance, meter (c) temperature, kelvin (d) time, second

1.78 *See Table 1.7.*

(a) mass, kilogram (b) distance, meter (c) temperature, kelvin (d) time, second

1.79 ***See Table 1.7.***

(a) 3.01 cm = 3.01×10^{-2} m = 30.1 mm = 3.01×10^{7} nm

(b) 50.0 cm^3 = 50.0×10^{-3} dm^3 = 50.0 mL = 5.00×10^{-2} L = 5.00×10^{-5} m^3

(c) 23.1 g = 2.31×10^{4} mg = 2.31×10^{-2} kg

(d) 72°F = 22°C = 295 K

1.80 ***See Table 1.7.***

(a) 45 x = 4.5×10^{4} ms = 0.75 minutes

(b) 650 nm = 6.50×10^{-5} cm = 6.50×10^{-7}m

(c) 30°C = 303 K = 86°F

(d) 2.00 L = 2.00×10^{3} cm^3 = 2.00×10^{-3} m^3

1.81 ***See Sections 1.4, 1.5, and Table 1.9.***

$$\text{? distance of 1500-meter race in mi} = 1,500\ \text{m} \times \frac{1\ \text{km}}{10^3\ \text{m}} \times \frac{1\ \text{mi}}{1.609\ \text{km}} = \mathbf{0.9323\ mi}$$

1.82 ***See Sections 1.4, 1.5, and Table 1.9.***

$$\text{? area in cm}^2 = (8.5\ \text{in} \times 11\ \text{in}) \times \left(\frac{2.54\ \text{cm}}{1\ \text{in}}\right)^2 = \mathbf{6.0 \times 10^2\ cm^2}$$

1.83 ***See Sections 1.4, 1.5, and Table 1.9.***

$$\text{? volume of wine in qt} = 12\ \text{bottles} \times \frac{750\ \text{mL}}{1\ \text{bottle}} \times \frac{1\ \text{L}}{10^3\ \text{mL}} \times \frac{1.057\ \text{qt}}{1\ \text{L}} = \mathbf{9.51\ qt}$$

Note: 12 bottles is taken to be an exact quantity.

1.84 ***See Sections 1.4, 1.5, and Table 1.9.***

$$\text{? speed limit in mi/hr} = \frac{100\ \text{km}}{\text{hr}} \times \frac{1\ \text{mi}}{1.609\ \text{km}} = \mathbf{62.2\ mi/hr}$$

$$\text{? speed limit in m/s} = \frac{100\ \text{km}}{\text{hr}} \times \frac{10^3\ \text{m}}{\text{km}} \times \frac{1\ \text{hr}}{60\ \text{min}} \times \frac{1\ \text{min}}{60\ \text{s}} = \mathbf{27.8\ m/s}$$

1.85 ***See Section 1.5.***

(a) $T_F = T_C\left[\frac{1.8°\ \text{F}}{1.0°\ \text{C}}\right] + 32°\ \text{F}$ yields $\mathbf{T_C = \left[\frac{1.0°\ C}{1.8°\ F}\right](T_F - 32°\ F)}$

(b) $T_K = T_C + 273.15$ and $T_F = T_C\left[\frac{1.8°\ \text{F}}{1.0°\ \text{C}}\right] + 32°\ \text{F}$ yield

$$T_F = (T_K - 273.15)\left[\frac{1.8^\circ F}{1.0^\circ C}\right] + 32^\circ F = \mathbf{T_K\left[\frac{1.8^\circ F}{1.0^\circ C}\right] - 459.67^\circ F}$$

1.86 ***See Section 1.5.***

(a) $T_K = T_C + 273.15$ and $T_C = \left[\frac{1.0^\circ C}{1.8^\circ F}\right](T_F - 32^\circ)$ yield $\mathbf{T_K = \left[\frac{1.0^\circ C}{1.8^\circ F}\right](T_F - 32^\circ F) + 273.15}$

(b) $\mathbf{T_F = T_c\left[\frac{1.8^\circ F}{1.0^\circ C}\right] + 32^\circ F}$

1.87 ***See Section 1.5.***

$T_K = T_C + 273.15$ yields $T_C = T_K - 273.15$, so $T_C = 4.21 - 273.15 = \mathbf{-268.94^\circ C}$

$T_F = (T_K - 273.15)\left[\frac{1.8^\circ F}{1.0^\circ C}\right] + 32^\circ F$ yields $T_F = (4.21 - 273.15)\left[\frac{1.8^\circ F}{1.0^\circ C}\right] + 32^\circ F = \mathbf{-452.09^\circ F}$

1.88 ***See Section 1.5.***

$T_F = T_c\left[\frac{1.8^\circ F}{1.0^\circ C}\right] + 32^\circ F = 80^\circ C\left[\frac{1.8^\circ F}{1.0^\circ C}\right] + 32^\circ F = 176^\circ F$

$T_K = T_c + 273.15 = 80 + 273.15 = \mathbf{353\ K}$

1.89 ***See Section 1.5.***

$T_F = T_c\left[\frac{1.8^\circ F}{1.0^\circ C}\right] + 32^\circ F = 801^\circ C\left[\frac{1.8^\circ F}{1.0^\circ C}\right] + 32^\circ F = \mathbf{1{,}474^\circ F}$

$T_K = T_c + 273.15 = 801 + 273.15 = \mathbf{1{,}074\ K}$

1.90 ***See Section 1.5.***

Let $y = T_F = T_c$ in $T_F = T_c\left[\frac{1.8^\circ F}{1.0^\circ C}\right] + 32^\circ F$ giving $y = y\left(\frac{1.8}{1.0}\right) + 32$.

This yields y = 1.8y + 32, - 0.8y = 32, and y= 40. Hence, **-40°F = -40°C.**

1.91 ***See Section 1.5.***

Volume of metal = volume of water displaced = 48.5 mL - 30.0 mL = 18.5 mL

Density of metal $= \frac{\text{mass}}{\text{volume}} = \frac{147.8\ g}{18.5\ mL} =$ **7.99 g/mL or 7.99 g/cm³**

1.92 ***See Section 1.5.***

Mass of liquid = 88.5 g - 52.1 g = 36.4 g

Density of liquid $= \frac{\text{mass}}{\text{volume}} = \frac{36.4\ g}{40.0\ mL} =$ **0.910 g/mL or 0.910 g/cm³**

1.93 *See Section 1.5.*

$$? \text{ volume of benzene in L} = 2.50 \text{ kg} \times \frac{10^3 \text{ g}}{\text{kg}} \times \frac{1 \text{ cm}^3}{0.879 \text{ g}} \times \frac{1 \text{ L}}{10^3 \text{ cm}^3} = \mathbf{2.84 \text{ L}}$$

1.94 *See Section 1.5.*

$$? \text{ volume of acetone} = 25.0 \text{ g} \times \frac{1 \text{ cm}^3}{0.791 \text{g}} = \mathbf{31.6 \text{ cm}^3 \text{ or } 31.6 \text{ mL}}$$

1.95 *See Section 1.5.*

$$\text{Volume of lead brick in cm}^3 = (8.50 \text{ in} \times 5.10 \text{ in} \times 3.20 \text{ in}) \times \left(\frac{2.54 \text{ cm}}{\text{in}}\right)^3 = 2.27 \times 10^3 \text{ cm}^3$$

$$? \text{ mass of lead brick in kg} = 2.27 \times 10^3 \text{ cm}^3 \times \frac{11.4 \text{ g}}{\text{cm}^3} \times \frac{1 \text{ kg}}{10^3 \text{ g}} = \mathbf{25.9 \text{ kg}}$$

1.96 *See Section 1.5.*

$$\text{Volume of copper sphere} = 3.73 \times 10^3 \text{ g} \times \frac{1 \text{ cm}^3}{8.92 \text{ g}} = 420 \text{ cm}^3$$

$$\text{Hence, } r^3 = \frac{3 \text{ V}}{4\Pi} = \frac{3 \times 420 \text{ cm}^3}{4 \times 3.1416} = 100 \text{ cm}^3 \text{ and } r = (100 \text{ cm}^3)^{1/3} = \mathbf{4.64 \text{ cm}}$$

1.97 *See Section 1.5.*

The volume of the cylinder can be calculated using $V_{cyl} = \Pi r^2 h$ giving

$$V_{cyl} = 3.1416 \times \left(\frac{105 \text{ mm}}{2} \times \frac{1 \text{ cm}}{10 \text{ mm}}\right)^2 \times 30 \text{ cm} = 2.6 \times 10^3 \text{ cm}^3$$

The mass of the cylinder can be calculated in grams from its volume and density giving

$$g_{cyl} = 2.6 \times 10^3 \text{ cm}^3 \times \frac{19.05 \text{ g}}{1 \text{ cm}^3} = 4.9 \times 10^4 \text{ g}$$

In terms of pounds the cylinder would weigh, $? \text{ lb} = 4.9 \times 10^4 \text{ g} \times \frac{1 \text{ lb}}{453.6 \text{ g}} = \mathbf{1.1 \times 10^2 \text{ lb}}$

Hence, it seems logical to use uranium for this type of bullet.

1.98 *See Section 1.5.*

(a) A ream of copy paper consisting of 500 sheets of paper has a height of approximately two inches. Assuming that a one-dollar bill has the thickness of a sheet of copy paper

$$? \text{ km for } 2.5 \times 10^{12} \text{ one dollar bills} = 2.5 \times 10^{12} \text{ dollar bills} \times \frac{1 \text{ sheet paper}}{1 \text{ dollar bill}} \times \frac{2.0 \text{ in}}{500 \text{ sheets paper}}$$

$$\times \frac{2.54 \text{ cm}}{1.0 \text{ in}} \times \frac{1 \text{ m}}{10^2 \text{ cm}} \times \frac{1 \text{ km}}{10^3 \text{ m}} = \mathbf{2.5 \times 10^5 \text{ km}}$$

(b) A ream of copy paper consisting of 500 sheets of paper weighs approximately five pounds. Assuming that it takes six one-dollar bills to cover the surface of one side of an 8.5 x 11 inch sheet of paper gives

$$? \text{ kg for } 2.5 \times 10^{12} \text{ one-dollar bills} = 2.5 \times 10^{12} \text{ dollar bills} \times \frac{1 \text{ sheet paper}}{6 \text{ dollar bills}} \times \frac{5.0 \text{ lb}}{500 \text{ sheets paper}}$$

$$\times \frac{453.6 \text{ g}}{1 \text{ lb}} \times \frac{1 \text{ kg}}{10^3 \text{ g}} = \mathbf{1.9 \times 10^9 \text{ kg}}$$

1.99 ***See Section 1.5.***

Since grams can be converted to volume using density and area = volume/thickness,

$$? \text{ volume in cm}^3 = 28.35 \text{ g} \times \frac{1 \text{ cm}^3}{19.3 \text{ g}} = 1.47 \text{ cm}^3 \text{ and}$$

$$? \text{ area in ft}^2 = \frac{1.47 \text{ cm}^3}{1.27 \times 10^{-5} \text{ cm}} \times \left(\frac{1 \text{ in}}{2.54 \text{ cm}}\right)^2 \times \left(\frac{1 \text{ ft}}{12 \text{ in}}\right)^2 = \mathbf{125 \text{ ft}^2}$$

1.100 ***See Section 1.5.***

$$? \text{ minutes} = 93{,}000{,}000 \text{ mi} \times \frac{1.609 \text{ km}}{1 \text{ mi}} \times \frac{10^3 \text{ m}}{1 \text{ km}} \times \frac{1 \text{ s}}{3.00 \times 10^8 \text{ m}} \times \frac{1 \text{ min}}{60 \text{ s}} = \mathbf{8.3 \text{ min}}$$

1.101 ***See Section 1.5.***

Volume of metal = volume of water displaced = 19.40 mL − 12.35 mL = 7.05 mL

$$\text{Density of metal} = \frac{\text{mass}}{\text{volume}} = \frac{134.412 \text{ g}}{7.05 \text{ mL}} = \mathbf{19.1 \text{ g/mL} \text{ or } 19.1 \text{g/cm}^3}$$

1.102 ***See Sections 1.4, 1.5.***

$$? \text{ mL A} = 9.9132 \text{ g A} \times \frac{1 \text{ mL A}}{0.98 \text{ g A}} = 10 \text{ mL A}$$

$$? \text{ mL B} = 9.9132 \text{ g B} \times \frac{1 \text{ mL B}}{1.03 \text{ g B}} = 9.62 \text{ mL B}$$

To obtain range of volumes based on uncertainties ± 1 in last digit of the densities,

$$? \text{ mL A max} = 9.9132 \text{ g A x } \frac{1 \text{ mL A}}{0.97 \text{ g A}} = 10 \text{ mL A}$$

$$? \text{ mL A min} = 9.9132 \text{ g A x } \frac{1 \text{ mL A}}{0.99 \text{ g A}} = 10 \text{ mL A}$$

$$? \text{ mL B max} = 9.9132 \text{ g B x } \frac{1 \text{ mL B}}{1.02 \text{ g B}} = 9.72 \text{ mL B}$$

$$? \text{ mL B min} = 9.9132 \text{ g B x } \frac{1 \text{ mL B}}{1.04 \text{ g B}} = 9.53 \text{ mL B}$$

The uncertainty in the density of A does not contribute to an uncertainty in the calculated volume of A having a mass of 9.9132 g because the result of the volume calculation is limited to just two significant figures. However, the uncertainty in the density of B causes the result of the calculated volume to be 9.62± 0.10 mL, since the density of B was known to three significant figures.

1.103 ***See Sections 1.4, 1.5.***

(a) Using a ruler to measure the distance between the maxima in the peaks, as estimated along the base line of the signal, and comparing the distances to the full width of the graph that represents 500 ns gives:

$$? \text{ speed of light from a} = \frac{312.5 \text{ ft} \times \frac{12 \text{ in}}{1 \text{ ft}} \times \frac{1 \text{ m}}{39.37 \text{ in}}}{317 \text{ ns} \times \frac{1 \text{ s}}{10^9 \text{ ns}}} = \mathbf{3.00 \times 10^8 \text{ m/s}}$$

and

$$? \text{ speed of light from b} = \frac{312.5 \text{ ft} \times \frac{12 \text{ in}}{1 \text{ ft}} \times \frac{1 \text{ m}}{39.37 \text{ in}}}{321 \text{ ns} \times \frac{1 \text{ s}}{10^9 \text{ ns}}} = \mathbf{2.97 \times 10^8 \text{ m/s}}$$

compared to the accepted value of 3.00 x 10^8 m/s. However, the uncertainty in the time that is indicated by one-half the width of the base of the peak is ± 5 s for graph a and ± 22 s for graph b. This corresponds to an uncertainty of 5 s out of 317 s for graph a, which is only 1.6%, and an uncertainty of 22 out of 321 s for graph b, which is a much higher 6.9%.

(b) The "noise level" in graph b is higher than in graph a, and this leads to worse precision. Also, the peak in graph b is broader, so the exact position of the maximum is harder to discern. Any improvements that lower the "noise level" or increase the signal to "noise" ratio and sharpen the peak will produce more precise results.

1.104 ***See Section 1.5.***

$$T_c \text{ for } 80.0°F = \left(\frac{1.0° \text{ C}}{1.8° \text{ F}}\right)(80.0°F - 32°F) = 26.7°C$$

$$T_c \text{ for } 215.0°F = \left(\frac{1.0° \text{ C}}{1.8° \text{ F}}\right)(215.0\text{ °F} - 32°F) = 101.7°C$$

$$\text{Thus, rate of change in °C per second} = \frac{(101.7° \text{ C} - 26.7° \text{ C})}{1 \text{ min}} \times \frac{1 \text{ min}}{60 \text{ s}} = \mathbf{1.25°C/s}$$

1.105 ***See Section 1.5.***

$$T_c = \left[\frac{1.0° \text{ C}}{1.8° \text{ F}}\right](T_F - 32° \text{ F}) = \left[\frac{1.0° \text{ C}}{1.8° \text{ F}}\right](98.6° \text{ F} - 32° \text{ F}) = \mathbf{37.0° \text{ C}}$$

$T_K = T_c + 273.15 = 37.0°C + 273.15 =$ **310.2 K**

1.106-1.109 ***Solutions for these Spreadsheet Exercises are on the Reger, Goode and Mercer Spreadsheet Solutions Diskette available from Saunders College Publishing.***

Chapter 2: Atoms, Molecules, and Ions

2.1 ***See Section 2.1.***

(a) Dalton's third postulate explains the Law of Constant Composition because it states that a given compound is always made up of the same types of atoms in the same ratios.

(b) Dalton's fourth postulate explains the Law of Conservation of Mass because the numbers and types of atoms do not change during a chemical reaction, only the way in which the atoms are joined together changes during chemical reactions.

2.2 ***See Section 2.1.***

(a) Dalton's fourth postulate explains the Law of Conservation of Mass.

(b) Dalton's third postulate explains the Law of Constant Composition.

2.3 ***See Section 2.5.***

8 S represents 8 atoms of sulfur, whereas S_8 represents molecules composed of 8 atoms of sulfur.

2.4 ***See Section 2.5.***

4 P represents 4 atoms of phosphorus, whereas P_4 represents molecules composed of 4 atoms of phosphorus.

2.5 ***See Sections 2.1, 2.5.***

Atom: smallest unit of an element that can enter into a chemical combination

Element: substance that cannot be decomposed into simpler substances by normal chemical processes

Molecule: combination of atoms joined tightly together so that they behave as a single particle

Compound: substance that can be decomposed into its constituent elements by chemical processes

Examples: Atoms and Elements: O, Fe and Xe

Molecules and Elements: H_2 and O_2

Molecules and Compounds: H_2O and CO_2
(Compounds composed of only nonmetals usually exist as molecules.)

Compounds: NaCl and $CaCO_3$
(Compounds composed of a metal and nonmetal or group of nonmetals usually exist as ionic compounds.)

2.6 ***See Section 2.2.***

The masses of protons and neutrons are approximately equal and approximately 1836 times greater than the mass of an electron. The protons and electrons have equal but opposite charges, and neutrons have no charge.

On a relative scale, we say the masses of protons and neutrons are each one and the electron has a mass of zero. Similarly, on a relative scale, we say the charge on a proton is +1 and the charge on an electron is -1.

2.7 *See Section 2.2.*

In Thomson's "plum pudding" model the positive charge and the mass were uniformly distributed, and the electrons were spread out to minimize electrostatic repulsions.

2.8 *See Section 2.2.*

The Rutherford experiment showed that the positive charge and nearly all of the mass of the atom are located in a central core, rather than being spread out as presumed by the Thomson model.

2.9 *See Section 2.2.*

Since the nuclei of aluminum atoms are much lighter than the nuclei of gold atoms and contain much less positive charge, there would be considerably less deflections through large angles with aluminum foil.

2.10 *See Section 2.2.*

The protons and neutrons are located in a central core that is called the nucleus. The electrons surround the nucleus and occupy most of the volume occupied by atoms.

2.11 *See Section 2.3.*

(a) Atomic number is the number of protons in the nucleus of an atom.

(b) Mass number is the total number of protons and neutrons in an atom.

(c) Isotopes are atoms of the same element that have different numbers of neutrons.

2.12 *See Section 2.3.*

(a) Atomic number is the number of protons in the nucleus of an atom.

(b) Mass number is the total number of protons and neutrons in an atom. Hence, the number of neutrons is equal to the mass number minus the atomic number.

(c) The chemical symbol of an element does not tell us directly the number of particles of any kind in atoms of that element. However, the chemical symbol is always accompanied by the atomic number of the element in the periodic table, and the atomic number is the number of protons in the nucleus of an atom.

2.13 *See Section 2.4.*

The percent abundances of the isotopes of carbon are not known to eight or nine significant digits, only the masses of the isotopes are known to eight or nine significant digits.

2.14 *See Section 2.5 and Figure 2.11.*

Each molecule of methane, CH_4, is composed of one carbon atom and four hydrogen atoms.

2.15 *See Section 2.5.*

Each molecule of dinitrogen tetroxide, N_2O_4, is composed of two nitrogen atoms and four oxygen atoms.

2.16 *See Sections 2.5, 2.6.*

Carbon monoxide, CO, exists as molecules composed of one carbon atom and one oxygen atom each. Cesium bromide, CsBr, exists as Cs^+ ions and Br^- ions present in a one-to-one ratio.

2.17 *See Sections 2.5, 2.6.*

Sulfur dioxide, SO_2, exists as molecules composed of one sulfur atom and two oxygen atoms each. Calcium chloride, $CaCl_2$, exists as Ca^{2+} ions and Cl^- ions present in one-to-two ratio.

2.18 *See Section 2.8.*

In naming acids corresponding to anions, if the name of the anion ends in *ate*, the *ate* ending is changed to *ic* and the word *acid* is added. If the name of the anion ends in *ite*, the *ite* ending is changed to *ous* and the word *acid* is added.

2.19 *See Sections 2.5, 2.6.*

Metals generally combine with nonmetals or groups of nonmetals to form ionic compounds, whereas nonmetals generally combine with nonmetals to form molecular compounds.

2.20 *See Section 2.7.*

Ionic compounds are generally hard crystalline solids which do not conduct electricity and have high melting points. However, ionic compounds do conduct electricity when melted or when dissolved in water.

Molecular compounds consisting of small molecules generally exist as low-melting solids, liquids or gases at room temperature and generally do not conduct electricity.

2.21 *See Section 2.7.*

The difference between the physical states of molecular and ionic compounds is due to the fact that the forces that hold one molecule to another are weaker than the strong electrostatic forces that hold ions of opposite charge together in ionic solids.

2.22 *See Section 2.7.*

The chemist should conclude the solid is an ionic compound because ionic compounds exist as crystalline solids having high melting points and conduct electricity when dissolved in water.

2.23 *See Section 2.8.*

The name nitrogen monoxide is a better name for NO than nitrogen oxide because nitrogen and oxygen form six nitrogen oxides.

2.24 *See Section 2.8.*

$CrCl_3$ is an ionic compound. Prefixes are used to name binary molecular compounds, not ionic compounds. The correct name for $CrCl_3$ is chromium(III) chloride.

2.25 *See Section 2.1.*

The ratio of the masses of S in combination with the fixed mass of 1.0 g O is $\frac{1.0}{0.67}$ or $\frac{3.0}{2.0}$ in accord with the Law of Multiple Proportions.

2.26 *See Section 2.1.*

The ratio of the masses of N in combination with the fixed mass of 1.00 g O is $\frac{1.75}{0.875}$ or $\frac{2.00}{1.00}$ in accord with the Law of Multiple Proportions.

2.27 *See Section 2.1.*

Assuming 100.0 g samples of each compound, the first contains $\frac{68.4 \text{ g Cr}}{31.6 \text{ g O}}$ or $\frac{2.16 \text{ g Cr}}{1.00 \text{ g O}}$ and the second contains $\frac{52.0 \text{ g Cr}}{48.0 \text{ g O}}$ or $\frac{1.08 \text{ g Cr}}{1.00 \text{ g O}}$. The ratio of the masses of Cr in combination with the fixed mass of 1.00 g O is $\frac{2.16}{1.08}$ or $\frac{2.00}{1.00}$ in accord with the Law of Multiple Proportions.

2.28 *See Section 2.1.*

Assuming 100.00 g samples of each compound, the first contains $\frac{17.82 \text{ g Cr}}{82.18 \text{ g Br}}$ or $\frac{0.2168 \text{ g Cr}}{1.000 \text{ g Br}}$ and the second contains $\frac{24.55 \text{ g Cr}}{75.45 \text{ g Br}}$ or $\frac{0.3254 \text{ g Cr}}{1.000 \text{ g Br}}$. The ratio of the masses of Cr in combination with the fixed mass of 1.000 g Br is $\frac{0.3254}{0.2168}$ or $\frac{1.500}{1.000}$ or $\frac{3.000}{2.000}$ in accord with the Law of Multiple Proportions.

2.29 *See Section 2.3 and Example 2.1.*

(a) $^{15}_{7}N$ (b) $^{70}_{31}Ga$ (c) $^{40}_{18}Ar$

2.30 *See Section 2.30.*

(a) $^{11}_{5}B$ (b) $^{55}_{25}Mn$ (c) $^{28}_{14}Si$

2.31 *See Section 2.3 and Example 2.1.*

(a) $^{79}_{33}As$: 33 protons, 46 neutrons, 33 electrons (b) $^{51}_{23}V$: 23 protons, 28 neutrons, 23 electron
(c) $^{128}_{52}Te$: 52 protons, 76 neutrons, 52 electrons

2.32 *See Section 2.3 and Examples 2.1, 2.2, 2.3.*

(a) $^{32}_{16}S$: 16 protons, 16 neutrons, 16 electrons (b) $^{24}_{12}Mg^{2+}$: 12 protons, 12 neutrons, 10 electrons
(c) $^{37}_{17}Cl^{-}$: 17 protons, 20 neutrons, 18 electrons

2.33 *See Section 2.3 and Example 2.2.*

(a) ${}^{16}_{8}O^{2-}$ (b) ${}^{79}_{34}Se^{2-}$ (c) ${}^{59}_{28}Ni^{2+}$

2.34 *See Section 2.3 and Example 2.2.*

(a) ${}^{9}_{4}Be^{2+}$ (b) ${}^{72}_{32}Ge^{2+}$ (c) ${}^{79}_{35}Br^{-}$

2.35 *See Section 2.3 and Examples 2.1, 2.2.*

(a) ${}^{25}_{12}Mg^{2+}$ (b) ${}^{27}_{13}Al^{3+}$ (c) ${}^{29}_{14}Si$ (d) ${}^{79}_{35}Br^{-}$

2.36 *See Section 2.3 and Examples 2.1, 2.2.*

(a) ${}^{51}_{23}V^{3+}$ (b) ${}^{127}_{53}I^{-}$ (c) ${}^{102}_{44}Ru^{3+}$ (d) ${}^{31}_{15}P$

2.37 *See Section 2.3 and Example 2.4.*

symbol	${}^{23}Na^{+}$	${}^{40}Ca^{+}$	${}^{81}Br^{-}$	${}^{128}Te^{2-}$
atomic number	11	20	35	52
mass number	23	40	81	128
charge	1+	2+	1-	2-
number protons	11	20	35	52
number electrons	10	18	36	54
number neutrons	12	20	46	76

2.38 *See Section 2.23 and Example 2.4.*

symbol	${}^{127}I^{-}$	${}^{16}O$	${}^{88}Sr^{2+}$	${}^{37}Cl^{-}$
atomic number	53	8	38	17
mass number	127	16	88	37
charge	1-	0	2+	1-
number protons	53	8	38	17
number electrons	54	8	36	18
number neutrons	74	8	50	20

2.39 *See Section 2.4 and Example 2.5.*

Multiply the mass of each isotope by its decimal fraction, and add these numbers together to obtain the atomic mass of the element.

Atomic mass = (0.6011)(68.93 u) + (0.3989)(70.925 u) = **69.72 u**
The element is gallium, **Ga.**

2.40 *See Section 2.4 and Example 2.5.*

Multiply the mass of each isotope by its decimal fraction, and add these numbers together to obtain the atomic mass of the element.

Atomic mass = (0.6917)(62.9396 u) + (0.3083)(64.9278 u) = **63.55 u**
The element is copper, **Cu.**

2.41 ***See Section 2.4 and Example 2.5.***

Multiply the mass of each isotope by its decimal fraction, and add these numbers together to obtain the atomic mass of the element.

Atomic mass of Rb = (0.7217)(84.912 u) + (0.2783)(86.909 u) = **85.47 u**

2.42 ***See Section 2.4 and Example 2.5.***

Multiply the mass of each isotope by its decimal fraction, and add these numbers together to obtain the atomic mass of the element.

Atomic mass of In = (0.957)(114.904 u) + (0.043)(112.904 u) = **115 u**

2.43 ***See Section 2.4 and Example 2.5.***

(a) Multiply the mass of each isotope by its decimal fraction, and add these numbers together to obtain the atomic mass of the element.

Atomic mass = (0.7899)(23.985 u) + (0.1000)(24.986 u) + (0.1101)(25.982 u) = **24.30 u**
The element is magnesium, **Mg.**

(b) ^{24}Mg , ^{25}Mg and ^{26}Mg

2.44 ***See Section 2.4 and Example 2.5.***

(a) Multiply the mass of each isotope by its decimal fraction, and add these numbers together to obtain the atomic mass of the element.

Atomic mass = (0.922)(27.977 u) + (0.0467)(28.976 u) + (0.0310)(29.974 u) = **28.1 μ**
The element is silicon, **Si.**

(b) ^{28}Si , ^{29}Si , and ^{30}Si

2.45 ***See Section 2.4 and Example 2.5.***

Let y be the percentage of atoms having a mass of 108.905 u and $\left(\frac{100}{92.9}\right)$y be the percentage of atoms having a mass of 106.905 u giving: $y + \left(\frac{100}{92.9}\right)y = 100.00\%$, 2.08 y = 100.00%, and y = 48.1%.

Atomic mass = (0.481)(108.905 u) + (0.519)(106.905 u) = **108 u**
The element is silver, **Ag.**

2.46 ***See Section 2.4 and Example 2.5.***

Let y be the percentage of atoms having a mass of 64.928 u and $\left(\frac{100.0}{44.58}\right)$ y be the percentage of atoms having a mass of 62.940 u giving: $y + \left(\frac{100.0}{44.58}\right)y = 100.00\%$, 3.243 y = 100.00%, and y = 30.83%.

Atomic mass = (0.3083)(64.928 u) + (0.6917)(62.940 u) = **63.55 u**
The element is copper, **Cu.**

2.47 ***See Section 2.4 and Example 2.5.***

(a) ^{121}Sb and ^{123}Sb

(b) Let y be the percentage of the atoms having a mass of 120.904 u and (100.00 - y) be the percentage of the atoms having a mass of 122.904 u giving: $\left(\frac{y}{100.00}\right)(120.904\text{ u}) + \left(\frac{100-y}{100.00}\right)(122.904\text{ u}) = 121.75\text{ u}$

(y)(120.904 u) + (100.00-y)(122.904 u) = (100.00)(121.75 u)

2.000y = 115 and y = 57.5

Hence, %^{121}Sb = **57.5%** and %^{123}Sb = 100.00% - 57.5% = **42.5%.**

2.48

(a) ^{79}Br and ^{81}Br

(b) Let y be the percentage of the atoms having a mass of 78.918 u and (100.00-y) be the percentage of the atoms having a mass of 80.916 u giving: $\left(\frac{y}{100.00}\right)(78.918\text{ u}) + \left(\frac{100.00-y}{100.00}\right)(80.916\text{ u}) = 79.904\text{ u}$

(y)(78.918 u) + (100.00-y)(80.916 u) = (100.00)(79.904 u)

1.998y = 101.2 and y = 50.65

Hence, % ^{79}Br = **50.65%** and % ^{81}Br = 100.00% - 50.65% = **49.35%.**

2.49 ***See Section 2.5 and Example 2.7.***

(a) Molecular mass for C_4H_6O:

4[C] x 12.01 = 48.04
6[H] x 1.01 = 6.06
1[O] x 16.00 = 16.00
70.10 u

(b) Molecular mass for $NOCl_2$:

1[N] x 14.01 = 14.01
1[O] x 16.00 = 16.00
2[Cl] x 35.45 = 70.90
100.91 u

(c) Molecular mass of N_2O_3:

2[N] x 14.01 = 28.02
3[O] x 16.00 = 48.00
76.02 u

2.50 ***See Section 2.5 and Example 2.7.***

(a) Molecular mass of P_4O_{10}:

4[P] x 30.97 =123.88
10[O] x 16.00 =160.00
283.88 u

(b) Molecular mass of C_6H_7N:

6[C] x 12.01 = 72.06
7[H] x 1.01 = 7.07
1[N] x 14.01 = 14.01
93.14 u

(c) Molecular mass of H_3PO_4:

3[H] x 1.01 = 3.03
1[P] x 30.97 = 3097
4[O] x 16.00 = 64.00
98.00 u

2.51 ***See Section 2.5 and Example 2.7.***

Molecular mass of $C_{14}H_{18}N_2O_5$:

14[C] x 12.01	=168.14	
18[H] x 1.01	= 18.18	
2[N] x 14.01	= 28.02	
5[O] x 16.00	= 80.00	
	294.34 u	

2.52 ***See Section 2.5 and Example 2.7.***

Molecular mass of B_4H_{10}:

4[B] x 10.81	= 43.24
10[H] x 1.01	= 10.10
	53.34 u

2.53 ***See Section 2.6 and Example 2.10.***

(a) Formula mass of K_2SO_4:

2[K] x 39.10	= 78.20
1[S] x 32.07	= 32.07
4[O] x 16.00	= 64.00
	174.27 u

(b) Formula mass of $AgNO_3$:

1[Ag] x 107.87	=107.87
1[N] x 14.01	= 14.01
3]O] x 16.00	= 48.00
	169.88 u

(c) Formula mass of NH_4Cl:

1[N] x 14.01	= 14.01
4[H] x 1.01	= 4.04
1[Cl] x 35.45	= 35.45
	53.50 u

2.54 ***See Section 2.6 and Example 2.10.***

(a) Formula mass of NaOH:

1[Na] x 22.99	= 22.99
1[O] x 16.00	= 16.00
1[H] x 1.01	= 1.01
	40.00 u

(b) Formula mass of K_2CO_3:

2[K] x 39.10	= 78.20
1[C] x 12.01	= 12.01
3[O] x 16.00	= 48.00
	138.2 1 u

(c) Formula mass of $Ca_3(PO_4)_2$:

3[Ca] x 40.08	=120.24
2[P] x 30.97	= 61.94
8[O] x 16.00	=128.00
	310.18 u

2.55 ***See Section 2.6 and Table 2.1.***

(a) I^- (b) Mg^{2+} (c) O^{2-} (d) Na^+

2.56 *See Section 2.6 and Table 2.1.*

(a) K^+ (b) Br^- (c) Ba^{2+} (d) S^{2-}

2.57 *See Section 2.6 and Example 2.8.*

(a) CaS (b) Mg_3N_2 (c) FeF_2

2.58 *See Section 2.6 and Example 2.8.*

(a) LiI (b) Cs_2O (c) YCl_3

2.59 *See Section 2.6, Table 2.1 and Example 2.8.*

(a) $CaCl_2$ (b) Rb_2S (c) Li_3N (d) Y_2Se_3

2.60 *See Section 2.6, Table 2.1, and Example 2.8.*

(a) MgF_2 (b) Na_2O (c) Sc_2Se_3 (d) Ba_3N_2

2.61 *See Table 2.2.*

(a) hydroxide ion, OH^- (b) chlorate ion, ClO_3^- (c) permanganate ion, MnO_4^-

2.62 *See Table 2.2.*

(a) chromate, CrO_4^{2-} (b) carbonate, CO_3^{2-} (c) sulfate, SO_4^{2-}

2.63 *See Table 2.2.*

(a) hydrogen sulfate, HSO_4^- (b) cyanide, CN^- (c) dihydrogen phosphate, $H_2PO_4^-$

2.64 *See Table 2.2.*

(a) perchlorate, ClO_4^- (b) sulfite, SO_3^{2-} (c) hydrogen carbonate, HCO_3^-

2.65 *See Section 2.6, Tables 2.1, 2.2, and Example 2.9.*

(a) magnesium nitrite, $Mg(NO_2)_2$
(b) lithium phosphate, Li_3PO_4
(c) barium cyanide, $Ba(CN)_2$
(d) ammonium sulfate, $(NH_4)_2SO_4$

2.66 *See Section 2.6, Tables 2.1, 2.2, and Example 2.9.*

(a) sodium nitrate, $NaNO_3$
(b) beryllium hydroxide, $Be(OH)_2$
(c) ammonium acetate, $NH_4CH_3CO_2$
(d) potassium sulfite, K_2SO_3

2.67 *See Section 2.6, Tables 2.1, 2.2, and Example 2.9.*

(a) strontium nitrate, $Sr(NO_3)_2$
(b) sodium dihydrogen phosphate, NaH_2PO_4
(c) potassium perchlorate, $KClO_4$
(d) lithium hydrogen sulfate, $LiHSO_4$

2.68 See Section 2.6, Tables 2.1, 2.2, and Example 2.9.

(a) barium hydrogen phosphate, $BaHPO_4$ (b) aluminum carbonate, $Al_2(CO_3)_3$
(c) potassium dichromate, $K_2Cr_2O_7$ (d) calcium hydrogen sulfite, $Ca(HSO_3)_2$

2.69 See Section 2.7.

(a) Solution contains Fe^{3+} and Cl^- ions and conducts electricity.
(b) Solution contains $CO(NH_2)_2$ molecules and does not conduct electricity.
(c) Solution contains NH_4^+ and Br^- ions and conducts electricity.
(d) Solution contains Na^+ and ClO_4^- ions and conducts electricity.
(e) Solution contains C_2H_5OH molecules and does not conduct electricity.

2.70 See Section 2.7.

(a) Solution contains Al^{3+} and Br^- ions and conducts electricity.
(b) Solution contains $C_2H_4(OH)_2$ molecules and does not conduct electricity.
(c) Solution contains Ca^{2+} and NO_3^- ions and conducts electricity.
(d) Solution contains NH_4^+ and SO_4^{2-} ions and conducts electricity.
(e) Solution contains K^+ and $Cr_2O_7^{2-}$ ions and conducts electricity.

2.71 See Section 2.8, Tables 2.1, 2.2, 2.3, and Example 2.11.

(a) LiI, lithium iodide (b) Mg_3N_2, magnesium nitride
(c) Na_3PO_4, sodium phosphate (d) $Ba(ClO_4)_2$, barium perchlorate

2.72 See Section 2.8 and Tables 2.1, 2.2, 2.3.

(a) NH_4Br, ammonium bromide (b) $BaCl_2$, barium chloride
(c) K_2O, potassium oxide (d) $Sr(NO_3)_2$, strontium nitrate

2.73 See Section 2.8, Tables 2.3, 2.4, and Example 2.13.

(a) $CoCl_3$, cobalt(III) chloride (b) $FeSO_4$, iron(II) sulfate (c) CuO, copper(II) oxide

2.74 See Section 2.8, Tables 2.2, 2.3, 2.4, and Example 2.13.

(a) $RhBr_2$, rhodium(II) bromide (b) CuCN, copper(I) cyanide (c) $V(NO_3)_3$, vanadium(III) nitrate

2.75 See Section 2.8, Tables 2.1, 2.3, 2.4, and Examples 2.12, 2.13.

(a) maganese(III) sulfide, Mn_2S_3 (b) iron(II) cyanide, $Fe(CN)_2$
(c) potassium sulfide, K_2S (d) mercury(II) chloride, $HgCl_2$

2.76 See Section 2.8, Tables 2.1, 2.2, 2.3, 2.4, and Examples 2.12, 2.13.

(a) calcium nitride, Ca_3N_2 (b) chromium(III) perchlorate, $Cr(ClO_4)_3$
(c) tin(II) fluoride, SnF_2 (d) potassium permanganate, $KMnO_4$

2.77 *See Section 2.8 and Tables 2.2, 2.3.*

(a) chloride ion: HCl, hydrochloric acid (b) nitrite ion: HNO_2, nitrous acid
(c) perchlorate ion: $HClO_4$, perchloric acid

2.78 *See Section 2.8 and Tables 2.2, 2.3.*

(a) cyanide ion: HCN, hydrocyanic acid (b) nitrate ion: HNO_3, nitric acid
(c) phosphate ion: H_3PO_4, phosphoric acid

2.79 *See Section 2.8, Table 2.5, and Example 2.14.*

(a) sulfur tetrafluoride, SF_4
(b) nitrogen trichloride, NCl_3
(c) dinitrogen pentoxide, N_2O_5
(d) chlorine trifluoride, ClF_3

2.80 *See Section 2.8 and Table 2.5.*

(a) sulfur difluoride, SF_2
(b) silicon tetrachloride, $SiCl_4$
(c) gallium trichloride, $GaCl_3$
(d) dinitrogen trioxide, N_2O_3

2.81 *See Section 2.8 and Tables 2.2, 2.3.*

(a) H_3PO_4 is related to the phosphate ion, PO_4^{3-}, and is phosphoric acid.
(b) H_2SO_3 is related to the sulfite ion, SO_3^{2-}, and is sulfurous acid.
(c) H_2Te is related to the telluride ion, Te^{2-}, and is hydrotelluric acid.

2.82 *See Section 2.8 and Tables 2.2, 2.3.*

(a) H_2CO_3 is related to the carbonate ion, CO_3^{2-}, and is carbonic acid.
(b) HBr is related to the bromide ion, Br^-, and is hydrobromic acid.
(c) HNO_2 is related to the nitrite ion, NO_2^-, and is nitrous acid.

2.83 *See Section 2.8, Table 2.5, and Example 2.14.*

(a) PBr_5, phosphorous pentabromide
(b) SeO_2, selenium dioxide
(c) B_2Cl_4, diboron tetrachloride
(d) S_2Cl_2, disulfur dichloride

2.84 *See Section 2.8, Table 2.5, and Example 2.15.*

(a) HI, hydrogen iodide
(b) NF_3, nitrogen trifluoride
(c) SO_2, sulfur dioxide
(d) N_2Cl_4, dinitrogen tetrachloride

2.85 *See Sections 2.8 and Tables 2.1, 2. 3.*

potassium chloride, KCl

2.86 *See Section 2.8 and Tables 2.3 and 2.4.*

MnO, manganese(II) oxide

2.88 ***See Section 2.8, Tables 2.2, 2.4, and Example 2.9.***

iron(III) sulfate, $Fe_2(SO_4)_3$

2.89 ***See Section 2.8 and Tables 2.1, 2.2.***

$Mg(OH)_2$, magnesium hydroxide

2.90 ***See Section 2.8 and Tables 2.1, 2.2.***

KNO_3, potassium nitrate; $(NH_4)_2CO_3$, ammonium carbonate

2.91 ***See Section 2.3 and Example 2.2.***

(a) ${}^{35}_{17}Cl^-$ (b) ${}^{39}_{19}K^+$

2.92 ***See Section 2.3***

(a) ${}^{39}_{18}Ar$ (must be Ar to have a mass number for an isotope that is close to its atomic mass)
(b) ${}^{40}_{20}Ca^{2+}$ (must be Ca, since the mass number of the isotope is close to the atomic mass of Ca but not that of any other alkaline earth metal)

2.93 ***See Section 2.8, Tables 2.1, 2.2, 2.3, 2.5, and Examples 2.11, 2.15.***

(a) NO, nitrogen monoxide, molecular
(b) $Y_2(SO_4)_3$, yttrium sulfate, ionic
(c) Na_2O, sodium oxide, ionic
(d) NBr_3, nitrogen tribromide, molecular

Note: Ionic compounds generally contain combinations of metals with nonmetals whereas molecular compounds generally contain just nonmetals.

2.94 ***See Section 2.8, Tables 2.3, 2.4, 2.5, and Examples 2.12, 2.14.***

(a) calcium phosphate, $Ca_3(PO_4)_2$, ionic
(b) germanium dioxide, GeO_2, ionic
(c) iron(III) sulfate, $Fe_2(SO_4)_3$, ionic
(d) phosphorous tribromide, PBr_3, molecular

2.95 ***See Section 2.3 and Example 2.4.***

symbol	${}^{70}Ga^{3+}$	${}^{103}Rh^{3+}$	${}^{114}In^{+}$	${}^{28}Si^{2-}$
atomic number	31	45	49	14
mass number	70	103	114	28
charge	3+	3+	1+	2-
number protons	31	45	49	14
number electrons	28	42	48	16
number neutrons	39	58	65	14

2.96 ***See Section 2.3 and Example 2.1.***

$^{239}_{94}Pu$: 94 protons, 145 neutrons, 94 electrons

2.97 ***See Sections 1.3, 2.5, 2.6.***

(a) calcium chloride, $CaCl_2$

(b) carbon dioxide, CO_2

(c) iron(III) oxide, Fe_2O_3

2.98 ***See Section 2.3 and Figure 2.7.***

$^{1}_{1}H$ (protium): 1 proton, 1 electron
$^{2}_{1}H$ (deuterium): 1 proton, 1 neutron, 1 electron
$^{3}_{1}H$ (tritium): 1 proton, 1 neutron, 1 electron

2.99 ***See Section 2.3 and Examples 2.1, 2.2.***

(a) $^{23}_{11}Na^{+}$

(b) $^{121}_{51}Sb^{3+}$

(c) $^{84}_{36}Kr$

2.100 ***See Section 2.8, Tables 2.1, 2.2, 2.3, 2.4, and Examples 2.12, 2.14.***

(a) sodium selenide, Na_2Se
(b) nickel (II) bromide, $NiBr_2$
(c) dinitrogen pentoxide, N_2O_5
(d) copper(II) sulfate, $CuSO_4$
(e) ammonium sulfite, $(NH_4)_2SO_3$

2.101 ***See Section 2.4 and Example 2.5.***

The atomic mass of Li is given by the sum (0.0742)(6.01512 u) + (0.9258)(7.01600 u) = 0.446 u + 6.495 u = 6.941 u which has four significant figures. Hencc, the apparent discrepancy between the three significant figures for the abundance of ^{6}Li and the four significant figures for the atomic mass of Li is due to the difference in the significant figure conventions for addition and subtraction compared to multiplication and division.

2.102, 2.103 ***Solutions for these Spreadsheet Exercises are on the Reger, Goode and Mercer Spreadsheet Solutions Diskette available from Saunders College Publishing.***

Chapter 3: Stoichiometry I

3.1 ***See Section 3.1.***

The word form of a chemical equation gives only the identities of the reactants and products in a chemical reaction. The symbol form of a chemical equation gives both the identities and relative amounts of reactants and products in a chemical reaction and is therefore useful for quantitative calculations for chemical reactions.

3.2 ***See Section 3.1.***

1. Assume the equation contains one formula unit of the most complicated substance, and bring the atoms of this substance into balance by adjusting the coefficients of the substances on the other side of the equation.

2. Continue by adjusting the coefficients of the reactants and products until the same numbers of each type of atom appear on both sides of the equation and the coefficients are given in terms of whole numbers.

3. Check to make sure your final equation contains the same numbers of atoms of each type on both sides.

3.3 ***See Section 3.2.***

The SI unit of amount of substance is the mole (abbreviated as mol). One mole is the amount of substance that contains as many entities as the number of atoms in exactly 12 grams of the carbon-12 (^{12}C) isotope of carbon.

3.4 ***See Section 3.2.***

The number of objects in one mole has been determined experimentally to be 6.022×10^{23} (when expressed to four significant figures) and is known as Avogadro's number.

3.5 ***See Section 3.2.***

The masses of molecules are expressed in terms of atomic mass units. The masses of moles are expressed in terms of grams, which is a more reasonable unit to use in the laboratory.

3.6 ***See Section 3.2 and Table 3.2.***

The units for atomic, molecular and formula masses are atomic mass units (μ), and the units for molar masses are grams per mole (g/mol).

3.7 ***See Section 3.2.***

$$\text{atoms} \xrightarrow{\text{Avogadro's number}} \text{moles}$$

where the number of atoms is multiplied by $\dfrac{1\ \text{mol}}{6.022 \times 10^{23}\ \text{atoms}}$.

3.8 ***See Section 3.2.***

$$\text{moles} \xrightarrow{\text{Avogadro's number}} \text{atoms}$$

where the number of moles is multiplied by $\dfrac{6.022 \times 10^{23}\ \text{atoms}}{1\ \text{mol}}$.

3.9 *See Section 3.3 and Figure 3.4.*

A small, carefully weighed sample of a compound is completely burned in a stream of oxygen. Excess oxygen is used to convert all of the carbon to carbon dioxide and all of the hydrogen to water. The carbon dioxide is collected in a trap that usually contains sodium hydroxide, and the water is collected in a trap that usually contains calcium chloride. The masses of carbon dioxide and water collected are obtained from the increases in the masses of the respective traps, and these masses are converted to the masses of C and H in the compound. The mass percentage of C in the compound is then calculated by dividing the mass of C by the mass of the sample and multiplying the result by 100%. The mass percentage of H is calculated in like manner using the mass of H.

3.10 *See Section 3.3, Figure 3.4, and Question 3.9.*

The masses of C and H are obtained as outlined in 3.9. These masses are subtracted from the total mass of sample burned to obtain the mass of O in the sample. The percentage O by mass is calculated using (mass of O/mass of sample) x 100%.

3.11 *See Section 3.3 and Example 3.12.*

The subscripts in a molecular formula are always a whole-number multiple of the subscripts in the empirical formula for the compound. To determine the value of this whole-number *n* using *n* = (molar mass of compound/ molar mass of empirical formula), the value of the molar mass of the compound must be known from a separate experiment. Once the value of *n* is the known, the subscripts in the empirical formula are each multiplied by *n* to obtain the molecular formula.

3.12 *See Section 3.4.*

One mol of N_2 reacts with two mol H_2 to form one mole N_2H_4.

3.13 *See Section 3.4 and Exercise 3.103 below.*

$$\text{mass } H_2 \xrightarrow{\text{molar mass } H_2} \text{moles } H_2 \xrightarrow{(\text{mol } N_2/\text{mol } H_2)\text{ ratio}} \text{moles of } N_2 \xrightarrow{\text{molar mass } N_2} \text{mass } N_2$$

Step 1: Multiply g H_2 by (1/molar mass H_2).
Step 2: Multiply mol H_2 by (mol N_2/mol H_2) ratio obtained from coefficients in balanced equation.
Step 3: Multiply mol N_2 by (molar mass N_2).

3.14 *See Section 3.4.*

"The reaction was carried out with the reactants present in stoichiometric amounts," means the reactants were present in the exact amounts necessary to react with one another and none was present in excess.

3.15 *See Section 3.4 and Exercise 3.123 below.*

The grams of product that can be obtained from each starting amount of reactant are calculated separately. The reactant which can produce the least amount of product is selected as the limiting reactant, because once that least amount of product is produced this particular reactant is completely consumed.

3.16 *See Section 3.4 and Exercise 3.123 below.*

The statement, "C_2H_4 is the limiting reactant and oxygen is present in excess in a combustion reaction," means the amount of carbon dioxide and water that can be produced by the reaction is limited by the amount of C_2H_4 present. It also means the C_2H_4 is completely consumed and an excess of oxygen remains when the reaction is complete.

3.17 ***See Section 3.1 and Examples 3.1, 3.2.***

(a) Unbalanced:	$C_5H_{12} + O_2 \rightarrow CO_2 + H_2O$	Start with C_5H_{12}.
Step 1:	$C_5H_{12} + O_2 \rightarrow \underline{5}CO_2 + \underline{6}H_2O$	Balances C, H.
Step 2 :	$C_5H_{12} + \underline{8}O_2 \rightarrow 5CO_2 + 6H_2O$	Balances O.

(b) Unbalanced:	$NH_3 + O_2 \rightarrow N_2 + H_2O$	Start with NH_3.
Step 1:	$\underline{2}NH_3 + O_2 \rightarrow N_2 + H_2O$	Balances N.
Step 2:	$2NH_3 + O_2 \rightarrow N_2 + \underline{3}H_2O$	Balances H.
Step 3:	$2NH_3 + \underline{\tfrac{3}{2}}O_2 \rightarrow N_2 + 3H_2O$	Balances O.
Step 4:	$\underline{4}NH_3 + \underline{3}O_2 \rightarrow \underline{2}N_2 + \underline{6}H_2O$	Gives whole number coefficients.

(c) Unbalanced:	$KOH + H_2SO_4 \rightarrow K_2SO_4 + H_2O$	Start with K_2SO_4.
Step 1:	$\underline{2}KOH + H_2SO_4 \rightarrow K_2SO_4 + H_2O$	Balances K.
Step 2:	$2KOH + H_2SO_4 \rightarrow K_2SO_4 + \underline{2}H_2O$	Balances O, H.

Note: S balance in starting equation is maintained through Steps 1 and 2.

3.18 ***See Section 3.1 and Examples 3.1, 3.2.***

(a) Unbalanced:	$Mg_3N_2 + H_2O \rightarrow NH_3 + Mg(OH)_2$	Start with Mg_3N_2.
Step 1:	$Mg_3N_2 + H_2O \rightarrow \underline{2}NH_3 + \underline{3}Mg(OH)_2$	Balances Mg, N.
Step 2:	$Mg_3N_2 + \underline{6}H_2O \rightarrow 2NH_3 + 3Mg(OH)_2$	Balances H, O.

(b) Unbalanced:	$Fe + O_2 \rightarrow Fe_2O_3$	Start with Fe.
Step 1:	$\underline{2}Fe + O_2 \rightarrow Fe_2O_3$	Balances Fe.
Step 2:	$2Fe + \underline{\tfrac{3}{2}}O_2 \rightarrow Fe_2O_3$	Balances O.
Step 3:	$\underline{4}Fe + \underline{3}O_2 \rightarrow \underline{2}Fe_2O_3$	Gives whole number coefficients.

(c) Unbalanced:	$Zn + H_3PO_4 \rightarrow H_2 + Zn_3(PO_4)_2$	Start with H_3PO_4.
Step 1:	$Zn + \underline{2}H_3PO_4 \rightarrow H_2 + Zn_3(PO_4)_2$	Balances PO_4 units.
Step 2:	$\underline{3}Zn + 2H_3PO_4 \rightarrow H_2 + Zn_3(PO_4)_2$	Balances Zn.
Step 3:	$3Zn + 2H_3PO_4 \rightarrow \underline{3}H_2 + Zn_3(PO_4)_2$	Balances H.

3.19 ***See Section 3.1 and Examples 3.1, 3.2.***

(a) Unbalanced:	$N_2H_4 + N_2O_4 \rightarrow N_2 + H_2O$	Start with N_2O_4.
Step 1:	$N_2H_4 + N_2O_4 \rightarrow N_2 + \underline{4}H_2O$	Balances O.
Step 2:	$\underline{2}N_2H_4 + N_2O_4 \rightarrow N_2 + 4H_2O$	Balances H.
Step 3:	$2N_2H_4 + N_2O_4 \rightarrow \underline{3}N_2 + 4H_2O$	Balances N.

(b) Unbalanced:	$F_2 + H_2O \rightarrow HF + O_2$	Start with H_2O.
Step 1:	$F_2 + \underline{2}H_2O \rightarrow HF + O_2$	Balances O.
Step 2:	$F_2 + 2H_2O \rightarrow \underline{4}HF + O_2$	Balances H.
Step 3:	$\underline{2}F_2 + 2H_2O \rightarrow 4HF + O_2$	Balances F.

(c) Unbalanced:	$Na_2O + H_2O \rightarrow NaOH$	Start with Na_2O.
Step 1:	$Na_2O + H_2O \rightarrow \underline{2}NaOH$	Balances Na, O, H.

3.20 ***See Section 3.1 and Examples 3.1, 3.2.***

(a) Unbalanced:	$N_2O + NH_3 \rightarrow N_2 + H_2O$	Start with NH_3.
Step 1	$N_2O + \underline{2}NH_3 \rightarrow N_2 + \underline{3}H_2O$	Balances H.
Step 2:	$\underline{3}N_2O + 2NH_3 \rightarrow N_2 + 3H_2O$	Balances O.
Step 3:	$3N_2O + 2NH_3 \rightarrow \underline{4}N_2 + 3H_2O$	Balances N.
(b) Unbalanced:	$Cr_2S_3 + HCl \rightarrow CrCl_3 + H_2S$	Start with Cr_2S_3.
Step 1:	$Cr_2S_3 + HCl \rightarrow \underline{2}CrCl_3 + \underline{3}H_2S$	Balances Cr, S.
Step 2:	$Cr_2S_3 + \underline{6}HCl \rightarrow 2CrCl_3 + 3H_2S$	Balances H, Cl.
(c) Unbalanced:	$Fe_2S_3 + O_2 \rightarrow Fe_2O_3 + SO_2$	Start with Fe_2S_3.
Step 1:	$Fe_2S_3 + O_2 \rightarrow Fe_2O_3 + \underline{3}SO_2$	Balances S.
Step 2:	$Fe_2S_3 + \underline{\frac{9}{2}}O_2 \rightarrow Fe_2O_3 + 3SO_2$	Balances O.
Step 3:	$\underline{2}Fe_2S_3 + \underline{9}O_2 \rightarrow \underline{2}Fe_2O_3 + \underline{6}SO_2$	Gives whole number coefficients.

Note: Fe balance in starting equation is maintained through Steps 1 and 2.

3.21 ***See Section 3.1 and Example 3.3.***

(a) Unbalanced:	$H_2SO_4 + NaOH \rightarrow H_2O + Na_2SO_4$	Start with Na_2SO_4.
Step 1:	$H_2SO_4 + \underline{2}NaOH \rightarrow H_2O + Na_2SO_4$	Balances Na.
Step 2:	$H_2SO_4 + 2NaOH \rightarrow \underline{2}H_2O + Na_2SO_4$	Balances O, H.
(b) Unbalanced:	$HCl + Ca(OH)_2 \rightarrow H_2O + CaCl_2$	Start with $CaCl_2$.
Step 1:	$\underline{2}HCl + Ca(OH)_2 \rightarrow H_2O + CaCl_2$	Balances Cl.
Step 2:	$2HCl + Ca(OH)_2 \rightarrow \underline{2}H_2O + CaCl_2$	Balances O, H.
(c) Balanced:	$HNO_3 + LiOH \rightarrow H_2O + LiNO_3$	

3.22 ***See Section 3.1 and Example 3.3.***

(a) Unbalanced:	$HF + Mg(OH)_2 \rightarrow H_2O + MgF_2$	Start with MgF_2.
Step 1:	$\underline{2}HF + Mg(OH)_2 \rightarrow H_2O + MgF_2$	Balances F.
Step 2:	$2HF + Mg(OH)_2 \rightarrow \underline{2}H_2O + MgF_2$	Balances H, O.
(b) Unbalanced:	$H_3PO_4 + NaOH \rightarrow H_2O + Na_3PO_4$	Start with Na_3PO_4.
Step 1:	$H_3PO_4 + \underline{3}NaOH \rightarrow H_2O + Na_3PO_4$	Balances Na.
Step 2:	$H_3PO_4 + 3NaOH \rightarrow \underline{3}H_2O + Na_3PO_4$	Balances N, O.
(c) Unbalanced:	$H_2SO_4 + Mg(OH)_2 \rightarrow H_2O + MgSO_4$	Start with H_2O.
Step 1:	$H_2SO_4 + Mg(OH)_2 \rightarrow \underline{2}H_2O + MgSO_4$	Balances H, O.

3.23 ***See Section 3.1 and Example 3.4.***

(a) Unbalanced:	$C_6H_{12} + O_2 \rightarrow CO_2 + H_2O$	Start with C_6H_{12}.
Step 1:	$C_6H_{12} + O_2 \rightarrow \underline{6}CO_2 + \underline{6}H_2O$	Balances C, H.
Step 2:	$C_6H_{12} + \underline{9}O_2 \rightarrow 6CO_2 + 6H_2O$	Balances O.
(b) Unbalanced:	$C_4H_8 + O_2 \rightarrow CO_2 + H_2O$	Start with C_4H_8.
Step 1:	$C_4H_8 + O_2 \rightarrow \underline{4}CO_2 + \underline{4}H_2O$	Balances C, H.
Step 2:	$C_4H_8 + \underline{6}O_2 \rightarrow 4CO_2 + 4H_2O$	Balances O.

(c)	Unbalanced:	$C_2H_4O + O_2 \rightarrow CO_2 + H_2O$	Start with C_2H_4O.
	Step 1:	$C_2H_4O + O_2 \rightarrow \underline{2}CO_2 + \underline{2}H_2O$	Balances C, H.
	Step 2:	$C_2H_4O + \underline{\tfrac{5}{2}}\, O_2 \rightarrow 2CO_2 + 2H_2O$	Balances O.
	Step 3:	$\underline{2}\, C_2H_4O + \underline{5}O_2 \rightarrow \underline{4}CO_2 + \underline{4}H_2O$	Gives whole number coefficients.
(d)	Unbalanced:	$C_4H_6O_2 + O_2 \rightarrow CO_2 + H_2O$	Start with $C_4H_6O_2$.
	Step 1:	$C_4H_6O_2 + O_2 \rightarrow \underline{4}CO_2 + \underline{3}H_2O$	Balances C, H.
	Step 2:	$C_4H_6O_2 + \underline{\tfrac{9}{2}}O_2 \rightarrow 4CO_2 + 3H_2O$	Balances O.
	Step 3:	$\underline{2}C_4H_6O_2 + \underline{9}O_2 \rightarrow \underline{8}CO_2 + \underline{6}H_2O$	Gives whole number coefficients.

3.24 *See Section 3.1 and Example 3.4.*

(a)	Unbalanced:	$C_8H_{12} + O_2 \rightarrow CO_2 + H_2O$	Start with C_8H_{12}.
	Step 1:	$C_8H_{12} + O_2 \rightarrow \underline{8}CO_2 + \underline{6}H_2O$	Balances C, H.
	Step 2:	$C_8H_{12} + \underline{11}O_2 \rightarrow 8CO_2 + 6H_2O$	Balances O.
(b)	Unbalanced:	$C_5H_{10} + O_2 \rightarrow CO_2 + H_2O$	Start with C_5H_{10}.
	Step 1:	$C_5H_{10} + O_2 \rightarrow \underline{5}CO_2 + \underline{5}H_2O$	Balances C, H.
	Step 2:	$C_5H_{10} + \underline{\tfrac{15}{2}}O_2 \rightarrow 5CO_2 + 5H_2O$	Balances O.
	Step 3:	$\underline{2}C_5H_{10} + \underline{15}O_2 \rightarrow \underline{10}CO_2 + \underline{10}H_2O$	Gives whole number coefficients.
(c)	Unbalanced:	$C_4H_8O + O_2 \rightarrow CO_2 + H_2O$	Start with C_4H_8O.
	Step 1:	$C_4H_8O + O_2 \rightarrow \underline{4}CO_2 + \underline{4}H_2O$	Balances C, H.
	Step 2:	$C_4H_8O + \underline{\tfrac{11}{2}}O_2 \rightarrow 4CO_2 + 4H_2O$	Balances O.
	Step 3:	$\underline{2}C_4H_8O + \underline{11}O_2 \rightarrow \underline{8}CO_2 + \underline{8}H_2O$	Gives whole number coefficients.
(d)	Unbalanced:	$C_5H_8O_2 + O_2 \rightarrow CO_2 + H_2O$	Start with $C_5H_8O_2$.
	Step 1:	$C_5H_8O_2 + O_2 \rightarrow \underline{5}CO_2 + \underline{4}H_2O$	Balances C, H.
	Step 2:	$C_5H_8O_2 + \underline{6}O_2 \rightarrow 5CO_2 + 4H_2O$	Balances O.

3.25 *See Section 3.1 and Examples 3.3, 3.4.*

(a)	Unbalanced:	$C_6H_{10} + O_2 \rightarrow CO_2 + H_2O$	Start with C_6H_{10}.
	Step 1:	$C_6H_{10} + O_2 \rightarrow \underline{6}CO_2 + \underline{5}H_2O$	Balances C, H.
	Step 2:	$C_6H_{10} + \underline{\tfrac{17}{2}}O_2 \rightarrow 6CO_2 + 5H_2O$	Balances O.
	Step 3:	$\underline{2}C_6H_{10} + \underline{17}O_2 \rightarrow \underline{12}CO_2 + \underline{10}H_2O$	Gives whole number coefficients.
(b)	Unbalanced:	$HNO_3 + Be(OH)_2 \rightarrow H_2O + Be(NO_3)_2$	Start with HNO_3.
	Step 1:	$\underline{2}HNO_3 + Be(OH)_2 \rightarrow H_2O + Be(NO_3)_2$	Balances NO_3 units.
	Step 2:	$2HNO_3 + Be(OH)_2 \rightarrow \underline{2}H_2O + Be(NO_3)_2$	Balances H, O.

3.26 *See Section 3.1 and Examples 3.3, 3.4.*

(a)	Unbalanced:	$C_8H_8 + O_2 \rightarrow CO_2 + H_2O$	Start with C_8H_8.
	Step 1:	$C_8H_8 + O_2 \rightarrow \underline{8}CO_2 + \underline{4}H_2O$	Balances C, H.
	Step 2:	$C_8H_8 + 10O_2 \rightarrow 8CO_2 + 4H_2O$	Balances O.
(b)	Unbalanced:	$H_2SO_4 + KOH \rightarrow H_2O + K_2SO_4$	Start with K_2SO_4.
	Step 1:	$H_2SO_4 + \underline{2}KOH \rightarrow H_2O + K_2SO_4$	Balances K.
	Step 2:	$H_2SO_4 + 2KOH \rightarrow \underline{2}H_2O + K_2SO_4$	Balances O, H.

3.27 *See Section 3.1 and Examples 3.1, 3.2.*

Unbalanced:	$CS_2 + O_2 \rightarrow CO_2 + SO_2$	Start with CS_2.
Step 1:	$CS_2 + O_2 \rightarrow CO_2 + \underline{2}SO_2$	Balances C,S.
Step 2:	$CS_2 + \underline{3}O_2 \rightarrow CO_2 + 2SO_2$	Balances O.

3.28 *See Section 3.1 and Example 3.4.*

Unbalanced:	$CH_3OC(CH_3)_3 + O_2 \rightarrow CO_2 + H_2O$	Start with $CH_3OC(CH_3)_3$.
Step 1:	$CH_3OC(CH_3)_3 + O_2 \rightarrow \underline{5}CO_2 + \underline{6}H_2O$	Balances C, H.
Step 2:	$CH_3OC(CH_3)_3 + \underline{\tfrac{15}{2}}O_2 \rightarrow 5CO_2 + 6H_2O$	Balances O.
Step 3:	$\underline{2}CH_3OC(CH_3)_3 + \underline{15}O_2 \rightarrow \underline{10}CO_2 + \underline{12}H_2O$	Gives whole number coefficients.

3.29 *See Section 3.1 and Example 3.4.*

Unbalanced:	$(CH_3)_2CO + O_2 \rightarrow CO_2 + H_2O$	Start with$(CH_3)_2CO$.
Step 1:	$(CH_3)_2CO + O_2 \rightarrow \underline{3}CO_2 + \underline{3}H_2O$	Balances C, H.
Step 2:	$(CH_3)_2CO + \underline{4}O_2 \rightarrow 3CO_2 + 3H_2O$	Balances O.

3.30 *See Section 3.1 and Example 3.3.*

Unbalanced:	$H_3PO_3 \rightarrow H_3PO_4 + PH_3$	Start with H_3PO_3 and H_3PO_4.
Step 1:	$\underline{4}H_3PO_3 \rightarrow \underline{3}H_3PO_4 + PH_3$	Balances O, P, H.

3.31 *See Section 3.1 and Examples 3.1, 3.2.*

Unbalanced:	$S_8 + Cl_2 \rightarrow S_2Cl_2$	Start with S_2Cl_2.
Step 1:	$S_8 + Cl_2 \rightarrow \underline{4}S_2Cl_2$	Balances S.
Step 2:	$S_8 + \underline{4}Cl_2 \rightarrow 4S_2Cl_2$	Balances Cl.

3.32 *See Section 3.1 and Examples 3.1, 3.2.*

Unbalanced:	$UO_2 + CCl_4 \rightarrow UCl_4 + COCl_2$	Start with CCl_4.
Step 1:	$UO_2 + \underline{2}CCl_4 \rightarrow UCl_4 + \underline{2}COCl_2$	Balances Cl, C, O.

3.33 *See Section 3.1 and Examples 3.1, 3.2.*

Unbalanced:	$N_2H_4 + O_2 \rightarrow NO_2 + H_2O$	Start with. N_2H_4.
Step 1:	$N_2H_4 + O_2 \rightarrow \underline{2}NO_2 + \underline{2}H_2O$	Balances N, H.
Step 2:	$N_2H_4 + \underline{3}O_2 \rightarrow 2NO_2 + 2H_2O$	Balances O.

Assigning an ON of +1 to H in N_2H_4 and H_2O and -2 to O in NO_2 and H_2O gives $N_2H_4 + 3O_2 \rightarrow 2NO_2 + 2H_2O$

ON: -2,+1   0   +4,-2   +1,-2

The ON of the N atoms increase, and the ON of the O atoms decrease.
The N atoms in N_2H_4 are oxidized, and the O atoms in O_2 are reduced.

3.34 *See Section 3.1 and Example 3.5.*

Unbalanced:	$Fe + O_2 \rightarrow Fe_2O_3$	Start with O_2.
Step 1:	$Fe + \underline{3}O_2 \rightarrow \underline{2}Fe_2O_3$	Balances O.
Step 2:	$\underline{4}Fe + 3O_2 \rightarrow 2Fe_2O_3$	Balances Fe.

Assigning an ON of -2 to oxygen in Fe_2O_3 gives $4Fe + 3O_2 \rightarrow 2Fe_2O_3$
ON: 0 0 +3,-2
The ON of the Fe atoms increase, and the ON of the O atoms decrease.
The Fe metal is oxidized, and the O atoms in O_2 are reduced.

3.35 ***See Section 3.1 and Example 3.5.***

Unbalanced:	$Zn + HCl \rightarrow ZnCl_2 + H_2$	Start with HCl.
Step 1:	$Zn + \underline{2}HCl \rightarrow ZnCl_2 + H_2$	Balances Cl, H.

Assigning an ON of +1 to H and -1 to Cl in HCl and $ZnCl_2$ gives $Zn + 2\,HCl \rightarrow ZnCl_2 + H_2$.
ON: 0 +1,-1 +2,-1 0
The ON of the Zn atoms increases, and the ON of the H atoms decreases.
The Zn metal is oxidized, and the H atoms in HCl are reduced.

3.36 ***See Section 3.1 and Example 3.5.***

Unbalanced:	$P_4(s) + O_2(g) \rightarrow P_4O_{10}(s)$	Start with O_2.
Step 1:	$P_4(s) + \underline{5}O_2(g) \rightarrow P_4O_{10}(s)$	Balances O.

Assigning an ON of -2 to O in P_4O_{10} gives $P_4(s) + 5O_2(g) \rightarrow P_4O_{10}(s)$
ON 0 0 +5,-2
The ON of the P atoms increases, and the ON of the O atoms decreases.
The P atoms in P_4 are oxidized, and the O atoms in O_2 are reduced.

3.37 ***See Section 3.1 and Example 3.5.***

Unbalanced:	$NO + NH_3 \rightarrow N_2 + H_2O$	Start with NH_3 and H_2O.
Step 1:	$NO + \underline{2}NH_3 \rightarrow N_2 + \underline{3}H_2O$	Balances H.
Step 2:	$\underline{3}NO + 2NH_3 \rightarrow N_2 + 3H_2O$	Balances O.
Step 3:	$3NO + 2NH_3 \rightarrow \underline{\frac{5}{2}}N_2 + 3H_2O$	Balances N.
Step 4:	$\underline{6}NO + \underline{4}NH_3 \rightarrow \underline{5}N_2 + \underline{6}H_2O$	Gives whole number coefficients.

Assigning an ON of +1 to H in NH_3 and H_2O and -2 to O in NO and H_2O gives
$6NO + 4NH_3 \rightarrow 5N_2 + 6H_2O$
ON: +2,-2 -3,+1 0 +1,-2
The ON of the N atoms in NH_3 increases, and the ON of the N atom in NO decreases.
The N atoms in NH_3 are oxidized, and the N atoms in NO are reduced.

3.38 ***See Section 3.1 and Example 3.5.***

Unbalanced:	$MnO_2 + HCl \rightarrow MnCl_2 + Cl_2 + H_2O$	Start with HCl.
Step 1:	$MnO_2 + \underline{4}HCl \rightarrow MnCl_2 + Cl_2 + H_2O$	Balances Cl.
Step 2:	$MnO_2 + 4HCl \rightarrow MnCl_2 + Cl_2 + \underline{2}H_2O$	Balances H, O.

Assigning an ON of -2 to O in MnO_2 and H_2O, an ON of +1 to H in HCl and H_2O and an ON of -1 to Cl in HCl and $MnCl_2$ gives $MnO_2 + 4HCl \rightarrow MnCl_2 + Cl_2 + 2H_2O$
ON: +4,-2 +1,-1 +2,-1 0 +1,-2
The ON of some of the Cl atoms in HCl increases, and the ON of the Mn atoms decreases.
The Cl atoms in HCl that undergo an increase in ON are oxidized, and the Mn atoms in MnO_2 are reduced.

3.39 ***See Section 3.2 and Example 3.6.***

(a) $? \text{ atoms Mg} = 1.44 \text{ mol Mg} \times \dfrac{6.022 \times 10^{23} \text{ atoms Mg}}{1 \text{ mol Mg}}$ = **8.67×10^{23} atoms Mg**

(b) $? \text{ atoms Ne} = 9.77 \text{ mol Ne} \times \dfrac{6.022 \times 10^{23} \text{ atoms Ne}}{1 \text{ mol Ne}}$ = **5.88×10^{24} atoms Ne**

(c) $? \text{ atoms Fe} = 0.099 \text{ mol Fe} \times \dfrac{6.022 \times 10^{23} \text{ atoms Fe}}{1 \text{ mol Fe}}$ = **6.0×10^{22} atoms Fe**

3.40 ***See Section 3.2 and Example 3.6.***

(a) $? \text{ atoms Xe} = 0.0778 \text{ mol Xe} \times \dfrac{6.022 \times 10^{23} \text{ atoms Xe}}{1 \text{ mol Xe}}$ = **4.69×10^{23} atoms Xe**

(b) $? \text{ atoms K} = 1.45 \text{ mol K} \times \dfrac{6.022 \times 10^{23} \text{ atoms K}}{1 \text{ mol K}}$ = **8.73×10^{23} atoms K**

(c) $? \text{ atoms Ti} = 55.8 \text{ mol Ti} \times \dfrac{6.022 \times 10^{23} \text{ atoms Ti}}{1 \text{ mol Ti}}$ = **3.36×10^{25} atoms Ti**

3.41 ***See Section 3.2 and Example 3.6.***

(a) $? \text{ molecules } H_2O = 99.2 \text{ mol } H_2O \times \dfrac{6.022 \times 10^{23} \text{ molecules } H_2O}{1 \text{ mol } H_2O}$ = **5.97×10^{25} molecules H_2O**

(b) $? \text{ molecules } N_2 = 1.22 \text{ mol } N_2 \times \dfrac{6.022 \times 10^{23} \text{ molecules } N_2}{1 \text{ mol } N_2}$ = **7.35×10^{25} molecules N_2**

(c) $? \text{ molecules } C_3H_6 = 22.9 \text{ mol } C_3H_6 \times \dfrac{6.022 \times 10^{23} \text{ molecules } C_3H_6}{1 \text{ mol } C_3H_6}$ = **1.38×10^{25} molecules C_3H_6**

(d) $? \text{ molecules } N_2O = 0.0022 \text{ mol } N_2O \times \dfrac{6.022 \times 10^{23} \text{ molecules } N_2O}{1 \text{ mol } N_2O}$ = **1.3×10^{21} molecules N_2O**

3.42 ***See Section 3.2 and Example 3.6.***

(a) $? \text{ molecules } Cl_2 = 0.223 \text{ mol } Cl_2 \times \dfrac{6.022 \times 10^{23} \text{ molecules } Cl_2}{1 \text{ mol } Cl_2}$ = **1.34×10^{23} molecules Cl_2**

(b) $? \text{ molecules } N_2H_4 = 14.7 \text{ mol } N_2H_4 \times \dfrac{6.022 \times 10^{23} \text{ molecules } N_2H_4}{1 \text{ mol } N_2H_4}$ = **8.85×10^{24} molecules N_2H_4**

(c) $? \text{ molecules } C_9H_{18} = 0.334 \text{ mol } C_9H_{18} \times \dfrac{6.022 \times 10^{23} \text{ molecules } C_9H_{18}}{1 \text{ mol } C_9H_{18}}$ = **2.01×10^{23} molecules C_9H_{18}**

(d) $? \text{ molecules } CO_2 = 1.22 \text{ mol } CO_2 \times \dfrac{6.022 \times 10^{23} \text{ molecules } CO_2}{1 \text{ mol } CO_2}$ = **7.35×10^{23} molecules CO_2**

3.43 ***See Section 3.2 and Example 3.6.***

(a) $? \text{ mol } O_2 = 3.44 \times 10^{24} \text{ molecules } O_2 \times \dfrac{1 \text{ mol } O_2}{6.022 \times 10^{23} \text{ molecules } O_2} = \mathbf{5.71 \text{ mol } O_2}$

(b) $? \text{ mol Na} = 1.11 \times 10^{22} \text{ atoms Na} \times \dfrac{1 \text{ mol Na}}{6.022 \times 10^{23} \text{ atoms Na}} = \mathbf{0.0184 \text{ mol Na}}$

(c) $? \text{ mol } C_2H_6 = 5.57 \times 10^{30} \text{ molecules } C_2H_6 \times \dfrac{1 \text{ mol } C_2H_6}{6.022 \times 10^{23} \text{ molecules } C_2H_6} = \mathbf{9.25 \times 10^6 \text{ mol } C_2H_6}$

(d) $? \text{ mol CO} = 1.66 \times 10^{24} \text{ molecules CO} \times \dfrac{1 \text{ mol CO}}{6.022 \times 10^{23} \text{ molecules CO}} = \mathbf{2.76 \text{ mol CO}}$

3.44 ***See Section 3.2 and Example 3.6.***

(a) $? \text{ mol } Br_2 = 1.33 \times 10^{26} \text{ molecules } Br_2 \times \dfrac{1 \text{ mol } Br_2}{6.022 \times 10^{23} \text{ molecules } Br_2} = \mathbf{221 \text{ mol } Br_2}$

(b) $? \text{ mol } C_5H_{12} = 7.71 \times 10^{26} \text{ molecules } C_5H_{12} \times \dfrac{1 \text{ mol } C_5H_{12}}{6.022 \times 10^{23} \text{ molecules } C_5H_{12}} = \mathbf{1.28 \times 10^3 \text{ mol } C_5H_{12}}$

(c) $? \text{ mol } B_2H_6 = 2.34 \times 10^{23} \text{ molecules } B_2H_6 \times \dfrac{1 \text{ mol } B_2H_6}{6.022 \times 10^{23} \text{ molecules } B_2H_6} = \mathbf{0.389 \text{ mol } B_2H_6}$

(d) $? \text{ mol Ne} = 7.76 \times 10^{23} \text{ atoms Ne} \times \dfrac{1 \text{ mol Ne}}{6.022 \times 10^{23} \text{ atoms Ne}} = \mathbf{1.29 \text{ mol Ne}}$

3.45 ***See Section 3.2 and Example 3.7.***

(a) Formula mass for NaOH:

1[Na] x 22.99 = 22.99
1[O] x 16.00 = 16.00
1[H] x 1.01 = 1.01
40.00 Hence, molar mass for NaOH is **40.00 g/mol.**

(b) Molecular mass for C_2H_4:

2[C] x 12.01 = 24.02
4[H] x 1.01 = 4.04
28.06 Hence, molar mass for C_2H_4 is **28.06 g/mol.**

(c) Formula mass for $Mg(OH)_2$:

1[Mg] x 24.30 = 24.30
2[O] x 16.00 = 32.00
2[H] x 1.01 = 2.02
58.32 Hence, molar mass for $Mg(OH)_2$ is **58.32 g/mol.**

3.46 ***See Section 3.2 and Example 3.7.***

(a) Molecular mass for N_2O_4:

2[N] x 14.01 = 28.02
4[O] x 16.00 = 64.00
92.02 Hence, molar mass for N_2O_4 is **92.02 g/mol.**

(b) Formula mass for Na_2SO_4:

1[Na] x 22.99 = 45.98
2[S] x 32.07 = 32.07
4[O] x 16.00 = 64.00
142.05 Hence, molar mass for Na_2SO_4 is **142.05 g/mol.**

(c) Molecular mass for $C_6H_{10}O_2$:

6[C] x 12.01	= 72.06		
10[H] x 1.01	= 10.10		
2[O] x 16.00	= 32.0		
	114.16	Hence, molar mass for $C_6H_{10}O_2$ is **114.16 g/mol.**	

3.47 ***See Section 3.2 and Example 3.7.***

(a) Formula mass for $ZnBr_2$:

1[Zn] x 65.39	= 65.39	
2[Br] x 79.90	=159.80	
	225.19	Hence, molar mass for $ZnBr_2$ is **225.19 g/mol.**

(b) Formula mass for K_2CrO_4:

2[K] x 39.10	= 78.20	
1[Cr] x 52.00	= 52.00	
4[O] x 16.00	= 64.00	
	194.20	Hence, molar mass for K_2CrO_4 is **194.20 g/mol.**

(c) Formula mass for BaS:

1[Ba] x 137.33	=137.33	
1[S] x 32.07	= 32.07	
	169.40	Hence, molar mass for BaS is **169.40 g/mol.**

3.48 ***See Section 3.2 and Example 3.7.***

(a) Molecular mass N_2O_2:

2[N] x 14.01	= 28.02	
2[O] x 16.00	= 32.00	
	60.02	Hence, molar mass for N_2O_2 is **60.02 g/mol.**

(b) Formula mass for $(NH_4)_2CO_3$:

2[N] x 14.01	= 28.02	
8[H] x 1.01	= 8.08	
1[C] x 12.01	= 12.01	
3[O] x 16.00	= 48.00	
	96.11	Hence, molar mass for $(NH_4)_2CO_3$ is **96.11 g/mol.**

(c) Molecular mass for $C_8H_{15}N$:

8[C] x 12.01	= 96.08	
15[H] x 1.01	= 15.15	
1[N] x 14.01	= 14.01	
	125.24	Hence, molar mass for $C_8H_{15}N$ is **125.24 g/mol.**

3.49 ***See Section 3.2 and Examples 3.6, 3.7, 3.8.***

(a) $? \text{ mol } SO_2 = 9.40 \text{ g } SO_2 \times \frac{1 \text{ mol } SO_2}{64.07 \text{ g } SO_2} = \mathbf{0.147 \text{ mol } SO_2}$

(b) $? \text{ g } AlCl_3 = 3.31 \text{ mol } AlCl_3 \times \frac{133.4 \text{ g } AlCl_3}{1 \text{ mol } AlCl_3} = \mathbf{4.41 \times 10^2 \text{ g } AlCl_3}$

(c) $? \text{ mol } H_2SO_4 = 1.12 \times 10^{23} \text{ molecules } H_2SO_4 \times \frac{1 \text{ mol } H_2SO_4}{6.022 \times 10^{23} \text{ molecules } H_2SO_4} = \mathbf{0.186 \text{ mol } H_2SO_4}$

3.50 ***See Section 3.2 and Examples 3.6, 3.7, 3.8.***

(a) $? \text{ mol Se} = 33.0 \text{ g Se} \times \frac{1 \text{ mol Se}}{78.96 \text{ g Se}} = \mathbf{0.418 \text{ mol Se}}$

(b) $? \text{ mol } NO_2 = 3.22 \times 10^{24} \text{ molecules } NO_2 \times \frac{1 \text{ mol } NO_2}{6.022 \times 10^{23} \text{ molecules } NO_2} = \mathbf{5.35 \text{ mol } NO_2}$

(c) $? \text{ g } H_2O = 5.57 \text{ mol } H_2O \times \frac{18.02 \text{ g } H_2O}{1 \text{ mol } H_2O} = \mathbf{100 \text{ g } H_2O}$

3.51 ***See Section 3.2 and Examples 3.7, 3.8.***

(a) $? \text{ mol } C_6H_6 = 14.3 \text{ g } C_6H_6 \times \frac{1 \text{ mol } C_6H_6}{78.12 \text{ g } C_6H_6} = \mathbf{0.183 \text{ mol } C_6H_6}$

(b) $? \text{ g } SiH_4 = 0.0535 \text{ mol } SiH_4 \times \frac{32.13 \text{ g } SiH_4}{1 \text{ mol } SiH_4} = \mathbf{1.72 \text{ g } SiH_4}$

(c) $? \text{ molecules } H_2O = 1.11 \text{ g } H_2O \times \frac{1 \text{ mol } H_2O}{18.02 \text{ g } H_2O} \times \frac{6.022 \times 10^{23} \text{ molecules } H_2O}{1 \text{ mol } H_2O} = \mathbf{3.71 \times 10^{22} \text{ molecules } H_2O}$

3.52 ***See Section 3.2 and Examples 3.7, 3.8.***

(a) $? \text{ g } CO_2 = 78.4 \text{ mol } CO_2 \times \frac{44.01 \text{ g } CO_2}{1 \text{ mol } CO_2} = \mathbf{3.45 \times 10^3 \text{ g } CO_2}$

(b) $? \text{ mol } AgNO_3 = 192 \text{ g } AgNO_3 \times \frac{1 \text{ mol } AgNO_3}{169.9 \text{ g } AgNO_3} = \mathbf{1.13 \text{ mol } AgNO_3}$

(c) $? \text{ molecules } CH_4 = 9.22 \text{ g } CH_4 \times \frac{1 \text{ mol } CH_4}{16.05 \text{ g } CH_4} \times \frac{6.022 \times 10^{23} \text{ molecules } CH_4}{1 \text{ mol } CH_4} = \mathbf{3.46 \times 10^{23} \text{ molecules } CH_4}$

3.53 ***See Section 3.2 and Example 3.8.***

(a) $? \text{ mol } K_2SO_4 = 2.2 \text{ g } K_2SO_4 \times \frac{1 \text{ mol } K_2SO_4}{174.3 \text{ g } K_2SO_4} = \mathbf{0.013 \text{ mol } K_2SO_4}$

(b) $? \text{ mol } C_8H_{12}N_4 = 6.4 \text{ g } C_8H_{12}N_4 \times \frac{1 \text{ mol } C_8H_{12}N_4}{164.0 \text{ g } C_8H_{12}N_4} = \mathbf{0.039 \text{ mol } C_8H_{12}N_4}$

(c) $? \text{ mol } Fe(C_5H_5)_2 = 7.13 \text{ g } Fe(C_5H_5)_2 \times \frac{1 \text{ mol } Fe(C_5H_5)_2}{186.05 \text{ g } Fe(C_5H_5)_2} = \mathbf{0.0383 \text{ mol } Fe(C_5H_5)_2}$

3.54 ***See Section 3.2 and Example 3.7.***

(a) $? \text{ g } N_2O_4 = 7.55 \text{ mol } N_2O_4 \times \frac{92.02 \text{ g } N_2O_4}{1 \text{ mol } N_2O_4} = \mathbf{695 \text{ g } N_2O_4}$

(b) $? \text{ g } CaCl_2 = 9.2 \text{ mol } CaCl_2 \times \frac{111.0 \text{ g } CaCl_2}{1 \text{ mol } CaCl_2} = \mathbf{1.0 \times 10^3 \text{ g } CaCl_2}$

(c) $? \text{ g } CO = 0.44 \text{ mol } CO \times \frac{28.0 \text{ g } CO}{1 \text{ mol } CO} = \mathbf{12 \text{ g } CO}$

3.55 *See Section 3.2 and Example 3.8.*

(a) $? \text{ mol } H_2O_2 = 48.0 \text{ g } H_2O_2 \times \frac{1 \text{ mol } H_2O_2}{34.02 \text{ g } H_2O_2} = \mathbf{1.41 \text{ mol } H_2O_2}$

(b) $? \text{ atoms O} = 1.41 \text{ mol } H_2O_2 \times \frac{2 \text{ mol atoms O}}{1 \text{ mol } H_2O_2} \times \frac{6.022 \times 10^{23} \text{ atoms O}}{1 \text{ mol atoms O}} = \mathbf{1.70 \times 10^{24} \text{ O atoms}}$

3.56 *See Section 3.2 and Example 3.8.*

(a) $? \text{ molecules } H_2 = 3.4 \text{ g } H_2 \times \frac{1 \text{ mol } H_2}{2.02 \text{ g } H_2} \times \frac{6.022 \times 10^{23} \text{ molecules } H_2}{1 \text{ mol } H_2} = \mathbf{1.0 \times 10^{24} \text{ molecules } H_2}$

(b) $? \text{ atoms H} = 1.0 \times 10^{24} \text{ molecules } H_2 \times \frac{2 \text{ atoms H}}{1 \text{ molecule } H_2} = \mathbf{2.0 \times 10^{24} \text{ atoms H}}$

3.57 *See Section 3.2 and Examples 3.6, 3.7, 3.8.*

(a) $? \text{ g } NO_2 = 3.50 \text{ mol } NO_2 \times \frac{46.01 \text{ g } NO_2}{1 \text{ mol } NO_2} = \mathbf{161 \text{ g } NO_2}$

(b) $? \text{ molecules } NO_2 = 3.50 \text{ mol } NO_2 \times \frac{6.022 \times 10^{23} \text{ molecules } NO_2}{1 \text{ mol } NO_2} = \mathbf{2.11 \times 10^{24} \text{ molecules } NO_2}$

(c) $? \text{ atoms N} = 2.11 \times 10^{24} \text{ molecules } NO_2 \times \frac{1 \text{ atom N}}{1 \text{ molecule } NO_2} = \mathbf{2.11 \times 10^{24} \text{ atoms}}$

$? \text{ atoms O} = 2.11 \times 10^{24} \text{ molecules } NO_2 \times \frac{2 \text{ atoms O}}{1 \text{ molecule } NO_2} = \mathbf{4.22 \times 10^{24} \text{ atoms}}$

3.58 *See Section 3.2 and Examples 3.6, 3.7, 3.8.*

(a) $? \text{ mol } SO_3 = 3.31 \text{ g } SO_3 \times \frac{1 \text{ mol } SO_3}{80.07 \text{ g } SO_3} = \mathbf{0.413 \text{ mol } SO_3}$

(b) $? \text{ molecules } SO_3 = 0.413 \text{ mol } SO_3 \times \frac{6.022 \times 10^{23} \text{ molecules } SO_3}{1 \text{ mol } SO_3} = \mathbf{2.49 \times 10^{23} \text{ molecules } SO_3}$

(c) $? \text{ atoms S} = 2.49 \times 10^{23} \text{ molecules } SO_3 \times \frac{1 \text{ atom S}}{1 \text{ molecule } SO_3} = \mathbf{2.49 \times 10^{23} \text{ atoms S}}$

$? \text{ atoms O} = 2.49 \times 10^{23} \text{ molecules } SO_3 \times \frac{3 \text{ atoms O}}{1 \text{ molecule } SO_3} = \mathbf{7.47 \times 10^{23} \text{ atoms O}}$

3.59 ***See Section 3.2 and Examples 3.6, 3.7, 3.8.***

(a) Molecular mass for $C_{22}H_{25}NO_6$:

22[C] x 12.01	=	264.22
25[H] x 1.01	=	25.25
1[N] x 14.01	=	14.01
6[O] x 16.00	=	96.00
		399.48

Hence, molar mass for $C_{22}H_{25}NO_6$ is **399.48 g/mol.**

(b) $? \text{ g } C_{22}H_{25}NO_6 = 3.2 \times 10^{22} \text{ molecules } C_{22}H_{25}NO_6 \times \frac{1 \text{ mol } C_{22}H_{25}NO_6}{6.022 \times 10^{23} \text{ molecules } C_{22}H_{25}NO_6}$

$\times \frac{399.48 \text{ g } C_{22}H_{25}NO_6}{1 \text{ mol } C_{22}H_{25}NO_6} = \mathbf{21 \text{ g } C_{22}H_{25}NO_6}$

(c) $? \text{ mol } C_{22}H_{25}NO_6 = 326 \text{ g } C_{22}H_{25}NO_6 \times \frac{1 \text{ mol } C_{22}H_{25}NO_6}{399.48 \text{ g } C_{22}H_{25}NO_6} = \mathbf{0.816 \text{ mol } C_{22}H_{25}NO_6}$

(d) $? \text{ atoms C} = 50 \text{ molecules } C_{22}H_{25}NO_6 \times \frac{22 \text{ atoms C}}{1 \text{ molecule } C_{22}H_{25}NO_6} = \mathbf{1.1 \times 10^3 \text{ atoms C}}$

3.60 ***See Section 3.2 and Examples 3.6, 3.7, 3.8.***

(a) Molecular mass for $Ni(CO)_4$:

1[Ni] x 58.69	=	58.69
4[C] x 12.01	=	48.04
4[O] x 16.00	=	64.00
		170.73

Hence, molar mass for $Ni(CO)_4$ is **170.73 g/mol.**

(b) $? \text{ mol } Ni(CO)_4 = 3.22 \text{ g } Ni(CO)_4 \times \frac{1 \text{ mol } Ni(CO)_4}{170.73 \text{ g } Ni(CO)_4} = \mathbf{1.89 \times 10^{-2} \text{ mol } Ni(CO)_4}$

(c) $? \text{ molecules } Ni(CO)_4 = 5.67 \text{ g } Ni(CO)_4 \times \frac{1 \text{ mol } Ni(CO)_4}{170.73 \text{ g } Ni(CO)_4} \times \frac{6.022 \times 10^{23} \text{ molecules } Ni(CO)_4}{1 \text{ mol } Ni(CO)_4}$

$= \mathbf{2.00 \times 10^{22} \text{ molecules } Ni(CO)_4}$

(d) $? \text{ atoms C} = 34 \text{ g } Ni(CO)_4 \times \frac{1 \text{ mol } Ni(CO)_4}{170.73 \text{ g } Ni(CO)_4} \times \frac{6.022 \times 10^{23} \text{ molecules } Ni(CO)_4}{1 \text{ mol } Ni(CO)_4} \times \frac{4 \text{ atom C}}{1 \text{ molecule } Ni(CO)_4}$

$= \mathbf{4.8 \times 10^{23} \text{ atoms C}}$

Note: $Ni(CO)_4$ is composed of a metal and nonmetals and would be predicted to be an ionic compound (2.6). However, the observation that it is a volatile compound rather than a crystalline solid suggests it is a molecular compound (2.7).

3.61 ***See Section 3.3 and Example 3.9.***

(a) Molecular mass for C_4H_8:

4[C] x 12.01	=	48.04
8[H] x 1.01	=	8.08
		56.12

$\% \text{C} = \frac{48.04 \text{ g C}}{56.12 \text{ g } C_4H_8} \times 100\% = \mathbf{85.60\% \text{ C}}$

$\% \text{H} = \frac{8.08 \text{ g H}}{56.12 \text{ g } C_4H_8} \times 100\% = \mathbf{14.40\% \text{ H}}$

Molar mass for C_4H_8 is 56.12 g/mol.

(b) Molecular mass for $C_3H_4N_2$: 3[C] x 12.01 =36.03 $\%C = \frac{36.03\ g\ C}{68.09\ g\ C_3H_4N_2} \times 100\% =$ **52.92% C**

4[H] x 1.01 = 4.04 $\%H = \frac{4.04\ g\ H}{68.09\ g\ C_3H_4N_2} \times 100\% =$ **5.93% H**

2[N] x 14.01 = <u>28.02</u> $\%N = \frac{28.02\ g\ N}{68.09\ g\ C_3H_4N_2} \times 100\% =$ **41.15% N**

68.09

Molar mass for $C_3H_4N_2$ is 68.09 g/mol.

(c) Formula mass for Fe_2O_3: 2[Fe] x 55.85 =111.70 $\%Fe = \frac{111.7\ g\ Fe}{159.70\ g\ Fe_2O_3} \times 100\% =$ **69.94% Fe**

3[O] x 16.0 = <u>48.00</u> $\%O = \frac{48.00\ g\ O}{159.70\ g\ Fe_2O_3} \times 100\% =$ **30.06% O**

159.70

Molar mass for Fe_2O_3 is 159.70 g/mol.

3.62 ***See Section 3.3 and Example 3.9.***

(a) Molecular mass for C_6H_{12}: 6[C] x 12.01 = 72.06 $\%C = \frac{72.06\ g\ C}{84.18\ g\ C_6H_{12}} \times 100\% =$ **85.60% C**

12[H] x 1.01 = <u>12.12</u> $\%H = \frac{12.12\ g\ H}{84.18\ g\ C_6H_{12}} \times 100\% =$ **14.40% H**

84.18

Molar mass for C_6H_{12} is 84.18 g/mol.

(b) Molecular mass for $C_5H_{12}O$: 5[C] x 12.01 = 60.05 $\%C = \frac{60.05\ g\ C}{88.17\ g\ C_5H_{12}O} \times 100\% =$ **68.11% C**

12[H] x 1.01 = 12.12 $\%H = \frac{12.12\ g\ H}{88.17\ g\ C_5H_{12}O} \times 100\% =$ **13.75% H**

1[O] x 16.00 = <u>16.00</u> $\%O = \frac{16.00\ g\ O}{88.17\ g\ C_5H_{12}O} \times 100\% =$ **18.15% O**

88.17

Molar mass for $C_5H_{12}O$ is 88.17 g/mol.

(c) Formula mass for $NiCl_2$: 1[Ni] x 58.69 = 58.69 $\%Ni = \frac{58.69\ g\ Ni}{129.59\ g\ NiCl_2} \times 100\% =$ **45.29% Ni**

2[Cl] x 35.45 = <u>70.90</u> $\%Cl = \frac{70.90\ g\ Cl}{129.59\ g\ NiCl_2} \times 100\% =$ **54.71% Cl**

129.59

Molar mass for $NiCl_2$ is 129.59 g/mol.

3.63 ***See Section 3.3 and Example 3.8.***

Molecular mass for C_3H_6O:

$3[C] \times 12.01 = 36.03$ $\quad \%C = \frac{36.03 \text{ g C}}{58.09 \text{ g } C_3H_6O} \times 100\% = \mathbf{62.02\%\ C}$

$6[H] \times 1.01 = 6.06$ $\quad \%H = \frac{6.06 \text{ g H}}{58.09 \text{ g } C_3H_6O} \times 100\% = \mathbf{10.43\%\ H}$

$1[O] \times 16.00 = \underline{16.00}$ $\quad \%O = \frac{16.00 \text{ g O}}{58.09 \text{ g } C_3H_6O} \times 100\ \% = \mathbf{27.54\%\ O}$

58.09

Molar mass for C_3H_6O is 58.09 g/mol.

3.64 ***See Section 3.3 and Example 3.8.***

Molecular mass for H_3PO_4:

$3[H] \times 1.01 = 3.03$ $\quad \%H = \frac{3.03 \text{ g H}}{98.01 \text{ g } H_3PO_4} \times 100\% = \mathbf{3.09\%\ H}$

$1[P] \times 30.97 = 30.97$ $\quad \%P = \frac{30.97 \text{ g P}}{98.00 \text{ g } H_3PO_4} \times 100\% = \mathbf{31.60\%\ P}$

$4[O] \times 16.00 = \underline{64.00}$ $\quad \%O = \frac{64.00 \text{ g O}}{98.00 \text{ g } H_3PO_4} \times 100\% = \mathbf{65.31\ \%\ O}$

98.00

Molar mass for H_3PO_4 is 98.00 g/mol.

3.65 ***See Section 2.8, 3.3, and Example 3.9.***

Formula mass for Na_2SO_4:

$2[Na] \times 22.99 = 45.98$ $\quad \%Na = \frac{45.98 \text{ g Na}}{142.05 \text{ g } Na_2SO_4} \times 100\% = \mathbf{32.37\%\ Na}$

$1[S] \times 32.07 = 32.07$ $\quad \%S = \frac{32.07 \text{ g S}}{142.05 \text{ g } Na_2SO_4} \times 100\% = \mathbf{22.58\%\ S}$

$4[O] \times 16.00 = \underline{64.00}$ $\quad \%O = \frac{64.00 \text{ g O}}{142.05 \text{ g } Na_2SO_4} \times 100\% = \mathbf{45.05\%\ O}$

142.05

Molar mass for Na_2SO_4 is 142.05 g/mol.

3.66 ***See Sections 2.8, 3.3, and Example 3.9.***

Formula mass for $MgCO_3$:

$1[Mg] \times 24.30 = 24.30$ $\quad \%Mg = \frac{24.30 \text{ g Mg}}{84.31 \text{ g } MgCO_3} \times 100\% = \mathbf{28.83\%\ Mg}$

$1[C] \times 12.01 = 12.01$ $\quad \%C = \frac{12.01 \text{ g C}}{84.31 \text{ g } MgCO_3} \times 100\% = \mathbf{14.25\%\ C}$

$3[O] \times 16.00 = \underline{48.00}$ $\quad \%O = \frac{48.00 \text{ g O}}{84.31 \text{ g } MgCO_3} \times 100\% = \mathbf{56.93\%\ O}$

84.31

Molar mass for $MgCO_3$ is 84.31 g/mol.

3.67 ***See Section3.3, and Example 3.9.***

(a)

Formula mass for $NaBH_4$:

1[Na] x 22.99 = 22.99

1[B] x 10.81 = 10.81

4[H] x 1.01 = 4.04

37.84

Molar mass for $NaBH_4$ is 37.84 g/mol.

(b)

$$\% Na = \frac{22.99 \text{ g Na}}{37.84 \text{ g NaBH}_4} \times 100\% = \mathbf{60.76\% \ Na}$$

$$\% B = \frac{10.81 \text{ g B}}{37.84 \text{ g NaBH}_4} \times 100\% = \mathbf{28.57\% \ B}$$

$$\% H = \frac{4.04 \text{ g H}}{37.84 \text{ g NaBH}_4} \times 100\% = \mathbf{10.68 \ \% \ H}$$

3.68 ***See Sections 2.8, 3.3, and Example 3.9.***

(a)

Formula mass for $CaCO_3$:

1[Ca] x 40.08 = 40.08

1[C] x 12.01 = 12.01

3[O] x 16.00 = 48.00

100.09

Molar mass for $CaCO_3$ is 100.09 g/mol.

(b)

$$\% Ca = \frac{40.08 \text{ g Ca}}{100.09 \text{ g CaCO}_3} \times 100\% = \mathbf{40.04\% \ Ca}$$

$$\% C = \frac{12.01 \text{ g C}}{100.09 \text{ g CaCO}_3} \times 100\% = \mathbf{12.00\% \ C}$$

$$\% O = \frac{48.00 \text{ g O}}{100.09 \text{ g CaCO}_3} \times 100\% = \mathbf{47.96\% \ O}$$

3.69 ***See Section 3.3 and Example 3.9.***

Formula mass for FeI_3:

1[Fe] x 55.85 = 55.85

3[I] x 126.90 =380.70

436.55

$$\% Fe = \frac{55.85 \text{ g Fe}}{436.55 \text{ g FeI}_3} \times 100\% = 12.79\% \text{ Fe}$$

$$\% I = \frac{380.78 \text{ g I}}{436.55 \text{ g FeI}_3} \times 100\% = 87.21\% \text{ I}$$

Molar mass for FeI_3 is 436.55 g/mol.

Analytical percent composition of 18.0% Fe and 82.0% I does not agree with percent composition of FeI_3. **Compound is not FeI_3.**

3.70 ***See Section 3.3 and Example 3.9.***

Formula mass for $C_6H_4(OH)Cl$:

6[C] x 12.01 = 72.06

5[H] x 1.01 = 5.05

1[O] x 16.00 = 16.00

1[Cl] x 35.45 = 35.45

128.56

$$\% C = \frac{72.06 \text{ g C}}{128.56 \text{ g C}_6\text{H}_4\text{(OH)Cl}} \times 100\% = 56.05\%\text{C}$$

$$\% H = \frac{5.05 \text{ g H}}{128.56 \text{ g C}_6\text{H}_4\text{(OH)Cl}} \times 100\% = 3.93\% \text{ H}$$

$$\% O = \frac{16.00 \text{ g O}}{128.56 \text{ g C}_6\text{H}_4\text{(OH)Cl}} \times 100\% = 12.45\%\text{O}$$

$$\%Cl = \frac{35.45 \text{ g Cl}}{128.56 \text{ g C}_6\text{H}_4\text{(OH)Cl}} \times 100\% = 27.57\%\text{Cl}$$

Molar mass for $C_6H_4(OH)Cl$ is 128.56 g/mol.

Analytical percent composition of 56.05% C, 3.93% H, and 27.57% Cl agrees with percent composition of $C_6H_4(OH)Cl$. **Compound is $C_6H_4(OH)Cl$.**

3.71 *See Section 3.3.*

(a) Molecular mass for CO:

1[C] x 12.01	=	12.01
1[O] x 16.0 0	=	16.00
		28.01

Hence, molar mass for CO is 28.01 g/mol.

$$? \text{ g C} = 4.9 \text{ g CO} \times \frac{12.01 \text{ g C}}{28.01 \text{ g CO}} = \mathbf{2.1 \text{ g C}}$$

(b) Molecular mass for C_3H_6:

3[C] x 12.01	=	36.03
6[H] x 1.01	=	6.06
		42.09

Hence, molar mass for C_3H_6 is 42.09 g/mol.

$$? \text{ g C} = 2.2 \text{ g } C_3H_6 \text{ x } \frac{36.03 \text{ g C}}{42.09 \text{ g } C_3H_6} = \mathbf{1.9 \text{ g C}}$$

(c) Molecular mass for C_2H_6O:

2[C] x 12.01	=	24.02
6[H] x 1.01	=	6.06
1[O] x 16.00	=	16.00
		46.08

Hence, molar mass for C_2H_6O is 46.08 g/mol.

$$? \text{ g C} = 9.33 \text{ g } C_2H_6O \times \frac{24.02 \text{ g C}}{46.08 \text{ g } C_2H_6O} = \mathbf{4.86 \text{ g C}}$$

Note: See direct method used in working 3.73.

3.72 *See Section 3.3.*

(a) Molecular mass for $C_4H_{10}O$:

4[C] x 12.01	=	48.04
10[H] x 1.01	=	10.10
1[O] x 16.00	=	16.00
		74.14

Hence, molar mass for $C_4H_{10}O$ is 74.14 g/mol

$$? \text{ g C} = 1.80 \text{ g } C_4H_{10}O \times \frac{48.04 \text{ g C}}{74.14 \text{ g } C_4H_{10}O} = \mathbf{1.17 \text{ g C}}$$

(b) Formula mass for Na_2CO_3:

2[Na] x 22.99	=	45.98
1[C] x 12.01	=	12.01
3[O] x 16.00	=	48.00
		105.99

Hence, molar mass for Na_2CO_3 is 105.99 g/mol

$$? \text{ g C} = 0.00223 \text{ g } Na_2CO_3 \times \frac{12.01 \text{ g C}}{105.99 \text{ g } Na_2CO_3} = \mathbf{2.53 \times 10^{-4} \text{ g C}}$$

(c) Molecular mass for $C_5H_{11}N$:

5[C] x 12.01	=	60.05
11[H] x 1.01	=	11.11
1[N] x 14.01	=	14.01
		85.17

Hence, molar mass for $C_5H_{11}N$ is 85.17 g/mol.

$$? \text{ g C} = 22.1 \text{ g } C_5H_{11}N \times \frac{60.05 \text{ g C}}{85.17 \text{ g } C_5H_{11}N} = \mathbf{15.6 \text{ g C}}$$

Note: See direct method used in working 3.73.

3.73 *See Section 3.3.*

(a) $$? \text{ g C} = 4.32 \text{ g } CO_2 \times \frac{12.01 \text{ g C}}{44.01 \text{ g } CO_2} = \mathbf{1.18 \text{ g C}}$$

(b) $$? \text{g C} = 2.21 \text{ g } C_2H_4 \times \frac{24.02 \text{ g C}}{28.06 \text{ g } C_2H_4} = \mathbf{1.89 \text{ g C}}$$

(c) $? \text{ g C} = 0.0443 \text{ g } CS_2 \times \frac{12.01 \text{ g C}}{76.15 \text{ g } CS_2} = \mathbf{0.00698 \text{ g C}}$

3.74 ***See Section 3.3.***

(a) $? \text{ g H} = 4.33 \text{ g } H_2O \times \frac{2.02 \text{ g H}}{18.02 \text{ g } H_2O} = \mathbf{0.485 \text{ g H}}$

(b) $? \text{ g H} = 1.22 \text{ g } C_2H_2 \times \frac{2.02 \text{ g H}}{26.04 \text{ g } C_2H_2} = \mathbf{0.0946 \text{ g H}}$

(c) $? \text{ g H} = 4.44 \text{ g } N_2H_4 \times \frac{4.04 \text{ g H}}{32.06 \text{ g } N_2H_4} = \mathbf{0.560 \text{ g H}}$

3.75 ***See Section 3.3 and Example 3.10.***

$? \text{ g C} = 1.80 \text{ g } CO_2 \times \frac{12.01 \text{ g C}}{44.01 \text{ g } CO_2} = 0.491 \text{ g C}$

$\% \text{C} = \frac{0.491 \text{ g C}}{1.070 \text{ g sample}} \times 100\% = \mathbf{45.9\% \text{ C}}$

$? \text{ g H} = 1.02 \text{ g } H_2O \times \frac{2.02 \text{ g H}}{18.02 \text{ g } H_2O} = 0.114 \text{ g H}$

$\% \text{H} = \frac{0.114 \text{ g H}}{1.070 \text{ g sample}} \times 100\% = \mathbf{10.7\% \text{ H}}$

$? \text{ g O} = \text{g sample} - \text{g C} - \text{g H}$

$= 1.070 \text{ g} - 0.491 \text{ g} - 0.114 \text{ g} = 0.465 \text{ g O}$

$\% \text{O} = \frac{0.465 \text{ g O}}{1.070 \text{ g sample}} \times 100\% = \mathbf{43.5\% \text{ O}}$

3.76 ***See Section 3.3 and Example 3.10.***

$? \text{ g C} = 4.06 \text{ g } CO_2 \times \frac{12.01 \text{ g C}}{44.01 \text{ g } CO_2} = 1.11 \text{ g C}$

$\% \text{C} = \frac{1.11 \text{ g C}}{2.770 \text{ g sample}} \times 100\% = \mathbf{40.1\% \text{ C}}$

$? \text{ g H} = 1.66 \text{ g } H_2O \times \frac{2.02 \text{ g H}}{18.02 \text{ g } H_2O} = 0.186 \text{ g H}$

$\% \text{H} = \frac{0.186 \text{ g H}}{2.770 \text{ g sample}} \times 100\% = \mathbf{6.7\% \text{ H}}$

$? \text{ g O} = \text{g sample} - \text{g C} - \text{g H}$

$= 2.770 \text{ g} - 1.11 \text{ g} - 0.186 \text{ g} = 1.47 \text{ g O}$

$\% \text{O} = \frac{1.47 \text{ g O}}{2.770 \text{ g sample}} \times 100\% = \mathbf{53.1\% \text{ O}}$

3.77 ***See Section 3.3 and Example 3.10.***

$? \text{ g C} = 5.06 \text{ g } CO_2 \times \frac{12.01 \text{ g C}}{44.01 \text{ g } CO_2} = 1.38 \text{ g C}$

$\% \text{C} = \frac{1.38 \text{ g C}}{3.11 \text{ g sample}} \times 100\% = \mathbf{44.4\% \text{ C}}$

$? \text{ g H} = 2.07 \text{ g } H_2O \times \frac{2.02 \text{ g H}}{18.02 \text{ g } H_2O} = 0.232 \text{ g H}$

$\% \text{H} = \frac{0.232 \text{g H}}{3.11 \text{ g sample}} \times 100\% = \mathbf{7.46\% \text{ H}}$

$? \text{ g N} = \text{g sample} - \text{g C} - \text{g H}$

$= 3.11 \text{ g} - 1.38 \text{ g} - 0.232 \text{ g} = 1.50 \text{ g N}$

$\% \text{N} = \frac{1.50 \text{ g N}}{3.11 \text{ g sample}} \times 100\% = \mathbf{48.2\% \text{ N}}$

3.78 *See Section 3.3.*

$$? \text{ g C} = 1.04 \text{ g CO}_2 \times \frac{12.01 \text{ g C}}{44.01 \text{ g CO}_2} = 0.284 \text{ g C} \qquad \% \text{C} = \frac{0.284 \text{ g C}}{0.513 \text{ g sample}} \times 100\% = \mathbf{55.4\% \ C}$$

$$? \text{ g H} = 0.704 \text{ g H}_2\text{O} \times \frac{2.02 \text{ g H}}{18.02 \text{ g H}_2\text{O}} = 0.0789 \text{ g H} \qquad \% \text{H} = \frac{0.0789 \text{ g H}}{0.513 \text{ g sample}} \times 100\% = \mathbf{15.4\% \ H}$$

$$? \text{ g N} = \text{g sample} - \text{g C} - \text{g H} \qquad \% \text{N} = \frac{0.150 \text{ g N}}{0.513 \text{ g sample}} \times 100\% = \mathbf{29.2 \ \% \ N}$$

$$= 0.513 \text{ g} - 0.284 \text{ g} - 0.0789 \text{ g} = 0.150 \text{ g N}$$

3.79 *See Section 3.3 and Example 3.11.*

$$? \text{ mol C} = 0.831 \text{ g C} \times \frac{1 \text{ mol C}}{12.011 \text{ g C}} = 0.0692 \text{ mol C} \qquad \text{relative mol C} = \frac{0.0692 \text{ mol C}}{0.0692} = 1.00 \text{ mol C}$$

$$? \text{ mol H} = 0.139 \text{ g H} \times \frac{1 \text{ mol H}}{1.008 \text{ g H}} = 0.138 \text{ mol H} \qquad \text{relative mol H} = \frac{0.138 \text{ mol H}}{0.0692} = 1.99 \text{ mol H}$$

The empirical formula is $\mathbf{CH_2}$.

3.80 *See Section 3.3 and Example 3.11.*

$$? \text{ mol C} = 0.80 \text{ g C} \times \frac{1 \text{ mol C}}{12.011 \text{ g C}} = 0.067 \text{ mol C} \qquad \text{relative mol C} = \frac{0.067 \text{ mol C}}{0.067} = 1.00 \text{ mol C}$$

$$? \text{ mol H} = 0.20 \text{ g H} \times \frac{1 \text{ mol H}}{1.008 \text{ g H}} = 0.20 \text{ mol H} \qquad \text{relative mol H} = \frac{0.20 \text{ mol H}}{0.067} = 3.0 \text{ mol H}$$

The empirical formula is $\mathbf{CH_3}$.

3.81 *See Section 3.3 and Example 3.11.*

$$? \text{ mol C} = 0.571 \text{ g C} \times \frac{1 \text{ mol C}}{12.011 \text{ g C}} = 0.0475 \text{ mol C} \qquad \text{relative mol C} = \frac{0.0475 \text{ mol C}}{0.0238} = 2.00 \text{ mol C}$$

$$? \text{ mol H} = 0.072 \text{ g H} \times \frac{1 \text{ mol H}}{1.008 \text{ g H}} = 0.071 \text{ mol H} \qquad \text{relative mol H} = \frac{0.071 \text{ mol H}}{0.0238} = 3.0 \text{ mol H}$$

$$? \text{ mol N} = 0.333 \text{ g N} \times \frac{1 \text{ mol N}}{14.007 \text{ g N}} = 0.0238 \text{ mol N} \qquad \text{relative mol N} = \frac{0.0238 \text{ mol N}}{0.0238} = 1.00 \text{ mol N}$$

The empirical formula is $\mathbf{C_2H_3N}$.

3.82 *See Section 3.3 and Example 3.11.*

$$? \text{ mol N} = 0.152 \text{ g N} \times \frac{1 \text{ mol N}}{14.007 \text{ g N}} = 0.0109 \text{ mol N} \qquad \text{relative mol N} = \frac{0.0109 \text{ mol N}}{0.0109} = 1.00 \text{ mol N}$$

$$? \text{ mol O} = 0.348 \text{ g O} \times \frac{1 \text{ mol O}}{15.9994 \text{ g O}} = 0.0218 \text{ mol O} \qquad \text{relative mol O} = \frac{0.0218 \text{ mol O}}{0.0109} = 2.00 \text{ mol O}$$

The empirical formula is $\mathbf{NO_2}$.

3.83 *See Section 3.3 and Example 3.11.*

Assume the sample has a mass of 100.00 g and therefore contains 44.06 g Fe and 55.93 g Cl.

$$? \text{ mol Fe} = 44.06 \text{ g Fe} \times \frac{1 \text{ mol Fe}}{55.847 \text{ g Fe}} = 0.7889 \text{ mol Fe} \quad \text{relative mol Fe} = \frac{0.7889 \text{ mol Fe}}{0.7889} = 1.00 \text{ mol Fe}$$

$$? \text{ mol Cl} = 55.94 \text{ g Cl} \times \frac{1 \text{ mol Cl}}{35.453 \text{ g Cl}} = 1.578 \text{ mol Cl} \quad \text{relative mol Cl} = \frac{1.578 \text{ mol Cl}}{0.7889} = 2.00 \text{ mol Cl}$$

The empirical formula is **$FeCl_2$**.

3.84 ***See Section 3.3 and Example 3.11.***

Assume the sample has a mass of 100.00 g and therefore contains 52.7 g Se and 47.3 g Cl.

$$? \text{ mol Se} = 52.7 \text{ g Se} \times \frac{1 \text{ mol Se}}{78.96 \text{ g Se}} = 0.667 \text{ mol Se} \quad \text{relative mol Se} = \frac{0.667 \text{ mol Se}}{0.667} = 1.00 \text{ mol Se}$$

$$? \text{ mol Cl} = 47.3 \text{ g Cl} \times \frac{1 \text{ mol Cl}}{35.453 \text{ g Cl}} = 1.33 \text{ mol Cl} \quad \text{relative mol Cl} = \frac{1.33 \text{ mol Cl}}{0.667} = 1.99 \text{ mol Cl}$$

The empirical formula is **$SeCl_2$.**

3.85 ***See Section 3.3 and Example 3.11.***

$$? \text{ mol Ti} = 0.252 \text{ g Ti} \times \frac{1 \text{ mol Ti}}{47.88 \text{ g Ti}} = 0.00526 \text{ mol Ti} \quad \text{relative mol Ti} = \frac{0.00526 \text{ mol Ti}}{0.00526} = 1.00 \text{ mol Ti}$$

$$? \text{ mol Cl} = 0.748 \text{ g Cl} \times \frac{1 \text{ mol Cl}}{35.453 \text{ g Cl}} = 0.0211 \text{ mol Cl} \quad \text{relative mol Cl} = \frac{0.0211 \text{ mol Cl}}{0.00526} = 4.01 \text{ mol Cl}$$

The empirical formula is **$TiCl_4$**.

3.86 ***See Section 3.3 and Example 3.11.***

$$? \text{ mol Cr} = 0.173 \text{ g Cr} \times \frac{1 \text{ mol Cr}}{51.996 \text{ g Cr}} = 0.00333 \text{ mol Cr} \quad \text{relative mol Cr} = \frac{0.00333 \text{ mol Cr}}{0.00333} = 1.00 \text{ mol Cr}$$

$$? \text{ mol O} = 0.160 \text{ g O} \times \frac{1 \text{ mol O}}{15.9994 \text{ g O}} = 0.0100 \text{ ml O} \quad \text{relative mol O} = \frac{0.0100 \text{ mol O}}{0.00333} = 3.00 \text{ mol O}$$

The empirical formula is **CrO_3.**

3.87 ***See Section 3.3 and Example 3.11.***

Assume the sample has a mass of 100.0 g and therefore contains 66.6 g C, 11.2 g H, and 22.2 g O.

$$? \text{ mol C} = 66.6 \text{ g C} \times \frac{1 \text{ mol C}}{12.011 \text{ g C}} = 5.54 \text{ mol C} \quad \text{relative mol C} = \frac{5.54 \text{ mol C}}{1.38} = 4.01 \text{ mol C}$$

$$? \text{ mol H} = 11.2 \text{ gH} \times \frac{1 \text{ mol H}}{1.008 \text{ gH}} = 11.1 \text{ mol H} \quad \text{relative mol H} = \frac{11.1 \text{ mol H}}{1.38} = 8.04 \text{ mol H}$$

$$? \text{ mol O} = 22.2 \text{ g O} \times \frac{1 \text{ mol O}}{15.999 \text{ g O}} = 1.38 \text{ mol O} \quad \text{relative mol O} = \frac{1.38 \text{ mol O}}{1.38} = 1.00 \text{ mol O}$$

The empirical formula is **C_4H_8O.**

3.88 ***See Section 3.3 and Example 3.11.***

Assume the sample has a mass of 100.00 g and therefore contains 26.89 g Ti, 67.44 g C and 5.67 g H.

$$? \text{ mol Ti} = 26.89 \text{ g Ti} \times \frac{1 \text{ mol Ti}}{47.88 \text{ g Ti}} = 0.5616 \text{ mol Ti} \quad \text{relative mol Ti} = \frac{0.5616 \text{ mol Ti}}{0.5616} = 1.000 \text{ mol Ti}$$

$$? \text{ mol C} = 67.44 \text{ g C} \times \frac{1 \text{ mol C}}{12.011 \text{ g C}} = 5.615 \text{ mol C} \quad \text{relative mol C} = \frac{5.615 \text{ mol C}}{0.5616} = 9.998 \text{ mol C}$$

$$? \text{ mol H} = 5.67 \text{ g H} \times \frac{1 \text{ mol H}}{1.008 \text{ g H}} = 5.625 \text{ mol H} \qquad \text{relative mol H} = \frac{5.625 \text{ mol H}}{0.5616} = 10.02 \text{ mol H}$$

The empirical formula is **$TiC_{10}H_{10}$.** The compound is $Ti(C_5H_5)_2$.

3.89 ***See Section 3.3 and Example 3.11.***

Assume the sample has a mass of 100.00 g and therefore contains 65.02 g Pt, 2.02 g H, 9.34 g N and 23.63 g Cl.

$$? \text{ mol Pt} = 65.02 \text{ g Pt} \times \frac{1 \text{ mol Pt}}{195.08 \text{ g Pt}} = 0.3333 \text{ mol Pt} \qquad \text{relative mol Pt} = \frac{0.3333 \text{ mol Pt}}{0.3333} = 1.00 \text{ mol Pt}$$

$$? \text{ mol H} = 2.02 \text{ g H} \times \frac{1 \text{ mol H}}{1.008 \text{ g H}} = 2.00 \text{ mol H} \qquad \text{relative mol H} = \frac{2.00 \text{ mol H}}{0.3333} = 6.00 \text{ mol H}$$

$$? \text{ mol N} = 9.34 \text{ g N} \times \frac{1 \text{ mol N}}{14.007 \text{ g N}} = 0.667 \text{ mol N} \qquad \text{relative mol N} = \frac{0.667 \text{ mol N}}{0.3333} = 2.00 \text{ mol N}$$

$$? \text{ mol Cl} = 23.63 \text{ g Cl} \times \frac{1 \text{ mol Cl}}{35.453 \text{ g Cl}} = 0.6666 \text{ mol Cl} \qquad \text{relative mol Cl} = \frac{0.6666 \text{ mol Cl}}{0.3333} = 2.00 \text{ mol Cl}$$

The empirical formula is **$PtH_6N_2Cl_2$**, and the compound is actually $Pt(NH_3)_2Cl_2$.

3.90 ***See Section 3.3 and Example 3.11.***

Assume the sample has a mass of 100.00 g and therefore contains 79.95 g C, 9.40 g H and 10.65 g O.

$$? \text{ mol C} = 79.95 \text{ g C} \times \frac{1 \text{ mol C}}{12.011 \text{ g C}} = 6.656 \text{ mol C} \qquad \text{relative mol C} = \frac{6.656 \text{ mol C}}{0.6656} = 10.00 \text{ mol C}$$

$$? \text{ mol H} = 9.40 \text{ g H} \times \frac{1 \text{ mol H}}{1.008 \text{ g H}} = 9.33 \text{ mol H} \qquad \text{relative mol H} = \frac{9.33 \text{ mol H}}{0.6656} = 14.0 \text{ mol H}$$

$$? \text{ mol O} = 10.65 \text{ g O} \times \frac{1 \text{ mol O}}{15.9994 \text{ g O}} = 0.6656 \text{ mol O} \qquad \text{relative mol O} = \frac{0.6656 \text{ mol O}}{0.6656} = 1.000 \text{ mol O}$$

The empirical formula is **$C_{10}H_{14}O$.**

3.91 ***See Section 3.3 and Example 3.11.***

$$? \text{ g C} = 3.800 \text{ g } CO_2 \times \frac{12.01 \text{ g C}}{44.01 \text{ g } CO_2} = 1.037 \text{ g C} \qquad ? \text{ g H} = 1.040 \text{ g } H_2O \times \frac{2.016 \text{ g H}}{18.02 \text{ g } H_2O} = 0.1163 \text{ g H}$$

$$? \text{ g O} = \text{g sample} - \text{g C} - \text{g H} = 2.074 \text{ g} - 1.037 \text{ g} - 0.1163 \text{ g} = 0.921 \text{ g O}$$

$$? \text{ mol C} = 1.037 \text{ g C} \times \frac{1 \text{ mol C}}{12.01 \text{ g C}} = 0.08634 \text{ mol C} \qquad \text{relative mol C} = \frac{0.08634 \text{ mol C}}{0.0576} = 1.50 \text{ mol C}$$

$$? \text{ mol H} = 0.1163 \text{ g H} \times \frac{1 \text{ mol H}}{1.008 \text{ g H}} = 0.1154 \text{ mol H} \qquad \text{relative mol H} = \frac{0.1154 \text{ mol H}}{0.0576} = 2.00 \text{ mol H}$$

$$? \text{ mol O} = 0.921 \text{ g O} \times \frac{1 \text{ mol O}}{15.9994 \text{ g O}} = 0.0576 \text{ mol O} \qquad \text{relative mol O} = \frac{0.0576 \text{ mol O}}{0.0576} = 1.00 \text{ mol O}$$

Multiplying each of these by the same smallest integer that gives whole numbers of relative moles of atoms for each yields:

relative mol C = 1.50 mol C x 2 = 3.00 mol C
relative mol H = 1.00 mol H x 2 = 4.00 mol H
relative mol O= 1.00 mol O x 2 = 2.00 mol O

The empirical formula is **$C_3H_4O_2$.**

3.92 ***See Section 3.3 and Example 3.11.***

$$? \text{ g Cl} = 5.166 \text{ g AgCl} \times \frac{35.453 \text{ g Cl}}{143.321 \text{ g AgCl}} = 1.278 \text{ g Cl}$$

$$? \text{ Ca} = \text{g sample} - \text{g Cl} = 2.000 \text{ g} - 1.278 \text{ g} = 0.722 \text{ g Ca}$$

$$? \text{ mol Ca} = 0.722 \text{ g Ca} \times \frac{1 \text{ mol Ca}}{40.078 \text{ g Ca}} = 0.0180 \text{ mol Ca} \qquad \text{relative mol Ca} = \frac{0.0180 \text{ mol Ca}}{0.0180} = 1.00 \text{ mol Ca}$$

$? \text{ mol Cl} = 1.278 \text{ g Cl} \times \frac{1 \text{ mol Cl}}{35.453 \text{ g Cl}} = 0.0360 \text{ mol Cl}$ relative mol Cl $= \frac{0.0360 \text{ mol Cl}}{0.0180} = 2.00 \text{ mol Cl}$

The empirical formula is **$CaCl_2$.**

Note: We should predict this based on our knowledge of ion charges from Section 2.6 and our knowledge that ionic compounds are represented by empirical formulas.

3.93 ***See Section 3.3 and Example 3.11.***

$? \text{ g C} = 2.60 \text{ g } CO_2 \times \frac{12.01 \text{ g C}}{44.01 \text{ g } CO_2} = 0.710 \text{ g C}$ $? \text{ g H} = 0.799 \text{ g } H_2O \times \frac{2.02 \text{ g H}}{18.02 \text{ g } H_2O} = 0.0896 \text{ g H}$

$? \text{ g N in 1.48 g sample} = 1.48 \text{ g sample} \times \frac{0.340 \text{ g N}}{2.43 \text{ g sample}} = 0.207 \text{ g N}$

$? \text{ O} = \text{g sample - g C - g H - g N} = 1.48 \text{ g} - 0.710 \text{ g C} - 0.0894 \text{ g H} - 0.207 \text{ g N} = 0.47 \text{ g O}$

$? \text{ mol C} = 0.710 \text{ g C} \times \frac{1 \text{ mol C}}{12.011 \text{ g C}} = 0.0591 \text{ mol C}$ relative mol C $= \frac{0.0591 \text{ mol C}}{0.029} = 2.0 \text{ mol C}$

$? \text{ mol H} = 0.0896 \text{ g H} \times \frac{1 \text{ mol H}}{1.008 \text{ g H}} = 0.0888 \text{ mol H}$ relative mol H $= \frac{0.0888 \text{ mol H}}{0.029} = 3.0 \text{ mol H}$

$? \text{ mol N} = 0.207 \text{ g N} \times \frac{1 \text{ mol N}}{14.01 \text{ g N}} = 0.0148 \text{ mol N}$ relative mol N $= \frac{0.0148 \text{ mol N}}{0.029} = 0.51 \text{ mol N}$

$? \text{ mol O} = 0.47 \text{ g O} \times \frac{1 \text{ mol O}}{15.999 \text{ g O}} = 0.029 \text{ mol O}$ relative mol O $= \frac{0.029 \text{ mol O}}{0.029} = 1.0 \text{ mol O}$

Multiplying each of these by the same smallest integer that gives whole numbers of relative moles of atoms for each yields:

relative mol C = 2.0 mol C x 2 = 4.0 mol C
relative mol H = 3.0 mol H x 2 = 6.0 mol H
relative mol N = 0.51 mol N x 2 = 1.0 mol N
relative mol O = 1.0 mol O x 2 = 2.0 mol O

The empirical formula is **$C_4H_6NO_2$.**

Note: An alternative method of solving this problem involves calculating % C from g C in 1.48 sample, % H from g H in 1.48 g sample, % N from g N in 2.43 g sample, % O from 100.00% - % C - % H - % N and working with a 100.00 g sample.

3.94 ***See Section 3.3 and Example 3.11.***

$? \text{ g C} = 3.94 \text{ g } CO_2 \times \frac{12.01 \text{ g C}}{44.01 \text{ g } CO_2} = 1.08 \text{ g C}$ $? \text{ g H} = 1.89 \text{ g } H_2O \times \frac{2.02 \text{ g H}}{18.02 \text{ g } H_2O} = 0.212 \text{ g H}$

$? \text{ g N in 2.18 g sample} = 2.18 \text{ g sample} \times \frac{0.235 \text{ g N}}{1.23 \text{ g sample}} = 0.417 \text{ g N}$

$? \text{ g} = \text{g sample - g C - g H - g N} = 2.18 \text{ g} - 1.08 \text{ g} - 0.212 \text{ g} - 0.417 \text{ g} = 0.47 \text{ g O}$

$? \text{ mol C} = 1.08 \text{ g C} \times \frac{1 \text{ mol C}}{12.011 \text{ g C}} = 0.0899 \text{ mol C}$ relative mol C $= \frac{0.0899 \text{ mol C}}{0.029} = 3.1 \text{ mol C}$

$? \text{ mol H} = 0.212 \text{ g H} \times \frac{1 \text{ mol H}}{1.008 \text{ g H}} = 0.210 \text{ mol H}$ relative mol H $= \frac{0.210 \text{ mol H}}{0.029} = 7.2 \text{ mol H}$

$? \text{ mol N} = 0.417 \text{ g N} \times \frac{1 \text{ mol N}}{14.01 \text{ g N}} = 0.0298 \text{ mol N}$ relative mol N $= \frac{0.0298 \text{ mol N}}{0.029} = 1.0 \text{ mol N}$

$? \text{ mol O} = 0.47 \text{ g O} \times \frac{1 \text{ mol O}}{15.999 \text{ g O}} = 0.029 \text{ mol O}$ relative mol O $= \frac{0.029 \text{ mol O}}{0.029} = 1.0 \text{ mol O}$

The empirical formula is **C_3H_7NO.**

Note: An alternative method of solving this problem involves calculating % C from g C in 2.18 g sample, % H from g H in 2.18 g sample, % N from g N in 1.23 g sample, % O from 100.00% - % C - % H - % N and working with a 100.00 sample.

3.95 *See Section 3.3 and Example 3.12.*

Formula mass for CH_2O:

1[C] x 12.0	=	12.0
2[H] x 1.0	=	2.0
1[O] x 16.0	=	16.0
		30.0

Hence, molar mass for CH_2O is 30.0 g/mol.

$$n = \frac{\text{molar mass compound}}{\text{molar mass } CH_2O} = \frac{90 \text{ g / mol}}{30.0 \text{ g / mol}} = 3$$

The molecular formula is $\mathbf{C_3H_6O_3}$.

3.96 *See Section 3.3 and Example 3.12.*

(a) Formula mass for HO:

1[H] x 1.0	=	1.0
1[O] x 16.0	=	16.0
		17.0

Hence, molar mass for HO is 17.0 g/mol.

$$n = \frac{\text{molar mass compound}}{\text{molar mass HO}} = \frac{34 \text{ g / mol}}{17.0 \text{ g / mol}} = 2$$

The molecular formula is $\mathbf{H_2O_2}$.

3.97 *See Section 3.3 and Example 3.12.*

(a) Formula mass for C_2H_4O:

2[C] x 12.0	=	24.0
4[H] x 1.0	=	4.0
1[O] x 16.0	=	16.0
		44.0

Hence, molar mass for C_2H_4O is 44.0 g/mol.

$$n = \frac{\text{molar mass compound}}{\text{molar mass } C_2H_4O} = \frac{132 \text{ g / mol}}{44.0 \text{ g / mol}} = 3$$

The molecular formula is $\mathbf{C_6H_{12}O_3}$.

(b) Formula mass for $C_3H_4NO_3$:

3[C] x 12.0	=	36.0
4[H] x 1.0	=	4.0
1[N] x 14.0	=	14.0
3[O] x 16.0	=	48.0
		102.0

Hence, molar mass for $C_3H_4NO_3$ is 102.0 g/mol.

$$n = \frac{\text{molar mass compound}}{\text{molar mass } C_3H_4NO_3} = \frac{408 \text{ g / mol}}{102.0 \text{ g / mol}} = 4$$

The molecular formula is $\mathbf{C_{12}H_{16}N_4O_{12}}$.

3.98 *See Section 3.3 and Example 3.12.*

(a) Formula mass for $C_5H_{10}O$:

5[C] x 12.0	=	60.0
10[H] x 1.0	=	10.0
1[O] x 16.0	=	16.0
		86.0

Hence, molar mass for $C_5H_{10}O$ is 86.0 g/mol.

$$n = \frac{\text{molar mass compound}}{\text{molar mass } C_5H_{10}O} = \frac{258 \text{ g / mol}}{86.0 \text{ g / mol}} = 3$$

The molecular formula is $\mathbf{C_{15}H_{30}O_3}$.

(b) Formula mass for PCl_3:

$$1[P] \times 31.0 = 31.0$$
$$3[Cl] \times 35.5 = 106.5$$
$$137.5$$

Hence, molar mass for PCl_3 is 137.5 g/mol.

$$n = \frac{\text{molar mass compound}}{\text{molar mass } PCl_3} = \frac{137.3 \text{ g/mol}}{137.5 \text{ g/mol}} = 1$$

The molecular formula is $\mathbf{PCl_3}$

3.99 ***See Section 3.3 and Example 3.12.***

Assume the sample has a mass of 100.0 g and therefore contains 62.0 g C, 10.4 g H and 27.5 g O.

$$? \text{ mol C} = 62.0 \text{ g C} \times \frac{1 \text{ mol C}}{12.011 \text{ g C}} = 5.16 \text{ mol C} \qquad \text{relative mol C} = \frac{5.16 \text{ mol C}}{1.72} = 3.00 \text{ mol C}$$

$$? \text{ mol H} = 10.4 \text{ g H} \times \frac{1 \text{ mol H}}{1.008 \text{ g H}} = 10.3 \text{ mol H} \qquad \text{relative mol H} = \frac{10.3 \text{ mol H}}{1.72} = 5.99 \text{ mol H}$$

$$? \text{ mol O} = 27.5 \text{ g O} \times \frac{1 \text{ mol O}}{15.999 \text{ g O}} = 1.72 \text{ mol O} \qquad \text{relative mol O} = \frac{1.72 \text{ mol O}}{1.72} = 1.00 \text{ mol O}$$

The empirical formula is C_3H_6O.

Formula mass for C_3H_6O:

$$3[C] \times 12.0 = 36.0$$
$$6[H] \times 1.0 = 6.0$$
$$1[O] \times 16.0 = 16.0$$
$$58.0$$

Hence, molar mass for C_3H_6O is 58.0 g/mol

$$n = \frac{\text{molar mass compound}}{\text{molar mass } C_3H_6O} = \frac{174 \text{ g/mol}}{58.0 \text{ g/mol}} = 3$$

The molecular formula is $\mathbf{C_9H_{18}O_3}$.

3.100 ***See Section 3.3 and Example 3.12.***

$$? \text{ g C} = 1.42 \text{ g } CO_2 \times \frac{12.01 \text{ g C}}{44.01 \text{ g } CO_2} = 0.388 \text{ g C} \qquad ? \text{ g H} = 0.338 \text{ g } H_2O \times \frac{2.02 \text{ g H}}{18.02 \text{ g } H_2O} = 0.0379 \text{ g H}$$

? g N = g sample - g C - g H = 0.500 g - 0.388 g - 0.0379 g = 0.074 g N

$$? \text{ mol C} = 0.388 \text{ g C} \times \frac{1 \text{ mol C}}{12.01 \text{ g C}} = 0.0323 \text{ mol C} \qquad \text{relative mol C} = \frac{0.0323 \text{ mol C}}{0.0053} = 6.1 \text{ mol C}$$

$$? \text{ mol H} = 0.0379 \text{ g H} \times \frac{1 \text{ mol H}}{1.008 \text{ g H}} = 0.0376 \text{ mol H} \qquad \text{relative mol H} = \frac{0.0376 \text{ mol H}}{0.0053} = 7.1 \text{ mol H}$$

$$? \text{ mol N} = 0.074 \text{ g N} \times \frac{1 \text{ mol N}}{14.01 \text{ g N}} = 0.0053 \text{ mol N} \qquad \text{relative mol N} = \frac{0.0053 \text{ mol N}}{0.0053} = 1.0 \text{ mol N}$$

The empirical formula is C_6H_7N:

$$6[C] \times 12.0 = 72.0$$
$$7[H] \times 1.0 = 7.0$$
$$1[N] \times 14.0 = 14.0$$
$$93.0$$

Hence, molar mass for C_6H_7N is 93.0 g/mol.

$$n = \frac{\text{molar mass compound}}{\text{molar mass } C_6H_7N} = \frac{186 \text{ g/mol}}{93.0 \text{ g/mol}} = 2$$

The molecular formula is $\mathbf{C_{12}H_{14}N_2}$.

3.101 ***See Section 3.3 and Example 3.12.***

Assume the sample has a mass of 100.0 g and therefore contains 40.0 g C, 6.71 g H and 53.3 g O.

$$? \text{ mol C} = 40.0 \text{ g C} \times \frac{1 \text{ mol C}}{12.011 \text{ g C}} = 3.33 \text{ mol C} \qquad \text{relative mol C} = \frac{3.33 \text{ mol C}}{3.33} = 1.00 \text{ mol C}$$

$$? \text{ mol H} = 6.71 \text{ g H} \times \frac{1 \text{ mol H}}{1.008 \text{ g H}} = 6.66 \text{ mol H} \qquad \text{relative mol H} = \frac{6.66 \text{ mol H}}{3.33} = 2.00 \text{ mol H}$$

$$? \text{ mol O} = 53.3 \text{ g O} \times \frac{1 \text{ mol O}}{15.999 \text{ g O}} = 3.33 \text{ mol O} \qquad \text{relative mol O} = \frac{3.33 \text{ mol O}}{3.33} = 1.00 \text{ mol O}$$

The empirical formula is CH_2O.

Formula mass for CH_2O:	1[C] x 12.0	=	12.0	
	2[H] x 1.0	=	2.0	
	1[O] x 16.0	=	16.0	
		=	30.0	Hence, molar mass for CH_2O is 30.0 g/mol.

$$n = \frac{\text{molar mass compound}}{\text{molar mass } CH_2O} = \frac{60 \text{ g/mol}}{30.0 \text{ g/mol}} = 2$$

The molecular formula for acetic acid is $\mathbf{C_2H_4O_2}$. This is usually written as CH_3COOH or CH_3CO_2H.

3.102 ***See Section 3.3 and Example 3.12.***

Assume the sample has a mass of 100.0 g and therefore contains 40.0 g C, 6.71 g H and 53.3 g O.

$$? \text{ mol C} = 40.0 \text{ g C} \times \frac{1 \text{ mol C}}{12.011 \text{ g C}} = 3.33 \text{ mol C} \qquad \text{relative mol C} = \frac{3.33 \text{ mol C}}{3.33} = 1.00 \text{ mol C}$$

$$? \text{ mol H} = 6.71 \text{ g H} \times \frac{1 \text{ mol H}}{1.008 \text{ g H}} = 6.66 \text{ mol H} \qquad \text{relative mol H} = \frac{6.66 \text{ mol H}}{3.33} = 2.00 \text{ mol H}$$

$$? \text{ mol O} = 53.3 \text{ g O} \times \frac{1 \text{ mol O}}{15.999 \text{ g O}} = 3.33 \text{ mol O} \qquad \text{relative mol O} = \frac{3.33 \text{ mol O}}{3.33} = 1.00 \text{ mol O}$$

The empirical formula is CH_2O.

Formula mass for CH_2O:	1[C] x 12.0	=	12.0	
	2[H] x 1.0	=	2.0	
	1[O] x 16.0	=	16.0	
		=	30.0	Hence, molar mass for CH_2O is 30.0 g/mol.

$$n = \frac{\text{molar mass compound}}{\text{molar mass } CH_2O} = \frac{180 \text{ g/mol}}{30.0 \text{ g/mol}} = 6$$

The molecular formula for fructose is $\mathbf{C_6H_{12}O_6}$.

3.103 ***See Sections 3.1, 3.4, and Examples 3.4, 3.13, 3.14.***

(a)

Unbalanced:	$C_3H_6 + O_2 \rightarrow CO_2 + H_2O$	Start with C_3H_6.
Step 1:	$C_3H_6 + O_2 \rightarrow \underline{3}CO_2 + \underline{3}H_2O$	Balances C, H.
Step 2:	$C_3H_6 + \underline{\frac{9}{2}}O_2 \rightarrow 3CO_2 + 3H_2O$	Balances O.
Step 3:	$\underline{2}C_3H_6 + \underline{9}O_2 \rightarrow \underline{6}CO_2 + \underline{6}H_2O$	Gives whole number coefficients.

(b) *Strategy:* $g\ C_3H_6 \rightarrow mol\ C_3H_6 \rightarrow mol\ CO_2 \rightarrow g\ CO_2$

$$? \text{ g } CO_2 = 2.45 \text{ g } C_3H_6 \times \frac{1 \text{ mol } C_3H_6}{42.0 \text{ g } C_3H_6} \times \frac{6 \text{ mol } CO_2}{2 \text{ mol } C_3H_6} \times \frac{44.0 \text{ g } CO_2}{1 \text{ mol } CO_2} = \mathbf{7.70 \text{ g } CO_2}$$

3.104 ***See Sections 3.2, 3.4, and Examples 3.4, 3.13, 3.14.***

(a)

Unbalanced:	$C_4H_8O + O_2 \rightarrow CO_2 + H_2O$	Start with C_4H_8O.
Step 1:	$C_4H_8O + O_2 \rightarrow \underline{4}CO_2 + \underline{4}H_2O$	Balances C, H.
Step 2:	$C_4H_8O + \underline{\frac{11}{2}}O_2 \rightarrow 4CO_2 + 4H_2O$	Balances O.

Step 3: $\underline{2}C_4H_8O + \underline{11}O_2 \rightarrow \underline{8}CO_2 + \underline{8}H_2O$ Gives whole number coefficients.

(b) *Strategy:* $g\ C_4H_8O \rightarrow mol\ C_4H_8O \rightarrow mol\ O_2 \rightarrow g\ O_2$

$$? \text{ g } O_2 = 5.33 \text{ g } C_4H_8O \times \frac{1 \text{ mol } C_4H_8O}{72.0 \text{ g } C_4H_8O} \times \frac{11 \text{ mol } O_2}{2 \text{ mol } C_4H_8O} \times \frac{32.0 \text{ g } O_2}{1 \text{ mol } O_2} = \mathbf{13.0 \text{ g } O_2}$$

3.105 ***See Sections 3.1 and 4 and Examples 3.1, 3.2, 3.13, 3.14.***

Unbalanced:	$P_4 + Cl_2 \rightarrow PCl_5$	Start with PCl_5.
Step 1:	$P_4 + Cl_2 \rightarrow \underline{4}PCl_5$	Balances P.
Step 2:	$P_4 + \underline{10}Cl_2 \rightarrow 4PCl_5$	Balances Cl.

Strategy: $g\ P_4 \rightarrow mol\ P_4 \rightarrow mol\ Cl_2 \rightarrow g\ Cl_2$

$$? \text{ g } Cl_2 = 0.567 \text{ g } P_4 \times \frac{1 \text{ mol } P_4}{123.9 \text{ g } P_4} \times \frac{10 \text{ mol } Cl_2}{1 \text{ mol } P_4} \times \frac{70.9 \text{ g } Cl_2}{1 \text{ mol } Cl_2} = \mathbf{3.24 \text{ g } Cl_2}$$

3.106 ***See Sections 3.1, 3.4, and Examples 3.1, 3.2, 3.13.***

Unbalanced:	$N_2 + H_2 \rightarrow NH_3$	Start with NH_3.
Step 1:	$N_2 + H_2 \rightarrow \underline{2}NH_3$	Balances N.
Step 2:	$N_2 + \underline{3}H_2 \rightarrow 2NH_3$	Balances H.

Strategy: $g\ N_2 \rightarrow mol\ N_2 \rightarrow mol\ NH_3 \rightarrow g\ NH_3$

$$? \text{ g } NH_3 = 5.33 \text{ g } N_2 \times \frac{1 \text{ mol } N_2}{28.0 \text{ g } N_2} \times \frac{2 \text{ mol } NH_3}{1 \text{ mol } N_2} \times \frac{17.0 \text{ g } NH_3}{1 \text{ mol } NH_3} = \mathbf{6.47 \text{ g } NH_3}$$

3.107 ***See Sections 3.1, 3.4, and Examples 3.1,3. 2, 3.13, 3.14.***

Unbalanced:	$Al + H_2SO_4 \rightarrow Al_2(SO_4)_3 + H_2$	Start with $Al_2(SO_4)_3$.
Step 1:	$Al + \underline{3}H_2SO_4 \rightarrow Al_2(SO_4)_3 + H_2$	Balances SO_4 units.
Step 2:	$\underline{2}Al + 3H_2SO_4 \rightarrow Al_2(SO_4)_3 + H_2$	Balances Al.
Step 3:	$2Al + 3H_2SO_4 \rightarrow Al_2(SO_4)_3 + \underline{3}H_2$	Balances H.

Strategy: $g\ H_2 \rightarrow mol\ H_2 \rightarrow mol\ Al \rightarrow g\ Al$

$$? \text{ g Al} = 13.2 \text{ g } H_2 \times \frac{1 \text{ mol } H_2}{2.02 \text{ g } H_2} \times \frac{2 \text{ mol Al}}{3 \text{ mol } H_2} \times \frac{27.0 \text{ g Al}}{1 \text{ mol Al}} = \mathbf{118 \text{ g Al}}$$

3.108 ***See Sections 3.1, 3.4, and Examples 3.1, 3.2, 3.13, 3.14.***

Balanced: $CaCO_3 \rightarrow CaO + CO_2$

Strategy: $g\ CaO \rightarrow mol\ CaO \rightarrow mol\ CaCO_3 \rightarrow g\ CaCO_3$

$$? \text{ g } CaCO_3 = 4.65 \text{ g CaO} \times \frac{1 \text{ mol CaO}}{56.1 \text{ g CaO}} \times \frac{1 \text{ mol } CaCO_3}{1 \text{ mol CaO}} \times \frac{100.1 \text{ g } CaCO_3}{1 \text{ mol } CaCO_3} = \mathbf{8.30 \text{ g } CaCO_3}$$

3.109 See Sections 3.1, 3.4, and Examples 3.1, 3.2, 3.13, 3.14.

Unbalanced:	$Li + O_2 \rightarrow Li_2O$	Start with Li_2O.
Step 1:	$Li + O_2 \rightarrow Li_2O$	Balances O.
Step 2:	$\underline{4}Li + O_2 \rightarrow Li_2O$	Balances Li.

Strategy: g Li → mol Li → mol Li_2O → g Li_2O

$$? \text{ g LiO} = 0.45 \text{ g Li} \times \frac{1 \text{ mol Li}}{6.9 \text{ g Li}} \times \frac{2 \text{ mol } Li_2O}{4 \text{ mol Li}} \times \frac{29.8\ Li_2O}{1 \text{ mol } Li_2O} = \mathbf{0.97\ g\ Li_2O}$$

3.110 See Sections 3.1, 3.4, and Examples 3.1, 3.2, 3.13, 3.14.

Balanced: $HCl + NaOH \rightarrow NaCl + H_2O$

Strategy: g H_2O → mol H_2O → mol NaCl → g NaCl

$$? \text{ g NaCl} = 78.2 \text{ g } H_2O \times \frac{1 \text{ mol } H_2O}{18.0 \text{ g } H_2O} \times \frac{1 \text{ mol NaCl}}{1 \text{ mol } H_2O} \times \frac{58.5 \text{ g NaCl}}{1 \text{ mol NaCl}} = \mathbf{254\ g\ NaCl}$$

3.111 See Sections 3.1, 3.4, and Examples 3.1, 3.2, 3.13, 3.14.

Unbalanced:	$Al + HCl \rightarrow AlCl_3 + H_2$	Start with $AlCl_3$.
Step 1:	$Al + \underline{3}HCl \rightarrow AlCl_3 + H_2$	Balances Cl.
Step 2:	$Al + 3HCl \rightarrow AlCl_3 + \underline{\tfrac{3}{2}}H_2$	Balances H.
Step 3:	$\underline{2}Al + \underline{6}HCl \rightarrow \underline{2}AlCl_3 + \underline{3}H_2$	Gives whole number coefficients.

Strategy: g Al → mol Al → mol $AlCl_3$ → g $AlCl_3$

$$? \text{ g } AlCl_3 = 3.3 \text{ g Al} \times \frac{1 \text{ mol Al}}{27.0 \text{ g Al}} \times \frac{2 \text{ mol } AlCl_3}{2 \text{ mol Al}} \times \frac{133.4 \text{ g } AlCl_3}{1 \text{ mol } AlCl_3} = 16.3 \text{ g } AlCl_3$$

$$\text{percent yield} = \frac{\text{actual yield}}{\text{theoretical yield}} \times 100\% = \frac{3.5 \text{ g } AlCl_3}{16.3 \text{ g } AlCl_3} \times 100\% = \mathbf{21\%}$$

3.112 See Sections 3.1, 3.4, and Examples 3.1, 3.2, 3.13, 3.14.

Unbalanced:	$CS_2 + Cl_2 \rightarrow CCl_4 + S_2Cl_2$	Start with Cl_2.
Step 1:	$CS_2 + \underline{3}Cl_2 \rightarrow CCl_4 + S_2Cl_2$	Balances Cl_2.

Strategy: g CS_2 → mol CS_2 → mol CCl_4 → g CCl_4

$$? \text{ g } CCl_4 = 43.1 \text{ g } CS_2 \times \frac{1 \text{ mol } CS_2}{76.2 \text{ g } CS_2} \times \frac{1 \text{ mol } CCl_4}{1 \text{ mol } CS_2} \times \frac{153.8 \text{ g } CCl_4}{1 \text{ mol } CCl_4} = 87.0 \text{ g } CCl_4$$

$$\text{percent yield } CCl_4 = \frac{\text{actual yield } CCl_4}{\text{theoretical yield } CCl_4} \times 100\% = \frac{45.2 \text{ g } CCl_4}{87.0 \text{ g } CCl_4} \times 100\% = \mathbf{52.0\%}$$

Strategy: g CS_2 → mol CS_2 → mol S_2Cl_2 → g S_2Cl_2

$$? \text{ g } S_2Cl_2 = 43.1 \text{ g } CS_2 \times \frac{1 \text{ mol } CS_2}{76.2 \text{ g } CS_2} \times \frac{1 \text{ mol } S_2Cl_2}{1 \text{ mol } CS_2} \times \frac{135.1 \text{ g } S_2Cl_2}{1 \text{ mol } S_2Cl_2} = 76.4 \text{ g } S_2Cl_2$$

$$\text{percent yield } S_2Cl_2 = \frac{\text{actual yield } S_2Cl_2}{\text{theoretical yield } S_2Cl_2} \times 100\% = \frac{41.3 \text{ g } S_2Cl_2}{76.4 \text{ g } S_2Cl_2} \times = \mathbf{54.1\%}$$

3.113 ***See Sections 3.1, 3.4, and Examples 3.1, 3.2, 3.13, 3.14.***

Balanced: $M + 2HCl \rightarrow MCl_2 + H_2$

Strategy: Let x equal molar mass of M in g M → mol M → mol H_2 → g H_2 set up.

$$? \text{ g } H_2 = 2.24 \text{ g M} \times \frac{1 \text{ mol M}}{x \text{ g M}} \times \frac{1 \text{ mol } H_2}{1 \text{ mol M}} \times \frac{2.02 \text{ g } H_2}{1 \text{ mol } H_2} = 0.0808 \text{ g } H_2$$

$$\text{So, } \frac{x \text{ g M}}{1 \text{ mol M}} = \frac{2.24 \text{ g M}}{0.0808 \text{ g } H_2} \times \frac{2.02 \text{ g } H_2}{1 \text{ mol } H_2} \times \frac{1 \text{ mol } H_2}{1 \text{ mol M}} = 56.0 \text{ g/mol}$$

Since Fe has a molar mass of 55.85 g/mol, the metal is **Fe.**

3.114 ***See Sections 3.1, 3.4, and Examples 3.1, 3.2, 3.13, 3.14.***

Balanced: $2 NaOH + X_2 \rightarrow NaX + NaXO + H_2O$

Strategy: Let y equal molar mass of X_2 in g X_2 → mol X_2 → mol NaX → g NaX set up.

$$? \text{ g NaX} = 3.11 \text{ g } X_2 \times \frac{1 \text{ mol } X_2}{y \text{ g } X_2} \times \frac{1 \text{ mol NaX}}{1 \text{ mol } X_2} \times \frac{(23.0 + 0.5 \text{ y}) \text{ g NaX}}{1 \text{ mole NaX}} = 2.00 \text{ g NaX}$$

Thus, (3.11)(23.0 + 0.5 y) = 2.00 y, 71.5 + 1.56 y = 2.00 y, and y = 162.
Since Br_2 has a molar mass of 159.8 g/mol, the halogen is **Br.**

3.115 ***See Sections 3.1, 3.5, and Examples 3.1,3.2, 3.16.***

Unbalanced:	$N_2 + H_2 \rightarrow NH_3$	Start with NH_3.
Step 1:	$N_2 + H_2 \rightarrow \underline{2}NH_3$	Balances N.
Step 2:	$N_2 + \underline{3}H_2 \rightarrow 2NH_3$	Balances H.

Strategy: g N_2 → mol N_2 → mol NH_3 → g NH_3

$$? \text{ g } NH_3 \text{ based on } N_2 = 14 \text{ g } N_2 \times \frac{1 \text{ mol } N_2}{28.0 \text{ g } N_2} \times \frac{2 \text{ mol } NH_3}{1 \text{ mol } N_2} \times \frac{17.0 \text{ g } NH_3}{1 \text{ mol } NH_3} = 17.0 \text{ g } NH_3$$

Strategy: g H_2 → mol H_2 → mol NH_3 → g NH_3

$$? \text{ g } NH_3 \text{ based on } H_2 = 1.0 \text{ g } H_2 \times \frac{1 \text{ mol } H_2}{2.0 \text{ g } H_2} \times \frac{2 \text{ mol } NH_3}{3 \text{ mol } H_2} \times \frac{17.0 \text{ g } NH_3}{1 \text{ mol } NH_3} = 5.7 \text{ g } NH_3$$

H_2 is the limiting reactant because it produces less NH_3. The maximum amount of NH_3 which can be produced from 14 g N_2 and 1.0 g H_2 is **5.7 g**.

3.116 ***See Section 3.5 and Example 3.16.***

Balanced: $Zn(s) + 2AgNO_3(aq) \rightarrow 2Ag(s) + Zn(NO_3)_2(aq)$

Strategy: $g\ Zn \rightarrow mol\ Zn \rightarrow mol\ Ag \rightarrow g\ Ag$

$$? \text{ g Ag based on Zn} = 3.22 \text{ g Zn} \times \frac{1 \text{ mol Zn}}{65.4 \text{ g Zn}} \times \frac{2 \text{ mol Ag}}{1 \text{ mol Zn}} \times \frac{107.9 \text{ g Ag}}{1 \text{ mol Ag}} = 10.6 \text{ g Ag}$$

Strategy: $g\ AgNO_3 \rightarrow mol\ AgNO_3 \rightarrow mol\ Ag \rightarrow g\ Ag$

$$? \text{ g Ag based on AgNO}_3 = 4.35 \text{ g AgNO}_3 \times \frac{1 \text{ mol AgNO}_3}{169.9 \text{ g AgNO}_3} \times \frac{2 \text{ mol Ag}}{2 \text{ mol AgNO}_3} \times \frac{107.9 \text{ g Ag}}{1 \text{ mol Ag}} = 2.76 \text{ g Ag}$$

$AgNO_3$ is the limiting reactant because it produces less Ag. The maximum amount of Ag which can be produced from 3.22 g Zn and 4.35 g $AgNO_3$ is **2.76 g.**

3.117 ***See Sections 3.1, 3.5, and Examples 3.1,3.2, 3.16.***

Unbalanced:	$CS_2 + O_2 \rightarrow CO_2 + SO_2$	Start with CS_2.
Step 1:	$CS_2 + O_2 \rightarrow CO_2 + \underline{2}SO_2$	Balances S.
Step 2:	$CS_2 + \underline{3}O_2 \rightarrow CO_2 + 2SO_2$	Balances O.

Strategy: $g\ CS_2 \rightarrow mol\ CS_2 \rightarrow mol\ CO_2 \rightarrow g\ CO_2$

$$? \text{ g CO}_2 \text{ based on CS}_2 = 3.12 \text{ g CS}_2 \times \frac{1 \text{ mol CS}_2}{76.15 \text{ g CS}_2} \times \frac{1 \text{ mol CO}_2}{1 \text{ mol CS}_2} \times \frac{44.01 \text{ g CO}_2}{1 \text{ mol CO}_2} = 1.80 \text{ g CO}_2$$

Strategy: $g\ O_2 \rightarrow mol\ O_2 \rightarrow mol\ CO_2 \rightarrow g\ CO_2$

$$? \text{ g CO}_2 \text{ based on O}_2 = 1.88 \text{ g O}_2 \times \frac{1 \text{ mol O}_2}{32.00 \text{ g O}_2} \times \frac{1 \text{ mol CO}_2}{3 \text{ mol O}_2} \times \frac{44.01 \text{ g CO}_2}{1 \text{ mol CO}_2} = 0.862 \text{ g CO}_2$$

O_2 is the limiting reactant because it produces less CO_2. The maximum amount of CO_2 that can be produced from 3.12 g CS_2 and 1.88 g O_2 is 0.862 g. Hence, the theoretical yield is **0.862 g CO_2.**

3.118 ***See Sections 3.1, 3.5, and Examples 3.1,3.2, 3.16.***

Unbalanced:	$P_4 + O_2 \rightarrow P_4O_{10}$	Start with P_4O_{10}.
Step 1:	$P_4 + \underline{5}O_2 \rightarrow P_4O_{10}$	Balances O.

Strategy: $g\ P_4 \rightarrow mol\ P_4 \rightarrow mol\ P_4O_{10} \rightarrow g\ P_4O_{10}$

$$? \text{ g P}_4\text{O}_{10} \text{ based on P}_4 = 2.2 \text{ g P}_4 \times \frac{1 \text{ mol P}_4}{124.0 \text{ g P}_4} \times \frac{1 \text{ mol P}_4\text{O}_{10}}{1 \text{ mol P}_4} \times \frac{284.0 \text{ g P}_4\text{O}_{10}}{1 \text{ mol P}_4\text{O}_{10}} = 5.0 \text{ g P}_4\text{O}_{10}$$

Strategy: $g\ O_2 \rightarrow mol\ O_2 \rightarrow mol\ P_4O_{10} \rightarrow g\ P_4O_{10}$

$$? \text{ g P}_4\text{O}_{10} \text{ based on O}_2 = 4.2 \text{ g O}_2 \times \frac{1 \text{ mol O}_2}{32.0 \text{ g O}_2} \times \frac{1 \text{ mol P}_4\text{O}_{10}}{5 \text{ mol O}_2} \times \frac{284.0 \text{ g P}_4\text{O}_{10}}{1 \text{ mol P}_4\text{O}_{10}} = 7.5 \text{ g P}_4\text{O}_{10}$$

P_4 is the limiting reactant because it produces less P_4O_{10}. The maximum amount of P_4O_{10} that can be produced from 2.2 g P_4 and 4.2 g O_2 is 5.0 g. Hence, the theoretical yield is **5.0 g P_4O_{10}.**

3.119 ***See Sections 3.1, 3.5, and Examples 3.1, 3.2, 3.17.***

Unbalanced:	$NO + O_2 \rightarrow NO_2$	Start with NO_2.
Step 1:	$NO + \frac{1}{2}O_2 \rightarrow NO_2$	Balances O.
Step 2:	$\underline{2}NO + O_2 \rightarrow \underline{2}NO_2$	Gives whole number coefficients.

Strategy: $g\ NO \rightarrow mol\ NO \rightarrow mol\ NO_2 \rightarrow g\ NO_2$

$$?\ g\ NO_2 \text{ based on } NO = 9.33\ g\ NO \times \frac{1\ mol\ NO}{30.0\ g\ NO} \times \frac{2\ mol\ NO_2}{2\ mol\ NO} \times \frac{46.0\ g\ NO_2}{1\ mol\ NO_2} = 14.3\ g\ NO_2$$

Strategy: $g\ O_2 \rightarrow mol\ O_2 \rightarrow mol\ NO_2 \rightarrow g\ NO_2$

$$?\ g\ NO_2 \text{ based on } O_2 = 9.66\ g\ O_2 \times \frac{1\ mol\ O_2}{32.0\ g\ O_2} \times \frac{2\ mol\ NO_2}{1\ mol\ O_2} \times \frac{46.0\ g\ NO_2}{1\ mol\ NO_2} = 27.8\ g\ NO_2$$

NO is the limiting reactant because it produces less NO_2. The maximum amount of NO_2 that can be produced from 9.33 g NO and 9.66 g O_2 is 14.3 g. Hence, the theoretical yield is **14.3 g NO_2.**

$$\text{percent yield} = \frac{\text{actual yield}}{\text{theoretical yield}} \times 100\% = \frac{10.1\ g\ NO_2}{14.3\ g\ NO_2} \times 100\% = \mathbf{70.6\%}.$$

3.120 ***See Sections 3.1, 3.5, and Examples 3.1, 3.2, 3.17.***

Unbalanced:	$P_4 + Cl_2 \rightarrow PCl_5$	Start with PCl_5
Step 1:	$P_4 + Cl_2 \rightarrow \underline{4}PCl_5$	Balances P.
Step 2:	$P_4 + \underline{10}Cl_2 \rightarrow 4PCl_5$	Balances Cl.

Strategy: $g\ P_4 \rightarrow mol\ P_4 \rightarrow mol\ PCl_5 \rightarrow g\ PCl_5$

$$?\ g\ PCl_5 \text{ based on } P_4 = 2.3\ g\ P_4 \times \frac{1\ mol\ P_4}{124.0\ g\ P_4} \times \frac{4\ mol\ PCl_5}{1\ mol\ PCl_5} \times \frac{208.2\ g\ PCl_5}{1\ mol\ PCl_5} = 15\ g\ PCl_5$$

Strategy: $g\ Cl_2 \rightarrow mol\ Cl_2 \rightarrow mol\ PCl_5 \rightarrow g\ PCl_5$

$$?\ g\ PCl_5 \text{ based on } Cl_2 = 7.0\ g\ Cl_2 \times \frac{1\ mol\ Cl_2}{70.9\ g\ Cl_2} \times \frac{4\ mol\ PCl_5}{10\ mol\ Cl_2} \times \frac{208.2\ g\ PCl_5}{1\ mol\ PCl_5} = 8.2\ g\ PCl_5$$

Cl_2 is the limiting reactant because it produces less PCl_5. The maximum amount of PCl_5 that can be produced from 2.3 g P_4 and 7.0 g Cl_2 is 8.2 g. Hence, the theoretical yield is **8.2 g PCl_5.**

$$\text{percent yield} = \frac{\text{actual yield}}{\text{theoretical yield}} \times 100\% = \frac{7.1\ g\ PCl_5}{8.2\ g\ PCl_5} \times 100\% = \mathbf{87\%}.$$

3.121 ***See Sections 3.1, 3.4, and Examples 3.4, 3.15.***

Use the balanced equation that was obtained in solving 3.103: $2C_3H_6 + 9O_2 \rightarrow 6CO_2 + 6H_2O$.

Strategy: $g\ C_3H_6 \rightarrow mol\ C_3H_6 \rightarrow mol\ H_2O \rightarrow g\ H_2O$

$$?\ g\ H_2O \text{ based on } C_3H_6 = 33.5\ g\ C_3H_6 \times \frac{1\ mol\ C_3H_6}{42.0\ g\ C_3H_6} \times \frac{6\ mol\ H_2O}{2\ mol\ C_3H_6} \times \frac{18.0\ g\ H_2O}{1\ mol\ H_2O} = 43.1\ g\ H_2O$$

Strategy: $g\ O_2 \rightarrow mol\ O_2 \rightarrow mol\ H_2O \rightarrow g\ H_2O$

$$? \text{ g } H_2O \text{ based on } O_2 = 127 \text{ g } O_2 \times \frac{1 \text{ mol } O_2}{32.0 \text{ g } O_2} \times \frac{6 \text{ mol } H_2O}{9 \text{ mol } O_2} \times \frac{18.0 \text{ g } H_2O}{1 \text{ mol } H_2O} = 47.6 \text{ g } H_2O$$

C_3H_6 is the limiting reactant because it produces less H_2O. The maximum amount of H_2O that can be produced from 33.5 g C_3H_6 and 127 g O_2 is 43.1 g. Hence, the theoretical yield is **43.1 g H_2O.**

$$\text{percent yield} = \frac{\text{actual yield}}{\text{theoretical yield}} \times 100\% = \frac{16.1 \text{ g } H_2O}{43.1 \text{ g } H_2O} \times 100\% = \mathbf{37.4\%}$$

3.122 ***See Sections 3.1, 3.5, and Examples 3.1, 3.2, 3.17.***

Unbalanced:	$H_2 + I_2 \rightarrow HI$	Start with HI.
Step 1:	$H_2 + I_2 \rightarrow \underline{2}HI$	Balances H, I.

Strategy: $g\ H_2 \rightarrow mol\ H_2 \rightarrow mol\ HI \rightarrow g\ HI$

$$? \text{ g HI based on } H_2 = 2.23 \text{ g } H_2 \times \frac{1 \text{ mol } H_2}{2.02 \text{ g } H_2} \times \frac{2 \text{ mol HI}}{1 \text{ mol } I_2} \times \frac{127.9 \text{ g HI}}{1 \text{ mol HI}} = 282 \text{ g HI}$$

Strategy: $g\ I_2 \rightarrow mol\ I_2 \rightarrow mol\ HI \rightarrow g\ HI$

$$\text{g HI based on } I_2 = 55.3 \text{ g } I_2 \times \frac{1 \text{ mol } I_2}{253.8 \text{ g } I_2} \times \frac{2 \text{ mol HI}}{1 \text{ mol } I_2} \times \frac{127.9 \text{ g HI}}{1 \text{ mol HI}} = 55.7 \text{ g HI}$$

I_2 is the limiting reactant because it produces less HI. The maximum amount of HI that can be produced from 2.33 g H_2 and 55.3 g I_2 is 55.7 g. Hence, the theoretical yield is **55.7 g HI.**

$$\text{percent yield} = \frac{\text{actual yield}}{\text{theoretical yield}} \times 100\% = \frac{43.1 \text{ g HI}}{55.7 \text{ g HI}} \times 100\% = \mathbf{77.4\%.}$$

3.123 ***See Sections 3.4, 3.5, and Example 3.17.***

(a) Balanced: $HNO_3 + NaOH \rightarrow H_2O + NaNO_3$

Strategy: $g\ HNO_3 \rightarrow mol\ HNO_3 \rightarrow mol\ NaNO_3 \rightarrow g\ NaNO_3$

$$? \text{ g } NaNO_3 \text{ based on } HNO_3 = 21.2 \text{ g } HNO_3 \times \frac{1 \text{ mol } HNO_3}{63.0 \text{ g } HNO_3} \times \frac{1 \text{ mol } NaNO_3}{1 \text{ mol } HNO_3} \times \frac{85.0 \text{ g } NaNO_3}{1 \text{ mol } NaNO_3} = 28.6 \ NaNO_3$$

Strategy: $g\ NaOH \rightarrow mol\ NaOH \rightarrow mol\ NaNO_3 \rightarrow g\ NaNO_3$

$$? \text{ g based on NaOH} = 23.1 \text{ g NaOH} \times \frac{1 \text{ mol NaOH}}{40.0 \text{ g NaOH}} \times \frac{1 \text{ mol } NaNO_3}{1 \text{ mol NaOH}} \times \frac{85.0 \text{ g } NaNO_3}{1 \text{ mol } NaNO_3} = 49.1 \text{ g } NaNO_3$$

HNO_3 is the limiting reactant because it produces less $NaNO_3$. The maximum amount of $NaNO_3$ which can be produced from 21.2 g HNO_3 and 23.1 NaOH is 28.6 g $NaNO_3$. This is the theorectical yield of the reaction.

$$\text{percent yield} = \frac{\text{actual yield}}{\text{theoretical yield}} \times 100\% = \frac{12.9 \text{ g } NaNO_3}{28.6 \text{ g } NaNO_3} \times 100\% = \mathbf{45.1\%}$$

(b) Since there is an excess of NaOH, some will remain at the end of the reaction.
Assuming complete reaction and conservation of mass,

initial g HNO_3 + initial g NaOH = g H_2O expected + g $NaNO_3$ expected + g excess NaOH

$$? \text{ g } H_2O \text{ based on } HNO_3 = 21.2 \text{ g } HNO_3 \times \frac{1 \text{ mol } HNO_3}{63.0 \text{ g } HNO_3} \times \frac{1 \text{ mol } H_2O}{1 \text{ mol } HNO_3} \times \frac{18.0 \text{ g } H_2O}{1 \text{ mol } H_2O} = 6.1 \text{ g } H_2O$$

g excess NaOH = initial g HNO_3 + initial g NaOH - g H_2O expected - g $NaNO_3$ expected
= 21.2 g + 23.1 g - 6.1 g - 28.6 g = **9.6 g NaOH**

Alternatively, the g NaOH required to produce 28.6 g $NaNO_3$ can be calculated using *g $NaNO_3$ → mol $NaNO_3$ → mol NaOH → g NaOH.* Then this amount can be subtracted from the initial amount of 23.1 g NaOH to determine the g excess NaOH.

$$? \text{ g NaOH for 28.6 } NaNO_3 = 28.6 \text{ g } NaNO_3 \times \frac{1 \text{ mol } NaNO_3}{85.0 \text{ g } NaNO_3} \times \frac{1 \text{ mol NaOH}}{1 \text{ mol } NaNO_3} \times \frac{40.0 \text{ g NaOH}}{1 \text{ mol NaOH}} = 13.5 \text{ g NaOH}$$

g excess NaOH = g NaOH available - g NaOH required = 23.1 g - 13.5 g = 9.6 g NaOH

3.124 ***See Sections 3.4, 3.5, and Example 3.17.***

(a) Balanced: $C_7H_6O_3 + C_4H_6O_3 \rightarrow C_9H_8O_4 + CH_3COOH$

Strategy: g $C_7H_6O_3$ → mol $C_7H_6O_3$ → mol $C_9H_8O_4$ → g $C_9H_8O_4$

$$? \text{ g } C_9H_8O_4 \text{ based on } C_7H_6O_3 = 4.6 \text{ g } C_7H_6O_3 \times \frac{1 \text{ mol } C_7H_6O_3}{138.0 \text{ g } C_7H_6O_3} \times \frac{1 \text{ mol } C_9H_8O_4}{1 \text{ mol } C_7H_6O_3} \times \frac{180.0 \text{ g } C_9H_8O_4}{1 \text{ mol } C_9H_8O_4}$$
$$= 6.0 \text{ g } C_9H_8O_4$$

Strategy: g $C_4H_6O_3$ → mol $C_4H_6O_3$ → mol $C_9H_8O_4$ → g $C_9H_8O_4$

$$? \text{ g } C_9H_8O_4 \text{ based on } C_4H_6O_3 = 4.6 \text{ g } C_4H_6O_3 \times \frac{1 \text{ mol } C_4H_6O_3}{102.0 \text{ g } C_4H_6O_3} \times \frac{1 \text{ mol } C_9H_8O_4}{1 \text{ mol } C_4H_6O_3} \times \frac{180.0 \text{ g } C_9H_8O_4}{1 \text{ mol } C_9H_8O_4}$$
$$= 8.1 \text{ g } C_9H_8O_4$$

$C_7H_6O_3$ is the limiting reactant because it produces less $C_9H_8O_4$. The maximum amount of $C_9H_8O_4$ that can be produced from 4.6 g $C_7H_6O_3$ and 4.6 g $C_4H_6O_3$ is 6.0 g. Hence, the theoretical yield is **6.0 g $C_9H_8O_4$.**

(b) Since there is an excess of $C_4H_6O_3$, some will remain at the end of the reaction. The g $C_4H_6O_3$ required to produce 6.0 g $C_9H_8O_4$ can be calculated using *g $C_9H_8O_4$ → mol $C_9H_8O_4$ → mol $C_4H_6O_3$ → g $C_4H_6O_3$.* Then this amount can be subtracted from the initial amount of 4.6 g $C_4H_6O_3$ to determine the g $C_4H_6O_3$ excess.

$$? \text{ g } C_4H_6O_3 \text{ required} = 6.0 \text{ g } C_9H_8O_4 \times \frac{1 \text{ mol } C_9H_8O_4}{180.0 \text{ g } C_9H_8O_4} \times \frac{1 \text{ mol } C_4H_6O_3}{1 \text{ mol } C_9H_8O_4} \times \frac{102.0 \text{ g } C_4H_6O_3}{1 \text{ mol } C_4H_6O_3} = 3.4 \text{ g } C_4H_6O_3$$

? g $C_4H_6O_3$ excess = g $C_4H_6O_3$ available - g $C_4H_6O_3$ required = 4.6 g - 3.4 g = 1.2 g $C_4H_6O_3$.

3.125 ***See Sections 3.1, 3.4, 3.5, and Examples 3.1, 3.2, 3.17.***

Unbalanced:	$NH_3 + O_2 \rightarrow NO + H_2O$	Start with NH_3 and H_2O.
Step 1:	$\underline{2}NH_3 + O_2 \rightarrow NO + \underline{3}H_2O$	Balances H.
Step 2:	$2NH_3 + O_2 \rightarrow \underline{2}NO + 3H_2O$	Balances N.

Step 3:	$2NH_3 + \frac{5}{2}O_2 \rightarrow 2NO + 3H_2O$	Balances O.
Step 4:	$\underline{4}NH_3 + \underline{5}O_2 \rightarrow \underline{4}NO + \underline{6}H_2O$	Gives whole number coefficients.

Strategy: ton NH_3 → lb NH_3 → g NH_3 → mol NH_3 → mol NO → g NO → lb NO → ton NO

$$? \text{ ton NO based on } NH_3 = 1.00 \text{ ton } NH_3 \times \frac{2{,}000 \text{ lb } NH_3}{1 \text{ ton } NH_3} \times \frac{453.6 \text{ g } NH_3}{1 \text{ lb } NH_3} \times \frac{1 \text{ mol } NH_3}{17.0 \text{ g } NH_3} \times \frac{4 \text{ mol NO}}{4 \text{ mol } NH_3}$$
$$\times \frac{30.0 \text{ g NO}}{1 \text{ mol NO}} \times \frac{1 \text{ lb NO}}{453.6 \text{ g NO}} \times \frac{1 \text{ ton NO}}{2{,}000 \text{ lb NO}} = 1.76 \text{ ton NO}$$

Strategy: ton O_2 → lb O_2 → g O_2 → mol O_2 → mol NO → g NO → lb NO → ton NO

$$? \text{ ton NO based on } O_2 = 1.00 \text{ ton } O_2 \times \frac{2{,}000 \text{ lb } O_2}{1 \text{ ton } O_2} \times \frac{453.6 \text{ g } O_2}{1 \text{ lb } O_2} \times \frac{1 \text{ mol } O_2}{32.0 \text{ g } O_2} \times \frac{4 \text{ mol NO}}{5 \text{ mol } O_2}$$
$$\times \frac{30.0 \text{ g NO}}{1 \text{ mol NO}} \times \frac{1 \text{ lb NO}}{453.6 \text{ g NO}} \times \frac{1 \text{ ton NO}}{2{,}000 \text{ lb NO}} = 0.750 \text{ ton NO}$$

O_2 is the limiting reactant because it produces less NO. The maximum amount of NO that can be produced from 1.00 ton N_2 and 1.00 ton O_2 is **0.750 ton NO.**

3.126 ***See Sections 3.1, 3.4, 3.5, Examples 3.1, 3.2, 3.17, and the Solution for Exercise 3.125.***

Unbalanced:	$NO + O_2 \rightarrow NO_2$	Start with NO_2.
Step 1:	$NO + \frac{1}{2}O_2 \rightarrow NO_2$	Balances O.
Step 2:	$\underline{2}NO + O_2 \rightarrow \underline{2}NO_2$	Gives whole number coefficients.

Strategy: ton NO → lb NO → g NO → mol NO → mol NO_2 → g NO_2 → lb NO_2 → ton NO_2

$$? \text{ ton } NO_2 = 0.75 \text{ ton NO} \times \frac{2{,}000 \text{ lb NO}}{1 \text{ ton NO}} \times \frac{453.6 \text{ g NO}}{1 \text{ lb NO}} \times \frac{1 \text{ mol NO}}{30.0 \text{ g NO}} \times \frac{2 \text{ mol } NO_2}{2 \text{ mol NO}}$$
$$\times \frac{46.0 \text{ g } NO_2}{1 \text{ mol } NO_2} \times \frac{1 \text{ lb } NO_2}{453.6 \text{ g } NO_2} \times \frac{1 \text{ ton } NO_2}{2{,}000 \text{ lb } NO_2} = \mathbf{1.2 \text{ ton } NO_2}.$$

3.127 ***See Sections 2.6, 3.3, and Example 3.9.***

Calcium and nitrogen would be expected to form Ca_3N_2.
Formula mass for Ca_3N_2:

3[Ca] x 40.1	=	120.3
2[N] x 28.0	=	28.0
		148.3

$$\% Ca = \frac{120.3 \text{ g Ca}}{148.3 \text{ g } Ca_3N_2} \times 100\% = \mathbf{81.1\% \ Ca}$$

$$\% N = \frac{28.0 \text{ g N}}{148.3 \text{ g } Ca_3N_2} \times 100\% = \mathbf{18.9\% \ N}$$

Molar mass for Ca_3N_2 is 148.3 g/mol.

3.128 ***See Sections 2.6, 2.8, 3.3, and Example 3.9.***

Iron(III) sulfate is $Fe_2(SO_4)_3$.
Formula mass for $Fe_2(SO_4)_3$:

2[Fe] x 55.8	=	111.6
3[S] x 32.1	=	96.3

$$\% Fe = \frac{111.6 \text{ g Fe}}{399.9 \text{ g } Fe_2(SO_4)_3} \times 100\% = \mathbf{27.9\% \ Fe}$$

$$\% S = \frac{96.3 \text{ g S}}{399.9 \text{ g } Fe_2(SO_4)_3} \times 100\% = \mathbf{24.1\% \ S}$$

$12[O] \times 16.0 = \underline{192.0}$ $\quad\quad$ $\%O = \frac{192.0 \text{ g O}}{399.9 \text{ g } Fe_2(SO_4)_3} \times 100\% = \mathbf{48.0\%\ O}$

399.9

Molar mass for $Fe_2(SO_4)_3$ is 399.9 g/mol.

3.129 ***See Sections 3.2 and 3.3.***

Strategy: tons Cu → tons $CuFeS_2$ → tons ore

$$? \text{ ton ore} = (0.990 \times 20.0 \text{ tons})Cu \times \frac{183.6 \text{ tons } CuFeS_2}{63.5 \text{ tons Cu}} \times \frac{100 \text{ tons ore}}{10 \text{ tons } CuFeS_2} = \mathbf{572\ tons\ ore}$$

3.130 ***See Sections 3.1, 3.2, 3.3, and Examples 3.1, 3.2, 3.13.***

(a)

Unbalanced:	$In_2S_3 + O_2 \rightarrow In_2O_3 + SO_2$	Start with In_2S_3.
Step 1:	$In_2S_3 + O_2 \rightarrow In_2O_3 + \underline{3}SO_2$	Balances S.
Step 2:	$In_2S_3 + \underline{\frac{9}{2}}O_2 \rightarrow In_2O_3 + 3SO_2$	Balances O.
Step 3:	$\underline{2}In_2S_3 + \underline{9}O_2 \rightarrow \underline{2}In_2O_3 + \underline{6}SO_2$	Gives whole number coefficients.
Unbalanced:	$In_2O_3 + CO \rightarrow In + CO_2$	Start with In_2O_3.
Step 1:	$In_2O_3 + CO \rightarrow \underline{2}In + CO_2$	Balances In.
Step 2:	$In_2O_3 + \underline{3}CO \rightarrow 2In + 3CO_2$	Balances O.

(b) *Strategy: kg In_2S_3 → g In_2S_3 → mol In_2S_3 → mol In_2S_3 → mol In → g In → kg In*

$$? \text{ kg In} = 35.7 \text{ kg } In_2S_3 \times \frac{10^3 \text{ g } In_2S_3}{1 \text{ kg } In_2S_3} \times \frac{1 \text{ mol } In_2S_3}{325.9 \text{ g } In_2S_3} \times \frac{2 \text{ mol } In_2O_3}{2 \text{ mol } In_2S_3} \times \frac{2 \text{ mol In}}{1 \text{ mol } In_2O_3}$$
$$\times \frac{114.8 \text{ g In}}{1 \text{ mol In}} \times \frac{1 \text{ kg In}}{10^3 \text{ g In}} = \mathbf{25.2\ kg\ In}$$

Alternatively, using the composition of In_2S_3 gives

$$? \text{ kg In} = 35.7 \text{ kg } In_2S_3 \times \frac{229.6 \text{ kg In}}{325.9 \text{ kg } In_2S_3} = \mathbf{25.2\ kg\ In}$$

3.131 ***See Sections 3.2 and 3.3.***

Assume the sample contains 100.0 g and therefore contains 25.5 g Cu, 12.8 g S, 57.7 g O and 4.0 g H.

Strategy: g Cu → mol Cu → mol $CuSO_4$

$$? \text{ mol } CuSO_4 = 25.5 \text{ g Cu} \times \frac{1 \text{ mol Cu}}{63.5 \text{ g Cu}} \times \frac{1 \text{ mol } CuSO_4}{1 \text{mol Cu}} = 0.402 \text{ mol } CuSO_4$$

Strategy: g H → mol H → mol H_2O

$$? \text{ mol } H_2O = 4.0 \text{ g H} \times \frac{1 \text{ mol H}}{1.008 \text{ g H}} \times \frac{1 \text{ mol } H_2O}{2 \text{ mol H}} = 2.0 \text{ mol } H_2O$$

$$\frac{\text{mol } H_2O}{\text{mol } CuSO_4} = \frac{2.0 \text{ mol}}{0.402 \text{ mol}} = 5.0$$

The value of x in $CuSO_4 \cdot x\ H_2O$ is **5.0.**

3.132 ***See Section 3.2 and Examples 3.6, 3.8.***

$$? \text{ atoms N} = 34.7 \text{ g } N_2O \times \frac{1 \text{ mol } N_2O}{44.0 \text{ g } N_2O} \times \frac{6.022 \times 10^{23} \text{ molecules } N_2O}{1 \text{ mol NO}} \times \frac{2 \text{ atoms N}}{1 \text{ molecule } N_2O} = \mathbf{9.50 \times 10^{23} \text{ atoms N}}$$

3.133 ***See Section 3.3 and Examples 3.12.***

Assume the sample contains 100.00 g and therefore contains 71.56 g C, 6.71 g H, 4.91 g N and 16.82 g O.

$$? \text{ mol C} = 71.56 \text{ g C} \times \frac{1 \text{ mol C}}{12.011 \text{ g C}} = 5.96 \text{ mol C} \qquad \text{relative mol C} = \frac{5.96 \text{ mol C}}{0.350} = 17.0 \text{ mol C}$$

$$? \text{ mol H} = 6.71 \text{ g H} \times \frac{1 \text{ mol H}}{1.008 \text{ g H}} = 6.66 \text{ mol H} \qquad \text{relative mol H} = \frac{6.66 \text{ mol H}}{0.350} = 19.0 \text{ mol H}$$

$$? \text{ mol N} = 4.91 \text{ g N} \times \frac{1 \text{ mol N}}{14.007 \text{ g N}} = 0.350 \text{ mol N} \qquad \text{relative mol N} = \frac{0.350 \text{ mol N}}{0.350} = 1.00 \text{ mol N}$$

$$? \text{ mol O} = 16.82 \text{ g O} \times \frac{1 \text{ mol O}}{15.999 \text{ g O}} = 1.05 \text{ mol O} \qquad \text{relative mol O} = \frac{1.05 \text{ mol O}}{0.350} = 3.00 \text{ mol O}$$

The empirical formula is $C_{17}H_{19}NO_3$.

Empirical formula mass for $C_{17}H_{19}NO_3$:

17[C] x 12.0	=	204.0
19[H] x 1.0	=	19.0
1[N] x 14.0	=	14.0
3[O] x 16.0	=	48.0
	=	285.0

The molar mass of the empirical formula for morphine is numerically equal to the molar mass of morphine. Hence, the molecular formula for morphine is $\mathbf{C_{17}H_{19}NO_3}$.

3.134 ***See Section 3.3 and Example 3.12.***

	Name	Empirical Formula	Molar Mass g/mol	Molecular Formula
a.	dimethyl sulfoxide	C_2H_6SO	78	$\underline{C_2H_6SO}$
b.	cyclopropane	$\underline{CH_2}$	$\underline{42}$	C_3H_6
c.	tryptamine	C_5H_6N	160	$\underline{C_{10}H_{12}N_2}$
d.	lactose	$\underline{C_{12}H_{22}O_{11}}$	$\underline{342}$	$C_{12}H_{22}O_{11}$

Comments:

a. The molar mass for dimethyl sulfide is numerically equal to the molar mass for the empirical formula. Hence, the molecular formula for dimethyl sulfoxide is the same as its empirical formula.
b. The molecular formula can be reduced to the empirical formula CH_2 by dividing each subscript in C_3H_6 by three.
c. The molar mass for tryptamine is two times the molar mass for the empirical formula. Hence, each subscript in the empirical formula must be multiplied by two to obtain the molecular formula.
d. The molecular formula cannot be reduced by dividing each subscript in $C_{12}H_{22}O_{11}$ by the same number.

3.135 ***See Sections 3.1, 3.4, and Examples 3.1, 3.2, 3.15..***

Unbalanced:	$NaWCl_6$	→	$Na_2WCl_6 + WCl_6$	Start with Na_2WCl_6.
Step 1:	$\underline{2}NaWCl_6$	→	$Na_2WCl_6 + WCl_6$	Balances Na, W, Cl.

Strategy: $g\ NaWCl_6 \rightarrow mol\ NaWCl_6 \rightarrow mol\ WCl_6 \rightarrow g\ WCl_6$

$$\text{percent yield} = \frac{\text{actual yield}}{\text{theorectical yield}} \times 100\% = \frac{1.52\text{ g WCl}_6}{2.67\text{ g WCl}_6} \times 100\% = \mathbf{56.9\%}$$

3.136 ***See Sections 3.4, 3.5, and Example 3.17.***

(a) Balanced: $K_2[PtCl_4] + C_2H_4 \rightarrow KPtCl_3(C_2H_4) + KCl$

Strategy: g $K_2[PtCl_4]$ → mol $K_2[PtCl_4]$ → mol $KPtCl_3(C_2H_4)$ → g $KPtCl_3(C_2H_4)$

$$? \text{ g K[PtCl}_3\text{(C}_2\text{H}_4\text{)]} = 45.8\text{ g K}_2\text{[PtCl}_4\text{]} \times \frac{1\text{ mol K}_2\text{[PtCl}_4\text{]}}{415.1\text{ g K}_2\text{[PtCl}_4\text{]}} \times \frac{1\text{ mol K[PtCl}_3\text{(C}_2\text{H}_4\text{)]}}{1\text{ mol K}_2\text{[PtCl}_4\text{]}}$$
$$\times \frac{368.6\text{ g K[PtCl}_3\text{(C}_2\text{H}_4\text{)]}}{1\text{ mol K[PtCl}_3\text{(C}_2\text{H}_4\text{)]}} = 40.7\text{ g K[PtCl}_3\text{(C}_2\text{H}_4\text{)]}$$

Stategy: g C_2H_4 → mol C_2H_4 → mol $K[PtCl_3(C_2H_4)]$ → g $K[PtCl_3(C_2H_4)]$

$$? \text{ g K[PtCl}_3\text{(C}_2\text{H}_4\text{)]} = 12.5\text{ g C}_2\text{H}_4 \times \frac{1\text{ mol C}_2\text{H}_4}{28.0\text{ g C}_2\text{H}_4} \times \frac{1\text{ mol K[PtCl}_3\text{(C}_2\text{H}_4\text{)]}}{1\text{ mol C}_2\text{H}_4}$$
$$\times \frac{368.6\text{ g K[PtCl}_3\text{(C}_2\text{H}_4\text{)]}}{1\text{ mol K[PtCl}_3\text{(C}_2\text{H}_4\text{)]}} = 164.6\text{ g K[PtCl}_3\text{(C}_2\text{H}_4\text{)]}$$

$K_2[PtCl_4]$ is the limiting reactant because it produces less $K[PtCl_3(C_2H_4)]$. The maximum amount of $K[PtCl_3(C_2H_4)]$ that can be produced from 45.8 g $K_2[PtCl_4]$ and 12.5 g C_2H_4 is **40.7 g $K[PtCl_3(C_2H_4)]$**.

(b) Since there is an excess of C_2H_4, some will remain at the end of the reaction. The g C_2H_4 required to produce 40.7 g $K[PtCl_3(C_2H_4)]$ can be calculated using *g $K[PtCl_3(C_2H_4)]$ → mol $K[PtCl_3(C_2H_4)]$ → mol C_2H_4 → g C_2H_4*. Then this amount can be subtracted from the initial amount of 12.5 g C_2H_4 to determine the g C_2H_4 excess.

$$? \text{ g C}_2\text{H}_4 \text{ required} = 40.7\text{ g K[PtCl}_3\text{(C}_2\text{H}_4\text{)]} \times \frac{1\text{ mol K[PtCl}_3\text{(C}_2\text{H}_4\text{)]}}{368.6\text{ g K[PtCl}_3\text{(C}_2\text{H}_4\text{)]}} \times \frac{1\text{ mol C}_2\text{H}_4}{1\text{ mol K[PtCl}_3\text{(C}_2\text{H}_4\text{)]}}$$
$$\times \frac{28.0\text{ g C}_2\text{H}_4}{1\text{ mol C}_2\text{H}_4} = 3.09\text{ g C}_2\text{H}_4$$

$? \text{ g } C_2H_4 \text{ excess} = \text{g } C_2H_4 \text{ available} - \text{g } C_2H_4 \text{ required} = 12.5\text{ g} - 3.09\text{ g} = \mathbf{9.4\ C_2H_4}$

3.137 ***See Section 3.4.***

$$\% \text{ Cu} = \frac{\text{g Cu}}{\text{g sample}} \times 100\% = \frac{0.306\text{ g Cu}}{1.20\text{ g sample}} \text{ x } 100\% = 0.255\%$$

$$x = \frac{\text{mol H}_2\text{O}}{\text{mol CuSO}_4}$$

$$? \text{ mol CuSO}_4 \text{ in sample} = 0.306\text{ g Cu} \times \frac{1\text{ mol Cu}}{63.55\text{ g Cu}} \times \frac{1\text{ mol CuSO}_4}{1\text{ mol Cu}} = 0.00482\text{ mol Cu}$$

$$? \text{ g CuSO}_4 \text{ in sample} = 0.00482\text{ mol CuSO}_4 \times \frac{159.62\text{ g CuSO}_4}{1\text{ mol CuSO}_4} = 0.769\text{ g CuSO}_4$$

$$? \text{ g H}_2\text{O in sample} = \text{g sample} - \text{g CuSO}_4 = 1.20\text{ g} - 0.769\text{ g} = 0.43\text{ g H}_2\text{O}$$

$$? \text{ mol H}_2\text{O in sample} = 0.43\text{ g H}_2\text{O} \times \frac{1\text{ mol H}_2\text{O}}{18.02\text{ g H}_2\text{O}} = 0.024\text{ mol H}_2\text{O}$$

$$\frac{\text{mol } H_2O}{\text{mol } CuSO_4} = \frac{0.024 \text{ mol}}{0.00482 \text{ mol}} = 5.0, \text{ so the value of x in } CuSO_4 \cdot xH_2O \text{ is 5.}$$

3.138 ***See Section 3.4.***

Balanced:	$N_2 + 3H_2$	→	$2NH_3$
Balanced:	$4NH_3 + 5O_2$	→	$4NO + 6H_2O$

Strategy: g N_2 → mol N_2 → mol NH_3 → mol NO → g NO

$$? \text{ g NO} = 100 \text{ g } N_2 \times \frac{1 \text{ mol } N_2}{28.0 \text{ g } N_2} \times \frac{2 \text{ mol } NH_3}{1 \text{ mol } N_2} \times \frac{4 \text{ mol NO}}{4 \text{ mol } NH_3} \times \frac{30.0 \text{ g NO}}{1 \text{ mol NO}} = \mathbf{214 \text{ g NO}}$$

3.139 ***See Sections 3.1, 3.5, and Examples 3.1, 3.2, 3.16.***

Unbalanced:	$SCl_2 + NaF$	→	$SF_4 + S_2Cl_2 + NaCl$	Start with NaF.
Step 1:	$SCl_2 + \underline{4}NaF$	→	$SF_4 + S_2Cl_2 + NaCl$	Balances F.
Step 2:	$SCl_2 + 4NaF$	→	$SF_4 + S_2Cl_2 + \underline{4}NaCl$	Balances Na..
Step 3:	$\underline{3}SCl_2 + 4NaF$	→	$SF_4 + S_2Cl_2 + 4NaCl$	Balances S, Cl.

Strategy: g SCl_2 → mol SCl_2 → mol SF_4 → g SF_4

$$? \text{ g } SF_4 \text{ based on } SCl_2 = 12.44 \text{ g } SCl_2 \times \frac{1 \text{ mol } SCl_2}{103.0 \text{ g } SCl_2} \times \frac{1 \text{ mol } SF_4}{3 \text{ mol } SCl_2} \times \frac{108.1 \text{ g } SF_4}{1 \text{ mol } SF_4} = 4.352 \text{ g } SF_4$$

Strategy: g NaF → mol NaF → mol SF_4 → g SF_4

$$? \text{ g } SF_4 \text{ based on NaF} = 10.11 \text{ g NaF} \times \frac{1 \text{ mol NaF}}{42.00 \text{ g NaF}} \times \frac{1 \text{ mol } SF_4}{4 \text{ mol NaF}} \times \frac{108.1 \text{ g } SF_4}{1 \text{ mol } SF_4} = 6.505 \text{ g } SF_4$$

SCl_2 is the limiting reactant because it produces less SF_4. The maximum amount of SF_4 that can be produced from 12.44 g SCl_2 and 10.11 g NaF is **4.352 g SF_4.**

3.140 ***See Sections 3.1, 3.3, and Example 3.12.***

Assume a sample of the liquid has a mass of 100.00 g and therefore contains 64.03 g C, 4.48 g H, and 31.49 g Cl.

$$? \text{ mol C} = 64.03 \text{ g C} \times \frac{1 \text{ mol C}}{12.011 \text{ g C}} = 5.331 \text{ mol C} \qquad \text{relative mol C} = \frac{5.331 \text{ mol C}}{0.888} = 6.00 \text{ mol C}$$

$$? \text{ mol H} = 4.48 \text{ g H} \times \frac{1 \text{ mol H}}{1.008 \text{ g H}} = 4.44 \text{ mol H} \qquad \text{relative mol H} = \frac{4.44 \text{ mol H}}{0.888} = 5.00 \text{ mol H}$$

$$? \text{ mol Cl} = 31.49 \text{ g Cl} \times \frac{1 \text{ mol Cl}}{35.453 \text{ g Cl}} = 0.888 \text{ mol Cl} \qquad \text{relative mol Cl} = \frac{0.888 \text{ mol Cl}}{0.888} = 1.00 \text{ mol Cl}$$

The empirical formula is C_6H_5Cl.
Formula mass for C_6H_5Cl:

6[C] x 12.0	=	72.0
5[H] x 1.0	=	5.0
1[Cl] x 35.5	=	35.5
		112.5

Hence, molar mass for C_6H_5Cl is 112.5 g/mol

$$n = \frac{\text{molar mass compound}}{\text{molar mass } C_6H_5Cl} = \frac{112.5 \text{ g/mol}}{112.5 \text{ g/mol}} = 1$$

The molecular formula is **C_6H_5Cl.**

Balanced:	$C_6H_6 + Cl_2$	→	$C_6H_5Cl + HCl$

3.141-3.143 ***Solutions for these Spreadsheet Exercises are on the Reger, Goode and Mercer Spreadsheet Solutions Diskette available from Saunders College Publishing.***

Chapter 4: Stoichiometry II

4.1 ***See Section 4.1.***

The sugar is the substance being dissolved and is the solute. The water is the substance doing the dissolving and is the solvent.

4.2 ***See Section 4.1.***

The ethanol is the liquid present in the lesser amount and is the solute. The gasoline is the liquid present in the greater amount and is the solvent.

4.3 ***See Section 4.1 and Table 4.1.***

Table 4.1 indicates Hg_2^{2+} forms a chloride precipitate and a sulfate precipitate, whereas Ca^{2+} does not. Hence, one can test for Hg_2^{2+} by adding NaCl or Na_2SO_4. If a precipitate forms, Hg_2^{2+} is present. If no precipitate forms, Hg_2^{2+} is absent, and Ca^{2+} is present. Table 4.1 also indicates Na_2CO_3 or Na_3PO_4 can be added to test for Ca^{2+} in the absence of Hg_2^{2+}. If Hg_2^{2+} is absent and a carbonate or phosphate precipitate is obtained, Ca^{2+} is present.

4.4 ***See Section 4.1 and Table 4.1.***

Table 4.1 indicates Ag^+ forms a chloride precipitate, whereas Ba^{2+} does not. Hence, one can test for Ag^+ by adding NaCl. If a precipitate forms, Ag^+ is present. If no precipitate forms, Ag^+ is absent, and Ba^{2+} is present. Table 4.1 also indicates Na_2CO_3 or Na_3PO_4 can be added to test for Ba^{2+} in the absence of Ag^+. If Ag^+ is absent and a carbonate or phosphate precipitate is obtained, Ba^{2+} is present.

4.5 ***See Section 4.1***

The strong acid nitric acid ionizes completely when dissolved in water. The weak acid propionic acid only ionizes slightly when dissolved in water. HNO_3 is a strong electrolyte, whereas CH_3CH_2COOH is a weak electrolyte.

4.6 ***See Section 4.2 and Figure 4.5.***

The appropriate amount of NaCl is weighed and placed into a volumetric flask that will hold the given volume of solute. Water is added to dissolve the NaCl solute. The flask is then filled with more water to the calibration mark.

4.7 ***See Sections 4.1, 4.2 and Figure 4.7.***

Li_2SO_4 is a soluble salt which dissociates completely into ions at low molar concentrations. Because there are two moles of lithium ions in every mole of Li_2SO_4, the molar concentration of the Li^+ ions in the solution is twice the molar concentration of the Li_2SO_4.

4.8 ***See Section 4.2 and Figure 4.8.***

The required amount of 12.1 *M* HCl is measured by pipet and placed into a volumetric flask of desired volume containing a reasonable amount of solvent. Then water is added to fill the volumetric flask to the calibration mark.

4.9 See Section 4.2 and Figure 4.8.

The volumetric flask of desired volume should be filled with a reasonable amount of water and cooled in an ice bath. The required amount of concentrated sulfuric acid should then be measured by pipet and added to the chilled water in the volumetric flask. Finally, water is added to fill the volumetric flask to the calibration mark. Note: The emphasis is on adding the concentrated acid to a reasonable amount of water. Cooling merely reduces the possibility of having too violent a reaction between the water and acid.

4.10 See Section 4.4 and Example 4.13.

$$\text{mass of } Na_2CO_3 \xrightarrow{\text{molar mass } Na_2CO_3} \text{mol } Na_2CO_3 \xrightarrow{(\text{mol } HNO_3/\text{mol } Na_2CO_3) \text{ ratio}} \text{mol } HNO_3 \xrightarrow{\text{L } HNO_3} \text{molarity } HNO_3$$

Step 1: Divide 1.00 g mass of Na_2CO_3 by molar mass of Na_2CO_3.
Step 2: Multiply mol Na_2CO_3 by ($\text{mol } HNO_3/\text{mol } Na_2CO_3$) ratio obtained from coefficients in balanced equation.
Step 3: Divide mol HNO_3 by liters of HNO_3 solution delivered from buret.

4.11 See Section 4.4 and Figure 4.9.

The titration of an acid with a base implies a given amount of acid solution is placed into a flask and the base solution is placed into a buret. An indicator is placed into a flask containing the acid solution and base is added from the buret until the indicator changes color. The point at which the indicator changes color is called the end point of the titration, since this is the point at which the titration ends. The indicator must be carefully selected to change color close to the equivalence point of the reaction, the point at which stoichiometrically equivalent amounts of acid and base have reacted. (Note that the concentration of either the acid or base solution must be known, so that the other can be determined.)

4.12 See Section 4.4 and Figure 4.10.

Assuming the Cl is present as Cl^-, it can be precipitated as AgCl by adding $AgNO_3$. The AgCl precipitate can be separated from solution by filtration, dried and weighed. The amount of Cl present in the sample can be obtained from the mass of AgCl precipitate using g Cl = g precipitate x (g Cl/g AgCl) and percent Cl can be calculated using % = (gCl/g sample) x 100%.

4.13 See Section 4.1, Table 4.1, and Example 4.1.

(a) BaI_2, (c) Na_2CO_3 and (d) $(NH_4)_2SO_4$ are classified as soluble salts and will dissolve in water. (b) $PbCl_2$ is classified as insoluble.

4.14 See Section 4.1, Table 4.1, and Example 4.1.

(b) $CaBr_2$, (c) KNO_3 and (d) $AgClO_4$ are classified as soluble salts and will dissolve in water. (a) Hg_2Cl_2 is classified as insoluble.

4.15 See Section 4.1, Table 4.1, and Example 4.1.

(a) $CaCl_2$, (b) $Ba(OH)_2$ and (c) $AgNO_3$ are classified as soluble compounds and will dissolve in water. (d) $CaCO_3$ is classified as insoluble.

4.16 See Section 4.1, Table 4.1, and Example 4.1.

(a) Na_3PO_4, (b) $(NH_4)_2CO_3$, (c) NH_4Cl and (d) $CaSO_4$ are all classified as soluble salts and will dissolve in water.

4.17 See Sections 3.1, 4.1, Table 4.1 and Examples 4.2, 4.3.

(a) The ions present from the reactants are $Na^+(aq)$, $OH^-(aq)$, $Mg^{2+}(aq)$ and $Cl^-(aq)$. The two possible products are NaCl and $Mg(OH)_2$, and Table 4.1 indicates $Mg(OH)_2$ is insoluble. The net ionic equation for the reaction is: $Mg^{2+}(aq) + 2OH^-(aq) \rightarrow Mg(OH)_2(s)$.

(b) The ions present from the reactants are $Na^+(aq)$, $NO_3^-(aq)$, $Mg^{2+}(aq)$ and $Br^-(aq)$. The two possible products are NaBr and $MgNO_3$, and Table 4.1 indicates both of these compounds are soluble. Hence, no precipitate is formed.

(c) Magnesium metal and hydrochoric acid react via the oxidation-reduction reaction $Mg(s) + 2HCl(aq) \rightarrow MgCl_2(aq) + H_2(g)$. The net ionic equation for the reaction is: $Mg(s) + 2H^+(aq) \rightarrow Mg^{2+}(aq) + H_2(g)$.

4.18 See Sections 3.1, 4.1, Table 4.1, and Examples 4.2, 4.3.

(a) The ions present from the reactants are $NH_4^+(aq)$, $CO_3^{2-}(aq)$, $Mg^{2+}(aq)$, and $Cl^-(aq)$. The two possible products are NH_4Cl and $MgCO_3$, and Table 4.1 indicates $MgCO_3$ is insoluble. The net ionic equation for the reaction is: $Mg^{2+}(aq) + CO_3^{2-}(aq) \rightarrow MgCO_3(s)$.

(b) HNO_3 and NaOH react via the acid-base reaction $HNO_3(aq) + NaOH(aq) \rightarrow H_2O(\ell) + NaNO_3(aq)$. The net ionic equation for the reaction is: $H^+(aq) + OH^-(aq) \rightarrow H_2O(\ell)$.

(c) The ions present from the reactants are $Be^{2+}(aq)$, $SO_4^{2-}(aq)$, $Na^+(aq)$, and $OH^-(aq)$. The two possible products are $Be(OH)_2$ and Na_2SO_4, and Table 4.1 indicates $Be(OH)_2$ is insoluble. The net ionic equation for the reaction is: $Be^{2+}(aq) + 2OH^-(aq) \rightarrow Be(OH)_2(s)$.

4.19 See Sections 3.1, 4.1, Table 4.1, and Examples 4.2, 4.3.

(a) HCl and $Ca(OH)_2$ react in the acid-base reaction $2HCl(aq) + Ca(OH)_2(aq) \rightarrow 2H_2O(\ell) + CaCl_2(aq)$. The net ionic equation for the reaction is: $H^+(aq) + OH^-(aq) \rightarrow H_2O(\ell)$.

(b) The ions from the reactants are $NH_4^+(aq)$, $Cl^-(aq)$, $Ag^+(aq)$, and $ClO_4^-(aq)$. The two possible products are NH_4ClO_4 and AgCl, and Table 4.1 indicates AgCl is insoluble. The net ionic equation for the reaction is: $Ag^+(aq) + Cl^-(aq) \rightarrow Ag(Cl(s)$.

(c) The ions present from the reactants are $Ba^{2+}(aq)$, $ClO_4^-(aq)$, $Na^+(aq)$, and $CO_3^{2-}(aq)$. The two possible products are $BaCO_3$ and $NaClO_4$, and Table 4.1 indicates $BaCO_3$ is insoluble. The net ionic equation for the reaction is: $Ba^{2+}(aq) + CO_3^{2-}(aq) \rightarrow BaCO_3(s)$.

4.20 See Sections 3.1, 4.1, Table 4.1, and Examples 4.2, 4.3.

(a) The ions present from the reactants are $K^+(aq)$, $Br^-(aq)$, $Ag^+(aq)$, and $NO_3^-(aq)$. The two possible products are KNO_3 and AgBr, and Table 4.1 indicates AgBr is insoluble. The net ionic equation for the reaction is: $Ag^+(aq) + Br^-(aq) \rightarrow AgBr(s)$.

(b) Ca and HNO_3 react via the oxidation-reduction reaction $Ca(s) + 2HNO_3(aq) \rightarrow Ca(NO_3)_2(aq) + H_2(g)$. The net ionic equation for the reaction is: $Ca(s) + 2H^+(aq) \rightarrow Ca^{2+}(aq) + H_2(g)$.

(c) The ions present from the reactants are $Li^+(aq)$, $OH^-(aq)$, $Fe^{3+}(aq)$, and $Cl^-(aq)$. The two possible products are LiCl and $Fe(OH)_3$, and Table 4.1 indicates $Fe(OH)_3$ is insoluble. The net ionic equation for the reaction is: $Fe^{3+}(aq) + 3OH^-(aq) \rightarrow Fe(OH)_3(s)$.

4.21 ***See Section 4.1, Table 4.1, and Examples 4.2, 4.3.***

The ions present from the reactants are $Ag^+(aq)$, NO_3^- (aq), $Ca^{2+}(aq)$, and $Cl^-(aq)$. The two possible products are AgCl and $Ca(NO_3)_2$, and Table 4.1 indicates AgCl is insoluble.

Overall: $2AgNO_3(aq) + CaCl_2(aq) \rightarrow 2AgCl(s) + Ca(NO_3)_2(aq)$

Complete ionic: $2Ag^+(aq) + 2NO_3^-(aq) + Ca^{2+}(aq) + 2Cl^-(aq) \rightarrow 2AgCl(s) + Ca^{2+}(aq) + 2NO_3^-(aq)$

Net ionic: $Ag^+(aq) + Cl^-(aq) \rightarrow AgCl(s)$

4.22 ***See Section 4.1, Table 4.1, and Examples 4.2, 4.3.***

The ions present from the reactants are $Co^{2+}(aq)$, $Br^-(aq)$, $Na^+(aq)$, and $OH^-(aq)$. The two possible products are $Co(OH)_2$ and NaBr, and Table 4.1 indicates $Co(OH)_2$ is insoluble.

Overall: $CoBr_2(aq) + 2NaOH(aq) \rightarrow Co(OH)_2(s) + 2\ NaBr(aq)$

Complete ionic: $Co^{2+}(aq) + 2Br^-(aq) + 2Na^+(aq) + 2OH^-(aq) \rightarrow Co(OH)_2(s) + 2Na^+(aq) + 2Br^-(aq)$

Net ionic: $Co^{2+}(aq) + 2OH^-(aq) \rightarrow Co(OH)_2(s)$

4.23 ***See Section 4.1, Table 4.1, and Examples 4.2, 4.3.***

The ions present from the reactants are NH_4^+ (aq), PO_4^{3-} (aq), $Ag^+(aq)$, and NO_3^- (aq). The two possible products are NH_4NO_3 and Ag_3PO_4, and Table 4.1 indicates Ag_3PO_4 is insoluble.

Overall: $(NH_4)_3PO_4(aq) + 3AgNO_3(aq) \rightarrow Ag_3PO_4(s) + 3NH_4NO_3(aq)$

Complete ionic: $3NH_4^+(aq) + PO_4^{3-}(aq) + 3Ag^+(aq) + 3NO_3^-(aq) \rightarrow Ag_3PO_4(s) + 3NH_4^+(aq) + 3NO_3^-(aq)$

Net ionic: $3Ag^+(aq) + PO_4^{3-}(aq) \rightarrow Ag_3PO_4(s)$

4.24 ***See Section 4.1, Table 4.1, Examples 4.2, 4.3.***

The ions present from the reactants are $Pb^{2+}(aq)$, $CH_3CO_2^-$ (aq), $Ba^{2+}(aq)$, and $Br^-(aq)$. The two possible products are $PbBr_2$ and $Ba(CH_3CO_2)_2$, and Table 4.1 indicates $PbBr_2$ is insoluble.

Overall: $Pb(CH_3CO_2)_2(aq) + BaBr_2(aq) \rightarrow PbBr_2(s) + Ba(CH_3CO_2)_2(aq)$

Complete ionic: $Pb^{2+}(aq) + 2CH_3CO_2^-(aq) + Ba^{2+}(aq) + 2Br^-(aq) \rightarrow PbBr_2(s) + Ba^{2+}(aq) + 2CH_3CO_2^-(aq)$

Net ionic: $Pb^{2+}(aq) + 2Br^-(aq) \rightarrow PbBr_2(s)$

4.25 ***See Section 4.1 and Table 4.1.***

Table 4.1 indicates only the chlorides of Ag^+, Pb^{2+}, and Hg_2^{2+} are insoluble. Hence, the solution must contain Pb^{2+}.

4.26 ***See Section 4.1 and Table 4.1.***

Table 4.1 indicates the hydroxides of both Ag^+ and Mg^{2+} are insoluble. However, Table 4.1 also indicates the bromide of Ag^+ is insoluble, whereas the bromide of Mg^{2+} is soluble. Hence, the solution must contain Mg^{2+} because it does not form an insoluble bromide.

4.27 *See Section 4.1 and Table 4.1.*

Table 4.1 indicates all carbonates, except those of the Group IA metals and NH_4^+, are insoluble. Hence, both Ca^{2+} and Ba^{2+} would be expected to form carbonate precipitates. However, Table 4.1 also indicates only the sulfates of Ba^{2+}, Pb^{2+}, Hg^{2+} and Hg_2^{2+} are insoluble. Since no sulfate precipitate forms, the solution cannot contain Ba^{2+}. It must therefore contain Ca^{2+} to give the carbonate precipitate.

4.28 *See Section 4.1 and Table 4.1.*

Table 4.1 indicates Pb^{2+} forms a sulfate precipitate, whereas Ca^{2+} does not. Hence, the solution must containPb^{2+}.

4.29 *See Section 4.2 and Example 4.4.*

$$\text{molarity} = \frac{\text{moles of solute}}{\text{liters of solution}}$$

$$? \text{ mol KOH} = 8.23 \text{ g KOH} \times \frac{1 \text{ mol KOH}}{56.1 \text{ g KOH}} = 0.147 \text{ mol KOH} \qquad ? \text{ L soln} = 250 \text{ mL} \times \frac{1 \text{ L}}{10^3 \text{ mL}} = 0.250 \text{ L soln}$$

$$? \text{ molarity} = \frac{0.147 \text{ mol KOH}}{.250 \text{ L soln}} = \mathbf{0.588\ \mathit{M}\ KOH}$$

See the alternative method used in working 4.31.

4.30 *See Section 4.2 and Example 4.4.*

$$\text{molarity} = \frac{\text{moles of solute}}{\text{liters of solution}}$$

$$? \text{ mol NaCl} = 23.1 \text{ g NaCl} \times \frac{1 \text{ mol NaCl}}{58.5 \text{ g NaCl}} = 0.395 \text{ mol NaCl} \qquad ? \text{ L soln} = 500 \text{ mL} \times \frac{1 \text{ L}}{10^3 \text{ mL}} = 0.500 \text{ L soln}$$

$$? \text{ molarity} = \frac{0.395 \text{ mol NaCl}}{0.500 \text{ L soln}} = \mathbf{0.790\ \mathit{M}\ NaCl}$$

Note: See the alternative method used in working 4.32.

4.31 *See Section 4.2 and Example 4.4.*

$$? M \text{ AgNO}_3 = \frac{1.44 \text{ g AgNO}_3}{1.00 \text{ L soln}} \times \frac{1 \text{ mol AgNO}_3}{169.9 \text{ g AgNO}_3} = \mathbf{8.48 \times 10^{-3}\ \mathit{M}\ AgNO_3}$$

Note: See the alternative method used in working 4.29.

4.32 *See Section 4.2 and Example 4.4.*

$$? M \text{ NaOH} = \frac{1.11 \text{ g NaOH}}{250 \text{ mL soln}} \times \frac{1 \text{ mol NaOH}}{40.0 \text{ g NaOH}} \times \frac{10^3 \text{ mL soln}}{1 \text{ L soln}} = \mathbf{0.111\ \mathit{M}\ NaOH}$$

Note: See the alternative method used in working 4.30.

4.33 *See Section 421 and Example 4.5.*

(a) $$? \text{ mol HNO}_3 = 0.033 \text{ L HNO}_3 \text{ soln} \times \frac{3.11 \text{ mol HNO}_3}{1 \text{ L HNO}_3 \text{ soln}} = \mathbf{0.10\ mol\ HNO_3}$$

(b) $$? \text{ mol HNO}_3 = 1.0 \text{ L HNO}_3 \text{ soln} \times \frac{3.2 \text{ mol HNO}_3}{1 \text{ L HNO}_3 \text{ soln}} = \mathbf{3.2\ mol\ HNO_3}$$

4.34 *See Section 4.2 and Example 4.5.*

(a) $? \text{ mol NaCl} = 0.22 \text{ L NaCl soln} \times \dfrac{1.2 \text{ mol NaCl}}{1 \text{ L NaCl soln}} = \mathbf{0.26 \text{ mol NaCl}}$

(b) $? \text{ mol AgNO}_3 = 0.500 \text{ L AgNO}_3 \text{ soln} \times \dfrac{0.22 \text{ mol AgNO}_3}{1 \text{ L AgNO}_3 \text{ soln}} = \mathbf{0.11 \text{ mol AgNO}_3}$

4.35 *See Section 4.2 and Example 4.5.*

(a) $? \text{ mol AgNO}_3 = 1.33 \text{ L AgNO}_3 \text{ soln} \times \dfrac{0.211 \text{ mol AgNO}_3}{1 \text{ L AgNO}_3 \text{ soln}} = \mathbf{0.281 \text{ mol AgNO}_3}$

(b) $? \text{ mol CaCl}_2 = 1.000 \text{ L CaCl}_2 \text{ soln} \times \dfrac{0.00113 \text{ mol CaCl}_2}{1 \text{ L CaCl}_2 \text{ soln}} = \mathbf{0.00113 \text{ mol CaCl}_2}$

4.36 *See Section 4.2 and Example 4.5.*

(a) $? \text{ mol NaBr} = 0.238 \text{ L NaBr soln} \times \dfrac{0.211 \text{ mol NaBr}}{1 \text{ L NaBr soln}} = \mathbf{0.0502 \text{ mol NaBr}}$

(b) $? \text{ mol NH}_4\text{Cl} = 1.2 \text{ L NH}_4\text{Cl soln} \times \dfrac{0.077 \text{ mol NH}_4\text{Cl}}{1 \text{ L NH}_4\text{Cl soln}} = \mathbf{0.092 \text{ mol NH}_4\text{Cl}}$

4.37 *See Section 4.2 and Examples 4.5, 4.6.*

(a) $? \text{ g HCl} = 3.13 \text{ L HCl soln} \times \dfrac{2.21 \text{ mol HCl}}{1 \text{ L HCl soln}} \times \dfrac{36.5 \text{ g HCl}}{1 \text{ mol HCl}} = \mathbf{252 \text{ g HCl}}$

(b) $? \text{ g KCl} = 1.5 \text{ L KCl soln} \times \dfrac{1.2 \text{ mol KCl}}{1 \text{ L KCl soln}} \times \dfrac{74.6 \text{ g KCl}}{1 \text{ mol KCl}} = \mathbf{1.3 \times 10^2 \text{ g KCl}}$

4.38 *See Section 4.2 and Examples 4.5, 4.6.*

(a) $? \text{ g KBr} = 0.113 \text{ L KBr soln} \times \dfrac{1.0 \text{ mol KBr}}{1 \text{ L KBr soln}} \times \dfrac{119.0 \text{ g KBr}}{1 \text{ mol KBr}} = \mathbf{13.4 \text{ g KBr}}$

(b) $? \text{ g KNO}_3 = 0.120 \text{ L KNO}_3 \text{ soln} \times \dfrac{2.11 \text{ mol KNO}_3}{1 \text{ L KNO}_3 \text{ soln}} \times \dfrac{101.1 \text{ g KNO}_3}{1 \text{ mol KNO}_3} = \mathbf{25.6 \text{ g KNO}_3}$

4.39 *See Section 4.2 and Example 4.6.*

$? \text{ g AgNO}_3 = 0.300 \text{ L AgNO}_3 \text{ soln} \times \dfrac{1.00 \text{ mol AgNO}_3}{1 \text{ L AgNO}_3 \text{ soln}} \times \dfrac{169.9 \text{ g AgNO}_3}{1 \text{ mol AgNO}_3} = \mathbf{51.0 \text{ g AgNO}_3}$

4.40 *See Section 4.2 and Example 4.6.*

$? \text{ g Na}_2\text{SO}_4 = 0.500 \text{ L Na}_2\text{SO}_4 \text{ soln} \times \dfrac{3.50 \text{ mol Na}_2\text{SO}_4}{1 \text{ L Na}_2\text{SO}_4 \text{ soln}} \times \dfrac{142.1 \text{ g Na}_2\text{SO}_4}{1 \text{ mol Na}_2\text{SO}_4} = \mathbf{249 \text{ g Na}_2\text{SO}_4}$

4.41 *See Section 4.2 and Example 4.6.*

$$? \text{ g BaCl}_2 = 1.00 \text{ L BaCl}_2 \text{ soln} \times \frac{0.100 \text{ mol BaCl}_2}{1 \text{ L BaCl}_2 \text{ soln}} \times \frac{208.2 \text{ g BaCl}_2}{1 \text{ mol BaCl}_2} = \mathbf{20.8 \text{ g BaCl}_2}$$

4.42 *See Section 4.2 and Example 4.6.*

$$? \text{ g Na}_2\text{SO}_4 = 0.400 \text{ L Na}_2\text{SO}_4 \text{ soln} \times \frac{2.50 \text{ mol Na}_2\text{SO}_4}{1 \text{ L Na}_2\text{SO}_4 \text{ soln}} \times \frac{142.1 \text{ g Na}_2\text{SO}_4}{1 \text{ mol Na}_2\text{SO}_4} = \mathbf{142 \text{ g Na}_2\text{SO}_4}$$

4.43 *See Section 4.2 and Examples 4.4, 4.7.*

$$? \, M \text{ SrCl}_2 = \frac{4.11 \text{ g SrCl}_2}{1.00 \text{ L soln}} \times \frac{1 \text{ mol SrCl}_2}{158.5 \text{ g SrCl}_2} = \mathbf{0.0259 \, \mathit{M} \text{ SrCl}_2}$$

$$? \, M \text{ Sr}^{2+} = \frac{2.59 \times 10^{-2} \text{ mol SrCl}_2}{1 \text{ L soln}} \times \frac{1 \text{ mol Sr}^{2+}}{1 \text{ mol SrCl}_2} = \mathbf{0.0259 \, \mathit{M} \text{ Sr}^{2+}}$$

$$? \, M \text{ Cl}^- = \frac{2.59 \times 10^{-2} \text{ mol SrCl}_2}{1 \text{ L soln}} \times \frac{2 \text{ mol Cl}^-}{1 \text{ mol SrCl}_2} = \mathbf{0.0518 \, \mathit{M} \text{ Cl}^-}$$

4.44 *See Section 4.2 and Examples 4.4, 4.7.*

$$? \, M \text{ Na}_2\text{SO}_4 = \frac{9.21 \text{ g Na}_2\text{SO}_4}{2.00 \text{ L soln}} \times \frac{1 \text{ mol Na}_2\text{SO}_4}{142.1 \text{ g Na}_2\text{SO}_4} = \mathbf{0.0324 \, \mathit{M} \text{ Na}_2\text{SO}_4}$$

$$? \, M \text{ Na}^+ = \frac{3.24 \times 10^{-2} \text{ g Na}_2\text{SO}_4}{1 \text{ L soln}} \times \frac{2 \text{ mol Na}_2}{1 \text{ mol Na}_2\text{SO}_4} = \mathbf{0.0648 \, \mathit{M} \text{ Na}^+}$$

$$? \, M \text{ SO}_4^{2-} = \frac{3.24 \times 10^{-2} \text{ mol Na}_2\text{SO}_4}{1 \text{ L soln}} \times \frac{1 \text{ mol SO}_4^{2-}}{1 \text{ mol Na}_2\text{SO}_4} = \mathbf{0.0324 \, \mathit{M} \text{ SO}_4^{2-}}$$

4.45 *See Section 4.2 and Examples 4.4, 4.7.*

$$? \, M \text{ Mg(NO}_3)_2 = \frac{21.5 \text{ g Mg(NO}_3)_2}{5.00 \text{ L soln}} \times \frac{1 \text{ mol Mg(NO}_3)_2}{148.3 \text{ g Mg(NO}_3)_2} = \mathbf{0.0290 \, \mathit{M} \text{ Mg(NO}_3)_2}$$

$$? \, M \text{ Mg}^{2+} = \frac{2.90 \times 10^{-2} \text{ mol Mg(NO}_3)_2}{1 \text{ L soln}} \times \frac{1 \text{ mol Mg}^{2+}}{1 \text{ mol Mg(NO}_3)_2} = \mathbf{0.0290 \, \mathit{M} \text{ Mg}^{2+}}$$

$$? \, M \text{ NO}_3^- = \frac{2.90 \times 10^{-2} \text{ mol Mg(NO}_3)_2}{1 \text{ L soln}} \times \frac{2 \text{ mol NO}_3^-}{1 \text{ mol Mg(NO}_3)_2} = \mathbf{0.0580 \, \mathit{M} \text{ NO}_3^-}$$

4.46 *See Section 4.2 and Examples 4.4, 4.7.*

$$? \, M \text{ Na}_2\text{CO}_3 = \frac{5.15 \text{ g Na}_2\text{CO}_3}{0.500 \text{ L soln}} \times \frac{1 \text{ mol Na}_2\text{CO}_3}{106.0 \text{ g Na}_2\text{CO}_3} = \mathbf{0.0972 \, \mathit{M} \text{ Na}_2\text{CO}_3}$$

$$? \, M \text{ Na}^+ = \frac{9.72 \times 10^{-2} \text{ mol Na}_2\text{CO}_3}{1 \text{ L soln}} \times \frac{2 \text{ mol Na}^+}{1 \text{ mol Na}_2\text{CO}_3} = \mathbf{0.194 \, \mathit{M} \text{ Na}^+}$$

$$? \, M \text{ CO}_3^{2-} = \frac{9.72 \times 10^{-2} \text{ mol Na}_2\text{CO}_3}{1 \text{ L soln}} \times \frac{1 \text{ mol CO}_3^{2-}}{1 \text{ mol Na}_2\text{CO}_3} = \mathbf{0.0972 \, \mathit{M} \text{ CO}_3^{2-}}$$

4.47 *See Section 4.2 and Example 4.6.*

$$? \text{ g KSCN} = 1.00 \text{ L KSCN soln} \times \frac{0.20 \text{ mol KSCN}}{1 \text{ L KSCN soln}} \times \frac{97.2 \text{ g KSCN}}{1 \text{ mol KSCN}} = \mathbf{19\ g\ KSCN}$$

4.48 *See Section 4.2 and Example 4.6.*

$$? \text{ g } KMnO_4 = 0.500 \text{ L } KMnO_4 \text{ soln} \times \frac{0.100 \text{ mol } KMnO_4}{1 \text{ L soln}} \times \frac{158.0 \text{ g } KMnO_4}{1 \text{ mol } KMnO_4} = \mathbf{7.90\ g\ KMnO_4}$$

4.49 *See Section 4.2 and Example 4.8.*

Solving $M(\text{con}) \times V(\text{con}) = M(\text{dil}) \times V(\text{dil})$ for $V(\text{con})$ gives

$$V(\text{con}) = \frac{M(\text{dil}) \times V(\text{dil})}{M(\text{con})} \qquad V(\text{con}) = \frac{0.45 M \times 2.5 \text{ L}}{2.3 M} = \mathbf{0.49\ L\ of\ 2.3\ \mathit{M}\ HCl}$$

4.50 *See Section 4.2 and Example 4.8.*

Solving $M(\text{con}) \times V(\text{con}) = M(\text{dil}) \times V(\text{dil})$ for $V(\text{con})$ gives

$$V(\text{con}) = \frac{M(\text{dil}) \times V(\text{dil})}{M(\text{con})} \qquad V(\text{con}) = \frac{2.35 M \text{ NaOH} \times 1.00 \text{ L}}{5.22 M} = \mathbf{0.450\ L\ of\ 5.22\ \mathit{M}\ NaOH}$$

4.51 *See Section 4.2 and example 4.8.*

Solving $M(\text{con}) \times V(\text{con}) = M(\text{dil}) \times V(\text{dil})$ for $V(\text{con})$ gives

$$V(\text{con}) = \frac{M(\text{dil}) \times V(\text{dil})}{M(\text{con})} \qquad V(\text{con}) = \frac{0.118 M\ Li_2CO_3 \times 2.00 \text{ L}}{2.11 M} = \mathbf{0.112\ L\ of\ 2.11\ \mathit{M}\ Li_2CO_3}$$

4.52 *See Section 4.2 and example 4.8.*

Solving $M(\text{con}) \times V(\text{con}) = M(\text{dil}) \times V(\text{dil})$ for $V(\text{con})$ gives

$$V(\text{con}) = \frac{M(\text{dil}) \times V(\text{dil})}{M(\text{con})} \qquad V(\text{con}) = \frac{0.113 M\ H_2SO_4 \times 1.00 \text{ L}}{5.00 M} = \mathbf{0.0226\ L\ of\ 5.00\ \mathit{M}\ H_2SO_4}$$

4.53 *See Section 4.2 and Example 4.9.*

Solving $M(\text{con}) \times V(\text{con}) = M(\text{dil}) \times V(\text{dil})$ for $M(\text{dil})$ gives

$$M(\text{dil}) = \frac{M(\text{con}) \times V(\text{con})}{V(\text{dil})} \qquad M(\text{dil}) = \frac{1.0\ M \times 0.050 \text{ L}}{2.0 \text{ L}} = \mathbf{0.025\ \mathit{M}\ C_6H_{12}O_6}$$

4.54 *See Section 4.2 and Example 4.9.*

Solving $M(\text{con}) \times V(\text{con}) = M(\text{dil}) \times V(\text{dil})$ for $M(\text{dil})$ gives

$$M(\text{dil}) = \frac{M(\text{con}) \times V(\text{con})}{V(\text{dil})} \qquad M(\text{dil}) = \frac{0.0153\ M\ Ca(OH)_2 \times 0.100 \text{ L}}{0.500 \text{ L}} = \mathbf{3.06 \times 10^{-3}\ \mathit{M}\ Ca(OH)_2}$$

4.55 *See Section 4.2 and Examples 4.4, 4.9.*

(a) $$? M \text{ NaOH} = \frac{3.56 \text{ g NaOH}}{2.0 \text{ L soln}} \times \frac{1 \text{ mol NaOH}}{40.0 \text{ g NaOH}} = \mathbf{0.044\ \mathit{M}\ NaOH}$$

Note: See the alternative method used in working 4.29.

(b) Solving M(con) x V(con) = M(dil) x V(dil) for M(dil) gives

$$M(\text{dil}) = \frac{M(\text{con}) \times V(\text{con})}{V(\text{dil})} \qquad M(\text{dil}) = \frac{1.4\ M \times 0.0250\ \text{L}}{2.0\ \text{L}} = \mathbf{0.018\ \mathit{M}\ NaOH}$$

4.56	***See Section 4.2 and Examples 4.4, 4.9.***

(a) $?\,M\ \text{NaNO}_3 = \frac{0.12\ \text{g NaNO}_3}{0.250\ \text{L soln}} \times \frac{1\ \text{mol NaNO}_3}{85.0\ \text{g NaNO}_3} = \mathbf{5.6 \times 10^{-3}\ \mathit{M}\ NaNO_3}$

(b) Solving M(con) x V(con) = M(dil) x V(dil) for M(dil) gives

$$M(\text{dil}) = \frac{M(\text{con}) \times V(\text{con})}{V(\text{dil})} \qquad M(\text{dil}) = \frac{0.42\ M \times 0.00075\ \text{L}}{0.250\ \text{L}} = \mathbf{1.3 \times 10^{-3}\ \mathit{M}\ NaOH}$$

4.57	***See Section 4.3 and Example 4.8.***

Solving M(con) x V(con) = M(dil) x V(dil) for M(dil) gives

$$M(\text{dil}) = \frac{M(\text{con}) \times V(\text{con})}{V(\text{dil})} \qquad V(\text{con}) = \frac{1.5\ M \times\ 2.0\ \text{L}}{5.0\ M} = \mathbf{0.60\ L\ of\ 5.0\ \mathit{M}\ NaOH}$$

Place 600 mL of 5.0 M NaOH into a reasonable amount of water in a 2.0 L volumetric flask. Add water to fill the volumetric flask to the calibration mark.

4.58	***See Section 4.2 and Examples 4.4, 4.9.***

$$?\,M\ \text{NaOH} = \frac{6.00\ \text{g NaOH}}{1.00\ \text{L soln}} \times \frac{1\ \text{mol NaOH}}{40.0\ \text{g NaOH}} = 0.150\ M\ \text{NaOH}$$

Solving M(con) x V(con) = M(dil) x V(dil) for M(dil) gives

$$M(\text{dil}) = \frac{M(\text{con}) \times V(\text{con})}{V(\text{dil})} \qquad M(dil) = \frac{1.50\ M \times 0.100\ \text{L}}{5.00\ \text{L}} = \mathbf{0.00300\ \mathit{M}\ NaOH}$$

Note: See the alternative method used in working 4.29.

4.59	***See Sections 3.1, 4.3, and Examples 3.1, 3.2, 4.10.***

Unbalanced:	$AgNO_3(aq) + CaCl_2(aq) \rightarrow AgCl(s) + Ca(NO_3)_2(aq)$	Start with $Ca(NO_3)_2$.
Step 1:	$\underline{2}AgNO_3(aq) + CaCl_2(aq) \rightarrow AgCl(s) + Ca(NO_3)_2(aq)$	Balances NO_3 units.
Step 2:	$2AgNO_3(aq) + CaCl_2(aq) \rightarrow 2AgCl(s) + Ca(NO_3)_2(aq)$	Balances Ag, Cl.

Strategy: L $AgNO_3$ soln → mol $AgNO_3$ → mol AgCl → g AgCl

$$?\ \text{g AgCl} = 0.00311\ \text{L AgNO}_3\ \text{soln} \times \frac{0.11\ \text{mol AgNO}_3}{1\ \text{L AgNO}_3\ \text{soln}} \times \frac{2\ \text{mol AgCl}}{2\ \text{mol AgNO}_3} \times \frac{143.4\ \text{g AgCl}}{1\ \text{mol AgCl}} = \mathbf{0.049\ g\ AgCl}$$

4.60	***See Sections 3.1, 4.3, and Examples 3.1, 3.2, 4.10.***

Balanced: $H_2SO_4(aq) + Ba(OH)_2(aq) \rightarrow 2H_2O(\ell) + BaSO_4(s)$

Strategy: L $Ba(OH)_2$ soln → mol $Ba(OH)_2$ → mol $BaSO_4(s)$ → g $BaSO_4(s)$

$$?\ \text{g BaSO}_4 = 0.0250\ \text{L Ba(OH)}_2\ \text{soln} \times \frac{0.11\ \text{mol Ba(OH)}_2}{1\ \text{L Ba(OH)}_2\ \text{soln}} \times \frac{1\ \text{mol BaSO}_4}{1\ \text{mol Ba(OH)}_2} \times \frac{233.4\ \text{g BaSO}_4}{1\ \text{mol BaSO}_4} = \mathbf{0.64\ g\ BaSO_4}$$

4.61 *See Sections 3.1, 4.3, and Examples 3.3, 4.10.*

Unbalanced:	$H_2SO_4(aq) + NaOH(s)$	→	$H_2O(\ell) + NaSO_4(s)$	Start with Na_2SO_4.
Step 1:	$H_2SO_4(aq) + \underline{2}NaOH(s)$	→	$H_2O(\ell) + Na_2SO_4(aq)$	Balances Na.
Step 2:	$H_2SO_4(aq) + 2NaOH(s)$	→	$2H_2O(\ell) + Na_2SO_4(aq)$	Balances H, O.

Strategy: L H_2SO_4 soln → mol H_2SO_4 → mol NaOH → g NaOH

$$? \text{ g NaOH} = 0.100 \text{ L } H_2SO_4 \text{ soln} \times \frac{3.13 \text{ mol } H_2SO_4}{1 \text{ L } H_2SO_4 \text{ soln}} \times \frac{2 \text{ mol NaCH}}{1 \text{ mol } H_2SO_4} \times \frac{40.0 \text{ g NaOH}}{1 \text{ mol NaOH}} = \mathbf{25.0 \text{ g NaOH}}$$

4.62 *See Sections 3.1, 4.3, and Examples 3.3, 4.10.*

Unbalanced:	$HCl(aq) + Ca(OH)_2(s)$	→	$H_2O(\ell) + CaCl_2(aq)$	Start with $CaCl_2$.
Step 1:	$\underline{2}HCl(aq) + Ca(OH)_2(s)$	→	$H_2O(\ell) + CaCl_2(aq)$	Balances Cl.
Step 2:	$2HCl(aq) + Ca(OH)_2(s)$	→	$\underline{2}H_2O(\ell) + CaCl_2(aq)$	Balances H, O.

Strategy: L HCl soln → mol HCl → mol $Ca(OH)_2$ → g $Ca(OH)_2$

$$? \text{ g } Ca(OH)_2 = 0.1000 \text{ L HCl soln} \frac{0.0922 \text{ mol HCl}}{1 \text{ L HCl soln}} \times \frac{1 \text{ mol } Ca(OH)_2}{2 \text{ mol HCl}} \times \frac{74.1 \text{ g } Ca(OH)_2}{1 \text{ mol } Ca(OH)_2} = \mathbf{0.342 \text{ g } Ca(OH)_2}$$

4.63 *See Sections 3.1, 4.3, and Examples 3.3, 4.11.*

Unbalanced:	$HNO_3(aq) + Ca(OH)_2(s)$	→	$H_2O(\ell) + Ca(NO_3)_2(aq)$	Start with $Ca(NO_3)_2$.
Step 1:	$\underline{2}\ HNO_3(aq) + Ca(OH)_2(s)$	→	$H_2O(\ell) + Ca(NO_3)_2(aq)$	Balances NO_3 units.
Step 2:	$2HNO_3(aq) + Ca(OH)_2(s)$	→	$\underline{2}H_2O(\ell) + Ca(NO_3)_2(aq)$	Balances H, O.

Strategy: g $Ca(OH)_2$ → mol $Ca(OH)_2$ → mol HNO_3 → L HNO_3

$$? \text{ L } HNO_3 \text{ soln} = 22 \text{ g } Ca(OH)_2 \times \frac{1 \text{ mol } Ca(OH)_2}{74.1 \text{ g } Ca(OH)_2} \times \frac{2 \text{ mol } HNO_3}{1 \text{ mol } Ca(OH)_2} \times \frac{1 \text{ L } HNO_3 \text{ soln}}{0.66 \text{ mol } HNO_3} = \mathbf{0.90 \text{ L } HNO_3 \text{ soln}}$$

4.64 *See Sections 3.1, 4.3, and Examples 3.3, 4.11.*

Unbalanced:	$HCl(aq) + Mg(OH)_2(s)$	→	$H_2O(\ell) + MgCl_2(aq)$	Start with $MgCl_2(aq)$.
Step 1:	$\underline{2}HCl(aq) + Mg(OH)_2(s)$	→	$H_2O(\ell) + MgCl_2(aq)$	Balances Cl.
Step 2:	$2HCl(aq) + Mg(OH)_2(s)$	→	$\underline{2}H_2O(\ell) + MgCl_2(aq)$	Balances H, O.

Strategy: g $Mg(OH)_2$ → mol $Mg(OH)_2$ → mol HCl → L HCl

$$? \text{ L HCl} = 2.5 \text{ g } Mg(OH)_2 \times \frac{1 \text{ mol } Mg(OH)_2}{58.3 \text{ g } Mg(OH)_2} \times \frac{2 \text{ mol HCl}}{1 \text{ mol } Mg(OH)_2} \times \frac{1 \text{ L HCl soln}}{0.22 \text{ mol HCl}} = \mathbf{0.39 \text{ L HCl soln}}$$

4.65 *See Sections 3.1, 4.4, and Examples 3.3, 4.12.*

Unbalanced:	$HCl(aq) + Ba(OH)_2(s)$	→	$H_2O(\ell) + BaCl_2(aq)$	Start with $BaCl_2$.
Step 1:	$\underline{2}HCl(aq) + Ba(OH)_2(s)$	→	$H_2O(\ell) + BaCl_2(aq)$	Balances Cl.
Step 2:	$2HCl(aq) + Ba(OH)_2(s)$	→	$\underline{2}H_2O(\ell) + BaCl_2(aq)$	Balances H, O.

Strategy: g $Ba(OH)_2$ → mol $Ba(OH)_2$ → mol HCl→ M HCl soln

$$? M \text{ HCl soln} = \frac{0.0298 \text{ mol HCl}}{0.135 \text{ L HCl soln}} = \mathbf{0.221\ \mathit{M}\ HCl}$$

4.66 *See Sections 3.1, 4.4, and Examples 3.3, 4.12.*

Unbalanced:	$H_2SO_4(aq) + NaOH(s)$	→	$2H_2O(\ell) + Na_2SO_4(aq)$	Start with Na_2SO_4.
Step 1:	$H_2SO_4(aq) + \underline{2}NaOH(s)$	→	$H_2O(\ell) + Na_2SO_4(aq)$	Balances Na.
Step 2:	$H_2SO_4(aq) + 2NaOH(s)$	→	$2H_2O(\ell) + Na_2SO_4(aq)$	Balances H, O.

Strategy: g NaOH (s)→ mol NaOH → mol H_2SO_4 → M H_2SO_4

$$? \text{ mol } H_2SO_4 = 0.155 \text{ g NaOH} \times \frac{1 \text{ mol NaOH}}{40.0 \text{ g NaOH}} \times \frac{1 \text{ mol } H_2SO_4}{2 \text{ mol NaOH}} = 0.00194 \text{ mol } H_2SO_4$$

$$? M\ H_2SO_4 = \frac{0.00194 \text{ mol } H_2SO_4}{0.00511 \text{ L } H_2SO_4 \text{ soln}} = \mathbf{0.380\ \mathit{M}\ H_2SO_4}$$

4.67 *See Sections 3.1, 3.5, 4.3, and Examples 3.3, 3.16, 4.10.*

Unbalanced:	$H_2SO_4 + Ba(OH)_2$	→	$H_2O + BaSO_4$	Start with $Ba(OH)_2$.
Step 1:	$H_2SO_4 + Ba(OH)_2$	→	$2H_2O + BaSO_4$	Balances H and O.

Strategy: L H_2SO_4 soln → mol H_2SO_4 → mol $BaSO_4$ → g $BaSO_4$

$$? \text{ g } BaSO_4 \text{ based on } H_2SO_4 = 0.355 \text{ L } H_2SO_4 \text{ soln} \times \frac{0.032 \text{ mol } H_2SO_4}{1 \text{ L } H_2SO_4 \text{ soln}}$$

$$\times \frac{1 \text{ mol } BaSO_4}{1 \text{ mol } H_2SO_4} \times \frac{233.4 \text{ g } BaSO_4}{1 \text{ mol } BaSO_4} = 2.6 \text{ g } BaSO_4$$

Strategy: L $Ba(OH)_2$ soln → mol $Ba(OH)_2$ → mol $BaSO_4$ → g $BaSO_4$

$$? \text{ g } BaSO_4 \text{ based on } Ba(OH)_2 = 0.266 \text{ L } Ba(OH)_2 \text{ soln} \times \frac{0.015 \text{ mol } Ba(OH)_2}{1 \text{ L } Ba(OH)_2 \text{ soln}}$$

$$\times \frac{1 \text{ mol } BaSO_4}{1 \text{ mol } Ba(OH)_2} \times \frac{233.4 \text{ g } BaSO_4}{1 \text{ mol } BaSO_4} = 0.93 \text{ g } BaSO_4$$

$Ba(OH)_2$ is the limiting reactant because it produces less $BaSO_4$. The maximum amount of $BaSO_4$ which can be produced from 355 mL of 0.032 *M* H_2SO_4 and 266 mL of 0.015 *M* $Ba(OH)_2$ is **0.93 g**.

4.68 *See Sections 3.1, 4.3, and Examples 3.1, 3.2, 4.10.*

Unbalanced:	$Mg(NO_3)_2(aq) + NaOH(aq) \rightarrow Mg(OH)_2(s) + NaNO_3(aq)$	Start with $Mg(NO_3)_2$
Step 1:	$Mg(NO_3)_2(aq) + NaOH(aq) \rightarrow Mg(OH)_2(s) + \underline{2}NaNO_3(aq)$	Balances NO_3 units.
Step 2:	$Mg(NO_3)_2(aq) + \underline{2}NaOH(aq) \rightarrow Mg(OH)_2(s) + 2NaNO_3(aq)$	Balances Na, OH units.

Strategy: L NaOH soln → mol NaOH → mol $Mg(OH)_2$ → g $Mg(OH)_2$

$$? \text{ g } Mg(OH)_2 = 1.2 \text{ L NaOH soln} \times \frac{5.5 \text{ mol NaOH}}{1 \text{ L NaOH sohn}} \times \frac{1 \text{ mol } Mg(OH)_2}{2 \text{ mol NaOH}} \times \frac{58.3 \text{ g } Mg(OH)_2}{1 \text{ mol } Mg(OH)_2} = \mathbf{192\ g\ Mg(OH)_2}$$

4.69 See Sections 3.1, 4.3 and Examples 3.1, 3.2, 4.10.

Unbalanced: $Pb(CH_3COO)_2(aq) + Na_2SO_4(aq) \rightarrow PbSO_4(s) + NaCH_3COO(aq)$ Start with $Pb(CH_3COO)_2$.
Step 1: $Pb(CH_3COO)_2(aq) + Na_2SO_4(aq) \rightarrow PbSO_4(s) + \underline{2}NaCH_3COO(aq)$ Balances CH_3COO units and Na.

Strategy: $L\ Pb(CH_3COO)_2\ soln \rightarrow mol\ Pb(CH_3COO)_2 \rightarrow mol\ PbSO_4 \rightarrow g\ PbSO_4$

$$? \text{ g PbSO}_4 = 0.0200 \text{ L Pb(CH}_3\text{COO)}_2 \text{ soln} \times \frac{1.11 \text{ mol Pb(CH}_3\text{COO)}_2}{1 \text{ L Pb(CH}_3\text{COO)}_2 \text{ soln}} \times \frac{1 \text{ mol PbSO}_4}{1 \text{ mol Pb(CH}_3\text{COO)}_2}$$
$$\times \frac{303.3 \text{ g PbSO}_4}{1 \text{ mol PbSO}_4} = \mathbf{6.73 \text{ g PbSO}_4}$$

4.70 See Sections 3.1, 4.3, and Examples 3.1, 3.2, 4.10.

Unbalanced: $Mg(NO_3)_2(aq) + NaOH(aq) \rightarrow Mg(OH)_2(s) + NaNO_3(aq)$ Start with $Mg(NO_3)_2$.
Step 1: $Mg(NO_3)_2(aq) + NaOH(aq) \rightarrow Mg(OH)_2(s) + 2NaNO_3(aq)$ Balances NO_3 units.
Step 2: $Mg(NO_3)_2(aq) + \underline{2}NaOH(aq) \rightarrow Mg(OH)_2(s) + 2NaNO_3(aq)$ Balances Na, OH units.

Strategy: $L\ Mg(NO_3)_2\ soln \rightarrow mol\ Mg(NO_3)_2 \rightarrow mol\ Mg(OH)_2 \rightarrow g\ Mg(OH)_2$

$$? \text{ g Mg(OH)}_2 = 0.1000 \text{ L Mg(NO}_3)_2 \text{ soln} \times \frac{1.545 \text{ mol Mg(NO}_3)_2}{1 \text{ L Ba(NO}_3)_2} \times \frac{1 \text{ mol Mg(OH)}_2}{1 \text{ mol Mg(NO}_3)_2} \times \frac{58.33 \text{ g Mg(OH)}_2}{1 \text{ mol Mg(OH)}_2}$$
$$= \mathbf{9.011 \text{ g Mg(OH)}_2}$$

4.71 See Section 4.1, 4.3, Table 4.1, and Examples 4.2, 4.3, 4.10.

The ions present from the reactants are $Ba^{2+}(aq)$, $Br^-(aq)$, $Na^+(aq)$, and $SO_4^{2-}(aq)$. The two possible products are $BaSO_4$ and NaBr, and Table 4.1 indicates $BaSO_4$ is insoluble.

Unbalanced: $BaBr_2(aq) + Na_2SO_4(aq) \rightarrow BaSO_4(s) + NaBr(aq)$ Start with $BaBr_2$.
Step 1: $BaBr_2(aq) + Na_2SO_4(aq) \rightarrow BaSO_4(s) + 2NaBr(aq)$ Balances Na, Br.

Strategy: $L\ Na_2SO_4\ soln \rightarrow mol\ Na_2SO_4 \rightarrow mol\ BaSO_4 \rightarrow g\ BaSO_4$

$$? \text{ g BaSO}_4 = 0.0100 \text{ L Na}_2\text{SO}_4 \text{ soln} \times \frac{2.10 \text{ mol Na}_2\text{SO}_4}{1 \text{ L Na}_2\text{SO}_4 \text{ soln}} \times \frac{1 \text{ mol BaSO}_4}{1 \text{ mol Na}_2\text{SO}_4} \times \frac{233.4 \text{ g BaSO}_4}{1 \text{ mol BaSO}_4} = \mathbf{4.90 \text{ g BaSO}_4}$$

4.72 See Sections 4.1, 4.3, Table 4.1, and Examples 4.2, 4.3, 4.10.

The ions present from the reactants are $Na^+(aq)$, $Cl^-(aq)$, $Ag^+(aq)$, and $NO_3^-(aq)$. The two possible products are $NaNO_3$ and AgCl, and Table 4.1 indicates AgCl is insoluble.

Balanced: $NaCl(aq) + AgNO_3(aq) \rightarrow AgCl(s) + NaNO_3(aq)$

Strategy: $L\ AgNO_3\ soln \rightarrow mol\ AgNO_3 \rightarrow mol\ AgCl \rightarrow g\ AgCl$

$$? \text{ g AgCl} = 0.0100 \text{ L AgNO}_3 \text{ soln} \times \frac{2.10 \text{ mol AgNO}_3}{1 \text{ L AgNO}_3 \text{ soln}} \times \frac{1 \text{ mol AgCl}}{1 \text{ mol AgNO}_3} \times \frac{143.4 \text{ g AgCl}}{1 \text{ mol AgCl}} = \mathbf{3.01 \text{ g AgCl}}$$

4.73 See Section 4.3 and Example 4.11.

Balanced: $AgNO_3(aq) + NaI(aq) \rightarrow AgI(s) + NaNO_3(aq)$

Strategy: L NaI soln → mol NaI → mol AgNO₃ → L AgNO₃ soln

$$? \text{ L AgNO}_3 = 0.1200 \text{ L NaI soln} \times \frac{1.200 \text{ mol NaI}}{1 \text{ L NaI soln}} \times \frac{1 \text{ mol AgNO}_3}{1 \text{ mol NaI}} \times \frac{1 \text{ L AgNO}_3 \text{ soln}}{1.212 \text{ mol AgNO}_3} = \mathbf{0.1188 \text{ L AgNO}_3 \text{ soln}}$$

4.74 ***See Section 4.3 and Example 4.11.***

Balanced: $K_2CO_3(aq) + CaCl_2(aq) \rightarrow 2KCl(aq) + CaCO_3(s)$

Strategy: L $CaCl_2$ soln → mol $CaCl_2$ → mol K_2CO_3 → L K_2CO_3 soln

$$? \text{ L K}_2\text{CO}_3 \text{ soln} = 0.0500 \text{ L CaCl}_2 \text{ soln} \times \frac{0.100 \text{ mol CaCl}_2}{1 \text{ L CaCl}_2 \text{ soln}} \times \frac{1 \text{ mol K}_2\text{CO}_3}{1 \text{ mol CaCl}_2} \times \frac{1 \text{ L K}_2\text{CO}_3 \text{ soln}}{0.112 \text{ mol K}_2\text{CO}_3}$$

$$= \mathbf{0.0446 \text{ L K}_2\text{CO}_3 \text{ soln}}$$

4.75 ***See Sections 4.1, 4.3, Table 4.1, and Examples 4.2, 4.10.***

The ions present from the reactants are $NH_4^+(aq)$, $SO_4^{2-}(aq)$, $Ba^{2+}(aq)$ and $Cl^-(aq)$. The two possible products are NH_4Cl and $BaSO_4$, and Table 4.1 indicates $BaSO_4$ is insoluble.

Unbalanced:	$(NH_4)_2SO_4(aq) + BaCl_2(aq)$	→	$BaSO_4(s) + NH_4Cl(aq)$	Start with $(NH_4)_2SO_4$.
Step 1:	$(NH_4)_2SO_4(aq) + BaCl_2(aq)$	→	$BaSO_4(s) + \underline{2}NH_4Cl(aq)$	Balances NH_4 units and Cl.

Strategy: L$(NH_4)_2SO_4$ soln → mol$(NH_4)_2SO_4$ → mol $BaSO_4$ → g $BaSO_4$

$$? \text{ g BaSO}_4 = 0.021 \text{ L (NH}_4)_2\text{SO}_4 \text{ soln} \times \frac{3.5 \text{ mol (NH}_4)_2\text{SO}_4}{1 \text{ L (NH}_4)_2\text{SO}_4 \text{ soln}} \times \frac{1 \text{ mol BaSO}_4}{1 \text{ mol (NH}_4)_2\text{SO}_4} \times \frac{233.4 \text{ g BaSO}_4}{1 \text{ mol BaSO}_4}$$

$$= \mathbf{17 \text{ g BaSO}_4}$$

4.76 ***See Sections 4.1, 4.3, Table 4.1, and Examples 4.2, 4.10.***

The ions present from the reactants are $Na^+(aq)$, $OH^-(aq)$, $Fe^{2+}(aq)$, and $Cl^-(aq)$. The two possible products are NaCl and $Fe(OH)_2$, and Table 4.1 indicates $Fe(OH)_2$ is insoluble.

Unbalanced:	$NaOH(aq) + FeCl_2(aq)$	→	$NaCl(aq) + Fe(OH)_2(s)$	Start with $Fe(OH)_2$.
Step 1:	$\underline{2}NaOH(aq) + FeCl_2(aq)$	→	$NaCl(aq) + Fe(OH)_2(s)$	Balances OH units.
Step 2:	$2NaOH(aq) + FeCl_2(aq)$	→	$\underline{2}NaCl(aq) + Fe(OH)_2(s)$	Balances Cl.

Strategy: L NaOH soln → mol NaOH → mol $Fe(OH)_2$ → g $Fe(OH)_2$

$$? \text{ g Fe(OH)}_2 = 0.220 \text{ L NaOH soln} \times \frac{1.22 \text{ mol NaOH}}{1 \text{ L NaOH soln}} \times \frac{1 \text{ mol Fe(OH)}_2}{2 \text{ mol NaOH}} \times \frac{89.87 \text{ g Fe(OH)}_2}{1 \text{ mol Fe(OH)}_2} = \mathbf{12.1 \text{ g Fe(OH)}_2}$$

4.77 ***See Sections 3.5, 4.1, 4.3, and Examples 3.16, 4.2, 4.10.***

The ions present from the reactants are $Ag^+(aq)$, $NO_3^-(aq)$, $Na^+(aq)$ and $Br^-(aq)$. The two possible products are AgBr and $NaNO_3$, and Table 4.1 indicates AgBr is insoluble.

Overall: $AgNO_3(aq) + NaBr(aq) \rightarrow AgBr(s) + NaNO_3(aq)$

Complete ionic: $Ag^+(aq) + NO_3^-(aq) + Na^+(aq) + Br^-(aq) \rightarrow AgBr(s) + Na^+(aq) + NO_3^-(aq)$

Net ionic: $Ag^+(aq) + Br^-(aq) \rightarrow AgBr(s)$

Strategy: L $AgNO_3$ soln → mol $AgNO_3$ → mol AgBr → gAgBr

$$? \text{ g AgBr based on } AgNO_3 = 0.345 \text{ L } AgNO_3 \text{ soln} \times \frac{0.330 \text{ mol } AgNO_3}{1 \text{ L } AgNO_3 \text{ soln}} \times \frac{1 \text{ mol AgBr}}{1 \text{ mol } AgNO_3} \times \frac{187.8 \text{ g AgBr}}{1 \text{ mol AgBr}}$$

$$= 21.4 \text{ g AgBr}$$

Stategy: LNaBr soln → mol NaBr → mol AgBr → g AgBr

$$? \text{ g AgBr based on NaBr} = 0.1000 \text{ L NaBr soln} \times \frac{1.30 \text{ mol NaBr}}{1 \text{ L NaBr soln}} \times \frac{1 \text{ mol AgBr}}{1 \text{ mol NaBr}} \times \frac{187.8 \text{ g AgBr}}{1 \text{ mol AgBr}} = 24.4 \text{ g AgBr}$$

$AgNO_3$ is the limiting reactant because it produces less AgBr. The maximum amount of AgBr which can be produced from 345 mL of 0.33 *M* $AgNO_3$ and 100 mL of 1.3 *M* NaBr is **21.4 g**.

4.78 *See Sections 3.5, 4.1, 4.3, and Examples 3.16, 4.2, 4.10.*

The ions present from the reactants are $Na^+(aq)$, $OH^-(aq)$, $Mg^{2+}(aq)$, and $Cl^-(aq)$. The two possible products are NaCl and $Mg(OH)_2$, and Table 4.1 indicates $Mg(OH)_2$ is insoluble.

Overall: $2NaOH(aq) + MgCl_2(aq) \rightarrow 2NaCl(aq) + Mg(OH)_2(s)$

Complete ionic: $2Na^+(aq) + 2OH^-(aq) + Mg^{2+}(aq) + 2Cl^-(aq) \rightarrow 2Na^+(aq) + 2Cl^-(aq) + Mg(OH)_2(s)$

Net ionic: $2OH^-(aq) + Mg^{2+}(aq) \rightarrow Mg(OH)_2(s)$

Strategy: L NaOH soln → mol NaOH → mol $Mg(OH)_2$ → g $Mg(OH)_2$

$$? \text{ g } Mg(OH)_2 \text{ based on NaOH} = 0.0500 \text{ L NaOH soln} \times \frac{3.30 \text{ mol NaOH}}{1 \text{ L NaOH soln}} \times \frac{1 \text{ mol } Mg(OH)_2}{2 \text{ mol NaOH}} \times \frac{58.3 \text{ g } Mg(OH)_2}{1 \text{ mol } Mg(OH)_2}$$

$$= 4.81 \text{ g } Mg(OH)_2$$

Strategy: L $MgCl_2$ → mol $MgCl_2$ → mol $Mg(OH)_2$ → g $Mg(OH)_2$

$$? \text{ g } Mg(OH)_2 \text{ based on } MgCl_2 = 0.0350 \text{ L } MgCl_2 \text{ soln} \times \frac{1.00 \text{ mol } MgCl_2}{1 \text{ L } MgCl_2 \text{ soln}} \times \frac{1 \text{ mol } Mg(OH)_2}{2 \text{ mol } MgCl_2} \times \frac{58.3 \text{ g } Mg(OH)_2}{1 \text{ mol } Mg(OH)_2}$$

$$= 2.04 \text{ g } Mg(OH)_2$$

$MgCl_2$ is the limiting reactant because it produces less $Mg(OH)_2$. The maximum amount of $Mg(OH)_2$ which can be produced from 50.0 mL of 3.30 *M* NaOH and 35.0 mL of 1.00 *M* $MgCl_2$ is **2.04 g $Mg(OH)_2$.**

4.79 *See Sections 3.1, 4.4, and Examples 3.3, 4.12.*

Unbalanced:	$HNO_3(aq) + Sr(OH)_2(aq) \rightarrow H_2O(\ell) + Sr(NO_3)_2(aq)$	Start with $Sr(NO_3)_2$.
Step 1:	$\underline{2}HNO_3(aq) + Sr(OH)_2(aq) \rightarrow H_2O(\ell) + Sr(NO_3)_2(aq)$	Balances NO_3 units.
Step 2:	$2HNO_3(aq) + Sr(OH)_2(aq) \rightarrow \underline{2}H_2O(\ell) + Sr(NO_3)_2(aq)$	Balances H, O.

Strategy: L $Sr(OH)_2$ soln → mol $Sr(OH)_2$ → mol HNO_3 → M HNO_3 soln

$$? \text{ mol } HNO_3 = 0.02240 \text{ L } Sr(OH)_2 \text{ soln} \times \frac{0.0229 \text{ mol } Sr(OH)_2}{1 \text{ L } Sr(OH)_2 \text{ soln}} \times \frac{2 \text{ mol } HNO_3}{1 \text{ mol } Sr(OH)_2} = 1.03 \times 10^{-3} \text{ mol } HNO_3$$

$$? M\ HNO_3 \text{ soln} = \frac{1.03 \times 10^{-3} \text{ mol } HNO_3}{0.05000 \text{ L } HNO_3 \text{ soln}} = \mathbf{2.05 \times 10^{-2}}\ \boldsymbol{M}\ \mathbf{HNO_3}$$

4.80 *See Sections 3.1, 4.4, and Examples 3.3, 4.12.*

Unbalanced:	$HCl(aq) + Ca(OH)_2(aq)$	$\rightarrow$	$H_2O(\ell) + CaCl_2(aq)$	Start with $CaCl_2$.
Step 1:	$\underline{2}HCl(aq) + Ca(OH)_2(aq)$	$\rightarrow$	$H_2O(\ell) + CaCl_2(aq)$	Balances Cl.
Step 2:	$2HCl(aq) + Ca(OH)_2(aq)$	$\rightarrow$	$\underline{2}H_2O(\ell) + CaCl_2(aq)$	Balances H, O.

Strategy: L HCl soln → mol HCl → mol Ca(OH)$_2$ → M Ca(OH)$_2$ soln

$$? \text{ mol } Ca(OH)_2 = 0.03340 \text{ L HCl soln} \times \frac{0.213 \text{ mol HCl}}{1 \text{ L HCl soln}} \times \frac{1 \text{ mol } Ca(OH)_2}{2 \text{ mol HCl}} = 3.56 \times 10^{-3} \text{ mol } Ca(OH)_2$$

$$? M \; Ca(OH)_2 \text{ soln} = \frac{3.56 \times 10^{-3} \text{ mol } Ca(OH)_2}{0.482 \text{ L } Ca(OH)_2 \text{ soln}} = \mathbf{7.39 \times 10^{-3}} \; \boldsymbol{M} \; \mathbf{Ca(OH)_2 \; soln}$$

4.81 *See Sections 3.1, 4.4, and Examples 3.3, 4.12.*

Balanced Overall Equation: $HCl(aq) + KOH(aq) \rightarrow H_2O(l) + KCl(aq)$

Strategy: L KOH soln → mol KOH → mol HCl → M HCl soln

$$? \text{ mol HCl} = 0.03340 \text{ L KOH soln} \times \frac{2.20 \text{ mol KOH}}{1 \text{ L KOH soln}} \times \frac{1 \text{ mol HCl}}{1 \text{ mol KOH}} = 0.0735 \text{ mol HCl}$$

$$? M \text{ HCl soln} = \frac{0.0735 \text{ mol HCl}}{0.1000 \text{ L HCl soln}} = \mathbf{0.735} \; \boldsymbol{M} \; \mathbf{HCl}$$

4.82 *See Sections 3.1, 4.4, and Examples 3.3, 4.12.*

Unbalanced:	$H_2SO_4(aq) + NaOH(s)$	$\rightarrow$	$2H_2O(\ell) + Na_2SO_4(aq)$	Start with Na_2SO_4.
Step 1:	$H_2SO_4(aq) + \underline{2}NaOH(s)$	$\rightarrow$	$H_2O(\ell) + Na_2SO_4(aq)$	Balances Na.
Step 2:	$H_2SO_4(aq) + 2NaOH(s)$	$\rightarrow$	$2H_2O(\ell) + Na_2SO_4(aq)$	Balances H, O.

Strategy: L NaOH soln → mol HaOH → mol H$_2$SO$_4$ → M H$_2$SO$_4$

$$? \text{ mol } H_2SO_4 = 0.00965 \text{ L NaOH soln} \times \frac{1.33 \text{ mol NaOH}}{1 \text{ L NaOH soln}} \times \frac{1 \text{ mol } H_2SO_4}{2 \text{ mol NaOH}} = 6.42 \times 10^{-3} \text{ mol } H_2SO_4$$

$$? M \; H_2SO_4 \text{ soln} = \frac{6.42 \times 10^{-3} \text{ mol } H_2SO_4}{0.0500 \text{ L } H_2SO_4 \text{ soln}} = \mathbf{0.128 \; M \; H_2SO_4}$$

4.83 *See Sections 3.1, 4.1, 4.3, 4.4, and Examples 3.1, 3.2, 3.3, 4.11.*

(a) Unbalanced:	$HNO_3(aq) + Ba(OH)_2(aq)$	$\rightarrow$	$H_2O(\ell) + Ba(NO_3)_2(aq)$	Start with $Ba(NO_3)_2$.
Step 1:	$\underline{2}HNO_3(aq) + Ba(OH)_2(aq)$	$\rightarrow$	$H_2O(\ell) + Ba(NO_3)_2(aq)$	Balances NO_3 units.
Step 2:	$2HNO_3(aq) + Ba(OH)_2(aq)$	$\rightarrow$	$\underline{2}H_2O(\ell) + Ba(NO_3)_2(aq)$	Balances H, O.

Strategy: L Ba(OH)$_2$ soln → mol Ba(OH)$_2$ → mol HNO$_3$ → L HNO$_3$ soln

$$? \text{ L } HNO_3 \text{ soln} = 0.0500 \text{ L } Ba(OH)_2 \text{ soln} \times \frac{0.033 \text{ mol } Ba(OH)_2}{1 \text{ L } Ba(OH)_2 \text{ soln}} \times \frac{2 \text{ mol } HNO_3}{1 \text{ mol } Ba(OH)_2} \times \frac{1 \text{ L } HNO_3 \text{ soln}}{0.223 \text{ mol } HNO_3}$$

$$= \mathbf{0.015 \; L \; HNO_3 \; soln}$$

(b) Unbalanced: $AgNO_3(aq) + CaCl_2(aq) \rightarrow AgCl(s) + Ca(NO_3)_2(aq)$ Start with $Ca(NO_3)_2$.
Step 1: $\underline{2}AgNO_3(aq) + CaCl_2(aq) \rightarrow AgCl(s) + Ca(NO_3)_2(aq)$ Balances NO_3 units.
Step 2: $2AgNO_3(aq) + CaCl_2(aq) \rightarrow \underline{2}AgCl(s) + Ca(NO_3)_2(aq)$ Balance Ag,Cl.

Strategy: L $CaCl_2$ soln → mol $CaCl_2$ → mol $AgNO_3$ → L $AgNO_3$

$$? \text{ L AgNO}_3 \text{ soln} = 0.01000 \text{ L CaCl}_2 \text{ soln} \times \frac{2.43 \text{ mol CaCl}_2}{1 \text{ L CaCl}_2 \text{ soln}} \times \frac{2 \text{ mol AgNO}_3}{1 \text{ mol CaCl}_2} \times \frac{1 \text{ L AgNO}_3 \text{ soln}}{1.13 \text{ mol AgNO}_3}$$

= **0.0430 L $AgNO_3$ soln**

4.84 ***See Sections 3.1, 4.1, 4.3, 4.4, and Examples 3.1, 3.2, 3.3, 4.11.***

Unbalanced: $H_2SO_4(aq) + NaOH(s) \rightarrow 2H_2O(\ell) + Na_2SO_4(aq)$ Start with Na_2SO_4.
Step 1: $H_2SO_4(aq) + \underline{2}NaOH(s) \rightarrow H_2O(\ell) + Na_2SO_4(aq)$ Balances Na.
Step 2: $H_2SO_4(aq) + 2NaOH(s) \rightarrow 2H_2O(\ell) + Na_2SO_4(aq)$ Balances H, O.

Strategy: L NaOH soln → mol NaOH → mol H_2SO_4 → L H_2SO_4 soln

$$? \text{ L H}_2\text{SO}_4 \text{ soln} = 0.02000 \text{ L NaOH soln} \times \frac{3.436 \text{ mol NaOH}}{1 \text{ L NaOH soln}} \times \frac{1 \text{ mol H}_2\text{SO}_4}{2 \text{ mol NaOH}} \times \frac{1 \text{ L H}_2\text{SO}_4 \text{ soln}}{2.112 \text{ mol H}_2\text{SO}_4}$$

= **0.01627 L H_2SO_4 soln**

(b) Unbalanced: $Pb(NO_3)_2(aq) + H_2SO_4(aq) \rightarrow PbSO_4(s) + HNO_3(aq)$ Start with $Pb(NO_3)_2$.
Step 1: $Pb(NO_3)_2(aq) + H_2SO_4(aq) \rightarrow PbSO_4(s) + 2HNO_3(aq)$ Balances NO_3 units, H.

Strategy: L H_2SO_4 soln → mol H_2SO_4 → mol $Pb(NO_3)_2$ → L $Pb(NO_3)_2$ soln

$$? \text{ L Pb(NO}_3)_2 \text{ soln} = 0.03000 \text{ L H}_2\text{SO}_4 \text{ soln} \times \frac{0.3030 \text{ mol H}_2\text{SO}_4}{1 \text{ L H}_2\text{SO}_4 \text{ soln}} \times \frac{1 \text{ mol Pb(NO}_3)_2}{1 \text{ mol H}_2\text{SO}_4} \times \frac{1 \text{ L Pb(NO}_3)_2 \text{ soln}}{1.787 \text{ mol Pb(NO}_3)_2}$$

= **5.087×10^{-3} L $Pb(NO_3)_2$ soln**

4.85 ***See Sections 3.1, 4.3, 4.4, and Examples 3.3, 4.12.***

Unbalanced: $CH_3COOH(aq) + Ba(OH)_2(aq) \rightarrow H_2O(\ell) + Ba(CH_3COO)_2(aq)$ Start with $Ba(CH_3COO)_2$.
Step 1: $\underline{2}CH_3COOH(aq) + Ba(OH)_2(aq) \rightarrow H_2O(\ell) + Ba(CH_3COO)_2(aq)$ Balances CH_3COO units.
Step 2: $2CH_3COOH(aq) + Ba(OH)_2(aq) \rightarrow \underline{2}H_2O(\ell) + Ba(CH_3COO)_2(aq)$ Balances H, O.

Strategy: L $Ba(OH)_2$ soln → mol $Ba(OH)_2$ → mol CJ_3COOH → M CH_3COOH

$$? \text{ mol CH}_3\text{COOH} = 0.00332 \text{ L Ba(OH)}_2 \text{ soln} \times \frac{0.0100 \text{ mol Ba(OH)}_2}{1 \text{ L Ba(OH)}_2 \text{ soln}} \times \frac{2 \text{ mol CH}_3\text{COOH}}{1 \text{ mol Ba(OH)}_2}$$

$= 6.64 \times 10^{-5}$ mol CH_3COOH

$$? M \text{ CH}_3\text{COOH} = \frac{6.64 \times 10^{-5} \text{ mol CH}_3\text{COOH}}{0.01000 \text{ L CH}_3\text{COOH soln}} = \mathbf{6.64 \times 10^{-3}\ \mathit{M}\ CH_3COOH\ soln}$$

4.86 ***See Sections 4.3. 4.4, and Example 4.13.***

Balanced: $KHC_8H_4O_2(s) + NaOH(aq) \rightarrow H_2O(\ell) + KNaC_8H_4O_2(aq)$

Strategy: g $KHC_8H_4O_2$ → mol $KHC_8H_4O_2$ → mol NaOH → M NaOH

$$? \text{ mol NaOH} = 0.0220 \text{ g KHC}_8\text{H}_4\text{O}_4 \times \frac{1 \text{ mol KHC}_8\text{H}_4\text{O}_2}{204.22 \text{ g KHC}_8\text{H}_4\text{O}_2} \times \frac{1 \text{ mol NaOH}}{1 \text{ mol KHC}_8\text{H}_4\text{O}_2} = 1.08 \times 10^{-4} \text{ mol NaOH}$$

$$? \, M \text{ NaOH} = \frac{1.08 \times 10^{-4} \text{ mol NaOH}}{0.03500 \text{ L NaOH soln}} = \mathbf{3.09 \times 10^{-3}} \, \boldsymbol{M} \textbf{ NaOH soln}$$

4.87 ***See Sections 3.1, 4.3, 4.4, and Examples 3.3, 4.12.***

Unbalanced:	$H_3C_6H_5O_7(aq) + KOH(aq) \rightarrow$	$H_2O(\ell) + K_3C_6H_5O_7(aq)$	Start with $K_3C_6H_5O_7$.
Step 1:	$H_3C_6H_5O_7(aq) + \underline{3}KOH(aq) \rightarrow$	$H_2O(\ell) + K_3C_6H_5O_7(aq)$	Balances K.
Step 2:	$H_3C_6H_5O_7(aq) + 3KOH(aq) \rightarrow$	$\underline{3}H_2O(\ell) + K_3C_6H_5O_7(aq)$	Balances H, O.

Strategy: L KOH soln→ mol KOH soln → mol $H_3C_6H_5O_7$ → M $H_3C_6H_5O_7$

$$? \text{ mol } H_3C_6H_5O_7 = 0.01510 \text{ L KOH soln} \times \frac{0.0100 \text{ mol KOH}}{1 \text{ L KOH soln}} \times \frac{1 \text{ mol } H_3C_6H_5O_7}{3 \text{ mol KOH}} = 5.03 \times 10^{-5} \text{ mol } H_3C_6H_5O_7$$

$$? \text{ mol } H_3C_6H_5O_7 = \frac{5.03 \times 10^{-5} \text{ mol } H_3C_6H_5O_7}{0.0300 \text{ L } H_3C_6H_5O_7 \text{ soln}} = \mathbf{1.68 \times 10^{-3}} \, \boldsymbol{M} \, \mathbf{H_3C_6H_5O_7} \textbf{ soln}$$

4.88 ***See Sections 3.1, 4.3, 4.4, and Examples 3.3, 4.12.***

Unbalanced:	$H_2C_2O_4(aq) + NaOH(aq) \rightarrow$	$H_2O(\ell) + Na_2C_2O_4(aq)$	Start with$Na_2C_2O_4(aq)$.
Step 1:	$H_2C_2O_4(aq) + \underline{2}NaOH(aq) \rightarrow$	$H_2O(\ell) + Na_2C_2O_4(aq)$	Balances Na.
Step 2:	$H_2C_2O_4(aq) + 2NaOH(aq) \rightarrow$	$\underline{2}H_2O(\ell) + Na_2C_2O_4(aq)$	Balances H,O..

Strategy: L NaOH soln→ mol NaOH soln → mol $H_2C_2O_4$ → M $H_2C_2O_4$ soln

$$? \text{ mol } H_2C_2O_4 = 0.02205 \text{ L NaOH soln} \times \frac{0.100 \text{ mol NaOH}}{1 \text{ L NaOH soln}} \times \frac{1 \text{ mol } H_2C_2O_4}{2 \text{ mol NaOH}} = 1.10 \times 10^{-3} \text{ mol } H_2C_2O_4$$

$$? \, M \, H_2C_2O_4 = \frac{1.10 \times 10^{-3} \text{ mol } H_2C_2O_4}{0.0100 \text{ L } H_2C_2O_4 \text{ soln}} = \mathbf{0.110} \, \boldsymbol{M} \, \mathbf{H_2C_2O_4} \textbf{ soln}$$

4.89 ***See Sections 3.1, 4.3, 4.4, and Examples 3.3, 4.12.***

Unbalanced:	$HCl + Ba(OH)_2 \rightarrow$	$H_2O + BaCl_2$	Start with $BaCl_2$.
Step 1:	$\underline{2}HCl + Ba(OH)_2 \rightarrow$	$H_2O + BaCl_2$	Balances Cl.
Step 2:	$2HCl + Ba(OH)_2 \rightarrow$	$\underline{2}H_2O + BaCl_2$	Balances H and O.

Strategy: L HCl→ mol HCl → mol $Ba(OH)_2$ → M $Ba(OH)_2$ soln

$$? \text{ mol Ba(OH)}_2 = \left[\left(0.075 \text{ L HCl soln} \times \frac{0.10 \text{ mol HCl}}{1 \text{ L HCl soln}}\right) + \left(0.035 \text{ L HCl soln} \times \frac{0.012 \text{ mol HCl}}{1 \text{ L HCl soln}}\right)\right] \times \frac{1 \text{ mol Ba(OH)}_2}{2 \text{ mol HCl}} = 4.0 \times 10^{-3} \text{ mol Ba(OH)}_2$$

$$? \, M \text{ Ba(OH)}_2 \text{ soln} = \frac{4.0 \times 10^{-3} \text{ mol Ba(OH)}_2}{0.125 \text{ L Ba(OH)}_2 \text{ soln}} = \mathbf{3.2 \times 10^{-2}} \, \boldsymbol{M} \, \mathbf{Ba(OH)_2}$$

4.90 ***See Sections 4.1, 4.2. 4.4, and Examples 4.2, 4.4, 4.12.***

The ions present from the reactants are $Ag^+(aq)$, NO_3^- (aq), $K^+(aq)$, and $Cl^-(aq)$. The two possible products are AgCl and KNO_3, and Table 4.1 indicates AgCl is insoluble.

Balanced: $AgNO_3(aq) + KCl(aq) \rightarrow AgCl(s) + KNO_3(aq)$

Strategy: $g\ KCl \rightarrow M\ KCl$ *and* $L\ KCl\ soln \rightarrow mol\ KCl \rightarrow mol\ AgNO_3 \rightarrow M\ AgNO_3$

$$?\ M\ KCl\ soln = \frac{14.2\ g\ KCl}{1.00\ L\ KCl\ soln} \times \frac{1\ mol\ KCl}{74.6\ g\ KCl} = 0.190\ M\ KCl\ soln$$

$$?\ mol\ AgNO_3 = 0.0250\ L\ KCl\ soln \times \frac{0.190\ mol\ KCl}{1\ L\ KCl\ soln} \times \frac{1\ mol\ AgNO_3}{1\ mol\ KCl} = 4.75 \times 10^{-3}\ mol\ AgNO_3$$

$$?\ M\ AgNO_3 = \frac{4.75 \times 10^{-3}\ mol\ AgNO_3}{0.0332\ L\ AgNO_3\ soln} = \mathbf{0.143\ M\ AgNO_3\ soln}$$

4.91 ***See Sections 4.1, 4.2, 4.4, and Examples 4.2, 4.13, 4.14.***

The ions present from the reactants are $K^+(aq)$, $CO_3^{2-}(aq)$, $Ca^{2+}(aq)$, and $Cl^-(aq)$. The two possible products are KCl and $CaCO_3$, and Table 4.1 indicates $CaCO_3$ is insoluble.

Unbalanced:	$K_2CO_3(aq) + CaCl_2(aq) \rightarrow KCl(aq) + CaCO_3(s)$	Start with $CaCl_2$.
Step 1:	$K_2CO_3(aq) + CaCl_2(aq) \rightarrow \underline{2}KCl(aq) + CaCO_3(s)$	Balances K, Cl.

Strategy: $g\ CaCO_3 \rightarrow mol\ CaCO_3 \rightarrow mol\ CaCl_2 \rightarrow M\ CaCl_2\ soln$

$$?\ mol\ CaCl_2 = 4.50\ g\ CaCO_3 \times \frac{1\ mol\ CaCO_3}{100.1\ g\ CaCO_3} \times \frac{1\ mol\ CaCl_2}{1\ mol\ CaCO_3} = 0.0450\ mol\ CaCl_2$$

$$?\ M\ CaCl_2\ soln = \frac{0.0450\ mol\ CaCl_2}{0.300\ L\ CaCl_2\ soln} = \mathbf{0.150\ M\ CaCl_2}$$

4.92 ***See Sections 4.1, 4.2, 4.4, and Examples 4.2, 4.13, 4.14.***

The ions present from the reactants are $Ag^+(aq)$, $NO_3^-(aq)$, $Mg^{2+}(aq)$, and $Cl^-(aq)$. The two possible products are AgCl and $Mg(NO_3)_2$, and Table 4.1 indicates AgCl is insoluble.

Unbalanced:	$AgNO_3(aq) + MgCl_2(aq) \rightarrow AgCl(s) + Mg(NO_3)_2(aq)$	Start with $Mg(NO_3)_2$.
Step 1:	$\underline{2}AgNO_3(aq) + MgCl_2(aq) \rightarrow AgCl(s) + Mg(NO_3)_2(aq)$	Balances NO_3 units.
Step 2:	$2AgNO_3(aq) + MgCl_2(aq) \rightarrow \underline{2}AgCl(s) + Mg(NO_3)_2(aq)$	Balances Ag, Cl.

Strategy: $g\ AgCl \rightarrow mol\ AgCl \rightarrow mol\ AgNO_3 \rightarrow M\ AgNO_3$

$$?\ mol\ AgNO_3 = 2.21\ g\ AgCl \times \frac{1\ mol\ AgCl}{143.35\ g\ AgCl} \times \frac{1\ mol\ AgNO_3}{1\ mol\ AgCl} = 0.0154\ mol\ AgNO_3$$

$$?\ M\ AgNO_3\ soln = \frac{0.0154\ mol\ AgNO_3}{0.246\ L\ AgNO_3\ soln} = \mathbf{0.0626\ M\ AgNO_3\ soln}$$

4.93 ***See Section 4.4 and Example 4.14.***

Strategy: $g\ CaCO_3 \rightarrow g\ Ca \rightarrow \%\ Ca$ *in compound.*

$$?\ g\ Ca = 1.22\ g\ CaCO_3 \times \frac{40.1\ g\ Ca}{100.1\ g\ CaCO_3} = 0.489\ g\ Ca$$

$$\%Ca = \frac{g\ Ca}{g\ compound} \times 100\% = \frac{0.489\ g\ Ca}{2.11\ g\ compound} \times 100\% = \mathbf{23.2\%\ Ca}$$

4.94 *See Section 4.4 and Example 4.14.*

Strategy: g AgCl → g Ag → % Ag in compound.

$$? \text{ g Ag} = 2.02 \text{ g AgCl} \times \frac{107.9 \text{ g Ag}}{143.4 \text{ g AgCl}} = 1.52 \text{ g Ag}$$

$$\% \text{ Ag} = \frac{\text{g Ag}}{\text{g compound}} \text{ x } 100\% = \frac{1.52 \text{ g Ag}}{3.13 \text{ g compound}} \text{ x } 100\% = \mathbf{48.6\% \text{ Ag}}$$

4.95 *See Section 4.4 and Example 4.14.*

Strategy: g AgCl → g Ag → % Ag in sterling silver.

$$? \text{ g Ag} = 0.435 \text{ g AgCl} \times \frac{107.9 \text{ g Ag}}{143.4 \text{ g AgCl}} = 0.327 \text{ g Ag}$$

$$\% \text{ Ag} = \frac{\text{g Ag}}{\text{g sample}} \text{ x } 100\% = \frac{0.327 \text{ g Ag}}{0.360 \text{ g compound}} \text{ x } 100\% = \mathbf{90.9\% \text{ Ag}}$$

4.96 *See Section 4.4 and Example 4.14.*

$$\% \text{ Cu} = \frac{\text{g Cu}}{\text{g sample}} \times 100\% = \frac{4.33 \text{ g Cu}}{15.5 \text{ g sample}} \times 100\% = \mathbf{27.9\% \text{ Cu.}}$$

4.97 *See Section 4.4 and Examples 4.13, 4.14.*

Balanced: $AgNO_3(aq) + NaCl(aq) \rightarrow AgCl(s) + NaNO_3(aq)$

Strategy: g AgCl → mol AgCl → mol NaCl → M NaCl soln

$$? \text{ mol NaCl} = 0.0112 \text{ g AgCl} \times \frac{1 \text{ mol AgCl}}{143.4 \text{ g AgCl}} \times \frac{1 \text{ mol NaCl}}{1 \text{ mol AgCl}} = 7.81 \text{ x } 10^{-5} \text{ mol NaCl}$$

$$? M \text{ NaCl} = \frac{7.81 \times 10^{-5} \text{ mol NaCl}}{0.020 \text{ L NaCl soln}} = \mathbf{3.9 \text{ x } 10^{-3}} \; \boldsymbol{M} \textbf{ NaCl soln}$$

4.98 *See Section 4.4 and Examples 4.13, 4.14.*

Unbalanced:	$K_2SO_4(aq) + BaCl_2(aq) \rightarrow KCl(aq) + BaSO_4(aq)$	Start with K_2SO_4
Step 1:	$K_2SO_4(aq) + BaCl_2(aq) \rightarrow \underline{2}KCl(aq) + BaSO_4(aq)$	Balances K, Cl.

Strategy: g $BaSO_4$ → mol $BaSO_4$ → mol K_2SO_4 → M K_2SO_4 soln

$$? \text{ mol } K_2SO_4 = 0.233 \text{ g } BaSO_4 \times \frac{1 \text{ mol } BaSO_4}{233.4 \text{ g } BaSO_4} \times \frac{1 \text{ mol } K_2SO_4}{1 \text{ mol } BaSO_4} = 9.98 \text{ x } 10^{-4} \text{ mol } K_2SO_4$$

$$? M \; K_2SO_4 \text{ soln} = \frac{9.98 \times 10^{-4} \text{ mol } K_2SO_4}{0.100 \text{ L } K_2SO_4 \text{ soln}} = \mathbf{9.98 \text{ x } 10^{-3}} \; \boldsymbol{M} \; \mathbf{K_2SO_4} \textbf{ soln}$$

4.99 *See Section 4.1 and Table 4.1.*

Table 4.1 indicates only $Be(OH)_2$ will precipitate.

4.100 *See Section 4.1 and Table 4.1.*

Table 4.1 indicates Hg^{2+} can be removed from water by adding Na_2CO_3 or Na_3PO_4 to form insoluble $HgCO_3$ or $Hg_3(PO_4)_2$.

4.101 *See Sections 4.1 and 2 and Examples 4.1, 4 and 7.*

$$? M \text{ original NaOH} = \frac{5.30 \text{ g NaOH}}{1.00 \text{ L NaOH soln}} \times \frac{1 \text{ mol NaOH}}{40.00 \text{ g NaOH}} = \frac{0.132 \text{ mol NaOH}}{\text{L NaOH soln}} = 0.132\, M \text{ NaOH}$$

Solving M(con) x V(con) = M(dil) x V(dil) for M(dil) gives

$$M(\text{dil}) = \frac{M(\text{con}) \times V(\text{con})}{V(\text{dil})} \qquad M(\text{dil}) = \frac{0.132\, M \times 0.1000 \text{ L}}{0.5000 \text{ L}} = 0.00264\, M \text{ NaOH(dilute)}$$

Unbalanced:	$H_2SO_4(aq) + NaOH(s)$	→	$2H_2O(\ell) + Na_2SO_4(aq)$	Start with Na_2SO_4.
Step 1:	$H_2SO_4(aq) + \underline{2}NaOH(s)$	→	$H_2O(\ell) + Na_2SO_4(aq)$	Balances Na.
Step 2:	$H_2SO_4(aq) + 2NaOH(s)$	→	$2H_2O(\ell) + Na_2SO_4(aq)$	Balances H, O.

Strategy: $L\ H_2SO_4$ *soln* → *mol* H_2SO_4 → *mol NaOH* → *L NaOH soln*

$$? \text{ L dil NaOH} = 0.0330 \text{ L } H_2SO_4 \text{ soln} \times \frac{0.0220 \text{ mol } H_2SO_4}{1 \text{ L } H_2SO_4 \text{ soln}} \times \frac{2 \text{ mol NaOH}}{1 \text{ mol } H_2SO_4} \times \frac{1 \text{ L dilute NaOH soln}}{0.0264 \text{ mol NaOH}}$$

$$= \mathbf{0.0550 \text{ L dilute NaOH soln}}$$

4.102 *See Section 4.2 and Examples 4.5, 4.6.*

Strategy: *L soln* → *mol* $NaHCO_3$ → *g* $NaHCO_3$

$$? \text{ g } NaHCO_3 = 0.0100 \text{ L soln} \times \frac{0.0332 \text{ mol } NaHCO_3}{1 \text{ L soln}} \times \frac{84.01 \text{ g } NaHCO_3}{1 \text{ mol } NaHCO_3} = 0.279 \text{ g } NaHCO_3$$

Strategy: *L soln* → *mol* K_2CO_3 → *g* K_2CO_3

$$? \text{ g } K_2CO_3 = 0.0100 \text{ L soln} \times \frac{0.222 \text{ mol } K_2CO_3}{1 \text{ L soln}} \times \frac{138.2 \text{ g } K_2CO_3}{1 \text{ mol } K_2CO_3} = 0.307 \text{ g } K_2CO_3$$

? mass solid = g $NaHCO_3$ + g K_2CO_3 = 0.279 g + 0.307 g = **0.586 g**

4.103 *See Section 4.2 and Examples 4.4, 4.7.*

Strategy: *L NaOH soln* → *mol NaOH* → *mol* OH^-
L $Ba(OH)_2$ *soln* → *mol* $Ba(OH)_2$ → *mol* OH^-
Total mol OH^- → M OH^- *soln*

$$? \text{ mol } OH^- \text{ from NaOH} = 0.2000 \text{ L NaOH soln} \times \frac{0.0123 \text{ mol NaOH}}{1 \text{ L NaOH soln}} \times \frac{1 \text{ mol } OH^-}{1 \text{ mol NaOH}} = 2.46 \times 10^{-3} \text{ mol } OH^-$$

$$? \text{ mol } OH^- \text{ from } Ba(OH)_2 = 0.200 \text{ L NaOH soln} \times \frac{0.0154 \text{ mol } Ba(OH)_2}{1 \text{ L } Ba(OH)_2 \text{ soln}} \times \frac{2 \text{ mol } OH^-}{1 \text{ mol } Ba(OH)_2}$$

$$= 6.16 \times 10^{-3} \text{ mol } OH^-.$$

Total mol OH^- = 2.46×10^{-3} mol OH^- + 6.16×10^{-3} mol OH^- = 8.62×10^{-3} mol OH^-
Total volume soln = 500.0 mL soln = 0.5000 L soln

$$? M\ OH^- = \frac{8.62 \times 10^{-3}\ \text{mol}\ OH^-}{0.5000\ \text{L soln}} = \mathbf{0.0172\ \mathit{M}\ OH^-\ soln}$$

4.104 ***See Section 4.2 and Examples 4.4, 4.7.***

Strategy: *L NaCl soln → mol NaCl → mol Cl^-*
L $CaCl_2$ soln → mol $CaCl_2$ → mol Cl^-
Total mol Cl^- → M Cl^- soln

$$? \text{mol}\ Cl^-\ \text{from NaCl} = 0.150\ \text{L NaCl soln} \times \frac{1.54\ \text{mol NaCl}}{1\ \text{L NaCl soln}} \times \frac{1\ \text{mol}\ Cl^-}{1\ \text{mol NaCl}} = 0.231\ \text{mol}\ Cl^-$$

$$? \text{mol}\ Cl^-\ \text{from}\ CaCl_2 = 0.2000\ \text{L}\ CaCl_2\ \text{soln} \times \frac{2.00\ \text{mol}\ CaCl_2}{1\ \text{L}\ CaCl_2\ \text{soln}} \times \frac{2\ \text{mol}\ Cl^-}{1\ \text{mol}\ CaCl_2} = 0.800\ \text{mol}\ Cl^-$$

Total mol Cl^- = 0.231 mol Cl^- + 0.800 mol Cl^- = 1.031 mol Cl^-.
Total volume soln = 500.0 mL soln = 0.5000 L soln

$$? M\ Cl^- = \frac{1.031\ \text{mol}\ Cl^-}{.5000\ \text{L soln}} = \mathbf{2.062\ \mathit{M}\ Cl^-}$$

4.105 ***See Sections 3.1, 4.2, 4.3, 4.4, and Examples 3.3, 4.5, 4.12.***

Unbalanced:	$H_2SO_4(aq) + NaOH(s)$	→	$2H_2O(\ell) + Na_2SO_4(aq)$	Start with Na_2SO_4.
Step 1:	$H_2SO_4(aq) + \underline{2}NaOH(s)$	→	$H_2O(\ell) + Na_2SO_4(aq)$	Balances Na.
Step 2:	$H_2SO_4(aq) + 2NaOH(s)$	→	$2H_2O(\ell) + Na_2SO_4(aq)$	Balances H, O.

Strategy: *L H_2SO_4 soln → mol H_2SO_4 → mol NaOH in 50.0 mL NaOH soln → mol NaOH in 1.00 L NaOH soln → g NaOH*

$$? \text{mol NaOH in 50.0 mL NaOH soln} = 0.0100\ \text{mL}\ H_2SO_4\ \text{soln} \times \frac{3.11\ \text{mol}\ H_2SO_4}{1\ \text{L}\ H_2SO_4\ \text{soln}} \times \frac{2\ \text{mol NaOH}}{1\ \text{mol}\ H_2SO_4}$$

$$= 0.0622\ \text{mol NaOH}$$

$$? \text{mol NaOH in 1.0 L NaOH soln} = \frac{0.0622\ \text{mol NaOH}}{0.0500\ \text{L NaOH soln}} \times 1.00\ \text{L NaOH soln} = 1.24\ \text{mol NaOH}$$

$$? \text{g NaOH for 1.0 L NaOH soln} = 1.24\ \text{mol NaOH} \times \frac{40.0\ \text{g NaOH}}{1\ \text{mol NaOH}} = \mathbf{49.6\ g\ NaOH}$$

4.106 ***See Sections 4.4, 4.12, though this is oxidation-reduction rather than acid-base.***

Balanced: $2Na_2S_2O_3(aq) + I_2(aq) \rightarrow Na_2S_4O_6(aq) + 2NaI(aq)$

Strategy: *L I_2 soln → mol I_2 → mol $Na_2S_2O_3$ → M $Na_2S_2O_3$ soln*

$$? \text{mol}\ Na_2S_2O_3 = 0.03030\ \text{L}\ I_2\ \text{soln} \times \frac{0.1120\ \text{mol}\ I_2}{1\ \text{L}\ I_2\ \text{soln}} \times \frac{2\ \text{mol}\ Na_2S_2O_3}{1\ \text{mol}\ I_2} = 6.787 \times 10^{-3}\ \text{mol}\ Na_2S_2O_3$$

$$? M\ Na_2S_2O_3 = \frac{6.787 \times 10^{-3}\ \text{mol}\ Na_2S_2O_3}{0.100\ \text{L}\ Na_2S_2O_3\ \text{soln}} = \mathbf{0.06787\ \mathit{M}\ Na_2S_2O_3\ soln}$$

4.107 ***See Section 4.3 and Example 4.11.***

Balanced: $3H_2SO_4(aq) + 3NaNO_2(aq) \rightarrow 2NO(g) + HNO_3(aq) + 3\,NaHSO_4(aq) + H_2O(\ell)$

Strategy: $g\ NO \rightarrow mol\ NO \rightarrow mol\ H_2SO_4 \rightarrow L\ H_2SO_4\ soln$

$$?\ L\ H_2SO_4\ soln = 2.44\ g\ NO \times \frac{1\ mol\ NO}{30.0\ g\ NO} \times \frac{3\ mol\ H_2SO_4}{2\ mol\ NO} \times \frac{1\ L\ H_2SO_4\ soln}{1.22\ mol\ H_2SO_4} = \mathbf{0.100\ L\ H_2SO_4\ soln}$$

4.108 ***See Sections 3.1, 4.3, and Examples 3.1, 3.2, 4.13, 4.14.***

Unbalanced:	$AgCl(s) + NH_3(aq) \rightarrow Ag(NH_3)_2^+(aq) + Cl^-(aq)$		Start with $Ag(NH_3)_2^+$.
Step 1:	$AgCl(s) + \underline{2}NH_3(aq) \rightarrow Ag(NH_3)_2^+(aq) + Cl^-(aq)$		Balances N, H.

Strategy: $g\ AgCl \rightarrow mol\ AgCl \rightarrow mol\ Ag(NH_3)_2^+ \rightarrow M\ Ag(NH_3)_2^+$

$$?\ mol\ Ag(NH_3)_2^+ = 0.022\ g\ AgCl \times \frac{1\ mol\ AgCl}{143.4\ g\ AgCl} \times \frac{1\ mol\ Ag(NH_3)_2^+}{1\ mol\ AgCl} = 1.53 \times 10^{-4}\ mol\ Ag(NH_3)_2^+$$

$$?\ M\ Ag(NH_3)_2^+ = \frac{1.53 \times 10^{-4}\ mol\ Ag(NH_3)_2^+}{0.150\ L\ soln} = \mathbf{1.02 \times 10^{-3}\ \mathit{M}\ Ag(NH_3)_2^+}$$

4.109 ***See Section 4.4 and Example 4.14.***

Strategy: $L\ NaCl \rightarrow mol\ NaCl \rightarrow g\ NaCl \rightarrow \%\ NaCl$

$$?\ g\ NaCl = 0.250\ L\ NaCl\ soln \times \frac{1.23\ mol\ NaCl}{1\ L\ NaCl\ soln} \times \frac{58.44\ g\ NaCl}{1\ mol\ NaCl} = 18.0\ g\ NaCl$$

$$\%\ Na = \frac{g\ NaCl}{g\ sample} \times 100\% = \frac{18.0\ g\ NaCl}{83.5\ g\ sample} \times 100\% = \mathbf{21.6\ \%\ NaCl}$$

4.110 ***See Sections 3.3, 4.4, and Example 3.11.***

Strategy: $g\ AgCl \rightarrow g\ Cl \rightarrow mol\ Cl$
$g\ sample - g\ Cl \rightarrow g\ Al \rightarrow mol\ Al$

$$?\ g\ Cl = 1.006\ g\ AgCl \times \frac{35.5\ g\ Cl}{143.4\ g\ AgCl} = 0.2490\ g\ Cl$$

$g\ Al = g\ sample - g\ Cl = 0.3120\ g - 0.2490\ g\ Cl = 0.0630\ g\ Al$

$$?\ mol\ Al = 0.0630\ g\ Al \times \frac{1\ mol\ Al}{27.0\ g\ Al} = 2.33 \times 10^{-3}\ mol\ Al \quad \text{relative mol Al} = \frac{2.33 \times 10^{-3}\ mol\ Al}{2.33 \times 10^{-3}} = 1.00\ mol\ Al$$

$$?\ mol\ Cl = 0.2490\ g\ Cl \times \frac{1\ mol\ Cl}{35.453\ g\ Cl} = 7.02 \times 10^{-3}\ mol\ Cl \quad \text{relative mol Cl} = \frac{7.02 \times 10^{-3}\ mol\ Cl}{2.33 \times 10^{-3}} = 3.01\ mol\ Cl$$

The empirical formula is $\mathbf{AlCl_3}$.

4.111 ***See Sections 4.3, 4.4, and Examples 4.10, 4.11.***

The products of the reaction of an acid and a strong base are water and a salt, in this case $BaSO_4$. Table 4.1 indicates $BaSO_4$ is insoluble. Hence,

Unbalanced: $H_2SO_4(aq) + Ba(OH)_2(aq) \rightarrow H_2O(\ell) + BaSO_4(s)$ Start with $Ba(OH)_2$.
Step 1: $H_2SO_4(aq) + Ba(OH)_2(aq) \rightarrow \underline{2}H_2O(\ell) + BaSO_4(s)$ Balances H, O.

Strategy: $L\ H_2SO_4\ soln \rightarrow mol\ H_2SO_4 \rightarrow mol\ BaSO_4 \rightarrow g\ BaSO_4$

$$?\ \text{g BaSO}_4 = 0.0334\ \text{L H}_2\text{SO}_4\ \text{soln} \times \frac{0.0554\ \text{mol H}_2\text{SO}_4}{1\ \text{L H}_2\text{SO}_4\ \text{soln}} \times \frac{1\ \text{mol BaSO}_4}{1\ \text{mol H}_2\text{SO}_4} \times \frac{233.4\ \text{g BaSO}_4}{1\ \text{mol BaSO}_4} = \mathbf{0.432\ g\ BaSO_4}$$

4.112 ***See Section 4.3, 4.4, and Example 4.10, though this is oxidation-reduction rather than precipitation.***

Balanced: $2NH_3(aq) + NaOCl(aq) \rightarrow N_2H_4(aq) + NaCl(aq) + H_2O(\ell)$

Strategy: $L\ NH_3(aq) \rightarrow mol\ NH_3 \rightarrow mol\ N_2H_4 \rightarrow g\ N_2H_4$

$$?\ \text{g N}_2\text{H}_4 = 0.0500\ \text{L NH}_3\ \text{soln} \times \frac{1.22\ \text{mol NH}_3}{1\ \text{L NH}_3\ \text{soln}} \times \frac{1\ \text{mol N}_2\text{H}_4}{2\ \text{mol NH}_3} \times \frac{32.06\ \text{g N}_2\text{H}_4}{1\ \text{mol N}_2\text{H}_4} = 0.979\ \text{g N}_2\text{H}_4$$

Strategy: $L\ NaOCl\ soln \rightarrow mol\ NaOCl \rightarrow mol\ N_2H_4 \rightarrow g\ N_2H_4$

$$?\ \text{g N}_2\text{H}_4 = 0.1000\ \text{L NaOCl soln} \times \frac{0.440\ \text{mol NaOCl}}{1\ \text{L NaOCl soln}} \times \frac{1\ \text{mol N}_2\text{H}_4}{1\ \text{mol NaOCl}} \times \frac{32.06\ \text{g N}_2\text{H}_4}{1\ \text{mol N}_2\text{H}_4} = 1.41\ \text{g N}_2\text{H}_4$$

NH_3 is the limiting reactant because it produces less N_2H_4. The maximum amount of N_2H_4 which can be produced from 50.0 mL of 1.22 M $NH_3(aq)$ and 100.0 mL of 0.440 M $NaOCl(aq)$ is 0.979 g. Hence, the theoretical yield of N_2H_4 is **0.979 g**.

4.113 ***See Section 4.4 and Example 4.14.***

Balanced: $SnF_2(aq) + 2Pb^{2+}(aq) + 2Cl^-(aq) \rightarrow 2PbClF(s) + Sn^{2+}(aq)$

Strategy: $g\ PbClF \rightarrow mol\ PbClF \rightarrow mol\ SnF_2 \rightarrow g\ SnF_2 \rightarrow \%\ SnF_2$

$$?\ \text{g SnF}_2 = 0.105\ \text{g PbClF} \times \frac{1\ \text{mol PbClF}}{261.6\ \text{g PbClF}} \times \frac{1\ \text{mol SnF}_2}{2\ \text{mol PbClF}} \times \frac{156.7\ \text{g SnF}_2}{1\ \text{mol SnF}_2} = 0.0314\ \text{g SnF}_2$$

$$\%\ \text{SnF}_2 = \frac{\text{g SnF}_2}{\text{g sample}} \times 100\% = \frac{0.0314\ \text{g SnF}_2}{10.50\ \text{g sample}} \times 100\% = \mathbf{0.299\%\ SnF_2}$$

4.114 ***See Section 4.4 and Example 4.14.***

Strategy: $g\ BaSO_4 \rightarrow g\ Ba \rightarrow \%\ Ba$ *in compound.*

$$?\ \text{g Ba} = 2.2\ \text{g BaSO}_4 \times \frac{137.3\ \text{g Ba}}{233.4\ \text{g BaSO}_4} = 1.3\ \text{g Ba}$$

$$\%\ \text{Ba} = \frac{\text{g Ba}}{\text{g compound}} \times 100\% = \frac{1.3\ \text{g Ba}}{2.3\ \text{g compound}} \times 100\% = \mathbf{57\%\ Ba}$$

4.115 ***See Section 4.2 and Examples 4.5, 4.6.***

Strategy: $L\ Ag^+\ soln \rightarrow mol\ Ag^+ \rightarrow mol\ AgCl \rightarrow g\ AgCl$

$$?\ \text{g AgCl} = 4.00\ \text{L Ag}^+\ \text{soln} \times \frac{0.0438\ \text{mol Ag}^+}{1\ \text{L Ag}^+\ \text{soln}} \times \frac{1\ \text{mol AgCl}}{1\ \text{mol Ag}^+} \times \frac{143.4\ \text{g AgCl}}{1\ \text{mol AgCl}} = \mathbf{25.1\ g\ AgCl}$$

4.116 ***Solutions for these Spreadsheet Exercises are on the Reger, Goode and Mercer Spreadsheet Solutions Diskette available from Saunders College Publishing.***

Chapter 5: Thermochemistry

5.1 *See Section 5.1.*

Since the energy of a substance depends on its physical state, the physical state of every substance must be included in a thermochemical equation.

5.2 *See Section 5.l.*

Kinetic energy is energy that matter possesses because it is in motion. Potential energy is energy that matter possesses because of its position or condition.

Compounds posses potential energy as a result of forces that hold the constituent atoms together in chemical combination. This form of potential energy is called chemical energy.

5.3 *See Sections 5.1, 5.4.*

The heat absorbed by the system at constant temperature and pressure during a reaction is called the enthalpy of reaction. The heat absorbed by the system during a reaction in which one mole of a single compound is formed from its elements is called the enthalpy of formation for that compound. Enthalpies of formation are usually quoted for forming one mole of a compound in its standard state from the most stable form of its elements in their standard states and are therefore called standard enthalpies of formation.

5.4 *See Section 5.1.*

The enthalpy of reaction equals the heat absorbed by the system when the reaction occurs at constant temperature and pressure. (This does not mean there can be no temperature change. Instead, this means when the products are allowed to reach the original temperature of the reactants.)

5.5 *See Section 5.1.*

The reaction of one mole of ethylene gas with three moles of oxygen gas to produce two moles of carbon dioxide gas and two moles of water liquid releases 1,411 kJ of energy as heat when the reaction is conducted at constant temperature and pressure. (This means when the products are allowed to cool to the original temperature of the reactants.)

5.6 *See Section 5.2.*

The insulation prevents any transfer of heat into or out of the calorimeter.

5.7 *See Section 5.2.*

The specific heat tells us the number of joules needed to raise the temperature of one gram of the contents by one degree Celsius and can therefore be used to calculate the enthalpy change for the reaction being conducted in the calorimeter.

5.8 *See Section 5.3.*

The (a) energy expended and (b) time expended depend on the pathway taken and are therefore path functions. The (c) change in altitude and (d) change in potential energy do not depend on the pathway taken (are path independent) and are therefore state functions.

5.9 *See Section 5.3.*

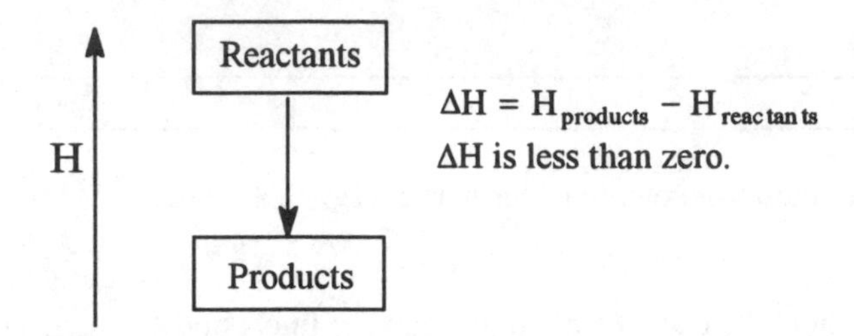

5.10 *See Section 5.3.*

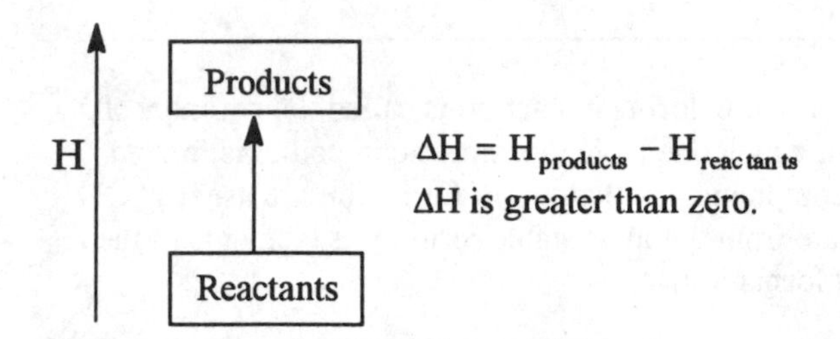

5.11 *See Section 5.4.*

Assuming the reaction is exothermic,

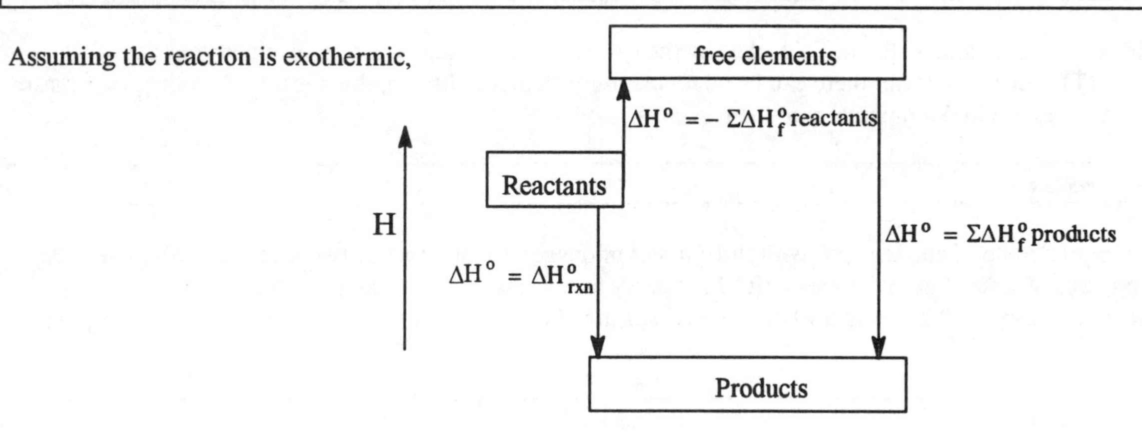

$\Delta H^\circ_{rxn} = -\Sigma\Delta H^\circ_f$ reactants + $\Sigma\Delta H^\circ_f$ products = $\Sigma\Delta H^\circ_f$ products - $\Sigma\Delta H^\circ_f$ reactants

5.12 *See Sections 5.3, 5.4.*

The relation between $\Delta H^\circ_{reaction}$ and enthalpies of formation that is given in Equation 5.13 can be derived using the equation for the thermite reaction. The thermite reaction was once used to weld broken railroad tracks back together and is illustrated in Figure 20.11 of the text. The equation for the thermite reaction is $2Al(s) + Fe_2O_3(s) \rightarrow 2Fe(s) + Al_2O_3(s)$, and the reaction can be considered to be the sum of two other reactions.

$2Al(s) + \frac{3}{2}O_2(g) \rightarrow Al_2O_3(s)$	ΔH°_1
$Fe_2O_3(s) \rightarrow 2Fe(s) + \frac{3}{2}O_2(g)$	ΔH°_2
$2Al(s) + Fe_2O_3(s) \rightarrow 2Fe(s) + Al_2O_3(s)$	ΔH°_3

From the definition of enthalpy of formation, $\Delta H_1^\circ = \Delta H_f^\circ Al_2O_3(s)$ and $\Delta H_2^\circ = -\Delta H_f^\circ Fe_2O_3(s)$.
Hence, applying Hess's law gives $\Delta H_3^\circ = \Delta H_1^\circ + \Delta H_2^\circ = \Delta H_f^\circ Al_2O_3(s) - \Delta H_f^\circ Fe_2O_3(s)$ and this is the same expression that is obtained by applying the general relation $\Delta H^\circ_{reaction} = \Sigma\, \Delta H_f^\circ$ products - $\Sigma\, \Delta H_f^\circ$ reactants to the reaction $2Al(s) + Fe_2O_3(s) \rightarrow 2Fe(s) + Al_2O_3(s)$ because $\Delta H_f^\circ Fe(s)$ and $\Delta H_f^\circ Al(s) = 0.$ kJ / mol.

5.13 *See Section 5.5.*

The ease of mining the coal, the sulfur content of the coal, the cost of labor to mine the coal, the distance between the mine and the location where the coal is to be used, and the profit margin to be realized affect the price of coal.

5.14 *See Section 5.5.*

The cost of crude oil, the cost of equipment to refine the crude oil, the cost of energy to refine the crude oil, and the extent to which the crude oil is refined affect the cost of gasoline, as does the cost of labor to refine the crude oil and the cost of adding performance enhancing additives.

5.15 *See Section 5.6.*

The primary cost of heating water with solar energy is the cost of the equipment. There is an additional cost associated with the need to have an electric backup for heating the water during early morning hours and cloudy days. When spread over the lifetime of the equipment, the cost is currently less expensive than the use of electricity but more expensive than the use of natural gas. This could change as energy costs shift in the future or as more efficient solar collectors are developed.

5.16 *See Section 5.5.*

Most of the sulfur in crude oil is removed in the refining stages, whereas coal refining methods are still in the research stages.

5.17 *See Sections 5.5, 5.6.*

These depend to some extent on the local climate. However, energy use can always be curtailed by turning off the lights in rooms that are not in use.

5.18 *See Sections 5.5, 5.6.*

These depend to some extent on the local climate. However, energy use can be curtailed by adding insulation, installing double-paned windows, turning the heating thermostat down slightly, turning the air conditioning thermostat up slightly, and always turning lights off in rooms not in use. Installing solar devices to curtail the use of electricity are generally not cost effective.

5.19 *See Section 5.6.*

Molar masses of fuels can vary widely. Hence, it is more important to know the amount of energy available per gram of fuel because the mass of the fuel must be moved with the vehicle.

5.20 *See Section 5.6.*

Costs associated with nuclear power generation include cost of design and manufacture of the facility, cost of mining and treating the uranium ore, cost of water to operate the cooling towers, cost of labor to run the facility, cost of complying with government regulations, and cost of storage of radioactive wastes. Only the costs of designing and manufacturing the facility are one-time expenditures.

5.21 *See Sections 5.5, 5.6.*

There are environmental concerns about the effects on the landscape of the terrain in which coal is mined. In addition, the burning of coal and oil contribute to air pollution, the most notable pollutant being carbon dioxide, which contributes to the greenhouse effect. On the other hand, there are environmental problems with the mining of uranium and the storage of radioactive waste with the use of nuclear power plants.

5.22 *See Sections 5.5, 5.6.*

(a) The use of ethanol would be favorable because there is a limited supply of petroleum, from which MTBE is made. It would also benefit the farming industry in the United States. However, the price of ethanol is probably greater than that of MTBE. Furthermore, ethanol is structurally similar to water and is likely to attract and add undesirable water molecules from the atmosphere to tanks of gasoline.

(b) The oxygen content per gram and the cost per gram should be considered to determine which compound places the most oxygen per dollar into the fuel to which it is added. The efficiency with which each compound removes the undesirable exhaust products should also be studied.

(c) The automobile industry is likely to favor the most efficient compound since it would want to avoid any adverse publicity concerning pollution and it is the consumer who will be purchasing the gasoline.

5.23 *See Section 5.1 and Example 5.1.*

(a) The negative sign for the enthalpy change means heat is given off by the system.

(b) $? \Delta H = 3.00 \text{ g C} \times \frac{1 \text{ mol C}}{12.0 \text{ g C}} \times \frac{-393.5 \text{ kJ}}{1 \text{ mol C}} = \mathbf{-98.4 \text{ kJ}}$

5.24 *See Section 5.1 and Example 5.1.*

(a) The negative sign for the enthalpy change means heat is given off by the system and the reaction is exothermic.

(b) $? \Delta H = 1.00 \text{ g } CH_4 \times \frac{1 \text{ mol } CH_4}{16.0 \text{ g } CH_4} \times \frac{-890 \text{ kJ}}{1 \text{ mol } CH_4} = \mathbf{-55.6 \text{ kJ}}$

5.25 *See Section 5.1 and Example 5.1.*

(a) The positive sign for the enthalpy change means heat is absorbed by the system and the reaction is endothermic.

(b) $? \Delta H = 6.00 \text{ g C} \times \frac{1 \text{ mol C}}{12.0 \text{ g C}} \times \frac{-131.3 \text{ kJ}}{1 \text{ mol C}} = \mathbf{+65.6 \text{ kJ}}$

5.26 *See Section 5.1 and Example 5.1.*

(a) $? \Delta H = 1.00 \text{ kg } SO_3 \times \frac{10^3 \text{ g } SO_3}{1 \text{ kg } SO_3} \times \frac{1 \text{ mol } SO_3}{80.07 \text{ g } SO_3} \times \frac{-98.9 \text{ kJ}}{1 \text{ mol } SO_3} = \mathbf{-1.24 \times 10^3 \text{ kJ}}$

(b) The negative sign for the enthalpy change means heat is given off by the system.

5.27 *See Section 5.1 and Example 5.1.*

$$? \Delta H = 2.20 \text{ g } N_2 \times \frac{1 \text{mol } N_2}{28.0 \text{ g } N_2} \times \frac{+180 \text{ kJ}}{1 \text{ mol } N_2} = \mathbf{+14.1 \text{ kJ}}$$

5.28 *See Section 5.1 and Example 5.1.*

$$? \Delta H = 10.0 \text{ g } O_2 \times \frac{1 \text{mol } O_2}{32.0 \text{ g } O_2} \times \frac{-572 \text{ kJ}}{1 \text{ mol } O_2} = \mathbf{-179 \text{ kJ}}$$

5.29 *See Section 5.1 and Example 5.1.*

$$? \Delta H = 12.2 \text{ g Al} \times \frac{1 \text{ mol Al}}{27.0 \text{ g Al}} \times \frac{-852 \text{ kJ}}{2 \text{ mol Al}} = \mathbf{-192 \text{ kJ}}$$

5.30 *See Section 5.1 and Example 5.5.*

$$? \Delta H = 33.2 \text{ g C} \times \frac{1 \text{ mol C}}{12.0 \text{ g C}} \times \frac{+227 \text{ kJ}}{2 \text{ mol C}} = \mathbf{+314 \text{ kJ}}$$

5.31 *See Section 5.1 and Example 5.11.*

(a) $C_8H_{18}(\ell) + \frac{25}{2}O_2(g) \rightarrow 8CO_2(g) + 9H_2O(\ell)$ $\Delta H = -5.46 \text{ X } 10^3 \text{ kJ}$

(b) $$? \Delta H = 10.0 \text{ g } C_8H_{18} \times \frac{1 \text{ mol } C_8H_{18}}{114.1 \text{ g } C_8H_{18}} \times \frac{-5.46 \text{ x } 10^3 \text{ kJ}}{1 \text{ mol } C_8H_{18}} = \mathbf{-479 \text{ kJ}}$$

5.32 *See Section 5.1, Example 5.1, and Solution for Exercise 5.31.*

(a) $CH_3OH(\ell) + \frac{3}{2}O_2(g) \rightarrow CO_2(g) + 2H_2O(\ell)$ $\Delta H = -727 \text{ kJ}$

(b) $$? \Delta H = 10.0 \text{ g } CH_3OH \times \frac{1 \text{ mol } CH_3OH}{32.0 \text{ g } CH_3OH} \times \frac{-727 \text{ kJ}}{1 \text{ mol } CH_3OH} = \mathbf{-227 \text{ kJ}}$$

(c) The combustion of 10.0 g $C_8H_{18}(\ell)$ releases 479 kJ and therefore releases more heat than the 227 kJ that is released by the combustion of 10.0 g $CH_3OH(\ell)$.

5.33 *See Section 5.1 and Example 5.1.*

$$? \Delta H = \frac{-95.44 \text{ kJ}}{2.00 \text{ g } C_8H_{18}} \times 228.3 \text{ g } C_8H_{18} \text{ for 2 moles } C_8H_{18} = \mathbf{-1.09 \text{ x } 10^4 \text{ kJ}}$$

5.34 *See Section 5.1 and Example 5.1.*

(a) Since heat is released, the reaction is exothermic and the sign of ΔH is negative.

(b) $$? \Delta H = \frac{-124 \text{ kJ}}{10.0 \text{ g } N_2O_4} \times 92.0 \text{ g } N_2O_4 \text{ for one mol } N_2O_4 = \mathbf{-1.14 \text{ x } 10^3 \text{ kJ}}$$

5.35 *See Section 5.1 and Example 5.1.*

$$? \Delta H = \frac{-5.65 \text{ kJ}}{3.00 \text{ g NO}} \times 60.0 \text{ g NO for 2 moles NO} = \mathbf{-1.13 \times 10^2 \text{ kJ}}$$

5.36 *See Section 5.1 and Example 5.1.*

(a) Since heat is released, the reaction is exothermic and the sign of ΔH is negative.

(b) $$? \Delta H = \frac{-13.56 \text{ kJ}}{5.00 \text{ g } NH_3} \times 34.0 \text{ g for 2 moles } NH_3 = \mathbf{-92.2 \text{ kJ}}$$

5.37 *See Section 5.1 and Example 5.1.*

$$? \Delta H = 2.00 \text{ g } C_2H_5OH \times \frac{1 \text{ mol } C_2H_5OH}{46.0 \text{ g } C_2H_5OH} \times \frac{-1{,}366.8 \text{ kJ}}{1 \text{ mol } C_2H_5OH} = \mathbf{-59.4 \text{ kJ}}$$

5.38 *See Section 5.1 and Example 5.2.*

$$? \text{ g } C_2H_5OH = 500 \text{ J} \times \frac{1 \text{ kJ}}{10^3 \text{ J}} \times \frac{1 \text{ mol } C_2H_5OH}{1{,}366.8 \text{ kJ}} \times \frac{46.0 \text{ g } C_2H_5OH}{1 \text{ mol } C_2H_5OH} = \mathbf{0.0168 \text{ g}}$$

5.39 *See Section 5.1 and Example 5.2.*

$$? \text{ g } CH_4 = 1.00 \text{ g } H_2O \times \frac{1 \text{ mol } H_2O}{18.0 \text{ g } H_2O} \times \frac{44.0 \text{ kJ}}{1 \text{ mol } H_2O} \times \frac{1 \text{ mol } CH_4}{890 \text{ kJ}} \times \frac{16.0 \text{ g } CH_4}{1 \text{ mol } CH_4} = \mathbf{0.0439 \text{ g } CH_4}$$

5.40 *See Section 5.1 and Example 5.2.*

$$? \text{ g } C_2H_5OH = 10.0 \text{ g } H_2O \times \frac{1 \text{ mol } H_2O}{18.0 \text{ g } H_2O} \times \frac{6.01 \text{ kJ}}{1 \text{ mol } H_2O} \times \frac{1 \text{ mol } C_2H_5OH}{-1{,}366.8 \text{ kJ}} \times \frac{46.0 \text{ g } C_2H_5OH}{1 \text{ mol } C_2H_5OH} = \mathbf{0.112 \text{ g } C_2H_5OH}$$

5.41 *See Section 5.2 and Example 5.3.*

Temperature change = $39.1^\circ C - 22.5^\circ C = 16.6^\circ C$ or 16.6 K

$$q = 300 \text{ g} \times 4.184 \times 10^{-3} \frac{\text{kJ}}{\text{g} \cdot \text{K}} \times 16.6 \text{ K} = \mathbf{20.8 \text{ kJ}}$$

5.42 *See Section 5.2 and Example 5.3.*

Temperature change = $44.1^\circ C - 20.2^\circ C = 23.9^\circ C$ or 23.9 K

$$q = 500 \text{ g} \times 2.419 \times 10^{-3} \frac{\text{kJ}}{\text{g} \cdot \text{K}} \times 23.9 \text{ K} = \mathbf{28.9 \text{ kJ}}$$

5.43 *See Section5.2 and Example 5.3.*

Temperature change = $34.2^\circ C - 22.5^\circ C = 11.7^\circ C$ or 11.7 K

$$q = 20.0 \text{ g} \times 0.900 \times 10^{-3} \frac{\text{kJ}}{\text{g} \cdot \text{K}} \times 11.7 \text{ K} = \mathbf{0.211 \text{ kJ}}$$

5.44 *See Section 5.2 and Example 5.3.*

Temperature change = 45.9°C - 21.4°C = 24.5°C or 24.5 K

$q = 500 \text{ g} \times 0.129 \times 10^{-3} \dfrac{\text{kJ}}{\text{g} \cdot \text{K}} \times 24.5 \text{ K} = \mathbf{1.58\ kJ}$

5.45 *See Section 5.2 and Example 5.4.*

Temperature change = 28.46°C - 23.50°C = 4.96°C

(a) $q \text{ water} = 40.0 \text{ g} \times 4.184 \dfrac{\text{J}}{\text{g} \cdot ^\circ\text{C}} \times 4.96^\circ\text{C} = \mathbf{830\ J}$

(b) q metal = -830 J, so **830 J** were given off by the metal.

(c) Temperature change = 28.46°C - 100.00°C = -71.54°C

$-830 \text{ J} = 50.0 \text{ g} \times C \times -71.45^\circ\text{C}$ and $C = \dfrac{-830 \text{ J}}{(50.0 \text{ g})(-71.45^\circ \text{C})} = \mathbf{0.232 \dfrac{J}{g \cdot ^\circ C}}$

5.46 *See Section 5.2 and Example 5.4.*

(a) Temperature change = 31.51°C - 25.00°C = 6.51°C

$q_{water} = 60.0 \text{ g} \times 4.184 \dfrac{\text{J}}{\text{g} \cdot ^\circ\text{C}} \times 6.51^\circ\text{C} = \mathbf{1.63 \times 10^3\ J}$

(b) $q_{metal} = -1.63 \times 10^3$J, so $\mathbf{1.63 \times 10^3}$ **J** were given off by the metal.

(c) Temperature change = 31.51°C - 100.00°C = -68.49°C

$-1.63 \times 10^3 \text{ J} = 50.0 \text{ g} \times C \times -68.49^\circ\text{C}$ and $C = \dfrac{-1.163 \times 10^3 \text{ J}}{(50.0 \text{ g})(-68.49^\circ \text{C})} = \mathbf{0.476 \dfrac{J}{g \cdot ^\circ C}}$

5.47 *See Section 5.2 and Example 5.4.*

(a) Temperature change = 28.76°C - 23.45°C = 5.31°C

$q_{water} = 21.50\text{g} \times 4.184 \dfrac{\text{J}}{\text{g} \cdot ^\circ\text{C}} \times 5.31^\circ\text{C} = \mathbf{478\ J}$

(b) q_{metal} = -478 J, so **478 J** were given off by the metal.

(c) Temperature change = 28.76°C - 100.00°C = -71.24°C

$-478 \text{ J} = 25.0 \text{ g} \times C \times -71.24^\circ\text{C}$ and $C = \dfrac{-478 \text{ J}}{(25.0 \text{ g})(-71.24^\circ \text{C})} = \mathbf{0.268 \dfrac{J}{g \cdot ^\circ C}}$

5.48 *See Section 5.2 and Example 5.4.*

(a) Temperature change = 28.37°C - 23.84°C = 4.53°C

$q_{water} = 43.58 \text{ g} \times 4.184 \dfrac{\text{J}}{\text{g} \cdot ^\circ\text{C}} \times 4.53^\circ\text{C} = \mathbf{826\ J}$

(b) q_{metal} = -826 J, so **826 J** were given off by the metal.

(c) Temperature change: 28.37°C - 95.72°C = -67.35°C

$-826\ J = 48.9\ g \times C \times -67.35°C$ and $C = \dfrac{-826\ J}{(48.9\ g)(-67.35°\ C)} = \mathbf{0.251 \dfrac{J}{g \cdot° C}}$

5.49 ***See Section 5.2 and Example 5.4.***

$q_{water} + q_{metal} = 0$, so $q_{water} = -q_{metal}$ and

$45.2\ g \times 4.184 \dfrac{J}{g \cdot° C} \times (T_F - 22.00)°C = -59.9\ g \times 0.385 \dfrac{J}{g \cdot° C} \times (T_F - 80.30)°C$

$189\ T_F - 4.16 \times 10^3 = -23.1\ T_F + 1.85 \times 10^3$

$212\ T_F = 6.01 \times 10^3$ and $\mathbf{T_F = 28.3°C}$

5.50 ***See Section 5.2 and Example 5.4.***

$q_{water} + q_{metal} = 0$, so $q_{water} = -q_{metal}$ and

$31.2\ g \times 4.184 \dfrac{J}{g \cdot° C} \times (T_F - 23.00)°C = -40.0\ g \times 0.900 \dfrac{J}{g \cdot° C} \times (T_F - 91.50)°C$

$131\ T_F - 3.00 \times 10^3 = -36\ T_F + 3.29 \times 10^3$

$167\ T_F = 6.29 \times 10^3$ and $\mathbf{T_F = 37.7°C}$

5.51 ***See Section 5.2 and Example 5.5.***

Temperature change = 17.1°C - 22.1°C = -5.0°C

$q_{surr} = 107\ g \times 4.184 \dfrac{J}{g \cdot° C} \times -5.0°C = -2.2 \times 10^3\ J$ or **- 2.2 kJ**

The decrease in the temperature of the solution indicates the process is endothermic.

5.52 ***See Section 5.2 and Example 5.5.***

Temperature change = 25.19°C - 23.78°C = 1.41°C

$q_{surr} = 150\ g \times 4.184 \dfrac{J}{g \cdot° C} \times 1.41°C = \mathbf{8.85 \times 10^2\ J}$

The increase in the temperature of the solution indicates the process is exothermic.

5.53 ***See Section 5.2 and Example 5.5.***

Temperature change = 23.1°C - 21.2°C = 1.9°C or 1.9

$q_{surr} = 300\ g \times 4.184 \dfrac{J}{g \cdot° C} \times 1.9°\ C = 2.4 \times 10^3\ J$ or **2.4 kJ**

The increase in the temperature of the solution indicates the process is **exothermic**.

5.54 ***See Section 5.2 and Example 5.5.***

Temperature change = 24.1°C - 22.3°C = 1.8°C

$q_{surr} = 100\ g \times 4.184 \dfrac{J}{g \cdot° C} \times 1.8°C = \mathbf{753\ J}$

The incease in the temperature of the mixture indicates the process is exothermic.

5.55 ***See Section 5.2 and Examples 5.5, 5.6.***

(a) $q_{surr} = 200 \text{ g} \times 4.184 \dfrac{\text{J}}{\text{g}\cdot^{\circ}\text{C}} \times 10.9^{\circ}\text{C} = \mathbf{9.12 \times 10^3\ J}$

The increase in the temperature of the solution indicates the process is exothermic.

(b) $? \text{ mol Mg} = 0.470 \text{ g Mg} \times \dfrac{1 \text{ mol Mg}}{24.31 \text{ g Mg}} = 0.0193 \text{ Mg}$

$\Delta H \text{ per mol Mg} = \dfrac{-9.12 \times 10^3 \text{ J}}{0.0193 \text{ mol Mg}} = -4.73 \times 10^5 \dfrac{\text{J}}{\text{mol Mg}} = \mathbf{-473\ kJ / mol\ Mg}$

$Mg(s) + 2HCl(aq) \rightarrow MgCl_2(aq) + H_2(g) \qquad \Delta H = -473 \text{ kJ}$

5.56 ***See Section 5.2 and Examples 5.5, 5.6.***

(a) $q_{surr} = 306 \text{ g} \times 4.18 \dfrac{\text{J}}{\text{g}\cdot^{\circ}\text{C}} \times 3.43^{\circ}\text{C} = \mathbf{4.39 \times 10^3\ J}$

The increase in the temperature of the solution indicates the process is exothermic.

(b) $? \text{ mol } CaCl_2 = 6.00 \text{ g } CaCl_2 \times \dfrac{1 \text{ mol } CaCl_2}{111.0 \text{ g } CaCl_2} = 0.0541 \text{ mol } CaCl_2$

$\Delta H \text{ per mol } CaCl_2 = \dfrac{-4.39 \times 10^3 \text{ J}}{0.0541 \text{ mol } CaCl_2} = -8.11 \times 10^4 \dfrac{\text{J}}{\text{mol } CaCl_2} = \mathbf{-81.1\ kJ / mol\ CaCl_2}$

5.57 ***See Section 5.2 and Example 5.5, 5.6.***

Temperature change = 24.28°C - 22.94°C = 1.34°C

(a) $q_{surr} = 250 \text{ g} \times 4.184 \dfrac{\text{J}}{\text{g}\cdot^{\circ}\text{C}} \times 1.34^{\circ}\text{C} = \mathbf{1.40 \times 10^3\ J}$

The increase in the temperature of the solution indicates the process is exothermic.

(b) $? \text{ mol AgI} = 2.93 \text{ g AgI} \times \dfrac{1 \text{ mol AgI}}{234.8 \text{ g AgI}} = 0.0125 \text{ mol AgI}$

$\Delta H \text{ per mol AgI} = \dfrac{-1.40 \times 10^3 \text{ J}}{0.0125 \text{ mol AgI}} = -1.12 \times 10^5 \dfrac{\text{J}}{\text{mol AgI}} = \mathbf{-112\ kJ/mol\ AgI}$

$Ag^+(aq) + I^-(aq) \rightarrow AgI(s) \qquad \Delta H = -112 \text{ kJ}$

5.58 ***See Section 5.2 and Examples 5.5, 5.6.***

(a) Temperature change = 25.49°C - 24.52°C = 0.97°C

$q_{surr} = 41.0 \text{ g} \times 4.18 \dfrac{\text{J}}{\text{g}\cdot^{\circ}\text{C}} \times 0.97^{\circ}\text{C} = \mathbf{1.7 \times 10^2\ J}$

The increase in the temperature of the solution indicates the process is exothermic.

(b) $? \text{ mol } ZnCO_3 = 1.00 \text{ g } ZnCO_3 = \times \dfrac{1 \text{ mol } ZnCO_3}{125.4 \text{ g } ZnCO_3} = 0.00797 \text{ mol } ZnCO_3$

$\Delta H \text{ per mol } ZnCO_3 = \dfrac{-1.7 \times 10^3 \text{ J}}{0.00797 \text{ mol } ZnCO_3} = 2.1 \times 10^4 \dfrac{\text{J}}{\text{mol } ZnCO_3} = \mathbf{-21\ kJ/mol\ ZnCO_3}$

$ZnCO_3(s) + 2HCl(aq) \rightarrow ZnCl_2(aq) + CO_2(g) \qquad \Delta H = -21 \text{ kJ}$

5.59 *See Section 5.3.*

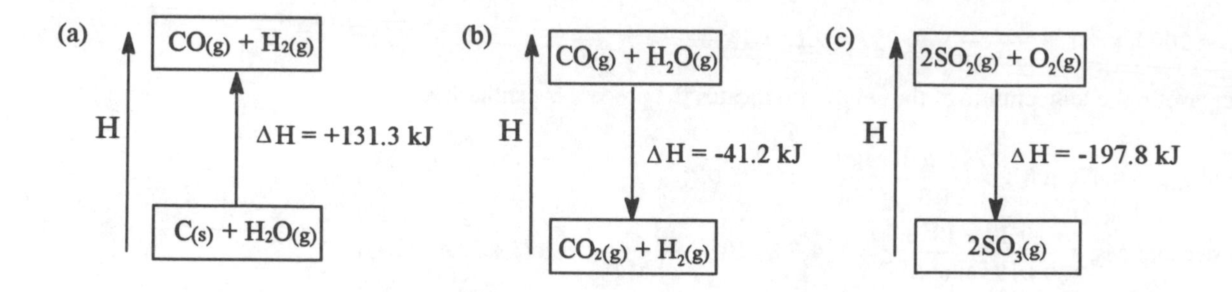

5.60 *See Section 5.3.*

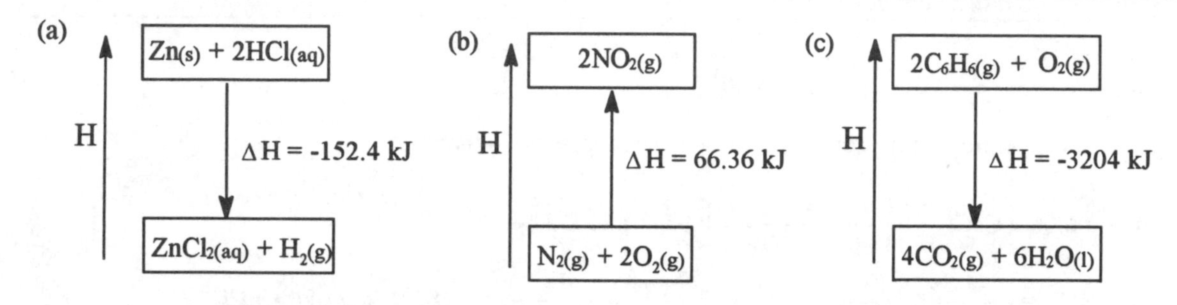

5.61 *See Section 5.3 and Examples 5.7, 5.8.*

The best approach for working Hess's Law problems involves finding one substance that appears in the desired final equation and in just one of the equations to be combined. Then use the latter equation to begin finding the algebraic combination of equations needed to obtain the desired final equation. In this case, C_2H_2, H_2 and C_2H_6 each appear separately in the given equations. Hence, any one of the given equations can be used to begin with. The desired combination is:

$$2CO_2(g) + 3H_2O(\ell) \rightarrow C_2H_6(g) + \tfrac{7}{2}O_2(g) \qquad \Delta H = +1560 \text{ kJ}$$
$$C_2H_2(g) + \tfrac{5}{2}O_2(g) \rightarrow 2CO_2(g) + H_2O(\ell) \qquad \Delta H = -1300 \text{ kJ}$$
$$2H_2(g) + O_2(g) \rightarrow 2H_2O(\ell) \qquad \Delta H = -572 \text{ kJ}$$
$$\overline{C_2H_2(g) + 2H_2(g) \rightarrow C_2H_6(g) \qquad \mathbf{\Delta H = -312\ kJ}}$$

The first equation was reversed to provide $C_2H_6(g)$ as a product in the appropriate numbers in the final equation, and the sign of the ΔH was changed. The second equation was multiplied by one-half to provide $C_2H_6(g)$ in appropriate numbers in the final equation, and the ΔH for the second equation was also multiplied by one-half. The third equation was multiplied by two to provide H_2 as a reactant in appropriate numbers in the final equation, and the ΔH was also multiplied by two. $H_2O(\ell)$ and $O_2(g)$ appear on both sides of the equation in equal numbers and are not shown in the final equation.

5.62 *See Section 5.3, Examples 5.7, 5.8, and the solution for Exercise 5.61.*

$$C_2H_4(g) + 3O_2(g) \rightarrow 2CO_2(g) + 2H_2O(\ell) \qquad \Delta H = -1411 \text{ kJ}$$
$$2CO_2(g) + 4H_2O(\ell) \rightarrow 2CH_4(g) + 4O_2(g) \qquad \Delta H = +1780 \text{ kJ}$$
$$2H_2(g) + O_2(g) \rightarrow 2H_2O(\ell) \qquad \Delta H = -572 \text{ kJ}$$
$$\overline{C_2H_4(g) + 2H_2(g) \rightarrow 2CH_4(g) \qquad \mathbf{\Delta H = -203\ kJ}}$$

5.63 ***See Section 5.3, Example 5.6, and the solution for Exercise 5.61.***

$$\begin{array}{rcll}
Zn(s) + 2HCl(aq) & \rightarrow & ZnCl_2(aq) + H_2(g) & \Delta H = -152.4\ kJ \\
ZnCl_2(aq) + H_2O(\ell) & \rightarrow & ZnO(s) + 2HCl(aq) & \Delta H = +90.2\ kJ \\
H_2(g) + \frac{1}{2}O_2(g) & \rightarrow & H_2O(\ell) & \Delta H = -285.8\ kJ \\
\hline
Zn(s) + \frac{1}{2}O_2(g) & \rightarrow & ZnO(s) & \mathbf{\Delta H = -348.0\ kJ}
\end{array}$$

5.64 ***See Section 5.3, Example 5.6, and the solution for Exercise 5.61.***

$$\begin{array}{rcll}
Mg(s) + 2HCl(aq) & \rightarrow & MgCl_2(aq) + H_2(g) & \Delta H = -462\ kJ \\
MgCl_2(aq) + H_2O(\ell) & \rightarrow & MgO(s) + 2HCl(aq) & \Delta H = +146\ kJ \\
H_2(g) + \frac{1}{2}O_2(g) & \rightarrow & H_2O(\ell) & \Delta H = -285.8\ kJ \\
\hline
Mg(s) + \frac{1}{2}O_2(g) & \rightarrow & MgO(s) & \mathbf{\Delta H = -602\ kJ}
\end{array}$$

5.65 ***See Section 5.3 and Example 5.6.***

$$\begin{array}{rcll}
C_2H_2(g) & \rightarrow & 2C(s) + H_2(g) & \Delta H = -227\ kJ \\
2C(s) + H_2(g) + 2Cl_2(g) & \rightarrow & C_2H_2Cl_4(\ell) & \Delta H = +130\ kJ \\
\hline
C_2H_2(g) + 2Cl_2(g) & \rightarrow & C_2H_2Cl_4(\ell) & \mathbf{\Delta H = -97\ kJ}
\end{array}$$

5.66 ***See Section 5.3 and Example 5.6.***

$$\begin{array}{rcll}
\frac{1}{2}H_2(g) + \frac{1}{2}Br_2(g) & \rightarrow & HBr(g) & \Delta H = -36\ kJ \\
H(g) & \rightarrow & \frac{1}{2}H_2(g) & \Delta H = -218\ kJ \\
Br(g) & \rightarrow & \frac{1}{2}Br_2(g) & \Delta H = -112\ kJ \\
\hline
H(g) + Br(g) & \rightarrow & HBr(g) & \Delta H = \mathbf{-366\ kJ}
\end{array}$$

5.67 ***See Section 5.3 and Example 5.6.***

$$\begin{array}{rcll}
Cu(s) + \frac{1}{2}Cl_2(g) & \rightarrow & CuCl(s) & \Delta H = -137.2\ kJ \\
CuCl(s) + \frac{1}{2}Cl_2(g) & \rightarrow & CuCl_2(s) & \Delta H = -82.9\ kJ \\
\hline
Cu(s) + Cl_2(g) & \rightarrow & CuCl_2(s) & \mathbf{\Delta H = -220.1\ kJ}
\end{array}$$

5.68 ***See Section 5.3 and Example 5.6.***

$$\begin{array}{rcll}
4Fe(s) + 6CO_2(g) & \rightarrow & 2Fe_2O_3(s) + 6CO(g) & \Delta H = 49.6\ kJ \\
6CO_2(g) + 3O_2(g) & \rightarrow & 6CO_2(g) & \Delta H = -1698\ kJ \\
\hline
4Fe(s) + 3O_2(g) & \rightarrow & 2Fe_2O_3(s) & \mathbf{\Delta H = -1648\ kJ}
\end{array}$$

5.69 ***See* Section 5.3.**

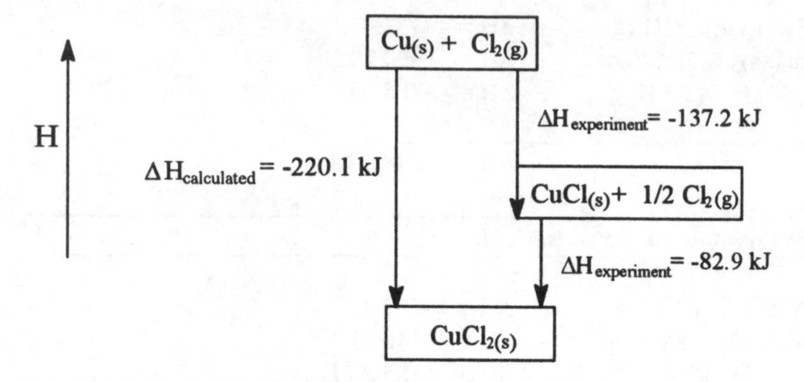

5.70 ***See Section 5.3.***

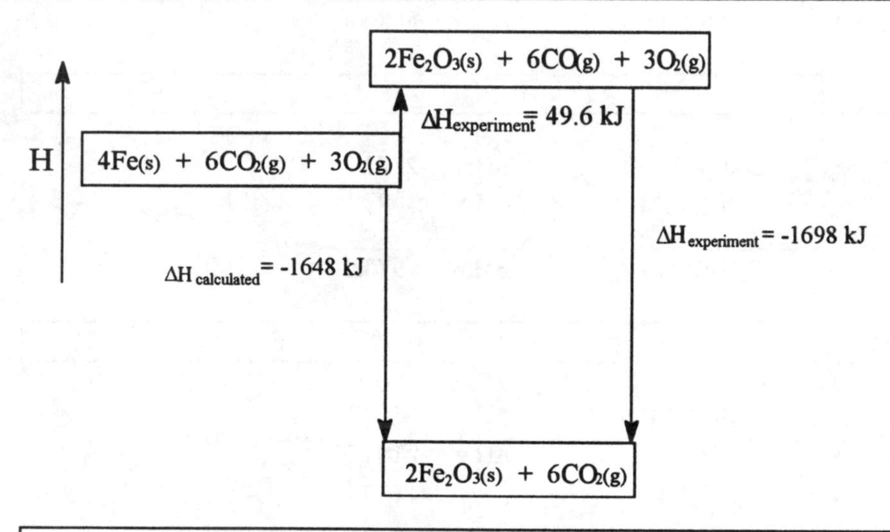

5.71 ***See Section 5.4.***

Thermochemical equations corresponding to standard enthalpies (heats) of formation involve forming one mole of a compound under standard conditions from the most stable forms of its constituent elements in their standard states. Hence,

(a) $\frac{1}{2}H_2(g) + \frac{1}{2}F_2(g) \rightarrow HF(g)$

(b) $H_2(g) + \frac{1}{8}S_8(s,\text{ rhombic}) + 2O_2(g) \rightarrow H_2SO_4(\ell)$ <u>or</u>
$H_2(g) + S(s,\text{ rhombic}) + 2O_2(g) \rightarrow H_2SO_4(\ell)$

(c) $Al(s) + \frac{3}{2}O_2(g) + \frac{3}{2}H_2(g) \rightarrow Al(OH)_3(s)$

(d) $Na(s) + K(s) + \frac{1}{8}S_8(s,\text{ rhombic}) + 2O_2(g) \rightarrow NaKSO_4(s)$ <u>or</u>
$Na(s) + K(s) + S(s,\text{ rhombic}) + 2O_2(g) \rightarrow NaKSO_4(s)$

5.72 *See Section 5.4.*

(a) $2C(graphite) + O_2(g) + 2H_2(g) \rightarrow CH_3COOH(\ell)$

(b) $\frac{3}{2}H_2(g) + \frac{1}{4}P_4(s) + 2O_2(g) \rightarrow H_3PO_4(\ell)$ or
$\frac{3}{2}H_2(g) + P(s) + 2O_2(g) \rightarrow H_3PO_4(\ell)$

(c) $C(graphite) \rightarrow C(diamond)$

(d) $Ca(s) + Cl_2(g) + 2H_2(g) + O_2(g) \rightarrow CaCl_2 \cdot 2H_2O(s)$

5.73 *See Section 5.4, Table 5.3, Example 5.9, and Appendix G.*

(a) For $C_6H_{12}O_6(s) \rightarrow 2CH_3CH_2OH(\ell) + 2CO_2(g)$,

$$\Delta H^\circ_{rxn} = \left[2\Delta H^\circ_f CH_3CH_2OH(\ell) + 2\Delta H^\circ_f CO_2(g)\right] - \left[\Delta H^\circ_f C_6H_{12}O(s)\right]$$

$$\Delta H^\circ_{rxn} = \left[\left(2\ \text{mol} \times -277.69\ \text{kJ}\cdot\text{mol}^{-1}\right) + \left(2\ \text{mol} \times -393.51\ \text{kJ}\cdot\text{mol}^{-1}\right)\right] - \left[1\ \text{mol} \times -1268\ \text{kJ}\cdot\text{mol}^{-1}\right] = \mathbf{-74\ kJ}$$

ΔH°_{rxn} is negative, and the reaction is exothermic.

(b) For $n-C_4H_{10}(g) + \frac{13}{2}O_2(g) \rightarrow 4CO_2(g) + 5H_2O(\ell)$

$$\Delta H^\circ_{rxn} = \left[4\Delta H^\circ_f CO_2(g) + 5\Delta H^\circ_f H_2O(\ell)\right] - \left[\Delta H^\circ_f n-C_4H_{10}(g) + \tfrac{13}{2}\Delta H^\circ_f O_2(g)\right]$$

$$\Delta H^\circ_{rxn} = \left[\left(4\ \text{mol} \times -393.51\ \text{kJ}\cdot\text{mol}^{-1}\right) + \left(5\ \text{mol} \times -285.83\ \text{kJ}\cdot\text{mol}^{-1}\right)\right]$$
$$-\left[\left(1\ \text{mol} \times -124.73\ \text{kJ}\cdot\text{mol}^{-1}\right) + \left(\tfrac{13}{2}\ \text{mol} \times 0.\ \text{kJ}\cdot\text{mol}^{-1}\right)\right] = \mathbf{-2878.46\ kJ}$$

ΔH°_{rxn} is negative, and the reaction is exothermic.

5.74 *See Section 5.4, Table 5.3, Example 5.9, and Appendix G.*

(a) For $6CO_2(g) + 6H_2O(\ell) \rightarrow C_6H_{12}O_6(s) + 6O_2(g)$,

$$\Delta H^\circ_{rxn} = \left[\Delta H^\circ_f C_6H_{12}O_6(s) + 6\Delta H^\circ_f O_2(g)\right] - \left[6\Delta H^\circ_f CO_2(g) + 6\Delta H^\circ_f H_2O(\ell)\right]$$

$$\Delta H^\circ_{rxn} = \left[\left(1\ \text{mol} \times -1268\ \text{kJ}\cdot\text{mol}^{-1}\right) + \left(6\ \text{mol} \times 0.\ \text{kJ}\cdot\text{mol}^{-1}\right)\right]$$
$$-\left[\left(6\ \text{mol} \times -393.51\ \text{kJ}\cdot\text{mol}^{-1}\right) + \left(6\ \text{mol} \times -285.83\ \text{kJ}\cdot\text{mol}^{-1}\right)\right] = \mathbf{2808\ kJ}$$

ΔH°_{rxn} is positive, and the reaction is endothermic.

(b) For $2Fe_2O_3(s) + 3C(s) \rightarrow 4Fe(s) + 3CO_2(g)$,

$$\Delta H^\circ_{rxn} = \left[4\Delta H^\circ_f Fe(s) + 3\Delta H^\circ_f CO_2(g)\right] - \left[2\Delta H^\circ_f Fe_2O_3(s) + 3\Delta H^\circ_f C(s)\right]$$

$$\Delta H^\circ_{rxn} = \left[\left(4\ \text{mol} \times 0.\ \text{kJ}\cdot\text{mol}^{-1}\right) + \left(3\ \text{mol} \times -393.51\ \text{kJ}\cdot\text{mol}^{-1}\right)\right]$$
$$-\left[\left(2\ \text{mol} \times -824.2\ \text{kJ}\cdot\text{mol}^{-1}\right) + \left(3\ \text{mol} \times 0.0\ \text{kJ}\cdot\text{mol}^{-1}\right)\right] = \mathbf{467.9\ kJ}$$

ΔH°_{rxn} is positive, and the reaction is endothermic.

5.75 *See Section 5.4, Table 5.3, Example 5.9, and Appendix G.*

(a) For $NaOH(s) + CO_2(g) \rightarrow NaHCO_3(s)$,

$$\Delta H^\circ_{rxn} = \left[\Delta H^\circ_f NaHCO_3(s)\right] - \left[\Delta H^\circ_f NaOH(s) + \Delta H^\circ_f CO_2(g)\right]$$

$$\Delta H^\circ_{rxn} = \left[\left(1 \text{ mol} \times -950.81 \text{ kJ}\cdot\text{mol}^{-1}\right)\right] - \left[\left(1 \text{ mol} \times -425.61 \text{ kJ}\cdot\text{mol}^{-1}\right) + \left(1 \text{ mol} \times -393.51 \text{ kJ}\cdot\text{mol}^{-1}\right)\right] = \mathbf{-131.69 \text{ kJ}}$$

ΔH°_{rxn} is negative , and the reaction is exothermic.

(b) For $2SO_2(g) + O_2(g) \rightarrow 2SO_3(g)$,

$$\Delta H^\circ_{rxn} = \left[2\Delta H^\circ_f SO_3(g)\right] - \left[2\Delta H^\circ_f SO_2(g) + \Delta H^\circ_f O_2(g)\right]$$

$$\Delta H^\circ_{rxn} = \left[\left(2 \text{ mol} \times -395.72 \text{ kJ}\cdot\text{mol}^{-1}\right)\right] - \left[\left(2 \text{ mol} \times -296.83 \text{ kJ}\cdot\text{mol}^{-1}\right) + \left(1 \text{ mol} \times 0. \text{ kJ}\cdot\text{mol}^{-1}\right)\right] = \mathbf{-197.78 \text{ kJ}}$$

ΔH°_{rxn} is negative, and the reaction is exothermic.

5.76 *See Section 5.4, Table 5.3, Example 5.9, and Appendix G.*

(a) For $NH_4NO_3(s) \rightarrow N_2O(g) + 2H_2O(\ell)$,

$$\Delta H^\circ_{rxn} = \left[\Delta H^\circ_f N_2O(g) + 2\Delta H^\circ_f H_2O(\ell)\right] - \left[\Delta H^\circ_f NH_4NO_3(s)\right]$$

$$\Delta H^\circ_{rxn} = \left[\left(1 \text{ mol} \times 82.05 \text{ kJ}\cdot\text{mol}^{-1}\right) + \left(2 \text{ mol} \times -285.83 \text{ kJ}\cdot\text{mol}^{-1}\right)\right] - \left[1 \text{ mol} \times -365.56 \text{ kJ}\cdot\text{mol}^{-1}\right] = \mathbf{-124.05 \text{ kJ}}$$

ΔH°_{rxn} is negative, and the reaction is exothermic.

(b) For $CO(g) + H_2O(\ell) \rightarrow HCOOH(\ell)$,

$$\Delta H^\circ_{rxn} = \left[\Delta H^\circ_f HCOOH(\ell)\right] - \left[\Delta H^\circ_f CO(g) + \Delta H^\circ_f H_2O_2(\ell)\right]$$

$$\Delta H^\circ_{rxn} = \left[\left(1 \text{ mol} \times 424.72 \text{ kJ}\cdot\text{mol}^{-1}\right)\right] - \left[\left(1 \text{ mol} \times -110.52 \text{ kJ}\cdot\text{mol}^{-1}\right) + \left(1 \text{ mol} \times -285.83 \text{ kJ}\cdot\text{mol}^{-1}\right)\right] = \mathbf{-28.37 \text{ kJ}}$$

ΔH°_{rxn} is negative, and the reaction is exothermic.

5.77 *See Section5.4, Table 5.3, Example 5.9, and Appendix G.*

(a) For C(graphite) $\rightarrow$ C(diamond),

$$\Delta H^\circ_{rxn} = \left[\Delta H^\circ_f C(\text{diamond})\right] - \left[\Delta H^\circ_f C(\text{graphite})\right]$$

$$\Delta H^\circ_{rxn} = \left[\left(1 \text{ mol} \times 1.895 \text{ kJ}\cdot\text{mol}^{-1}\right)\right] - \left[1 \text{ mol} \times 0. \text{ kJ}\cdot\text{mol}^{-1}\right] = \mathbf{1.895 \text{ kJ}}$$

ΔH°_{rxn} is positive, and the reaction is endothermic.

(b) For $H_2O(\ell) \rightarrow H_2O(g)$,

$$\Delta H^\circ_{rxn} = \left[\Delta H^\circ_f H_2O(g)\right] - \left[\Delta H^\circ_f H_2O(l)\right]$$

$$\Delta H^\circ_{rxn} = \left[\left(1 \text{ mol} \times -241.82 \text{ kJ}\cdot\text{mol}^{-1}\right)\right] - \left[\left(1 \text{ mol} \times -285.83 \text{ kJ}\cdot\text{mol}^{-1}\right)\right] = \mathbf{44.01 \text{ kJ}}$$

ΔH°_{rxn} is positive, and the reaction is endothermic.

5.78 *See Section5.4, Table 5.3, Example 5.9, and Appendix G.*

(a) For $Br_2(g) \rightarrow Br_2(\ell)$,

$$\Delta H^\circ_{rxn} = \left[\Delta H^\circ_f Br_2(\ell)\right] - \left[\Delta H^\circ_f Br_2(g)\right]$$

$$\Delta H^\circ_{rxn} = \left[\left(1 \text{ mol} \times 0. \text{ kJ}\cdot\text{mol}^{-1}\right)\right] - \left[1 \text{ mol} \times 30.91. \text{ kJ}\cdot\text{mol}^{-1}\right] = \mathbf{-30.91 \text{ kJ}}$$

ΔH°_{rxn} is negative, and the reaction is exothermic.

(b) For $AlCl_3(s) + 3H_2O(\ell) \rightarrow Al(OH)_3(s) + 3HCl(g)$,

$$\Delta H^\circ_{rxn} = \left[\Delta H^\circ_f Al(OH)_3(s) + 3\Delta H^\circ_f HCl(g)\right] - \left[\Delta H^\circ_f AlCl_3(s) + 3\Delta H^\circ_f H_2O(\ell)\right]$$

$$= \left[\left(1 \text{ mol} \times -1287 \text{ kJ} \cdot \text{mol}^{-1}\right) + \left(3 \text{ mol} \times -92.31 \text{ kJ} \cdot \text{mol}^{-1}\right)\right]$$

$$- \left[\left(1 \text{ mol} \times -704.2 \text{ kJ} \cdot \text{mol}^{-1}\right) + \left(3 \text{ mol} \times -285.83 \text{ kJ} \cdot \text{mol}^{-1}\right)\right] = \mathbf{-2 \text{ kJ}}$$

ΔH°_{rxn} is negative, and the reaction is exothermic.

5.79 ***See Section 5.4, Table 5.3, Example 5.9, and Appendix G.***

For $C_6H_{12}O_6(s) + 6O_2(g) \rightarrow 6CO_2(g) + 6H_2O(\ell)$

$$\Delta H^\circ_{rxn} = \left[6\Delta H^\circ_f CO_2(g) + 6\Delta H^\circ_f H_2O(\ell)\right] - \left[\Delta H^\circ_f C_6H_{12}O_6(s) + 6\Delta H^\circ_f O_2(g)\right]$$

$$\Delta H^\circ_{rxn} = \left[\left(6 \text{ mol} \times -393.51 \text{ kJ} \cdot \text{mol}^{-1}\right) + \left(6 \text{ mol} \times -285.83 \text{ kJ} \cdot \text{mol}^{-1}\right)\right]$$

$$- \left[\left(1 \text{ mol} \times -1268 \text{ kJ} \cdot \text{mol}^{-1}\right) + \left(6 \text{ mol} \times 0. \text{ kJ} \cdot \text{mol}^{-1}\right)\right] = -2808 \text{ kJ}$$

For 28 g $C_6H_{12}O_6(s)$,

$$? \Delta H^\circ = 28 \text{ g } C_6H_{12}O_6(s) \times \frac{1 \text{ mol } C_6H_{12}O_6(s)}{180.0 \text{ g } C_6H_{12}O_6(s)} \times \frac{-2808 \text{ kJ}}{1 \text{ mol } C_6H_{12}O_6(s)} = \mathbf{-4.4 \times 10^2 \text{ kJ}}$$

Thus, 4.4×10^2 kJ will be evolved as heat when 28.0 g glucose is burned.

5.80 ***See Section 5.4, Table 5.3, Example 5.9, and Appendix G.***

For $C_6H_6(\ell) + \frac{15}{2}O_2(g) \rightarrow 6CO_2(g) + 3H_2O(\ell)$,

$$\Delta H^\circ_{rxn} = \left[6\Delta H^\circ_f CO_2(g) + 3\Delta H^\circ_f H_2O(\ell)\right] - \left[\Delta H^\circ_f C_6H_6(\ell) + \tfrac{15}{2}\Delta H^\circ_f O_2(g)\right]$$

$$\Delta H^\circ_{rxn} = \left[\left(6 \text{ mol} \times -393.51 \text{ kJ} \cdot \text{mol}^{-1}\right) + \left(3 \text{ mol} \times -285.83 \text{ kJ} \cdot \text{mol}^{-1}\right)\right]$$

$$- \left[\left(1 \text{ mol} \times 49.03 \text{ kJ} \cdot \text{mol}^{-1}\right) + \left(\tfrac{15}{2} \text{ mol} \times 0. \text{ kJ} \cdot \text{mol}^{-1}\right)\right] = \mathbf{-3267.58 \text{ kJ}}$$

For 0.1045 g $C_6H_6(\ell)$,

$$? \Delta H^\circ = 0.1045 \text{ g } C_6H_6(\ell) \times \frac{1 \text{ mol } C_6H_6(\ell)}{78.0 \text{ g } C_6H_6(\ell)} \times \frac{-3267.58 \text{ kJ}}{1 \text{ mol } C_6H_6(\ell)} = \mathbf{-4.4 \text{ kJ}}$$

Thus, 4.4 kJ will be evolved as heat when 0.1045 g benzene is burned.

5.81 ***See Section 5.4 and Example 5.10.***

(a) $C_8H_{18}(\ell) + \frac{25}{2}O_2(g) \rightarrow 8CO_2(g) + 9H_2O(\ell)$ $\quad\quad \Delta H = -5456.6$ kJ

(b) $$\Delta H^\circ_{rxn} = \left[8\Delta H^\circ_f CO_2(g) + 9\Delta H^\circ_f H_2O(\ell)\right] - \left[\Delta H^\circ_f C_8H_{18}(\ell) + \tfrac{25}{2}\Delta H^\circ_f O_2(g)\right]$$

$$-5456.6 \text{ kJ} = \left[\left(8 \text{ mol} \times -393.51 \text{ kJ} \cdot \text{mol}^{-1}\right) + \left(9 \text{ mol} \times -285.83 \text{ kJ} \cdot \text{mol}^{-1}\right)\right]$$

$$- \left[\left(1 \text{ mol} \times \Delta H^\circ_f C_8H_{18}(\ell)\right) + \left(\tfrac{25}{2} \text{ mol} \times 0. \text{ kJ} \cdot \text{mol}^{-1}\right)\right]$$

$\Delta H^\circ_f C_8H_{18}(\ell) = \mathbf{-264.0 \text{ kJ} \cdot \text{mol}^{-1}}$

5.82 *See Section 5.4 and Example 5.10.*

(a) $C_{12}H_{26}(\ell) + \frac{37}{2}O_2(g) \rightarrow 12CO_2(g) + 13H_2O(\ell)$ $\Delta H = -8080.1$ kJ

(b) $\Delta H^\circ_{rxn} = \left[12\Delta H^\circ_f CO_2(g) + 13\Delta H^\circ_f H_2O(\ell)\right] - \left[\Delta H^\circ_f C_{12}H_{26}(\ell) + \frac{37}{2}\Delta H^\circ_f O_2(g)\right]$

$-8080.1 \text{ kJ} = \left[\left(12 \text{ mol} \times -393.51 \text{ kJ} \cdot \text{mol}^{-1}\right) + \left(13 \text{ mol} \times -285.83 \text{ kJ} \cdot \text{mol}^{-1}\right)\right]$

$-\left[\left(1 \text{ mol} \times \Delta H^\circ_f C_{12}H_{26}(\ell)\right) + \left(\frac{37}{2} \text{ mol} \times 0. \text{ kJ} \cdot \text{mol}^{-1}\right)\right]$

$\Delta H^\circ_f C_{12}H_{26}(\ell) = \mathbf{-357.81\ kJ \cdot mol^{-1}}$

5.83 *See Section 5.5.*

No. The cost also depends on the efficiency with which the heat energy is tranferred to the water.

5.84 *See Section 5.4, Table 5.3, Example 5.9, and Appendix G.*

(a) For $SO_2(g) + \frac{1}{2}O_2(g) \rightarrow SO_3(g)$,

$\Delta H^\circ_{rxn} = \left[\Delta H^\circ_f SO_3(g)\right] - \left[\Delta H^\circ_f SO_2(g) + \frac{1}{2}\Delta H^\circ_{rxn} O_2(g)\right]$

$\Delta H^\circ_{rxn} = \left[\left(1 \text{ mol} \times -395.72 \text{ kJ} \cdot \text{mol}^{-1}\right) + \left(1 \text{ mol} \times -296.83 \text{ kJ} \cdot \text{mol}^{-1}\right)\right] + \left[\frac{1}{2} \text{ mol} \times 0. \text{ kJ} \cdot \text{mol}^{-1}\right] = \mathbf{-98.89\ kJ}$

Sulfur trioxide, formed by combustion of sulfur-containing fossil fuels and reaction of the resultant sulfur dioxide with oxygen, reacts with water to form sulfuric acid, one of the major components of acid rain.

5.85 *See Section 5.6.*

(a) ? metric tons coal supplying equal energy $= 8.1 \times 10^7 \text{ kJ} \times \frac{1.0 \text{ kg coal}}{2.8 \times 10^4 \text{ kJ}} \times \frac{10^3 \text{ g coal}}{1 \text{ kg coal}} \times \frac{1 \text{ metric ton coal}}{10^6 \text{ g coal}}$

$= \mathbf{2.9\ metric\ tons\ coal}$

(b) $? \text{metric tons } SO_2 = 2.9 \text{ metric tons coal} \times \frac{0.90 \text{ metric ton S}}{100.00 \text{ metric tons coal}} \times \frac{64.0 \text{ metric tons } SO_2}{32.0 \text{ metric tons S}}$

$\times \frac{10^6 \text{ g } SO_2}{1 \text{ metric ton } SO_2} \times \frac{1 \text{ kg } SO_2}{10^3 \text{ g } SO_2} = 52 \text{ kg } SO_2$

(c) The production of SO_2 by the burning of coal creates an immediate problem in terms of producing solid $CaSO_3$ to be stored or disposed of if the SO_2 is removed by scrubbing or acid rain as H_2SO_4 if the SO_2 is not removed by scrubbing. On the other hand, the storage of radioactive waste presents a longterm storage problem.

5.86 *See Section 5.5.*

$? \text{kJ per gram } CH_4 = \frac{890 \text{ kJ}}{1 \text{ mol } CH_4} \times \frac{1 \text{ mol } CH_4}{16.0 \text{ g } CH_4} = \mathbf{55.7\ kJ/g\ CH_4}$

$? \text{kJ per gram } C_8H_{18} = \frac{5466 \text{ J}}{1 \text{ mol } C_8H_{18}} \times \frac{1 \text{ mol } C_8H_{18}}{114.0 \text{ g } C_8H_{18}} = \mathbf{47.95\ kJ/g\ C_8H_{18}}$

CH_4 produces more energy per gram of compound burned than C_8H_{18}. The difference is 7.8 kJ/g.

5.87 See Section 5.3.

$$? \text{ mol Ca} = 1.20 \text{ g Ca} \times \frac{1 \text{ mol} \cdot \text{Ca}}{40.1 \text{ g Ca}} = 0.0299 \text{ mol Ca}$$

$$? \Delta H \text{ per mole Ca} = \frac{-16.25 \text{ kJ}}{0.0299 \text{ mol Ca}} = \mathbf{-543 \text{ kJ/mol Ca}}$$

$Ca(s) + 2\,HCl(aq) \rightarrow CaCl_2(aq) + H_2(g) \qquad \Delta H = -543$ kJ

5.88 See Section 5.3, Examples 5.7, 5.8, and the solution for Exercise 5.61.

Starting with	$C_6H_{14}(\ell) + \frac{19}{2}O_2(g)$	$\rightarrow$	$6CO_2(g) + 7H_2O(\ell)$	$\Delta H = -4159.5$ kJ
	$C_6H_{14}(g) + \frac{19}{2}O_2(g)$	$\rightarrow$	$6CO_2(g) + 7H_2O(\ell)$	$\Delta H = -4191.1$ kJ
yields	$C_6H_{14}(\ell) + \frac{19}{2}O_2(g)$	$\rightarrow$	$6CO_2(g) + 7H_2O(\ell)$	$\Delta H = -4159.5$ kJ
	$6CO_2(g) + 7H_2O(\ell)$	$\rightarrow$	$C_6H_{14}(g) + \frac{19}{2}O_2(g)$	$\Delta H = -4191.1$ kJ

$C_6H_{14}(\ell) \rightarrow C_6H_{14}(g) \qquad$ **$\Delta H = 31.6$ kJ**

5.89 See Section 5.3, Examples 5.7, 5.8, and the solution for Exercise 5.61.

Temperature change = 34.7°C - 23.5°C = 11.2°C

$$677 \text{ J} = 25.0 \text{ g x C x } 11.2°\text{C} = \frac{677 \text{ J}}{(25.0 \text{ g})(11.2°\text{ C})} = \mathbf{2.42 \frac{J}{g \cdot ° C}}$$

5.90 See Section 5.4 and Example 5.10.

For $C_2H_2(g) + 2Cl_2(g) \rightarrow C_2H_2Cl_4(\ell)$,

$$\Delta H^\circ_{rxn} = \left[\Delta H^\circ_f\ C_2H_2Cl_4(\ell)\right] - \left[\Delta H^\circ_f C_2H_2(g)\ +\ 2\Delta H^\circ_f Cl_2(g)\right]$$

$$-97 \text{ kJ} = \left[\left(1 \text{ mol} \times \Delta H^\circ_f C_2H_2Cl_4(\ell)\right)\right] - \left[(1 \text{ mol} \times 226.73 \text{ kJ} \cdot \text{mol}^{-1}) + (2 \text{ mol} \times 0.\ \text{kJ} \cdot \text{mol}^{-1})\right]$$

$\Delta H^\circ_f\ C_2H_2Cl_4(\ell) = \mathbf{-324\ kJ \cdot mol^{-1}}$

5.91 See Section 5.1 and Example 5.1.

(a) Let y = g CO(g) and (10.0 -y) = g $H_2(g)$ present in the sample of water gas. Since mol CO = mol H_2 in the mixture,

$$y \text{ g CO} \times \frac{1 \text{ mol CO}}{28.0 \text{ g CO}} = (10.0\ -y) \text{ g } H_2 \times \frac{1 \text{ mol } H_2}{2.01 \text{ g } H_2}$$, 2.01 y = 28.0 (10.0 -y), 30.0 y = 280 and y = 9.33.

$$? \text{ mol CO} = 9.33 \text{ g CO} \times \frac{1 \text{ mol CO}}{28.0 \text{ g CO}} = \mathbf{0.333 \text{ mol CO}}$$

$$? \text{ mol } H_2 = (10.0 - 9.33) \text{ g } H_2 \times \frac{1 \text{ mol } H_2}{2.01 \text{ g } H_2} = 0.33 \text{ mol } H_2$$, but since mol H_2 = mol CO, **mol H_2 = 0.333 mol.**

(b) Using	$2CO(g) + O_2(g)$	$\rightarrow$	$2CO_2(g)$	$\Delta H = -566$ kJ
	$2H_2(g) + O_2(g)$	$\rightarrow$	$2H_2O(\ell)$	$\Delta H = -571.7$ kJ

$$? \Delta H \text{ for CO reacting} = 0.333 \text{ mol CO} \times \frac{-566 \text{ kJ}}{2 \text{ mol CO}} = -94.2 \text{ kJ}$$

$$? \Delta H \text{ for } H_2 \text{ reacting} = 0.333 \text{ mol } H_2 \times \frac{-571.7 \text{ kJ}}{2 \text{ mol } H_2} = -95.2 \text{ kJ}$$

Total ΔH = -94.2 kJ + -95.2 kJ = **-189.4 kJ**

Hence, burning 10.0 g water gas in air releases 189.4 kJ.

5.92 ***See Section 5.4 and Example 5.10.***

(a) $C_5H_5(\ell) + \frac{25}{4}O_2(g) \rightarrow 5CO_2(g) + \frac{5}{2}H_2O(\ell)$ $\Delta H° = -3288.9$ kJ

(b) $\Delta H^{\circ}_{rxn} = \left[5\Delta H^{\circ}_{f}CO_2(g) + \frac{5}{2}\Delta H^{\circ}_{f}H_2O(\ell)\right] - \left[\Delta H^{\circ}_{f}C_5H_5(\ell) + \frac{25}{4}\Delta H^{\circ}_{f}O_2(g)\right]$

$-3288.9 \text{ kJ} = \left[\left(5 \text{ mol} \times -393.51 \text{ kJ}\cdot\text{mol}^{-1}\right) + \left(\frac{5}{2}\text{mol} \times -285.83 \text{ kJ}\cdot\text{mol}^{-1}\right)\right]$

$- \left[\left(1 \text{ mol} \times \Delta H^{\circ}_{f}C_5H_5(\ell)\right) + \left(\frac{25}{4} \text{ mol} \times 0. \text{ kJ}\cdot\text{mol}^{-1}\right)\right]$

$\Delta H^{\circ}_{f}C_5H_5(\ell) = \mathbf{-606.78 \text{ kJ}\cdot\text{mol}^{-1}}$

5.93 ***See Section 5.4 and Example 5.10.***

(a) $CH_3COC_2H_5(\ell) + \frac{11}{2}O_2(g) \rightarrow 4CO_2(g) + 4H_2O(\ell)$ $\Delta H° = -2444.1$ kJ

(b) $\Delta H^{\circ}_{rxn} = \left[4\Delta H^{\circ}_{f}CO_2(g) + 4\Delta H^{\circ}_{f}H_2O(\ell)\right] - \left[\Delta H^{\circ}_{f}CH_3COC_2H_5(\ell) + \frac{11}{2}\Delta H^{\circ}_{f}O_2(g)\right]$

$-2444.1 \text{ kJ} = \left[\left(4 \text{ mol} \times -393.51 \text{ kJ}\cdot\text{mol}^{-1}\right) + \left(4\text{mol} \times -285.83 \text{ kJ}\cdot\text{mol}^{-1}\right)\right]$

$- \left[\left(1 \text{ mol} \times \Delta H^{\circ}_{f}CH_3COC_2H_5(\ell)\right) + \left(\frac{11}{2} \text{ mol} \times 0. \text{ kJ}\cdot\text{mol}^{-1}\right)\right]$

$\Delta H^{\circ}_{f}CH_3COC_2H_5(\ell) = \mathbf{-273.3 \text{ kJ}\cdot\text{mol}^{-1}}$

5.94 ***See Sections 5.1, 5.4, and Examples 5.1, 5.9.***

$C_2H_2(g) + \frac{5}{2}O_2(g) \rightarrow 2CO_2(g) + H_2O(\ell)$

$\Delta H^{\circ}_{rxn} = \left[2\Delta H^{\circ}_{f}CO_2(g) + \Delta H^{\circ}_{f}H_2O(\ell)\right] - \left[\Delta H^{\circ}_{f}C_2H_2(g) + \frac{5}{2}\Delta H^{\circ}_{f}O_2(g)\right]$

$\Delta H^{\circ}_{rxn} = \left[\left(2 \text{ mol} \times -393.51 \text{ kJ}\cdot\text{mol}^{-1}\right) + \left(1\text{mol} \times -285.83 \text{ kJ}\cdot\text{mol}^{-1}\right)\right]$

$- \left[\left(1 \text{ mol} \times 226.73 \text{ kJ}\cdot\text{mol}^{-1}\right) + \left(\frac{5}{2} \text{ mol} \times 0. \text{ kJ}\cdot\text{mol}^{-1}\right)\right]$

$= -1299.58 \text{ kJ}$

$?\Delta H° = 1.00 \text{ g } C_2H_2 \times \frac{1 \text{ mol } C_2H_2}{26.0 \text{ g } C_2H_2} \times \frac{-1299.58 \text{ kJ}}{1 \text{ mol } C_2H_2} = \mathbf{-50.0 \text{ kJ}}$

5.95 ***See Sections 5.1, 5.4, and Examples 5.1, 5.9.***

(a) $C_2H_2(g) + 5N_2O(g) \rightarrow 2CO_2(g) + H_2O(\ell) + 5N_2(g)$

(b) $\Delta H^{\circ}_{rxn} = \left[2\Delta H^{\circ}_{f}CO_2(g) + \Delta H^{\circ}_{rxn}H_2O(\ell) + 5\Delta H^{\circ}_{rxn}N_2(g)\right] - \left[\Delta H^{\circ}_{f}C_2H_2(g) + 5\Delta H^{\circ}_{f}N_2O(g)\right]$

$\Delta H^{\circ}_{rxn} = \left[\left(2 \text{ mol} \times -393.51 \text{ kJ}\cdot\text{mol}^{-1}\right) + \left(1\text{mol} \times -285.83 \text{ kJ}\cdot\text{mol}^{-1}\right) + \left(5 \text{ mol} \times 0. \text{ kJ}\cdot\text{mol}^{-1}\right)\right]$

$- \left[\left(1 \text{ mol} \times 226.73 \text{ kJ}\cdot\text{mol}^{-1}\right) + \left(5 \text{ mol} \times 82.05 \text{ kJ}\cdot\text{mol}^{-1}\right)\right] = -1709.83 \text{ kJ}$

$?\Delta H° = 65 \text{ g } C_2H_2 \times \frac{1 \text{ mol } C_2H_2}{26.0 \text{ g } C_2H_2} \times \frac{-1709.83 \text{ kJ}}{1 \text{ mol } C_2H_2} = \mathbf{-4.3 \times 10^3 \text{ kJ}}$

5.96 ***See Section 5.1 and Example 5.2.***

$? \text{ g } C_2H_4 = 3420 \text{ kJ} \times \frac{1 \text{ mol } C_2H_4}{1410.1 \text{ kJ}} \times \frac{28.06 \text{ g } C_2H_4}{1 \text{ mol } C_2H_4} = \mathbf{68.06 \text{ g } C_2H_4}$

5.97 *See Sections 3.3, 5.4, and Examples 3.11, 3.12, 5.10.*

(a) Assume the sample has a mass of 100.00 g and therefore contains 82.7 g C and 17.3 g H.

$$? \text{ mol C} = 82.7 \text{ g C} \times \frac{1 \text{ mol C}}{12.01 \text{ g C}} = 6.89 \text{ mol C} \qquad \text{relative mol C} = \frac{6.89 \text{ mol C}}{6.89} = 1.00 \text{ mol C}$$

$$? \text{ mol H} = 17.3 \text{ g H} \times \frac{1 \text{ mol H}}{1.008 \text{ g H}} = 17.2 \text{ mol H} \qquad \text{relative mol H} = \frac{17.2 \text{ mol H}}{6.89} = 2.50 \text{ mol H}$$

Multiplying each of these by the smallest iteger that gives whole numbers of relative moles of atoms for each yields:

relative mol C = 1.00 mol C x 2 = 2.00 mol C
relative mol H = 2.50 mol H x 2 = 5.00 mol H

The empirical formula is $\mathbf{C_2H_5}$.

(b) $$n = \frac{\text{molar mass compound}}{\text{molar mass } C_2H_5} = \frac{60 \text{ g/mol}}{29 \text{ g/mol}} = 2$$

The molecular formula is $\mathbf{C_4H_{10}}$.

(c) For $C_4H_{10}(?) + \frac{13}{2}O_2(g) \rightarrow 4CO_2(g) + 5H_2O(\ell)$,

$$\Delta H^\circ_{rxn} = \left[4\Delta H^\circ_f CO_2(g) + 5\Delta H^\circ_f H_2O(\ell)\right] - \left[\Delta H^\circ_f C_4H_{10}(?) + \tfrac{13}{2}\Delta H^\circ_f O_2(g)\right]$$

$$\Delta H^\circ_{rxn} = \frac{-49.53 \text{ kJ}}{1.000 \text{ g } C_4H_{10}} \times \frac{58.12 \text{ g } C_4H_{10}}{1 \text{ mol } C_4H_{10}} = -2879 \text{ kJ}$$

$$-2879 \text{ kJ} = \left[\left(4 \text{ mol} \times -393.51 \text{ kJ} \cdot \text{mol}^{-1}\right) + \left(5\text{mol} \times -285.83 \text{ kJ} \cdot \text{mol}^{-1}\right)\right]$$
$$- \left[\left(1 \text{ mol} \times \Delta H^\circ_f C_4H_{10}(?)\right) + \left(\tfrac{13}{2} \text{ mol} \times 0. \text{ kJ} \cdot \text{mol}^{-1}\right)\right]$$

$$\Delta H^\circ_f\ C_4H_{10}(?) = \mathbf{-124\ kJ \cdot mol^{-1}}$$

(d) The compound is n-butane, **n-C_4H_{10}**.

5.98 *See Section 3.3 and Examples 3.11, 3.12.*

(a) Assume the sample has a mass of 100.00 g and therefore contains 85.6 g C and 14.4 g H.

$$? \text{ mol C} = 85.6 \text{ g C} \times \frac{1 \text{ mol C}}{12.011 \text{ g C}} = 7.13 \text{ mol C} \qquad \text{relative mol C} = \frac{7.13 \text{ mol C}}{7.13} = 1.00 \text{ mol C}$$

$$? \text{ mol H} = 14.4 \text{ g H} \times \frac{1 \text{ mol H}}{1.008 \text{ g H}} = 14.3 \text{ mol H} \qquad \text{relative mol H} = \frac{14.3 \text{ mol H}}{7.13} = 2.00 \text{ mol H}$$

The empirical formula is $\mathbf{CH_2}$.

(b) ?ΔH vap for one mol = 88 x 322 = $2.8 \times 10^4 \text{ J} \cdot \text{mol}^{-1}$

$$? \text{ molar mass} = \frac{2.8 \times 10^4 \text{ J}}{1 \text{ mol}} \times \frac{1 \text{ g}}{389 \text{ J}} = 72 \text{ g/mol}$$

(c) $$n = \frac{\text{molar mass compound}}{\text{molar mass } CH_2} = \frac{72 \text{ g/mol}}{14 \text{ g/mol}} = 5$$

The molecular formula is $\mathbf{C_5H_{10}}$.

5.99 *See Section 3.3 and Examples 3.11, 3.12.*

(a) Assume the sample has a mass of 100.00 g and therefore contains 54.5 g C, 9.15 g H, and 100.00 -54.4 - 9.15 = 36.4 g O.

$$? \text{ mol C} = 54.5 \text{ g C} \times \frac{1 \text{ mol C}}{12.011 \text{ g C}} = 4.54 \text{ mol C} \qquad \text{relative mol C} = \frac{4.54 \text{ mol C}}{2.28} = 1.99 \text{ mol C}$$

$$? \text{ mol H} = 9.15 \text{ g H} \times \frac{1 \text{ mol H}}{1.008 \text{ g H}} = 9.08 \text{ mol H} \qquad \text{relative mol H} = \frac{9.08 \text{ mol H}}{2.28} = 3.98 \text{ mol H}$$

$$? \text{ mol O} = 36.4 \text{ g O} \times \frac{1 \text{ mol O}}{15.994 \text{ g O}} = 2.28 \text{ mol O} \qquad \text{relative mol O} = \frac{2.28 \text{ mol O}}{2.28} = 1.00 \text{ mol O}$$

The empirical formula is **C_2H_4O.**

(b) $?\Delta H_{vap}$ for one mole = 88 x 374 = 3.3 x 10^4 J · mol^{-1}

$$? \text{ molar mass} = \frac{3.3 \times 10^4 \text{ J}}{1 \text{ mol}} \times \frac{1 \text{ g}}{388 \text{ J}} = 85 \text{ g / mol}$$

(c) $$n = \frac{\text{molar mass compound}}{\text{molar mass } C_4H_8O} = \frac{85 \text{ g / mol}}{44 \text{ g / mol}} = 2$$

The molecular formula is **$C_4H_8O_2$.**

5.100 *See Section 5.2 and Example 5.3.*

(a) $$? \$ \text{ for one mol Au} = \frac{197.0 \text{ g Au}}{1 \text{ mol Au}} \times \frac{1 \text{ oz Au}}{31.10 \text{ g Au}} \times \frac{\$383}{1 \text{ oz Au}} = \mathbf{2.43 \times 10^3 \ \$ / mol\ Au}$$

(b) $$? \text{ g Au} = \$5{,}000 \times \frac{1 \text{ mol Au}}{\$2.43 \times 10^3} \times \frac{197.0 \text{ g Au}}{1 \text{ mol Au}} = 405 \text{ g Au}$$

Temperature change: 99°C - 15°C = 84°C

$$q = 405 \text{ g} \times 0.129 \frac{\text{J}}{\text{g} \cdot {}^\circ\text{C}} \times 84^\circ\text{C} = 4.4 \times 10^3 \text{ J} = \mathbf{4.4 \text{ kJ}}$$

5.101 *See Section 3.3 and Example 3.11.*

(a) $$? \text{ approximate molar mass of metal} = \frac{25 \text{ J}}{\text{mol} \cdot {}^\circ\text{C}} \times \frac{1 \text{ g} \cdot {}^\circ\text{C}}{0.24 \text{ J}} = \mathbf{104 \text{ g / mol}}$$

(b) Assume the sample has a mass of 100.00 g and therefore contains 93.10 g metal M and 6.90 g O.

$$? \text{ mol M} = 93.10 \text{ g M} \times \frac{1 \text{ mol M}}{104 \text{ g M}} = 0.895 \text{ mol M} \qquad \text{relative mol M} = \frac{0895 \text{ mol M}}{0.431} = 2.08 \text{ mol M}$$

$$? \text{ mol O} = 6.90 \text{ g O} \times \frac{1 \text{ mol O}}{16.0 \text{ g O}} = 0.431 \text{ mol O} \qquad \text{relative mol O} = \frac{0.431 \text{ mol O}}{0.431} = 1.00 \text{ mol O}$$

The empirical formula is **M_2O.**

(c) $$? \text{ mass of M combined with 16.0 g O in } M_2O = 16.0 \text{ g O} \times \frac{93.10 \text{ g M}}{6.90 \text{ g O}} = 216 \text{ g M}$$

$$? \text{ molar mass M} = \frac{216 \text{ g M}}{1 \text{ mol } M_2O} \times \frac{1 \text{ mol } M_2O}{2 \text{ mol M}} = \mathbf{108 \text{ g / mol M}}$$

The metal is **Ag.**

5.102 *See Section 3.3 and Example 3.11.*

(a) ? approximate molar mass of metal = $\frac{25\ \text{J}}{\text{mol}\cdot{}^\circ\text{C}} \times \frac{1\ \text{g}\cdot{}^\circ\text{C}}{0.46\ \text{J}} = \mathbf{54\ g/mol}$

(b) Assume the sample has a mass of 100.00 g and therefore contains 32.8 g M and 67.2 g Cl.

? mol M = 32.8 g M $\times \frac{1\ \text{mol M}}{54\ \text{g M}}$ = 0.61 mol M

relative mol M = $\frac{0.61\ \text{mol M}}{0.61}$ = 1.0 mol M

? mol Cl = 67.2 g Cl $\times \frac{1\ \text{mol Cl}}{35.453\ \text{g Cl}}$ = 1.90 mol Cl

relative mol Cl = $\frac{1.90\ \text{mol Cl}}{0.61}$ = 3.1 mol Cl

The empirical formula is **MCl_3.**

(c) ? mass of M combined with 106.4 g Cl (3 x 35.453) in MCl_3 = 106.4 g Cl $\times \frac{32.8\ \text{g M}}{657.2\ \text{g Cl}}$ = 51.9 g M

? molar mass M = $\frac{51.9\ \text{g M}}{1\ \text{mol MCl}_3} \times \frac{1\ \text{mol MCl}_3}{1\ \text{mol M}} = \mathbf{51.9\ g/mol\ M}$

The metal is **$CrCl_3$.**

5.103 *Solutions for Spreadsheet Exercises are on the Reger, Goode and Mercer Spreadsheet Solutions Diskette available from Saunders College Publishing.*

Chapter 6: The Gaseous State

6.1 ***See Section 6.1.***

Gases and liquids take the shape of their containers, but a liquid has a definite volume and does not expand to fill the container.

6.2 ***See Section 6.1.***

For most substances, density decreases from solid to liquid to gas as the substance expands. However, water is an important exception to the solid-to-liquid density trend because its solid form is less dense than its liquid form. This is why ice floats on liquid water and why lakes freeze from the top down instead of from the bottom up.

6.3 ***See Section 6.1 and Figure 6.4.***

In a barometer, the column of mercury creates a pressure that is equal to the prevailing atmospheric pressure. In a manometer with an open end, the difference between the heights of the two mercury surfaces measures the difference between the pressure of the gas in the container and the prevailing atmospheric pressure.

6.4 ***See Section 6.1 and Figure 6.4.***

The drawing should look like Figure 6a with the height of the column of mercury on the left being equal to P_{atm}- 200 torr, where P_{atm} is the prevailing atmospheric pressure.

6.5 ***See Section 6.1 and Table 6.1.***

One atmosphere of pressure (1 atm) is the normal pressure exerted by the atmosphere at sea level. It is equivalent to 760 mm Hg = 760 torr = 14.7 psi = 101.325 kPa.

6.6 ***See Section 6.2.***

(a) Volume decreases as the pressure increases in accord with Boyle's Law.

(b) Volume increases as the temperature increases in accord with Charles's Law.

(c) Volume increases as the amount increases in accord with Avogadro's Law.

6.7 ***See Section 6.2.***

A graph of volume versus temperature for a given mass of a gas at constant pressure can be extrapolated to intersect the temperature at -273°C, which is taken to be 0 kelvins and the basis for the absolute temperature scale. The temperature of absolute zero is not measured, since all gases liquify before that temperature is reached.

6.8 ***See Section 6.2.***

According to Avogadro's Law, one mole of any gas contains the same number of molecules and therefore exerts the same pressure as one mole of any other gas at the same temperature. The molar mass of nitrogen, N_2, is 28.0 g/mol, and the molar mass of oxygen, O_2, is 32.0 g/mol, so the oxygen will weigh more than the nitrogen.

6.9 *See Section 6.6.*

Dalton's Law of Partial Pressures recognizes that each gas in a mixture of nonreacting gases exerts a pressure that is the same as though it occupied the container by itself. Hence, knowing the mass of each gas in the container enables one to calculate the pressure due to each gas using the Ideal Gas Law. Then $P_T = P_1 + P_2 + P_3$ can be used to calculate the total pressure of the gases.

6.10 *See Section 6.6.*

The concentration term "molarity" refers to the moles of solute per liter of solution. The concentration term "mole fraction" refers to the moles of the component of interest per total moles present. Since mixtures of gases can be homogeneous mixtures which are solutions, one could use the term "molarity" in conjuction with gases. However, "molarity" would not be as useful as "mole fraction," because properties like the partial pressure of a gas depend on what fraction of the total moles of gases present is the gas of interest.

6.11 *See Section 6.7.*

The four assumptions/postulates of the kinetic molecular theory are:

1. Gases consist of small particles that are in constant and random motion.
2. There are no attractive or repulsive forces between particles, so collisions between molecules are elastic, meaning there is no loss in the total kinetic energy when the particles collide.
3. The gas particles are very small in size compared to the distances between them, so the volume occupied by the gas particles is generally small compared to the volume of the container.
4. The average kinetic energy of a collection of gas particles is proportional to the temperature on the Kelvin scale.

6.12 *See Section 6.7.*

The kinetic molecular theory assumes that the pressure exerted by a gas comes from the collisions of the individual gas particles with the walls of the container.

6.13 *See Section 6.7 and Figure 6.16.*

A Maxwell-Boltzman distribution curve shows the number of particles that have a given speed (y axis) versus the speed (x axis). See Figure 6.16.

6.14 *See Section 6.8.*

Diffusion is the mixing of gas particles due to motion. Effusion is the passage of gas particles through a small hole into an evacuated space.

6.15 *See Sction 6.9.*

As the pressure of a gas increases, the gas particles are forced closer together making the attractive forces between the particles more important. Hence, the particles are attracted to one another and do not hit the walls of the container as hard, so the pressure is reduced below that predicted by the Ideal Gas Law. In addition, the volume occupied by the gas particles becomes significant compared to the volume of the container. This works to make the pressure higher than predicted by the Ideal Gas Law.

6.16 *See Section 6.9 and Figure 6.21.*

(a) If the attractive forces between gas particles are strong, the ratio PV/nRT is less than one.

(b) At very high pressures, the reduced volume factor usually becomes more important than the reduced pressure factor, so the ratio PV/nRT becomes greater than one.
Notice, however, that Figure 6.21 indicates the attractive forces between H_2 molecules are never strong enough to cause PV/nRT to be less than one.

6.17 ***See Section 6.1, Table 6.1, and Example 6.1.***

(a) $? \text{ atm} = 334 \text{ torr} \times \frac{1 \text{ atm}}{760 \text{ torr}} = \mathbf{0.439 \text{ atm}}$

(b) $? \text{ atm} = 3944 \text{ Pa} \times \frac{1 \text{ kPa}}{10^3 \text{ Pa}} \times \frac{1 \text{ atm}}{101.325 \text{ kPa}} = \mathbf{0.03892 \text{ atm}}$

(c) $? \text{ torr} = 2.4 \text{ atm} \times \frac{760 \text{ torr}}{1 \text{ atm}} = \mathbf{1.8 \times 10^3 \text{ torr}}$

6.18 ***See Section 6.1, Table 6.1, and Example 6.1.***

(a) $? \text{ torr} = 3.2 \text{ atm} \times \frac{760 \text{ torr}}{1 \text{ atm}} = \mathbf{2.4 \times 10^3 \text{ torr}}$

(b) $? \text{ kPa} = 54.9 \text{ atm} \times \frac{101.325 \text{ kPa}}{1 \text{ atm}} = \mathbf{5.56 \times 10^3 \text{ kPa}}$

(c) $? \text{ atm} = 356 \text{ torr} \times \frac{1 \text{ atm}}{760 \text{ torr}} = \mathbf{0.468 \text{ atm}}$

6.19 ***See Section 6.1 and recall $T_K = T_C + 273$.***

(a) $T_K = 45 + 273 = \mathbf{318 \text{ K}}$ (b) $T_K = -28 + 273 = \mathbf{245 \text{ K}}$ (c) $T_K = 230 + 273 = \mathbf{503 \text{ K}}$

6.20 ***See Section 6.1 and recall $T_K = T_C + 273$, so $T_C = T_K - 273$.***

(a) $T_C = 344 - 273 = \mathbf{71°C}$ (b) $T_C = 122 - 273 = \mathbf{-151°C}$ (c) $T_C = 1537 - 273 = \mathbf{1264°C}$

6.21 ***See Section 6.3.***

State One:
$P_1 = 1.02 \text{ atm}$
$T_1 = 39 + 273 = 312 \text{ K}$

State Two:
$P_2 = ?$
$T_2 = 499 + 273 = 772 \text{ K}$

Solving $\frac{P_1}{T_1} = \frac{P_2}{T_2}$ for P_2 gives $P_2 = P_1 \times \frac{T_2}{T_1}$ $P_2 = 1.02 \text{ atm} \times \frac{772 \text{ K}}{312 \text{ K}} = \mathbf{2.52 \text{ atm}}$

Since the temperature increase occurred at constant volume, an increase in pressure is expected.

6.22 ***See Section 6.3 and Example 6.2.***

State One:
$V_1 = 455 \text{ mL}$
$P_1 = 0.330 \text{ atm}$

State Two:
$V_2 = 9.36 \times 10^3 \text{ mL}$
$P_2 = ?$

Solving $P_1V_1 = P_2V_2$ for P_2 gives $P_2 = P_1 \times \frac{V_1}{V_2}$ $P_2 = 0.330 \text{ atm} \times \frac{455 \text{ mL}}{9.36 \times 10^3 \text{ mL}} = \mathbf{0.0160 \text{ atm}}$

Since the volume increase occurred at constant temperature, a decrease in pressure is expected.

6.23 ***See Section 6.3 and Example 6.3.***

State One:
$V_1 = 39.6$ mL
$T_1 = 27 + 273 = 300$ K

State Two:
$V_2 = ?$
$T_2 = 127 + 273 = 400$ K

Solving $\frac{V_1}{T_1} = \frac{V_2}{T_2}$ for V_2 gives $V_2 = V_1 \times \frac{T_2}{T_1}$ $\quad V_2 = 39.6 \text{ mL} \times \frac{400 \text{ K}}{300 \text{ K}} = \mathbf{52.8\ mL}$

Since the temperature increase occurred at constant pressure, an increase in volume is expected.

6.24 ***See Section 6.3.***

State One:
$n_1 = 3.2$ mol
$V_1 = 34$ L

State Two:
$n_2 = 5.3$ mol
$V_2 = ?$

Solving $PV_1 = n_1RT$ and $PV_2 = n_2RT$ for $\frac{RT}{P}$ gives $\frac{V_1}{n_1} = \frac{RT}{P} = \frac{V_2}{n_2}$.

Solving $\frac{V_1}{n_1} = \frac{V_2}{n_2}$ for V_2 gives $V_2 = V_1 \times \frac{n_2}{n_1}$ $\quad V_2 = 34 \text{ L} \times \frac{5.3 \text{ mol}}{3.2 \text{ mol}} = \mathbf{56\ L}$

Since the number of moles increased at constant termperature and pressure, an increase in volume is expected.

6.25 ***See Section 6.3, Example 6.2.***

State One:
$P_1 = 399$ torr
$V_1 = 166$ mL

State Two:
$P_2 = 1$ atm or 760 torr
$V_2 = ?$

Solving $P_1 V_1 = P_2 V_2$ for V_2 gives $V_2 = V_1 \times \frac{P_1}{P_2}$ $\quad V_2 = 166 \text{ mL} \times \frac{399 \text{ torr}}{760 \text{ torr}} = \mathbf{87.2\ mL}$

Since the pressure increase occurred at constant temperature, a decrease in volume is expected.

6.26 ***See Section 6.3 and Example 6.4.***

State One:
$P_1 = 745$ torr
$V_1 = 2.33$ L
$T_1 = 27 + 273 = 300$ K

State Two:
$P_2 = ?$
$V_2 = 1.22$ L
$T_2 = 100 + 273 = 373$ K

Solving $\frac{P_1V_1}{T_1} = \frac{P_2V_2}{T_2}$ for P_2 gives $P_2 = P_1 \times \frac{V_1}{V_2} \times \frac{T_2}{T_1}$ $\quad P_2 = 745 \text{ torr} \times \frac{2.33 \text{ L}}{1.22 \text{ L}} \times \frac{373 \text{ K}}{300 \text{ K}} = 1.77 \times 10^3$ torr

Converting to atmospheres gives $P_2 = 1.77 \times 10^3 \text{ torr} \times \frac{1 \text{ atm}}{760 \text{ torr}} = \mathbf{2.33\ atm}$.

The decrease in volume and increase in temperature both lead to an increase in pressure.

6.27 ***See Section 6.3 and Example 6.4.***

State One:
$P_1 = 2.11$ atm
$V_1 = 900$ mL
$T_1 = 0 + 273 = 273$ K

State Two:
$P_2 = 4.33$ atm
$V_2 = ?$
$T_2 = 22 + 273 = 295$ K

Solving $\frac{P_1V_1}{T_1} = \frac{P_2V_2}{T_2}$ for V_2 gives $V_2 = V_1 \times \frac{P_1}{P_2} \times \frac{T_2}{T_1}$ $\quad V_2 = 900 \text{ mL} \times \frac{2.11 \text{ atm}}{4.33 \text{ atm}} \times \frac{295 \text{ K}}{273 \text{ K}} = \mathbf{474\ mL}$

6.28 See Section 6.3.

State One: $P_1 = 345$ torr, $T_1 = 358$ K

State Two: $P_2 = 938$ torr, $T_2 = ?$

Solving $\frac{P_1}{T_1} = \frac{P_2}{T_2}$ for T_2 gives $T_2 = T_1 \times \frac{P_2}{P_1}$ $\quad T_2 = 358\text{ K} \times \frac{938\text{ torr}}{345\text{ torr}} = \mathbf{973\ K}$

Since the pressure increase occurred at constant volume, an increase in temperature is expected.

6.29 See Section 6.3.

State One: $n_1 = 77.4 \times 10^6$ mol, $V_1 = 4.50 \times 10^4\ ft^3$, $T_1 = -5 + 273 = 268$ K

State Two: $n_2 = 72.1 \times 10^6$ mol, $V_2 = ?$, $T_2 = +7 + 273 = 280$ K

Solving $P_1 V_1 = n_1RT_1$ for R and $P_2 V_2 = n_2RT_2$ for R gives $\frac{P_1V_1}{n_1T_1} = R = \frac{P_2V_2}{n_2T_2}$

At constant pressure, this yields $\frac{V_1}{n_1T_1} = \frac{V_2}{n_2T_2}$ which gives $V_2 = V_1 \times \frac{n_2}{n_1} \times \frac{T_2}{T_1}$

$$V_2 = 4.50 \times 10^4\ ft^3 \times \frac{72.1 \times 10^6\ mol}{77.4 \times 10^6\ mol} \times \frac{280\ K}{268\ K} = \mathbf{4.38 \times 10^4\ ft^3}$$

6.30 See Section 6.3.

State One: $P_1 = 1.54$ atm, $V_1 = 1.33$ L, $T_1 = 22 + 273 = 295$ K

State Two: $P_2 = 2.13$ atm, $V_2 = 1.89$ L, $T_2 = ?$

Solving $\frac{P_1V_1}{T_1} = \frac{P_2V_2}{T_2}$ for T_2 gives $T_2 = T_1 \times \frac{P_2}{P_1} \times \frac{V_2}{V_1}$ $\quad T_2 = 295\text{ K} \times \frac{2.13\text{ atm}}{1.54\text{ atm}} \times \frac{1.89\text{ L}}{1.33\text{ L}} = 580\text{ K}$

Hence, $T_2 = 580 - 273 = $ **307°C**.

6.31 See Section 6.3 and Example 6.2.

The key to solving this problem is recognizing that PV = nRT and that nRT remains constant.

Hence, (P_1 cylinder)(V cylinder) = (P_2 cylinder)(V cylinder) + (P balloons)(V balloons)

(10.3 atm)(30 L) = (9.22 atm)(30 L) + (1.02 atm)(50 V balloon)

(1.02 atm)(50 V balloon) = (10.3 atm)(30 L) - (9.22 atm)(30 L)

$$\text{V balloon} = \frac{(10.3\text{ atm})(30\text{L}) - (9.22\text{ atm})(30\text{L})}{(1.02\text{ atm})(50)} = \mathbf{0.64\ L}$$

6.32 See Section 6.3 and Example 6.2.

The key to solving this problem is recognizing that PV = nRT and that nRT remains constant.

Hence, (P_1 cylinder)(V cylinder) = (P_2 cylinder)(V cylinder) + (P balloon)(V balloon)

(20.3 atm)(40 L) = (P_2 cylinder)(40 L) + (1.03 atm)(105 L)

(P_2 cylinder)(40 L) = (20.3 atm)(40 L) - (1.03 atm)(105 L)

$$P_2\text{ cylinder} = \frac{(20.3\text{ atm})(40\text{L}) - (1.03\text{ atm})(105\text{ L})}{(40\text{ L})} = \mathbf{17.6\ atm}$$

6.33 *See Section 6.4.*

Known Quantities: $P = 230\ \text{torr} \times \dfrac{1\ \text{atm}}{760\ \text{torr}} = 0.303\ \text{atm}$ $V = 3.22\ \text{L}$ $T = 33 + 273 = 306\ \text{K}$

Solving PV = nRT for n gives $n = \dfrac{PV}{RT}$ $n = \dfrac{(0.303\ \text{atm})(3.22\ \text{L})}{\left(0.0821 \dfrac{\text{L} \cdot \text{atm}}{\text{mol} \cdot \text{K}}\right)(306\ \text{K})} = \mathbf{0.0388\ mol\ Ar}$

6.34 *See Section 6.4.*

Known Quantities: $n = 2.49\ \text{mol}$ $P = 2.44\ \text{atm}$ $V = 24.0\ \text{L}$

Solving PV = nRT for T gives $T = \dfrac{PV}{nR}$ $T = \dfrac{(2.44\ \text{atm})(24.0\text{L})}{(2.49\ \text{mol})\left(0.0821 \dfrac{\text{L} \cdot \text{atm}}{\text{mol} \cdot \text{K}}\right)} = 286\ \text{K}$

$T_C = T_K - 273 = 286 - 273 = \mathbf{13°C}$

6.35 *See Section 6.4.*

Known Quantities: $P = 100\ \text{atm}$ $V = 3.00\ \text{L}$ $T = 27 + 273 = 300\ \text{K}$

Solving PV = nRT for n gives $n = \dfrac{PV}{RT}$ $n = \dfrac{(100\ \text{atm})(3.00\ \text{L})}{\left(0.0821 \dfrac{\text{L} \cdot \text{atm}}{\text{mol} \cdot \text{K}}\right)(300\ \text{K})} = \mathbf{12.2\ mol}$

6.36 *See Section 6.4 and Example 6.5.*

Known Quantities: $n = 0.322\ \text{g}\ N_2 \times \dfrac{1\ \text{mol}\ N_2}{28.0\ \text{g}\ N_2} = 0.0115\ \text{mol}\ N_2$ $V = 0.300\ \text{L}$

$T = 24 + 273 = 297\ \text{K}$

Solving PV = nRT for P gives $P = \dfrac{nRT}{V}$ $P = \dfrac{(0.0115\ \text{mol})\left(0.0821 \dfrac{\text{L} \cdot \text{atm}}{\text{mol} \cdot \text{K}}\right)(297\ \text{K})}{0.300\ \text{L}} = \mathbf{0.934\ atm}$

6.37 *See Section 6.4 and Table 6.1.*

Known Quantities: $n = 82.3\ \text{mol}$ $P = 1.01 \times 10^5\ \text{Pa} \times \dfrac{1\ \text{mmHg}}{133.3\ \text{Pa}} \times \dfrac{1\ \text{atm}}{760\ \text{mmHg}} = 0.997\ \text{atm}$

$T = 25 + 273 = 298\ \text{K}$

Solving PV = nRT for V gives $V = \dfrac{nRT}{P}$ $V = \dfrac{(82.3\ \text{mol})\left(0.0821 \dfrac{\text{L} \cdot \text{atm}}{\text{mol} \cdot \text{K}}\right)(298\ \text{K})}{(0.997\ \text{atm})} = \mathbf{2.02 \times 10^3\ L\ H_2}$

6.38 *See Section 6.4.*

Known Quantities: $n = 1.33\ \text{mol}$ $P = 1.21\ \text{atm}$ $V = 22.1\ \text{L}$

Solving PV = nRT for T gives $T = \dfrac{PV}{nR}$ $T = \dfrac{(1.21\ \text{atm})(22.1\ \text{L}))}{(1.33\ \text{mol})\left(0.0821 \dfrac{\text{L} \cdot \text{atm}}{\text{mol} \cdot \text{K}}\right)} = \mathbf{245\ K}$

6.39 *See Section 6.4.*

Known Quantities: $n = 1.44 \text{ g } CO_2 \times \dfrac{1 \text{ mol } CO_2}{44.0 \text{ g } CO_2} = 0.0327 \text{ mol } CO_2$ $V = 2.33 \text{ L}$

$T = 211 + 273 = 484 \text{ K}$

Solving PV = nRT for P gives $P = \dfrac{nRT}{V}$ $P = \dfrac{(0.0327 \text{ mol})\left(0.0821 \dfrac{\text{L} \cdot \text{atm}}{\text{mol} \cdot \text{K}}\right)(484 \text{ K})}{(2.33 \text{ L})} = \mathbf{0.558 \text{ atm}}$

6.40 *See Section 6.4.*

Known Quantities: $P = 1.22 \text{ atm}$ $V = 0.200 \text{ L}$ $T = 27 + 273 = 300 \text{ K}$

Solving PV = nRT for n gives $n = \dfrac{PV}{RT}$ $n = \dfrac{(1.22 \text{ atm})(0.200 \text{ L})}{\left(0.0821 \dfrac{\text{L} \cdot \text{atm}}{\text{mol} \cdot \text{K}}\right)(300 \text{ K})} = 9.91 \times 10^{-3} \text{ mol } SO_2$

$? \text{ g } SO_2 = 9.91 \times 10^{-3} \text{ mol } SO_2 \times \dfrac{64.1 \text{ g } SO_2}{1 \text{ mol } SO_2} = \mathbf{0.635 \text{ g } SO_2}$

6.41 *See Section 6.4.*

Known Quantities: $P = 1.13 \text{ atm}$ $V = 33.2 \text{ L}$ $T = 122 + 273 = 395 \text{ K}$

Solving PV = nRT for n gives $n = \dfrac{PV}{RT}$ $n = \dfrac{(1.13 \text{ atm})(33.2 \text{ L})}{\left(0.0821 \dfrac{\text{L} \cdot \text{atm}}{\text{mol} \cdot \text{K}}\right)(395 \text{ K})} = 1.16 \text{ mol } N_2$

$? \text{ molecules } N_2 = 1.16 \text{ mol } N_2 \times \dfrac{6.022 \times 10^{23} \text{ molecules } N_2}{1 \text{ mol } N_2} = \mathbf{6.99 \times 10^{23} \text{ molecules } N_2}$

6.42 *See Section 6.4.*

Known Quantities: $n = 2.35 \times 10^{25} \text{ molecules } H_2O \times \dfrac{1 \text{ mol } H_2O}{6.022 \times 10^{23} \text{ } H_2O \text{ molecules}} = 39.0 \text{ mol } H_2O$

$P = 0.173 \text{ atm}$ $T = 229 + 273 = 502 \text{ K}$

Solving PV = nRT for V gives $V = \dfrac{nRT}{P}$ $V = \dfrac{(39.0 \text{ mol})\left(0.0821 \dfrac{\text{L} \cdot \text{atm}}{\text{mol} \cdot \text{K}}\right)(502 \text{ K})}{(0.173 \text{ atm})} = \mathbf{9.29 \times 10^3 \text{ L}}$

6.43 *See Section 6.4 and Example 6.6.*

Known Quantities: $m = 0.550 \text{ g}$ $P = 744 \text{ torr} \times \dfrac{1 \text{ atm}}{760 \text{ torr}} = 0.979 \text{ atm}$

$V = 258 \text{ mL} \times \dfrac{1 \text{ L}}{10^3 \text{ mL}} = 0.258 \text{ L}$ $T = 22 + 273 = 295 \text{ K}$

Solving PV = nRT for n gives $n = \dfrac{PV}{RT}$ $n = \dfrac{(0.979 \text{ atm})(0.258 \text{ L})}{\left(0.0821 \dfrac{\text{L} \cdot \text{atm}}{\text{mol} \cdot \text{K}}\right)(295 \text{ K})} = 0.0104 \text{ mol}$

and using $\mathcal{M} = \dfrac{m}{n}$ yields $\mathcal{M} = \dfrac{0.550 \text{ g}}{0.0104 \text{ mol}} = \mathbf{52.9 \text{ g/mol}}$

6.44 *See Section 6.4 and Example 6.6.*

Known Quantities: $m = 0.165 \text{ g}$ $\quad P = 1.22 \text{ atm}$

$V = 34.8 \text{ mL} \times \dfrac{1 \text{ L}}{10^3 \text{ mL}} = 0.0348 \text{ L}$ $\quad T = 50 + 273 = 323 \text{ K}$

Solving PV = nRT for n gives $n = \dfrac{PV}{RT}$ $\quad n = \dfrac{(1.22 \text{ atm})(0.0348 \text{ L})}{\left(0.0821 \dfrac{\text{L} \cdot \text{atm}}{\text{mol} \cdot \text{K}}\right)(323 \text{ K})} = 0.00160 \text{ mol}$

and using $\mathcal{M} = \dfrac{m}{n}$ yields $\mathcal{M} = \dfrac{0.165 \text{ g}}{0.00160 \text{ mol}} = \mathbf{103 \text{ g / mol}}$

6.45 *See Section 6.4 and Example 6.6.*

Known Quantities: $m = 0.121 \text{ g}$ $\quad P = 740 \text{ torr} \times \dfrac{1 \text{ atm}}{760 \text{ torr}} = 0.974 \text{ atm}$

$V = 21.0 \text{ mL} \times \dfrac{1 \text{ L}}{10^3 \text{ mL}} = 0.0210 \text{ L}$ $\quad T = 29 + 273 = 302 \text{ K}$

Solving PV = nRT for n gives $n = \dfrac{PV}{RT}$ $\quad n = \dfrac{(0.0974 \text{ atm})(0.0210 \text{ L})}{\left(0.0821 \dfrac{\text{L} \cdot \text{atm}}{\text{mol} \cdot \text{K}}\right)(302 \text{ K})} = 8.25 \times 10^{-4} \text{ mol}$

and using $\mathcal{M} = \dfrac{m}{n}$ yields $\mathcal{M} = \dfrac{0.121 \text{ g}}{8.25 \times 10^{-4} \text{ mol}} = \mathbf{147 \text{ g / mol}}$

6.46 *See Section 6.4 and Example 6.6.*

Known Quantities: $m = 0.335 \text{ g}$ $\quad P = 1.02 \text{ atm}$

$V = 32.6 \text{ mL} \times \dfrac{1 \text{ L}}{10^3 \text{ mL}} = 0.0326 \text{ L}$ $\quad T = 27 + 273 = 300 \text{ K}$

Solving PV = nRT for n gives $n = \dfrac{PV}{RT}$ $\quad n = \dfrac{(1.02 \text{ atm})(0.0326 \text{ L})}{\left(0.0821 \dfrac{\text{L} \cdot \text{atm}}{\text{mol} \cdot \text{K}}\right)(300 \text{ K})} = 1.35 \times 10^{-3} \text{ mol}$

and using $\mathcal{M} = \dfrac{m}{n}$ yields $\mathcal{M} = \dfrac{0.335 \text{ g}}{1.35 \times 10^{-3} \text{ mol}} = \mathbf{248 \text{ g / mol}}$

6.47 *See Section 6.4 and Example 6.7.*

Known Quantities: mass of 1 mol He = 4.00 g $\quad P = 10.00 \text{ atm}$ $\quad T = 0 + 273 = 273 \text{ K}$

Solving PV = nRT for V gives $V = \dfrac{nRT}{P}$ $\quad V = \dfrac{(1 \text{ mol})\left(0.0821 \dfrac{\text{L} \cdot \text{atm}}{\text{mol} \cdot \text{K}}\right)(273 \text{ K})}{(10.00 \text{ atm})} = 2.24 \text{ L}$

and using $d = \dfrac{m}{V}$ yields $d = \dfrac{4.00 \text{ g}}{2.24 \text{ L}} = \mathbf{1.79 \text{ g/L}}$

6.48 *See Section 6.4 and Example 6.7.*

Known Quantities: mass of one mole O_2 = 32.0 g $\quad P = 0.100 \text{ atm}$ $\quad T = 35 + 273 = 308 \text{ K}$

Solving PV = nRT for V gives $V = \dfrac{nRT}{P}$ $\quad V = \dfrac{(1 \text{ mol})\left(0.0821 \dfrac{\text{L} \cdot \text{atm}}{\text{mol} \cdot \text{K}}\right)(308\text{K})}{(0.100 \text{ atm})} = 253 \text{ L}$

and using $d = \dfrac{m}{V}$ yields $d = \dfrac{32.0 \text{ g}}{253 \text{ L}} = \mathbf{0.126 \text{ g/L}}$

6.49 *See Section 6.4 and Example 6.7.*

Known Quantities: mass of one mole CO_2 = 44.0 g   P = 1.00 atm
T = 27 + 273 = 300 K

Solving PV = nRT for V gives $V = \frac{nRT}{P}$

$$V = \frac{(1\ \text{mol})\left(0.0821\ \frac{\text{L}\cdot\text{atm}}{\text{mol}\cdot\text{K}}\right)(300\ \text{K})}{(1.00\ \text{atm})} = 24.6\ \text{L}$$

and using $d = \frac{m}{V}$ yields $d = \frac{44.0\ \text{g}}{24.6\ \text{L}} = \mathbf{1.79\ g/L}$

6.50 *See Section 6.4 and Example 6.7.*

Known Quantities: mass of one mole C_2H_6 = 30.0 g   P = 0.55 atm
T = 100 + 273 = 373 K

Solving PV = nRT for V gives $V = \frac{nRT}{P}$

$$V = \frac{(1\ \text{mol})\left(0.0821\ \frac{\text{L}\cdot\text{atm})}{\text{mol}\cdot\text{K}}\right)(373\ \text{K})}{(0.55\ \text{atm})} = 55.7\ \text{L}$$

and using $d = \frac{m}{V}$ yields $d = \frac{30.0\ \text{g}}{55.7\ \text{L}} = \mathbf{0.539\ g/L}$

6.51 *See Section 6.5.*

Unbalanced: $Li(s) + H_2O(\ell) \rightarrow LiOH(aq) + H_2(g)$ Start with H_2, since Li and O are balanced..
Step 1: $Li(s) + H_2O(\ell) \rightarrow LiOH(aq) + \underline{\tfrac{1}{2}}H_2(g)$ Balances H.
Step 2: $\underline{2}Li(s) + \underline{2}H_2O(\ell) \rightarrow \underline{2}LiOH(aq) + H_2(g)$ Gives whole number coefficients.

Strategy: g Li → mol Li → mol H_2 → L H_2

$$?\ \text{mol H}_2 = 0.0223\ \text{g Li} \times \frac{1\ \text{mol Li}}{6.94\ \text{g Li}} \times \frac{1\ \text{mol H}_2}{2\ \text{mol Li}} = 0.00161\ \text{mol H}_2$$

Known Quantities: n = 0.00161 mol H_2 P = 1.33 atm   T = 33 + 273 = 306 K

Solving PV = nRT for V gives $V = \frac{nRT}{P}$

$$V = \frac{(0.00161\ \text{mol})\left(0.0821\frac{\text{L}\cdot\text{atm}}{\text{mol}\cdot\text{K}}\right)(306\ \text{K})}{(1.33\ \text{atm})} = \mathbf{0.0304\ L\ H_2}$$

6.52 *See Section 6.5.*

Unbalanced: $Mg(s) + HCl(aq) \rightarrow MgCl_2(aq) + H_2(g)$ Start with $MgCl_2$.
Step 1: $Mg(s) + \underline{2}HCl(aq) \rightarrow MgCl_2(aq) + H_2(g)$ Balances H, Cl.

Strategy: g Mg → mol Mg → mol H_2 → L H_2

$$?\ \text{mol H}_2 = 2.23\ \text{g Mg} \times \frac{1\ \text{mol Mg}}{24.3\ \text{g Mg}} \times \frac{1\ \text{mol H}_2}{1\ \text{mol Mg}} = 0.0918\ \text{mol H}_2$$

Known Quantities: n = 0.0918 mol H_2 $P = 744\ \text{torr} \times \frac{1\ \text{atm}}{760\ \text{torr}} = 0.979\ \text{atm}$
T = 26 + 273 = 299 K

Solving PV = nRT for V gives $V = \frac{nRT}{P}$

$$V = \frac{(0.0918\ \text{mol})\left(0.0821\ \frac{\text{L}\cdot\text{atm}}{\text{mol}\cdot\text{K}}\right)(299\ \text{K})}{(0.979\ \text{atm})} = \mathbf{2.30\ L\ H_2}$$

6.53 *See Section 6.5.*

Unbalanced:	$KClO_3(s)$	→	$KCl(s) + O_2(g)$	Start with $KClO_3$.
Step 1:	$\underline{2}KClO_3(s)$	→	$KCl(s) + \underline{3}O_2(g)$	Balances O.
Step 2:	$2KClO_3(s)$	→	$\underline{2}KCl(s) + 3O_2(g)$	Balances K and Cl.

Strategy: $g\ KClO_3 \rightarrow mol\ KClO_3 \rightarrow mol\ O_2 \rightarrow L\ O_2$

$$? \text{ mol } O_2 = 4.42 \text{ g } KClO_3 \times \frac{1 \text{ mol } KClO_3}{122.6 \text{ g } KClO_3} \times \frac{3 \text{ mol } O_2}{2 \text{ mol } KClO_3} = 0.0541 \text{ mol } O_2$$

Known Quantities: $n = 0.0541 \text{ mol } O_2$ $P = 760 \text{ torr} = 1.00 \text{ atm}$ $T = 23 + 273 = 296 \text{ K}$

Solving PV = nRT for V gives $V = \frac{nRT}{P}$ $V = \frac{(0.0541 \text{ mol})\left(0.0821 \frac{L \cdot atm}{mol \cdot K}\right)(296 \text{ K})}{(1.00 \text{ atm})} = \mathbf{1.31\ L\ O_2}$

6.54 *See Section 6.5.*

Unbalanced:	$H_2(g) + O_2(g)$	→	$H_2O(\ell)$	Start with H_2O.
Step 1:	$H_2(g) + O_2(g)$	→	$\underline{2}H_2O(\ell)$	Balances O.
Step 2:	$\underline{2}H_2(g) + O_2(g)$	→	$2H_2O(\ell)$	Balances H.

Strategy: $gH_2O \rightarrow mol\ H_2O \rightarrow mol\ O_2 \rightarrow L\ O_2$

$$? \text{ mol } O_2 = 4.22 \text{ g } H_2O \times \frac{1 \text{ mol } H_2O}{18.0 \text{ g } H_2O} \times \frac{1 \text{ mol } O_2}{2 \text{ mol } H_2O} = 0.117 \text{ mol } O_2$$

Known Quantities: $n = 0.117 \text{ mol } O_2$ $P = 0.993 \text{ atm}$ $T = 30 + 273 = 303 \text{ K}$

Solving PV = nRT for V gives $V = \frac{nRT}{P}$ $V = \frac{(0.117 \text{ mol})\left(0.0821 \frac{L \cdot atm}{mol \cdot K}\right)(303 \text{ K})}{(0.993 \text{ atm})} = \mathbf{2.93\ L\ O_2}$

6.55 *See Section 6.5 and Example 6.8.*

Balanced: $Zn(s) + H_2SO_4(aq) \rightarrow ZnSO_4(aq) + H_2(g)$

Strategy: $g\ Zn \rightarrow mol\ Zn \rightarrow mol\ H_2$

$$? \text{ mol } H_2 \text{ based on Zn} = 1.33 \text{ g Zn} \times \frac{1 \text{ mol Zn}}{65.4 \text{ g Zn}} \times \frac{1 \text{ mol } H_2}{1 \text{ mol Zn}} = 0.0203 \text{ mol } H_2$$

Strategy: $L\ H_2SO_4\ soln \rightarrow mol\ H_2SO_4 \rightarrow mol\ H_2$

$$? \text{ mol } H_2 \text{ based on } H_2SO_4 = 0.300 \text{ L } H_2SO_4 \text{ soln} \times \frac{2.33 \text{ mol } H_2SO_4}{1 \text{ L } H_2SO_4 \text{ soln}} \times \frac{1 \text{ mol } H_2}{1 \text{ mol } H_2SO_4} = 0.699 \text{ mol } H_2$$

Zn is the limiting reactant, since it yields less H_2.

Known Quantities: $n = 0.0203 \text{ mol } H_2$ $P = 1.12 \text{ atm}$ $T = 25 + 273 = 298 \text{ K}$

Solving PV = nRT for V gives $V = \frac{nRT}{P}$ $V = \frac{(0.0203 \text{ mol})\left(0.0821 \frac{L \cdot atm}{mol \cdot K}\right)(298 \text{ K})}{(1.12 \text{ atm})} = \mathbf{0.443\ L\ H_2}$

6.56 ***See Section 6.5 and Example 6.8.***

Unbalanced:	$H_2(g) + O_2(g)$	$\rightarrow$	$H_2O(\ell)$	Start with H_2O.
Step 1:	$H_2(g) + O_2(g)$	$\rightarrow$	$\underline{2}H_2O(\ell)$	Balances O.
Step 2:	$\underline{2}H_2(g) + O_2(g)$	$\rightarrow$	$2H_2O(\ell)$	Balances H.

Strategy: mol H_2 → mol H_2O → g H_2O

Known Quantities: $P = 734 \text{ torr} \times \dfrac{1 \text{ atm}}{760 \text{ torr}} = 0.966 \text{ atm}$ $V = 2.44 \text{ L}$ $T = 27 + 273 = 300 \text{ K}$

Solving PV = nRT for n gives $n = \dfrac{PV}{RT}$ $n\ H_2 = \dfrac{(0.966 \text{ atm})(2.44 \text{ L})}{\left(0.0821 \dfrac{\text{L}\cdot\text{atm}}{\text{mol}\cdot\text{K}}\right)(300 \text{ K})} = 0.0957 \text{ mol } H_2$

$? \text{ g } H_2O \text{ based on } H_2 = 0.0957 \text{ mol } H_2 \times \dfrac{2 \text{ mol } H_2O}{2 \text{ mol } H_2} \times \dfrac{18.0 \text{ g } H_2O}{1 \text{ mol } H_2O} = \mathbf{1.72\ g\ H_2O}$

Strategy: mol O_2 → mol H_2O → g H_2O

Known Quantities: $P = 734 \text{ torr} \times \dfrac{1 \text{ atm}}{760 \text{ torr}} = 0.966 \text{ atm}$ $V = 3.11 \text{ L}$ $T = 27 + 273 = 300 \text{ K}$

Solving PV = nRT for n gives $n = \dfrac{PV}{RT}$ $n_{O_2} = \dfrac{(0.966 \text{ atm})(3.11 \text{ L})}{\left(0.0821 \dfrac{\text{L}\cdot\text{atm}}{\text{mol}\cdot\text{K}}\right)(300 \text{ K})} = 0.126 \text{ mol } O_2$

$? \text{ g } H_2O \text{ based on } O_2 = 0.126 \text{ mol } O_2 \times \dfrac{2 \text{ mol } H_2O}{1 \text{ mol} O_2} \times \dfrac{18.0 \text{ g } H_2O}{1 \text{ mol } H_2O} = \mathbf{4.54\ g\ H_2O}$

H_2 is the limiting reactant because it produces less H_2O. The maximum amount of H_2O that can be produced from the 2.44 L H_2 and 3.11 L O_2 is **1.72 g.**

6.57 ***See Section 6.5.***

Unbalanced:	$NaN_3(s)$	$\rightarrow$	$Na(s) + N_2(g)$	Start with NaN_3.
Step 1:	$\underline{2}NaN_3(s)$	$\rightarrow$	$Na(s) + \underline{3}N_2(g)$	Balances N.
Step 2:	$2NaN_3(s)$	$\rightarrow$	$\underline{2}Na(s) + 3N_2(g)$	Balances Na.

Strategy: g NaN_3 → mol NaN_3 → mol N_2 → L N_2

$? \text{ mol } N_2 = 1.88 \text{ g NaN}_3 \times \dfrac{1 \text{ mol NaN}_3}{65.0 \text{ g NaN}_3} \times \dfrac{3 \text{ mol } N_2}{2 \text{ mol NaN}_3} = 0.0434 \text{ mol } N_2$

Known Quantities: $n = 0.0434 \text{ mol } N_2$ $P = 755 \text{ torr} \times \dfrac{1 \text{ atm}}{760 \text{ torr}} = 0.993 \text{ atm}$ $T = 24 + 273 = 297 \text{ K}$

Solving PV = nRT for V gives $V = \dfrac{nRT}{P}$ $V = \dfrac{(0.0434 \text{ mol})\left(0.0821 \dfrac{\text{L}\cdot\text{atm}}{\text{mol}\cdot\text{K}}\right)(297 \text{ K})}{0.993 \text{ atm}} = \mathbf{1.07\ L\ N_2}$

6.58 ***See Section 6.5 and Example 6.9.***

Unbalanced:	$H_2(g) + O_2(g)$	$\rightarrow$	$H_2O(\ell)$	Start with H_2O.
Step 1:	$H_2(g) + O_2(g)$	$\rightarrow$	$\underline{2}H_2O(\ell)$	Balances O.
Step 2:	$\underline{2}H_2(g) + O_2(g)$	$\rightarrow$	$2H_2O(\ell)$	Balances H.

$? \text{ L } O_2 = 3.22 \text{ L } H_2 \times \dfrac{1 \text{ L } O_2}{2 \text{ L } H_2} = \mathbf{1.61\ L\ H_2}$

6.59 See Section 6.5 and Example 6.9.

Unbalanced:	$H_2S(g) + O_2(g) \rightarrow$	$SO_2(g) + H_2O$	Start with O_2, since H and S are balanced.
Step 1:	$H_2S(g) + \underline{3/2}O_2(g) \rightarrow$	$SO_2 + H_2O$	Balances O.
Step 2:	$\underline{2}H_2S(g) + \underline{3}O_2(g) \rightarrow$	$\underline{2}SO_2(g) + \underline{2}H_2O$	Gives whole number coefficients.

$? \text{ L } SO_2 = 2.44 \text{ L } H_2S \times \frac{2 \text{ L } SO_2}{2 \text{ L } H_2S} = \mathbf{2.44 \text{ L } SO_2}$

$? \text{ L } O_2 = 2.44 \text{ L } H_2S \times \frac{3 \text{ L } O_2}{2 \text{ L } H_2S} = \mathbf{3.66 \text{ L } O_2}$

6.60 See Section 6.6 and Example 6.9.

Unbalanced:	$CH_4(g) + O_2(g) \rightarrow$	$CO_2(g) + H_2O(\ell)$	Start with CH_4.
Step 1:	$CH_4(g) + O_2(g) \rightarrow$	$CO_2(g) + \underline{2}H_2O(\ell)$	Balances H.
Step 2:	$CH_4(g) + \underline{2}O_2(g) \rightarrow$	$CO_2(g) + 2H_2O(\ell)$	Balances O.

$? \text{ L } CO_2 = 2.00 \times 10^3 \text{ L } CH_4 \times \frac{1 \text{ L } CO_2(g)}{1 \text{ L } CH_4(g)} = \mathbf{2.00 \times 10^3 \text{ L } CH_4}$

$? \text{ L } O_2 = 2.00 \times 10^3 \text{ L } CH_4 \times \frac{2 \text{ L } O_2}{1 \text{ L } CH_4} = \mathbf{4.00 \times 10^3 \text{ L } O_2}$

6.61 See Section 6.6.

$P_T = P_{H_2} + P_{Ar} = 1.22 \text{ atm} + 4.33 \text{ atm} = \mathbf{5.55 \text{ atm}}$

6.62 See Section 6.6.

Solving $P_T = P_{Ar} + P_{Ne}$ for P_{Ar} gives $P_{Ar} = P_T - P_{Ne} = 500 \text{ torr} - 235 \text{ torr} = \mathbf{265 \text{ torr.}}$

6.63 See Sections 6.3, 6.4, 6.6.

The partial pressure of the gas already in the 3.11 L container, gas 1, will remain the same. The partial pressure of the gas that is added to the 3.11 L container, gas 2, can be calculated using Boyle's Law or the Ideal Gas Law. The approach of Boyle's Law will be used here.

State One:	State Two:
$P_1 = 2.55 \text{ atm}$	$P_2 = ?$
$V_1 = 2.11 \text{ L}$	$V_2 = 3.11 \text{ L}$
$T_1 = 27 + 273 = 300 \text{ K}$	$T_2 = 27 + 273 = 300 \text{ K}$

Solving $\frac{P_1 V_1}{T_1} = \frac{P_2 V_2}{T_2}$ for P_2 at a constant temperature gives $P_2 = P_1 \times \frac{V_1}{V_2}$ $\quad P_2 = 2.55 \text{ atm} \times \frac{2.11 \text{ L}}{3.11 \text{ L}} = 1.73 \text{ atm}$

Hence, the total pressure of the mixture of gases 1 and 2 in the 3.11 L container is given by $P_T = P_{gas} + P_{gas\ 2}$ and is $P_T = 4.33 \text{ atm} + 1.73 \text{ atm} = \mathbf{6.06 \text{ atm.}}$

Note: The alternative approach involves using PV = nRT to calculate the moles of gas in the 2.11 L container. The pressure that would be exerted by this same number of moles of gas in the 3.11 L container is then calculated by using PV = nRT a second time. The pressure exerted by this second gas is added to the pressure exerted by the first gas to obtain the total pressure.

6.64 *See Sections 6.3, 6.4, 6.6.*

The partial pressure of neon in the 10-L cylinder can be calculated using Boyle's Law.

State One: $P_1 = 3.22$ atm, $V_1 = 4.53$ L

State Two: $P_2 = ?$, $V_2 = 10.0$ L

Solving $P_1V_1 = P_2V_2$ for P_2 gives $P_2 = P_1 \times \frac{V_1}{V_2}$ $\quad P_2 = 3.22\text{ atm} \times \frac{4.53\text{ L}}{10.0\text{ L}} = \mathbf{1.46\ atm}$

Solving $P_T = P_{Ar} + P_{Ne}$ for P_{Ar} gives $P_{Ar} = P_T - P_{Ne} = 5.32\text{ atm} - 1.46\text{ atm} = \mathbf{3.86\ atm}$.

6.65 *See Section 6.6 and Example 6.10.*

Known Quantities: $n_{O_2} = 0.322$ mol $\quad n_{N_2} = 1.53$ mol $\quad n_T = 0.322 + 1.53 = 1.85$

$V = 3.22$ L $\quad T = 100 + 273 = 373$ K

Solving $PV = nRT$ for P_T gives $P_T = \frac{n_T RT}{V}$ $\quad P_T = \frac{(1.85\text{ mol})\left(0.0821\frac{\text{L}\cdot\text{atm}}{\text{mol}\cdot\text{K}}\right)(373\text{ K})}{3.22\text{ L}} = \mathbf{17.6\ atm}$

6.66 *See Section 6.6 and Example 6.11.*

$n_T = 3.22\text{ mol } O_2 + 4.53\text{ mol } N_2 = 7.75\text{ mol}$

$P_{O_2} = \chi_{O_2} P_T = \frac{3.22\text{ mol}}{7.75\text{ mol}} \times 7.32\text{ atm} = \mathbf{3.04\ atm}$

6.67 *See Sections 6.4, 6.6 and Example 6.5.*

Known Quantities: $n_{H_2} = 10.5\text{ g } H_2 \times \frac{1\text{ mol } H_2}{2.016\text{ g } H_2} = 5.21\text{ mol } H_2$ $\quad V = 30.0$ L $\quad T = 120 + 273 = 393$ K

$P_{Ar} = 1.53$ atm

Solving $PV = nRT$ for P gives $P_{H_2} = \frac{nRT}{V}$ $\quad P_{H_2} = \frac{(5.21\text{ mol})\left(0.0821\frac{\text{L}\cdot\text{atm}}{\text{mol}\cdot\text{K}}\right)(393\text{ K})}{30.0\text{ L}} = \mathbf{5.60\ atm}$

Hence, $P_T = P_{H_2} + P_{Ar} = 5.60\text{ atm} + 1.53\text{ atm} = \mathbf{7.13\ atm}$

6.68 *See Sections 6.4, 6.6 and Example 6.5.*

Known Quantities: $n_{Ne} = 5.34\text{ g Ne} \times \frac{1\text{ mol Ne}}{20.2\text{ g Ne}} = 0.264\text{ mol Ne}$

$n_{Ar} = 1.22\text{ g Ar} \times \frac{1\text{ mol Ar}}{39.9\text{ g Ar}} = 0.0306\text{ mol Ar}$

$V = 5.00$ L $\quad T = 30 + 273 = 303$ K

Solving $PV = nRT$ for P gives $P_{Ne} = \frac{n_{Ne}RT}{V}$ $\quad P_{Ne} = \frac{(0.264\text{ mol})\left(0.0821\frac{\text{L}\cdot\text{atm}}{\text{mol}\cdot\text{K}}\right)(303\text{K})}{(5.00\text{ L})} = \mathbf{1.31\ atm}$

and $P_{Ar} = \frac{n_{Ar}RT}{V}$ $\quad P_{Ar} = \frac{(0.0306\text{ mol})\left(0.0821\frac{\text{L}\cdot\text{atm}}{\text{mol}\cdot\text{K}}\right)(303\text{ K})}{(5.00\text{ L})} = \mathbf{0.152\ atm}$

Hence, $P_T = P_{Ne} + P_{Ar} = 1.31\text{ atm} + 0.152\text{ atm} = \mathbf{1.46\ atm}$

6.69   *See Section 6.6 and Example 6.11.*

Known Quantities: $n_{H_2} = 0.220$ mol $n_{N_2} = 0.432$ mol $P_T = 5.22$ atm

Hence, $\chi_{H_2} = \dfrac{0.220 \text{ mol}}{0.652 \text{ mol}} = 0.337$ and $P_{H_2} = \chi_{H_2} P_T = (0.337)(5.22 \text{ atm}) = \mathbf{1.76 \text{ atm}}$

6.70   *See Section 6.6 and Example 6.11.*

Known Quantities: $n_{Ne} = 3.11$ mol $n_{Ar} = 1.02$ mol $P_T = 209$ torr

Hence, $\chi_{Ne} = \dfrac{3.11 \text{ mol}}{4.13 \text{ mol}} = 0.753$ and $P_{Ne} = \chi_{Ne} P_T = (0.753)(209 \text{ torr}) = \mathbf{157 \text{ torr}}$

6.71   *See Section 6.6 and Example 6.11.*

Known Quantities: $n_{Ne} = 0.22$ mol $n_{N_2} = 0.33$ mol $n_{O_2} = 0.22$

$P_T = 2.6$ atm

$$\chi_{Ne} = \frac{0.22 \text{ mol}}{0.77 \text{ mol}} = 0.29 \text{ and } P_{Ne} = \chi_{Ne} P_T = (0.29)(2.6 \text{ atm}) = \mathbf{0.75 \text{ atm}}$$

$$\chi_{N_2} = \frac{0.33 \text{ mol}}{0.77 \text{ mol}} = 0.43 \text{ and } P_{N_2} = \chi_{N_2} P_T = (0.43)(2.6 \text{ atm}) = \mathbf{1.1 \text{ atm}}$$

$$\chi_{O_2} = \frac{0.22 \text{ mol}}{0.77 \text{ mol}} = 0.29 \text{ and } P_{O_2} = \chi_{O_2} P_T = (0.29)(2.6 \text{ atm}) = \mathbf{0.75 \text{ atm}}$$

6.72   *See Section 6.6 and Example 6.11.*

Known Quantities:

$$n_{Ne} = 2.3 \text{ g Ne} \times \frac{1 \text{ mol Ne}}{20.2 \text{ g Ne}} = 0.11 \text{ mol Ne}$$

$$n_{Xe} = 0.33 \text{ g Xe} \times \frac{1 \text{ mol Xe}}{131.3 \text{ g Xe}} = 0.0025 \text{ mol Xe}$$

$$n_{Ar} = 1.1 \text{ g Ar} \times \frac{1 \text{ mol Ar}}{39.9 \text{ g Ar}} = 0.028 \text{ mol Ar}$$

$P_T = 2.6$ atm

$$\chi_{Ne} = \frac{0.11 \text{ mol}}{0.14 \text{ mol}} = 0.78 \text{ and } P_{Ne} = \chi_{Ne} P_T = (0.78)(2.6 \text{ atm}) = \mathbf{2.0 \text{ atm}}$$

$$\chi_{Xe} = \frac{0.0025 \text{ mol}}{0.14 \text{ mol}} = 0.018 \text{ and } P_{Xe} = \chi_{Xe} P_T = (0.018)(2.6 \text{ atm}) = \mathbf{0.047 \text{ atm}}$$

$$\chi_{Ar} = \frac{0.028 \text{ mol}}{0.14 \text{ mol}} = 0.20 \text{ and } P_{Ar} = \chi_{Ar} P_T = (0.20)(2.6 \text{ atm}) = \mathbf{0.52 \text{ atm}}$$

6.73   *See Section 6.6, Table 6.3 and Example 6.12.*

Solving $P_T = P_{O_2} + P_{H_2O}$ for P_{O_2} gives $P_{O_2} = P_T - P_{H_2O}$ At 26°C, $P_{O_2} = 755 \text{ torr} - 25.21 \text{ torr} = \mathbf{730 \text{ torr}}$

6.74   *See Section 6.6, Table 6.3,and Example 6.12.*

$P_T = P_{H_2} + P_{H_2O}$ P_{H_2O} is given in Table 6.3, so P_{H_2} must be calculated.

Known Quantities: $n_{H_2} = 0.0311$ mol $V = 1.00$ L $T = 25 + 273 = 298$ K

Solving PV = nRT for P gives $P_{H_2} = \dfrac{n_{H_2} RT}{V}$ $P_{H_2} = \dfrac{(0.0311 \text{ mol})\left(0.0821 \dfrac{\text{L} \cdot \text{atm}}{\text{mol} \cdot \text{K}}\right)(298 \text{ K})}{1.00 \text{ L}} = \mathbf{0.761 \text{ atm}}$

$? P_{H_2}$ in torr $= 0.761 \text{ atm} \times \frac{760 \text{ torr}}{1 \text{ atm}} = \mathbf{578 \text{ torr}}$

$P_T = P_{H_2} + P_{H_2O} = 578 \text{ torr} + 23.77 \text{ torr} = \mathbf{602 \text{ torr}}$

6.75 ***See Sections 6.5, 6.6.***

Balanced: $Zn + H_2SO_4 \rightarrow ZnSO_4 + H_2$

Strategy: $gZn \rightarrow mol\ Zn \rightarrow mol\ H_2 \rightarrow L\ H_2$

$? \text{ mol } H_2 = 0.113 \text{ g Zn} \times \frac{1 \text{ mol Zn}}{65.4 \text{ g Zn}} \times \frac{1 \text{ mol } H_2}{1 \text{ mol Zn}} = 1.73 \times 10^{-3} \text{ mol } H_2$

Known Quantities: $n_{H_2} = 1.73 \times 10^{-3}$ mol $P_{H_2} = P_{atm} - P_{H_2O} = 750 \text{ torr} - 22.39 \text{ torr} = 728 \text{ torr}$

$P_{H_2} = 728 \text{ torr} \times \frac{1 \text{ atm}}{760 \text{ torr}} = 0.958 \text{ atm}$ $T = 24 + 273 = 297 \text{ K}$

Solving PV = nRT for V gives $V_{H_2} = \frac{n_{H_2}RT}{P_{H_2}}$

$$V_{H_2} = \frac{(1.73 \times 10^{-3} \text{ mol})\left(0.0821 \frac{L \cdot atm}{mol \cdot K}\right)(297 \text{ K})}{(0.958 \text{ atm})} = \mathbf{0.0440 \text{ L } H_2}$$

6.76 ***See Section 6.5, 6.6.***

Unbalanced: $Na(s) + H_2O(\ell) \rightarrow NaOH(aq) + H_2(g)$ Start with H.

Step 1: $\underline{2}Na(s) + \underline{2}H_2O(\ell) \rightarrow \underline{2}NaOH(aq) + H_2(g)$ Balances Na, O, H.

Strategy: $L\ H_2 \rightarrow mol\ H_2 \rightarrow mol\ Na \rightarrow g\ Na$

Known Quantities: $P_{H_2} = P_{atm} - P_{H_2O} = 755 \text{ torr} - 19.84 \text{ torr} = 735 \text{ torr}$

P_{H_2} in atm $= 735 \text{ torr} \times \frac{1 \text{ atm}}{760 \text{ torr}} = 0.967 \text{ atm}$

$V = 499 \text{ mL} \times \frac{1 \text{ L}}{10^3 \text{ mL}} = 0.499 \text{ L}$ $T = 22 + 273 = 295 \text{ K}$

Solving PV = nRT for n gives $n_{H_2} = \frac{P_{H_2}V}{RT}$

$$n_{H_2} = \frac{(0.967 \text{ atm})(0.499 \text{ L})}{\left(0.0821 \frac{L \cdot atm}{mol \cdot K}\right)(295 \text{ K})} = 0.0199 \text{ mol } H_2$$

$? \text{ g Na} = 0.0199 \text{ mol } H_2 \times \frac{2 \text{ mol Na}}{1 \text{ mol } H_2} \times \frac{23.0 \text{ g Na}}{1 \text{ mol Na}} = \mathbf{0.915 \text{ g Na}}$

6.77 ***See Section 6.7.***

Since μ_{rms} increases with decreasing molar mass ($\mathcal{M}$), we obtain the following order of increasing rms speeds at a given temperature: $\mathbf{O_2 < N_2 < Ne}$.

6.78 ***See Section 6.7.***

Since μ_{rms} increases with decreasing molar mass ($\mathcal{M}$), we obtain the following order of increasing rms speeds at a given temperature: $\mathbf{SO_2 < Ar < H_2}$.

6.79 *See Section 6.7.*

Since μ_{rms} increases with decreasing molar mass ($\mathcal{M}$) and increases with increasing absolute temperature, we obtain the following order of increasing rms speeds: **Ar at 25°C < Ne at 25°C < Ne at 100°C**.

6.80 *See Section 6.7.*

Since μ_{rms} increases with decreasing molar mass ($\mathcal{M}$) and increases with increasing absolute temperature, we obtain the following order of increasing rms speeds: **Ar at 0°C < Ne at 50°C < He at 100°C**.

6.81 *See Section 6.7 and Example 6.13.*

For Ne atoms at 100°C, $$\mu_{rms} = \sqrt{\frac{3RT}{\mathcal{M}}} = \sqrt{\frac{(3)(8.314\ kg \cdot m^2 \cdot s^{-2} \cdot mol^{-1} \cdot K^{-1})(373\ K)}{0.02018\ kg \cdot mol^{-1}}} = \mathbf{679\ m/s}$$

6.82 *See Section 6.7 and Example 6.13.*

For SO_2 molecules at 127°C, $$\mu_{rms} = \sqrt{\frac{3RT}{\mathcal{M}}} = \sqrt{\frac{(3)(8.314\ kg \cdot m^2 \cdot s^{-2} \cdot mol^{-1} \cdot K^{-1})(400\ K)}{0.06410\ kg \cdot mol^{-1}}} = \mathbf{395\ m/s}$$

If the temperature on the Kelvin scale is doubled, the μ_{rms} increases by a factor of $\sqrt{2}$.
At 800 K, $\mu_{rms}\ SO_2 = (\sqrt{2})(395\ m/s) = \mathbf{559\ m/s.}$

6.83 *See Section 6.7 and Example 6.13.*

Solving $\mu_{rms} = \sqrt{\frac{3RT}{\mathcal{M}}}$ for $\mathcal{M}$ gives

$$\mathcal{M} = \frac{3RT}{(\mu_{rms})^2} = \frac{(3)(8.314\ kg \cdot m^2 \cdot s^{-2} \cdot mol^{-1} \cdot K^{-1})(301\ K)}{(518\ m \cdot s^{-1})^2} = 0.0280\ kg \cdot mol^{-1} = \mathbf{28.0\ g/mol}$$

6.84 *See Section 6.7.*

Solving $\mu_{rms} = \sqrt{\frac{3RT}{\mathcal{M}}}$ for T gives $T = \frac{\mathcal{M}(\mu_{rms})^2}{3R}$

$$T = \frac{(0.02018\ kg \cdot mol^{-1})(700\ m \cdot s^{-1})^2}{(3)(8.314\ kg \cdot m^2 \cdot s^{-2} \cdot mol^{-1} \cdot K^{-1})} = \mathbf{396\ K}$$

6.85 *See Section 6.8.*

$$\frac{\text{rate of effusion of He}}{\text{rate of effusion of Ne}} = \sqrt{\frac{\mathcal{M}_{Ne}}{\mathcal{M}_{He}}} = \sqrt{\frac{20.18}{4.003}} = \sqrt{5.041} = \mathbf{2.245}$$

We should expect He to effuse faster than Ne since He has a smaller molar mass than Ne.

6.86 *See Section 6.8.*

$$\frac{\text{rate of effusion of } CO_2}{\text{rate of effusion of } CH_4} = \sqrt{\frac{\mathcal{M}_{CH_4}}{\mathcal{M}_{CO_2}}} = \sqrt{\frac{16.0}{44.0}} = \sqrt{0.364} = \mathbf{0.603}$$

We should expect CO_2 to effuse slower than CH_4 since CO_2 has a larger molar mass than CH_4.

6.87 ***See Section 6.8.***

$$\frac{\text{rate of effusion of He}}{\text{rate of effusion of Ar}} = \sqrt{\frac{\mathcal{M}_{Ar}}{\mathcal{M}_{He}}} = \sqrt{\frac{39.9}{4.00}} = \sqrt{9.98} = \mathbf{3.16}$$

We should expect He to effuse faster than Ar since He has a smaller molar mass than Ar.

6.88 ***See Section 6.8.***

$$\frac{\text{rate of effusion of } N_2}{\text{rate of effusion of } SO_2} = \sqrt{\frac{\mathcal{M}_{SO_2}}{\mathcal{M}_{N_2}}} = \sqrt{\frac{64.1}{28.0}} = \sqrt{2.29} = \mathbf{1.51}$$

We should expect N_2 to effuse faster than SO_2 since N_2 has a smaller molar mass than SO_2.

6.89 ***See Section 6.8 and Example 6.14.***

$$\frac{t_x}{t_{H_2}} = \sqrt{\frac{\mathcal{M}_x}{\mathcal{M}_{H_2}}} \text{ yields } \mathcal{M}_x = \mathcal{M}_{H_2} \times \frac{(t_x)^2}{(t_{H_2})^2} = 2.016 \text{ g / mol} \times \frac{(9.12 \text{ min})^2}{(1.20 \text{ min})^2} = \mathbf{116 \text{ g / mol}}$$

6.90 ***See Section 6.8 and Example 6.14.***

$$\frac{t_x}{t_{O_2}} = \sqrt{\frac{\mathcal{M}_x}{\mathcal{M}_{O_2}}} \text{ yields } \mathcal{M}_x = \mathcal{M}_{O_2} \times \frac{(t_x)^2}{(t_{O_2})^2} = 32.0 \text{ g / mol} \times \frac{(8.3 \text{ min})^2}{(5.2 \text{ min})^2} = \mathbf{81.5 \text{ g / mol}}$$

6.91 ***See Section 6.8 and Example 6.14.***

$$\frac{t_x}{t_{N_2}} = \sqrt{\frac{\mathcal{M}_x}{\mathcal{M}_{N_2}}} \text{ yields } \mathcal{M}_x = \mathcal{M}_{N_2} \times \frac{(t_x)^2}{(t_{N_2})^2} = 28.0 \text{ g / mol} \times \frac{(163 \text{ s})^2}{(103 \text{ s})^2} = \mathbf{70.1 \text{ g / mol}}$$

6.92 ***See Section 6.8 and Example 6.14.***

$$\frac{t_x}{t_{O_2}} = \sqrt{\frac{\mathcal{M}_x}{\mathcal{M}_{O_2}}} \text{ yields } \mathcal{M}_x = \mathcal{M}_{O_2} \times \frac{(t_x)^2}{(t_{O_2})^2} = 32.0 \text{ g / mol} \times \frac{(90 \text{ s})^2}{(136 \text{ s})^2} = \mathbf{14.0 \text{ g / mol}}$$

6.93 ***See Section 6.9 and Example 6.15..***

(a) Oxygen gas at 30°C is more likely to follow the Ideal Gas Law than oxygen gas at -150°C. Deviations caused by the actual size of the gas particles and the attractive forces become less important in gas samples at higher temperatures, since the gas particles are further apart and have higher rms speeds. In addition, -150°C is near the boiling point/condensation point of oxygen (-183°C). Deviations caused by the actual size of the gas particles and the attractive forces between the particles are likely to be high near the boiling point/condensation point of a substance, the temperature at which the gaseous form of the substance becomes a liquid.

(b) Nitrogen gas is more likely to follow the Ideal Gas Law at -100°C than xenon at -100°C, because -100°C is closer to the boiling point of xenon (-107°C) than that of nitrogen (-196°C). Deviations caused by the actual size of the gas particles and the attractive forces between the particles are likely to be high near the boiling point/ condensation point of a substance, the temperature at which the gaseous form of the substance becomes a liquid.

(c) Argon gas at 1 atm pressure is more likely to follow the Ideal Gas Law than argon gas at 50 atm pressure. At high pressures, the gas particles are forced closer together causing the volumes of the gas particles and attractive forces between particles to become significant.

6.94 ***See Section 6.9 and Example 6.15.***

(a) Oxygen gas at 25°C and 1 atmosphere is more likely to follow the Ideal Gas Law than sulfur dioxide gas at 25°C and 1 atmosphere. The much lower boiling point of O_2 (-183°C) than SO_2(-10°C) indicates the attractive forces are lower between O_2 molecules than between SO_2 molecules.

(b) Nitrogen gas at 100°C is more likely to follow the Ideal Gas Law than nitrogen gas at -150°C. Deviations caused by the actual size of the gas particles and the attractive forces between the particles are likely to be high near the boiling point/condensation point of a substance, the temperature at which the gaseous form of the substance becomes a liquid.

(c) Argon gas at 1 atm pressure is more likely to follow the Ideal Gas Law than argon gas at 200 atm. At high presures, the gas particles are forced closer together causing the volumes of the gas particles and attractive forces between particles to become significant.

6.95 ***See Section 6.9 and Example 6.15.***

(a) Carbon dioxide gas at 0.05 atm is more likely to follow the Ideal Gas Law than carbon dioxide gas at 10 atm. At high pressures, the gas particles are forced closer together causing the volumes of the gas particles and attractive forces between particles to become significant.

(b) Neon gas is more likely to follow the Ideal Gas Law at -20°C than propane at -20°C, because -20°C is closer to the boiling point of propane (-45°C) than that of neon (-246°C). Deviations caused by the actual size of the gas particles and the attractive forces between the particles are likely to be high near the boiling point/condensation point of a substance, the temperature at which the gaseous form of the substance becomes a liquid.

(c) Sulfur dioxide gas at 50°C is more likely to follow the Ideal Gas Law than sulfur dioxide gas at 0°C. Deviations caused by the actual size of the gas particles and the attractive forces become less important in gas samples at higher temperatures, since the gas particles are further apart and have higher rms speeds. In addition, 0°C is near the boiling point/condensation point of oxygen (-10°C). Deviations caused by the actual size of the gas particles and the attractive forces between the particles are likely to be high near the boiling point/condensation point of a substance, the temperature at which the gaseous form of the substance becomes a liquid.

6.96 ***See Section 6.9 and Example 6.15.***

(a) Nitrogen at 25°C is more likely to follow the Ideal Gas Law than butane at 25°C. The much lower boiling point of N_2(-196°C) than C_4H_{10}(-1°C) indicates the attractive forces are lower between N_2 molecules than between C_4H_{10} molecules.

(b) Oxygen gas at 0.50 atm is more likely to follow the Ideal Gas Law than oxygen gas at 150 atm. At high pressures the gas particles are forced closer together causing the volumes of the gas particles and attractive forces between particles to become significant.

(c) Argon gas at 10°C is more likely to follow the Ideal Gas Law than argon gas at -160°C. Deviations caused by the actual size of the gas particles and the attractive forces become less important in gas samples at higher temperatures, since the gas particles are further apart and have higher rms speeds. In addition, -160°C is near the boiling point/condensation point of argon (-186°C). Deviations caused by the actual size of the gas particles and the attractive forces between the particles are likely to be high near the boiling point/condensation point of a substance, the temperature at which the gaseous form of the substance becomes a liquid.

6.97 *See Section 6.9, Table 6.4, and Example 6.16.*

Solving the Ideal Gas Law equation for pressure gives $P = \frac{nRT}{V}$

$$P = \frac{(10.2\ \text{mol})\left(0.0821\frac{L \cdot atm}{mol \cdot K}\right)(803\ \text{K})}{3.23\ \text{L}} = \mathbf{208\ atm}$$

Solving the van der Waals equation for pressure gives $P = \frac{nRT}{V - bn} - \frac{an^2}{V^2}$

$$P = \frac{(10.2\ \text{mol})\left(0.0821\frac{\text{L} \cdot \text{atm}}{\text{mol} \cdot \text{K}}\right)(803\ \text{K})}{3.23\ \text{L} - \left(0.0322\ \text{L} \cdot \text{mol}^{-1}\right)(10.2\ \text{mol})} - \frac{\left(1.34\ \text{atm} \cdot \text{L}^2 \cdot \text{mol}^{-2}\right)(10.2\ \text{mol})^2}{(3.23\ \text{L})^2} = \mathbf{218\ atm}$$

The pressure calculated using the Ideal Gas Law is less than that calculated using the van der Waals equation. Under these conditions, the correction for the actual size of the Ar atoms (bn) dominates the correction for attractive forces between Ar atoms ($-an^2/V^2$), since the former leads to an increase in pressure and the latter to a decrease in pressure.

6.98 *See Section 6.9, Table 6.4, and Example 6.16.*

Solving the Ideal Gas Law equation for pressure gives $P = \frac{nRT}{V}$

$$P = \frac{(1.55\ \text{mol})\left(0.0821\frac{L \cdot atm}{mol \cdot K}\right)(803\ \text{K})}{3.23\ \text{L}} = \mathbf{31.6\ atm}$$

Solving the van der Waals equation for pressure gives $P = \frac{nRT}{V - bn} - \frac{an^2}{V^2}$

$$P = \frac{(1.55\ \text{mol})\left(0.0821\frac{\text{L} \cdot \text{atm}}{\text{mol} \cdot \text{K}}\right)(803\ \text{K})}{3.23\ \text{L} - \left(0.0391\ \text{L} \cdot \text{mol}^{-1}\right)(1.55\ \text{mol})} - \frac{\left(1.39\ \text{atm} \cdot \text{L}^2 \cdot \text{mol}^{-2}\right)(1.55\ \text{mol})^2}{(3.23\ \text{L})^2} = \mathbf{31.9\ atm}$$

The pressure calculated using the Ideal Gas Law is less than that calculated using the van der Waals equation. Under these conditions, the correction for the actual size of the N_2 molecules dominates the correction for attractive forces between N_2 molecules ($-an^2/V^2$), since the former leads to an increase in pressure and the latter to a decrease in pressure.

6.99 *See Section 6.9, Table 6.4, and Example 6.16.*

Solving the Ideal Gas Law equation for pressure gives $P = \frac{nRT}{V}$

$$P = \frac{(13.9\ \text{mol})\left(0.0821\frac{\text{L} \cdot \text{atm}}{\text{mol} \cdot \text{K}}\right)(693\ \text{K})}{4.73\ \text{L}} = \mathbf{167\ atm}$$

Solving the van der Waals equation for pressure gives $P = \frac{nRT}{V - bn} - \frac{an^2}{V^2}$

$$P = \frac{(13.9\ \text{mol})\left(0.0821\frac{\text{L} \cdot \text{atm}}{\text{mol} \cdot \text{K}}\right)(693\ \text{K})}{4.73\ \text{L} - \left(0.0171\ \text{L} \cdot \text{mol}^{-1}\right)(13.9\ \text{mol})} - \frac{\left(0.211\ \text{atm} \cdot \text{L}^2 \cdot \text{mol}^{-2}\right)(13.9\ \text{mol})^2}{(4.73\ \text{L})^2} = \mathbf{174\ atm}$$

The pressure calculated using the Ideal Gas Law is less than that calculated using the van der Waals equation. Under these conditions, the correction for the actual size of the Ne atoms (bn) dominates the correction for attractive forces between Ne atoms ($-an^2/V^2$), since the former leads to an increase in pressure and the latter to a decrease in pressure.

6.100 ***See Section 6.9, Table 6.4, and Example 6.16.***

Solving the Ideal Gas Law equation for pressure gives $P = \dfrac{nRT}{V}$

$$P = \frac{(6.75\ \text{mol})\left(0.0821\dfrac{\text{L}\cdot\text{atm}}{\text{mol}\cdot\text{K}}\right)(713\ \text{K})}{4.93\ \text{L}} = \mathbf{80.1\ atm}$$

Solving the van der Waals equation for pressure gives $P = \dfrac{nRT}{V - bn} - \dfrac{an^2}{V^2}$

$$P = \frac{(6.75\ \text{mol})\left(0.0821\dfrac{\text{L}\cdot\text{atm}}{\text{mol}\cdot\text{K}}\right)(713\ \text{K})}{4.93\ \text{L} - \left(0.0428\ \text{L}\cdot\text{mol}^{-1}\right)(6.75\ \text{mol})} - \frac{\left(2.25\ \text{atm}\cdot\text{L}^2\cdot\text{mol}^{-2}\right)(6.75\ \text{mol})^2}{(4.93\ \text{L})^2} = \mathbf{80.9\ atm}$$

The pressure calculated using the Ideal Gas Law is greater than that calculated using the van der Waals equation. Under these conditions, the correction for the actual size of the CH_4 molecules (bn) is less than the correction for attractive forces between CH_4 molecules ($-an^2/V^2$), since the former leads to an increase in pressure and the latter to a decrease in pressure.

6.101 ***See Section 6.3.***

A tire gauge measures the difference in pressure between the pressure of the gas in the tire and atmospheric pressure: $P_{gauge} = P_{gas} - P_{atm}$. Hence, we have $P_{gas} = P_{gauge} + P_{atm}$ and

State One:
$P_1 = 32\ \text{psi} + 15\ \text{psi} = 47\ \text{psi}$
$T_1 = \left(\dfrac{1.0^\circ\text{C}}{1.8^\circ\text{F}}\right)(90 - 32)^\circ\text{F} + 273 = 305\ \text{K}$

State Two:
$P_2 = ?$
$T_2 = \left(\dfrac{1.0^\circ\text{C}}{1.8^\circ\text{F}}\right)(32 - 32)^\circ\text{F} + 273 = 273\ \text{K}$

Solving $\dfrac{P_1}{T_1} = \dfrac{P_2}{T_2}$ for P_2 gives $P_2 = P_1 \times \dfrac{T_2}{T_1}$ $\qquad P_2 = 47\ \text{psi} \times \dfrac{273\ \text{K}}{305\ \text{K}} = 42\ \text{psi}$

This means the pressure of the gas in the tire decreases by **5 psi** and $P_{gauge} = P_{gas} - P_{atm} = 42\ \text{psi} - 15\ \text{psi} = 27\ \text{psi}$.

6.102 ***See Section 6.3.***

State One:
$P_1 = 764$ torr
$T_1 = -20 + 273 = 253$

State Two:
$P_2 = ?$
$T_2 = 22 + 273 = 295$ K

Solving $\dfrac{P_1}{T_1} = \dfrac{P_2}{T_2}$ for P_2 gives $P_2 = P_1 \times \dfrac{T_2}{T_1}$ $\qquad P_2 = 764\ \text{torr} \times \dfrac{295\ \text{K}}{253\ \text{K}} = \mathbf{891\ torr}$

Since the temperature increase occurred at constant volume, an increase in pressure is expected.

6.103 ***See Section 6.4 and Example 6.5.***

Known Quantities: $\quad n_{O_2} = 0.24\ \text{kg}\ O_2 \times \dfrac{10^3\ \text{g}\ O_2}{\text{kg}\ O_2} \times \dfrac{1\ \text{mol}\ O_2}{32.0\ \text{g}\ O_2} = 7.5\ \text{mol}\ O_2 \qquad V = 2.8\ \text{L}$

$T = 20 + 273 = 293\ \text{K}$

Solving PV = nRT for P gives $P = \frac{nRT}{V}$ $\qquad P = \frac{(7.5\text{ mol})\left(0.0821\frac{\text{L}\cdot\text{atm}}{\text{mol}\cdot\text{K}}\right)(293\text{ K})}{2.8\text{ L}} = \mathbf{64\ atm}$

6.104 *See Sections 3.3, 6.4, and Examples 3.19, 6.6.*

Known Quantities: $\quad m = 1.26\text{ g} \qquad P = 744\text{ torr} \times \frac{1\text{ atm}}{760\text{ torr}} = 0.979\text{ atm}$

$V = 544\text{ mL} \times \frac{1\text{ L}}{10^3\text{ mL}} = 0.544\text{ L} \qquad T = 27 + 273 = 300\text{ K}$

Solving PV = nRT for n gives $n = \frac{PV}{RT}$ $\qquad n = \frac{(0.979\text{ atm})(0.544\text{ L})}{\left(0.0821\frac{\text{L}\cdot\text{atm}}{\text{mol}\cdot\text{K}}\right)(300\text{ K})} = 0.0216\text{ mol}$

Using $\mathcal{M} = \frac{m}{n}$ yields $\mathcal{M} = \frac{1.26\text{ g}}{0.0216\text{ mol}} = \mathbf{58.3\ g/mol}$, and $n = \frac{\text{molar mass compound}}{\text{molar mass } C_2H_5} = \frac{58.3\text{ g / mol}}{29.0\text{ g / mol}} = 2.$

Hence, the molecular formula is $\mathbf{C_4H_{10}}$.

6.105 See Sections 3.1, 3.4, 6.4 and Examples 3.4, 6.6.

Unbalanced:	$C_{56}H_{108}O_6 + O_2 \rightarrow CO_2 + H_2O$	Start with $C_{56}H_{108}O_6$.
Step 1:	$C_{56}H_{108}O_6 + O_2 \rightarrow \underline{56}CO_2 + \underline{54}H_2O$	Balances C and H.
Step 2:	$C_{56}H_{108}O_6 + \underline{80O_2} \rightarrow 56CO_2 + 54H_2O$	Balances O.

Strategy: $lb\ C_{56}H_{108}O_6 \rightarrow g\ C_{56}H_{108}O_6 \rightarrow mol\ C_{56}H_{108}O_6 \rightarrow mol\ O_2 \rightarrow L\ O_2$

$$?\text{ mol } O_2 = 5.0\text{ lb } C_{56}H_{108}O_6 \times \frac{453.6\text{ g } C_{56}H_{108}O_6}{1\text{ lb } C_{56}H_{108}O_6} \times \frac{1\text{ mol } C_{56}H_{108}O_6}{876.9\text{ g } C_{56}H_{108}O_6}$$

$$\times \frac{80\text{ mol } O_2}{1\text{ mol } C_{56}H_{108}O_6} = 2.1 \times 10^2\text{ mol } O_2$$

Known Quantities: $\quad n_{O_2} = 2.1 \times 10^2\text{ mol} \qquad P = 1.00\text{ atm} \qquad T = 22 + 273 = 295\text{ K}$

Solving PV = nRT for V gives $V = \frac{nRT}{P}$ $\qquad V = \frac{(2.1 \times 10^2\text{ mol})\left(0.0821\frac{\text{L}\cdot\text{atm}}{\text{mol}\cdot\text{K}}\right)(295\text{ K})}{1.00\text{ atm}} = \mathbf{5.1 \times 10^3\ L\ O_2}$

6.106 *See Sections 6.2, 6.6, and Example 6.2.*

State One:	State Two:
For Ar:	
$P_1 = 2.00$ atm	$P_2 = ?$
$V_1 = 50.0$ mL	$V_2 = 50.0\text{ mL} + 250\text{ mL} + 25.0\text{ mL} + 22.0\text{ mL} = 347\text{ mL}$
For Ne:	
$P_1 = 1.00$ atm	$P_2 = ?$
$V_1 = 250$ mL	$V_2 = 347$ mL
For H_2:	
$P_1 = 5.00$ atm	$P_2 = ?$
$V_1 = 25.0$ mL	$V_2 = 347$

Solving $P_1V_1 = P_2V_2$ for P_2 gives $P_2 = P_1 \times \frac{V_1}{V_2}$

For Ar: $P_{Ar} = 2.00\text{ atm} \times \frac{50.0\text{ mL}}{347\text{ mL}} = \mathbf{0.288\ atm}$

For Ne: $P_{Ne} = 1.00 \text{ atm} \times \frac{250 \text{ mL}}{347 \text{ mL}} = \mathbf{0.720 \text{ atm}}$

For H_2: $P_{H_2} = 5.00 \text{ atm} \times \frac{25.0 \text{ mL}}{347 \text{ mL}} = \mathbf{0.360 \text{ atm}}$

Hence, $P_T = P_{Ar} + P_{Ne} + P_{H_2} = 0.288 \text{ atm} + 0.720 \text{ atm} + 0.360 \text{ atm} = \mathbf{1.368 \text{ atm.}}$

6.107 ***See Sections 3.1, 3.5, 6.4, 6. 5, and Examples 3.1 3.2, 3.16.***

Unbalanced:	$H_2 + O_2$	$\rightarrow$	H_2O	Start with H_2O.
Step 1:	$H_2 + O_2$	$\rightarrow$	$\underline{2}H_2O$	Balances O.
Step 2:	$\underline{2}H_2 + O_2$	$\rightarrow$	$2H_2O$	Balances H.

Strategy: $L\ O_2 \rightarrow mol\ O_2 \rightarrow mol\ H_2O \rightarrow g\ H_2O$
$L\ H_2 \rightarrow mol\ H_2 \rightarrow mol\ H_2O \rightarrow g\ H_2O$

Known Quantities: $P = 1.22 \text{ atm}$ $V_{O_2} = 4.33 \text{ L}$ $V_{H_2} = 6.77 \text{ L}$ $T = 27 + 273 = 300 \text{ K}$

Solving PV = nRT for n gives $n_{O_2} = \frac{PV_{O_2}}{RT}$ $n_{O_2} = \frac{(1.22 \text{ atm})(4.33 \text{ L})}{\left(0.0821 \frac{\text{L}\cdot\text{atm}}{\text{mol}\cdot\text{K}}\right)(300 \text{ K})} = 0.214 \text{ mol } O_2$

and $n_{H_2} = \frac{PV_{H_2}}{RT}$ $n_{H_2} = \frac{(1.22 \text{ atm})(6.77 \text{ L})}{\left(0.0821 \frac{\text{L}\cdot\text{atm}}{\text{mol}\cdot\text{K}}\right)(300 \text{ K})} = 0.335 \text{ mol } H_2$

$? \text{ g } H_2O \text{ based on } O_2 = 0.214 \text{ mol } O_2 \times \frac{2 \text{ mol } H_2O}{1 \text{ mol } O_2} \times \frac{18.02 \text{ g } H_2O}{1 \text{ mol } H_2O} = 7.71 \text{ g } H_2O$

$? \text{ g } H_2O \text{ based on } H_2 = 0.335 \text{ mol } H_2 \times \frac{2 \text{ mol } H_2O}{2 \text{ mol } H_2} \times \frac{18.02 \text{ g } H_2O}{1 \text{ mol } H_2O} = \mathbf{6.04 \text{ g } H_2O}$

H_2 is the limiting reactant since it gives less grams of H_2O.

6.108 ***See Section 6.7 and Example 6.13.***

(a) For H_2 molecules at STP, $\mu_{rms} = \sqrt{\frac{3RT}{M}} = \sqrt{\frac{3\left(8.314 \frac{\text{kg}\cdot\text{m}^2}{\text{s}^2\cdot\text{mol}\cdot\text{K}}\right)(273 \text{ K})}{0.002016 \text{ kg}\cdot\text{mol}^{-1}}} = \mathbf{1.84 \times 10^3 \text{ m/s}}$

For N_2 at STP, $\mu_{rms} = \sqrt{\frac{3RT}{M}} = \sqrt{\frac{(3)\left(8.314 \frac{\text{kg}\cdot\text{m}^2}{\text{s}^2\cdot\text{mol}\cdot\text{K}}\right)(273 \text{ K})}{0.02802 \text{ kg}\cdot\text{mol}^{-1}}} = \mathbf{493 \text{ m/s}}$

(b) For H_2 molecules at STP,

$\overline{KE} = \frac{1}{2}m\overline{u^2} = \left(\frac{1}{2}\right)\left(\frac{0.00202 \text{ kg}\cdot\text{mol}^{-1}}{6.022 \times 10^{23} \text{ molecules}\cdot\text{mol}^{-1}}\right)(1.84 \times 10^3 \text{ m}\cdot\text{s}^{-1})^2 = \mathbf{5.68 \times 10^{-21} \text{ kg}\cdot\text{m}^2\cdot\text{s}^{-2}}$

For N_2 molecules at STP,

$\overline{KE} = \frac{1}{2}m\overline{u^2} = \frac{1}{2}\left(\frac{0.0280 \text{ kg}\cdot\text{mol}^{-1}}{6.022 \times 10^{23} \text{ molecules}\cdot\text{mol}^{-1}}\right)(493 \text{ m}\cdot\text{s}^{-1})^2 = \mathbf{5.65 \times 10^{-21} \text{ kg}\cdot\text{m}^2\cdot\text{s}^{-2}}$

6.109 See Sections 3.4, 6.4, 6.5.

Balanced: $2LiOH(s) + CO_2(g) \rightarrow Li_2CO_3(s) + H_2O(l)$

Strategy: L CO2 → mol CO_2 → mol LiOH → g LiOH

Known Quantities for CO_2: P = 1.00 atm V = 400 L T = 24 + 273 = 297 K

Solving PV = nRT for n gives $n_{CO_2} = \frac{PV_{CO_2}}{RT}$ $n_{CO_2} = \frac{(1.00\text{ atm})(400\text{ L})}{\left(0.0821\frac{L\cdot atm}{mol\cdot K}\right)(297\text{ K})} = 16.4\text{ mol }CO_2$

$$? \text{ g LiOH} = 16.4\text{ mol }CO_2 \times \frac{2\text{ mol LiOH}}{1\text{ mol }CO_2} \times \frac{23.95\text{ g LiOH}}{1\text{ mol LiOH}} = \mathbf{786\text{ g LiO}}$$

6.110 See Section 6.9, Table 6.4, and Example 6.16.

(a) Solving the van der Waals equation for pressure gives $P = \frac{nRT}{V-bn} - \frac{an^2}{V^2}$

$$P_{H_2} = \frac{(30.33\text{ mol})\left(0.0821\frac{L\cdot atm}{mol\cdot K}\right)(513\text{ K})}{2.44\text{ L} - (0.0266\text{ L}\cdot mol^{-1})(30.33\text{ mol})} - \frac{(0.244\text{ atm}\cdot L^2\cdot mol^{-2})(30.33\text{ mol})^2}{(2.44\text{ L})^2} = \mathbf{744\text{ atm}}$$

(b) Solving the van der Waals equation for pressure gives $P = \frac{nRT}{V-bn} - \frac{an^2}{V^2}$

$$P_{H_2} = \frac{(30.33\text{ mol})\left(0.0821\frac{L\cdot atm}{mol\cdot K}\right)(513\text{ K})}{2.44\text{ L} - (0.0428\text{ L}\cdot mol^{-1})(30.33\text{ mol})} - \frac{(2.25\text{ atm}\cdot L^2\cdot mol^{-2})(30.33\text{ mol})^2}{(2.44\text{ L})^2} = \mathbf{771\text{ atm}}$$

(c) The greater van der Waals pressure for CH_4 is due to the larger size of CH_4 molecules leading to a smaller effective volume for CH_4 than H_2 (1.14 L vs. 1.63 L).

6.111 See Sections 6.4, 6.5.

Balanced: $C_3H_4O_3(aq) \rightarrow CO_2(g) + C_2H_4O(aq)$

Strategy: g $C_3H_4O_3$ → mol $C_3H_4O_3$ → mol CO_2 → L CO_2

$$? \text{ mol }CO_2 = 0.113\text{ g }C_3H_4O_3 \times \frac{1\text{ mol }C_3H_4O_3}{88.0\text{ g }C_3H_4O_3} \times \frac{1\text{ mol }CO_2}{1\text{ mol }C_3H_4O_3} = 1.28 \times 10^{-3}\text{ mol }CO_2$$

Known Quantities: $n_{CO_2} = 1.28 \times 10^{-3}$ mol $P = 755\text{ torr} \times \frac{1\text{ atm}}{760\text{ torr}} = 0.993\text{ atm}$

T = 25 + 273 = 298 K

Solving PV = nRT for V gives $V = \frac{nRT}{P}$

$$V = \frac{(1.28 \times 10^{-3}\text{ mol})\left(0.0821\frac{L\cdot atm}{mol\cdot K}\right)(298\text{ K})}{(0.993\text{ atm})} = \mathbf{0.0315\text{ L }CO_2}$$

6.112 *See Sections 3.1, 5.1, 6.4.*

Unbalanced:	$B_2H_6(g) + O_2(g)$	→	$B_2O_3(s) + H_2O(\ell)$	Start with B_2H_6.
Step 1:	$B_2H_6(g) + O_2(g)$	→	$B_2O_3(s) + \underline{3}H_2O(\ell)$	Balances H.
Step 2:	$B_2H_6(g) + 3O_2(g)$	→	$B_2O_3(s) + 3H_2O(\ell)$	Balances O.

Strategy: L B_2H_6 → mol B_2H_6 → kJ/mol B_2H_6

Known Quantities: $P = 744 \text{ torr} \times \frac{1 \text{ atm}}{760 \text{ torr}} = 0.979 \text{ atm}$ $V = 2.329 \text{ L}$ $T = 120.0 + 273 = 393$

Solving PV = nRT for n gives $n = \frac{PV}{RT}$ $n_{B_2H_6} = \frac{(0.979 \text{ atm})(2.329 \text{ L})}{\left(0.0821 \frac{\text{L}\cdot\text{atm}}{\text{mol}\cdot\text{K}}\right)(393 \text{ K})} = \mathbf{0.0707\ mol\ B_2H_6}$

$? \Delta H \text{ per mol } B_2H_6 = \frac{-143.9 \text{ kJ}}{0.0707 \text{ mol } B_2H_6} = \mathbf{-2.04 \times 10^3\ kJ/mol\ B_2H_6}$

6.113 *See Section 6.5 and Example 6.8.*

Unbalanced:	$H_2(g) + Cl_2(g)$	→	$HCl(g)$	Start with HCl.
Step 1:	$H_2(g) + Cl_2(g)$	→	$2HCl(g)$	Balances H, Cl.

Strategy: g H_2 → mol H_2 → mol HCl

$? \text{ mol HCl} = 2.66 \text{ g } H_2 \times \frac{1 \text{ mol } H_2}{2.016 \text{ g } H_2} \times \frac{2 \text{ mol HCl}}{1 \text{ mol } H_2} = 2.64 \text{ mol HCl}$

Strategy: g Cl_2 → mol Cl_2 → mol HCl

$? \text{ mol HCl} = 4.88 \text{ g } Cl_2 \times \frac{1 \text{ mol } Cl_2}{70.90 \text{ g } Cl_2} \times \frac{2 \text{ mol HCl}}{1 \text{ mol } Cl_2} = 0.138 \text{ mol HCl}$

Cl_2 is the limiting reactant because it produces less moles of HCl.

$? \text{ mol } H_2 \text{ present} = 2.66 \text{ g } H_2 \times \frac{1 \text{ mol } H_2}{2.016 \text{ g } H_2} = 1.32 \text{ mol } H_2$

$? \text{ mol } Cl_2 \text{ present} = 4.88 \text{ g } Cl_2 \times \frac{1 \text{ mol } Cl_2}{70.90 \text{ g } Cl_2} = 0.0688 \text{ mol } Cl_2$

$? \text{ mol } H_2 \text{ reacting} = 0.0688 \text{ mol } Cl_2 \times \frac{1 \text{ mol } H_2}{1 \text{ mol } Cl_2} = 0.0688 \text{ mol } H_2$

? mol H_2 in excess = mol H_2 present - mol H_2 reacting = 1.32 mol - 0.0688 mol = 1.25 mol H_2

Known Quantities: $n_{total} = n_{HCl}$ formed + n_{H_2} in excess = 0.138 mol + 1.25 mol = 1.39 mol

V = 10.0 L T = 111 + 273 = 384 K

Solving PV = nRT for P gives $P = \frac{nRT}{V}$ $P = \frac{(1.39 \text{ mol})\left(0.0821 \frac{\text{L}\cdot\text{atm}}{\text{mol}\cdot\text{K}}\right)(384 \text{ K})}{(10.0 \text{ L})} = \mathbf{4.38\ atm}$

6.114 *See Sections 3.3, 6.4.*

Strategy: L H_2O → mol H_2O → mol H

Known Quantities: $P = 755 \text{ torr} \times \frac{1 \text{ atm}}{760 \text{ torr}} = 0.0993 \text{ atm}$ $V = 1.23 \text{ L}$ $T = 180 + 273 = 453 \text{ K}$

Solving PV = nRT for n gives $n = \frac{PV}{RT}$ $\quad n_{H_2O} = \frac{(0.993\ \text{atm})(1.23\ \text{L})}{\left(0.0821\frac{\text{L}\cdot\text{atm}}{\text{mol}\cdot\text{K}}\right)(453\ \text{K})} = 0.0328\ \text{mol}\ H_2O$

$? \text{ mol H} = 0.0328 \text{ mol } H_2O \times \frac{2 \text{ mol H}}{1 \text{ mol } H_2O} = 0.0656 \text{ mol H}$

Strategy: L CO_2 → mol CO_2 → mol C

Known Quantities: $P = 755 \text{ torr} \times \frac{1 \text{ atm}}{760 \text{ torr}} = \times \frac{1 \text{ atm}}{760 \text{ torr}} = 0.993 \text{ atm}$ V = 0.984 L $\quad$ T = 180 + 273 = 453 K

Solving PV = nRT for n gives $n = \frac{PV}{RT}$ $\quad n_{CO_2} = \frac{(0.993\ \text{atm})(0.984\ \text{L})}{\left(0.0821\frac{\text{L}\cdot\text{atm}}{\text{mol}\cdot\text{K}}\right)(453\ \text{K})} = 0.0263\ \text{mol}\ CO_2$

$? \text{ mol C} = 0.0263 \text{ mol } CO_2 \times \frac{1 \text{ mol C}}{1 \text{ mol } CO_2} = 0.0263 \text{ mol C}$

$\text{Relative mol C} = \frac{0.0263 \text{ mol C}}{0.0263} = 1.00 \text{ mol C} \qquad \text{Relative mol H} = \frac{0.0656 \text{ mol H}}{0.0263} = 2.49 \text{ mol H}$

Multiplying each of these by the same smallest integer that gives whole numbers of relative mols of atoms for each element yields: relative mol C = 1.00 mol C x 2 = 2.00 mol C

relative mol H = 2.49 mol H x 2 = 4.98 mol H

The empirical formula is **C_2H_5**.

6.115 ***See Sections 3.3, 6.4.***

Strategy: L CO_2 → mol CO_2 → g C → % C

Known Quantities: m sample = 4.33 g $\quad$ P = 0.99 atm $\quad$ V = 2.20 L $\quad$ T = 27 + 273 = 300 K

Solving PV = nRT for n gives $n = \frac{PV}{RT}$ $\quad n_{CO_2} = \frac{(0.99\ \text{atm})(2.20\ \text{L})}{\left(0.0821\ \frac{\text{L}\cdot\text{atm}}{\text{mol}\cdot\text{K}}\right)(300\ \text{K})} = 0.088\ \text{mol}\ CO_2$

$? \text{ g C} = 0.088 \text{ mol } CO_2 \times \frac{1 \text{ mol C}}{1 \text{ mol } CO_2} \times \frac{12.0 \text{ g C}}{1 \text{ mol C}} = 1.06 \text{ g C}$

$? \% \text{ C} = \frac{\text{g C}}{\text{g sample}} \times 100\% = \frac{1.06 \text{ g C}}{4.33 \text{ g}} \times 100\% = \mathbf{24.5\% \ C}$

6.116 ***See Section 3.1, 4.3, 6.4.***

Unbalanced:	$H_2O_2(aq)$	→	$O_2(g) + H_2O(\ell)$	Start with O atoms.
Step 1:	$\underline{2}H_2O_2(aq)$	→	$O_2(g) + \underline{2}H_2O(\ell)$	Balances H, O.

Strategy: L $H_2O_2(aq)$ → mol H_2O_2 → mol O_2 → L O_2

$? \text{ mol } O_2 = 0.1000 \text{ L } H_2O_2 \text{ soln} \times \frac{0.88 \text{ mol } H_2O_2}{1 \text{ L } H_2O_2 \text{ soln}} \times \frac{1 \text{ mol } O_2}{2 \text{ mol } H_2O_2} = 0.044 \text{ mol } O_2$

Known Quantities: $n_{O_2} = 0.044$ mol $\quad$ P = 0.971 atm $\quad$ T = 22 + 273 = 293 K

Solving PV = nRT for V gives $V = \frac{nRT}{P}$ $\quad V = \frac{(0.044\ \text{mol})\left(0.0821\frac{\text{L}\cdot\text{atm}}{\text{mol}\cdot\text{K}}\right)(293\ \text{K})}{(0.971\ \text{atm})} = \mathbf{1.1\ L\ O_2}$

6.116-6.119 ***Solutions for these Spreadsheet Exercises are on the Reger, Goode and Mercer Spreadsheet Solutions Diskette available from Saunders College Publishing.***

Chapter 7: Electrons in Atoms

7.1 *See Section 7.1.*

The speed, frequency and wavelength are the same, but the amplitude of the light wave from the second source is twice the amplitude of the light wave from the first source since it is twice as bright.

The energy of the photons and the total energy of the light from the two sources is the same since the energy is directly proportional to the frequency of the wave and independent of the amplitude.

7.2 *See Section 7.1.*

The product of wavelength times frequency, $\lambda\nu$, is equal to the speed of light. Since both waves travel at the same speed, the second one with the longer wavelength (560 nm vs. 720 nm) must have the lower frequency.

7.3 *See Section 7.2.*

With absorption spectra, electrons are promoted from a given energy level, usually the ground state with $n = 1$, to given excited state energy levels with $n > 1$. With emission spectra, electrons in excited states can return to lower level excited states or to the ground state. This gives rise to more lines in emission spectra than absorption spectra.

7.4 *See Section 7.2.*

Bohr assumed the electron of the hydrogen atom moved in fixed circular orbits around the nucleus corresponding to fixed energies for the electron. Bohr also assumed the electron could only have certain values of angular momentum.

7.5 *See Section 7.3.*

deBroglie assumed the circumference of a Bohr orbit was equal to a whole-number multiple of wavelengths so that a standing wave is produced and angular momentum is quantized.

7.6 *See Section 7.3.*

The square of the wave function, ψ^2, gives the probability of finding the electron at any point in space.

7.7 *See Section 7.4.*

The fourth quantum number, the spin quantum number m_s, does not affect the energy of the electron in the hydrogen atom.

7.8 *See Section 7.8.*

(a) The principal quantum number, n, is related to the average distance of the electron from the nucleus.

(b) The angular momentum quantum number, ℓ, is related to the shape that an orbital can have.

(c) The magnetic quantum number, m_ℓ, is related to the orientation in space of the orbital.

7.9 *See Section 7.5 and Figure 7.19.*

A 2s electron in carbon experiences a higher effective nuclear charge than a 2p electron in carbon because it penetrates the electron density of the filled 1s shell more than does the 2p electron.

7.10 *See Section 7.5.*

The interelectronic repulsions by inner electrons contributes to the overall phenomenon called electron shielding. The greater the repulsions the greater the shielding effect.

7.11 *See Section 7.5 and Figure 7.19.*

The nuclear charge for Be is +4 since there are four protons in the nucleus of a beryllium atom. The effective nuclear charge is less than four due to interelectronic repulsions. In addition, the effective nuclear charge experienced by electrons in the 2s subshell is lessened by the screening of nuclear charge caused by electrons in the 1s shell being closer to the nucleus than those in the 2s subshell. Hence, the effective nuclear charge acting on 1s electrons is greater than that acting on 2s electrons in beryllium.

7.12 *See Section 7.5 and Figure 7.19.*

Penetration refers to the location of some region of probability of an outer subshell within the location occupied primarily by electrons in inner shells and subshells. We say the probability pattern of the outer subshell penetrates the region occupied by the probability patterns of the inner shells and subshells, meaning it penetrates toward the nucleus and in so doing experiences a higher effective nuclear charge.

Within any given principal shell, penetration decreases with increasing value of ℓ . This is why the subshells in any given principal shell increase in energy in the order of increasing value of ℓ .

7.13 *See Section 7.3 and The Closer View.*

The equation relating wavelength and mass for matter is $\lambda = h/mv$. The more massive the object, the smaller its wavelength. For an electron the uncertainty in position is large compared to the size of the atom; for a baseball the uncertainty is negligible compared to the size of the ball.

7.14 *See Sections 7.4, 7.5, and Figures 7.16, 7.20.*

In the case of the hydrogen atom all the subshells of a given principal shell are equal in energy. In the case of multielectron atoms there is a difference in energy between the subshells of a given shell, and there is even some overlap between the subshells of differing principal shells, as with the 4s and 3d.

7.15 *See Section 7.5 and Figure 7.20.*

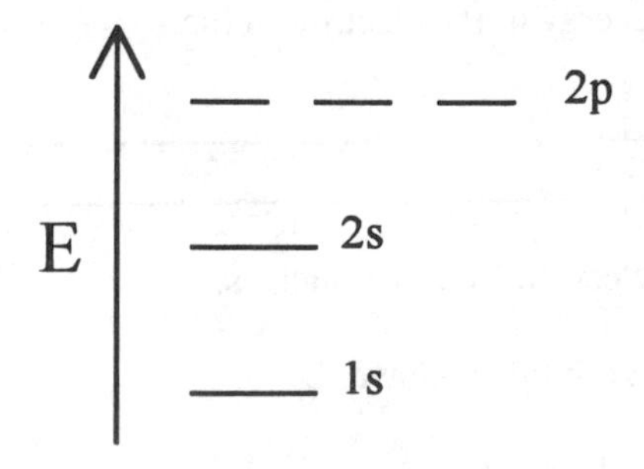

7.16 *See Section 7.6.*

The Pauli exclusion principle states: no two electrons in the same atom can have the same set of all four quantum numbers. Since n, ℓ, and m_ℓ are required to specify an orbital and m_s can only be $+\frac{1}{2}$ or $-\frac{1}{2}$, only two electrons can occupy the same orbital, provided they are of opposite spin.

7.17 *See Section 7.6, especially the orbital diagrams for C and O.*

Hund's rule states: in filling degenerate orbitals (orbitals with idential energies), one electron occupies each orbital separately with like spins before two electrons are placed in any of these orbitals and pairing occurs.

Carbon atoms contain two electrons in the 2p subshell and oxygen atoms contain four electrons in the 2p subshell. Following Hund's rule and using orbital diagrams, it can be shown that each configuration gives two unpaired electrons in the 2p subshell.

7.18 *See Sections 7.4, 7.6, and Table 7.1.*

The ℓ value associated with the 3p subshell is 1, and there are three allowed values for m_ℓ (-1, 0, and +1) when $\ell = 1$, meaning there are 3 orbitals in the 3p subshell. Since the Pauli exclusion principle limits the number of electrons to two per orbital, provided they are of opposite spin, the 3p subshell can contain a total of 6 electrons.

7.19 *See Section 7.1, Figure 7.4, and Example 7.1.*

Solving $\lambda\nu = c$ for ν gives $\nu = \dfrac{c}{\lambda}$ $\qquad \nu = \dfrac{3.00 \times 10^{8}\ \text{m} \cdot \text{s}^{-1}}{200\ \text{nm}} \times \dfrac{1\ \text{nm}}{10^{-4}\ \text{m}} = \mathbf{1.50 \times 10^{15}\ s^{-1}}$

The wavelength of 200 nm indicates the radiation is in the **near-ultraviolet** portion of the electromagnetic spectrum.

7.20 *See Section 7.1, Figure 7.4, and Example 7.1.*

Solving $\lambda\nu = c$ for ν gives $\nu = \dfrac{c}{\lambda}$ $\qquad \nu = \dfrac{3.00 \times 10^{8}\ \text{m} \cdot \text{s}^{-1}}{400\ \text{nm}} \times \dfrac{1\ \text{nm}}{10^{-9}\ \text{m}} = 7.5 \times 10^{14}\ \text{s}^{-1}$

The wavelength of 400 nm indicates the radiation is in the violet end of the **visible** protion of the electromagnetic spectrum.

7.21 *See Section 7.1 and Figure 7.4.*

Solving $\lambda\nu = c$ for λ gives $\lambda = \dfrac{c}{\nu}$, and recognizing that 1 Hz = 1 s^{-1} yields

$$\lambda = \frac{3.00 \times 10^{8}\ \text{m} \cdot \text{s}^{-1}}{6.00 \times 10^{13}\ \text{s}^{-1}} = \mathbf{5.00 \times 10^{-6}\ m}$$

This wavelength is equivalent to 5.00×10^3 nm. The radiation is in the **infrared** region of the electromagnetic spectrum.

7.22 *See Section 7.1 and Figure 7.4.*

Solving $\lambda\nu = c$ for λ gives $\lambda = \dfrac{c}{\nu}$, and recognizing that 1 Hz = 1 s^{-1} yields

$$\lambda = \frac{3.00 \times 10^{8}\ \text{m} \cdot \text{s}^{-1}}{1.50 \times 10^{15}\ \text{s}^{-1}} = \mathbf{2.00 \times 10^{-7}\ m}$$

This wavelength is equivalent to 200 nm. The radiation is in the **ultraviolet** portion of the electromagnetic spectrum.

7.23 See Section 7.1.

Solving $\lambda\nu = c$ for λ gives $\lambda = \frac{c}{\nu}$, and recognizing that 1 Hz = 1 s^{-1} yields

$$\lambda = \frac{3.00 \times 10^8 \ m \cdot s^{-1}}{580 \ ks^{-1}} \times \frac{1 \ ks^{-1}}{10^3 \ s^{-1}} = \mathbf{517 \ m}$$

7.24 See Section 7.1.

Solving $\lambda\nu = c$ for λ gives $\lambda = \frac{c}{\nu}$, and recognizing that 1 Hz = 1 s^{-1} yields

$$\lambda = \frac{3.00 \times 10^8 \ m \cdot s^{-1}}{101.3 \ Ms^{-1}} \times \frac{1 \ Ms^{-1}}{10^6 \ s^{-1}} = \mathbf{2.96 \ m}$$

7.25 See Section 7.1.

Solving $\lambda\nu = c$ for λ gives $\lambda = \frac{c}{\nu}$ and recognizing that 1 Hz = 1 s^{-1} yields

$$\lambda = \frac{3.00 \times 10^8 \ m \cdot s^{-1}}{2.47 \times 10^{20} \ s^{-1}} \times \frac{10^{12} \ pm}{m} = \mathbf{1.21 \ pm}$$

7.26 See Section 7.1.

Solving $\lambda\nu = c$ for λ gives $\lambda = \frac{c}{\nu}$ and recognizing that 1 Hz = 1 s^{-1} yields

$$\lambda = \frac{3.00 \times 10^8 \ m \cdot s^{-1}}{1.44 \times 10^{20} \ Hz} \times \frac{10^{12} \ pm}{m} = \mathbf{2.08 \ pm}$$

7.27 See Section 7.1 and Example 7.1.

Solving $\lambda\nu = c$ for ν gives $\nu = \frac{c}{\lambda}$ $\qquad \nu = \frac{3.00 \times 10^8 \ m \cdot s^{-1}}{21 \ m} = 1.4 \times 10^7 \ s^{-1} = \mathbf{1.4 \times 10^7 \ Hz}$

7.28 See Section 7.1 and Example 7.1.

Solving $\lambda\nu = c$ for ν gives $\nu = \frac{c}{\lambda}$ $\qquad \nu = \frac{3.00 \times 10^8 \ m \cdot s^{-1}}{10 \ m} = 3.00 \times 10^7 \ s^{-1} = \mathbf{3.00 \times 10^7 \ Hz}$

7.29 See Section 7.1 and Examples 7.1, 7.2.

(a) Solving $\lambda\nu = c$ for ν gives $\nu = \frac{c}{\lambda}$ $\qquad \nu = \frac{3.00 \times 10^8 \ m \cdot s^{-1}}{488 \ nm} \times \frac{10^9 \ nm}{1 \ m} = 6.15 \times 10^{14} \ s^{-1}$

Hence, $E = h\nu = (6.63 \times 10^{-34} \ J \cdot s)(6.15 \times 10^{14} \ s^{-1}) = \mathbf{4.08 \times 10^{-19} \ J.}$

(b) $? \ photons = \frac{1 \ photon}{4.08 \times 10^{-19} \ J} \times 1.00 \frac{J}{s} = \mathbf{2.45 \times 10^{18} \ photons/s}$

7.30 ***See Section 7.1 and Example 7.2.***

(a) Solving $\lambda\nu = c$ for λ gives $\lambda = \frac{c}{\nu}$, and recognizing that 1 Hz = $1 s^{-1}$ yields

$$\lambda = \frac{3.00 \times 10^{8}\ m \cdot s^{-1}}{9.83 \times 10^{14}\ s^{-1}} = 3.05 \times 10^{-7}\ m \text{ or } \mathbf{305\ nm}$$

(b) $E = h\nu = (6.63 \times 10^{-34}\ J \cdot s)\,(9.83 \times 10^{14}\ s^{-1}) = \mathbf{6.52 \times 10^{-19}\ J}$

(c) ? E for one mol of photons = $6.52 \times 10^{-19}\ \frac{J}{photon} \times \frac{6.022 \times 10^{23}\ photons}{1\ mol} \times \frac{1\ kJ}{10^{3}\ J} = \mathbf{393\ kJ/mol}$

7.31 ***See Section 7.1 and Example 7.2.***

$E = h\nu = (6.63 \times 10^{-34}\ J \cdot s)(3.70 \times 10^{15}\ s^{-1}) = \mathbf{2.45 \times 10^{-18}\ J}$

? E for one mol photons = $2.45 \times 10^{-18}\ \frac{J}{photon} \times 6.022 \times 10^{23}\ \frac{photons}{mol} \times \frac{1\ kJ}{10^{3}\ J} = \mathbf{1.48 \times 10^{3}\ kJ/mol}$

7.32 ***See Section 7.1 and Example 7.2.***

$E = h\nu = (6.63 \times 10^{-34}\ J \cdot s)(2.50 \times 10^{14}\ s^{-1}) = 1.66 \times 10^{-19}\ J$

? E for one mol photons = $1.66 \times 10^{-19}\ \frac{J}{photon} \times 6.022 \times 10^{23}\ \frac{photons}{mol} \times \frac{1\ kJ}{10^{3}\ J} = \mathbf{100\ kJ/mol}$

7.33 ***See Section 7.1 and Examples 7.1, 7.2.***

Solving $\lambda\nu = c$ for ν gives $\nu = \frac{c}{\lambda}$ $\qquad \nu = \frac{3.00 \times 10^{8}\ m \cdot s^{-1}}{589\ nm} \times \frac{10^{9}\ nm}{1\ m} = 5.09 \times 10^{14}\ s^{-1}$

Hence, $E = h\nu\,(6.63 \times 10^{-34}\ J \cdot s)(5.09 \times 10^{14}\ s^{-1}) = \mathbf{3.37 \times 10^{-19}\ J.}$

7.34 ***See Section 7.1 and Examples 7.1, 7.2.***

Solving $\lambda\nu = c$ for ν gives $\nu = \frac{c}{\lambda}$ $\qquad \nu = \frac{3.00 \times 10^{8}\ m \cdot s^{-1}}{640\ nm} \times \frac{10^{9}\ nm}{1\ m} = 4.69 \times 10^{14}\ s^{-1}$

Hence, $E = h\nu\,(6.63 \times 10^{-34}\ J \cdot s)(4.69 \times 10^{14}\ s^{-1}) = \mathbf{3.11 \times 10^{-19}\ J.}$

7.35 ***See Section 7.1 and Figure 7.4.***

(a) Solving $\lambda\nu = c$ for λ gives $\lambda = \frac{c}{\nu}$ $\qquad \lambda = \frac{3.00 \times 10^{8}\ m \cdot s^{-1}}{1.16 \times 10^{15}\ s^{-1}} = \mathbf{2.59 \times 10^{-7}\ m}$

This wavelength is equivalent to 259 nm. The radiation is in the **ultraviolet** region of the electromagnetic spectrum.

(b) Light in the visible region of the electromagnetic spectrum (400-700 nm) has a longer wavelength. Hence, it does not have sufficient energy to dislodge electrons from carbon.

7.36 ***See Section 7.1 and Example 7.1.***

Solving $\lambda\nu = c$ for ν gives $\nu = \frac{c}{\lambda}$ $\qquad \nu_o = \frac{3.00 \times 10^{8}\ m \cdot s^{-1}}{544\ nm} \times \frac{10^{9}\ nm}{1\ m} = \mathbf{5.51 \times 10^{14}\ s^{-1}}$

7.37 *See Section 7.2.*

$$? \text{ electrons} = 1.0 \times 10^{-3} \frac{\text{coul}}{\text{s}} \times \frac{1 \text{ electron}}{1.602 \times 10^{-19} \text{ coul}} = \mathbf{6.2 \times 10^{15}\ electrons/\ s}$$

An electron is dislodged from a metal by collision with a single photon, having an energy of $h\nu$, provided that energy exceeds the threshold energy. Hence, 6.2×10^{15} photons must be absorbed per second to produce 6.2×10^{15} electrons per second.

7.38 *See Section 7.1.*

$$? \text{ microamps} = 2.50 \times 10^{13} \frac{\text{electrons}}{\text{s}} \times \frac{1.602 \times 10^{-19} \text{ coul}}{1 \text{ electron}} \times \frac{10^6\ \mu\text{coul}}{1 \text{ coul}} = 4.00 \frac{\mu\text{coul}}{\text{s}} = 4.00\ \mu\text{amps}$$

An electron is dislodged from a metal by collision with a single photon, having an energy of $h\nu$, provided that energy exceeds the threshold energy. Hence, 2.50×10^{13} photons must be absorbed per second to produce 2.50×10^{13} electrons per second.

7.39 *See Section 7.2 and Example 7.3.*

$$\frac{1}{\lambda} = R_h\left(\frac{1}{n_A^2} - \frac{1}{n_B^2}\right) = 1.097 \times 10^7\ \text{m}^{-1}\left(\frac{1}{1^2} - \frac{1}{3^2}\right) = 9.751 \times 10^6\ \text{m}^{-1}$$

Solving for λ yields $\lambda = \dfrac{1}{9.751 \times 10^6\ \text{m}^{-1}} = \mathbf{1.026 \times 10^{-7}\ m}$

This wavelength is equivalent to 103 nm. The radiation is in the **ultraviolet** portion of the electromagnetic spectrum.

7.40 *See Section 7.2 and Example 7.3.*

$$\frac{1}{\lambda} = R_h\left(\frac{1}{n_A^2} - \frac{1}{n_B^2}\right) = 1.097 \times 10^7\ \text{m}^{-1}\left(\frac{1}{2^2} - \frac{1}{5^2}\right) = 2.304 \times 10^6\ \text{m}^{-1}$$

Solving for λ yields $\lambda = \dfrac{1}{2.304 \times 10^6\ \text{m}^{-1}} = \mathbf{4.340 \times 10^{-7}\ m}$

This wavelength is equivalent to 434 nm. The radiation is in the **visible** portion of the electromagnetic spectrum.

7.41 *See Section 7.3 and Example 7.4.*

Solving $p = mv = \dfrac{h}{\lambda}$ for λ gives $\lambda = \dfrac{h}{mv}$ $\quad \lambda = \dfrac{6.63 \times 10^{-34}\ \text{kg}\cdot\text{m}^2\cdot\text{s}^{-1}}{(1.67 \times 10^{-27}\ \text{kg})(1.7 \times 10^2\ \text{m}\cdot\text{s}^{-1})} \times \dfrac{10^9\ \text{nm}}{1\ \text{m}} = \mathbf{2.3\ nm}$

7.42 *See Section 7.3 and Example 7.4.*

Solving $p = mv = \dfrac{h}{\lambda}$ for λ gives $\lambda = \dfrac{h}{mv}$ $\quad \lambda = \dfrac{6.63 \times 10^{-34}\ \text{kg}\cdot\text{m}^2\cdot\text{s}^{-1}}{(9.11 \times 10^{-31}\ \text{kg})(2.9 \times 10^5\ \text{m}\cdot\text{s}^{-1})} \times \dfrac{10^9\ \text{nm}}{1\ \text{m}} = \mathbf{2.51\ nm}$

7.43 *See Section 7.3 and Example 7.4.*

Solving $p = mv = \dfrac{h}{\lambda}$ for λ gives $\lambda = \dfrac{h}{mv}$ $\quad \lambda = \dfrac{6.63 \times 10^{-34}\ \text{kg}\cdot\text{m}^2\cdot\text{s}^{-1}}{(9.11 \times 10^{-31}\ \text{kg})(2.19 \times 10^6\ \text{m}\cdot\text{s}^{-1})} = 3.32 \times 10^{-10}\ \text{m}$

This wavelength is equivalent to **332 pm**.

The circumference of the first Bohr orbit is given by $c = 2\pi r = (2)(3.143)(52.9 \text{ pm}) = \mathbf{332 \text{ pm}}$.
Hence, the wavelength of an electron traveling in the first Bohr orbit is exactly equal to the circumference of the first Bohr orbit.

7.44 ***See Section 7.3 and Example 7.4.***

Solving $p = mv = \dfrac{h}{\lambda}$ for λ gives $\lambda = \dfrac{h}{mv}$ $\quad \lambda = \dfrac{6.63 \times 10^{-34} \text{ kg}\cdot\text{m}^2\cdot\text{s}^{-1}}{(9.11 \times 10^{-31} \text{ kg})(1.094 \times 10^6 \text{ m}\cdot\text{s}^{-1})} = 6.65 \times 10^{-10} \text{ m}$

This wavelength is equivalent to **665 pm**.
The circumference of the first Bohr orbit is given by $c = 2\pi r = (2)(3.143)(212 \text{ pm}) = \mathbf{1.33 \times 10^3 \text{ pm}}$.
Hence, the wavelength of an electron traveling in the second Bohr orbit is exactly equal to one-half the circumference of the second Bohr orbit.

7.45 ***See Section 7.3 and Example 7.4.***

Solving $p = mv = \dfrac{h}{\lambda}$ for λ gives $\lambda = \dfrac{h}{mv}$

(a) $\lambda = \dfrac{6.63 \times 10^{-34} \text{ kg}\cdot\text{m}^2\cdot\text{s}^{-1}}{(0.100 \text{ kg})(40.0 \text{ m}\cdot\text{s}^{-1})} = \mathbf{1.66 \times 10^{-34} \text{ m}}$

(b) $\lambda = \dfrac{6.63 \times 10^{-34} \text{ kg}\cdot\text{m}^2\cdot\text{s}^{-1}}{(753 \text{ kg})(24.6 \text{ m}\cdot\text{s}^{-1})} = \mathbf{3.58 \times 10^{-38} \text{ m}}$

(c) $\lambda = \dfrac{6.63 \times 10^{-34} \text{ kg}\cdot\text{m}^2\cdot\text{s}^{-1}}{(1.67 \times 10^{-27} \text{ kg})(2.70 \times 10^3 \text{ m}\cdot\text{s}^{-1})} = \mathbf{1.47 \times 10^{-10} \text{ m}}$

7.46 ***See Section 7.4 and Example 7.4.***

Solving $p = mv = \dfrac{h}{\lambda}$ for λ gives $\lambda = \dfrac{h}{mv}$

(a) $\lambda = \dfrac{6.63 \times 10^{-34} \text{ kg}\cdot\text{m}^2\cdot\text{s}^{-1}}{(68 \text{ kg})(10 \text{ m}\cdot\text{s}^{-1})} = \mathbf{9.8 \times 10^{-35} \text{ m}}$

(b) $\lambda = \dfrac{6.63 \times 10^{-34} \text{ kg}\cdot\text{m}^2\cdot\text{s}^{-1}}{(0.0500 \text{ kg})(44.7 \text{ m}\cdot\text{s}^{-1})} = \mathbf{2.97 \times 10^{-34} \text{ m}}$

(c) $\lambda = \dfrac{6.63 \times 10^{-34} \text{ kg}\cdot\text{m}^2\cdot\text{s}^{-1}}{(9.11 \times 10^{-31} \text{ kg})(1.2 \times 10^5 \text{ m}\cdot\text{s}^{-1})} = \mathbf{6.06 \times 10^{-9} \text{ m}}$

7.47 ***See Section 7.4.***

Solving $p = mv = \dfrac{h}{\lambda}$ for v gives $v = \dfrac{h}{m\lambda}$ $\quad v = \dfrac{6.63 \times 10^{-34} \text{ kg}\cdot\text{m}^2\cdot\text{s}^{-1}}{(1.67 \times 10^{-27} \text{ kg})(0.150 \text{ nm})} \times \dfrac{10^9 \text{ nm}}{\text{m}} = \mathbf{2.65 \times 10^3 \text{ m/s}}$

7.48 *See Section 7.4.*

Solving $p = mv = \frac{h}{\lambda}$ for v gives $v = \frac{h}{m\lambda}$ $\quad v = \frac{6.63 \times 10^{-34}\ kg \cdot m^2 \cdot s^{-1}}{(9.11 \times 10^{-31}\ kg)(1.00\ nm)} \times \frac{10^9\ nm}{m} = \mathbf{7.28 \times 10^5\ m/s}$

7.49 *See Section 7.4 and Example 7.5.*

(a) $n = 6$ and $\ell = 1$ are associated with the 6p subshell.
(b) $n = 3$ and $\ell = 0$ are associated with the 3s subshell.
(c) $n = 5$ and $\ell = 2$ are associated with the 5d subshell.
(d) $n = 4$ and $\ell = 0$ are associated with the 4s subshell.
(e) Since ℓ must be at least one less than n, a value of $\ell = 3$ is not possible when $n = 2$.

7.50 *See Section 7.4 and Example 7.5.*

(a) $n = 5$ and $\ell = 1$ are associated with the 5s subshell.
(b) Since ℓ must be at least one less than n, a value of $\ell = 1$ is not possible when $n = 1$.
(c) $n = 3$ and $\ell = 2$ are associated with the 3d subshell.
(d) $n = 4$ and $\ell = 3$ are associated with the 4f subshell.
(e) $n = 7$ and $\ell = 0$ are associated with the 7s subshell.

7.51 *See Section 7.4.*

(a) 3p corresponds to $n = 3$ and $\ell = 1$.
(b) 5d corresponds to $n = 5$ and $\ell = 2$.
(c) 7s corresponds to $n = 7$ and $\ell = 0$.
(d) 4f corresponds to $n = 4$ and $\ell = 3$.
(e) 2s corresponds to $n = 2$ and $\ell = 0$.

7.52 *See Section 7.4.*

(a) 3d corresponds to $n = 3$ and $\ell = 2$.
(b) 5p corresponds to $n = 5$ and $\ell = 1$.
(c) 6s corresponds to $n = 6$ and $\ell = 0$.
(d) 5f corresponds to $n = 5$ and $\ell = 3$.
(e) 1s corresponds to $n = 1$ and $\ell = 0$.

7.53 *See Section 7.4.*

(a) $n = 2$ and $\ell = 1$ are associated with the 2p subshell.
(b) Since ℓ must be at least one less than n, a value of $\ell = 2$ is not possible when $n = 2$.
(c) $n = 3$ and $\ell = 0$ are associated with the 3s subshell.
(d) Since allowed values for m_ℓ are all whole numbers from $-\ell$ to $+\ell$, a value m_ℓ=1 is not possible when $\ell = 0$.
(e) $n = 3$ and $\ell = 2$ are associated with the 3d subshell.
(f) $n = 5$ and $\ell = 0$ are associated with the 5s subshell.

7.54 *See Section 7.4.*

(a) Since ℓ must be at least one less than n, a value of $\ell = 1$ is not possible when $n = 1$.
(b) $n = 4$ and $\ell = 2$ are associated with the 4d subshell.
(c) $n = 2$ and $\ell = 0$ are associated with the 2s subshell.

d) Since allowed values for m_ℓ are all whole numbers from $-\ell$ to $+\ell$, a value of $m_\ell = 1$ is not possible when $\ell = 0$.
(e) $n = 3$ and $\ell = 2$ are associated with the 3d subshell.
(f) $n = 6$ and $\ell = 0$ are associated with the 6s subshell.

7.55 ***See Section 7.4 and Table 7.1.***

(a) There are 4 subshells in the $n = 4$ shell ($\ell = 0,1,2,3$).
(b) There are 5 orbitals in 3d subshell ($\ell = 2$ and $m_\ell = 2,-1,0,+1,+2$).
(c) The maximum value of ℓ for $n = 3$ is 2 ($\ell = 0,1,...(n-1)$max).
(d) For the 3p subshell, $n = 3$, $\ell = 1$ and $m_\ell = -1,0,+1$.

7.56 ***See Section 7.4 and Table 7.1.***

(a) There are 3 subshells in the $n = 3$ shell ($\ell = 0,1,2$).
(b) There are 3 orbitals in the 4p subshell ($\ell = 1$ and $m_\ell = -1,0+1$).
(c) The maximum value of ℓ for $n = 4$ is 3 ($\ell = 0,1,...(n-1)$max).
(d) For the 3d subshell, $n = 3$, $\ell = 2$ and $m_\ell = -2,-1,0,+1,+2$.

7.57 ***See Section 7.4 and Figures 7.12, 7.14, 7.15.***

(a) $n = 2$ and $\ell = 0$ are associated with the 2s orbital, which is spherical. — 2s

(b) The $3p_x$ is "dumbbell" shaped along the x axis. — $3p_x$

(c) The $4d_{xy}$ is 45° off the x and y axes in the plane formed by the intersection of the x and y axes. — $4d_{xy}$

(d) $n = 2$ and $\ell = 1$ are associated with the 2p subshell. 2p orbitals are "dumbbell" shaped and are located along the x, y and z axes. — 2p

(e) $n = 1$ and $\ell = 0$ are associated with the 1s orbital, which is spherical. — 1s

7.58 ***See Section 7.4 and Figures 7.12, 7.14, 7.15.***

(a) $n = 3$ and $\ell = 0$ are associated with the 3s orbital, which is spherical. — 3s

(b) The $3d_{xy}$ is 45° off the x and y axes in the plane formed by the intersection of the x and y axes. — $3d_{xy}$

(c) The $4p_y$ is "dumbbell" shaped along the y axis. — $4p_y$

(d) $n = 3$ and $\ell = 1$ are associated with the 3p subshell. 3p orbitals are "dumbbell" shaped and are located along the x, y, and z axes.

3p

(e) $n = 2$ and $\ell = 0$ are associated with the 2s orbital, which is spherical.

2s

7.59 *See Section 7.4 and Figure 7.13.*

(a) The probability of finding a p_y electron is highest on either side of the nucleus along the y axis.

(b) The probability of finding a p_y electron is zero anywhere in the xz plane including at the nucleus along the y axis.

7.60 *See Section 7.4 and Figure 7.13.*

(a) The probability of finding a p_z electron is highest on either side of the nucleus along the z axis.

(b) The probability of finding a p_z electron is zero anywhere in the xy plane including at the nucleus along the z axis.

7.61 *See Section 7.4 and Figure 7.12.*

"s" orbitals have spherical contours about the nucleus.

7.62 *See Section 7.4 and Figure 7.12.*

$n = 3$ and $\ell = 0$ are associated with the 3s orbital, which is spherical.

3s

7.63 *See Section 7.4 and Figure 7.14.*

$n = 3$ and $\ell = 2$ are associated with the 3d subshell.

3d or

7.64 *See Section 7.4 and Figure 7.14.*

"p" orbitals are "dumbbell" shaped.

7.65 *See Section 7.4 and Figure 7.16.*

$2s < 3s = 3p_x = 3p_y < 4s = 4p_z = 4d_{xy}$

7.66 *See Section 7.4 and Figure 7.16.*

$1s < 3s = 3p_y < 4s = 4p_x = 4d_{xy} = 4d_{yz}$

7.67 *See Section 7.4 and Example 7.7.*

Substituting in $\frac{1}{\lambda} = Z^2 R_h \left(\frac{1}{n_A^2} - \frac{1}{n_B^2} \right)$ and noting that Z = 3 for Li^{2+} gives

$$\frac{1}{\lambda} = (3)^2 (1.097 \text{ x } 10^7 \text{ m}^{-1}) \left(\frac{1}{1^2} - \frac{1}{3^2} \right) = (3)^2 (1.097 \text{ x } 10^7 \text{ m}^{-1}) \left(\frac{3^2 - 1^2}{1^2 \times 3^2} \right) = 8.776 \text{ x } 10^7 \text{ m}^{-1}$$

Solving for λ yields $\lambda = \frac{1}{8.776 \text{ x } 10^7 \text{ m}^{-1}} = \mathbf{1.139 \text{ x } 10^{-8} \text{ m}}$

This wavelength is equivalent to 11.39 nm. The radiation is in the **far ultraviolet** region of the electromagnetic spectrum.

7.68 *See Section 7.4 and Example 7.7.*

The wavelength of light absorbed when an electron moves from the n = 3 to n = 4 Bohr orbits of He^+ is equal to the wavelength of light that would be emitted as an electron moves from the n = 4 to n = 3 Bohr orbits of He^+.

Substituting in $\frac{1}{\lambda} = Z^2 R_h \left(\frac{1}{n_A^2} - \frac{1}{n_B^2} \right)$ and noting that Z = 2 for He^+ gives

$$\frac{1}{\lambda} = (2)^2 (1.097 \text{ x } 10^7 \text{ m}^{-1}) \left(\frac{1}{3^2} - \frac{1}{4^2} \right) = (2)^2 (1.097 \text{ x } 10^7 \text{ m}^{-1}) \left(\frac{4^2 - 3^2}{3^2 \times 4^2} \right) = 2.133 \text{ x } 10^6 \text{ m}^{-1}$$

Solving for λ yields $\lambda = \frac{1}{2.133 \text{ x } 10^6 \text{ m}^{-1}} = \mathbf{4.688 \text{ x } 10^{-7} \text{ m}}$

This wavelength is equivalent to 468.8 nm.

7.69 *See Section 7.5 and Figure 7.20.*

(a) 2s < 2p < 3p < 3d < 5p
(b) 1s < 2s < 2p < 3s < 3d < 4d
(c) 1s < 2s < 2p < 3s < 3p < 3d < 4p

7.70 *See Section 7.5 and Figure 7.20.*

(a) $2s < 3s < 3p_x = 3p_y < 4s < 4p_z < 4d_{xy}$
(b) $1s < 3p_x = 3p_y < 4s < 3d_{xy}$
(c) $2s < 3p_x < 4s < 3d_{xz} < 5s$

7.71 *See Section 7.6.*

(a) C $1s^2 2s^2 2p^2$ O $1s^2 2s^2 2p^4$
(b) N $1s^2 2s^2 2p^3$
(c) Li $1s^2 2s^1$ Be $1s^2 2s^2$

7.72 *See Section 7.6.*

(a) He $1s^2$ Be $1s^2 2s^2$ Ne $1s^2 2s^2 2p^6$
(b) N $1s^2 2s^2 2p^3$
(c) H $1s^1$ Li $1s^2 2s^1$ B $1s^2 2s^2 2p^1$ F $1s^2 2s^2 2p^5$

7.73 *See Section 7.6.*

(a) C $1s^2 2s^2 2p^2$ ↑↓ (1s) ↑↓ (2s) ↑ ↑ _ (2p)

O $1s^2 2s^2 2p^4$ ↑↓ (1s) ↑↓ (2s) ↑↓ ↑ ↑ (2p)

(b) N $1s^2 2s^2 2p^3$ ↑↓ (1s) ↑↓ (2s) ↑ ↑ ↑ (2p)

(c) Li $1s^2 2s^1$ ↑↓ (1s) ↑ (2s)

Be $1s^2 2s^2$ ↑↓ (1s) ↑↓ (2s)

7.74 *See Section 7.6.*

(a) He $1s^2$ ↑↓ (1s)

Be $1s^2 2s^2$ ↑↓ (1s) ↑↓ (2s)

Ne $1s^2 2s^2 2p^6$ ↑↓ (1s) ↑↓ (2s) ↑↓ ↑↓ ↑↓ (2p)

(b) N $1s^2 2s^2 2p^3$ ↑↓ (1s) ↑↓ (2s) ↑ ↑ ↑ (2p)

(c) H $1s^1$ ↑ (1s)

Li $1s^2 2s^1$ ↑↓ (1s) ↑ (2s)

B $1s^2 2s^2 2p^1$ ↑↓ (1s) ↑↓ (2s) ↑ _ _ (2p)

F $1s^2 2s^2 2p^5$ ↑↓ (1s) ↑↓ (2s) ↑↓ ↑↓ ↑ (2p)

7.75 *See Section 7.6.*

Be and Ne; see the solution for 7.74a or the orbital diagrams given in the text for B, C, Ne, and Be.

7.76 *See Section 7.6..*

(a) Li $1s^2 2s^1$ ↑↓ (1s) ↑ (2s) 1 unpaired

(b) He $1s^2$ ↑↓ (1s) 0 unpaired

(c) F $1s^2 2s^2 2p^5$ ↑↓ (1s) ↑↓ (2s) ↑↓ ↑↓ ↑ (2p) 1 unpaired

(d) B $1s^2 2s^2 2p^1$ ↑↓ (1s) ↑↓ (2s) ↑ (2p) 1 unpaired

7.77 *See Section 7.6.*

(a) He, 1s (b) Be, 2s (c) C, 2p (d) F, 2p

7.78 *See Section 7.6.*

(a) H, 1s (b) Li, 2s (c) N, 2p (d) Ne, 2p

7.79 *See Sections 7.4, 7.6, and Figure 7.20.*

Energy level diagram for C:

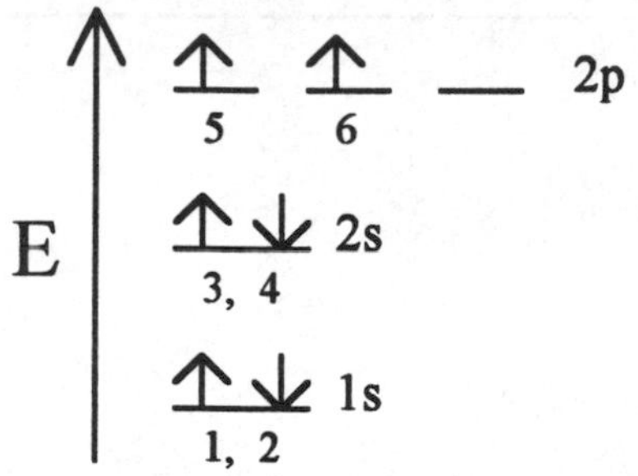

E ↑
- 2p: ↑ (5) ↑ (6) __
- 2s: ↑↓ (3, 4)
- 1s: ↑↓ (1, 2)

Quantum numbers for electrons in C:

electron	n	ℓ	m_ℓ	m_s
1	1	0	0	+1 / 2
2	1	0	0	−1 / 2
3	2	0	0	+1 / 2
4	2	0	0	−1 / 2
5	2	1	−1	+1 / 2
6	2	1	0	+1 / 2

The first electron placed in an orbital was arbitrarily assigned a m_s value of +1/2. The m_l values chosen for the 2p electrons were arbitrarily chosen to be -1 and 0 in keeping with $m_l = -\ell$ to $+\ell$.

7.80 *See Sections 7.4, 7.6 , and Figure 7.20.*

Energy level diagram for N:

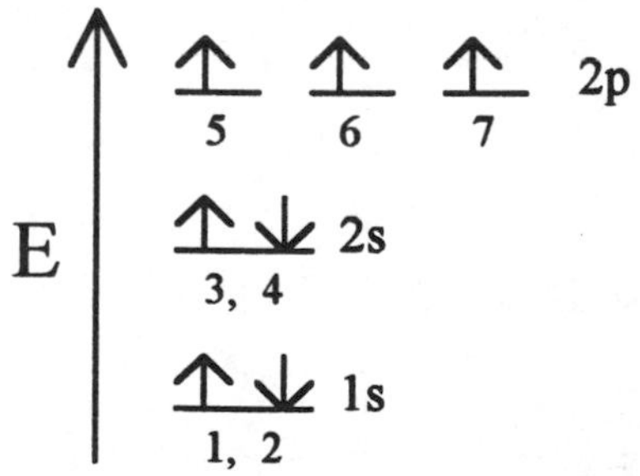

E ↑
- 2p: ↑ (5) ↑ (6) ↑ (7)
- 2s: ↑↓ (3, 4)
- 1s: ↑↓ (1, 2)

Quantum numbers for electrons in N:

electron	n	ℓ	m_ℓ	m_s
1	1	0	0	+1 / 2
2	1	0	0	−1 / 2
3	2	0	0	+1 / 2
4	2	0	0	−1 / 2
5	2	1	−1	+1 / 2
6	2	1	0	+1 / 2
7	2	1	1	+1 / 2

The first electron placed in an orbital was arbitrarily assigned a m_s value of +1/2. The m_ℓ values chosen for the 2p electrons were arbitrarily assigned in the order -1, 0, and 1 in keeping with $m_\ell = -\ell$ to $+\ell$.

7.81 *See Sections 7.4, 7.6.*

Li $1s^2 2s^1$ The highest energy electron has $n = 2$, $\ell = 0$, $m_\ell = 0$, and $m_s = +1/2$, with the choice of assigning +1/2 for m_s being arbitrary.

7.82	*See Sections 7.4, 7.6.*

B $1s^2\ 2s^2\ 2p^1$ The highest energy electron has $n = 2$, $\ell = 1$, $m_\ell = -1$, and $m_s = +1/2$. The m_ℓ value for the 2p electron was arbitrarily chosen to be -1 in keeping with $m_\ell = -\ell$ to $+\ell$. The 2p electron also was arbitrarily assigned an m_s value of +1/2.

7.83	*See Section 7.6 and Table 7.2.*

(a) Ten electrons can occupy the 3d subshell [2 $(2\ell + 1)$ with $\ell = 2$].

(b) Two electrons can occupy the 4s subshell [2 $(2\ell + 1)$ with $\ell = 0$].

(c) Eight electrons can occupy the 2nd principal shell ($2n^2$ with $n = 2$).

(d) Fifty electrons can occupy the 5th principal shell ($2n^2$ with $n = 5$).

7.84	*See Section 7.6 and Table 7.2.*

(a) Six electrons can occupy the 3p subshell [2 $(2\ell + 1)$ with $\ell = 1$].

(b) Ten electrons can occupy the 4d subshell [2 $(2\ell + 1)$ with $\ell = 2$].

(c) Thirty two electrons can occupy the 4th principal shell ($2n^2$ with $n = 4$).

(d) Eighteen electrons can occupy the 3rd principal shell ($2n^2$ with $n = 3$).

7.85	*See Section 7.6.*

(a) ↑↓ (1s) ↑↓ (2s) ↑↓ ↑ ↑ (2p)   corresponds to the ground state for O, $1s^2\ 2s^2\ 2p^4$.

(b) ↑↓ (1s) ↑ (2s) ↑ __ __ (2p)   corresponds to an excited state of Be.

↑↓ (1s) ↑↓ (2s)   correspond to the ground state of Be, $1s^2\ 2s^2$.

(c) ↑↓ (1s) ↑↓ (2s) ↑↓ ↑ __ (2p)   corresponds to an excited state of N.

↑↓ (1s) ↑↓ (2s) ↑ ↑ ↑ (2p)   corresponds to the ground state of N, $1s^2\ 2s^2\ 2p^3$.

(d) ↑ (1s) ↑↓ (2s)   corresponds to an excited state of Li.

↑↓ (1s) ↑ (2s)   corresponds to the ground state of Li, $1s^2\ 2s^1$.

7.86 *See Section 7.6.*

(a) ↑↓ (1s) ↑ (2s) ↑ ↑ ↑ (2p) corresponds to an excited state of C.

↑↓ (1s) ↑↓ (2s) ↑ ↑ _ (2p) corresponds to the ground state of C, $1s^2\ 2s^2\ 2p^2$.

(b) ↑↓ (1s) ↑↓ (2s) corresponds to the ground state for Be, $1s^2\ 2s^2$.

(c) ↑ (1s) ↑↓ (2s) ↑↓ _ _ (2p) corresponds to an excited state for B.

↑↓ (1s) ↑↓ (2s) ↑ _ _ (2p) corresponds to the ground state of B, $1s^2\ 2s^2\ 2p^1$.

(d) ↑↓ (1s) ↑↓ (2s) ↑↓ ↑↓ ↑ (2p) corresponds to the ground state for F, $1s^2\ 2s^2\ 2p^5$.

7.87 *See Section 7.1.*

Solving $\lambda\nu$ = speed for λ gives $\lambda = \frac{\text{speed}}{\nu}$, and recognizing that $1\ \text{Hz} = 1\text{s}^{-1}$ yields $\lambda = \frac{344\ \text{m}\cdot\text{s}^{-1}}{512\ \text{s}^{-1}} = \mathbf{0.672\ m.}$

7.88 *See Section 7.1 and Example 7.2.*

The energy of the photon that is absorbed in the photoelectric effect ($h\nu_{photon}$) is equal to the threshold energy of the metal ($h\nu_o$) plus the kinetic energy of the electron that is ejected (KE). Hence,

$h\nu_{photon} = h\nu_o + KE$, $h\nu_o = h\nu_{photon} - KE$, and $\nu_o = \nu_{photon} - \frac{KE}{h}$, where

$$\nu_{photon} = \frac{c}{\lambda} = \frac{3.00 \times 10^8\ \text{m}\cdot\text{s}^{-1}}{400\ \text{nm}} \times \frac{10^9\ \text{nm}}{\text{m}} = 7.50 \times 10^{14}\ \text{s}^{-1}.$$

This gives $\nu_o = 7.5 \times 10^{14}\ \text{s}^{-1} - \frac{1.38 \times 10^{-19}\ \text{J}}{6.63 \times 10^{-34}\ \text{J}\cdot\text{s}} = \mathbf{5.41 \times 10^{14}\ s^{-1}}.$

7.89 *See Section 7.1 and Examples 7.1, 7.2.*

Solving $\lambda\nu = c$ for ν gives $\nu = \frac{c}{\lambda}$ $\quad \nu = \frac{3.00 \times 10^8\ \text{m}\cdot\text{s}^{-1}}{493\ \text{nm}} \times \frac{10^9\ \text{nm}}{1\ \text{m}} = 6.09 \times 10^{14}\ \text{s}^{-1}$

Hence, $E = h\nu = (6.63 \times 10^{-34}\ \text{J}\cdot\text{s})(6.09 \times 10^{14}\ \text{s}^{-1}) = \mathbf{4.04 \times 10^{-19}\ J}$.

7.90 *See Section 7.1 and Examples 7.1, 7.2*

Solving $\lambda\nu = c$ for ν gives $\nu = \frac{c}{\lambda}$ $\quad \nu = \frac{3.00 \times 10^8\ \text{m}\cdot\text{s}^{-1}}{641\ \text{nm}} \times \frac{10^9\ \text{nm}}{1\ \text{m}} = 4.68 \times 10^{14}\ \text{s}^{-1}$

Hence, $E = h\nu = (6.63 \times 10^{-34}\ \text{J}\cdot\text{s})(4.68 \times 10^{14}\ \text{s}^{-1}) = \mathbf{3.10 \times 10^{-19}\ J}$.

7.91 *See Section 7.2 and Example 7.3.*

The Rydberg equation is $\frac{1}{\lambda} = R_h\left(\frac{1}{n_A^2} - \frac{1}{n_B^2}\right)$.

The two lowest energy lines in the Paschen series ($n_A = 3$) involve transitions from $n_B = 4$ and $n_B = 5$ to $n_A = 3$.
Substituting $n_A = 3$ and $n_B = 4$ into the Rydberg equation gives

$$\frac{1}{\lambda} = \left(1.097 \times 10^7 \text{ m}^{-1}\right)\left(\frac{1}{3^2} - \frac{1}{4^2}\right) = \left(1.097 \times 10^7 \text{ m}^{-1}\right)\left(\frac{4^2 - 3^2}{3^2 \times 4^2}\right) = 5.333 \times 10^5 \text{ m}^{-1}$$

Solving for λ in nm yields $\lambda = \frac{1}{5.333 \times 10^5 \text{ m}^{-1}} \times \frac{10^9 \text{ nm}}{1 \text{ m}} = \mathbf{1.875 \times 10^3 \text{ nm}}$

Substituting $n_A = 3$ and $n_B = 5$ into the Rydberg equation gives

$$\frac{1}{\lambda} = \left(1.097 \times 10^7 \text{ m}^{-1}\right)\left(\frac{1}{3^2} - \frac{1}{5^2}\right) = \left(1.097 \times 10^7 \text{ m}^{-1}\right)\left(\frac{5^2 - 3^2}{3^2 \times 5^2}\right) = 7.801 \times 10^5 \text{ m}^{-1}$$

Solving for λ in nm yields $\lambda = \frac{1}{7.801 \times 10^5 \text{ m}^{-1}} \times \frac{10^9 \text{ nm}}{1 \text{ m}} = \mathbf{1.282 \times 10^3 \text{ nm}}$

7.92 *See Section 7.2 and Example 7.3.*

The Rydberg equation is $\frac{1}{\lambda} = R_h\left(\frac{1}{n_A^2} - \frac{1}{n_B^2}\right)$.

The two lowest energy lines in the Lyman series ($n_A = 1$) involve transitions from $n_B = 2$ and $n_B = 3$ to $n_A = 1$.
Substituting $n_A = 1$ and $n_B = 2$ into the Rydberg equation gives

$$\frac{1}{\lambda} = \left(1.097 \times 10^7 \text{ m}^{-1}\right)\left(\frac{1}{1^2} - \frac{1}{2^2}\right) = \left(1.097 \times 10^7 \text{ m}^{-1}\right)\left(\frac{2^2 - 1^2}{1^2 \times 2^2}\right) = 8.228 \times 10^6 \text{ m}^{-1}$$

Solving for λ in nm yields $\lambda = \frac{1}{8.228 \times 10^6 \text{ m}^{-1}} \times \frac{10^9 \text{ nm}}{1 \text{ m}} = \mathbf{121.5 \text{ nm}}$

Substituting $n_A = 1$ and $n_B = 3$ into the Rydberg equation gives

$$\frac{1}{\lambda} = \left(1.097 \times 10^7 \text{ m}^{-1}\right)\left(\frac{1}{1^2} - \frac{1}{3^2}\right) = \left(1.097 \times 10^7 \text{ m}^{-1}\right)\left(\frac{3^2 - 1^2}{1^2 \times 3^2}\right) = 9.751 \times 10^6 \text{ m}^{-1}$$

Solving for λ in nm yields $\lambda = \frac{1}{9.751 \times 10^6 \text{ m}^{-1}} \times \frac{10^9 \text{ nm}}{1 \text{ m}} = \mathbf{102.6 \text{ nm}}$

7.93 *See Insights into Chemistry: A Closer View.*

Recognizing that the momentum of the automobile is equal to the product of mass times velocity gives

$$momentum, mv = 650 \text{ kg} \times 55\frac{\text{mi}}{\text{hr}} \times \frac{1.609 \text{ km}}{1 \text{ mi}} \times \frac{10^3 \text{ m}}{1 \text{ km}} \times \frac{1 \text{ hr}}{3600 \text{ s}} = 1.6 \times 10^4 \text{ kg}\cdot\text{m}\cdot\text{s}^{-1}$$

and $\Delta mv = 0.1 \times 10^4 \text{ kg}\cdot\text{m}^2\cdot\text{s}^{-1}$.

Solving $\Delta x \Delta mv = 1.06 \times 10^{-34}$ kg·m^2·s^{-1} for Δx gives $\Delta x = \frac{1.06 \times 10^{-34} \text{ kg}\cdot\text{m}^2\cdot\text{s}^{-1}}{\Delta mv}$

and substituting for Δmv gives $\Delta x = \frac{1.06 \times 10^{-34} \text{ kg}\cdot\text{m}^2\cdot\text{s}^{-1}}{0.1 \times 10^4 \text{ kg}\cdot\text{m}\cdot\text{s}^{-1}} = \mathbf{1 \times 10^{-37} \text{ m}}$

This amount of uncertainty in position is insignificant.

7.94 *See Section 7.4.*

The three lowest potential energy states correspond to n = 1, 2, and 3 because the electron is closest to the positive nuclear charge.

$$E_1 = \frac{-2.18 \times 10^{-18}\ J}{1^2} = \mathbf{-2.18 \times 10^{-18}\ J} \qquad E_2 = \frac{-2.18 \times 10^{-18}\ J}{2^2} = \mathbf{-5.45 \times 10^{-19}\ J}$$

$$E_3 = \frac{-2.18 \times 10^{-18}\ J}{3^2} = \mathbf{-2.42 \times 10^{-19}\ J}$$

Notice that the potential energy of an electron in an atom is always negative. The reference point of infinite separation between the nucleus and the electron corresponds to zero potential energy. As the distance between the nucleus and the electron increases, the potential energy becomes less negative corresponding to higher energy states.

7.95 *See Section 7.4.*

$$E_1 = \frac{2.18 \times 10^{-18}}{n^2}\frac{J}{atom} \times \frac{6.022 \times 10^{23}\ atoms}{1\ mol} \times \frac{1\ kJ}{10^3\ J} = \frac{\mathbf{-1313}}{\boldsymbol{n^2}}\ \mathbf{kJ / mol}$$

7.96 *See Section 7.4.*

The three lowest (potential) energy states correspond to n = 1, 2, and 3 because the electron is closest to the positive nuclear charge.

$$E_1 = \frac{-1312}{1^2}\ kJ/mol = \mathbf{-1312\ kJ/mol} \qquad E_2 = \frac{-1312}{2^2}\ kJ/mol = \mathbf{-328\ kJ/mol}$$

$$E_3 = \frac{-1312}{3^2}\ kJ/mol = \mathbf{-146\ kJ}$$

Notice that the potential energy of an electron in an atom is always negative. The reference point of infinite separation between the nucleus and the electron corresponds to zero potential energy. As the distance between the nucleus and the electron increases, the potential energy becomes less negative corresponding to higher energy states.

7.97 *See Section 7.6.*

(a) Li $1s^2 2s^1$

↑↓	↑			
1s	2s			

(b) F $1s^2\ 2s^2\ 2p^5$

↑↓	↑↓	↑↓	↑↓	↑
1s	2s		2p	

(c) O $1s^2\ 2s^2\ 2p^4$

↑↓	↑↓	↑↓	↑	↑
1s	2s		2p	

7.98 *See Section 7.6.*

(a) Be $1s^2 2s^2$

↑↓	↑↓			
1s	2s			

(b) B $1s^2 2s^2 2p^1$

↑↓	↑↓	↑	__	__
1s	2s		2p	

(c) Ne $1s^2 2s^2 2p^6$

↑↓	↑↓	↑↓	↑↓	↑↓
1s	2s		2p	

7.99 *See Sections 7.1, 7.2.*

The Rydberg equation is $\frac{1}{\lambda} = R_h\left(\frac{1}{n_A^2} - \frac{1}{n_B^2}\right)$.

(a) Substituting $n_A = 2$ and $n_B = 3$ into the Rydberg equation gives

$$\frac{1}{\lambda} = \left(1.097 \text{ x } 10^7 \text{ m}^{-1}\right)\left(\frac{1}{2^2} - \frac{1}{3^2}\right) = \left(1.097 \text{ x } 10^7 \text{ m}^{-1}\right)\left(\frac{3^2 - 2^2}{2^2 \times 3^2}\right) = 1.524 \text{ x } 10^6 \text{ m}^{-1}$$

Substituting $\frac{1}{\lambda}$ in $E = h\nu = \frac{hc}{\lambda} = hc\frac{1}{\lambda}$ yields

$$E = \left(6.63 \text{ x } 10^{-34} \text{ J}\cdot\text{s}\right)\left(3.00 \text{ x } 10^8 \text{ m}\cdot\text{s}^{-1}\right)\left(1.524 \text{ x } 10^6 \text{ m}^{-1}\right) = \mathbf{3.03 \text{ x } 10^{-19} \text{ J}}$$

Substituting $n_A = 32$ and $n_B = 33$ into the Rydberg equation gives

$$\frac{1}{\lambda} = \left(1.097 \text{ x } 10^7 \text{ m}^{-1}\right)\left(\frac{1}{32^2} - \frac{1}{33^2}\right) = \left(1.097 \text{ x } 10^7 \text{ m}^{-1}\right)\left(\frac{33^2 - 32^2}{32^2 \times 33^2}\right) = 6.394 \text{ x } 10^2 \text{ m}^{-1}$$

Substituting $\frac{1}{\lambda}$ in $E = h\nu = \frac{hc}{\lambda} = hc\frac{1}{\lambda}$ yields

$$E = \left(6.63 \text{ x } 10^{-34} \text{ J}\cdot\text{s}\right)\left(3.00 \text{ x } 10^8 \text{ m}\cdot\text{s}^{-1}\right)\left(6.394 \text{ x } 10^2 \text{ m}^{-1}\right) = \mathbf{1.27 \text{ x } 10^{-22} \text{ J}}$$

The difference between the $n = 2$ and $n = 3$ levels is much larger than the energy difference between the $n = 32$ and $n = 33$ levels.

(b) The largest energy difference that can be observed for the hydrogen atom is the difference between the $n = 1$ level and the $n = \infty$ level. Sustituting $n_A = 1$ and $n_B = \infty$ into the Rydberg equation gives

$$\frac{1}{\lambda} = \left(1.097 \text{ x } 10^7 \text{ m}^{-1}\right)\left(\frac{1}{1^2} - \frac{1}{\infty^2}\right) = 1.097 \text{ x } 10^7 \text{ m}^{-1}$$

Substituting $\frac{1}{\lambda}$ in $E = h\nu = \frac{hc}{\lambda} = hc\frac{1}{\lambda}$ yields

$$E = \left(6.63 \text{ x } 10^{-34} \text{ J}\cdot\text{s}\right)\left(3.00 \text{ x } 10^8 \text{ m}\cdot\text{s}^{-1}\right)\left(1.097 \text{ x } 10^7 \text{ m}^{-1}\right) = \mathbf{2.18 \text{ x } 10^{-18} \text{ J}}$$

(c) There is a limit to the energy difference, since the $n = \infty$ level corresponds to the electron being infinitely far from the nucleus.

(d) The energy associated with removing an electron from an atom in the gas phase is known as the ionization energy of the atom; see Section 8.3. The energy that was calculated in part c corresponds to removing the electron from one atom of H. The value that is commonly tabulated for the ionization energy of H corresponds to the energy that is needed to remove the electron from one mole of H atoms, namely

$$\left(2.18 \text{ x } 10^{-18} \frac{\text{J}}{\text{atom}}\right)\left(6.022 \text{ x } 10^{23} \frac{\text{atoms}}{\text{mol}}\right) = 1.314 \text{ x } 10^6 \frac{\text{J}}{\text{mol}} \text{ or } \mathbf{1314 \frac{kJ}{mol}}$$

7.100 *See Section 7.1 and Example 7.2.*

$$E = h\nu = \frac{hc}{\lambda} \qquad E = \frac{\left(6.63 \times 10^{-34} \text{ J}\cdot\text{s}\right)\left(3.00 \times 10^8 \text{ m}\cdot\text{s}^{-1}\right)}{(589 \text{ nm})} \times \frac{10^9 \text{ nm}}{1 \text{ m}} = 3.38 \times 10^{-19} \text{ J}$$

$$? \text{ power} = \frac{3.38 \times 10^{-19} \text{ J}}{1 \text{ photon}} \times \frac{1004 \text{ photons}}{10 \text{ s}} = 3.39 \times 10^{-17} \frac{\text{J}}{\text{S}} = \mathbf{3.39 \text{ x } 10^{-17} \text{ watts}}$$

7.101 *See Section 7.6.*

(a) ↑↓ (1s) ↑↓ (2s) ↑ __ __ (2p) corresponds to the ground state for B, $1s^2\,2s^2\,2p^1$.

(b) ↑↓ (1s) ↑↓ (2s) ↑ ↑ _ (2p) corresponds to the ground state for C, $1s^2\ 2s^2\ 2p^2$.

(c) ↑↓ (1s) ↑ (2s) ↑ ↑ ↑ (2p) corresponds to an excited state for C.

↑↓ (1s) ↑↓ (2s) ↑ ↑ _ (2p) corresponds to the ground state for C, $1s^2\ 2s^2\ 2p^2$.

(d) ↑↓ (1s) ↑↓ (2s) ↑↑↓ ↑ _ (2p) violates the Pauli exclusion principal and is an impossible state for O.

↑↓ (1s) ↑↓ (2s) ↑↓ ↑ ↑ (2p) corresponds to the ground state for O, $1s^2\ 2s^2\ 2p^4$.

7.102 ***See Section 7.6.***

(a) ↑ (1s) ↑↓ (2s) ↑ _ _ (2p) corresponds to an excited state for Be.

↑↓ (1s) ↑↓ (2s) corresponds to the ground state for Be, $1s^2\ 2s^2$.

(b) ↑↓ (1s) ↑↓ (2s) ↑↓ _ _ (2p) corresponds to an excited state for C.

↑↓ (1s) ↑↓ (2s) ↑ ↑ _ (2p) corresponds to the ground state for C, $1s^2\ 2s^2\ 2p^2$.

(c) ↑↓ (1s) ↑↑ (2s) ↑ ↑ _ (2p) two electrons of like spin in 2s and therefore corresponds to an impossible configuration for C.

(d) ↑↓ (1s) ↑↓ (2s) ↑ ↑ ↑ (2p) corresponds to the ground state for N, $1s^2\ 2s^2\ 2p^3$.

7.103 ***See Sections 2.2, 6.7, 7.3, and Examples 6.13, 7.4.***

(a) Solving $p = mv = \frac{h}{\lambda}$ for v gives $v = \frac{h}{m\lambda}$

$$v = \frac{6.63 \times 10^{-34}\ \text{kg} \cdot \text{m}^2 \cdot \text{s}^{-1}}{(9.11 \times 10^{-31}\ \text{kg})\left(100\ \text{pm} \times \frac{10^{-12}\ \text{m}}{1\ \text{pm}}\right)} = \mathbf{7.28 \times 10^{6}\ m \cdot s^{-1}}$$

(b) Solving $p = mv = \frac{h}{\lambda}$ for v gives $v = \frac{h}{m\lambda}$

$$v = \frac{6.63 \times 10^{-34}\ \text{kg} \cdot \text{m}^2 \cdot \text{s}^{-1}}{(1.67 \times 10^{-27}\ \text{kg})\left(100\ \text{pm} \times \frac{10^{-12}\ \text{m}}{1\ \text{pm}}\right)} = \mathbf{3.97 \times 10^{3}\ m \cdot s^{-1}}$$

For neutrons at 300 K,

$$\mu_{rms} = \sqrt{\frac{3RT}{M}} = \sqrt{\frac{(3)(8.314\ \text{kg} \cdot \text{m}^2 \cdot \text{s}^{-2} \cdot \text{mol}^{-1} \cdot \text{K}^{-1})(300\ \text{K})}{\left[\left(1.675 \times 10^{-27}\ \frac{kg}{neutron}\right)\left(6.022 \times 10^{23}\ \frac{neutrons}{mol}\right)\right]}} = \mathbf{2.73 \times 10^{3}\ m \cdot s^{-1}}$$

A velocity higher than the rms speed at 300 K is needed to obtain a wavelength of 100 pm.

7.104 *See Insights into Chemistry: A Closer View*

Solving $p = mv = \frac{h}{\lambda}$ for λ gives $\lambda = \frac{h}{mv}$.

For the 100 mi/hr fastball, $\lambda = \frac{6.63 \times 10^{-34}\ \text{kg} \cdot \text{m}^2 \cdot \text{s}^{-1}}{(.0220\ \text{kg})\left(100\frac{\text{mi}}{\text{hr}}\right)} \times \frac{1\ \text{mi}}{1.609\ \text{km}} \times \frac{1\ \text{km}}{10^3\ \text{m}} \times \frac{3600\ \text{s}}{1\ \text{hr}} = \mathbf{6.74 \times 10^{-34}\ m}$

For the 80 mi/hr curveball, $\lambda = \frac{6.63 \times 10^{-34}\ \text{kg} \cdot \text{m}^2 \cdot \text{s}^{-1}}{(.0220\ \text{kg})\left(80\frac{\text{mi}}{\text{hr}}\right)} \times \frac{1\ \text{mi}}{1.609\ \text{km}} \times \frac{1\ \text{km}}{10^3\ \text{m}} \times \frac{3600\ \text{s}}{1\ \text{hr}} = \mathbf{8.42 \times 10^{-34}\ m}$

The uncertainty in the position of the 100 mi/hr fastball is ($\frac{1}{2}$)(6.74×10^{-34} m) = 3.37×10^{-34} m, and the uncertainty in the position of the 80 mi/hr curveball is ($\frac{1}{2}$)(8.42×10^{-34} m) = 4.21×10^{-34} m. Hence, the uncertainty in the position of the fastball is calculated to be less than that for the curveball. However, the uncertainty in the position of both the fastball and the curveball is insignificant. So this cannot be the reason why batters frequently miss a curveball.

Chapter 8: Periodic Trends

8.1 ***See Section 8.1 and Figure 8.1.***

The periodic table can be subdivided into sections based on the subshell being filled. The two columns on the left are the *s*-block elements, the six columns on the right are the *p*-block elements, the ten columns in the middle are the *d*-block elements, and the *f*-block elements are at the bottom. The period number gives the value of *n* for the *s* and *p* subshells, the period number minus one gives the value of *n* for the *d* subshells, and the period number minus two gives the value of *n* for the *f* subshells.

8.2 ***See Section 8.1.***

An isoelectronic series is a group of atoms and ions that have the same number of electrons. The species S^{2-}, Cl^-, Ar, K^+, and Ca^{2+} are isoelectronic - they all have 18 electrons and the electron configuration $1s^2\ 2s^2\ 2p^6\ 3s^2\ 3p^6$.

8.3 ***See Section 8.1.***

Valence electrons are electrons occupying the valence orbitals of an atom, those orbitals of the highest occupied principal level and those orbitals of partially filled subshells of lower principal quantum number. All of the other electrons of the atom are usually referred to as core electrons. Valence electrons are used in chemical bonding and reactivity, whereas core electrons are not directly involved in chemical bonding and reactivity.

8.4 ***See Sections 7.5, 8.1.***

The energy of the 4s subshell is lower than that of the 3d subshell because of the greater penetration by the 4*s* orbital electrons into the region occupied by the first and second principal shells.

8.5 ***See Sections 7.4, 8.1.***

The first shell can only consist of the ls subshell because the maximum value of ℓ for any given *n* is $(n - 1)$, in this case $\ell_{max} = 0$. Hence, the first shell can only contain two electrons, provided they are of opposite spins. On the other hand, the second shell can consist of the 2s and the 2p subshells, which can contain a total of eight electrons, two in the 2s and six in the 2p.

8.6 ***See Section 8.1 and Figure 8.1.***

The fourth period is the first to contain 18 elements because it is the first period in which the *d* orbitals are being filled. The subshells that are filled going across the fourth period are the 4s, 3d, and 4p.

8.7 ***See Sections 7.5, 8.1.***

The electrons in the 2s orbital are closer to the nucleus on the average in oxygen than beryllium because atoms of oxygen contain 8 protons at the nucleus, whereas atoms of beryllium only contain 4 protons.

8.8 ***See Section 8.2 and Figure 8.3.***

The atomic radius of fluorine can be taken to be one-half of the known F-F bond length, namely $\frac{1}{2}$(143 pm), and the atomic radius of chlorine can be taken to be one-half of the Cl-Cl bond length, namely $\frac{1}{2}$(198 pm). The radii of F and Cl can then be summed to predict a bond length of $\frac{1}{2}$(143 pm) + $\frac{1}{2}$(198 pm) = 170 pm for the bond length in ClF.

8.9	*See Section 8.2 and Figure 8.8.*

The drawing should show size decreasing across the periods and increasing from top to bottom within groups, as in Figure 8.8. Size decreases with increasing effective nuclear charge going across a period and increases with the addition of shells of higher *n* value going down a group. The latter occurs in spite of an increase in effective nuclear charge going down a group.

8.10	*See Sections 7.5, 8.2.*

Atoms of carbon have fewer protons and a lower effective nuclear charge than atoms of oxygen. Electrons occupying the same subshell occupy the same general region of space and do not shield each other effectively.

8.11	*See Section 8.2 and Figures 8.1, 8.8.*

Atoms of sulfur are larger than atoms of oxygen because the valence electrons of oxygen are located in the second shell and those of sulfur are located in the third shell.

8.12	*See Section 8.3.*

Ionization energy is defined as the energy that is required to remove an electon from a gaseous atom or ion in its electronic ground state. In the case of Li this is energy corresponding to: Li(g) → Li^+(g) + e^-.

Electron configurations: Li [He]$2s^1$ Li^+ [He]

8.13	*See Section 8.3 and Figure 8.13.*

The drawing should show a general increase in ionization energy from Li to Ne with discontinuities at B and O. Ionization energies generally increase as effective nuclear charge increases and size decreases across a period. The drop at B is due to the fact that the electron is being removed from the less penetrating and inherently higher energy 2p subshell rather than the 2s. The drop at O is due to the fact that pairing is occurring in the 2p subshell for the first time at O. The repulsion associated with the pairing of electrons in O offsets the increase in effective nuclear charge and decrease in size from N to O. However, the normal trend resumes from O to F to Ne because the increase in effective nuclear charge and decrease in size from O to F to Ne dominates the effects of pairing.

8.14	*See Section 8.3.*

The drop in ionization energy from Mg to Al is due to the fact that the electron of Al is being removed from the less penetrating and inherently higher energy 3p subshell rather than the 3s subshell.

8.15	*See Section 8.3.*

The electron that is removed from Na is its outermost $3s^1$ electron, whereas the electron that is removed from Li is its outermost $2s^1$ electron. The fact that the electron that is being removed from Na is in a subshell of higher *n* value and is further from the nucelus than that of Li is the reason why the first ionization energy of sodium is lower than of lithium. The reason it is only slightly lower is that the size increase from lithium to sodium is somewhat offset by an increase in effective nuclear charge from lithium to sodium.

8.16	*See Section 8.3.*

The increase in effective nuclear charge from Mn to Co is less than that from Ga to As leading to less change in ionization energy from Mn to Co. The effective nuclear charge changes less for the transition metals because the increase in the nuclear charge is almost cancelled by the shielding of the valence *ns* electrons by the added (*n* - 1)d electrons.

8.17 *See Section 8.3 and Example 8.7.*

An increase in ionization energy is to be expected as successive electrons are removed from an atom because there is an increase in excess nuclear charge (cation charge) with each electron removed. The large increase from the second to third ionization of magnesium is due to the fact that the removal of the third electron from magnesium ($1s^2\ 2s^2\ 2p^6\ 3s^2$) involves the removal of a core electron.

8.18 *See Section 8.3 and Example 8.7.*

An increase in ionization energy is to be expected as successive electrons are removed from an atom because there is an increase in excess nuclear charge (cation charge) with each electron removed. A large increase is to be expected when the fourth electron is removed from B ($1s^2\ 2s^2\ 2p^1$) because that would involve the removal of a core electron.

8.19 *See Section 8.3.*

The first ionization energy of magnesium is greater than that of sodium because there is an increase in effective nuclear charge and decrease in size going across a period in the periodic table. However, the electron that is removed from aluminum is in the less penetrating and inherently higher energy 3p subshell, so there is a drop in ionization energy from magnesium to aluminum.

8.20 *See Sections 8.2, 8.3.*

Each successive element in the same principal level has an increase in nuclear charge and an increase in effective nuclear charge because electrons added to the same subshell do not shield each other effectively. This causes the outermost electrons of silicon to be drawn closer to the nucleus than those of aluminum and to also be more difficult to remove during the ionization process.

8.21 *See Section 8.4.*

The electron affinity of an element is the energy change that accompanies the addition of an electron to a gaseous atom to form an anion.

8.22 *See Section 8.4.*

The members of the halogen group, Group VIIA (17) have the most favorable, most negative, electron affinities of members of their respective periods. This is due to the fact that effective nuclear charge increases and size decreases going across a period. The effective nuclear charge and size trends continue with the noble gases, but these elements have completed subshells and no tendency to acquire another electron.

8.23 *See Section 8.4 and Figure 8.17.*

Electron affinities generally become less favorable with increasing size going down a group and do so in spite of the increase in effective nuclear charge going down a group. However, the addition of an electron to the small members of the second period causes sufficient electron-electron repulsion to cause them to have less favorable electron affinity values than the corresponding group members of the third period. Hence, for most groups the electron affinity becomes more favorable from the second to third periods and then less favorable from the third to subsequent periods.

8.24 *See Sections 8.3, 8.4.*

The effective nuclear charge for Be is greater than that for Li, and Be is smaller than Li. This causes the first ionization energy for Be to be greater than that for Li and might lead us to predict that Be should also have the greater electron affinity. However, the added electron completes the 2s subshell for Li and must be added to the inherently less penetrating and higher energy 2p subshell of Be ($1s^2$ $2s^2$) causing Li to have a greater electron affinity than Be. Indeed, the electron affinity of Li is exothermic and that of Be is endothermic.

8.25 *See Section 8.5.*

The metallic elements in Group IA are soft, silver-colored metals (Cs is golden) that have low melting points and low ionization energies. They emit light of a characteristic color when their compounds are placed in a flame.

8.26 *See Section 8.5.*

The alkaline earth metals have a silvery-white appearance and are harder than the alkali metals but softer than most other metals. They also have higher melting points and ionization energies than the alkali metals but lower than those of most other metals. In addition, they emit light of a characteristic color when their compounds are placed in a flame.

8.27 *See Section 8.5.*

The halogens exist as diatomic molecules. Under standard conditions, F_2 is a pale light yellow gas, chlorine a deeper greenish-yellow gas, bromine a deep red liquid, and iodine a shiny violet-black solid.

8.28 *See Section 8.5.*

(a) Reactivity increases going down Group IA(1).
(b) Reactivity increases going down Group IIA (2).
(c) Reactivity decreases going down Group VIIA (17).

8.29 *See Section 8.5 and Figure 8.18.*

Compounds containing lithium and sodium can be put in fireworks to give red and yellow colors, respectively.

8.30 *See Section 8.5.*

Fluorine is the most reactive of all the nonmetals. This means it is easily reduced and difficult to obtain as a free or uncombined element by oxidation.

8.31 *See Section 8.1, Figure 8.1, Table 8.1, and Example 8.1.*

(a) Ge $[Ar]4s^2 3s^{10} 4p^2$ [Ar] ↑↓ (4s) ↑↓ ↑↓ ↑↓ ↑↓ ↑↓ (3d) ↑ ↑ _ (4p)

(b) S $[Ne]3s^2 3p^4$ [Ne] ↑↓ (3s) ↑↓ ↑ ↑ (3p)

(c) Rb $[Kr]5s^1$ [Kr] ↑ (5s)

8.32 *See Section 8.1, Figure 8.1, Table 8.1, and Example 8.1.*

(a) Ar [Ne]$3s^23p^6$ [Ne] ↑↓ (3s) ↑↓ ↑↓ ↑↓ (3p)

(b) Br [Ar]$4s^23d^{10}4p^5$ [Ar] ↑↓ (4s) ↑↓ ↑↓ ↑↓ ↑↓ ↑↓ (3d) ↑↓ ↑↓ ↑ (4p)

(c) K [Ar]$4s^1$ [Ar] ↑ (4s)

8.33 *See Section 8.1, Figure 8.1, Table 8.1, and Example 8.1.*

(a) C [He]$2s^22p^2$ [He] ↑↓ (2s) ↑ ↑ _ (2p)

(b) Cl [Ne]$3s^23p^5$ [Ne] ↑↓ (3s) ↑↓ ↑↓ ↑ (3p)

(c) Na [Ne]$3s^1$ [Ne] ↑ (3s)

8.34 *See Section 8.1, Figure 8.1, Table 8.1, and Example 8.1.*

(a) Mg [Ne]$3s^2$ [Ne] ↑↓ (3s)

(b) Si [Ne]$3s^23p^2$ [Ne] ↑↓ (3s) ↑ ↑ _ (3p)

(c) B [He]$2s^22p^1$ [He] ↑↓ (2s) ↑ _ _ (2p)

8.35 *See Section 8.1 and Figure 8.1.*

(a) The element which has 13 total electrons and is the first element in the p-block in the third period ($3p^1$) is Al.

(b) The element which has 11 total electrons and is the first element in the s-block in the third period ($3s^1$) is Na.

(c) The element which has 25 total electrons and is the fifth element in the 3d-block ($3d^5$) is Mn.

8.36 *See Section 8.1 and Figure 8.1.*

(a) The element which has 15 total electrons and is the third element in the p-block in the third period ($3p^3$) is P.

(b) The element which has 12 total electrons and is the second element in the s-block in the third period ($3s^2$) is Mg.

(c) The element which has 27 total electrons and is the seventh element in the 3d-block ($3d^7$) is Co.

8.37	*See Section 8.1 and Figure 8.1.*

(a) The element which has 14 total electrons and is the second element in the p-block in the third period ($3p^2$) is Si.

(b) The element which has 6 total electrons and is the second element in the p-block in the second period ($2p^2$) is C.

(c) The element which has 14 total electrons and is the first element in the s-block in the fourth period ($4s^1$) is K.

8.38	*See Section 8.1 and Figure 8.1.*

(a) The element which has 7 total electrons and is the third element in the p-block in the second period ($2p^3$) is N.

(b) The element which has 18 total electrons and is the sixth element in the p-block in the third period ($3p^6$) is Ar.

(c) The element which has 28 total electrons and is the eigth element in the 3d-block ($3d^8$) is Ni.

8.39	*See Section 8.1 and Figure 8.1.*

The element which is a member of the halogen group (VIIA, 17) and a member of the fourth period is Br.

8.40	*See Section 8.1 and Figure 8.1.*

The element which is a member of the alkali metal group (IA, 1) and a member of the fifth period is Rb.

8.41	*See Section 8.1 and Figure 8.1.*

The element which is a member of the alkaline earth group (IIA, 2) and a member of the third period is Mg.

8.42	*See Section 8.1 and Figure 8.1.*

The element which is a member of the oxygen group (VIA, 16) and a member of the fifth period is Te.

8.43	*See Section 8.1, Figure 8.1, Table 8.1, and Example 8.1.*

(a) Ca [Ar]$4s^2$ — [Ar] ↑↓ (4s)

(b) Tc [Kr]$5s^2 4d^5$ — [Kr] ↑↓ (5s) ↑ ↑ ↑ ↑ ↑ (4d)

(c) In [Kr]$5s^2 4d^{10} 5p^1$ — [Kr] ↑↓ (5s) ↑↓ ↑↓ ↑↓ ↑↓ ↑↓ (4d) ↑ _ _ (5p)

8.44	*See Section 8.1, Figure 8.1, Table 8.1, and Example 8.1.*

(a) Si [Ne]$3s^2 3p^2$ — [Ne] ↑↓ (3s) ↑ ↑ _ (3p)

(b) Ni [Ar]$4s^2 3d^8$ — [Ar] ↑↓ (4s) ↑↓ ↑↓ ↑↓ ↑ ↑ (3d)

(c) Sr $[Kr]5s^2$ [Kr] ↑↓
5s

8.45 *See Section 8.1, Figure 8.1, Table 8.1, and Example 8.1.*

(a) As $[Ar]4s^23d^{10}4p^3$ [Ar] ↑↓ ↑↓ ↑↓ ↑↓ ↑↓ ↑↓ ↑ ↑ ↑
4s 3d 4p

(b) Fe $[Ar]4s^23d^6$ [Ar] ↑↓ ↑↓ ↑ ↑ ↑ ↑
4s 3d

(c) Ba $[Xe]6s^2$ [Xe] ↑↓
6s

8.46 *See Section 8.1, Figure 8.1, Table 8.1, and Example 8.1.*

(a) O $[He]2s^22p^4$ [He] ↑↓ ↑↓ ↑ ↑
2s 2p

(b) V $[Ar]4s^23d^3$ [Ar] ↑↓ ↑ ↑ ↑ __ __
4s 3d

(c) I $[Kr]5s^24d^{10}5p^5$ [Kr] ↑↓ ↑↓ ↑↓ ↑↓ ↑↓ ↑↓ ↑↓ ↑↓ ↑
5s 4d 5p

8.47 *See Section 8.1, Figure 8.1, Table 8.1, and Example 8.1.*

(a) Y $[Kr]5s^24d^1$ [Kr] ↑↓ ↑ __ __ __ __ 1 unpaired
5s 4d

(b) Se $[Ar]5s^24d^{10}5p^4$ [Ar] ↑↓ ↑↓ ↑↓ ↑↓ ↑↓ ↑↓ ↑↓ ↑ ↑ 2 unpaired
5s 4d 5p

(c) Cd $[Kr]5s^24d^{10}$ [Kr] ↑↓ ↑↓ ↑↓ ↑↓ ↑↓ ↑↓ 0 unpaired
5s 4d

8.48 *See Section 8.1, Figure 8.1, Table 8.1, and Example 8.1.*

(a) Fe $[Ar]4s^23d^6$ [Ar] ↑↓ ↑↓ ↑ ↑ ↑ ↑ 4 unpaired
4s 3d

(b) S $[Ne]3s^23p^4$ [Ne] ↑↓ ↑↓ ↑ ↑ 2 unpaired
3s 3p

(c) Tl $[Xe]6s^24f^{14}5d^{10}6p^1$
[Xe] ↑↓ ↑↓ ↑↓ ↑↓ ↑↓ ↑↓ ↑↓ ↑↓ ↑↓ ↑↓ ↑↓ ↑↓ ↑↓ ↑ 1 unpaired
6s 4f 5d 6p

8.49 *See Section 8.1, Figure 8.1, Table 8.1, and Example 8.1.*

(a) Sc $[Ar]4s^23d^1$ [Ar] ↑↓ ↑ __ __ __ __ 1 unpaired
4s 3d

(b) F $[He]2s^2 2p^5$ — [He] ↑↓ (2s) ↑↓ ↑↓ ↑ (2p) — 1 unpaired

(c) Ti $[Ar]4s^2 3d^2$ — [Ar] ↑↓ (4s) ↑ ↑ _ _ _ (3d) — 2 unpaired

8.50 ***See Section 8.1, Figure 8.1, Table 8.1, and Example 8.1.***

(a) Be $[He]2s^2$ — [He] ↑↓ (2s) — 0 unpaired

(b) He $1s^2$ — ↑↓ (1s) — 0 unpaired

(c) Zr $[Kr]5s^2 4d^2$ — [Kr] ↑↓ (5s) ↑ ↑ _ _ _ (4d) — 2 unpaired

8.51 ***See Section 8.1, Figure 8.1, Table 8.1, and Examples 8.1, 8.2.***

(a) Al $[Ne]3s^2 3p^1$ Valence electrons $3s^2 3p^1$

(b) Cs $[Xe]6s^1$ Valence electron $6s^1$

(c) As $[Ar]4s^2 3d^{10} 4p^3$ Valence electrons $4s^2 4p^3$

8.52 ***See Section 8.1, Figure 8.1, Table 8.1, and Examples 8.1, 8.2.***

(a) Zr $[Kr]5s^2 4d^2$ Valence electrons $5s^2 4d^2$

(b) Ca $[Ar]4s^2$ Valence electrons $4s^2$

(c) I $[Kr]5s^2 4d^{10} 5p^5$ Valence electrons $5s^2 5p^5$

8.53 ***See Section 8.1, Figure 8.1, Table 8.1, and Examples 8.1, 8.2.***

(a) Be $[He]2s^2$ Valence electrons $2s^2$

(b) Br $[Ar]4s^2 3d^{10} 4p^5$ Valence electrons $4s^2 4p^5$

(c) In $[Kr]5s^2 4d^{10} 5p^1$ Valence electrons $5s^2 5p^1$

8.54 ***See Section 8.1, Figure 8.1, Table 8.1, and Examples 8.1, 8.2.***

(a) Co $[Ar]4s^2 3d^7$ Valence electrons $4s^2 3d^7$

(b) Pb $[Xe]6s^2 4f^{14} 5d^{10} 6p^2$ Valence electrons $6s^2 6p^2$

(c) S $[Ne]3s^2 3p^4$ Valence electrons $3s^2 3p^4$

8.55 ***See Section 8.1 and Figure 8.1.***

A valence shell configuration of ns^2 corresponds to Group IIA (2).

8.56 *See Section 8.1 and Figure 8.1.*

A valence shell configuration of ns^2np^4 corresponds to Group VIA (16).

8.57 *See Section 8.1 and Figure 8.1.*

(a) Group IA, ns^1 (b) Group IVA, ns^2np^2 (c) Group VIIA, ns^2np^5

8.58 *See Section 8.1 and Figure 8.1.*

(a) Group IIA, ns^2 (b) Group IIIA, ns^2np^1 (c) Group VIA, ns^2np^4

8.59 *See Section 8.1, Figure 8.1 and Example 8.3.*

(a) S^{2-} has 18 electrons, two more than S: $[Ne]3s^23p^6$ or [Ar].

(b) Mn^{2+} has 23 electrons, two fewer than Mn: $[Ar]3d^5$.
Note: The electrons of highest *n* value, the $4s^2$ electrons, are removed first in forming Mn^{2+} from Mn.

(c) Ge^{2+} has 30 electrons, two fewer than Ge: $[Ar]4s^23d^{10}$.
Note: The electrons of highest ℓ value, the $4p^2$ electrons, are removed in forming Ge^{2+} from Ge, not the $4s^2$ electrons.

8.60 *See Section 8.1, Figure 8.1 and Example 8.3.*

(a) Y^{3+} has 36 electrons, three fewer than Y: [Kr].

(b) Br^- has 36 electrons, one more than Br: $[Ar]4s^23d^{10}4p^6$ or [Kr].

(c) Rh^{2+} has 43 electrons, two fewer than Rh: $[Kr]4d^7$.
Note: The electron of highest *n* value, the $5s^1$ electron, is removed first in forming Rh^{2+} from Rh.

8.61 *See Section 8.1, Figure 8.1 and Example 8.3.*

(a) P^{2-} has 17 electrons, two more than P: $[Ne]3s^23p^5$.

(b) Fe^{2+} has 24 electrons, two fewer than Fe: $[Ar]3d^6$.
Note: The electrons of highest *n* value, the $4s^2$ electrons, are removed in forming Fe^{2+} from Fe.

(c) Co^{3+} has 24 electrons, three fewer than Co: $[Ar]3d^6$.
Note: The electrons of highest *n* value, the $4s^2$ electrons, are removed in forming Co^{3+} from Co.

8.62 *See Section 8.1, Figure 8.1 and Example 8.3.*

(a) Ga^{2+} has 29 electrons, two fewer than Ga: $[Ar]4s^13d^{10}$
Note: The electron of highest ℓ value, the $4p^1$ electron, is removed first in forming Ga^{2+} from Ga. Then one of the $4s^2$ electrons is removed because it has a higher *n* value than a $3d^{10}$ electron.

(b) Se^{2-} has 36 electrons, two more than Se: $[Ar]4s^23d^{10}4p^6$ or [Kr].

(c) Ru^{2+} has 42 electrons, two fewer than Ru: $[Kr]4d^6$.
Note: The electron of highest *n* value, the $5s^1$ electron, is removed first in forming Ru^{2+} from Ru.

8.63 *See Section 8.1, Figure 8.l, Table 8.1, and Example 8.3.*

(a) The 1+ cation would be formed from an element having a total of 14 electrons and a $3s^23p^2$ valence configuration. The cation is Si^+.

(b) The 1+ cation would be formed from an element having a total of 12 electrons and a $3s^2$ valence configuration. The cation is Mg^+.

(c) The 1+ cation would be formed from an element having a total of 31 electrons and a $4s^24p^1$ valence configuration. The cation is Ga^+.

8.64 *See Section 8.1, Figure 8.l, Table 8.1, and Example 8.3.*

(a) The 2- anion would be formed from an element having a total of 34 electrons and a $4s^24p^4$ valence configuration. The anion is Se^{2-}.

(b) The 2- anion would be formed from an element having a total of 7 electrons and a $2s^22p^3$ valence configuration. The anion is N^{2-}.

(c) The 2- anion would be formed from an element having a total of 16 electrons and a $3s^23p^4$ valence configuration. The anion is S^{2-}.

8.65 *See Section 8.1, Figure 8.l, Table 8.1, and Example 8.3.*

(a) The 1- anion would be formed from an element having a total of 17 electrons and a $3s^23p^5$ valence configuration. The anion is Cl^-.

(b) The 1- anion would be formed from an element having a total of 8 electrons and a $2s^22p^4$ valence configuration. The anion is O^-.

(c) The 1- anion would be formed from an element having a total of 34 electrons and a $4s^24p^4$ valence configuration. The anion is Se^-.

8.66 *See Section 8.1, Figure 8.l, Table 8.1, and Example 8.3.*

(a) The 2+ cation would be formed from an element having a total of 33 electrons and a $4s^24p^3$ valence configuration. The cation is As^{2+}.

(b) The 2+ cation would be formed from an element having a total of 7 electrons and a $2s^22p^3$ valence configuration. The cation is N^{2+}.

(c) The 2+ cation would be formed from an element having a total of 15 electrons and a $3s^23p^3$ valence configuration. The cation is P^{2+}.

8.67 *See Section 8.1, Figure 8.1, Table 8.1, and Examples 8.1, 8.3.*

(a) Y^{3+} [Kr] 0 unpaired electrons

(b) Ni^{2+} $[Ar]3d^8$ [Ar] ↑↓ ↑↓ ↑↓ ↑ ↑ (3d) 2 unpaired electrons

(c) Cl^- [Ar] 0 unpaired electrons

8.68 See Section 8.1, Figure 8.1, Table 8.1, and Examples 8.1, 8.3.

(a) Co^{3+} [Ar]$3d^6$ [Ar] ↑↓ ↑ ↑ ↑ ↑ (3d) 4 unpaired electrons

(b) Sn^{2+} [Kr]$5s^24d^{10}$ 0 unpaired electrons

(c) Ru^{2+} [Kr]$4d^6$ [Kr] ↑↓ ↑ ↑ ↑ ↑ (4d) 4 unpaired electrons

8.69 See Section 8.1, Figure 8.1, Table 8.1, and Examples 8.1, 8.3.

(a) Co^{2+} [Ar]$3d^7$ [Ar] ↑↓ ↑↓ ↑ ↑ ↑ (3d) 3 unpaired electrons

(b) Ca^{2+} [Ar] 0 unpaired electrons

(c) S^- [Ne]$3s^23p^5$ [Ne] ↑↓ (3s) ↑↓ ↑↓ ↑ (3p) 1 unpaired electron

8.70 See Section 8.1, Figure 8.1, Table 8.1, and Examples 8.1, 8.3.

(a) V^{3+} [Ar]$3d^2$ [Ar] ↑ ↑ _ _ _ (3d) 2 unpaired electrons

(b) Sb^{2+} [Kr]$5s^24d^{10}5p^1$ 1 unpaired electron

(c) Pd^{2+} [Kr]$4d^8$ [Kr] ↑↓ ↑↓ ↑↓ ↑ ↑ (4d) 2 unpaired electrons

8.71 See Section 8.1, Figure 8.1, Table 8.1, and Example 8.3.

(a) Fe^{3+} [Ar]$3d^5$ (b) Cr^{3+} [Ar]$3d^3$

Note: The electrons of highest *n* value, the 4s electrons, are removed first in forming these cations from their respective neutral atoms.

8.72 See Section 8.1, Figure 8.1, Table 8.1, and Example 8.3.

The 3+ cation would be formed from an element having a total of 44 electrons. The element is Ru.

8.73 See Section 8.1 and Example 8.4.

Se has 34 electrons and the electron configuration [Ar]$4s^23d^{10}4p^4$. As^- and Br^+ would have the same number of electrons as Se and also have identical electron configurations.

8.74 See Section 8.1 and Example 8.4.

Kr has 36 electrons. Se^{2-}, Br^-, Rb^+, and Sr^{2+} have the same number of electrons as Kr and also have the same electron configuration as Kr.

8.75 See Section 8.1, Table 8.1, and Examples 8.1, 8.3.

Li^{3+}, B^{3+}, and N^{3+} would have no unpaired electrons. Li^{3+} would have zero electrons, B^{3+} would have $1s^2$ and N^{3+} would have $1s^22s^2$.

8.76 ***See Section 8.1, Table 8.1, and Examples 8.1, 8.3.***

Only nonmetals commonly form anions, and only O^{2-} from the second period would have no unpaired electrons as an anion with a 2^- charge. O^{2-} is $[He]2s^22p^6$ or [Ne].

8.77 ***See Section 8.1, Table 8.1, and Examples 8.1, 8.3.***

The maximum number of unpaired electrons that is possible for the d orbitals is 5. The transition metal in the fourth period which forms a 2+ ion having 5 unpaired electrons is Mn.

Mn^{2+} $[Ar]3d^5$ ↑ ↑ ↑ ↑ ↑
3d

8.78 ***See Section 8.1, Table 8.1, and Examples 8.1, 8.3.***

The maximum number of unpaired electrons that is possible for the d orbitals is 5. The transition metal in the fourth period which forms a 3+ ion having 5 unpaired electrons is Fe.

Fe^{3+} $[Ar]3d^5$ ↑ ↑ ↑ ↑ ↑
3d

8.79 ***See Section 8.2, Figures 8.4, 8.5, 8.6, 8.7, and Example 8.4.***

(a) Na is larger than Na^+. Cations are normally smaller than their corresponding neutral atoms, because the same numbers of protons are exerting an attractive force on fewer electrons and there are less electron-electron repulsions in the cations. However, in this case it is also important to realize that the formation of Na^+ from Na involves the removal of the only electron in the outermost shell of Na. Hence, Na^+ is smaller than Na.

(b) O^{2-} is larger than F^-. O^{2-} and F^- both have 10 electrons and are therefore isoelectronic. O^{2-} has fewer protons attracting the same number of electrons as F^- and therefore has a lower effective nuclear charge.

(c) Ni^{2+} is larger than Ni^{3+}. Ni^{3+} has the same number of protons exerting an attractive force on fewer electrons and also has less electron-electron repulsions. Hence, Ni^{3+} has a higher effective nuclear charge than Ni^{2+} and is smaller than Ni^{2+}.

8.80 ***See Section 8.2, Figures 8.4, 8.5, 8.6, 8.7, and Example 8.4.***

(a) Se^{2-} is larger than Br^-. Se^{2-} and Br^- both have 36 electrons and are therefore isoelectronic. Se^{2-} has fewer protons attracting the same number of electrons as Br^- and therefore has a lower effective nuclear charge.

(b) Ru^{2+} is larger than Ru^{3+}. Ru^{3+} has the same number of protons exerting an attractive force on fewer electrons and also has less electron-electron repulsions. Hence, Ru^{3+} has a higher effective nuclear charge than Ru^{2+} and is smaller than Ru^{2+}.

(c) Mg is larger than Mg^{2+}. Cations are normally smaller than their corresponding neutral atoms, because the same numbers of protons are exerting an attractive force on fewer electrons and there are less electron-electron repulsions in the cations. However, in this case it is also important to realize that the formation of Mg^{+2} from Mg involves the removal of the only electrons in the outermost shell of Mg. Hence, Mg^{2+} is smaller than Mg.

8.81 ***See Section 8.2, Figures 8.4, 8.5, 8.6, 8.7,8.8, and Examples 8.4, 8.5.***

(a) Na > Mg Size generally decreases as effective nuclear charge increases from left to right in a given period.

(b) B > O Size generally decreases as effective nuclear charge increases from left to right in a given period.

(c) $Be^{2+} > Be^{3+}$ Be^{3+} has the same number of protons exerting an attractive force on fewer electrons and also has less electron-electron repulsions. Hence, Be^{3+} has a higher effective nuclear charge than Be^{2+} and is smaller than Be^{2+}.

8.82 ***See Section 8.2, Figures 8.4, 8.5, 8.6, 8.7, 8.8, and Examples 8.4, 8.5.***

(a) Na > Li Size generally increases as more shells are added going down a group in spite of the increase in effective nuclear charge.

(b) P > O Size generally decreases as effective nuclear charge increases from left to right in a given period and generally increases as more shells are added going down a given group. The increase in size associated with going down groups usually dominates the decrease in size associated with going across periods when the atoms are relatively close together in the periodic table.

(c) $Rb > Rb^{+}$ Cations are normally smaller than their corresponding neutral atoms, because the same numbers of protons are exerting an attractive force on fewer electrons and there are less electron-electron repulsions in the cations. However, in this case it is also important to realize that the formation of Rb^{+} from Rb involves the removal of the only electron in the outermost shell of Rb. Hence, Rb^{+} is smaller than Rb.

8.83 ***See Section 8.2, Figure 8.8, and Example 8.5.***

(a) O < B < Li Size generally decreases as effective nuclear charge increases from left to right in a given period.

(b) N < C < Si Size generally decreases as effective nuclear charge increases from left to right in a given period and generally increases as more shells are added going down a given group.

(c) S < As < Sn Size generally decreases as effective nuclear charge increases from left to right in a given period and generally increases as more shells are added going down a given group. The increase in size associated with going down groups usually dominates the decrease in size associated with going across periods when the atoms are relatively close together in the periodic table.

8.84 ***See Section 8.2, Figure 8.8, and Example 8.5.***

(a) Be < Li < Na Size generally decreases as effective nuclear charge increases from left to right in a given period and generally increases as more shells are added going down a given group.

(b) F < N < P Size generally decreases as effective nuclear charge increases from left to right in a given period and generally increases as more shells are added going down a given group.

(c) O < I < Sn Size generally decreases as effective nuclear charge increases from left to right in a given period and generally increases as more shells are added going down a given group.

8.85 ***See Section 8.2, Figures 8.4, 8.8, and Examples 8.4, 8.5.***

(a) Be^{2+} < Be < Li Cations are normally smaller than their corresponding neutral atoms, because the same numbers of protons are exerting an attractive force on fewer electrons and there are less electron-electron repulsions in the cations. In addition, size generally decreases as effective nuclear charge increases from left to right in a given period.

(b) Cl < S < S^{2-} Size generally decreases as effective charge increases from left to right in a given period. In addition, anions are normally larger than their corresponding neutral atoms because the same number of protons are exerting a less effective attractive force on more electrons and there are more electron-electron repulsions in anions.

(c) O < S < Si   Size generally increases as more shells are added going down a given group and generally decreases as effective nuclear charge increases from left to right in a given period.

8.86	*See Section 8.2, Figures 8.4, 8.8, and Examples 8.4, 8.5.*

(a) $F < F^- < O^{2-}$ Anions are normally larger than their corresponding neutral atoms because the same number of protons are exerting a less effective attractive force on more electrons and there are more electron-electron repulsions in anions. In addition, O^{2-} and F^- both have 10 electrons and are therefore isoelectronic. O^{2-} has fewer protons attracting the same number of electrons as F^- and therefore has a lower effective nuclear charge.

(b) Al^{3+} < Mg < Na   Size generally decreases as effective nuclear charge increases from left to right in a given period and generally increases as more shells are added going down a given group. In addition, cations are normally smaller than their corresponding neutral atoms, because the same numbers of protons are exerting an attractive force on fewer electrons and there are less electron-electron repulsions in the cations.

(c) N < P < Si   Size generally increases as more shells are added going down a group. In addition, size generally decreases as effective nuclear charge increases from left to right in a given period

8.87	*See Section 8.2, Figure 8.7, and Example 8.5.*

Li has the largest atomic radius of elements in the second period. Size generally decreases as effective nuclear charge increases from left to right in a given period.

8.88	*See Section 8.2, Figure 8.7, and Example 8.5.*

C has the smallest atomic radius of elements in Group IVA. Size generally increases as more shells are added going down a group in spite of the increase in effective nuclear charge.

8.89	*See Section 8.2, Figure 8.7, and Example 8.5.*

The element having the electron configuration $1s^22s^22p^63s^23p^3$ (P) is larger than the element having the electron configuration $1s^22s^22p^5$ (Cl). Size generally decreases as effective nuclear charge increases from left to right in a given period.

8.90	*See Section 8.2, Figure 8.7, and Example 8.5.*

The element having the electron configuration $1s^22s^22p^2$ (C) is larger than the element having the electron configuration $1s^22s^22p^4$ (O). Size generally decreases as effective nuclear charge increases from left to right in a given period.

8.91	*See Section 8.3, Figures 8.13, 8.14, and Example 8.6.*

(a) Cl has a higher ionization energy than Si. Cl has a higher effective nuclear charge than Si and a smaller size than Si. Hence, Cl exerts a greater attractive force on its outer electrons than Si.

(b) Na has a higher ionization energy than Rb. The electron that is removed from Rb is in an orbital that is larger in size and therefore further from the nucleus. This increase in size dominates the increase in effective nuclear charge that occurs from Na to Rb.

(c) F^- would have a higher ionization energy than O^{2-}. F^- and O^{2-} are isoelectronic. However, F^- has more protons attracting the same number of electrons and therefore has a higher effective nuclear charge and smaller size than O^{2-}.

8.92 *See Section 8.3, Figures 8.13, 8.14, and Example 8.6.*

(a) F has a higher ionization energy than N. F has a higher effective nuclear charge than N and a smaller size than N. Hence, F exerts a greater attractive force on its outer electrons than N.

(b) Mg^{2+} would have a higher ionization energy than Na^+. Mg^{2+} and Na^+ are isoelectonic. However, Mg^{2+} has more protons attracting the same number of electrons and therefore has a higher effective nuclear charge and smaller size than Na^+.

(c) Si has a higher ionization energy than K. Ionization energies generally increase with increasing effective nuclear charge and decreasing size going across a given period. In addition, ionization energies generally decrease with increasing size going down a group. Hence, Si has a higher ionization energy than K for several reasons.

8.93 *See Section 8.3, Figures 8.12, 8.13, and Example 8.6.*

(a) Cl has a higher ionization energy than Ge. Ionization energies generally increase with increasing effective nuclear charge and decreasing size going across a given period. Ionization energies generally decrease with increasing size going down a given group. Hence, Cl has a higher ionization energy than Ge for several reasons.

(b) F has a higher ionization energy than B. Ionization energies generally increase with increasing effective nuclear charge and decreasing size going across a given period.

(c) Al^{3+} has a higher ionization energy than Na^+. Al^{3+} and Na^+ are isoelectronic. However, Al^{3+} has more protons attracting the same number of electrons and therefore has a higher effective nuclear charge and smaller size than Na^+.

8.94 *See Section 8.3, Figures 8.12, 8.13, and Example 8.6.*

(a) I has a higher ionization energy than K. Ionization energies generally increase with increasing effective nuclear charge and decreasing size going across a given period. In addition, ionization energies decrease with increasing size going down a given group in the periodic table. The increase from K to Br dominates the decrease from Br to I, so I has a higher ionization energy than K.

(b) Al^+ has a higher ionization than Al. Al^+ has the same number of protons as Al attracting one less electron and therefore has a higher effective nuclear charge and smaller size than Al.

(c) Ar has a higher ionization energy than Cl^-. Ar and Cl^- are isoelectronic. However, Ar has one more proton attracting the same number of electrons and therefore has a higher effective nuclear charge and smaller size than Cl^-

8.95 *See Section 8.3, Figures 8.12, 8.13, and Example 8.6.*

(a) $O^{2-} < O < F$ This order is one of increasing effective nuclear charge and decreasing size.

(b) $Si < C < N$ This order is one of increasing effective nuclear charge and decreasing size.
Note: See solution for 8.83b.

(c) $Sr < Ru < Te$ This order is one of increasing effective nuclear charge and decreasing size.

8.96 *See Section 8.3, Figures 8.12, 8.13, and Example 8.6.*

(a) $Se^{2-} < S < O$ This order is one of decreasing size which is the dominate factor influencing ionization energies within a given group.

(b) Fe < Br < F	This order is one of decreasing size.

(c) Cl^- < Cl < F	This order is one of decreasing size.

8.97 ***See Section 8.3, Figures 8.12, 8.13, and Example 8.6.***

(a) N^{3-} < N < Ne	This order is one of increasing effective nuclear charge and decreasing size.

(b) Si < P < Cl	This order is one of increasing effective nuclear charge and decreasing size

(c) Ga < Se < O	This order is one of decreasing size.

8.98 ***See Section 8.3, Figures 8.12, 8.13, and Example 8.6.***

(a) K < Se < S	This order is one of decreasing size.

(b) Cr < As < O	This order is one of decreasing size.

(c) O^- < O < F	This order is one of decreasing size.

8.99 ***See Section 8.3, Figures 8.12, 8.13, and Example 8.6.***

Al^+ has a higher effective nuclear charge and smaller size than Mg^+ and therefore has a higher ionization energy.
Note: This comparison is analogous to the comparison of the first ionization energies of Na and Mg.

8.100 ***See Section 8.3, Figures 8.12, 8.13, and Example 8.6.***

Cl^+ has a higher effective nuclear charge and smaller size than P^+ and therefore has a higher ionization energy.
Note: This comparison is analogous to the comparison of the first ionization energies of Si and S.

8.101 ***See Section 8.3, Figures 8.12, 8.13, and Examples 8.6, 8.7.***

Although Al^{2+} has a higher effective nuclear charge and smaller size than Mg^{2+}, Mg^{2+} has a higher third ionization energy than Al^{2+}. This is due to the fact that removing the third electron from magnesium involves removing a core electron from the 2p subshell, and it is more difficult to remove core electrons than valence electrons.

8.102 ***See Section 8.2 and Figure 8.8.***

Although removing the second electron from O and the third electron from F both involve removing an electron from a $[He]2s^22p^3$ electron configuration, F^{2+} has a higher ionization energy than O^+. This is due to the fact that F^{2+} has a higher effective nuclear charge and smaller size than O^+.

8.103 ***See Section 8.4 and Figure 8.17.***

(a) Cl has a more favorable (more negative) electron affinity than S. Effective nuclear charge increases and size decreases from left to right in the periodic table causing Cl to exert a greater attractive force on an added electron than S.

(b) O has a more favorable (more negative) electron affinity than N. Effective nuclear charge increases and size decreases from left to right in the periodic table causing O to exert a greater attractive force on an added electron than N.

(c) F has a more favorable (more negative) electron affinity than S because F is a much smaller atom than S.

8.104 *See Section 8.4 and Figure 8.17.*

(a) Cl has a more favorable (more negative) electron affinity than S. Effective nuclear charge increases and size decreases from left to right in the periodic table causing Cl to exert a greater attractive force on an added electron than S.

(b) S has a more favorable (more negative) electron affinity than P. Effective nuclear charge increases and size decreases from left to right in the periodic table causing S to exert a greater attractive force on an added electron than P.

(c) Br has a more favorable (more negative) electron affinity than As. Effective nuclear charge increases and size decreases from left to right in the periodic table causing Br to exert a greater attractive force on an added electron than As.

8.105 *See Section 8.4 and Figure 8.17.*

(a) F has a more favorable (more negative) electron affinity than O. Effective nuclear charge increases and size decreases from left to right in the periodic table causing F to exert a greater attractive force on an added electron than O.

(b) Cl has a more favorable (more negative) electron affinity than P. Effective nuclear charge increases and size decreases from left to right in the periodic table causing Cl to exert a greater attractive force on an added electron than P.

(c) Br has a more favorable (more negative) electron affinity than Se. Effective nuclear charge increases and size decreases from left to right in the periodic table causing Br to exert a greater attractive force on an added electron than Se.

8.106 *See Section 8.4 and Figure 8.17.*

(a) Br has a more favorable (more negative) electron affinity than Te because Br is a much smaller atom than Te.

(b) O has a more favorable (more negative) electron affinity than B. Effective nuclear charge increases and size decreases from left to right in the periodic table causing O to exert a greater attractive force on an added electron than B.

(c) Se has a more favorable (more negative) electron affinity than In. Effective nuclear charge increases and size decreases from left to right in the periodic table causing Se to exert a greater attractive force on an added electron than In.

8.107 *See Section 8.5.*

(a) $2Na(s) + O_2(g) \rightarrow Na_2O_2(s)$ (principal product)

(b) $Na(s) + N_2(g) \rightarrow$ no reaction

(c) $2Na(s) + Cl_2(g) \rightarrow 2NaCl(s)$

(d) $2Na(s) + 2H_2O(\ell) \rightarrow 2NaOH(aq) + H_2(g)$

8.108 *See Section 8.5.*

(a) $4Li(s) + O_2(g) \rightarrow 2Li_2O(s)$

(b) $6Li(s) + N_2(g) \rightarrow 2Li_3N(s)$
(c) $2Li(s) + Cl_2(g) \rightarrow 2LiCl(s)$
(d) $2Li(s) + 2\ H_2O\ (\ell\ ) \rightarrow 2LiOH(aq) + H_2(g)$

8.109 ***See Section 8.5.***

(a) $Ba(s) + O_2(g) \rightarrow BaO_2(s)$

(b) $Ba(s) + 2H_2O(\ell\ ) \rightarrow Ba(OH)_2(aq) + H_2(g)$

8.110 ***See Section 8.5.***

(a) $2Ca(s) + O_2(g) \rightarrow 2CaO(s)$

(b) $Ca(s) + 2H_2O(\ell\ ) \rightarrow Ca(OH)_2(aq) + H_2(g)$

8.111 ***See Section 8.2, 8.3, 8.4.***

Going from potassium to calcium size decreases, ionization energy increases and electron afinity becomes less favorable. The latter is due to the fact that the electrons of calcium occupy completed subshells.

8.112 ***See Section 8.1, Figure 8.1, and Example 8.3.***

Ti has 22 electrons and is [Ar] $4s^2 3d^2$. Hence, the energy level diagram for Ti is:

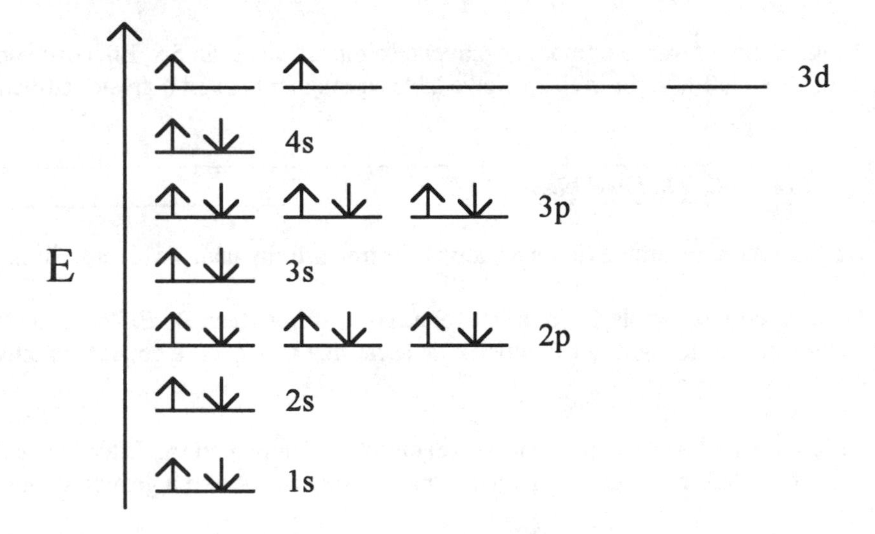

8.113 ***See Section 8.1, Figure 8.1, and Example 8.3.***

Ca^{2+} has 18 electrons, two fewer than Ca, and is [Ar]. Hence, the energy level diagram for Ca^{2+} is:

E

↑↓ ↑↓ ↑↓ 3p

↑↓ 3s

↑↓ ↑↓ ↑↓ 2p

↑↓ 2s

↑↓ 1s

8.114 *See Section 8.1, Figure 8.1, Table 8.1, and Example 8.l.*

(a) $1s^1 2s^2 3s^1$ is an excited state for an element having 5 electrons, an excited state for B which has $1s^2 2s^2 2p^1$ as a ground state electron configuration.

(b) $1s^2 2s^2 2p^6 3s^2$ is the ground state for an element having 12 electrons, the ground state for Mg.

(c) $1s^2 2s^2 2p^6 3p^1$ is an excited state for an element having 11 electrons, an excited state for Na which has $1s^2 2s^2 2p^6 3s^1$ as a ground state electron configuration.

8.115 *See Sections 8.1, 8.2, and Figures 8.1, 8.8.*

Electron configuration	Element	Atomic radius (pm)
$1s^2 2s^2 2p^6 3s^2$	Mg	160
$[Ar]\ 4s^2$	Ca	231
$1s^2 2s^2 2p^6 3s^2 3p^5$	Cl	99

Size generally decreases going across the periodic table and generally increases going down the periodic table.

8.116 *See Section 8.l, Figure 8.1, Table 8.1, and Example 8.3.*

Cu $[Ar]\ 4s^1 3d^{10}$ Cu^{2+} $[Ar]\ 3d^9$
No. The electron configuration of Cu^{2+} would be predicted to be $[Ar]\ 3d^9$ whether starting with $[Ar]\ 4s^2 3d^9$ or $[Ar]\ 4s^1 3d^{10}$ because electrons are removed from the subshell of highest *n* value first.

8.117 *See Section 8.l, Figure 8.1, Table 8.1, and Example 8.3.*

Pd $[Kr]\ 4d^{10}$ Pd^{2+} $[Kr]\ 4d^8$
No. The electron configuration of Pd^{2+} would be predicted to be $[Kr]\ 4d^8$ starting with $[Ar]\ 5s^2 4d^8$ or $[Ar]\ 4d^{10}$ because electrons are removed from the subshell of highest *n* value first.

8.118 *See Section 8.1 and Example 8.3.*

The 2+ cation having the electron configuration $[Ar]\ 3d^4$ would be formed from an element having 24 total electrons. The element is Cr.

8.119 *See Section 8.1, Figure 8.1, Table 8.1, and Example 8.1, 8.3.*

Cr $[Ar]\ 4s^1 3d^5$ Cr^{4+} $[Ar]$ ↑ ↑ _ _ _ (3d) 2 unpaired electrons

8.120 *See Sections 8.1, 8.2, 8.3, and Figures 8.8, 8.13, 8.14, 8.17.*

(a) Bi is the largest member of Group VA. Size generally increases as more shells are added going down a group.

(b) C has the largest ionization energy of members of Group VA. Ionization energy generally decreases as size increases going down a group.

(c) Bi has the "lowest" electron affinity of members of Group VA. Electron affinity generally becomes less favorable as size increases going down a group.

8.121 ***See Sections 3.2, 3.4, 6.5, 8.5..***

Unbalanced: NaCl $\xrightarrow{\text{elect}}$ Na + Cl_2 Start with Cl_2.

Step 1: _2_NaCl $\xrightarrow{\text{elect}}$ _2_Na + Cl_2. Balances Na, Cl.

Strategy: g NaCl → mol NaCl → mol Cl_2 → L Cl_2 at STP

$$? \text{ L Cl}_2 \text{ at STP} = 2.44 \text{ g NaCl} \times \frac{1 \text{ mol NaCl}}{58.45 \text{ g NaCl}} \times \frac{1 \text{ mol Cl}_2}{2 \text{ mol NaCl}} \times \frac{22.4 \text{ L Cl}_2}{1 \text{ mol Cl}_2} = \mathbf{0.468 \text{ L Cl}_2}$$

8.122 ***See Sections 8.1, 8.2, and Figures 8.1, 8.8.***

Electron configuration	Element
$1s^2\,2s^2\,2p^1$	B
$1s^2\,2s^2\,2p^4$	O
$1s^2\,2s^2\,2p^5$	F

(a) B has the largest size. Size generally decreases as effective nuclear increases from left to right in a given period.

(b) B has the smallest ionization energy. B has the lowest effective nuclear charge and largest size.

(c) F has the greatest electron affinity. F has the highest effective nuclear charge and smallest size.

8.123 ***See Section 8.1, Table 8.1, and Example 8.3.***

Fe [Ar] $4s^2\,3d^6$
Fe^{2+} has 24 electrons, two fewer than Fe: [Ar]$3d^6$
Fe^{3+} has 23 electrons, three fewer than Fe: [Ar]$3d^5$

8.124 ***See Section 8.1 and Example 8.4.***

Ga [Ar] $4s^2\,3d^{10}\,4p^1$ Ga^{4+} [Ar] $3d^9$
Mn [Ar] $4s^2\,3d^5$ Mn^{4+} [Ar] $3d^3$
Ga^{4+} is not a stable species. Forming Ga^{4+} requires removal of a 3d electron which is from a completed inner subshell, an electron which is not a valence electron. On the other hand, the 3d electrons of Mn are considered to be valence elctrons because the 3d is only partially filled.

8.125 ***See Sections 8.2, 8.3.***

The effective nuclear charge increases with the increasing number of protons in the isoelectronic series S^{2-}, K^+ and Ca^{2+} causing the following trends:

Size: $Ca^{2+} < K^+ < S^{2-}$ IE: $S^{2-} < K^+ < Ca^{2+}$

8.126 ***See Section 8.1, 8.2, 8.3.***

(a) Size: O < C < Li (b) First Ionization energy: Li < C < O
Effective nuclear charge increases from left to right in a given period causing the size to decrease and ionization energy to increase.

(c) Second ionization energy: C < O < Li
Removing a second electron from Li involves removing an electron from an inner shell, a core electron.

(d) Unpaired electrons: Li < C equal to O

Li	[He] $2s^1$	[He]	$\underline{\uparrow}$ (2s)		1 unpaired electron
C	[He] $2s^2\,2p^2$	[He]	$\underline{\uparrow\downarrow}$ (2s)	$\underline{\uparrow}\ \underline{\uparrow}\ \underline{\ \ }$ (2p)	2 unpaired electrons
O	[He] $2s^2\,2p^4$	[O]	$\underline{\uparrow\downarrow}$ (2s)	$\underline{\uparrow\downarrow}\ \underline{\uparrow}\ \underline{\uparrow}$ (2p)	2 unpaired electrons

8.127 ***See Sections 6.4, 8.1.***

The given information can be used to determine the molar mass and identity of the gas.
Assuming a volume of 1.00 L of gas gives a mass of 1.62 g of gas, and the Ideal Gas can be used to calculate the moles of gas involved.

Known Quantities: m = 1.62 g P = 1.00 atm V = 1.00 L T = 27 + 273 = 300 K

Solving PV = nRT for n gives $n = \frac{PV}{RT}$ $n = \frac{(1.00\ \text{atm})(1.00\ \text{L})}{\left(0.0821\frac{\text{L}\cdot\text{atm}}{\text{mol}\cdot\text{K}}\right)(300\ \text{K})} = 0.0406\ \text{mol}$

$\mathcal{M} = \frac{m}{n} = \frac{1.62\ \text{g}}{0.0406\ \text{mol}} = 39.9$ g/mol, so the noble gas is Ar.

Ar [Ne] $3s^2\,3p^6$ [Ne] $\underline{\uparrow\downarrow}$ (3s) $\underline{\uparrow\downarrow}\ \underline{\uparrow\downarrow}\ \underline{\uparrow\downarrow}$ (3p)

8.128 ***See Sections 3.1, 3.4, 8.1.***

The given information can be used to determine the molar mass and identity of the gas.

Unbalanced:	$M(s) + O_2(g)$	→	$M_2O(s)$	Start with M_2O.
Step 1:	$\underline{2}M(s) + O_2(g)$	→	$M_2O(s)$	Balances M.
Step2:	$2M(s) + \underline{\tfrac{1}{2}}\ O_2(g)$	→	$M_2O(s)$	Balances O.
Step 3:	$4M(s) + O_2(g)$	→	$2\ M_2O(s)$	Gives whole number coefficients.

Strategy: g O_2 → mol O_2 → mol M → molar mass M

$$? \text{ mol M} = 1.41\ \text{g } O_2 \times \frac{1\ \text{mol } O_2}{32.0\ \text{g } O_2} \times \frac{4\ \text{mol M}}{1\ \text{mol } O_2} = 0.176\ \text{mol}$$

$\mathcal{M} = \frac{m}{n} = \frac{1.22\ \text{g}}{0.176\ \text{mol}} = 6.93$ g/mol, so the akali metal is Li. Li [He] $2s^1$

Note: The mass of metal oxide formed is equal to the sum of the masses of the metal and oxygen present. This tells us the metal and oxygen are present in stoichiometric amounts, and the problem is not a limiting reactant problem.

8.129 ***See Sections 8.1, 3.1, 3.4.***

M [Ne] $3s^1$, so M is Na.

Unbalanced:	$Na(s) + H_2O(\ell)$	→	$NaOH(aq) + H_2(g)$	Start with H.
Step 1:	$Na(s) + \underline{2}\ H_2O(\ell)$	→	$\underline{2}\ NaOH(aq) + H_2(g)$	Balances H, O.
Step 2:	$\underline{2}\ Na(s) + 2H_2O(\ell)$	→	$2\ NaOH(aq) + H_2(g)$	Balances Na.

Strategy: g Na → mol Na → mol H_2O → g H_2O

$$? \text{ g } H_2O = 2.34 \text{ g Na} \times \frac{1 \text{ mol Na}}{23.0 \text{ g Na}} \times \frac{2 \text{ mol } H_2O}{2 \text{ mol Na}} \times \frac{18.0 \text{ g } H_2O}{1 \text{ mol } H_2O} = \mathbf{1.83 \text{ g } H_2O}$$

8.130 ***See Sections 8.1, 3.3.***

$1s^2 2s^2 2p^6 3s^2 3p^6 3d^5$ corresponds to the M^{3+} cation of a metal having 26 total electrons as a neutral atom. The metal is Fe, and the compound is $FeCl_3$.

$$\% \text{ Fe} = \frac{55.83 \text{ g Fe}}{162.2 \text{ g } FeCl_3} \times 100\% = \mathbf{34.4\% \text{ Fe}}$$

8.131 ***See Sections 8.1, 3.1, 3.5.***

$1s^2 2s^2 2p^6 3s^2 3p^6 3d^6$ corresponds to the electron configuration of Fe^{2+}. Hence, the metal halide is $FeCl_2$.

Balanced: $Fe(s) + Cl_2(g) \rightarrow FeCl_2(s)$

Strategy: $g\ FeCl_2 \rightarrow mol\ FeCl_2 \rightarrow mol\ Fe \rightarrow g\ Fe$

$$? \text{ g Fe} = 7.88 \text{ g } FeCl_2 \times \frac{1 \text{ mol } FeCl_2}{126.8 \text{ g } FeCl_2} \times \frac{1 \text{ mol Fe}}{1 \text{ mol } FeCl_2} \times \frac{55.85 \text{ g Fe}}{1 \text{ mol Fe}} = \mathbf{3.47 \text{ g Fe}}$$

Strategy: $g\ FeCl_2 \rightarrow mol\ FeCl_2 \rightarrow mol\ Cl_2 \rightarrow g\ Cl_2$

$$? \text{ g } Cl_2 = 7.88 \text{ g } FeCl_2 \times \frac{1 \text{ mol } FeCl_2}{126.8 \text{ g } FeCl_2} \times \frac{1 \text{ mol } Cl_2}{1 \text{ mol } FeCl_2} \times \frac{70.9 \text{ g } Cl_2}{1 \text{ mol } Cl_2} = \mathbf{4.41 \text{ g } Cl_2}$$

Alternatively, $g\ Cl_2 = g\ FeCl_2 - g\ Fe = 7.88\ g - 3.47\ g = 4.41\ g\ Cl_2$.

8.132 ***See Section 8.5 and Figure 8.1.***

M in MF has a charge of 1+. The second period metal which forms a 1+ cation in combination with fluorine is lithium.

M′ in $M'F_2$ has a charge of 2+. The second period metal which forms a 2+ cation in combination with fluorine is beryllium.

Li^+ is larger than Be^{2+}. These cations are isoelectronic and Li^+ has less protons altractiving the same number of electrons as Be^{2+}.

8.133 ***See Sections 6.4, 8.1.***

The given information can be used to determine the molar mass and identity of the diatomic molecular gas. Assuming a volume of 1.00 L of gas gives a mass of 1.14 g of gas, and the Ideal Gas Law can be used to calculate the moles of gas involved.

Known Quantities: m = 1.14 g P = 1.00 atm V = 1.00 L T = 27 + 273 = 300 K

Solving PV = nRT for n gives $n = \frac{PV}{RT}$ $n = \frac{(1.00 \text{ atm})(1.00 \text{ L})}{\left(0.0821 \frac{\text{L} \cdot \text{atm}}{\text{mol} \cdot \text{K}}\right)(300 \text{ K})} = 0.0406 \text{ mol}$

$\mathcal{M} = \frac{m}{n} = \frac{1.14 \text{ g}}{0.0406 \text{ mol}} = 28.1$ g /mol, so the diatomic element is N.

N [Ne] $2s^2 2p^3$ [Ne] $\underset{2s}{\underline{\uparrow\downarrow}}$ $\underset{2p}{\underline{\uparrow}\ \underline{\uparrow}\ \underline{\uparrow}}$

Chapter 9: Chemical Bonds

9.1 *See Section 9.1.*

A Lewis electron-dot symbol for an atom consists of the symbol for the element surrounded by one dot for each valence electron of the atom. A Lewis electron-dot symbol for a monatomic cation or anion consists of the symbol for the element surrounded by one dot for each electron in the valence shell of the ion and a superscript on the right that shows the charge on the ion.

9.2 *See Section 9.2 and Example 9.2.*

(a) $\cdot\mathrm{Ba}\cdot + 2\ :\overset{\cdot\cdot}{\underset{\cdot}{\mathrm{Br}}}: \rightarrow \mathrm{Ba}^{2+} + 2\ :\overset{\cdot\cdot}{\underset{\cdot\cdot}{\mathrm{Br}}}:^{-}$

(b) $2\ \mathrm{K}\cdot + :\overset{\cdot}{\underset{\cdot}{\mathrm{S}}}: \rightarrow 2\ \mathrm{K}^{+} + :\overset{\cdot\cdot}{\underset{\cdot\cdot}{\mathrm{S}}}:^{2-}$

9.3 *See Section 9.2 and Example 9.2.*

(a) $\cdot\mathrm{Be}\cdot + :\overset{\cdot}{\underset{\cdot}{\mathrm{O}}}: \rightarrow \mathrm{Be}^{2+} + :\overset{\cdot\cdot}{\underset{\cdot\cdot}{\mathrm{O}}}:^{2-}$

(b) $\cdot\overset{\cdot}{\mathrm{Y}}\cdot + 3\ :\overset{\cdot\cdot}{\underset{\cdot}{\mathrm{Cl}}}: \rightarrow \mathrm{Y}^{3+} + 3\ :\overset{\cdot\cdot}{\underset{\cdot\cdot}{\mathrm{Cl}}}:^{-}$

9.4 *See Section 9.2.*

The lattice energy is the energy that is required to separate one mole of an ionic crystalline solid into the isolated gaseous ions. It is proportional to the product of the charge of the cation times the charge of the anion and inversely proportional to the distance between the charges, which is usually taken to be equal to the sum of the radii of the cation and anion. However, the total number of attractions between the cations and anions and the repulsions between the cations and the repulsions between the anions also affect the lattice energy.

Al_2O_3 is an example of an ionic compound that has a high lattice energy. The Al^{3+} and O^{2-} cations and anions have high ion charges and small radii leading to strong coulombic attractions.

9.5 *See Section 9.2 and Example 9.3.*

(a) The lattice energy of NaI is greater than that of KI because the smaller cation size of Na^+ causes the coulombic attractions to be stronger in NaI than in KI.

(b) The lattice energy of $MgCl_2$ greater than that of NaCl because the 2+ and 1- coulombic attractions in $MgCl_2$ are stronger than the 1+ and 1- coulombic attractions in NaCl and Mg^{2+} is smaller than Na^+.

9.6 *See Section 9.2 and Example 9.3.*

(a) The lattice energy of LiCl is greater than that of LiBr because the smaller anion size of Cl^- causes the coulombic attractions to be stronger in LiCl than LiBr.

(b) The lattice energy of Na_2O is greater than that of NaF because Na_2O has 1+ and 2- coulombic attractions compared to 1+ and 1- coulombic attractions for NaF. The effect of increased charges in Na_2O dominates the increase in lattice energy in NaF caused by the smaller anion size of F^-.

9.7 *See Sections 8.3, 9.2.*

The energy required to remove a second electron from Na, a core electron, is too high to be repaid by the greater lattice energy for 2+ and 1- coulombic attractions for NaF_2 compared to the 1+ and 1- coulombic attractions for NaF.

9.8 *See Sections 9.2, 9.3.*

Covalent bonding involves the sharing of electrons, whereas ionic bonding involves the transfer of electrons to form cations and anions that are held together by coulombic attractions.

9.9 *See Section 9.3.*

The octet rule states that each atom in a molecule shares electrons until it is surrounded by eight valence electrons. It is obeyed best by members of the second period because their valence shell is completed by eight electrons.

9.10 *See Section 9.3.*

A correct Lewis structure must show the correct connectivity of the atoms in the molecule or ion, the correct total number of valence electrons for the molecule or ion and an acceptable distribution of the total number of valence electrons.

9.11 *See Section 9.4.*

HCl is a polar molecule because Cl is more electronegative than H, whereas Cl_2 is composed of atoms of identical electronegativity and is nonpolar. The H atoms in HCl are partially positive and will be oriented toward the negatively charged plate, while the Cl atoms are partially negative and will be oriented toward the positively charged plate. On the other hand, the nonpolar Cl_2 molecules show no tendency to align themselves with the oppositely charged plates.

9.12 *See Section 9.4.*

A polar bond is a covalent bond in which the bonding electrons are not shared equally by the two bonded atoms. Polar bonds are formed when atoms having differing electonegativies form covalent bonds with each other.

9.13 *See Section 9.4 and Figure 9.9.*

Fluorine has the highest electronegativity. Electronegativity increases in all directions toward fluorine. Hence, electronegativity increases going from bottom to top within groups and going from left to right across periods.

9.14 *See Sections 8.3, 9.4, and Figures 8.13, 8.14, 9.9.*

Ionization energy and electronegativity both increase from bottom to top within groups and going from left to right across periods. However, discontinuities in ionization energies occur at Groups IIIA and VIA when going across the groups of the representative elements. Furthermore, the halogens, rather than the noble gases, have the highest electronegativities in their respective periods. The latter is due to the fact that the noble gases have completed subshells, and the halogens do not.

9.15 *See Section 9.5.*

Formal charges are charges that are assigned to atoms in Lewis structures by assuming the shared electrons are divided equally between the bonded atoms and comparing the number of assigned electrons to the number of valence electrons of the atoms. If an atom is assigned less electrons than its number of valence electrons, it is assigned a positive formal charge. If an atom is assigned more electrons than its number of valence electrons, it is assigned a negative formal charge. Hence, the formal charge of an atom can be calculated using the equation: formal charge = (number of valence electrons of atom) - (number of lone pair electrons) - $\frac{1}{2}$ number of shared electrons).

9.16 *See Section 9.6.*

Resonance structures are structures that differ only in the distribution of the valence electrons in representing the bonding in a species and generally correspond to equally acceptable Lewis structures. However, no resonance structure actually exists; the correct structure is an average or hybrid of the contributing resonance structures. We draw resonance structures when we cannot draw a single Lewis structure to adequately portray the bonding in molecule or ion. In most cases this involves the inability to use Lewis structures to portray bonds that are known to be somewhere between single and double bond character from experimental measurements of bond lengths and strengths.

9.17 *See Sections 9.5, 9.6, and Example 9.12.*

The most stable Lewis structures are usually those which 1) place the lowest formal charges on the atoms, 2) avoid placing like formal charges on adjacent atoms, 3) place negative formal charges on the more electronegative atoms, and 4) place opposite formal charges on adjacent atoms. These guidelines can be used to determine whether nonequivalent resonance structures are equally probable or even feasible.

9.18 *See Section 9.7.*

An electron deficient molecule is one that does not have eight electrons around the central atom in its correct Lewis structure. Electron deficient molecules usually have central atoms of elements from Groups IA, IIA, or IIIA and readily participate in reactions in which the central atom can acquire an octet. Hence, electron deficient molecules are usually quite reactive.

9.19 *See Section 9.7.*

Odd electron molecules are called radicals. They are usually quite reactive and often dimerize to form even electron species that obey the octet rule.

9.20 *See Section 9.7.*

Electron rich compounds have the general formula YF_n where Y = P, As, S, Se, Te, Cl, Br, I, or Xe, and $n > 3$ for P and As, $n> 2$ for S, Se, and Te, $n > 1$ for Cl, Br and I and $n > 0$ for Xe. In some caes, a fluorine atom can be replaced by another halogen or an oxygen atom. Examples include XeF_2, ClF_3, $TeCl_4$, XeO_4, PCl_5, and SF_6. In each case the central atom is the electron rich atom.

9.21 *See Section 9.8.*

Bond dissociation energy refers to the energy required to break one mole of the bonds being considered. The equation for using bond dissociation energies to calculate approximate heats of reactions is:

$$\Delta H_{reaction} = \Sigma(\text{bond energies of bonds broken}) - \Sigma(\text{bond energies of bonds formed})$$

9.22 *See Section 9.8.*

The bond dissociation energy for HCl can be determined directly and exactly by dissociating diatomic HCl molecules. However, the C-C bond always occurs in compounds in which the carbon atoms are bonded to other atoms, because carbon atoms form four bonds to obtain a noble gas configuration. This makes it impossible to directly measure the C-C bond energy, so an average value based on measurements for more than compound is given in tables of bond dissociation energies.

9.23 *See Section 9.1, Figure 9.1, and Example 9.1.*

(a) $Na\cdot$ (b) $:\ddot{\underset{\cdot}{F}}:$ (c) $:\ddot{\underset{\cdot\cdot}{O}}:^{2-}$ (d) Mg^{2+}

9.24 *See Section 9.1, Figure 9.1, and Example 9.1.*

(a) $:\dot{\underset{\cdot}{S}}:$ (b) $:\ddot{\underset{\cdot\cdot}{I}}:^{-}$ (c) $\cdot Be\cdot$ (d) $\cdot Ga^{2+}$

9.25 *See Section 9.1, Figure 9.1, and Example 9.1.*

(a) $Li\cdot$ (b) $:\ddot{\underset{\cdot\cdot}{S}}:^{2-}$ (c) $\cdot Mg\cdot$ (d) $:\ddot{\underset{\cdot}{Br}}:$

9.26 *See Section 9.1, Figure 9.1, and Example 9.1.*

(a) $K\cdot$ (b) $\cdot\ddot{\underset{\cdot}{N}}\cdot$ (c) $\cdot\dot{B}\cdot$ (d) $:\ddot{\underset{\cdot\cdot}{F}}:^{-}$

9.27 *See Section 9.1 and Examples 9.2, 9.3.*

(a) Lithium having one valence electron forms Li^+ and oxygen having six valence electrons forms O^{2-} as they react to form Li_2O. Similarly, sodium having one valence electron forms Na^+ and sulfur having six valence electrons forms S^{2-} as they react to form Na_2S. Since the smaller sizes of Li^+ and O^{2+} compared to Na^+ and S^{2-} lead to stronger coulombic attractions, Li_2O has the greater lattice energy.

(b) Potassium having one valence electron forms K^+ and chlorine having seven valence electrons forms Cl^- as they react to form KCl. Similarly, magnesium having two valence electrons forms Mg^{2+} and fluorine having seven valence electrons forms F^- as they react to form MgF_2. Since Mg^{2+} has a higher charge than K^+ and is smaller than K^+ and since F^- is smaller than Cl^-, MgF_2 has stronger coulombic attractions than KCl and the greater lattice energy.

9.28 *See Section 9.1 and Examples 9.2, 9.3.*

(a) Potassium having one valence electron forms K^+ and sulfur having six valence electrons forms S^{2-} as they react to form K_2S. Similarly, potassium having one valence electron forms K^+ and chlorine having seven valence electrons forms Cl^- as they react to form KCl. Since S^{2-} has a more negative anion charge than Cl^-, K_2S has stronger coulombic attractions than KCl in spite of the larger size of S^{2-} compared to Cl^-. Hence, K_2S has the greater lattice energy.

(b) Lithium having one valence electron forms Li^+ and fluorine having seven valence electrons forms F^- as they react to form LiF. Rubidium having one valence electron forms Rb^+ and chlorine having seven valence electrons forms Cl^- as they react to form RbCl. Since the smaller sizes of Li^+ and F^- compared to Rb^+ and Cl^- lead to stronger coulombic attractions, LiF has the greater lattice energy.

9.29 *See Section 9.1 and Example 9.3.*

The smaller size of Cl^- compared to I^- leads to stronger coulombic attractions and a greater lattice energy for LiCl than LiI.

9.30 *See Section 9.1 and Example 9.3.*

The smaller size of Ca^{2+} compared to Ba^{2+} leads to stronger coulombic attractions and a greater lattice energy for CaO than BaO.

9.31 *See Section 9.2 and Example 9.2.*

(a) KBr < NaBr < NaCl The cation charges are all 1+, and the anion charges are all 1-. However, $r_{Cl^-} < r_{Br^-}$ and $r_{Na^+} < r_{K^+}$ leading to the given order of increasing coulombic attractions and greater lattice energies.

(b) $CaCl_2$ < CaO < MgO The cation charges are all 1+. However O^{2-} has an anion charge of 2- compared to the anion charge of 1- for Cl^- and $r_{O^{2-}} < r_{Cl^-}$ leading to stronger coulombic attractions and greater lattice energies for the oxides. In addition, $r_{Mg^{2+}} < r_{Ca^{2+}}$ leading to stronger coulombic attractions and a greater lattice energy for MgO than CaO.

(c) LiF < BeF_2 < BeO The cation charges of 2+ for beryllium and 1+ for lithium lead to stronger coulombic attractions and greater lattice energies for the beryllium compounds than the lithium compound. In addition, the more negative anion charge of O^{2-} compared to F^- leads to stronger coulombic attractions and a greater lattice energy for BeO than BeF_2 in spite of the fact O^{2-} is larger than F^-.

9.32 *See Section 9.2 and Example 9.2.*

(a) NaCl < LiCl < $BeCl_2$ The decreasing cation size in the order Na^+, Li^+, Be^{2+} leads to stronger coulombic attractions and greater lattice energies in the order NaCl, LiCl, $BeCl_2$. More importantly, $BeCl_2$ has the strongest coulombic attractions and greatest lattice energy due to the 2+ cation charge of Be^{2+} compared to the 1+ cation charge of Na^+ and Li^+.

(b) $MgBr_2$ < MgF_2 < MgO The more negative anion charge of O^{2-} compared to F^- and Br^- leads to the strongest coulombic attractions and greatest lattice energy for MgO in spite of the fact that O^{2-} is larger than F^-. In addition, the smaller size of F^- compared to Br^- leads to stronger coulombic attractions and a greater lattice energy for MgF_2 than $MgBr_2$.

(c) $MgCl_2$ < $BeCl_2$ < BeO The more negative anion charge of O^{2-} compared to Cl^- leads to the strongest coulombic attractions and greatest lattice energy for BeO. In addition, the smaller size of Be^{2+} compared to Mg^{2+} leads to stronger coulombic attractions and a greater lattice energy for $BeCl_2$ than $MgCl_2$.

9.33 *See Section 9.3 and Figure 9.1.*

(a) CF_4 Carbon has four valence electrons and each fluorine has seven valence electrons. Carbon obtains an octet by forming four covalent bonds whereas each fluorine obtains an octet by forming one covalent bond.

(b) NI_3 Nitrogen has five valence electrons and each iodine has seven valence electrons. Nitrogen obtains an octet by forming three covalent bonds whereas each iodine obtains an octet by forming one covalent bond.

9.34 *See Section 9.3 and Figure 9.1.*

(a) SiH_4 Silicon has four valence electrons and each hydrogen has one valence electron. Silicon obtains an octet by forming four covalent bonds, whereas each hydrogen obtains the noble gas configuration of helium by forming one covalent bond.

(b) NCl_3 Nitrogen has five valence electrons and each chlorine has seven valence electrons. Nitrogen obtains an octet by forming three covalent bonds, whereas each chlorine obtains an octet by forming one covalent bond.

(c) OF_2 Oxygen has six valence electrons and each fluorine has seven valence electrons. Oxygen obtains an octet by forming two covalent bonds, whereas each fluorine obtains an octet by forming one covalent bond.

9.35 *See Section 9.3 and Examples 9.4, 9.5, 9.6.*

(a) H_2S

H—S—H

H—S—H (with two lone pairs on S)

Total valence electrons $= [2 \times 1(H) + 1 \times 6(S)] = 8$.
Four electrons remain after assigning two single bonds, and four unshared electrons are needed to give each atom a noble gas configuration (4 for S). Eight electrons are used in writing the Lewis structure in which bonding pairs of electrons are represented by dashes and lone pairs of electrons are represented by a pair of dots.

(b) H_2CO

H—C—O, with H bonded to C

H—C=O, with H bonded to C (two lone pairs on O)

Total valence electrons $= [2 \times 1(H) + 1 \times 4(C) + 1 \times 6(O)] = 12$.
Six electrons remain after assigning three single bonds, and eight unshared electrons are needed to give each atom a noble gas configuration (2 for C and 6 for O). Hence, two electrons (8-6) must be used to form an additional bond. The additional bond must be between C and O, since H obtains a noble gas configuration by forming just one bond. Twelve electrons are used in writing the Lewis structure in which bonding pairs of electrons are represented by dashes and lone pairs of electrons are represented by a pair of dots.

(c) PF_3

F—P—F, with F bonded to P

:F—P—F: with F bonded to P (three lone pairs on each F, one lone pair on P)

Total valence electrons $= [1 \times 5(P) + 3 \times 7(F)] = 26$.
Twenty electrons remain after assigning three single bonds, and twenty unshared electrons are needed to give each atom a noble gas configuration (2 for P and 6 for each F). Twenty six electrons are used in writing the Lewis structure in which bonding pairs of electrons are represented by dashes and lone pairs of electrons are represented by a pair of dots.

9.36 See Section 9.3 and Examples 9.4, 9.5, 9.6.

(a) $SiCl_4$

Total valence electrons $= [1 \times 4(\text{Si}) + 4 \times 7(\text{Cl})] = 32$.
Twenty four electrons remain after assigning four single bonds, and twenty four unshared electrons are needed to give each atom a noble gas configuration (6 for each Cl). Thirty two electrons are used in writing the Lewis structure in which bonding pairs of electrons are represented by dashes and lone pairs of electrons are represented by a pair of dots.

(b) SF_2

Total valence electrons $= [1 \times 6(\text{Si}) + 2 \times 7(\text{F})] = 20$.
Sixteen electrons remain after assigning two single bonds, and sixteen unshared electrons are needed to give each atom a noble gas configuration (4 for S and 6 for each F). Twenty electrons are used in writing the Lewis structure in which bonding pairs of electrons are represented by dashes and lone pairs of electrons are represented by a pair of dots.

(c) CS_2

Total valence electrons $= [1 \times 4(\text{C}) + 2 \times 6(\text{S})] = 16$.
Twelve electrons remain after assigning two single bonds, and sixteen unshared electrons are needed to give each atom a noble gas configuration. Hence, four electrons (16-12) must be used to form two additional bonds. The most likely structure involves a second bond to each S. Sixteen electrons are used in writing the Lewis structure in which bonding pairs of electrons are represented by dashes and lone pairs of electrons are represented by a pair of dots.

9.37 See Section 9.3.

(a) The H-S bonds are single bonds having a bond order of 1.

(b) The H-C bonds are single bonds having a bond order of 1. The C=O double bond has a bond order of 2.

(c) The F-P bonds are single bonds having a bond order of 1.

9.38 See Section 9.3.

(a) The Cl-Si bonds are single bonds having a bond order of 1.

(b) The F-S bonds are single bonds having a bond order of 1.

(c) The S=C bonds are double bonds having a bond order of 2.

9.39 *See Section 9.3 and Examples 9.4, 9.5, 9.6.*

(a) AsH_3

Total valence electrons $= [1 \times 5(\text{As}) + 3 \times 1(\text{H})] = 8$.

```
H—As—H
   |
   H

   ••
H—As—H
   |
   H
```

Two electrons remain after assigning three single bonds, and two unshared electrons are needed to give each atom a noble gas configuration (2 for As). Eight electrons are used in writing the Lewis structure.

(b) ClF

Total valence electrons = [1x7(Cl)+1x7(F)] = 14.

```
Cl—F

 ••    ••
:Cl—F:
 ••    ••
```

Twelve electrons remain after assigning one single bond, and twelve unshared electrons are needed to give each atom a noble gas configuration (6 for Cl and 6 for F). Fourteen electrons are used in writing the Lewis structure.

(c) CF_3OH

Total valence electrons $= [1 \times 4(\text{C}) + 3 \times 7(\text{F}) + 1 \times 6(\text{O}) + 1 \times 1(\text{H})] = 32$.

```
    F
    |
F—C—O—H
    |
    F

    ••
   :F:
 ••  |   ••
:F—C—O—H
 ••  |   ••
   :F:
    ••
```

Twenty two electrons remain after assigning five single bonds, and twenty two electrons are needed to give each atom a noble gas configuration (6 for each F and 4 for O). Thirty two electrons are used in writing the Lewis structure.

9.40 *See Section 9.3 and Examples 9.4, 9.5, 9.6.*

(a) CN^-

Total valence electrons $= [1 \times 4(\text{C}) + 1 \times 5(\text{N}) + 1(\text{charge})] = 10$.

```
C—N⁻
```

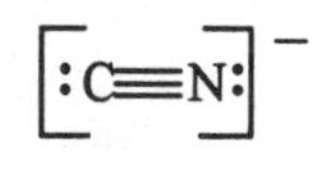

Eight electrons remain after assigning one single bond, and twelve electrons are needed to give each atom a noble gas configuration (6 for C and 6 for N). Hence, four electrons (12-8) must be used to form two additonal bonds. Ten electrons are used in writing the Lewis structure.

(b) C_2H_6

Total valence electrons $= [2 \times 4(\text{C}) + 6 \times 1(\text{H})] = 14$.

```
   H  H
   |  |
H—C—C—H
   |  |
   H  H
```

All fourteen electrons are used in assigning seven signle bonds, and each atom has a noble gas configuration. Fourteen electrons are used in writing the Lewis structure.

(c) N_2H_4

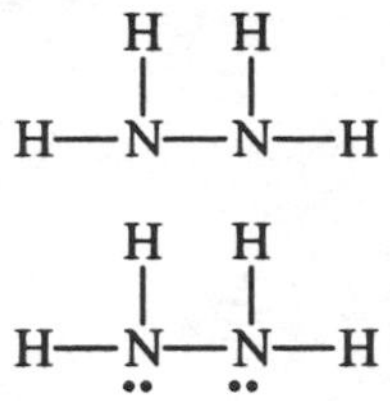

```
   H   H
   |   |
H—N—N—H

   H   H
   |   |
H—N—N—H
   ••  ••
```

Total valence electrons $=[2 \times 5(N) + 4 \times 1(H)] = 14$.
Four electrons remain after assigning five single bonds, and four electrons are needed to give each atom a noble gas configuration (2 for each N). Fourteen electrons are used in writing the Lewis structure.

9.41	*See Section 9.3 and Examples 9.4, 9.5, 9.6.*

(a) CCl_4

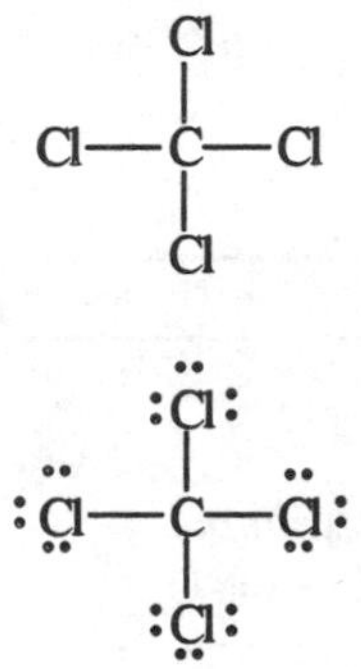

Total valence electrons $=[1 \times 4(C) + 4 \times 7(Cl)] = 32$.
Twenty four electrons remain after assigning four single bonds, and twenty four electrons are needed to give each atom a noble gas configuration (6 for each Cl). Thirty two electrons are used in writing the Lewis structure.

(b) NO^+

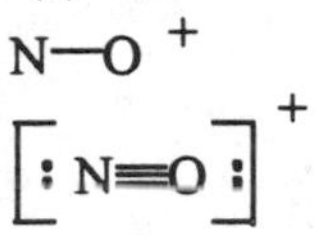

N—O +

[: N≡O :] +

Total valence electrons $= [1 \times 5(N) + 1 \times 6(O) - 1(\text{charge})] = 10$.
Eight electrons remain after assigning a single bond, and twelve unshared electrons are needed to give each atom a noble gas configuration (6 for N and 6 for O). Hence, four electrons (12-8) must be used to form two additional bonds between N and O making the bond a triple bond. Ten electrons are used in writing the Lewis structure.

(c) BCl_4^-

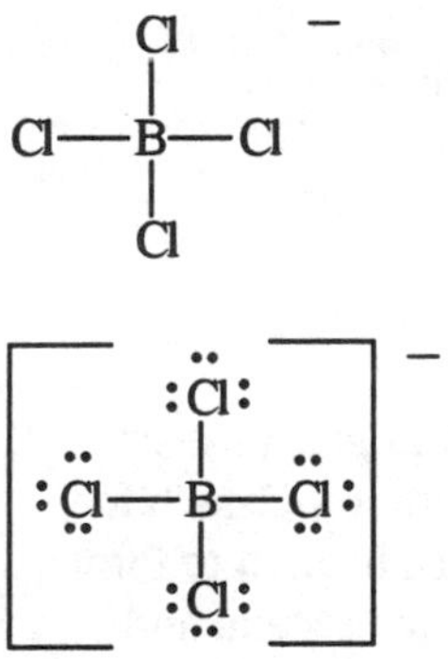

Total valence electrons $=[1 \times 3(B) + 4 \times 7(Cl) + 1(\text{charge})] = 32$.
Twenty four electons remain after assigning four single bonds, and twenty four electons are needed to give each atom a noble gas configuration (6 for eachCl). Thirty two electrons are used in writing the Lewis structure.

9.42	*See Section 9.3 and Examples 9.4, 9.5, 9.6.*

(a) NO^-

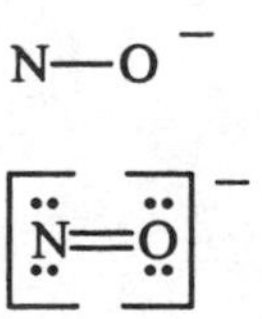

N—O −

[N=O] −

Total valence electrons $=[1 \times 5(N) + 1 \times 6(O) + 1(\text{charge})] = 12$.
Ten electrons remain after assigning one single bond, and twelve unshared electrons are needed to give each atom a noble gas configuration. Hence, two electrons (12-10) must be used to form one additional bond. Twelve electrons are used in writing the Lewis structure.

(b) C_3H_8

```
   H   H   H
   |   |   |
H—C—C—C—H
   |   |   |
   H   H   H
```

Total valence elctrons $=[3 \times 4(C) + 8 \times 1(H)] = 20$.
No electrons remain after assigning 10 single bonds, and each atom has a noble gas configuration. Twenty electrons are used in writing the Lewis structure.

(c) ClO^-

```
Cl—O   −
```

```
 ..   ..  −
[:Cl—O:]
 ..   ..
```

Total valence electrons $=[1 \times 7(Cl) + 1 \times 6(O) + 1(charge)] = 14$.
Twelve electrons remain after assigning one single bond, and twelve electrons are needed to give eadh atom a noble gas configuration (6 for Cl and 6 for O). Fourteen electrons are used in writing the Lewis structure.

9.43 *See Section 9.3 and Examples 9.4, 9.5, 9.6.*

(a) CH_3CHO

```
   H   O
   |   |
H—C—C—H
   |
   H
```

```
   H  :O:
   |   ||
H—C—C—H
   |
   H
```

Total valence electrons $= [2 \times 4(C) + 4 \times 1(H) + 1 \times 6(O)] = 18$.
Six electrons remain after assigning six single bonds, and eight unshared electrons are needed to give each atom a noble gas configuration (6 for O and 2 for second C). Hence, two electrons (8-6) must be used to form an additional bond. The additional bond must be between C and O, since H attains a noble gas configuration by forming just one bond. Eighteen electrons are used in writing the Lewis structure.

(b) NH_2OH

```
H—N—O—H
   |
   H
```

```
   ..  ..
H—N—O—H
   |   ..
   H
```

Total valence electrons $= [1 \times 5(N) + 3 \times 1(H) + 1 \times 6(O)] = 14$.
Six electrons remain after assigning four single bonds, and six unshared electrons are needed to give each atom a noble gas configuration (2 for N and 4 for O). Fourteen electrons are used in writing the Lewis structure.

(c) CH_3CHCH_2

```
   H   H   H
   |   |   |
H—C—C—C—H
   |
   H
```

```
   H   H   H
   |   |   |
H—C—C=C—H
   |
   H
```

Total valence electrons $= [3 \times 4(C) + 6 \times 1(H)] = 18$.
Two electrons remain after assigning eight single bonds, and four unshared electrons are needed to give each atom a noble gas configuration (2 each for second and third C atoms). Hence, two electrons (4-2) must be used to form an additional bond. The additional bond must be between the second and third C atoms, since the first C atom has an octet and H attains a noble gas configuration by forming just one bond. Eighteen electrons are used in writing the Lewis structure.

(d) CH_2CCH_2

```
H—C—C—C—H
  |   |
  H   H

H—C=C=C—H
  |   |
  H   H
```

Total valence electrons = $[3 \times 4(C) + 4 \times 1(H)] = 16$.
Four electrons remain after assigning six single bonds, and eight unshared electrons are needed to give each atom a noble gas configuration (2 for each C on the ends and 4 for the C in the middle). Hence, four electrons (8-4) must be used to form two additional bonds. The additional bonds must be formed between the carbon atoms, since hydrogen attains a noble gas configuration by forming just one bond. Fourteen electrons are used in writing the Lewis structure.

9.44 ***See Section 9.3 and Examples 9.4, 9.5, 9.6.***

(a) O_2F_2

```
F—O—O—F

 ..  ..  ..  ..
:F—O—O—F:
 ..  ..  ..  ..
```

Total valence electrons = $[2 \times 6(O) + 2 \times 7(F)] = 26$.
Twenty electrons remain after assigning three single bonds, and twenty electrons are needed to give each atom a noble gas configuration (6 for each F and 4 for each O). Twenty six electrons are used in writing the Lewis structure.

(b) HCO_2H

```
    O
    |
H—C—O—H

   :O:
    ||  ..
H—C—O—H
        ..
```

Total valence electrons = $[2 \times 1(H) + 1 \times 4(C) + 2 \times 6(O)] = 18$.
Ten electrons remain after assigning four single bonds, and twelve electrons are needed to give each atom a noble gas configuration (6 for the top O, 2 for C, and 4 for the O on the right). Hence, two electrons (12-10) must be used to form an additional bond. Hydrogen attains a noble gas configuration by forming just one bond, and the oxygen atom on the right has attained a noble gas configuration by forming two bonds as usual. Hence, an additional bond is formed between the carbon and top oxygen atom. Eighteen electrons are used in writing the Lewis structure.

(c) CH_2CCl_2

```
    H  Cl
    |  |
H—C—C—Cl

        ..
    H  :Cl:
    |   |    ..
H—C=C—Cl:
             ..
```

Total valence electrons= $[2 \times 4(C) + 2 \times 1(H) + 2 \times 7(Cl)] = 24$.
Fourteen electrons remain after assigning five single bonds, and sixteen electrons are needed to give each atom a noble gas configuration (2 for each C and 6 for each Cl). Hence, two electrons (16-14) must be used to form an additional bond. The additional bond is formed by the carbon atoms to give each carbon atom a noble gas configuration.

(d) CH_3NHCH_3

```
    H  H  H
    |  |  |
H—C—N—C—H
    |     |
    H     H

    H  H  H
    |  |  |
H—C—N—C—H
    |  ..  |
    H     H
```

Total valence electrons = [2x4(C)+1x5(N)+7x1(H)] = 20.
Eighteen electrons are used in assigning nine single bonds, and two electrons are needed to give each atom a noble gas configuration (2 for N). Twenty electrons are used in writing the Lewis structure.

9.45 *See Section 9.3 and Examples 9.4, 9.5, 9.6.*

(a) C_2H_2

H—C—C—H

H—C≡C—H

Total valence electrons $= [2 \times 4(C) + 2 \times 1(H)] = 10$.

Four electrons remain after assigning three single bonds, and eight unshared electrons are needed to give each atom a noble gas configuration (4 for each C). Hence, four electrons (8-4) must be used to form two additional bonds. The additional bonds must be formed between the carbon atoms, since hydrogen attains a noble gas configuration by forming just one bond. Ten electrons are used in writing the Lewis structure.

(b) HOCl

H—O—Cl

H—Ö—C̈l:

Total valence electrons $= [1 \times 1(H) + 6 \times 1(O) + 7 \times 1(Cl)] = 14$.

Ten electrons remain after assigning two single bonds, and ten unshared electrons are needed to give each atom a noble gas configuration (4 for O and 6 for Cl). Ten electrons are used in writing the Lewis structure.

(c) CH_2CHCN

H—C(H)—C(H)—C—N

H—C(H)=C(H)—C≡N:

Total valence electrons $= [3 \times 4(C) + 3 \times 1(H) + 1 \times 5(N)] = 20$.

Eight electrons remain after assigning six single bonds, and fourteen unshared electrons are needed to give each atom a noble gas configuration (2 each for first and second C, 4 for third C and 6 for N). Hence, six electrons (14-8) must be used to form three additional bonds. One additional bond is placed between the first and second carbon atoms to give these atoms a noble gas configuration, and two additional bonds are placed between the third carbon atom and the nitrogen atom. Twenty electrons are used in writing the Lewis structure.

(d) H_2O_2

H—O—O—H

H—Ö—Ö—H

Total valence electrons $= [2 \times 1(H) + 2 \times 6(O)] = 14$.

Eight electrons remain after assigning three single bonds, and eight unshared electrons are needed to give each atom a noble gas configuration (4 for each O). Fourteen electrons are used in writing the Lewis structure.

9.46 *See Section 9.3 and Examples 9.4, 9.5, 9.6.*

(a) BF_4^-

F—B(F)(F)—F (−)

[:F̤̈:—B(:F̤̈:)(:F̤̈:)—:F̤̈:]⁻

Total valence electrons $= [1 \times 3(B) + 4 \times 7(F) + 1(\text{charge})] = 32$.

Twenty four electrons remain after assigning four single bonds, and twenty four electrons are needed to give each atom a noble gas configuration (6 for each F). Thirty two electrons are used in writing the Lewis structure.

(b) $CH_2CHC(O)H$

```
    H   H   O
    |   |   |
H—C—C—C—H

    H   H  :O:
    |   |   ||
H—C=C—C—H
```

Total valence electrons = $[3 \times 4(C) + 1 \times 6(O) + 4 \times 1(H)] = 22$.
Eight electrons remain after assigning seven single bonds, and twelve electrons are needed to give each atom a noble gas configuration. Hence, four electrons (12-8) must be used to form two additional bonds. One additional bond is formed between the first and second carbon atoms and another is formed between the third carbon atom and the oxygen atom. Twenty two electrons are used in writing the Lewis structure.

(c) C_2F_4

```
    F   F
    |   |
F—C—C—F

    ..  ..
   :F: :F:
 ..  |   |   ..
:F—C=C—F:
 ..          ..
```

Total valence electrons = $[2 \times 4(C) + 4 \times 7(F)] = 36$.
Twenty six electons remain after assigning six single bonds, and twenty eight electrons are needed to give each atom a noble gas configuration. Hence, two electrons (28-26) must be used to form an additional bond. The additional bond is formed between the carbon atoms to give each carbon atom four bonds and each atom a noble gas configuration. Thirty six electrons are used in writing the Lewis structure.

(d) CH_2CHCCH

```
    H   H
    |   |
H—C—C—C—C—H

    H   H
    |   |
H—C=C—C≡C—H
```

Total valence electrons =$[4 \times 4(C) + 4 \times 1(H)] = 20$.
Six electrons remain after assigning seven single bonds, and twelve electrons are needed to give each atom a noble gas configuration. Hence, six electrons (12-6) must be used to form three additional bonds. One additional bond is formed between the carbon atoms on the left and two additional bonds are formed between the carbon atoms on the right to give each atom a noble gas configuration. Twenty electrons are used in writing the Lewis structure.

9.47 ***See Section 9.4 and Figure 9.9.***

(a) Br (b) Cl (c) N

Electronegativity values decrease down the periodic table and also to the left across the table.

9.48 ***See Section 9.4 and Figure 9.9.***

(a) O (b) Ge (c) Br

Electronegativity values decrease down the periodic table and also to the left across the table.

9.49 ***See Section 9.4 and Figure 9.9.***

(a) I < Br < Cl (b) Ca < Ga < Br (c) K < Ge < O

Electronegativity values decrease down the periodic table and also to the left across the table.

9.50 ***See Section 9.4 and Figure 9.9.***

(a) Si < S < F (b) Li < P < O (c) Ga < B < S

Electronegativity values decrease down the periodic table and also to the left across the table.

9.51 *See Section 9.4, Figure 9.8, and Example 9.7.*

(a) The order of increasing electronegativity is C < N < O (2.5 < 3.0 < 3.5). The C-O bond is more polar than the N-O bond due to the greater difference in electronegativity. The polarity is $\overset{\longmapsto}{\text{C}-\text{O}}$.

(b) The order of increasing electronegativity is Ge = Si < C (1.9 = 1.9 < 2.5). The Ge-C bond is more polar than the Si-Ge bond due to the greater difference in electronegativity. The polarity is $\overset{\longmapsto}{\text{Ge}-\text{C}}$.

(c) The order of increasing electronegativity is H < S < O (2.1 < 2.5 < 3.5). The H-O bond is more polar than the H-S bond due to the greater difference in electronegativity. The polarity is $\overset{\longmapsto}{\text{H}-\text{O}}$.

(d) The order of increasing electronegativity is Si < B < C (1.8 < 2.0 < 2.5). The B-C bond is more polar than the B-Si bond due to the greater difference in electronegativity. The polarity is $\overset{\longmapsto}{\text{B}-\text{C}}$.

9.52 *See Section 9.4, Figure 9.8, and Example 9.7.*

(a) The order of increasing electronegativity is Br < N < F (2.8 < 3.0 < 4.0). The F-Br bond is more polar than the N-F bond due to the greater difference in electronegatvity. The polarity is $\overset{\longleftarrow\!\shortmid}{\text{F}-\text{Br}}$

(b) The order of increasing electronegativity is P < S < N < F (2.1 < 2.5 < 3.0 < 4.0). The F-N bond is more polar than the S-P bond due to the greater difference in electronegativity. The polarity is $\overset{\longleftarrow\!\shortmid}{\text{F}-\text{N}}$.

(c) The order of increasing electronegativity is Se < S < Br < Cl (2.4 < 2.5 < 2.8 < 3.0). The Cl-Br bond is slightly more polar than the S-Se bond due to the slightly greater difference in electronegativity.

The polarity is $\overset{\longleftarrow\!\shortmid}{\text{Cl}-\text{Br}}$

(d) The order of increasing electronegativity is B < C < N < F (2.0 < 2.5 < 3.0 < 4.0). The B-F is more polar than the C-N bond due to the greater difference in electronegativity. The polarity is $\overset{\longmapsto}{\text{B}-\text{F}}$.

9.53 *See Section 9.4 and Example 9.7.*

N_2 has identical nuclei bonded together and is therefore nonpolar. BrF is more polar than ClF due to the greater difference in electronegativity (1.2 vs. 1.0).

9.54 *See Section 9.4 and Example 9.7.*

O_2 has identical nuclei bonded together and is therefore nonpolar. ICl is more polar than BrCl due to the greater difference in electronegativity (0.5 vs. 0.2).

9.55 ***See Section 9.5 and Examples 9.8, 9.9, 9.10.***

H_2CO, COH_2

Total valence electrons = $[1 \times 4(\text{C}) + 1 \times 6(\text{O}) + 2 \times 1(\text{H})] = 12.$

Six electrons remain after assigning three single bonds in either conntictivity. Eight unshared electrons are needed to give each atom a noble gas configuration (2 for C and 6 for O in A, and 6 for C and 2 for O in B). Hence, two electrons (8-6) must be used to form an additional bond beyond carbon and oxygen. The appropriate Lewis structures associated with the given connectivities are:

A: H—C(—H)=Ö: B: :C=Ö(—H)—H

Twelve electrons are used in writing the Lewis structures.

Formal Charges:	A	B
H	$1 - 0 - (\frac{1}{2})(2) = 0$	$1 - 0 - (\frac{1}{2})2 = 0$
C	$4 - 0 - (\frac{1}{2})(8) = 0$	$4 - 4 - (\frac{1}{2})(4) = -2$
O	$6 - 4 - (\frac{1}{2})(4) = 0$	$6 - 0 - (\frac{1}{2})8 = +2$

When these formal charges are included in the Lewis structures, we obtain:

A: H—C(—H)=Ö: B: ⊖⊖:C=O(⊕⊕)(—H)—H

Structure A has the smaller formal charges and is threfore the more favorable arrangement.

9.56 ***See Section 9.5 and Examples 9.8, 9.9, 9.10.***

HNO, HON

Total valence electrons= $[1 \times 1(\text{H}) + 1 \times 5(\text{N}) + 1 \times 6(\text{O})] = 12.$

Eight electrons remain after assigning two single bonds in either connectivity. Ten unshared electrons are needed to give each atom a noble gas configuartion. Hence, two electrons (10-8) must be used to form one additional bond between nitrogen and oxygen. The appropriate Lewis structure associated with the given connectivities are:

A: H—N̈=Ö: B: H—Ö=N̈:

Twelve electrons are used in writing the Lewis structures.

Formal Charges:	A	B
H	$1 - 0 - (\frac{1}{2})(2) = 0$	$1 - 0 - (\frac{1}{2})(2) = 0$
N	$5 - 2 - (\frac{1}{2})(6) = 0$	$5 - 4 - (\frac{1}{2})(4) = -1$
O	$6 - 4 - (\frac{1}{2})(4) = 0$	$6 - 2 - (\frac{1}{2})(6) = +1$

When these formal charges are included in the Lewis structures, we obtain:

A: $H-\ddot{N}=\ddot{O}:$ (no formal charges)

B: $H-\ddot{O}^{\oplus}=\ddot{N}:^{\ominus}$

Structure A has the smaller formal charges and is therefore the more favorable arrangement.

9.57 *See Section 9.5 and Examples 8.8,9.9, 9.10.*

H_2CNH, $HCNH_2$

Total valence electrons $= \left[3\times 1(H)+1\times 4(C)+1\times 5(N)\right]=12.$

Four electrons remain after assigning four single bonds in either connectivity. Six unshared electrons are needed to give each atom a noble gas configuration. Hence, two electrons (6-4) must be used to form one additional bond between carbon and nitrogen. The appropriate Lewis structures associated with the given connectivities are:

A: $H-C(H)=\ddot{N}-H$

B: $H-\ddot{C}=N(H)-H$

Formal Charges:	A	B
H	$1-0-\left(\frac{1}{2}\right)(2)=0$	$1-0-\left(\frac{1}{2}\right)(2)=0$
C	$4-0-\left(\frac{1}{2}\right)(8)=0$	$4-2-\left(\frac{1}{2}\right)(6)=-1$
N	$5-2-\left(\frac{1}{2}\right)(6)=0$	$5-0-\left(\frac{1}{2}\right)(8)=+1$

When these formal charges are included in the Lewis structures, we obtain:

A: $H-C(H)=\ddot{N}-H$ (no formal charges)

B: $H-\ddot{C}^{\ominus}=N^{\oplus}(H)-H$

Structure A has the smaller formal charges and is therefore the more favorable arrangement.

9.58 *See Section 9.5 and Examples 9.8, 9.9, 9.10, 9.12.*

SCN^-, CSN^-

Total valence electrons $= [1 \times 6(\mathrm{S}) + 1 \times 4(\mathrm{C}) + 1 \times 5(\mathrm{N}) = 1(\mathrm{charge})] = 16$.

Twelve electrons remain after assigning two single bonds in either connectivity. Sixteen unshared electrons are needed to give each atom a noble gas configuration. Hence, four electrons (16-12) must be used to form two additional bonds. This gives three possible resonance forms having the SCN^- connectivity:

A	B	C
$[\mathrm{S{=}C{=}N}]^-$	$[\mathrm{S{\equiv}C{-}N}]^-$	$[\mathrm{S{-}C{\equiv}N}]^-$

Formal Charges:

	A	B	C
S	$6 - 4 - (\frac{1}{2})(4) = 0$	$6 - 2 - (\frac{1}{2})(6) = +1$	$6 - 6 - (\frac{1}{2})(2) = -1$
C	$4 - 0 - (\frac{1}{2})(8) = 0$	$4 - 0 - (\frac{1}{2})(8) = 0$	$4 - 0 - (\frac{1}{2})(8) = 0$
N	$5 - 4 - (\frac{1}{2})(4) = -1$	$5 - 6 - (\frac{1}{2})(2) = -2$	$5 - 2 - (\frac{1}{2})(6) = 0$

When these formal charges are included in the Lewis structures, we obtain:

A	B	C
$\mathrm{S{=}C{=}N}^{\ominus}$	$\mathrm{S}^{\oplus}\mathrm{{\equiv}C{-}N}^{\ominus\ominus}$	$\mathrm{S}^{\ominus}\mathrm{{-}C{\equiv}N}$

Structures A and C have lower formal charges than B and are favored over Structure B. Structure A is favored over Structure C, because it places the negative charge on the more electronegative N atom.

There also are three possible resonance forms having the CSN^- connectivity:

A′	B′	C′
$[\mathrm{C{=}S{=}N}]^-$	$[\mathrm{C{\equiv}S{-}N}]^-$	$[\mathrm{C{-}S{\equiv}N}]^-$

Formal Charges:

	A′	B′	C′
C	$4 - 4 - (\frac{1}{2})(4) = -2$	$4 - 2 - (\frac{1}{2})(6) = -1$	$4 - 6 - (\frac{1}{2})(2) = -3$
S	$6 - 0 - (\frac{1}{2})(8) = +2$	$6 - 0 - (\frac{1}{2})(8) = +2$	$6 - 0 - (\frac{1}{2})(8) = +2$
N	$5 - 4 - (\frac{1}{2})(4) = -1$	$5 - 6 - (\frac{1}{2})(2) = -2$	$5 - 2 - (\frac{1}{2})(6) = 0$

When these formal charges are included in the Lewis structures, we obtain:

A′	B′	C′
$^{\ominus\ominus}\mathrm{C{=}S}^{\oplus\oplus}\mathrm{{=}N}^{\ominus}$	$^{\ominus}\mathrm{C{\equiv}S}^{\oplus\oplus}\mathrm{{-}N}^{\ominus\ominus}$	$^{\ominus\ominus\ominus}\mathrm{C{-}S}^{\oplus\oplus}\mathrm{{\equiv}N}$

Hence, the structures having CSN^- connectivity have higher formal charges than those having SCN^- connectivity and are not as favorable. Structure A above has formal charges which are comparable to those of Structure C but has the advantage of placing the negative formal charge on the more electronegative N atom. Hence, Structure A is favored.

9.59 *See Section 9.6 and Example 9.12.*

(a) NO_2^-, O-N-O

Total valence electrons $= [1\times 5(\text{N}) + 2\times 6(\text{O}) + 1(\text{charge})] = 18$.

Fourteen electrons remain after assigning two single bonds, and sixteen unshared electrons are needed to give each atom a noble gas configuration (6 for each O and 4 for N). Hence, two electrons (16-14) must be used to form one additional bond. This gives the following as possible resonance forms:

A: $[\,:\ddot{\underset{..}{O}}-\ddot{N}=\ddot{\underset{..}{O}}\,]^-$

B: $[\,\ddot{\underset{..}{O}}=\ddot{N}-\ddot{\underset{..}{O}}:\,]^-$

Formal Charges:

	A	B
O left	$6-6-\left(\frac{1}{2}\right)(2)=-$	$6-4-\left(\frac{1}{2}\right)(4)=0$
N	$5-2-\left(\frac{1}{2}\right)(6)=0$	$5-2-\left(\frac{1}{2}\right)(6)=0$
O right	$6-4-\left(\frac{1}{2}\right)(4)=0$	$6-6-\left(\frac{1}{2}\right)(2)=-$

When these formal charges are included in the Lewis structures, we obtain

A: $\ominus:\ddot{\underset{..}{O}}-\ddot{N}=\ddot{\underset{..}{O}}$

B: $\ddot{\underset{..}{O}}=\ddot{N}-\ddot{\underset{..}{O}}:\ominus$

Structures A and B have equal formal charges and are therefore equally important.

(b) ClCN, Cl-C-N

Total valence electrons $= [1\times 7(\text{Cl}) + 1\times 4(\text{C}) + 1\times 5(\text{N})] = 16$.

Twelve electrons remain after assigning two single bonds, and sixteen electrons are needed to give each atom a noble gas configuration (6 for Cl, 4 for C and 6 for N).

Hence, four electrons (16-12) must be used to form two additional bonds. This gives the following as possible resonance forms:

A: $\ddot{\underset{..}{Cl}}=C=\ddot{\underset{..}{N}}$

B: $:\ddot{\underset{..}{Cl}}-C\equiv N:$

C: $\ddot{Cl}\equiv C-\ddot{\underset{..}{N}}:$

Formal Charges:

	A	B	C
Cl	$7-4-\left(\frac{1}{2}\right)(4)=+1$	$7-6-\left(\frac{1}{2}\right)(2)=0$	$7-2-\left(\frac{1}{2}\right)(6)=+2$
C	$4-0-\left(\frac{1}{2}\right)(8)=0$	$4-\left(\frac{1}{2}\right)(8)=0$	$4-\left(\frac{1}{2}\right)(8)=0$
N	$5-4-\left(\frac{1}{2}\right)(4)=-1$	$5-2-\left(\frac{1}{2}\right)(6)=0$	$5-6-\left(\frac{1}{2}\right)(2)=-2$

When these formal charges are included in the Lewis structures, we obtain

A: $\oplus\,\ddot{\underset{..}{Cl}}=C=\ddot{\underset{..}{N}}\,\ominus$

B: $:\ddot{\underset{..}{Cl}}-C\equiv N:$

C: $\oplus\oplus\ddot{Cl}\equiv C-\ddot{\underset{..}{N}}:\ominus\ominus$

Structure B has the lowest formal charges and is therefore most important. Structure C has high formal charges and isn't very important.

9.60	*See Section 9.6 and Example 9.12*

(a) NH_2NO_2,

H—N—N—O
    | |
    H   O

Total valence electrons $= [2 \times 1(H) + 2 \times 5(N) + 2 \times 6(O)] = 24$.
Fourteen electrons remain after assigning five single bonds, and sixteen unshared electrons are needed to give each atom a noble gas configuration (2 for each N and 6 for each O). Hence, two electrons (16-14) must be used to form an additional bond. The additional bond could occur between the nitogen atoms or between the nitrogen atom on the right and either oxygen atom giving three resonance forms, two of which would be equivalent. These resonance forms are:

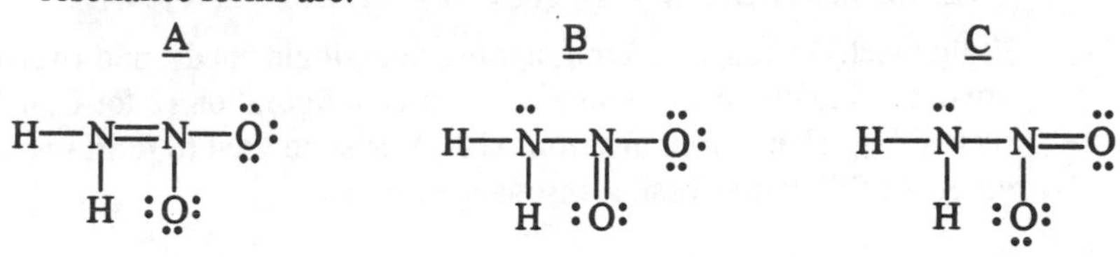

Formal charges:

	A	B	C
H	$1 - 0 - (\frac{1}{2})(2) = 0$	$1 - 0 - (\frac{1}{2})(2) = 0$	$1 - 0 - (\frac{1}{2})(2) = 0$
N left (first)	$5 - 0 - (\frac{1}{2})(8) = +1$	$5 - 2 - (\frac{1}{2})(6) = 0$	$5 - 2 - (\frac{1}{2})(6) = 0$
N right (second)	$5 - 0 - (\frac{1}{2})(8) = +1$	$5 - 0 - (\frac{1}{2})(8) = +1$	$5 - 0 - (\frac{1}{2})(8) = +1$
O bottom	$6 - 6 - (\frac{1}{2})(2) = -1$	$6 - 4 - (\frac{1}{2})(4) = 0$	$6 - 6 - (\frac{1}{2})(2) = -1$
O right	$6 - 6 - (\frac{1}{2})(2) = -1$	$6 - 6 - (\frac{1}{2})(2) = -1$	$6 - 4 - (\frac{1}{2})(4) = 0$

When these formal charges are included in the Lewis structures, we obtain:

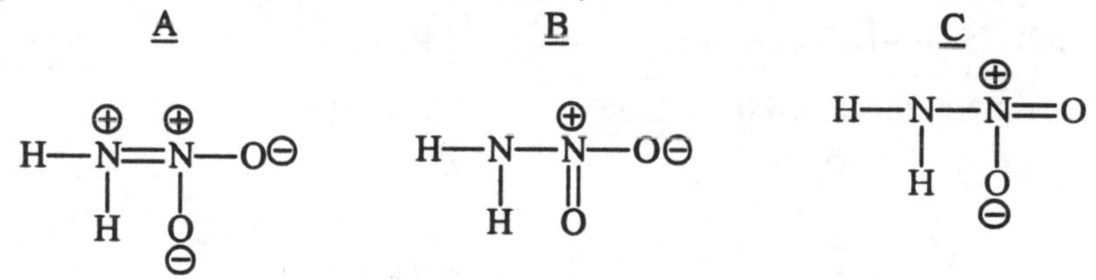

Structure A has higher formal charges and places like charges on adjacent atoms, so it isn't very important. Structures B and C have the same formal charges and are equally important.

(b) HN_3, H-N-N-N
Total valence electrons $= [1 \times 1(H) + 3 \times 5(N)] = 16$.
Ten electrons remain after assigning three single bonds, and fourteen unshared electrons are needed to give each atom a noble gas configuration (4 for each of the middle N atoms and 6 for the end N atom). Hence, four electrons (14-10) must be used to form additional bonds. This leads to three possible resonance forms:

A	B	C
H—N̈=N=N̈:	H—N≡N—N̈:	H—N̈—N≡N:

Formal charges:

	A	B	C
H	$1 - 0 - (\frac{1}{2})(2) = 0$	$1 - 0 - (\frac{1}{2})(2) = 0$	$1 - 0 - (\frac{1}{2})(2) = 0$
N (left)	$5 - 2 - (\frac{1}{2})(6) = 0$	$5 - 0 - (\frac{1}{2})(8) = +1$	$5 - 4 - (\frac{1}{2})(4) = -1$
N (center)	$5 - 0 - (\frac{1}{2})(8) = +1$	$5 - 0 - (\frac{1}{2})(8) = +1$	$5 - 0 - (\frac{1}{2})(8) = +1$
N (right)	$5 - 4 - (\frac{1}{2})(4) = -1$	$5 - 6 - (\frac{1}{2})(2) = -2$	$5 - 2 - (\frac{1}{2})(6) = 0$

Structures A and C have the lowest formal charges and are equally important. Structure B places like formal charges on adjacent atoms and has higher overall formal charges. Hence, Structure B isn't very important.

When these formal charges are included in the Lewis structures, we obtain:

A: $H-\ddot{N}=N=\ddot{N}:$ with $\oplus$ on the central N and $\ominus$ on the terminal N

B: $H-N\equiv N-\ddot{N}:^{\ominus\ominus}$ with $\oplus$ on the first N and $\oplus$ on the central N

C: $H-\ddot{N}-N\equiv N:$ with $\ominus$ on the first N and $\oplus$ on the central N

9.61 ***See Section 9.6 and Example 9.12.***

(a) CH_2N_2, $H-C(H)-N-N$

Total valence electrons $= [1\times 4(C)+2\times 1(H)+2\times 5(N)] = 16$.
Eight electrons remain after assigning four single bonds, and twelve unshared electrons are needed to give each atom a noble gas configuration (2 for C, 4 for first N and 6 for second N). Hence, four electrons (12-8) must be used to form two additional bonds. This gives the following possible resonance forms:

A: $H-C(H)=N=\ddot{N}:$

B: $H-\ddot{C}(H)-N\equiv N:$

Formal Charges:

	A	B
H	$1-0-(\frac{1}{2})(2)=0$	$1-0-(\frac{1}{2})(2)=0$
C	$4-0-(\frac{1}{2})(8)=0$	$4-2-(\frac{1}{2})(6)=-1$
N left (first)	$5-0-(\frac{1}{2})(8)=+1$	$5-0-(\frac{1}{2})(8)=+1$
N right (second)	$5-4-(\frac{1}{2})(4)=-1$	$5-2-(\frac{1}{2})(6)=0$

When these formal charges are included in the Lewis structures, we obtain

A: $H-C(H)=N=\ddot{N}:$ with $\oplus$ on the first N and $\ominus$ on the second N

B: $H-\ddot{C}(H)-N\equiv N:$ with $\ominus$ on C and $\oplus$ on the first N

Structure A is favored, since it places the -1 formal charge on the more electronegative N atom and not the less electronegative C atom.

(b) OCN^-

$[O—C—N]^-$

Total valence electrons $= [1 \times 6(O) + 1 \times 4(C) + 1 \times 5(N) + 1(\text{charge})] = 16.$ Twelve electrons remain after assigning two single bonds, and sixteen unshared electrons are needed to give each atom a noble gas configuration (6 for O, 4 for C and 6 for N). Hence four electrons (16-12) must be used to form two additional bonds. This gives the following possible resonance forms:

A	B	C
$[O{=}C{=}N:]^-$	$[:O{\equiv}C—N:]^-$	$[:O—C{\equiv}N:]^-$

Formal Charges:

	A	B	C
O	$6-4-\left(\frac{1}{2}\right)(4)=0$	$6-2-\left(\frac{1}{2}\right)(6)=+1$	$6-6-\left(\frac{1}{2}\right)(2)=-1$
C	$4-0-\left(\frac{1}{2}\right)(8)=0$	$4-0-\left(\frac{1}{2}\right)(8)=0$	$4-0-\left(\frac{1}{2}\right)(8)=0$
N	$5-4-\left(\frac{1}{2}\right)(4)=-1$	$5-6-\left(\frac{1}{2}\right)(2)=-2$	$5-2-\left(\frac{1}{2}\right)(6)=0$

When these formal charges are included in the Lewis structures, we obtain

A	B	C
$O{=}C{=}N:$ (⊖ on N)	$:O{\equiv}C—N:$ (⊕ on O, ⊖⊖ on N)	$:O—C{\equiv}N:$ (⊖ on O)

Structure C is favored, since it places the -1 formal charge on the more electronegative O atom and not the less electronegative N atom.

9.62 ***See Section 9.6 and Example 9.12.***

(a) $CH_3C(O)NO_2$

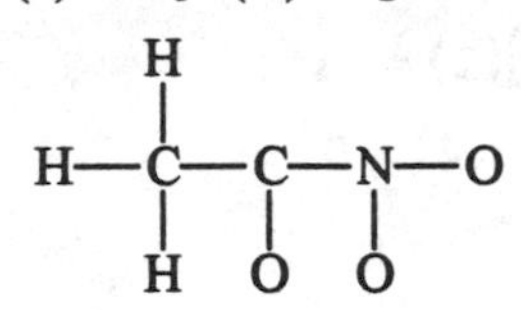

Total valence electrons$= [2 \times 4(C) + 3 \times 1(H) + 1 \times 5(N) + 3 \times 6(O)] = 34.$ Eighteen electrons remain after assigning eight single bonds, and twenty two unshared electrons are needed to give each atom a noble gas configuration (2 for C, 2 for N, and 6 for each O). Hence, four electrons (22-18) must be used to form two additional bonds. This gives the following possible resonance forms:

A	B
$H_3C—C({=}O:)—N({=}O:)—O:$ (N—O single on right)	$H_3C—C({=}O:)—N(—O:)={O}:$ (N=O double on right)

Formal Charges:

	A	B
H	$1-0-\left(\frac{1}{2}\right)(2)=0$	$1-0-\left(\frac{1}{2}\right)(2)=0$
C (left)	$4-0-\left(\frac{1}{2}\right)(8)=0$	$4-0-\left(\frac{1}{2}\right)(8)=0$
C (right)	$4-0-\left(\frac{1}{2}\right)(8)=0$	$4-0-\left(\frac{1}{2}\right)(8)=0$
O (on C)	$6-4-\left(\frac{1}{2}\right)(4)=0$	$6-4-\left(\frac{1}{2}\right)(4)=0$
N	$5-0-\left(\frac{1}{2}\right)(8)=+1$	$5-0-\left(\frac{1}{2}\right)(8)=+1$
O (over N)	$6-4-\left(\frac{1}{2}\right)(4)=0$	$6-6-\left(\frac{1}{2}\right)(2)=-1$
O (right)	$6-6-\left(\frac{1}{2}\right)(2)=-1$	$6-4-\left(\frac{1}{2}\right)(4)=0$

When these formal charges are included in the Lewis structures, we obtain:

A　　B

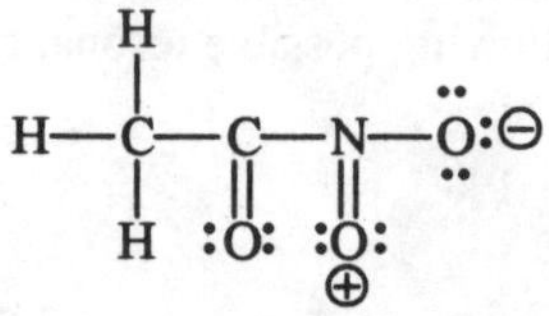

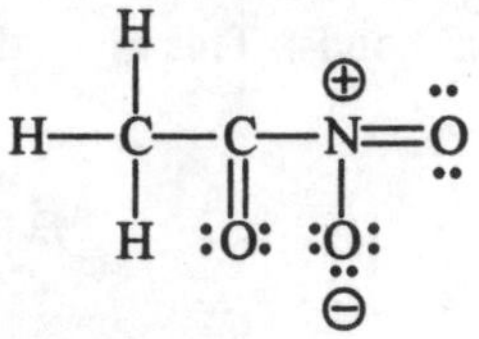

Structures A and B are equivalent and are equally important.

(b) NO_3^-

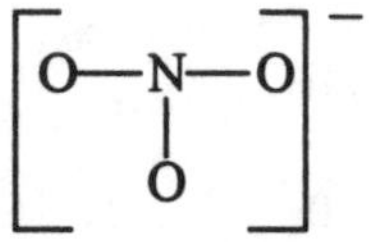

$[O—N(—O)—O]^-$

Total valence electrons $= [1 \times 5(N) + 3 \times 6(O) + 1(\text{charge})] = 24$.
Eighteen electrons remain after assigning three single bonds, and twenty unshared electrons are needed to give each atom a noble gas configuration. Hence, two electrons (20-18) must be used to form an additional bond. This gives the following possible resonance forms:

A　　B 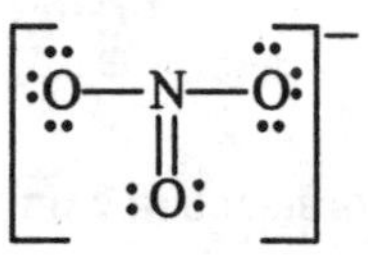　　C

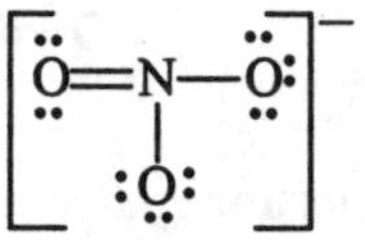

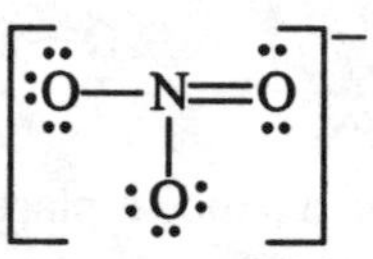

Formal Charges:

	A	B	C
O (left)	$6-4-(\frac{1}{2})(4)=0$	$6-6-(\frac{1}{2})(2)=-1$	$6-6-(\frac{1}{2})(2)=-1$
N	$5-0-(\frac{1}{2})(8)=+1$	$5-0-(\frac{1}{2})(8)=+1$	$5-0-(\frac{1}{2})(8)=+1$
O (down)	$6-6-(\frac{1}{2})(2)=-1$	$6-4-(\frac{1}{2})(4)=0$	$6-6-(\frac{1}{2})(2)=-1$
O (right)	$6-6-(\frac{1}{2})(2)=-1$	$6-6-(\frac{1}{2})(2)=-1$	$6-4-(\frac{1}{2})(4)=0$

When these formal charges are included in the Lewis structures, we obtain:

A　　B 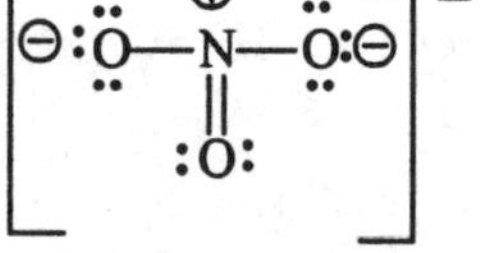　　C

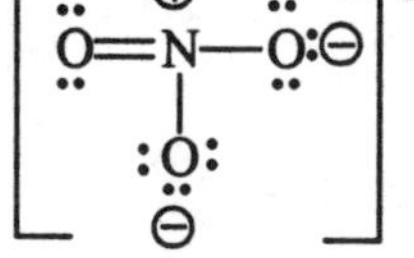

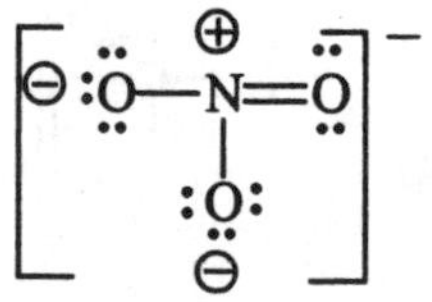

Structures A, B and C are equivalent and are equally important.

9.63	***See Section 9.6 and Example 9.12.***

(a) N_3^-

$[N-N-N]^-$

Total valence electrons $= [3 \times 5(N) + 1(\text{charge})] = 16$.

Twelve electrons remain after assigning two single bonds and sixteen unshared electrons are needed to give each atom a noble gas configuration (6 for left and right N and 4 for middle N). Hence, four electrons (16-12) must be used to form two additional bonds. This gives the following possible resonance forms:

A	B	C
$[:\ddot{N}=N=\ddot{N}:]^-$	$[:N\equiv N-\ddot{\underset{..}{N}}:]^-$	$[:\ddot{\underset{..}{N}}-N\equiv N:]^-$

Formal Charges:

	A	B	C
N left	$5-4-(\frac{1}{2})(4)=-1$	$5-2-(\frac{1}{2})(6)=0$	$5-6-(\frac{1}{2})(2)=-2$
N middle	$5-0-(\frac{1}{2})(8)=+1$	$5-0-(\frac{1}{2})(8)=+1$	$5-0-(\frac{1}{2})(8)=+1$
N right	$5-4-(\frac{1}{2})(4)=-1$	$5-6-(\frac{1}{2})(2)=-2$	$5-2-(\frac{1}{2})(6)=0$

When these formal charges are included in the Lewis structures, we obtain

A	B	C
$:\ddot{N}=N=\ddot{N}:$ (⊖ ⊕ ⊖)	$:N\equiv N-\ddot{\underset{..}{N}}:$ (⊕ ⊖⊖)	$:\ddot{\underset{..}{N}}-N\equiv N:$ (⊖⊖ ⊕)

Structure A is favored, since it avoids having a -2 formal charge on one atom.

(b) CO_3^{2-}

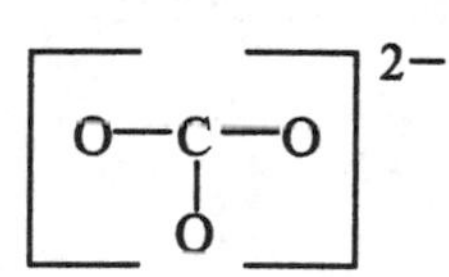

Total valence electrons $= [1 \times 4(C) + 3 \times 6(O) + 2(\text{charge})] = 24$.

Eighteen electrons remain after assigning three single bonds, and twenty unshared electrons are needed to give each atom a noble gas configuration (6 for each O and 2 for C). Hence, two electrons (20-18) must be used to form one additional bond. This gives the following possible resonance forms:

A	B	C
$[O=C(-O:)-O:]^{2-}$	$[:O-C(=O)-O:]^{2-}$	$[:O-C(-O:)=O]^{2-}$

Formal Charges:

	A	B	C
O left	$6-4-(\frac{1}{2})(4)=0$	$6-6-(\frac{1}{2})(2)=-1$	$6-6-(\frac{1}{2})(2)=-1$
C	$4-0-(\frac{1}{2})(8)=0$	$4-0-(\frac{1}{2})(8)=0$	$4-0-(\frac{1}{2})(8)=0$
O middle	$6-6-(\frac{1}{2})(2)=-1$	$6-4-(\frac{1}{2})(4)=0$	$6-6-(\frac{1}{2})(2)=-1$
O right	$6-6-(\frac{1}{2})(2)=-1$	$6-6-(\frac{1}{2})(2)=-1$	$6-4-(\frac{1}{2})(4)=0$

When these formal charges are included in the Lewis structures, we obtain

A B C

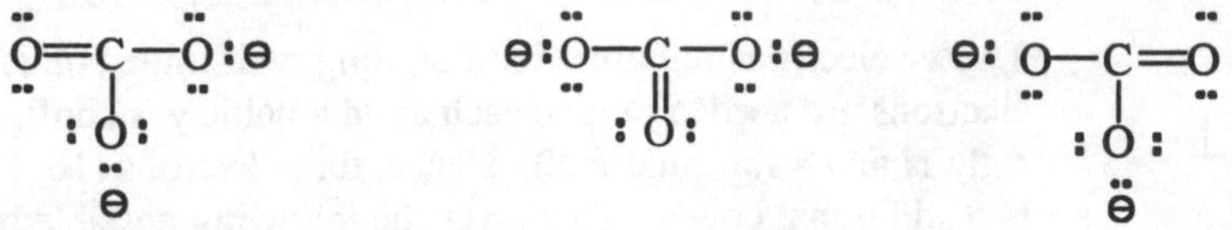

Structures A, B, and C have equal formal charges and are therefore equally important.

9.64 ***See Section 9.6 and Example 9.12.***

(a) HCO_2^-

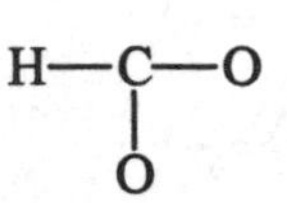

Total valence electrons = $\left[1\times 1(\text{H}) + 1\times 4(\text{C}) + 2\times 6(\text{O}) + 1(\text{charge})\right] = 18$.

Twelve electrons remain after assigning three single bonds, and fourteen unshared electrons are needed to give each atom a noble gas configuration (2 for C and 6 for each O). Hence, two electrons (14-12) must be used to form one additional bond. This gives the following possible resonance forms:

A B

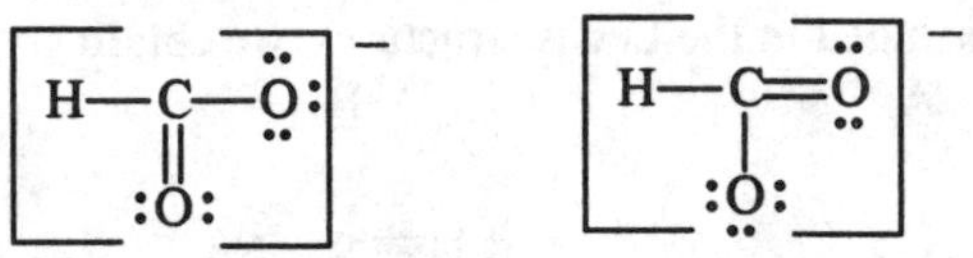

Formal Charges:

	A	B
H	$1-0-(\frac{1}{2})(2)=0$	$1-0-(\frac{1}{2})(2)=0$
C	$4-0-(\frac{1}{2})(8)=0$	$4-0-(\frac{1}{2})(8)=0$
O (down)	$6-4-(\frac{1}{2})(4)=0$	$6-6-(\frac{1}{2})(2)=-1$
O (right)	$6-6-(\frac{1}{2})(2)=-1$	$6-4-(\frac{1}{2})(4)=0$

When these formal charges are included in the Lewis structures, we obtain:

A B

Structures A and B are equivalent and are equally important.

(b) HNO_2, H-O-N-O

Total valence electrons = $[1 \times 1(\mathrm{H}) + 1 \times 5(\mathrm{N}) + 2 \times 6(\mathrm{O})] = 18$.

Twelve electrons remain after assigning three single bonds, and fourteen unshared electrons are needed to give each atom a noble gas configuration (4 for O on the left, 4 for N, and 6 for O on the right). Hence, two electrons (14-12) must be used to form one additional bond. This gives the following possible resonance forms:

A	B
H—O=N—O	H—O—N=O

Formal Charges:

	A	B
H	$1-0-(\frac{1}{2})(2)=0$	$1-0-(\frac{1}{2})(2)=0$
O (left)	$6-2-(\frac{1}{2})(6)=+1$	$6-4-(\frac{1}{2})(4)=0$
N	$5-2-(\frac{1}{2})(6)=0$	$5-2-(\frac{1}{2})(6)=0$
O (right)	$6-6-(\frac{1}{2})(2)=-1$	$6-4-(\frac{1}{2})(4)=0$

Structure B has the lower formal charges and is important. Structure A has higher formal charges and places an unlikely positive formal charge on oxygen and isn't very important.

9.65 ***See Section 9.6 and Example 9.11.***

$C_6H_5CH_3$

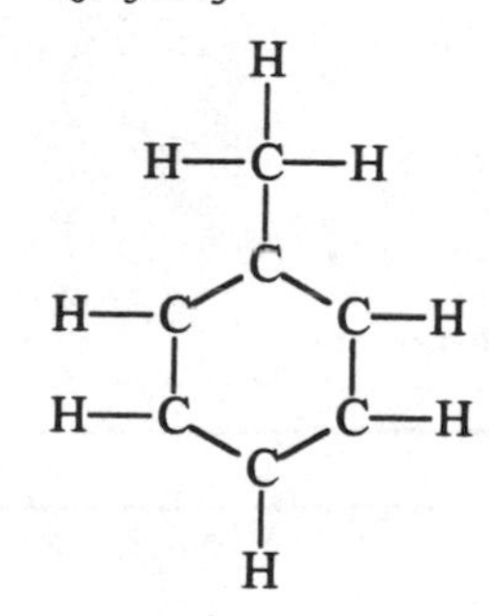

Total valence electrons = $[7 \times 4(\mathrm{C}) + 8 \times 1(\mathrm{H})] = 36$.

Six electrons remain after assigning fifteen single bonds, and twelve unshared electrons are needed to give each atom a noble gas configuration (2 for each C member of the cyclic ring). Hence, six electrons (12-6) must be used to form three additional bonds. This gives the following resonance possibilities:

A B

Formal Charges:

	A	B
H	$1-0-(\frac{1}{2})(2)=0$	$1-0-(\frac{1}{2})(2)=0$
C (top)	$4-0-(\frac{1}{2})(8)=0$	$4-0-(\frac{1}{2})(8)=0$
C (ring)	$4-0-(\frac{1}{2})(8)=0$	$4-0-(\frac{1}{2})(8)=0$

Structures A and B have formal charges of zero and are equivalent and equally important.

9.66 *See Section 9.6 and Example 9.11.*

C_6H_5Cl

Total valence electrons = $[6 \times 4(C) + 5 \times 1(H) + 7 \times 1(Cl)] = 36$.

Twelve electrons remain after assigning twelve single bonds, and eighteen unshared electrons are needed to give each atom a noble gas configuration (6 for Cl and 2 for each O). Hence, six electrons (18-12) must be used to form three additional bonds. This gives the following possible resonance forms:

A B

Formal Charges:

	A	B
Cl	$7-6-(\frac{1}{2})(2)=0$	$7-6-(\frac{1}{2})(2)=0$
H	$1-0-(\frac{1}{2})(2)=0$	$1-0-(\frac{1}{2})(2)=0$
C	$4-0-(\frac{1}{2})(8)=0$	$4-0-(\frac{1}{2})(8)=0$

Structures A and B have formal charges of zero and are equivalent and equally important.

9.67 *See Section 9.6 and Example 9.12.*

CH_3NCO

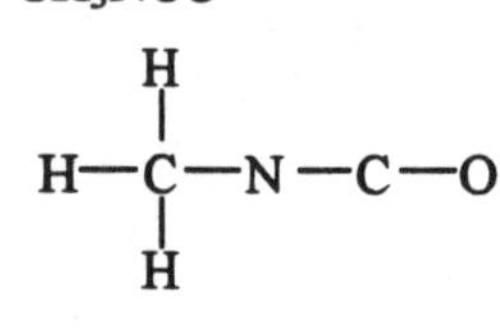

Total valence electrons $= [2 \times 4(C) + 3 \times 1(H) + 1 \times 5(N) + 1 \times 6(O)] = 22$.

Ten electrons remain after assigning six single bonds, and fourteen unshared electrons are needed to give each atom a noble gas configuration (4 for N, 4 for second C and 6 for O). Hence, four electrons(14-10) must be used to form two additional bonds. This gives the following as possible resonance forms:

The first structure is likely to be the most important and the last structure is likely to be the least important, since it places a positive formal charge on the most electronegative element in the molecule.

9.68 ***See Section 9.6 and Example 9.12.***

$C_2O_4^{2-}$

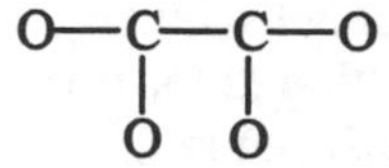

Total valence electons = $[2 \times 4(C) + 4 \times 6(O) + 2(\text{charge})] = 34$.

Twenty four electrons remain after assigning five single bonds, and twenty eight unshared electrons are needed to give each atom a noble gas configuration. Hence, four electrons (28-24) must be used to form two additional bonds. This gives the following possible resonance forms:

A B C D

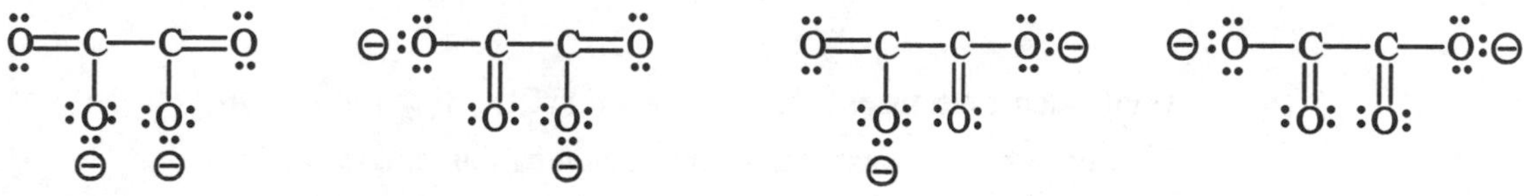

The structures have equivalent formal charges and are equally important.

9.69 ***See Section 9.7 and Examples 9.13, 9.14.***

(a) SeF_6

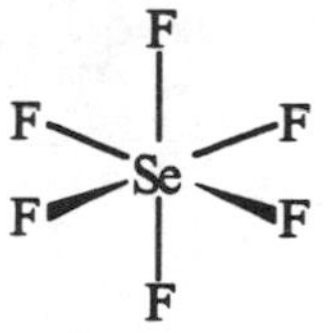

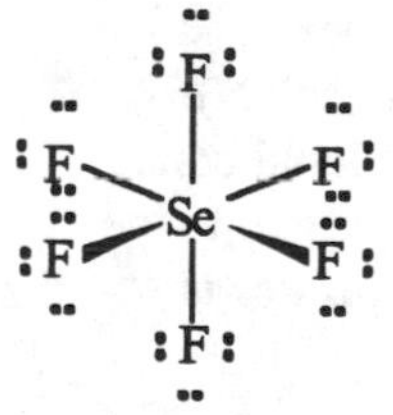

Total valence electrons $= [1 \times 6(Se) + 6 \times 7(F)] = 48$.

Thirty six electrons remain after assigning six single bonds, and thirty six unshared electrons are needed to give each atom a noble gas configuration (6 for each F). Forty eight electrons are used in writing the **electron rich** Lewis structure.

(b) BBr_3

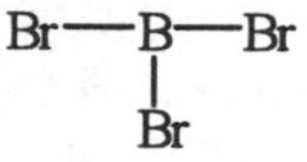

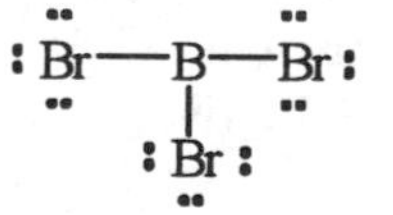

Total valence electrons $= [1 \times 3(B) + 3 \times 7(Br)] = 24$.

Eighteen electrons remain after assigning three single bonds, and eighteen unshared electrons are needed to give each atom a noble gas configuration (6 for each Br). Eighteen electrons are shown in the **electron deficient** Lewis structure.

(c) NO_2

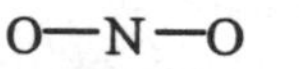

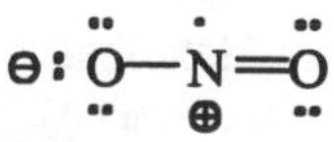

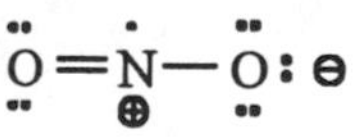

Total valence electrons $= [1 \times 5(N) + 2 \times 6(O)] = 17$.

Thirteen electrons remain after assigning two single bonds, and sixteen unshared electrons are needed to give each atom a noble gas configuration (6 for each O and 4 for N). Hence, three electrons (16-13) are available to form additional bonds. Two of these are used to form one additional bond, and the odd electron is assigned to N. Seventeen electrons are used in writing the **odd electron** Lewis structure.

9.70 *See Section 9.7 and Examples 9.13, 9.14.*

(a) IF_3

F—I—F
 |
 F

Total valence electrons = $[1 \times 7(\text{I}) + 3 \times 7(\text{F})] = 28$.

Twenty two electrons remain after assigning three single bonds, and only twenty unshared electrons are needed to give each atom a noble gas configuration (2 for I and 6 for each F). Hence, two additional electrons (22-20) must be assigned to iodine, the central atom. Twenty eight electrons are used in writng the **electron rich** Lewis structure.

(b) ICl_4^-

Total valence electrons = $[1 \times 7(\text{I}) + 4 \times 7(\text{Cl}) + 1(\text{charge})] = 36$.

Twenty eight electrons remain after assigning four single bonds, and only twenty four unshared electrons are needed to give each atom a noble gas configuration (6 for each Cl). Hence, four additional electrons (28-24) must be assigned to iodine, the central atom. Thirty six electrons are used in writing the **electron rich** Lewis structure.

(c) N_2^+

N—N $^+$

Total valence electrons = $[2 \times 5(\text{N}) - 1(\text{charge})] = 9$.

Seven electrons remain after assigning one single bond, and twelve unshared electrons are needed to give each atom a noble gas configuration (6 for each N). Hence, four of the seven remaining electrons are used to form two additional bonds, two are assigned to one nitrogen atom and the remaining electron is assigned to the other nitrogen atom. Nine electrons are used in writing the **odd electron** Lewis structure.

9.71 *See Section 9.7 and Examples 9.13, 9.14.*

(a) XeF_2

F—Xe—F

Total valence electrons = $[1 \times 8(\text{Xe}) + 2 \times 7(\text{F})] = 22$.

Twenty electrons remain after assigning two single bonds, and only sixteen unshared electrons are needed to give each atom a noble gas configuration (6 for each F and 4 for Xe). Hence, four additional electrons (20-16) must be assigned to xenon, the central atom. Twenty two electrons are used in writing the **electron rich** Lewis structure.

(b) $BeCl_2$

Cl—Be—Cl

Total valence elctrons = $[1 \times 2(\text{Be}) + 2 \times 7(\text{Cl})]16$.

Twelve electrons remain after assigning two single bonds, and sixteen unshared electrons are needed to give each atom a noble gas configuration (4 for Be and 6 for each Cl). Hence, it would appear the four electrons (16-12) must be used to form additional bonds, but beryllium is an element that is known to form electron deficient compounds. Hence, sixteen electrons are used in writing a single-bonded **electron deficient** Lewis structure.

(c) XeO_2F_4

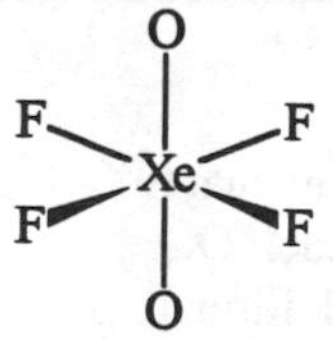

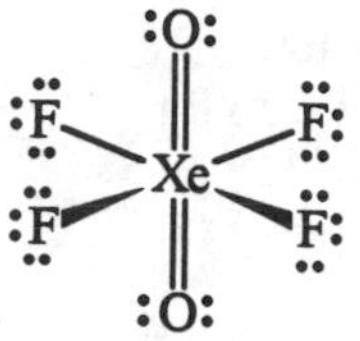

Total valence electrons = $[1 \times 8(\text{Xe}) + 2 \times 6(\text{O}) + 4 \times 7(\text{F})] = 48.$
Thirty six electrons remain after assigning six single bonds, and thirty six unshared electrons are needed to give each atom a noble gas configuration (6 for each O and 6 for each F). Forty eight electrons are used in writing the **electron rich** Lewis structure that leads to the lowest formal charges..

9.72	*See Section 9.7 and Examples 9.13, 9.14.*

(a) BI_3

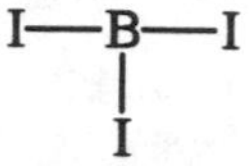

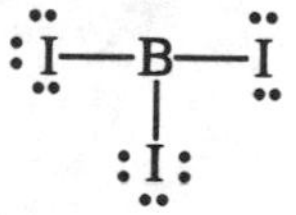

I—B—I with I below B

Total valence electrons = $[1 \times 3(\text{B}) + 3 \times 7(\text{I})] = 24.$
Eighteen electrons remain after assigning three single bonds, and eighteen unshared electrons are needed to give each atom a noble gas configuration (6 for each I). Eighteen electrons are shown in the **electron deficient** Lewis structure.

(b) IF_5

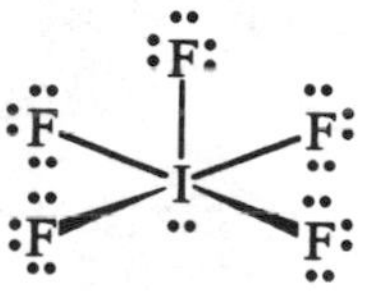

Total valence electrons = $[1 \times 7(\text{I}) + 5 \times 7(\text{F})] = 42.$
Thirty two electrons remain after assigning five single bonds, and only thirty unshared electrons are needed to give each atom a noble gas configuration (6 for each F). Hence, two additional electrons (32-30) must be added to iodine, the central atom. Forty two electrons are used in writing the **electron rich** Lewis structure.

(c) HN_2

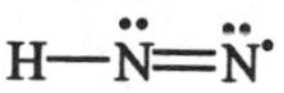

H—N—N

H—N=N

Total valence electrons = $[1 \times 1(\text{H}) + 2 \times 5(\text{N})] = 11.$
Seven electrons remain after assigning two single bonds, and ten electrons are needed to give each atom a noble gas configuration (4 for N on the left and 6 for N on the right). Hence, two of the seven remaining electrons are assigned to the nitrogen atom on the left, two are used to form an additional bond between the nitrogen atoms and three are assigned to the nitrogen atom on the right because this arrangement comes closest to giving each atom a noble gas configuration. Eleven electrons are used in writing the **odd electron** Lewis structure.

9.73 ***See Section 9.7 and Example 9.15.***

(a) SeO_2

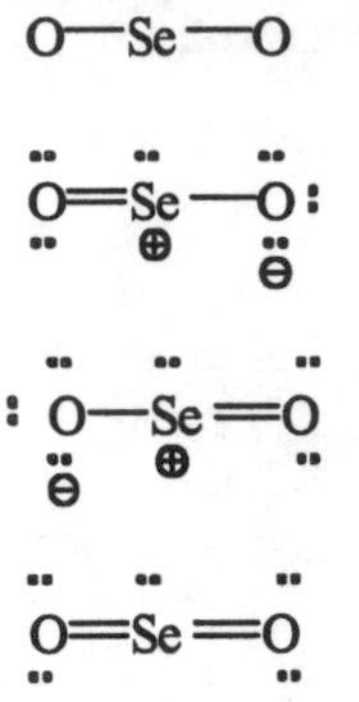

O—Se—O

Total valence electrons $= [1 \times 6(\text{Se}) + 2 \times 6(\text{O})] = 18$.

Fourteen electrons remain after assigning two single bonds, and sixteen unshared electrons are needed to give each atom a noble gas configuration (6 for each O and 4 for Se). Hence, two electrons (16-14) must be used to form one additional bond. This gives two equally important resonance forms which obey the octet rule. However, eighteen electrons can also be used to draw a Lewis structure which shows an expanded valence shell for Se and has formal charges of zero for all atoms. This is likely to be the most important Lewis structure.

(b) SO_3

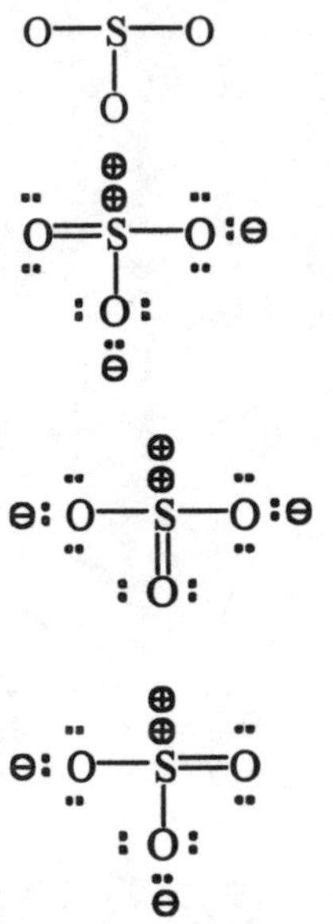

Total valence electrons $= [1 \times 6(\text{S}) + 3 \times 6(\text{O})] = 24$.

Eighteen electrons remain after assigning three single bonds, and twenty unshared electrons are needed to give each atom a noble gas configuration (6 for each O and 2 for S). Hence, two electrons must be used to form an additional bond. This gives three equally important resonance forms which obey the octet rule. However, twenty four electrons can also be used to draw three resonance forms having two double bonds and one form having three double bonds. All of these involve an expanded valence shell for S and reduced formal charges. However, only the last structure has formal charges of zero for all atoms and is therefore likely to be the most important Lewis structure.

9.74 ***See Section 9.7 and Example 9.15.***

(a) $HClO_4$

Total valence electrons = $[1 \times 1(H) + 1 \times 7(Cl) + 4 \times 6(O)] = 32$

Twenty two electrons remain after assigning four single bonds, and twenty two electrons are needed to give each atom a noble gas configuration (4 for O on left, and 6 for the other O atoms). Thirty two electrons can be used in writing a Lewis structure which obeys the octet rule. However, thirty two electrons can also be used to write a Lewis structure involving an expanded valence shell for chlorine and zero formal charges for all the atoms. This is likely to be the more important Lewis structure.

(b) H_3PO_4

Total valence electrons = $[3 \times 1(H) + 1 \times 5(P) + 4 \times 6(O)] = 32$.

Eighteen electrons remain after assigning seven single bonds, and eighteen electrons are needed to give each atom a noble gas configuration (6 for top O and 4 for other O atoms). Thirty two electrons can be used in writing a Lewis structure which obeys the octet rule. However, thirty two electrons can also be used to write a Lewis structure involving an expanded octet for phosphorus and zero formal charges for all the atoms. This is likely to be the more important Lewis structure.

9.75 ***See Section 9.7 and Example 9.15.***

(a) ClO_2^-, $[O-Cl-O]^-$

Total valence electrons = $[1 \times 7(Cl) + 2 \times 6(O) + 1(charge)] = 20$.

Sixteen electrons remain after assigning two single bonds, and sixteen unshared electrons are needed to give each atom a noble gas configuration (6 for each O and 4 for C). Sixteen electrons can be used to draw a Lewis structure which obeys the octet rule. However, sixteen electrons can also be used to draw three resonance forms which have an expanded valence shell for Cl and reduced formal charges. However, the last one shown places the -1 formal charge on the less electronegative atom and isn't likely to be important. The forms with just one double bond are therefore likely to be the most important resonance forms.

(b) HSO_3^-

Total valence electrons $= [1 \times 1(H) + 1 \times 6(S) + 3 \times 6(O) + 1(\text{charge})] = 26.$

[O—S—O—H]⁻ with O below S

Eighteen electrons remain after assigning four single bonds, and eighteen unshared electrons are needed to give each atom a noble gas configuration (6 for each terminal O, 2 for S and 4 for other O). Eighteen electrons can be used to draw a Lewis structure which obeys the octet rule. However, eighteen electrons can also be used to draw three resonance forms which have an expanded valence shell for S. Two of these have reduced formal charges and are likely to be the most important resonance forms.

O=S—O—H (O below S)   O—S—O—H (O=S below)   O—S=O—H (O below S)

9.76 *See Section 9.7 and Example 9.15.*

(a) SO_3^{2-}

Total valence electrons $= [1 \times 6(S) + 3 \times 6(O) + 2(\text{charge})] = 26.$

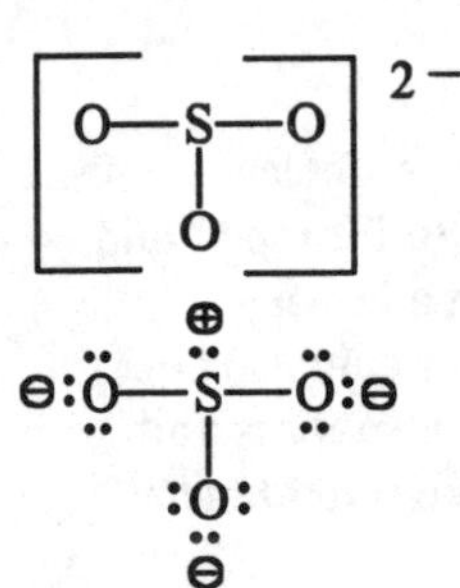

Twenty electrons remain after assigning three single bonds, and twenty electrons are needed to give each atom a noble gas configuration (2 for S and 6 for each O). Twenty six electrons can be used to write a Lewis structure which obeys the octet rule. However, twenty six electrons can also be used to draw three resonance forms involving an expanded valence shell for sulfur and lower formal charges.

O=S—O (O below S)   O—S=O (O below S)   O—S—O (O=S below)

These structures are equivalent and are likely to be the most important Lewis structures.

(b) ClO_4^-

Total valence electrons = $[1 \times 7(Cl) + 4 \times 6(O) + 1(charge)] = 32$.

Twenty four electrons remain after assigning four single bonds, and twenty four unshared electrons are needed to give each atom a noble gas configuration (6 for each O). Thirty two electrons can be used write a Lewis structure which obeys the octet rule. However, thirty two electrons can also be used to draw four resonance forms involving an expanded octet for chlorine and reduced formal charges.

These structures are equivalent and are likely to be the most important Lewis structures.

9.77 ***See Section 9.8 and Table 9.4.***

H_2CNH

Total valence electrons = $[3 \times 1(H) + 1 \times 4(C) + 1 \times 5(N)] = 12$.

Four electrons remain after assigning four single bonds, and six electrons are needed to give each atom a noble gas configuration (2 for C and 4 for N). Hence, two electrons (6-4) must be used to form one additional bond between the carbon and nitrogen atoms. Twelve electrons are used in writing the Lewis structure.

H_3CNH_2

Total valence electrons = $[5 \times 1(H) + 1 \times 4(C) + 1 \times 5(N)] = 14$.

Two electrons remain after assigning six single bonds, and two unshared electrons are needed to give each atom a noble gas configuration (2 for N). Fourteen electrons are used in writing the Lewis structure.

The carbon-nitrogen bond in H_2CNH is a double bond that is shorter and stronger than the carbon-nitrogen single bond in H_3CNH_2 because it places more electron density between the positively chaged carbon and nitrogen nuclei.

9.78 *See Section 9.8 and Table 9.4.*

HNNH

Total valence electrons = $[2 \times 1(H) + 2 \times 5(N)] = 12$.

H—N—N—H

Six electrons remain after assigning three single bonds, and eight unshared electrons are needed to give each atom a noble gas configuration (4 for each N atom). Hence, two electrons (8-6) must be used to form one additional bond between the nitrogen atoms. Twelve electrons are used in writing the Lewis structure.

$H-\ddot{N}=\ddot{N}-H$

H_2NNH_2

Total valence electrons = $[4 \times 1(H) + 2 \times 5(N)] = 14$.

Four electrons remain after assigning five single bonds, and four unshared electrons are needed to give each atom a noble gas configuration (2 for each N). Fourteen electrons are used in writing the Lewis structure.

```
H—N—N—H
  |  |
  H  H
```

```
   ••  ••
H—N——N—H
  |   |
  H   H
```

The nitogen-nitrogen bond in HNNH is a double bond that is shorter and stronger than the nitrogen-nitrogen single bond in H_2NNH_2 because it places more electron density between the positively charged nitrogen nuclei.

9.79 *See Section 9.8 and Table 9.4.*

(a) $NH_3(g) \rightarrow N(g) + 3H(g)$ $\qquad \Delta H = 3D_{N-H}$

$$\Delta H = (3\text{ mol})\left(389\frac{\text{kJ}}{\text{mol}}\right) = \mathbf{1167\ kJ}$$

(b) $CH_3OH(g) \rightarrow C(g) + 4H(g) + O(g)$ $\qquad \Delta H = 3D_{C-H} + D_{C-O} + D_{O-H}$

$$\Delta H = (3\text{ mol})\left(414\frac{\text{kJ}}{\text{mol}}\right) + (1\text{ mol})\left(351\frac{\text{kJ}}{\text{mol}}\right) + (1\text{ mol})\left(463\frac{\text{kJ}}{\text{mol}}\right) = \mathbf{2056\ kJ}$$

Note: Lewis strutctures can be used to determine what bonds are broken.

9.80 *See Section 9.8 and Table 9.4.*

(a) $CH_2CF_2(g) \rightarrow 2C(g) + 2H(g) + 2F(g)$ $\qquad \Delta H = 2D_{C-H} + D_{C=C} + 2D_{C-F}$

$$\Delta H = (2\text{ mol})\left(414\frac{\text{kJ}}{\text{mol}}\right) + (1\text{ mol})\left(611\frac{\text{kJ}}{\text{mol}}\right) + (2\text{ mol})\left(439\frac{\text{kJ}}{\text{mol}}\right) = \mathbf{2317\ kJ}$$

(b) $N_2H_4(g) \rightarrow 2N(g) + 4H(g)$ $\qquad \Delta H = D_{N-N} + 4D_{N-H}$

$$\Delta H = (1\text{ mol})\left(163\frac{\text{kJ}}{\text{mol}}\right) + (4\text{ mol})\left(389\frac{\text{kJ}}{\text{mol}}\right) = \mathbf{1719\ kJ}$$

Note: Lewis structures can be used to determine what bonds are broken.

9.81 *See Section 9.8, Table 9.4, Example 9.16.*

(a) $2H_2(g) + O_2(g) \rightarrow 2H_2O(g)$

$$\Delta H_{rxn} = [2D_{H-H} + D_{O=O}] - [4D_{O-H}]$$

$$\Delta H_{rxn} = \left[(2\text{ mol})\left(436\frac{\text{kJ}}{\text{mol}}\right) + (1\text{ mol})\left(498\frac{\text{kJ}}{\text{mol}}\right)\right] - \left[(4\text{ mol})\left(463\frac{\text{kJ}}{\text{mol}}\right)\right] = \mathbf{-482\ kJ}$$

(b) $2CO(g) + O_2(g) \rightarrow 2CO_2(g)$

$$\Delta H_{rxn} = [2D_{C\equiv O} + D_{O=O}] - [4D_{C=O}]$$

$$\Delta H_{rxn} = \left[(2\text{ mol})\left(1{,}072\frac{\text{kJ}}{\text{mol}}\right) + (1\text{ mol})\left(498\frac{\text{kJ}}{\text{mol}}\right)\right] - \left[(4\text{ mol})\left(799\frac{\text{kJ}}{\text{mol}}\right)\right] = \mathbf{-554\ kJ}$$

Note: Lewis structures can be used to determine what bonds are broken and what bonds are formed.

9.82 *See Section 9.8, Table 9.4, Example 9.16.*

(a) $CH_3OH(g) + \frac{3}{2}O_2(g) \rightarrow CO_2(g) + 2H_2O(g)$

$$\Delta H_{rxn} = [3D_{C-H} + D_{C-O} + D_{O-H} + \tfrac{3}{2}D_{O=O}] - [2D_{C=O} + 4D_{O-H}]$$

$$= \left[(3\text{ mol})\left(414\frac{\text{kJ}}{\text{mol}}\right) + (1\text{ mol})\left(351\frac{\text{kJ}}{\text{mol}}\right) + (1\text{ mol})\left(463\frac{\text{kJ}}{\text{mol}}\right) + (\tfrac{3}{2}\text{ mol})\left(498\frac{\text{kJ}}{\text{mol}}\right)\right]$$

$$- \left[(2\text{ mol})\left(799\frac{\text{kJ}}{\text{mol}}\right) + (4\text{ mol})\left(463\frac{\text{kJ}}{\text{mol}}\right)\right] = \mathbf{--647\ kJ}$$

(b) $CO(g) + 2H_2(g) \rightarrow CH_3OH(g)$

$$\Delta H_{rxn} = [D_{C\equiv O} + 2D_{H-H}] - [3D_{C-H} + D_{C-O} + D_{O-H}]$$

$$= \left[(1\text{ mol})\left(1072\frac{\text{kJ}}{\text{mol}}\right) + (2\text{ mol})\left(436\frac{\text{kJ}}{\text{mol}}\right)\right]$$

$$- \left[(3\text{ mol})\left(414\frac{\text{kJ}}{\text{mol}}\right) + (1\text{ mol})\left(351\frac{\text{kJ}}{\text{mol}}\right) + (1\text{ mol})\left(463\frac{\text{kJ}}{\text{mol}}\right)\right] = \mathbf{-112\ kJ}$$

Note: Lewis structures can be used to determine what bonds are broken and what bonds are formed.

9.83 ***See Section 9.8, Table 9.4, and Example 9.16.***

(a) $C_2H_4(g) + 3O_2(g) \rightarrow 2CO_2(g) + 2H_2O(g)$

$$\Delta H_{rxn} = [4D_{C-H} + D_{C=C} + 3D_{O=O}] - [4D_{C=O} + 4D_{O-H}]$$

$$\Delta H_{rxn} = \left[(4\ mol)\left(414\frac{kJ}{mol}\right) + (1\ mol)\left(611\frac{kJ}{mol}\right) + (3\ mol)\left(498\frac{kJ}{mol}\right)\right] - \left[(4\ mol)\left(799\frac{kJ}{mol}\right) + (4\ mol)\left(463\frac{kJ}{mol}\right)\right] = \mathbf{-1.29 \times 10^3\ kJ}$$

(b) $H_2CO(g) + H_2(g) \rightarrow H_3COH\ (CH_3OH)$

$$\Delta H_{rxn} = [2D_{C-H} + D_{C=O} + D_{H-H}] - [3D_{C-H} + D_{C-O} + D_{O-H}]$$

$$\Delta H_{rxn} = \left[(2\ mol)\left(414\frac{kJ}{mol}\right) + (1\ mol)\left(799\frac{kJ}{mol}\right) + (1\ mol)\left(436\frac{kJ}{mol}\right)\right] - \left[(3\ mol)\left(414\frac{kJ}{mol}\right) + (1\ mol)\left(351\frac{kJ}{mol}\right) + (1mol)\left(463\frac{kJ}{mol}\right)\right] = \mathbf{7\ kJ}$$

Note: Lewis strutctures can be used to determine what bonds are broken and what bonds are formed.

9.84 ***SeeSection 9.8, Table 9.4, andExample 9.16.***

(a) $2H_2(g) + C_2H_2(g) \rightarrow C_2H_6(g)$

$$\Delta H_{rxn} = [2D_{H-H} + D_{C\equiv C} + 2D_{C-H}] - [D_{C-C} + 6D_{C-H}]$$

$$\Delta H_{rxn} = \left[(2\ mol)\left(436\frac{kJ}{mol}\right) + (1\ mol)\left(837\frac{kJ}{mol}\right) + (2\ mol)\left(414\frac{kJ}{mol}\right)\right] - \left[(1\ mol)\left(348\frac{kJ}{mol}\right) + (6\ mol)\left(414\frac{kJ}{mol}\right)\right] = \mathbf{-295\ kJ}$$

(b) $\frac{1}{2}N_2(g) + \frac{3}{2}H_2(g) \rightarrow NH_3(g)$

$$\Delta H = [\tfrac{1}{2}D_{N\equiv N} + \tfrac{3}{2}D_{H-H}] - [3D_{N-H}]$$

$$= \left[(\tfrac{1}{2}\ mol)\left(946\frac{kJ}{mol}\right) + (\tfrac{3}{2}\ mol)\left(436\frac{kJ}{mol}\right)\right] - \left[(3\ mol)\left(389\frac{kJ}{mol}\right)\right] = \mathbf{-40\ kJ}$$

Note: Lewis structures can be used to determine what bonds are broken and what bonds are formed.

9.85 ***See Section 9.6 and Example 9.12.***

$H_2NCH_2CO_2H$

```
    H  O
    |  |
H—N—C—C—O—H
  |  |
  H  H
```

Total valence electrons = $[5 \times 1(H) + 1 \times 5(N) + 2 \times 4(C) + 2 \times 6(O)] = 30.$
Twelve electrons remain after assigning nine single bonds, and fourteen unshared electrons are needed to give each atom a noble gas configuration (2 for N, 2 for C (right), 6 for O (top), and 4 for O (right)). Hence, two electrons (14-12) must be used to form one additional bond. This gives the following possible resonance forms:

A

```
      H :O:
  ..  |  ||  ..
H—N—C—C—O—H
  |  |       ..
  H  H
```

B

```
      H  ..
  ..  | :O:
H—N—C—C=O—H
  |  |    ..
  H  H
```

Formal Charges:

	A	B
H	$1\text{-}0\text{-}\frac{1}{2}(2) = 0$	$1\text{-}0\text{-}\frac{1}{2}(2) = 0$
N	$5\text{-}2\text{-}\frac{1}{2}(6) = 0$	$5\text{-}2\text{-}\frac{1}{2}(6) = 0$
C (left)	$4\text{-}0\text{-}\frac{1}{2}(8) = 0$	$4\text{-}0\text{-}\frac{1}{2}(8) = 0$
C (right)	$4\text{-}0\text{-}\frac{1}{2}(8) = 0$	$4\text{-}0\text{-}\frac{1}{2}(8) = 0$
O (top)	$6\text{-}4\text{-}\frac{1}{2}(4) = 0$	$6\text{-}6\text{-}\frac{1}{2}(2) = -1$
O (right)	$6\text{-}4\text{-}\frac{1}{2}(4) = 0$	$6\text{-}2\text{-}\frac{1}{2}(6) = +1$

Hence, Structure A has lower formal charges and is favored over Structure B.

9.86 ***See Section 9.3.***

(a)

```
[    ..    ]-
[   :F:    ]
[ ..  |  .. ]
[:F—E—F:]
[ ..  |  .. ]
[   :F:    ]
[    ..    ]
```

Total valence electrons = $[1 \times ?(E) + 4 \times 7(F) + 1(\text{charge})] = 32.$

? = 3 and E = B, so EF_4^- is BF_4^-.

(b)

```
[    ..    ]+
[   :F:    ]
[ ..  |  .. ]
[:F—E—F:]
[ ..  |  .. ]
[   :F:    ]
[    ..    ]
```

Total valence electrons = $[1 \times ?(E) + 4 \times 7(F) - 1(\text{charge})] = 32.$

? = 5 and E = N, so EF_4^+ is NF_4^+.

9.87 *See Section 9.3, 9.7, and Examples 9.4, 9.5,9.6, 9.15.*

S_2N_2

Total valence electrons $= [2 \times 6(S) + 2 \times 5(N)] = 22$. Fourteen electrons remain after assigning four single bonds, and sixteen unshared electrons are needed to give each atom a noble gas configuration. Hence, two electrons (16-14) must be used to form one additional bond. This bond can exist in any one of the four bonding positions. Twenty two electrons are used drawing the Lewis structure which obeys the octet rule. However, twenty two electrons can also be used to draw a Lewis strucutre which shows an expanded valence shell for sulfur and has formal charges of zero for all atoms.

9.88 *See Section 9.7 and Example 9.15.*

$[H_2P_2O_4]^{2-}$

Total valence electrons $= [2 \times 1(H) + 2 \times 5(P) + 4 \times 6(O) + 2(\text{charge})] = 38$

A B

Structure A which obeys the octet rule places the negative formal charges on the more electronegative oxygen atoms and is therefore favored over the electron rich Structure B that places the negative formal charges on the less electronegative phosphorus atoms.

9.89 *See Section 9.7 and Example 9.15.*

SF_4CH_2

Total valence electrons $= [1 \times 6(S) + 4 \times 7(F) + 1 \times 4(C) + 2 \times 1(H)] = 40$.
Twenty six electrons remain after assigning seven single bonds, and twenty six unshared electrons are needed to give each attached atom a noble gas configuration (6 for each F and 2 for C). Twenty six electrons can be used to draw a single bonded structure. However, twenty six electrons can also be used to draw a strucutre with one double bond and reduced formal charges. This structure is also compatible with the unusually short C-S bond distance of 155 pm.

Note: The S atom already has an expanded valence shell with the single bonded structure. However, electron withdrawal from S by the highly electronegative fluorine atoms probably contributes to forming a C-S double bond.

9.90 *See Section 9.6.*

N_2O_5

O—N—O—N—O (each N also bonded to a lower O)

:Ö—N—Ö—N—Ö: with :O: double-bonded below each N

Total valence electrons = $[2 \times 5(N) + 5 \times 6(O)] = 40.$

Twenty eight electrons remain after assigning six single bonds, and thirty two unshared electrons are needed to give each atom a noble gas configuration (2 for each N, 4 for middle O, and 6 for other O atoms). Hence, four electrons (32-28) must be used to form two additional bonds, one from each nitrogen to a terminal oxygen. Forty electrons are used in writing the Lewis structure, which is one of four possible resonance forms.

NO_2^+

$[O—N—O]^+$

Total valence electrons = $[1 \times 5(N) + 2 \times 6(O) - 1(\text{charge})] = 16.$

Twelve electrons remain after assigning two single bonds, and sixteen unshared electrons are needed to give each atom a noble gas configuration (4 for N and 6 for each O). Hence, four electrons (16-12) must be used to form two additional bonds. This gives the following resonance possiblities:

$[\ddot{O}{=}N{=}\ddot{O}]^+ \quad [O{\equiv}N{-}\ddot{O}:]^+ \quad [:\ddot{O}{-}N{\equiv}O]^+$

$[NO_3]^-$

$[O—N(—O)—O]^-$

Total valence electrons = $[1 \times 5(N) + 3 \times 6(O) + 1(\text{charge})] = 24.$

Eighteen electrons remain after assigning three single bonds, and twenty unshared electrons are needed to give each atom a noble gas configuration (2 for N and 6 for each O). Hence, two electrons (20-18) must be used to form one additional bond. This gives the following resonance possibilities:

$[\ddot{O}{=}N(-\ddot{O}:){-}\ddot{O}:]^- \quad [:\ddot{O}{-}N({=}\ddot{O}:){-}\ddot{O}:]^- \quad [:\ddot{O}{-}N(-\ddot{O}:){=}\ddot{O}]^-$

9.91 *See Section 9.6 and Example 9.12.*

N_2O Total valence electrons = $[2 \times 5(N) + 1 \times 6(O)] = 16.$

Twelve electrons remain after assigning two single bonds in either connectivity, and sixteen electrons are needed to give each atom a noble gas configuration (6 for each terminal atom and 4 for the central atom). Hence, four electrons (16-12) must be used to form two additional bonds. This gives the following as possible resonance forms:

	A	B	C
	$:N{\equiv}N{-}\ddot{O}:$	$:\ddot{N}{=}N{=}\ddot{O}:$	$:\ddot{N}{-}N{\equiv}O:$
Formal Charges:			
N left	$5-2-\left(\frac{1}{2}\right)(6)=0$	$5-4-\left(\frac{1}{2}\right)(4)=-1$	$5-6-\left(\frac{1}{2}\right)(2)=-2$
N right	$5-0-\left(\frac{1}{2}\right)(8)=+1$	$5-0-\left(\frac{1}{2}\right)(8)=+1$	$5-0-\left(\frac{1}{2}\right)(8)=+1$
O	$6-6-\left(\frac{1}{2}\right)(2)=-$	$6-4-\left(\frac{1}{2}\right)(4)=0$	$6-2-\left(\frac{1}{2}\right)(6)=+1$

	D	E	F
	$:\ddot{N}=\ddot{O}=\ddot{N}:$	$:N\equiv O-\ddot{N}:$	$:\ddot{N}-O\equiv N:$
N left	$5-4-\left(\frac{1}{2}\right)(4)=-1$	$5-2-\left(\frac{1}{2}\right)(6)=0$	$5-6-\left(\frac{1}{2}\right)(2)=-2$
N right	$5-4-\left(\frac{1}{2}\right)(4)=-1$	$5-6-\left(\frac{1}{2}\right)(2)=-2$	$5-2-\left(\frac{1}{2}\right)(6)=0$
O	$6-0-\left(\frac{1}{2}\right)(8)=+2$	$6-0-\left(\frac{1}{2}\right)(8)=+2$	$6-0-\left(\frac{1}{2}\right)(8)=+2$

When these formal charges are included in the Lewis structures, we obtain

A: $:N\equiv N^{\oplus}-\ddot{O}:^{\ominus}$ B: $:\ddot{N}^{\ominus}=N^{\oplus}=\ddot{O}:$ C: $^{2\ominus}:\ddot{N}-N^{\oplus}\equiv O^{\oplus}:$

D: $:\ddot{N}^{\ominus}=O^{2\oplus}=\ddot{N}:^{\ominus}$ E: $:N\equiv O^{2\oplus}-\ddot{N}:^{2\ominus}$ F: $^{2\ominus}:\ddot{N}-O^{2\oplus}\equiv N:$

Structures A and B have lower formal charges and also avoid having a positive formal charge on the more electronegative O atom, as occurs in C, D, E and F. Structure A has its negative formal charge on the most electronegative element in the molecule and is therefore likely to be the most important resonance form. These considerations indicate a N-N-O connectivity is more likely than a N-O-N connectivity.

9.92 ***See Sections 9.3, 9.8, Table 9.4, and Examples 9.4, 9.5, 9.6.***

CH_3NH_2

```
    H   H
    |   |
H — C — N — H
    |   ..
    H
```

Total valence electrons $= [1 \times 4(C) + 1 \times 5(N) + 5 \times 1(H)] = 14$.
Two electrons remain after assigning six single bonds, and two unshared electrons are needed to give each atom a noble gas configuration (2 for N). Fourteen electrons are used in writing the Lewis structure.

$CH_3NH_2 \rightarrow C(g) + N(g) + 5H(g)$ $\Delta H_{rxn} = 3D_{C\text{-}H} + D_{C\text{-}N} + 2D_{N\text{-}H}$

$$\Delta H_{rxn} = (3\text{ mol})\left(414\frac{\text{kJ}}{\text{mol}}\right) + (1\text{ mol})\left(293\frac{\text{kJ}}{\text{mol}}\right) + (2\text{ mol})\left(389\frac{\text{kJ}}{\text{mol}}\right) = \mathbf{2313\ kJ}$$

9.93 ***See Sections 9.3, 9.4, Figure 9.8, and Examples 9.4, 9.5, 9.6, 9.7.***

BrNO

Br—N—O

$:\ddot{\underset{..}{Br}}-\ddot{N}=\ddot{\underset{..}{O}}$

Total valence electrons $= [1 \times 7(Br) + 1 \times 5(N) + 1 \times 6(O)] = 18$.
Fourteen electrons remain after assigning two single bonds, and sixteen electrons are needed to give each atom a noble gas configuration (6 for Br, 4 for N and 6 for O). Hence, two electrons (16-14) must be used to form an additional bond between N and O. Eighteen electrons are used in writing the Lewis structure.

According to Figure 9.8, the electronegativity difference between N and O is greater than that between N and Br. Hence, the N-O bond should be the more polar bond. However, caution should be used in assuming the values given in Figure 9.8 are applicable to multiply bonded atoms.

9.94 *See Section 9.8, Table 9.4, and Example 9.16.*

$CO(g) + Cl_2(g) \rightarrow Cl_2CO(g)$

$$\Delta H_{rxn} = [C_{C\equiv O} + D_{Cl\text{-}Cl}] - [2D_{C\text{-}Cl} + D_{C=O}]$$

$$= \left[(1\text{ mol})\left(1072\frac{\text{kJ}}{\text{mol}}\right) + (1\text{ mol})\left(242\frac{\text{kJ}}{\text{mol}}\right)\right] - \left[(2\text{ mol})\left(328\frac{\text{kJ}}{\text{mol}}\right) + (1\text{ mol})\left(799\frac{\text{kJ}}{\text{mol}}\right)\right] = \mathbf{-141\ kJ}$$

Note: Lewis structures can be used to determine what bonds are broken and what bonds are formed.

9.95 *See Section 9.8, Table 9.4. and Example 9.16.*

$HCN(g) + 2H_2(g) \rightarrow H_3CNH_2(g)$

$$\Delta H_{rxn} = [D_{C-H} + D_{C\equiv N} + 2D_{H-H}] - [3D_{C-H} + D_{C-N} + 2D_{N-H}]$$

$$\Delta H_{rxn} = \left[(1\text{ mol})\left(414\frac{\text{kJ}}{\text{mol}}\right) + (1\text{ mol})\left(891\frac{\text{kJ}}{\text{mol}}\right) + (2\text{ mol})\left(436\frac{\text{kJ}}{\text{mol}}\right)\right]$$

$$- \left[(3\text{ mol})\left(414\frac{\text{kJ}}{\text{mol}}\right) + (1\text{ mol})\left(293\frac{\text{kJ}}{\text{mol}}\right) + (2\text{ mol})\left(389\frac{\text{kJ}}{\text{mol}}\right)\right] = \mathbf{-136\ kJ}$$

Note: Lewis structures can be used to determine what bonds are broken and what bonds are formed.

9.96 *See Section 9.2 and Example 9.3.*

All three compounds contain M^+ and X^- cations and anions. Hence, the smaller the ions the stronger the coulombic attractions and the greater the lattice energy. This gives: **CsI = 582 kJ/mol, KCl = 692 kJ/mol and LiF = 1004 kJ/mol.**

9.97 *See Sections 9.3, 9.5, 9.7.*

N_2O

N—N—O

$:N\equiv N^{\oplus}—\ddot{O}:^{\ominus}$

Total valence electrons = $[2 \times 5(N) + 1 \times 6(O)] = 16$.
Twelve electrons remain after assigning two single bonds, and sixteen unshared electrons are needed to give each atom a noble gas configuration (6 for left N, 4 for right N, and 6 for O). Hence, four electrons (16-12) must used to form two additional bonds. A Lewis structure having a triple bond between the nitrogen atoms is favored because it places the negative formal charge on the more elcctronegative oxygen atom. Sixteen electrons are used in writing the Lewis structure.

NO_2

O—N—O

Total valence electrons = $[1 \times 5(N) + 2 \times 6(O)] = 17$
Thirteen electrons remain after assigning two single bonds, and sixteen unshared electrons are needed to give each atom a noble gas configuration (4 for N and 6 for each O). Hence, two electrons are used to form one additional bond and one electron is assigned to nitrogen. This gives the following resonance possibilities:

$^{\ominus}:\ddot{O}—\dot{N}^{\oplus}=\ddot{O}$ $\quad$ $\ddot{O}=\dot{N}^{\oplus}—\ddot{O}:^{\ominus}$

The N-O bond in N_2O is a single bond and that in NO_2 is intermediate between a single and double bond. Hence, the latter places more electron density between the positively charged nitrogen and oxygen nuclei leading to a shorter and stronger N-O bond in NO_2 than N_2O.

NO_2 contains an unpaired electron.

9.98 *See Section 9.7 and Examples 9.14, 9.15.*

$XeF_6 + H_2O \rightarrow XeOF_4 + 2HF$

XeF_6:

Total valence electrons $= [1 \times 8(\text{Xe}) + 6 \times 7(\text{F})] = 50.$

$XeOF_4$:

Total valence electrons $= [1 \times 8(\text{Xe}) + 1 \times 6(\text{O}) + 4 \times 7(\text{F})] = 42.$

9.99 *See Sections 3.5, 9.7, and Examples 3.16, 9.14.*

(a) $3ClF_3 + U \rightarrow UF_6 + 3ClF$
$2ClF_3 + Pu \rightarrow PuF_4 + 2ClF$

(b) ClF_3

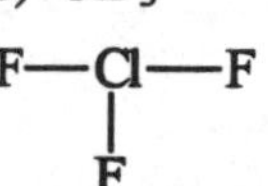

Total valence electrons $= [1 \times 7(\text{Cl}) + 3 \times 7(\text{F})] =$ **28.**
Twenty two electrons remain after assigning three single bonds, and twenty electrons are needed to give each atom a noble gas configuration (2 for Cl and 6 for each F). Hence, two additional electrons must be added to the chlorine atom, the central atom. Twenty eight electrons are used in writing the **electron rich** Lewis structure for ClF_3.

(c) *Strategy:* $g\ UF_6 \rightarrow mol\ UF_6 \rightarrow mol\ U \rightarrow g\ U$

$$?\ \text{g U} = 43.5\ \text{g UF}_6 \times \frac{1\ \text{mol UF}_6}{352.0\ \text{g UF}_6} \times \frac{1\ \text{mol U}}{1\ \text{mol UF}_6} \times \frac{238.0\ \text{g U}}{1\ \text{mol U}} = 29.4\ \text{g U}$$

Strategy: $g\ PuF_4 \rightarrow mol\ PuF_4 \rightarrow mol\ Pu \rightarrow g\ Pu$

$$?\ \text{g Pu} = 22.1\ \text{g PuF}_4 \times \frac{1\ \text{mol PuF}_4}{320.0\ \text{g PuF}_4} \times \frac{1\ \text{mol Pu}}{1\ \text{mol PuF}_4} \times \frac{244\ \text{g Pu}}{1\ \text{mol Pu}} = 16.9\ \text{g Pu}$$

Mass of mixture = 29.4 g U + 16.9 g Pu = **46.3 g.**

9.100 ***See Sections 5.1, 6.4, 9.8, and Examples 5.1, 6.6, 9.16.***

$CH_4(g) + 2O_2(g) \rightarrow CO_2(g) + 2H_2O(g)$

$$\Delta H_{rxn} = [4\ D_{C\text{-}H} + 2\ D_{O=O}] - [2\ D_{C=O} + 4\ D_{O\text{-}H}]$$
$$= \left[(4\ mol)\left(414\frac{kJ}{mol}\right) + (2\ mol)\left(498\frac{kJ}{mol}\right)\right] - \left[(2\ mol)\left(799\frac{kJ}{mol}\right) + (4\ mol)\left(463\frac{kJ}{mol}\right)\right] = -798\ kJ$$

$$?\ g\ CH_4 = 2044\ kJ \times \frac{1\ mol\ CH_4}{798\ kJ} \times \frac{16.05\ g\ CH_4}{1\ mol\ CH_4} = \mathbf{41.1\ g\ CH_4}$$

Known Quantities: $n_{CH_4} = 41.0\ g\ CH_4 \times \frac{1\ mol\ CH_4}{16.0\ g\ CH_4} = 2.56\ mol$

$P_{atm} = 1.44\ atm$ $T = 27 + 273 = 300\ K$

Solving PV = nRT for V gives $V = \frac{nRT}{P}$. $V = \frac{(2.56\ mol)\left(0.0821\frac{L \cdot atm}{mol \cdot K}\right)(300\ K)}{1.44\ atm} = \mathbf{43.8\ L\ CH_4}$

Chapter 10: Molecular Structure

10.1 ***See Section 10.1.***

The main premise of the VSEPR model is that the elctron pairs within the valence shell of an atom repel each other and determine the molecular geometry of the molecule or ion of interest.

10.2 ***See Section 10.1 and Figure 10.1***

The electron-pair arrangement about a central atom in a molecule or ion can be determined by using the Lewis structure. The number of lone pairs about a central atom plus the number of atoms bonded to it is equal to the steric number of the atom, and the steric number can be used with Figure 10.1 to determine the electron-pair arrangement.

10.3 ***See Section 10.1.***

Formaldehyde, H_2CO, is an example of a trigonal planar molecule in which the carbon forms four bonds, one of which is part of a double bond to the oxygen atom. Phosgene, Cl_2CO, is another example of a trigonal planar molecule in which carbon forms four bonds.

10.4 ***See Section 10.1.***

The VSEPR model states that the order of importance of repulsions within the valence shell of an atom is *lp-lp* > *lp-bp* > *bp-bp*. The actual positions of atoms in molecules and ions are those which minimize the repulsions in the valence shell of the central atom and cause these various kinds of repulsions to becomc equal. In water, having two lone pairs and two bonded atoms about the oxygen atom, this occurs at an H-O-H bond angle of 104.5° compared to the 109.5° of a perfectly tetrahedral arrangement.

10.5 ***See Section 10.1.***

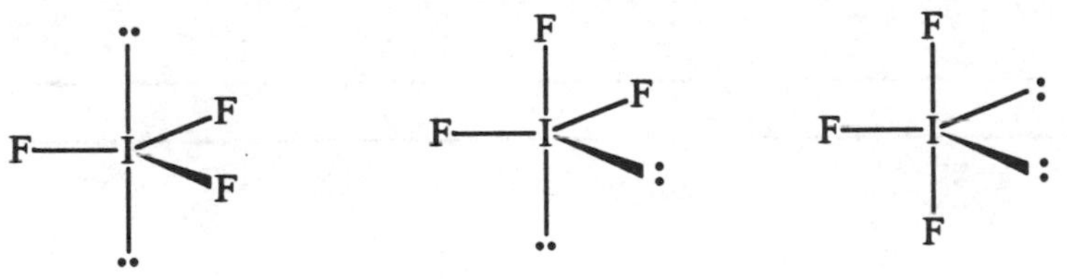

F
F—I
F
F
F—I
F
F
F—I
F

The VSEPR model states that the order of importance of repulsions within the valence shell of an atom is *lp-lp* > *lp-bp* > *bp-bp*. In addition, repulsions diminish as the angle between the electron pairs increases from 90° to 120° to 180°. Hence, lone pairs always occupy equatorial rather than axial positions in trigonal bipyramidal arrangements because there are only two interactions at 90° for equatorial positions in trigonal bipyramidal positions compared to three interactions at 90° for axial positions. The structure on the right is favored and is observed experimentally.

10.6 *See Section 10.1.*

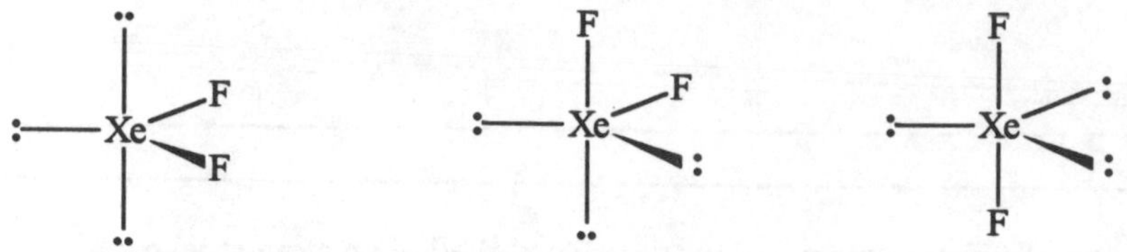

[Structures: three XeF_2 trigonal bipyramidal arrangements — (left) two F equatorial with three lone pairs, one equatorial and two axial; (middle) F axial and F equatorial; (right) two F axial with three lone pairs equatorial.]

The VSEPR model states that the order of importance of repulsions within the valence shell of an atom is *lp-lp* > *lp-bp* > *bp-bp*. In addition, repulsions diminish as the angle between the electron pairs increases from 90° to 120° to 180°. Hence, lone pairs always occupy equatorial rather than axial positions in trigonal bipyramidal arrangements because there are only two interactions at 90° for equatorial positions in trigonal bipyramidal positions compared to three interactions at 90° for axial positions. The structure on the right is favored and is observed experimentally.

10.7 *See Section 10.2.*

Any molecule with a totally symmetrical arrangement of atoms and lone pairs of electrons is nonpolar. Hence, any of the electron-pair arrangements shown in Figure 10.1 having like atoms attached to a central atom that is different has polar bonds but is nonpolar overall. For example, $BeCl_2(g)$ as Cl—Be—Cl is nonpolar because the bond dipoles cancel each other. This always occurs when the molecule is totally symmetrical.

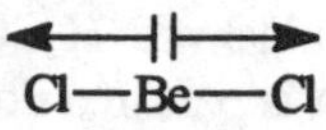

10.8 *See Section 10.2.*

SF_6 exists in a totally symmetrical octahedral arrangement. Each sulfur-fluorine dipole is exactly cancelled by a sulfur-fluorine dipole pointing in the opposite direction. Hence, it has polar bonds but is nonpolar overall.

10.9 *See Section 10.3.*

Valence bond theory describes bonds as being formed by atoms sharing valence electrons in overlapping valence orbitals. These overlaps are caused by the attraction between the nuclear charge of one of the bonded atoms and the electron cloud of the other atom and vice versa.

10.10 *See Section 10.3 and Example 10.6.*

H		↑ (1s)	
I	[Kr]	↑↓ (5s)	↑↓ ↑↓ ↑ (5p)

The partially filled ls orbital of H overlaps with the partially filled 5p orbital of I to form the bond in HI.

10.11 *See Section 10.3 and Example 10.6.*

Cl	[Ne]	↑↓ (3s)	↑↓ ↑↓ ↑ (3p)
F	[He]	↑↓ (2s)	↑↓ ↑↓ ↑ (2p)

The partially filled 3p orbital of Cl overlaps with the partially filled 2p orbital of F to form the bond in ClF.

10.12 *See Section 10.3.*

There are just two unpaired electrons in the valence shell of a carbon atom suggesting the simplest carbon-hydrogen compound should be CH_2. Instead, it is CH_4 and all of the bonds are equivalent and the experimental bond angle is 109.5°, the angle associated with a tetrahedral arrangement and sp^3 hybrids. Hence, sp^3 hybridization for carbon can account for both the number of bonds formed and the experimental bond angle for the simplest carbon-hydrogen compound, CH_4, whereas the use of the atomic orbitals that are used for isolated carbon atoms (the Aufbau order orbitals) cannot account for these observations using valence bond theory.

10.13 *See Sections 10.1, 10.3.*

BCl_3 SN for B = 3, trigonal planar electron-pair arrangement, sp^2 hybrids for B.

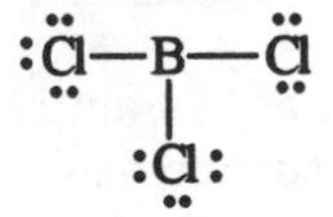

BCl_4^- SN for B = 4, tetrahedral electron-pair arrangement, sp^3 hybrids for B.

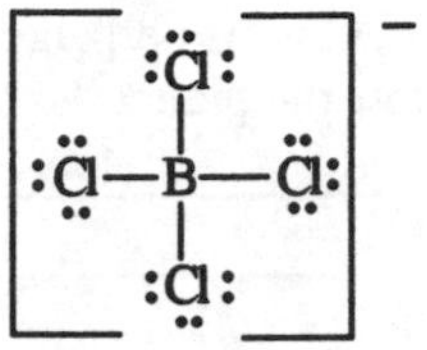

10.14 *See Sections 10.1, 10.3.*

$SbCl_5$ SN for Sb = 5, trigonal bipyramidal electron-pair arrangement, sp^3d hybrids for Sb.

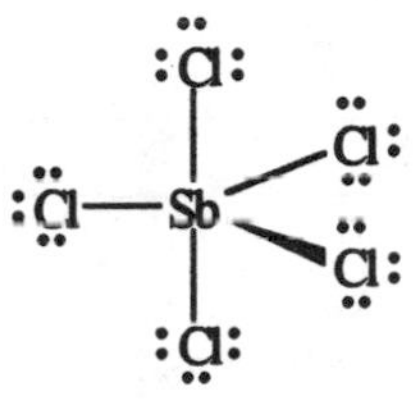

$SbCl_6^-$ SN for Sb = 6, octahedral electron-pair arrangement, sp^3d^2 hybrids for Sb.

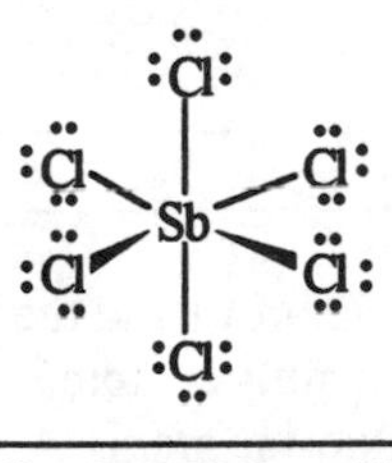

10.15 *See Section 10.3 and Table 10.1.*

H:Cl: Hybrid orbitals are predicted by looking at Lewis structures, observing the steric number, predicting the electron-pair arrangement and selecting the corresponding hybrid orbitals for the central atom. However, we can only determine the positions of attached atoms and are unable to determine the positions of lone pairs. Hence, any prediction of use of hybrid atomic orbitals by Cl in HCl would be a matter of pure conjecture that could not be verified by experiment.

10.16 *See Section 10.3 and Table 10.1.*

Angle	Hybrids
180°	sp
120°	sp^2
109.5°	sp^3

10.17 *See Section 10.4 and Figures 10.22, 10.24.*

A sigma (σ) bond is a bond in which the shared pair of electrons is symmetric about the axis joining the two nuclei of the bonded atoms. A pi (π) bond is a bond that places electron density above and below the line joining the bonded atoms and can be formed by the sideways overlap of p orbitals.

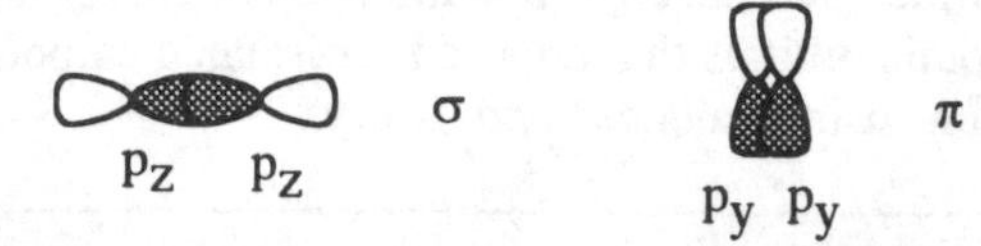

10.18 *See Section 10.3 and Figures 10.23, 10.24, 10.25.*

H:C::C:H
H H

According to valence bond theory, each C atom is predicted to use sp^2 hybrids that are 120° apart in the same plane. The C-H bonds are formed by overlap of sp^2 orbitals of C containing one electron and 1s orbitals of H containing one electron giving an overall planar molecule. The carbon-carbon double bond consists of a sigma bond formed by overlap of sp^2 orbitals containing one electron each and a pi bond formed by sideways overlap of 2p atomic orbitals containing one electron each. The pi bond is perpendicular to the plane of the atoms with one-half of the pi bond above the plane and the other half below the plane.

10.19 *See Section 10.5 and Figures 10.35, 10.36.*

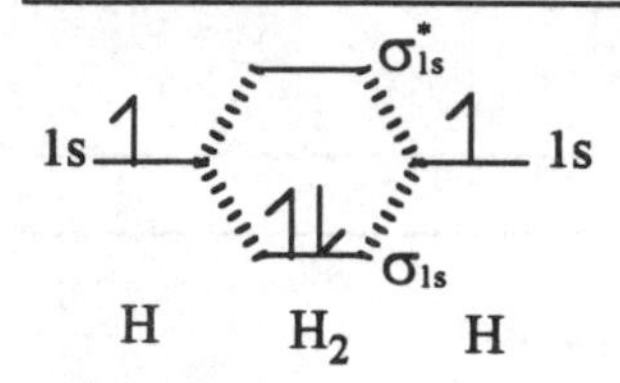

10.20 *See Section 10.5 and Figures 10.38, 10.39.*

10.21 *See Section 10.6 and Figure 10.48.*

According to molecular orbital theory, there is a weak interaction between the 2s orbital of Li and the 2p orbital of F pointing toward Li resulting in a sigma bonding molecular orbital. This orbital is mainly centered around the fluorine atom, and the valence electron of Li is nearly completely transferred to the fluorine atom. According to the ionic bonding description, the valence electron of lithium is completely transferred from Li to F giving Li^+ and F^-. Hence, the two descriptions differ by a very small percentage in terms of electron transfer from Li to F.

10.22 *See Section 10.6 and Figure 10.49.*

The three p orbitals that are perpendicular to the O_3 plane interact to form the delocalized pi bonding molecular orbital that is shown in Figure 10.49.

10.23 *See Section 10.1 and Figure 10.1.*

(a) SN = 3, trigonal planar (b) SN = 4, tetrahedral (c) SN = 4, tetrahedral
(d) SN = 5, trigonal bipyramidal

10.24 *See Section 10.1 and Figure 10.1.*

(a) SN = 3, trigonal planar (b) SN = 5, trigonal bipyramidal (c) SN = 6, octahedral
(d) SN = 6, octahedral

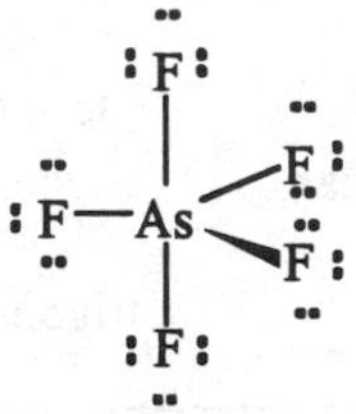

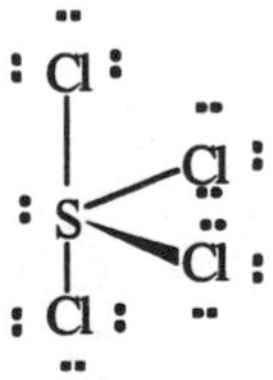

10.25 *See Section 10.1, Figure 10.6, and Example 10.1.*

(a) CF_4	(b) CS_2	(c) AsF_5	(d) CF_2O	(e) NH_4^+
SN = 4	SN = 2	SN = 5	SN = 3	SN = 4
tetrahedral	linear	trigonal bipyramidal	trigonal planar	tetrahedral

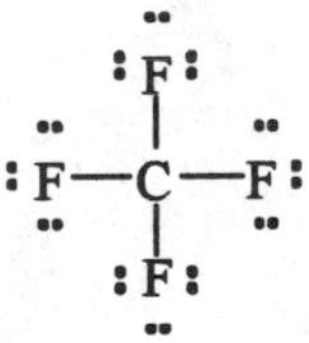

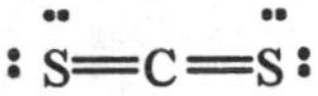

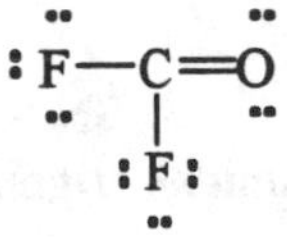

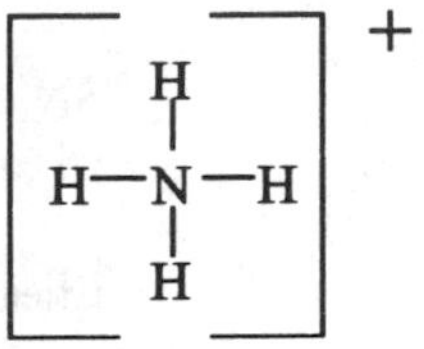

10.26 *See Section 10.1, Figure 10.6, and Example 10.1.*

(a) $BeF_2(g)$	(b) SF_6	(c) SiH_4	(d) FCN	(e) BeF_3^-
SN = 3	SN = 6	SN = 4	SN = 2	SN = 3
linear	octahedral	tetrahedral	linear	trigonal planar

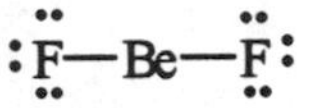

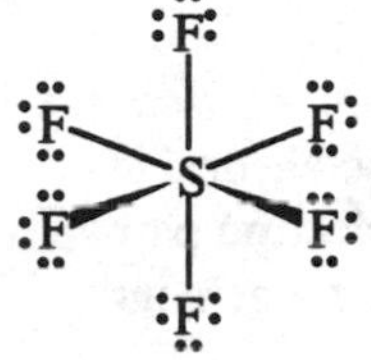

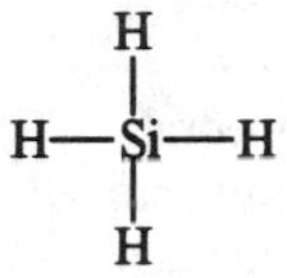

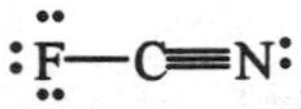

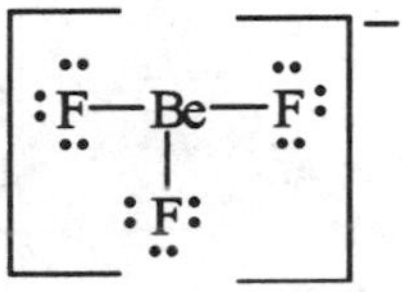

10.27 *See Section 10.1. Figures 10.1, 10.6, and Examples 10.1, 10.2..*

(a) SeO_2	(b) N_2O	(c) H_3O^+	(d) IF_5	(e) SCl_4
electron rich				
SN = 3	SN = 2	SN = 4	SN = 6	SN = 5
trigonal planar	linear	tetrahedral	octahedral	trigonal bipyramidal
bent	linear	trigonal pyramidal	square pyramidal	see-saw

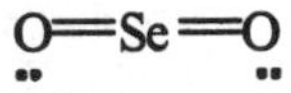

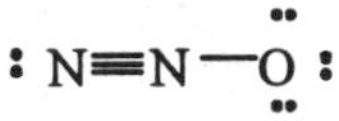

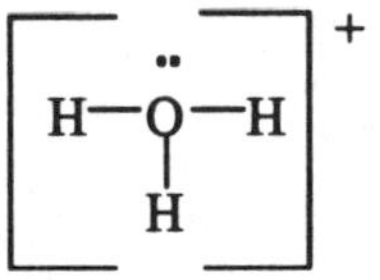

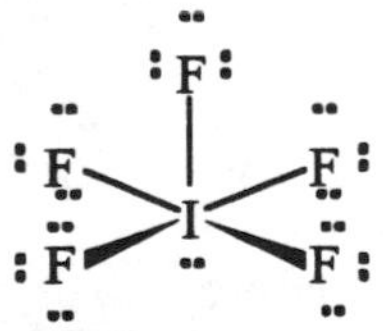

10.28 ***See Section 10.1. Figures 10.1, 10.6, and Examples 10.1, 10.2..***

(a) XeO_2	(b) I_3^-	(c) NO_2^-	(d) PCl_5	(e) $AlCl_3(g)$
SN = 4	SN = 5	SN = 3	SN = 5	SN = 3
tetrahedral	trigonal bipyramidal	trigonal planar	trigonal bipyramidal	trigonal planar
bent	linear	bent	trigonal bipyramidal	trigonal planar

10.29 ***See Section 10.1, Figures 10.1, 10.6, and Examples 10.1, 10.2..***

(a) BCl_3	NCl_3	(b) OF_2	SF_6
SN = 3	SN = 4	SN = 4	SN = 6
trigonal planar electron pair arrangement	tetrahedral electron pair arrangement	tetrahedral electron pair arrangement	octahedral electron pair arrangement
120° bond angles	109° bond angles	109° bond angles	90°&180° bond angles

NCl_3 has smaller bond angles than BCl_3.

SF_6 has smaller bond angles than OF_2.

10.30 ***See Section 10.1, Figures 10.1, 10.6, and Examples 10.1, 10.2..***

(a) SO_4^{2-} | $AlBr_3(g)$ | (b) CCl_4 | $BeI_2(g)$

SN = 4
tetrahedral electron
pair arrangement
109° bond angles

SN = 3
trigonal planar electron
pair arrangement
120° bond angles

SO_4^{2-} has smaller bond angles than $AlBr_3(g)$.

SN = 4
tetrahedral electron
pair arrangement
109° bond angles

SN = 2
linear electron
pair arrangement
180° bond angles

CCl_4 has smaller bond angles than $BeI_2(g)$.

10.31 ***See Section 10.1, Figures 10.1, 10.6, and Examples 10.1, 10.2..***

(a) Cl_2NH | NH_4^+ | (b) SF_2 | IF_4^-

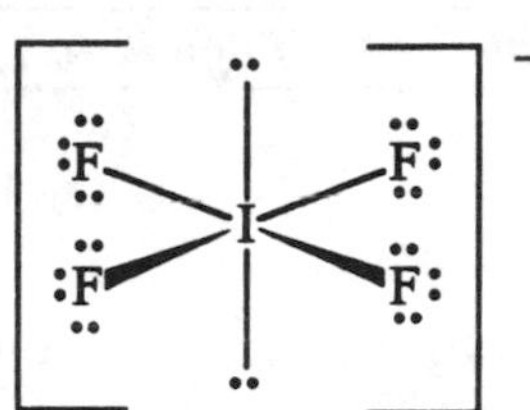

SN = 4
tetrahedral electron
pair arrangement
109° bond angles

SN = 4
tetrahedral electron
pair arrangement
109° bond angles

Cl_2NH has slightly smaller bond angles due to the lone pair on the N.

SN = 4
tetrahedral electron
pair arrangement
109° bond angles

SN = 6
octahedral electron
pair arrangement
90° & 180° bond angles

IF_4^- has smaller bond angles than SF_2.

10.32 ***See Section 10.1, Figures 10.1, 10.6, and Examples 10.1, 10.2..***

(a) BF_3 | $AsCl_4^+$ | (b) $CS_2(g)$ | $AsCl_3$

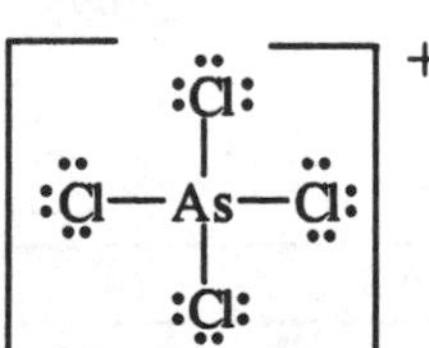

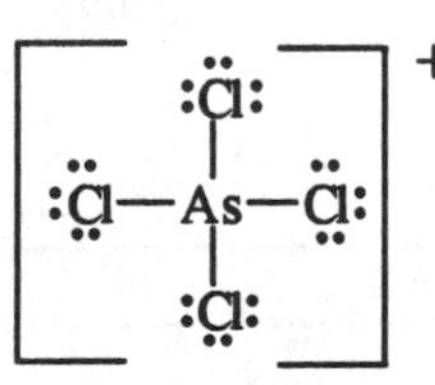

SN = 3
trigonal eletron
pair arrangement
120° bond angles

SN = 4
tetrahedral electron
pair arrangement
109° bond angles

$AsCl_4^+$ has smaller bond angles than BF_3.

SN = 2
linear electron
pair arrangement
180° bond angles

SN = 4
tetrahedral electron
pair arrangement
109° bond angles

$AsCl_3$ has smaller bond angles than $CS_2(g)$.

10.33 ***See Section 10.1, Figures 10.1, 10.6, and Examples 10.2, 10.3.***

(a) H_3CCCH

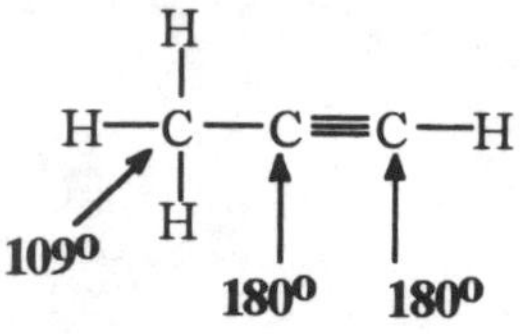

(b) Br_2CCH_2

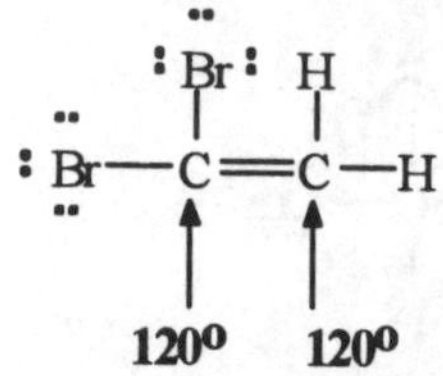

(c) H_3CNH_2

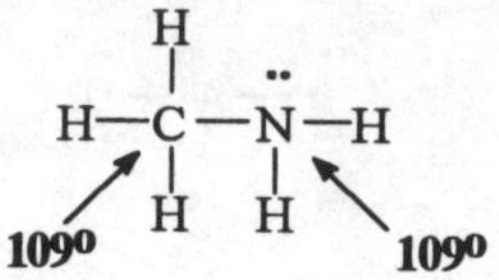

10.34 ***See Section 10.1, Figures 10.1, 10.6, and Examples 10.2, 10.3.***

(a) $ClC(O)NH_2$

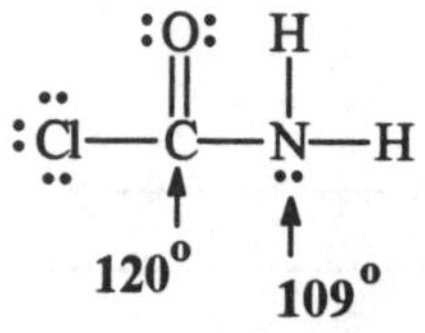

(b) $HOCH_2CH_2OH$

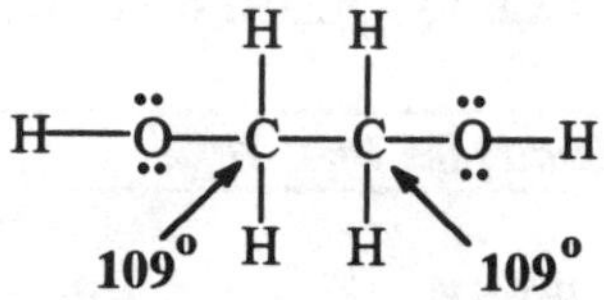

(c) NCCN

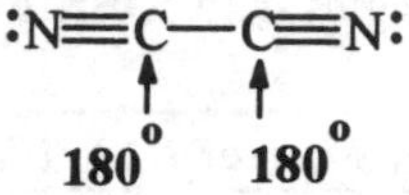

10.35 ***See Section 10.1, Figures 10.1, 10.6, and Examples 10.2, 10.3.***

(a) SO_2

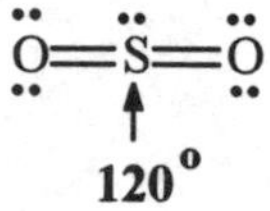

electron rich

(b) H_3CCN

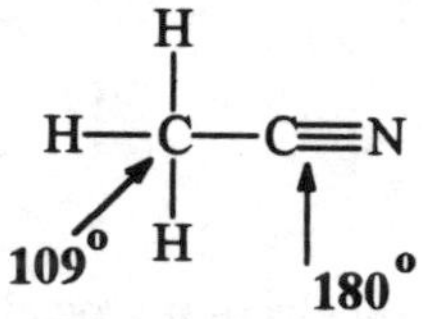

(c) SCN^-

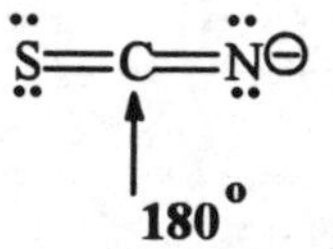

10.36 ***See Section 10.1, Figures 10.1, 10.6, and Examples 10.2, 10.3.***

(a) SeF_4

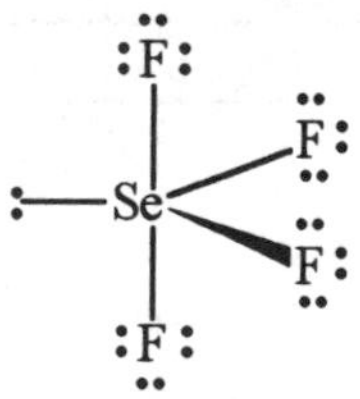

$90^\circ, 120^\circ, 180^\circ$

(b) SCl_2

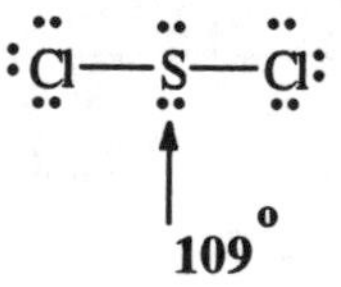

(c) GeH_4

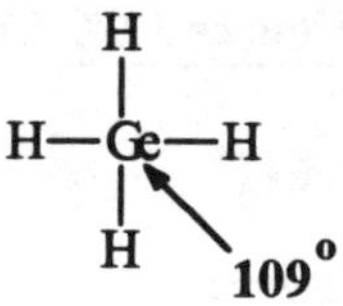

10.37 ***See Section 10.1, Figures 10.1, 10.6, and Examples 10.2, 10.3.***

(a)

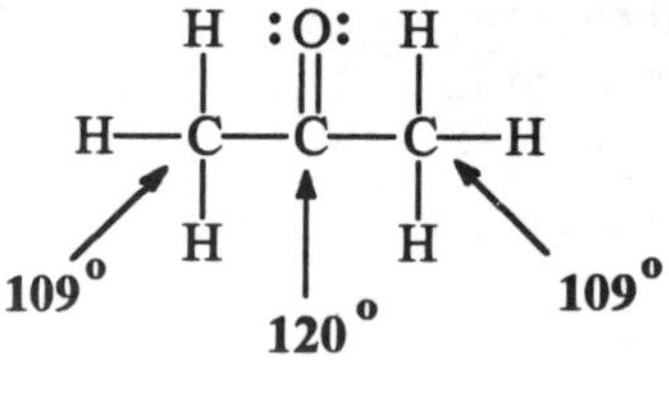

(b)

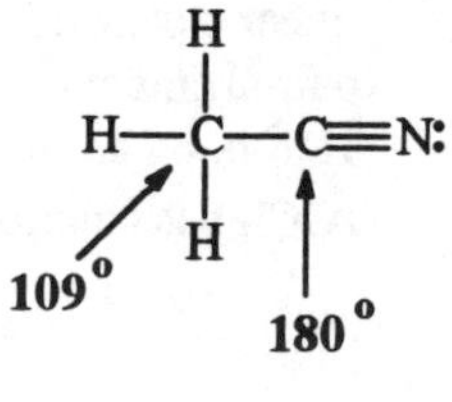

10.38 ***See Section 10.1, Figures 10.1, 10.6, and Examples 10.2, 10.3.***

(a)

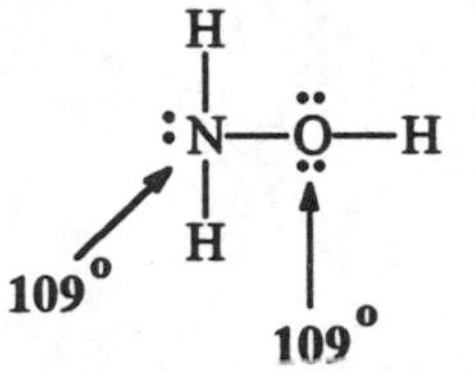

H
:N—Ö—H (O has two lone pairs)
H

109° (at N) 109° (at O)

(b)

H H H
H—C—N—C—H (N has one lone pair)
H H

109° (at each C) 109° (at N)

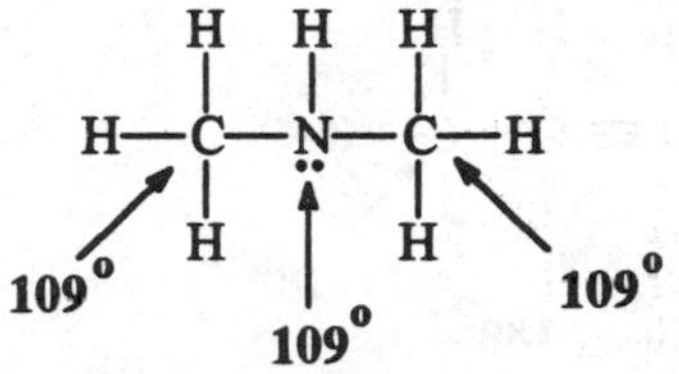

10.39 ***See Section 10.1, Figures 10.1, 10.6, and Examples 10.2, 10.3.***

(a)

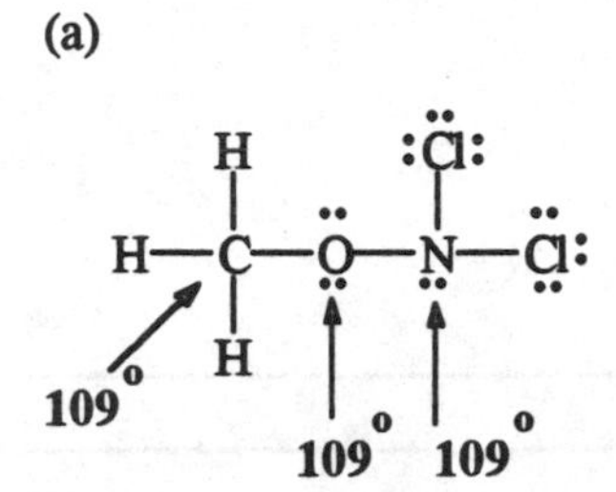

(b)

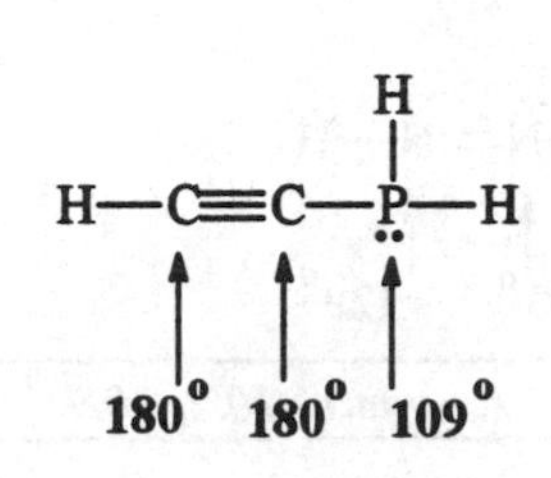

10.40 ***See Section 10.1, Figures 10.1, 10.6, and Examples 10.2, 10.3.***

(a)

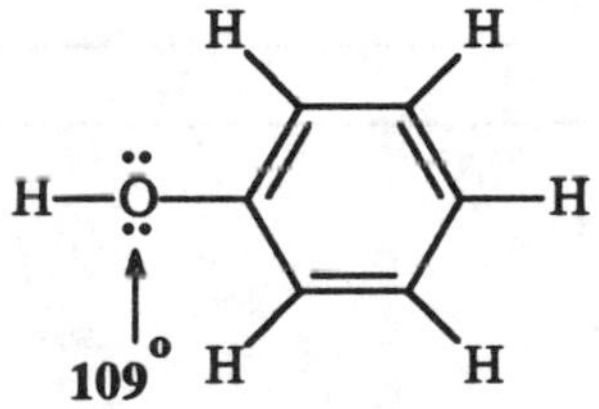

H—Ö— (benzene ring) with four H atoms on the ring and one more H drawn on the right (—H)

109° (at O)

120° at each C

(b)

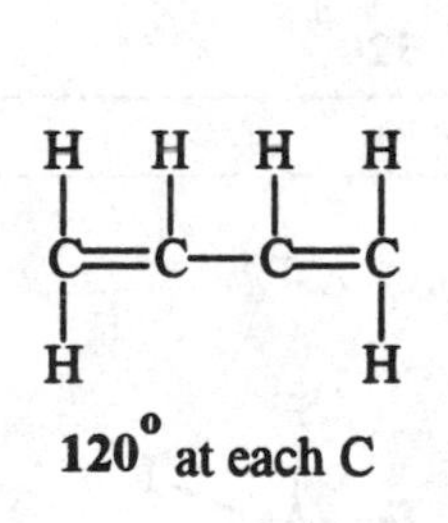

10.41 ***See Section 10.1, Figures 10.1, 10.6, and Examples 10.2, 10.3.***

(a) C_2H_4O

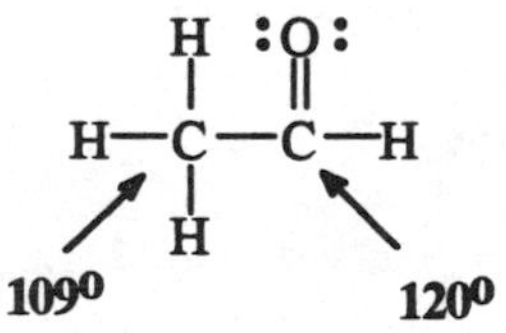

H :O:
H—C—C—H (C=O double bond)
H

109° (at the first C) 120° (at the C bonded to O)

(b) C_3H_6O

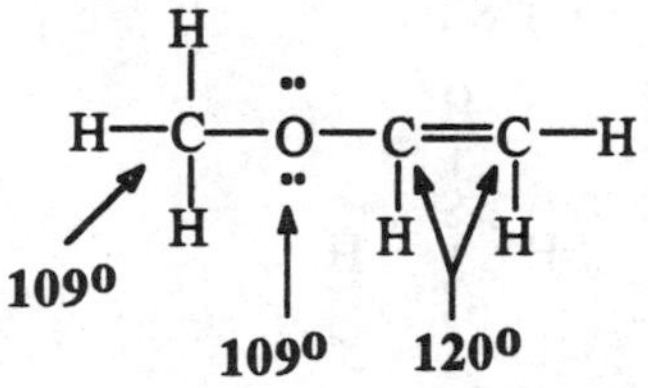

H
H—C—Ö—C=C—H
H H H

109° (at C) 109° (at O) 120° (at each C=C carbon)

(c) XeF_2

:F—Xe—F: (each F and Xe with lone pairs)

180°

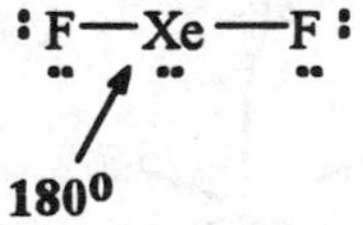

10.42 ***See Section 10.1, Figures 10.1, 10.6, and Examples 10.2, 10.3.***

(a)

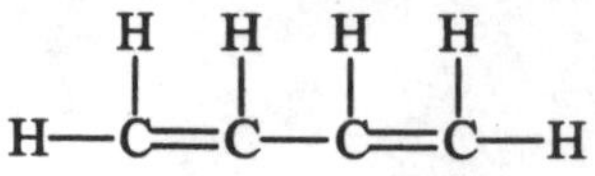

H—C(H)=C(H)—C(H)=C(H)—H

120° at each C

(b)

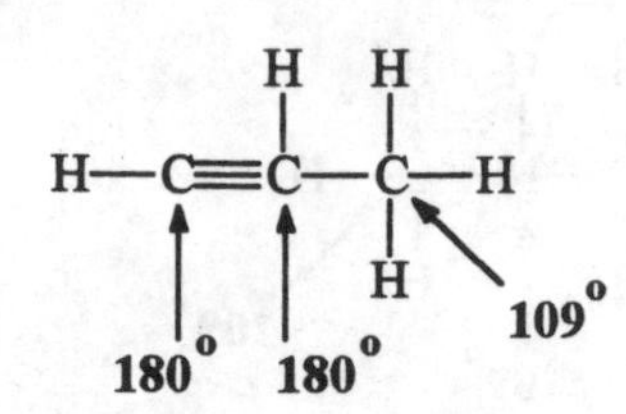

(c)

:Cl—P—Cl: with :Cl: below P

109°

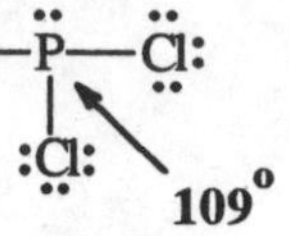

10.43 ***See Section 10.1, Figures 10.1, 10.6, and Examples 10.2, 10.3.***

(a)

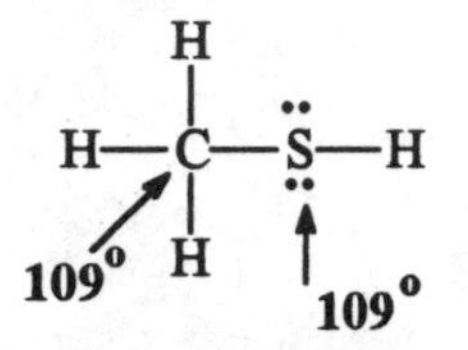

(b)

H—N=N—H

120° 120°

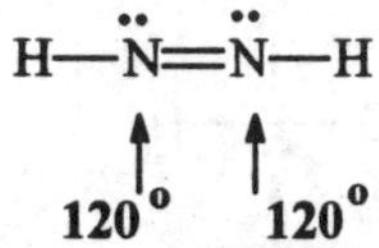

10.44 ***See Section 10.1, Figures 10.1, 10.6, and Examples 10.2, 10.3.***

(a)

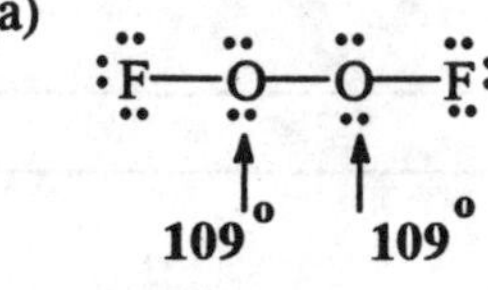

(b)

:Cl: :Cl:
:Cl—C=C—Cl:

120° 120°

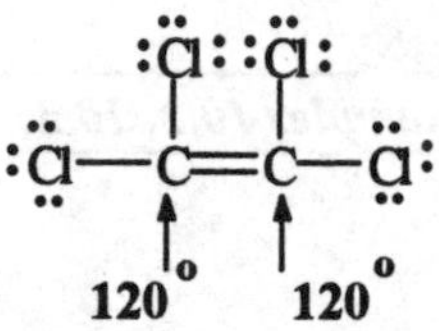

10.45 ***See Section 10.2 and Example 10.4.***

(a) CF_4

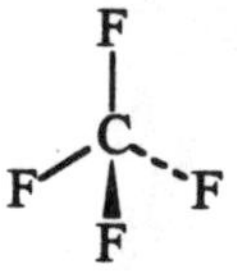

symmetrical
nonpolar

(b) CS_2

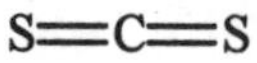

S=C=S

symmetrical
nonpolar

(c) AsF_5

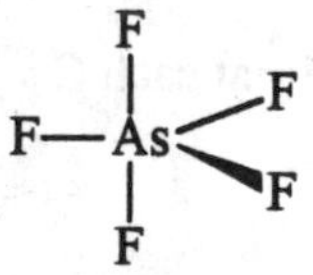

symmetrical
nonpolar

(d) F_2CO

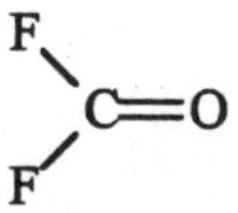

unsymmetrical
polar

10.46 ***See Section 10.2 and Example 10.4.***

(a) BeF_2

F—Be—F

symmetrical
nonpolar

(b) SF_6

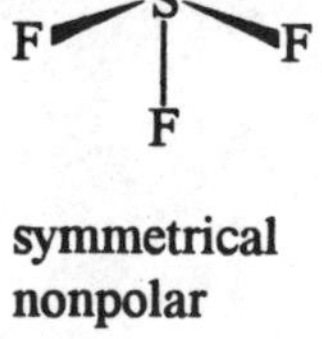

symmetrical
nonpolar

(c) SiH_4

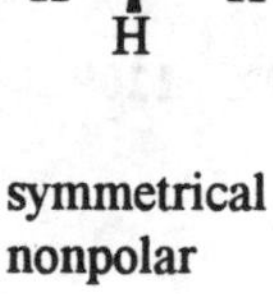

symmetrical
nonpolar

(d) FCN

F—C≡N

unsymmetrical
polar

10.47 *See Section 10.2 and Example 10.4.*

(a) SeO_2

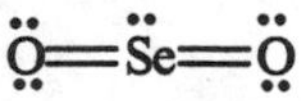

O=Se=O

electron rich

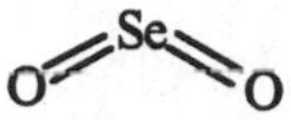

unsymmetrical
polar

(b) N_2O

:N≡N—O:

N≡N—O

unsymmetrical
polar

(c) SCl_4

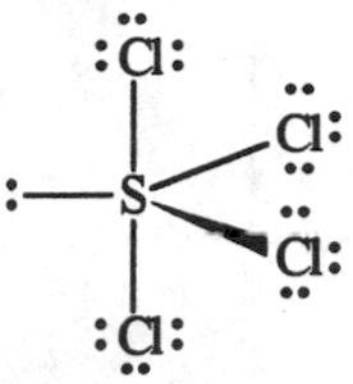

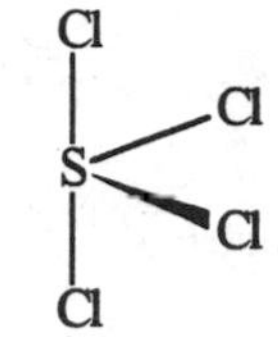

unsymmetrical
polar

10.48 *See Section 10.2 and Example 10.4.*

(a) SF_2

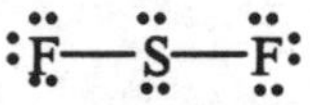

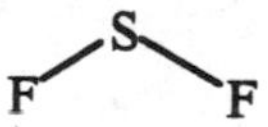

:F—S—F:

unsymmetrical
polar

(b) PCl_5

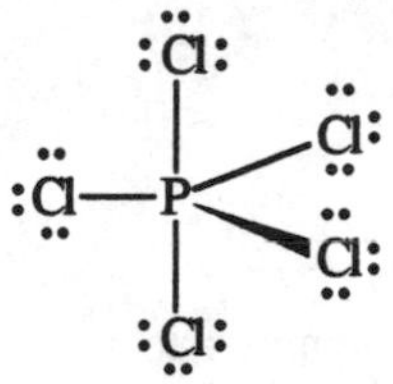

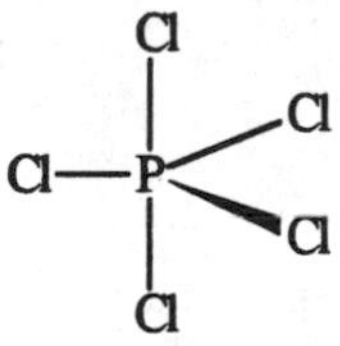

symmetrical
nonpolar

(c) $AlCl_3(g)$

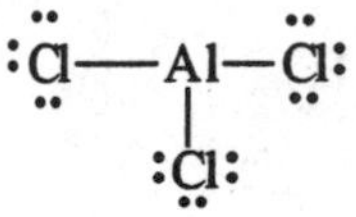

electron deficient

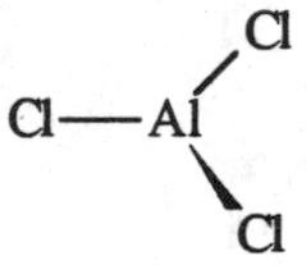

symmetrical
nonpolar

10.49 *See Section 10.2 and Example 10.4.*

(a) HCN

H—C≡N

unsymmetrical
polar

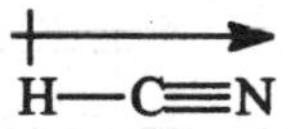

H—C≡N

(b) I_2

I—I

symmetrical
nonpolar

(c) NO

N=O

unsymmetrical
polar

N=O

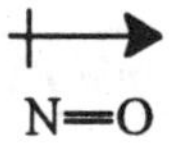

10.50 *See Section 10.2, Example 10.4, and Solution for Exercise10.47.*

(a) SiH_4

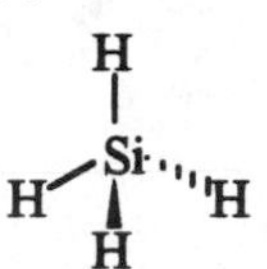

symmetrical
nonpolar

(b) PCl_3

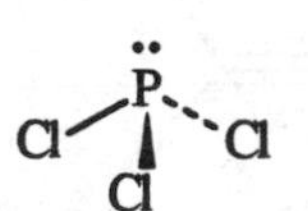

unsymmetrical
polar

(c) IF_5

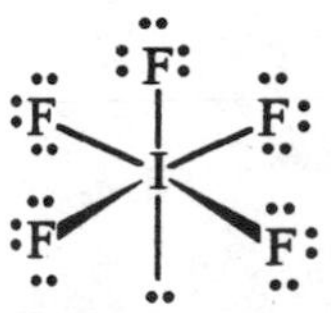

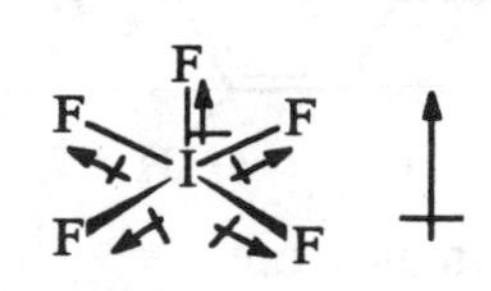

unsymmetrical
polar

10.51 ***See Section 10.2 and Example 10.4.***

(a) NF_3

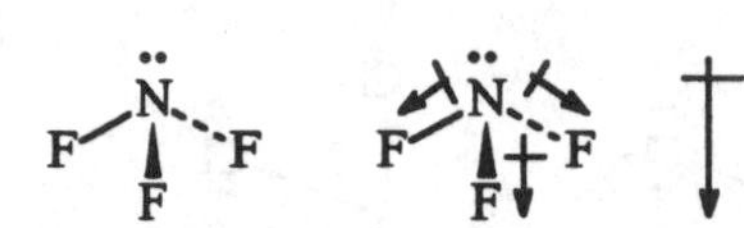

unsymmetrical
polar

(b) CBr_4

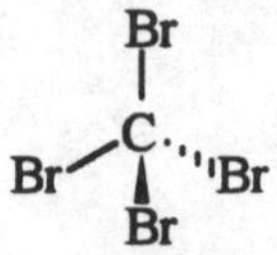

symmetrical
nonpolar

(c) BeI_2

I—Be—I

symmetrical
nonpolar

10.52 ***See Section 10.2 and Example 10.4.***

(a) BCl_3

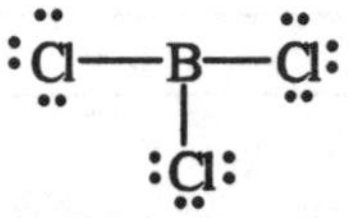

electron deficient

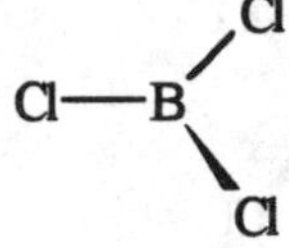

symmetrical
nonpolar

(b) OF_2

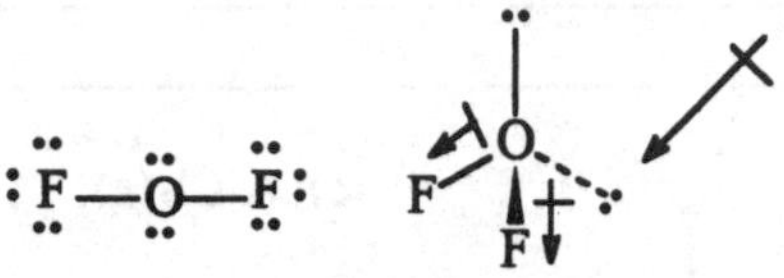

unysmmetrical
polar

(c) SF_6

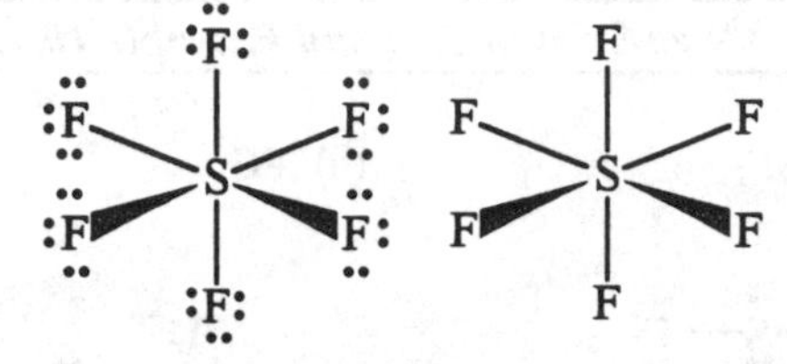

symmetrical
nonpolar

10.53 ***See Section 10.2 and Example 10.4.***

(a) F - C ≡ C - F
symmetrical
nonpolar
bond dipoles cancel

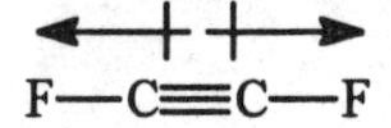

(b) H - C ≡ C - F
unsymmetrical
polar
bond dipoles do not cancel

H—C≡C—F

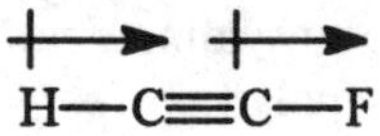

10.54 ***See Section 10.2 and Example 10.4.***

(a)

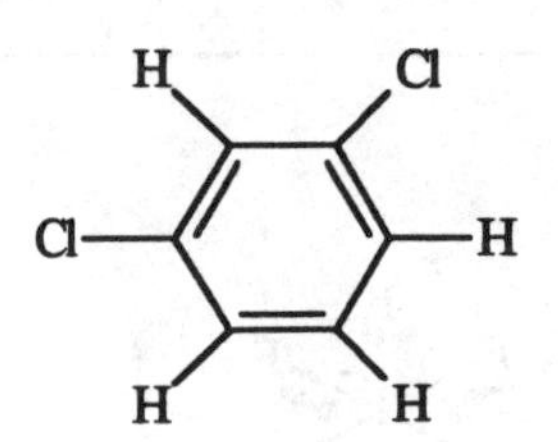

unsymmetrical
polar
C-Cl bond dipoles do not cancel C-H bond dipoles.

(b)

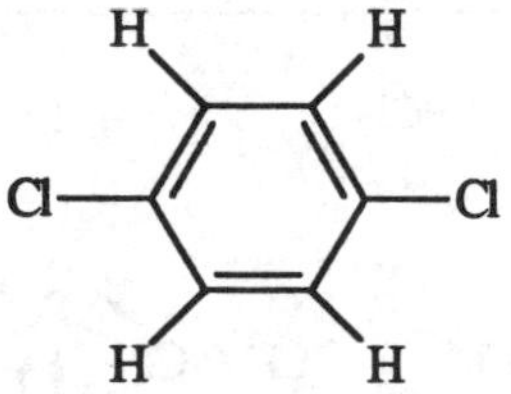

symmetrical
nonpolar
All bond dipoles cancel.

10.55 ***See Section 10.3.***

(a) 120°, sp^2 and sp^3d hybrids (b) 90°, sp^3d and sp^3d^2 hybrids (c) 180°, sp, sp^3d and sp^3d^2 hybrids

10.56 ***See Section 10.3.***

(a) tetrahedral, sp^3 (b) trigonal bipyramidal, sp^3d (c) octahedral, sp^3d^2

10.57 ***See Section 10.3 and Example 10.7.***

(a) CF_4	(b) $SbCl_6^-$	(c) AsF_5	(d) SiH_4	(e) NH_4^+
SN = 4	SN = 6	SN = 5	SN = 4	SN = 4
tetrahedral	octahedral	trigonal bipyramidal	tetrahedral	tetrahedral
sp^3 for C	sp^3d^2 for Sb	sp^3d for As	sp^3 for Si	sp^3 for N

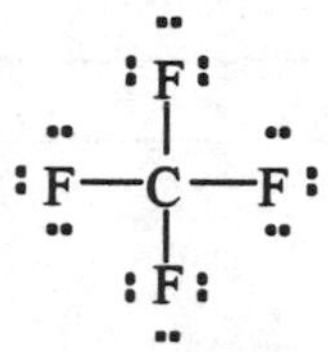

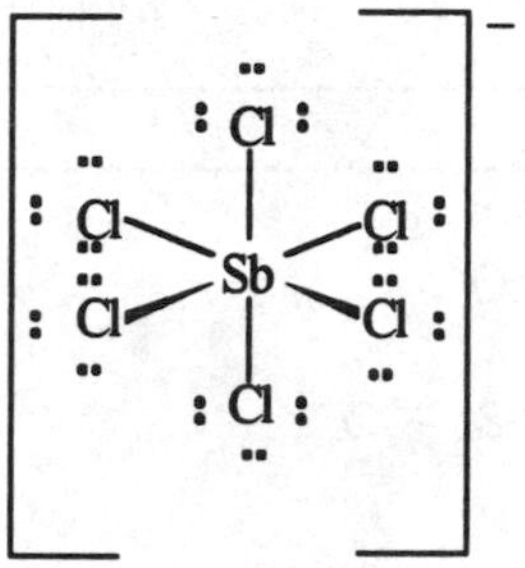

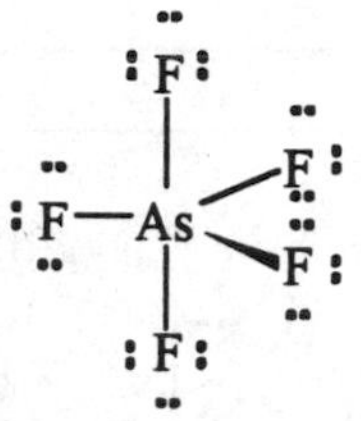

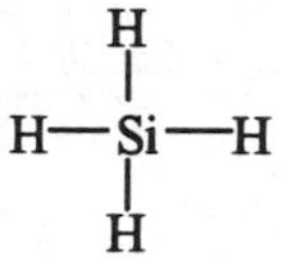

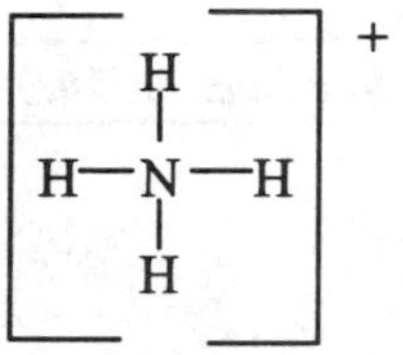

10.58 ***See Section 10.3 and Example 10.7.***

(a) NF_3

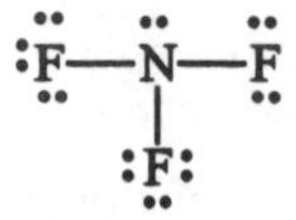

SN = 4
tetrahedral

sp^3 for N

(b) SCl_2

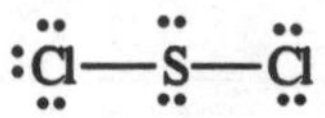

SN = 4
tetrahedral

sp^3 for S

(c) H_3O^+

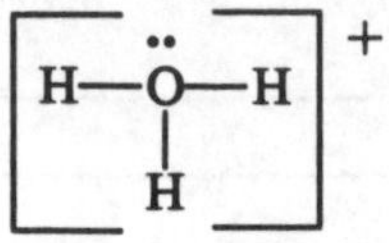

SN = 4
tetrahedral

sp^3 for O

(d) IF_5

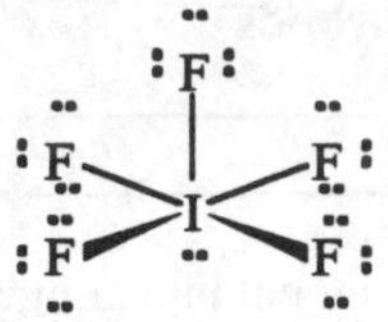

SN = 6
octahedral

sp^3d^2 for I

(e) SCl_4

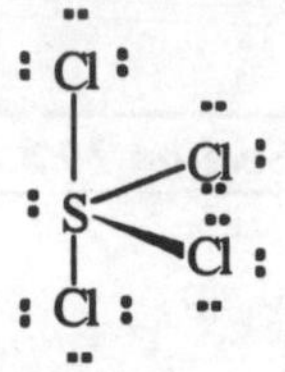

SN = 5
trigonal bipyramidal
sp^3d for S

10.59 ***See Section 10.3 and Examples 10.7, 10.8.***

(a) N_2O

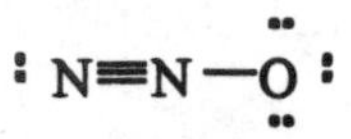

SN = 2
linear

sp for central N

(b) $SnCl_2(g)$

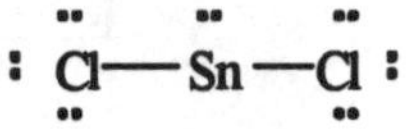

SN = 3
trigonal planar

sp^2 for Sn

(c) I_3^-

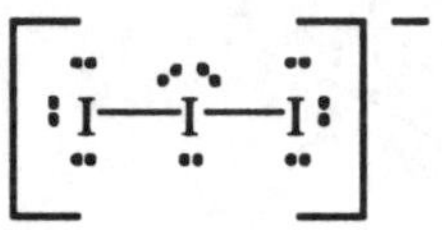

SN = 5
trigonal bipyramidal
sp^3d for central I

(d) SeO_2

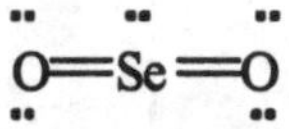

SN = 3
trigonal planar

sp^2 for Se

10.60 ***See Section 10.3 and Examples 10.7, 10.8.***

(a) ClF_3

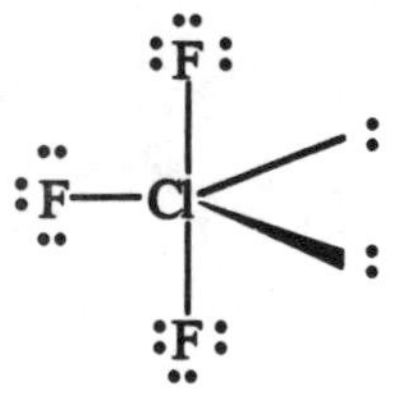

SN = 5
trigonal bipyramidal
sp^3d for Cl

(b) BBr_3

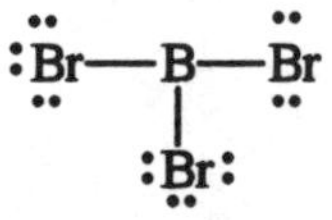

SN = 3
trigonal planar

sp^2 for B

(c) $BeF_2(g)$

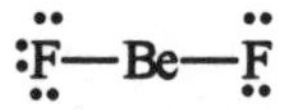

SN = 2
linear

sp for Be

(d) ONCl

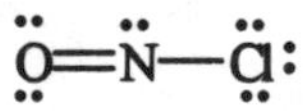

SN = 3
trigonal planar

sp^2 for N

10.61 *See Section 10.3 and Examples 10.7, 10.8.*

(a) CO_3^{2-}

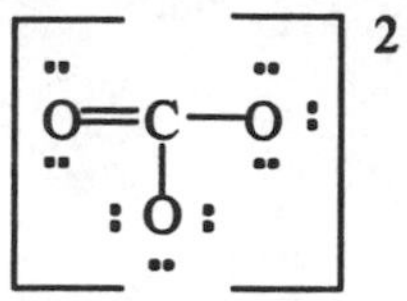

```
 ..     ..
[O=C—O:]2
 ..  |  ..
    :O:
     ..
```

SN = 3
trigonal planar
sp^2 for C

(b) CH_2F_2

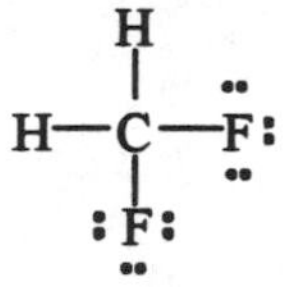

```
     H
     |   ..
  H—C—F:
     |   ..
    :F:
     ..
```

SN = 4
tetrahedral
sp^3 for C

(c) H_2CO

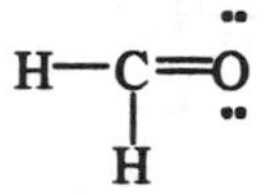

```
        ..
  H—C=O
     |  ..
     H
```

SN = 3
trigonal planar
sp^2 for C

10.62 *See Section 10.3 and Examples 10.7, 10.8.*

(a) C_2H_6

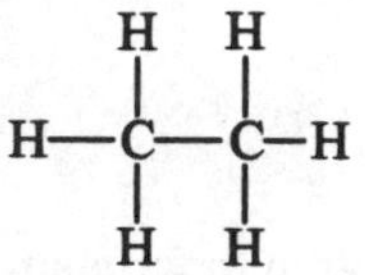

```
    H  H
    |  |
 H—C—C—H
    |  |
    H  H
```

SNs = 4, 4
tetrahdral about
each C atom
sp^3 for each C

(b) C_2H_4

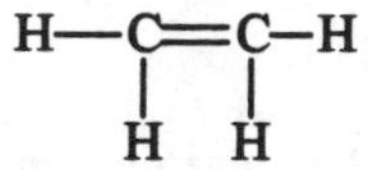

```
 H—C=C—H
    |  |
    H  H
```

SNs = 3, 3
trigonal planar
about each C atom
sp^2 for each C

(c) CBr_4

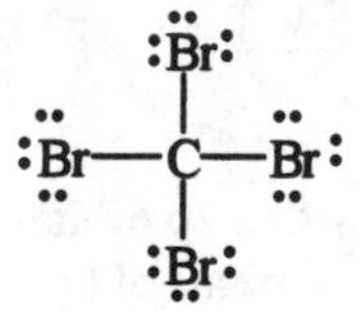

```
        ..
       :Br:
  ..    |    ..
 :Br—C—Br:
  ..    |    ..
       :Br:
        ..
```

SN = 4
tetrahedral

sp^3 for C

10.63 *See Section 10.3 and Examples 10.7, 10.8.*

(a) H_3O^+

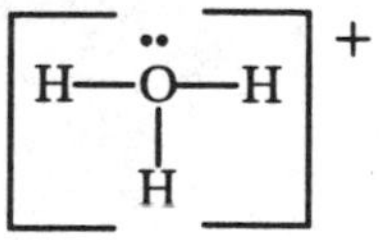

```
     ..
[ H—O—H ]+
     |
     H
```

SN = 4
tetrahedral
sp^3 for O

(b) H_3COH

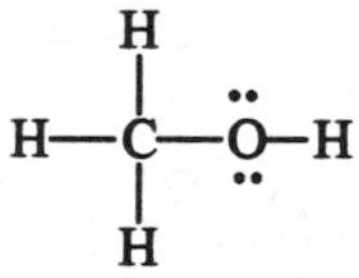

```
     H
     |  ..
  H—C—O—H
     |  ..
     H
```

SN = 4
tetrahedral
sp^3 for O

(c) Cl_2O

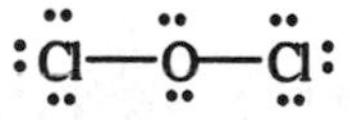

```
  ..   ..   ..
 :Cl—O—Cl:
  ..   ..   ..
```

SN = 4
tetrahedral
sp^3 for O

10.64 *See Section 10.3 and Examples 10.7, 10.8.*

(a) $HNCl_2$

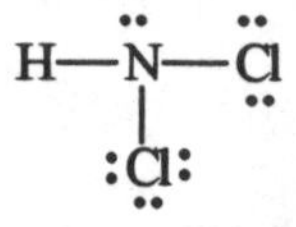

H—N—Cl
|
Cl

SN = 4
tetrahedral

sp^3 for N

(b) NO_3^-

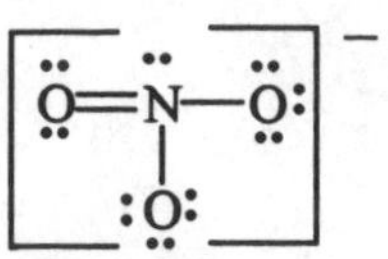

[O=N—O
|
O]$^-$

SN = 3
trigonal planar

sp^2 for N

(c) N_2H_2

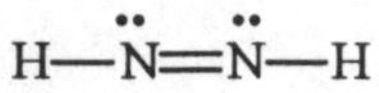

H—N=N—H

SNs = 3, 3
trigonal planar
about each N atom

sp^2 for each N

10.65 *See Section 10.3 and Examples 10.7, 10.8.*

(a) OF_2 — SN = 4, tetrahedral

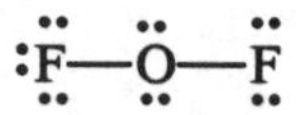

:F—O—F:

sp^3 for O and [He] ↑↓ (2s) ↑↓ ↑↓ ↑ (2p) for F.

A sp^3 orbital from O containing one electron overlaps with a 2p orbital from F containing one electron to form an O-F bond in OF_2. The lone pairs of electrons are in sp^3 orbitals of O.

(b) NH_3 — SN = 4, tetrahedral

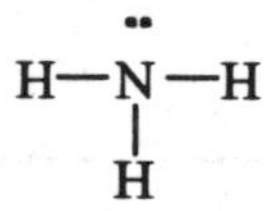

H—N—H
|
H

sp^3 for N and ↑ (1s) for H.

A sp^3 orbital from N containing one electron overlaps with a 1s orbital of H containing one electron to form a N-H bond in NH_3. The lone pair of electrons is in a sp^3 orbital of N.

(c) BCl_3 — SN = 3, trigonal planar

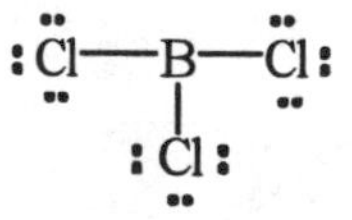

:Cl—B—Cl:
|
:Cl:

sp^2 for B and [Ne] ↑↓ (3s) ↑↓ ↑↓ ↑ (3p) for Cl.

A sp^2 orbital from B containing one electron overlaps with a 3p orbital of Cl containing one electron to form a B-Cl bond in BCl_3.

10.66 ***See Section 10.3 and Examples 10.7, 10.8.***

(a) IO_4^- SN = 4, tetrahedral

sp^3 for I and [He] ↑↓ (2s) ↑↓ ↑ ↑ (2p) for O.

A sp^3 orbital from I overlaps with a 2p orbital of O to form an I-O bond in IO_4^-. There are no lone pairs on I.

(b) $AlF_3(g)$ SN = 3, trigonal planar

sp^2 for Al and [He] ↑↓ (2s) ↑↓ ↑↓ ↑ (2p) for F.

A sp^2 orbital from Al containing one electron overlaps with a 2p orbital of F containing one electron to form an Al-F bond in $AlF_3(g)$. There are no lone pairs on Al.

(c) NO_2^- SN = 3, trigonal planar

sp^2 for N and [He] ↑↓ (2s) ↑↓ ↑ ↑ (2p) for O.

A sp^2 orbital from N overlaps with a 2p orbital of O to form a N-O bond in NO_2^-. The lone pair on N occupies an sp^2 orbital.

10.67 ***See Section 10.3 and Example 10.7.***

SeF_4 SN = 5, trigonal bipyramidal

sp^3d for Se and [He] ↑↓ (2s) ↑↓ ↑↓ ↑ (2p) for F.

A sp^3d orbital from Se containing one electron overlaps with a 2p orbital of F containing one electron to form a Se-F bond in SeF_4. The lone pair of electrons is in a sp^3d orbital of Se.

10.68 ***See Section 10.3 and Examples 10.7, 10.8.***

NOHO For N: SN = 3, trigonal planar, sp^2.

H—O—N=O For O: SN = 4, tetrahedral, sp^3.

10.69 ***See Section 10.3 and Examples 10.7, 10.8.***

(a)

```
  H     :Cl:
  |      |
H—C—O—N—Cl:
  |
  H
```

For C: SN = 4, tetrahedral, sp^3.

For O: SN = 4, tetrahedral, sp^3.

For N: SN = 4, tetrahedral, sp^3.

(b)

```
          H
          |
H—C≡C—P—H
```

For both C: SN = 2, linear, sp.

For P: SN = 4, tetrahedral, sp^3.

10.70 ***See Section 10.3 and Examples 10.7, 10.8.***

(a)

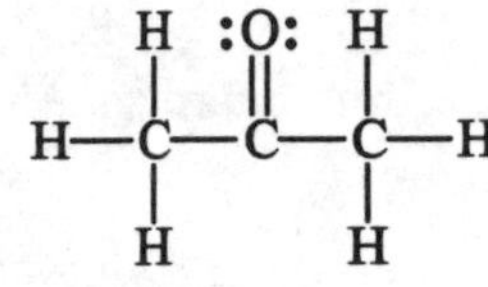

For end C atoms: SN = 4, tetrahedral, sp^3.

For center C atom: SN = 3, trigonal planar, sp^2.

(b)

```
  H
  |
H—C—C≡N
  |
  H
```

For left C: SN = 4, tetrahedral, sp^3.

For right C: SN = 2, linear, sp.

10.71 ***See Section 10.4, Figures 10.24-31.***

(a) 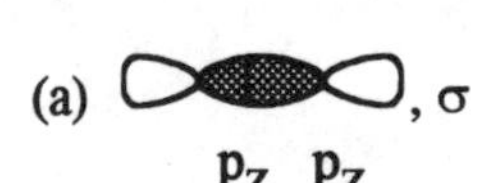, σ

p_z p_z

(b) 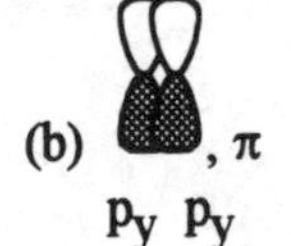, π

p_y p_y

(c) , σ

sp_z p_z

10.72 ***See Section 10.4, Figures 10.24-31.***

(a) 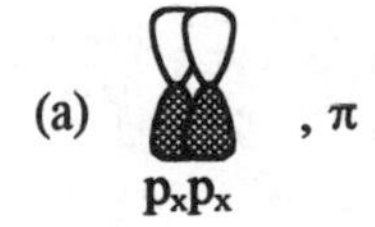, π

p_xp_x

(b), σ

s p_z

(c), σ

sp^2 s

10.73 ***See Sections 10.3, 10.4, and Examples 10.8, 10.9.***

H_3CCN

$$H-\overset{H}{\underset{H}{C_1}}-C_2\equiv N:$$

SN C_1 = 4, tetrahedral, sp^3

SN C_2 = 2, linear, sp

Bond	Orbital Overlaps	Bond Type
C-H	sp^3-1s	σ
C_1-C_2	sp^3-sp	σ
$C_2\equiv N$	sp-p_Z	σ
	p_Y-p_Y	π
	p_X-p_X	π

There are a total of five σ bonds and two π bonds inH_3CCN.

10.74 ***See Sections 10.3, 10.4, and Examples 10.8, 10.9.***

H, H, $C_1=C_2$, H, C_3, H, H, H

SN C_1 = 3, trigonal planar, sp^2

Sn C_2 = 3, trigonal planar, sp^2

SN C_3 = 4, tetrahedral, sp^3

Bond	Orbital Overlaps	Bond Type
C_1-H	sp^2-1s	σ
C_2-H	sp^2-1s	σ
C_3-H	sp^3-1s	σ
C_1-C_2	sp^2-sp^2	σ
	p-p	π
C_2-C_3	sp^2-sp^3	σ

10.75 ***See Sections 10.3, 10.4, and Examples 10.8, 10.9.***

H, H, C=N, :, H

SN C = 3, trigonal planar, sp^2

SN N = 3, trigonal planar, sp^2

Bond	Orbital Overlaps	Bond Type
C-H	sp^2-1s	σ
N-H	sp^2-1s	σ
C-N	sp^2-sp^2	σ
	p-p	π

Bond overlaps are similar to those shown for C_2H_4 in Figure 10.25.

10.76 ***See Sections 10.3, 10.4, and Examples 10.8, 10.9.***

:N≡N:

SN N = 2, linear, sp

Bond	Orbital Overlaps	Bond Type
N-N	sp_z-sp_z	σ
	P_x-P_x	π
	p_y-p_y	π

Bond overlaps are similar to those shown for carbon-carbon bond of C_2H_2 in Figures 10.30a and 10.30b.

10.77 *See Sections 10.3, 10.4, and Examples 10.7, 10.8, 10.9..*

(a)

```
   H  H
    \ /
     C    H
   /   \ /
H—C     C—H
  ||    |
H—C     C—H
   \   / \
     C    H
    / \
   H   H
```

The single bonded carbon atoms are tetrahedral and sp^3.

The double bonded carbon atoms are trigonal planar and sp^2.

(b)

```
:Cl:
     \
      C=O:
     /
:Cl:
```

The C is trigonal planar and sp^2.

(c)

```
         H  :O:
H        |   ||    ..
  \      |   ||
  :N — C1 — C2 — O — H
  /      |         ..
H        H
```

The N is tetrahedral and sp^3.

The C_1 is tetrahedral and sp^3.

The C_2 is trigonal planar and sp^2.
The O of C_2-O-H is tetrahedral and sp^3.

10.78 *See Sections 10.3, 10.4, and Examples 10.7, 10.8, 10.9..*

(a) $\ddot{O}=C=\ddot{O}$

For C: SN = 2, linear, sp.

(b)

```
     H
     |
H — C1 — C2≡C3 — H
     |
     H
```

For C_1: SN = 4, tetrahedral, sp^3.

For C_2: SN = 2, linear, sp.

For C_3: SN = 2, linear, sp.

(c)

```
     H   :O:
     |    ||
H — C1 — C2 — H
     |
     H
```

For C_1: SN = 4, tetrahedral, sp^3.

For C_2: SN = 3, trigonal planar, sp^2.

10.79 *See Sections 10.3, 10.4, and Examples 10.7, 10.8, 10.9.*

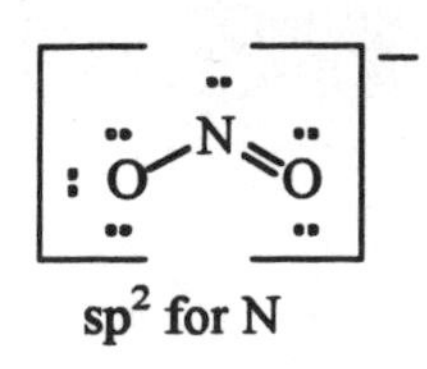

sp^2 for N

```
[ O=N—O: ]⁻
```

sp^2 for N

Note: A top view of a trigonal planar arrangement is shown here for convenience. A side view is often shown like that shown in 10.77b.

10.80 ***See Sections 10.3, 10.4, and Examples 10.7, 10.8, 10.9.***

$[:\ddot{N}{=}N{=}\ddot{N}:]^-$ — sp for central N

$[:N{\equiv}N{-}\ddot{\underset{..}{N}}:]^-$ — sp for central N

$[:\underset{..}{\ddot{N}}{-}N{\equiv}N:]^-$ — sp for central N

10.81 ***See Sections 10.3, 10.4, and Examples 10.7, 10.8, 10.9.***

(a)

```
   H  H
   |  |
H—C—N:
   |  |
   H  H
```

C is tetrahedral and sp^3.

N is tetrahedral and sp^3.

(b)

```
   H
   |
H—C1—C2≡C3—H
   |
   H
```

C_1 is tetrahedral and sp^3.

C_2 and C_3 are linear and sp.

10.82 ***See Sections 10.3, 10.4, and Examples 10.7, 10.8, 10.9.***

(a)

```
   H  :O:
   |   ||
H-C1—C2——H
   |
   H
```

C_1 is tetrahedral and sp^3.

C_2 is trigonal planar and sp^2.

(b)

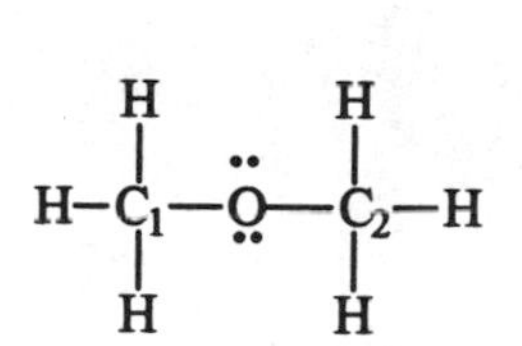

C_1 and C_3 are tetrahedral and sp^3.

O is tetrahedral and sp^3.

10.83 ***See Sections 10.3, 10.4, and Examples 10.7, 10.8, 10.9.***

H_2CCHCN

```
H—C1=C2—C3≡N:
   |   |
   H   H
```

C_1 is trigonal planar and sp^2.

C_2 is trigonal planar and sp^2.

C_3 is linear and sp.

10.84 ***See Sections 10.3, 10.4, and Examples 10.7, 10.8, 10.9.***

C_2F_4

C_1 and C_2 are trigonal planar and sp^2.

10.85 ***See Section 10.5, Figure 10.37, and Example 10.10.***

For He_2^{2+}, the electron configuration is $(\sigma_{1s})^2$, there are no unpaired electrons, the bond order is $\frac{1}{2}[2-0] = 1.0$, and it is predicted to be stable.

He^+ He_2^{2+} He^+

10.86 ***See Section 10.5, Figure 10.37, and Example 10.10.***

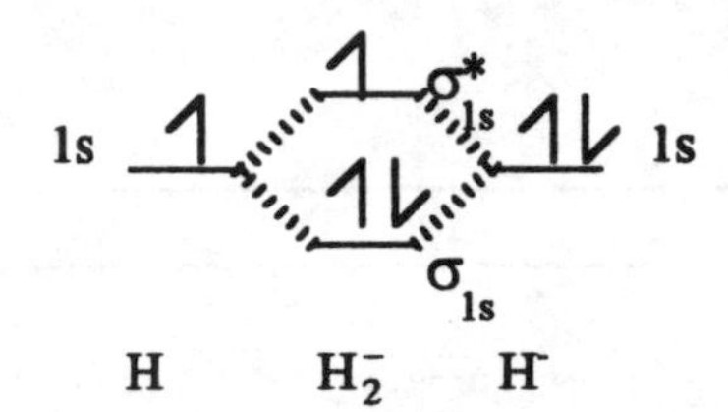

For H_2^-, the electron configuration is $(\sigma_{1s})^2(\sigma_{1s}^*)^1$, there is one unpaired electron, the bond order is $\frac{1}{2}[2-1] = 0.5$, and it is predicted to be stable.

10.87 ***See Section 10.5, Figure 10.37, and Example 10.10.***

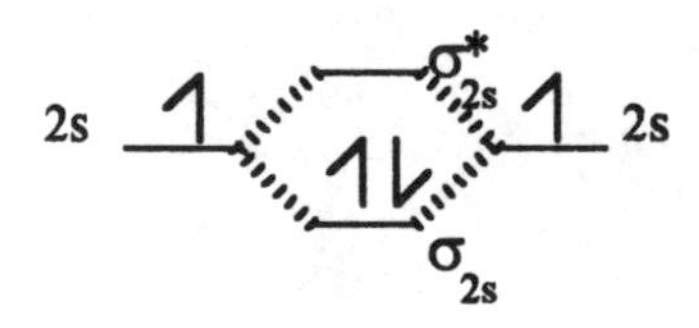

For Li_2, the electron configuration is $(\sigma_{2s})^2$, there are no unpaired electrons, the bond order is $\frac{1}{2}[2-0] = 1$, and it is predicted to be stable.

Li Li_2 Li

10.88 ***See Section 10.5, Figure 10.40, and Example 10.11.***

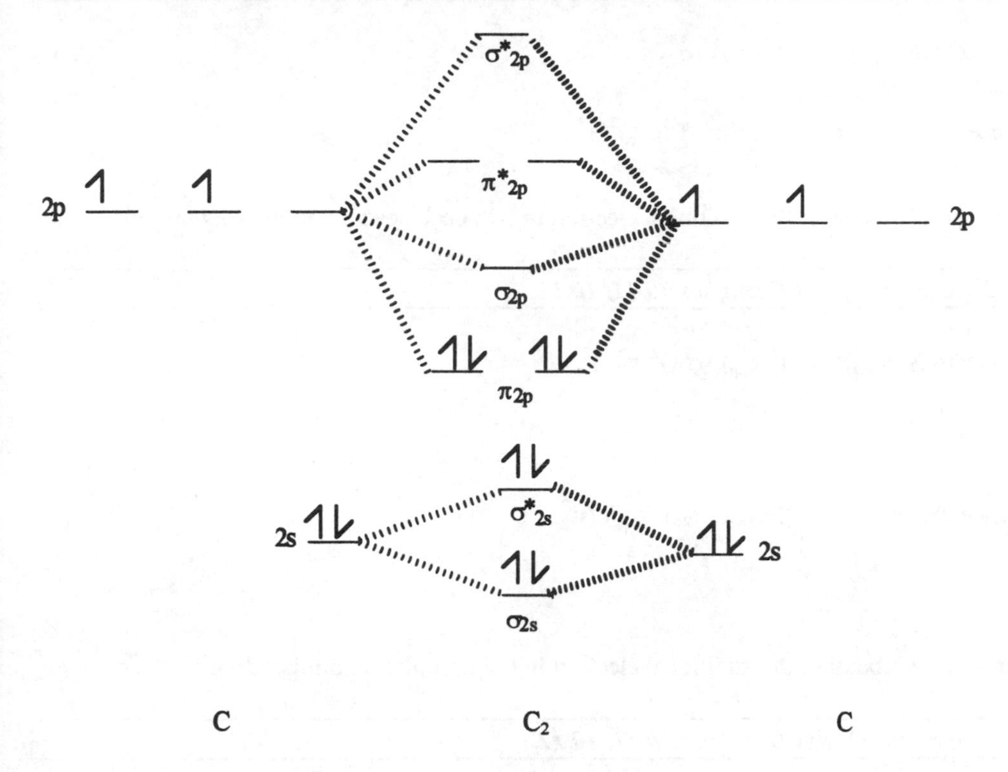

For C_2, the electron configuration is $(\sigma_{2s})^2$ $(\sigma^*_{2s})^2$ $(\pi_{2p})^4$, there are no unpaired electrons, the bond order is $\frac{1}{2}[6\text{-}2] = 2$, and it is predicted to be stable.

10.89 ***See Section 10.5, Figure 10.40, and Examples 10.11, 10.12.***

(a) C_2^+ has $(4 + 4 - 1) = 7$ valence electrons : $(\sigma_{2s})^2(\sigma^*_{2s})^2(\pi_{2p})^3$.
The bond order is $\frac{1}{2}[5 - 2] = 1.5$, and there is one unpaired electron in a π_{2p} orbital.

(b) N_2^- has $(5 + 5 + 1) = 11$ valence electrons: $(\sigma_{2s})^2(\sigma^*_{2s})^2(\pi_{2p})^4(\sigma_{2p})^2(\pi^*_{2p})^1$.
The bond order is $\frac{1}{2}[8 - 3] = 2.5$, and there is one unpaired electron in a π^*_{2p} orbital.

(c) Be_2^- has $(2 + 2 + 1) = 5$ valence electrons: $(\sigma_{2s})^2(\sigma^*_{2s})^2(\pi_{2p})^1$.
The bond order is $\frac{1}{2}[3 - 2] = 0.5$, and there is one unpaired electron in a π_{2p} orbital.

10.90 ***See Section 10.5, Figure 10.40, and Examples 10.11, 10.12.***

(a) O_2^+ has $(6+6-1) = 11$ valence electrons: $(\sigma_{2s})^2(\sigma^*_{2s})^2(\pi_{2p})^4(\sigma_{2p})^2(\pi^*_{2p})^1$.
The bond order is $\frac{1}{2}[8\text{-}3] = 2.5$, and there is one unpaired electron in a π^*_{2p} orbital.

(b) Li_2^+ has $(1+1-1) = 1$ valence electron: $(\sigma_{2s})^1$
The bond order is $\frac{1}{2}[1\text{-}0] = 0.5$, and there is one unpaired electron in a σ_{2s} orbital.

(c) C_2^- has $(4+4+1) = 9$ valence electrons: $(\sigma_{2s})^2(\sigma^*_{2s})^2(\pi_{2p})^4(\sigma_{2p})^1$.
The bond order is $\frac{1}{2}[7\text{-}2] = 2.5$, and there is one unpaired electron in a σ_{2p} orbital.

10.91 *See Section 10.5, Figure 10.40, and Examples 10.11, 10.12.*

N_2 has (5+5) + 10 valence electrons: $(\sigma_{2s})^2(\sigma^*_{2s})^2(\pi_{2p})^4(\sigma_{2p})^2$.
The bond order is $\frac{1}{2}$[8-2] = 3.0.
N_2^- has (5+5+1) = 11 valence elctrons: $(\sigma_{2s})^2(\sigma^*_{2s})^2(\pi_{2p})^4(\sigma_{2p})^2(\pi^*_{2p})^1$.
The bond order is $\frac{1}{2}$[8-3] = 2.5.
N_2 has a higher bond order than N_2^- because the additional electron in N_2^- occupies an antibonding orbital.

10.92 *See Section 10.5, Figure 10.40, and Examples 10.11, 10.12.*

O_2 has (6+6) = 12 valence electons: $(\sigma_{2s})^2(\sigma^*_{2s})^2(\pi_{2p})^4(\sigma_{2p})^2(\pi^*_{2p})^2$.

The bond order is $\frac{1}{2}$[8-4] = 2.0.

O_2^- has (6+6+1) = 13 valence electrons: $(\sigma_{2s})^2(\sigma^*_{2p})^2(\pi_{2p})^4(\sigma_{2p})^2(\pi^*_{2p})^3$.

The bond order is $\frac{1}{2}$[8-5] = 1.5

O_2 has a higher bond order than O_2^- because the additional electron in O_2^- occupies an antibonding electron.

10.93 *See Section 10.5, Figure 10.40, and Examples 10.11, 10.12.*

(a) B_2 has (3+3) = 6 valence electrons: $(\sigma_{2s})^2(\sigma^*_{2p})^2(\pi_{2p})^2$.
The bond order is $\frac{1}{2}$[4-2] = 1.0.
B_2^- has (3+3+1) = 7 valence electrons: $(\sigma_{2s})^2(\sigma^*_{2p})^2(\pi_{2p})^3$.
The bond order is $\frac{1}{2}$[5-2] = 1.5.
B_2^- has a higher bond order and stronger bond than B_2 because the additional electron in B_2^- occupies a bonding orbital.

(b) C_2^- has (4+4+1) = 9 valence electrons: $(\sigma_{2s})^2(\sigma^*_{2s})^2(\pi_{2p})^4(\sigma_{2p})^1$.
The bond order is $\frac{1}{2}$[7-2] = 2.5.
C_2^+ has (4+4-1) = 7 valence electrons: $(\sigma_{2s})^2(\sigma^*_{2s})^2(\pi_{2p})^3$.
The bond order is $\frac{1}{2}$[5-2] = 1.5.
C_2^- has a higher bond order and stronger bond than C_2^+. Forming C_2^- from C_2 involves adding an electron to a bonding orbital, whereas forming C_2^+ from C_2 involves removing an electron from a bonding orbital.

(c) O_2^{2+} has (6+6-2) = 10 valence electrons: $(\sigma_{2s})^2(\sigma^*_{2s})^2(\pi_{2p})^4(\sigma_{2p})^2$.
The bond order is $\frac{1}{2}$[8-2] = 3.0.
O_2 has (6+6) = 12 valence electrons: $(\sigma_{2s})^2(\sigma^*_{2s})^2(\pi_{2p})^4(\sigma_{2p})^2(\pi^*_{2p})^2$.
The bond order is $\frac{1}{2}$[8-4] = 2.0.
O_2^{2+} has a higher bond order and stronger bond than O_2 because two electrons are removed from antibonding orbitals of O_2 to form O_2^{2+}.

10.94 See Section 10.5, Figure 10.40, and Examples 10.11, 10.12.

(a) F_2 has (7+7) = 14 valence electrons: $(\sigma_{2s})^2(\sigma^*_{2s})^2(\pi_{2p})^4(\sigma_{2p})^2(\pi^*_{2p})^4$.
The bond order is $\frac{1}{2}$[8-6] = 1.0.
F_2^- has (7+7+1) = 15 valence electrons: $(\sigma_{2s})^2(\sigma^*_{2s})^2(\pi_{2p})^4(\sigma_{2p})^2(\pi^*_{2p})^4(\sigma^*_{2p})^1$.
The bond order is $\frac{1}{2}$[8-7] = 0.5.
F_2 has a higher bond order and stronger bond than F_2^- because the additional electron in F_2^- occupies an antibonding orbital.

(b) O_2^- has (6+6+1) = 13 valence electrons: $(\sigma_{2s})^2(\sigma^*_{2s})^2(\pi_{2p})^4(\sigma_{2p})^2(\pi^*_{2p})^3$.
The bond order is $\frac{1}{2}$[8-5] = 1.5.
O_2^+ has (6+6-1) = 11 valence electrons: $(\sigma_{2s})^2(\sigma^*_{2s})^2(\pi_{2p})^4(\sigma_{2p})^2(\pi^*_{2p})^1$.
The bond order is $\frac{1}{2}$[8-3] = 2.5.
O_2^+ has a higher bond order and stronger bond than O_2^-. Forming O_2^+ from O_2 involves removing an antibonding electron, whereas forming O_2^- from O_2 involves adding an electron to an antibonding orbital.

(c) C_2^{2+} has (4+4-2) = 6 valence electrons: $(\sigma_{2s})^2(\sigma^*_{2s})^2(\pi_{2p})^2$.
The bond order is $\frac{1}{2}$[4-2] = 1.0.
C_2 has (4+4) = 8 valence electrons: $(\sigma_{2s})^2(\sigma^*_{2s})^2(\pi_{2p})^4$.
The bond order is $\frac{1}{2}$[6-2] = 2.0.
C_2 has a higher bond order and stronger bond than C_2^{2+} because forming C_2^{2+} from C_2 involves removing two bonding electrons.

10.95 See Section 10.5.

(a) O_2^- and F_2^+ have 13 valence electrons. A bond order of $\frac{1}{2}[8-5]=1.5$ indicates these species should be stable.

(b) C_2^{2-}, N_2, and O_2^{2+} have 10 valence electrons. A bond order of $\frac{1}{2}[8-2]=3.0$ indicates these species should be stable.

(c) Li_2^{2-}, Be_2, and B_2^{2+} have 4 valence electrons. A bond order of $\frac{1}{2}[2-2]=0$ indicates these species would not be stable.

10.96 See Section 10.5.

(a) C_2^- and N_2^+ have 9 valence electrons. A bond order of $\frac{1}{2}$[7-2] = 2.5 indicates these species should be stable.

(b) C_2 and N_2^{2+} have 8 valence electrons. A bond order of $\frac{1}{2}$[6-2] = 2.0 indicates these species should be stable.

(c) Li_2^- and Be_2^+ have 3 valence electrons. A bond order of $\frac{1}{2}$[2-1] = 0.5 indicates these species should be stable.

10.97 *See Section 10.5.*

(a) N_2 and CO have 10 valence electrons and $(\sigma_{2s})^2(\sigma^*_{2s})^2(\pi_{2p})^4(\sigma_{2p})^2$.

(b) B_2 and BeC have 6 valence electrons and $(\sigma_{2s})^2(\sigma^*_{2s})^2(\pi_{2p})^2$.

10.98 *See Section 10.5.*

(a) O_2 and NF have 12 valence electrons and $(\sigma_{2s})^2(\sigma^*_{2s})^2(\pi_{2p})^4(\sigma_{2p})^2(\pi^*_{2p})^2$.

(b) C_2 and BN have 8 valence electrons and $(\sigma_{2s})^2(\sigma^*_{2s})^2(\pi_{2p})^4$.

10.99 *See Section 10.5.*

Specie	Number of valence electrons	Electron Configuration	Bond Order	Number of unpaired electrons
(a) CN	9	$(\sigma_{2s})^2 (\sigma^*_{2s})^2 (\pi_{2p})^4 (\sigma_{2p})^1$	$\frac{1}{2}[7-2]=2.5$	1 in σ_{2p}
(b) CO^-	11	$(\sigma_{2s})^2 (\sigma^*_{2s})^2 (\pi_{2p})^4 (\sigma_{2p})^2 (\pi^*_{2p})^1$	$\frac{1}{2}[8-3]=2.5$	1 in π^*_{2p}
(c) BeB^-	6	$(\sigma_{2s})^2 (\sigma^*_{2s})^2 (\pi_{2p})^2$	$\frac{1}{2}[4-2]=1.0$	2 in π_{2p}
(d) BC^+	6	$(\sigma_{2s})^2 (\sigma^*_{2s})^2 (\pi_{2p})^2$	$\frac{1}{2}[4-2]=1.0$	2 in π_{2p}

10.100 *See Section 10.5.*

Specie	Number of valence electrons	Eelctron Configuration	Bond Order	Number of unpaired electrons
(a) $LiBe^+$	2	$(\sigma_{2s})^2$	$\frac{1}{2}[2-0]=1.0$	none
(b) CO^+	9	$(\sigma_{2s})^2 (\sigma^*_{2s})^2 (\pi_{2p})^4 (\sigma_{2p})^1$	$\frac{1}{2}[7-2]=2.5$	1 in σ_{2p}
(c) CN^-	10	$(\sigma_{2s})^2 (\sigma^*_{2s})^2 (\pi_{2p})^4 (\sigma_{2p})^2$	$\frac{1}{2}[8-2]=3.0$	none
(d) OF	13	$(\sigma_{2s})^2 (\sigma^*_{2s})^2 (\pi_{2p})^4 (\sigma_{2p})^2 (\pi^*_{2p})^3$	$\frac{1}{2}[8-5]=1.5$	1 in π^*_{2p}

10.101 ***See Section 10.5 and Figure 10.48.***

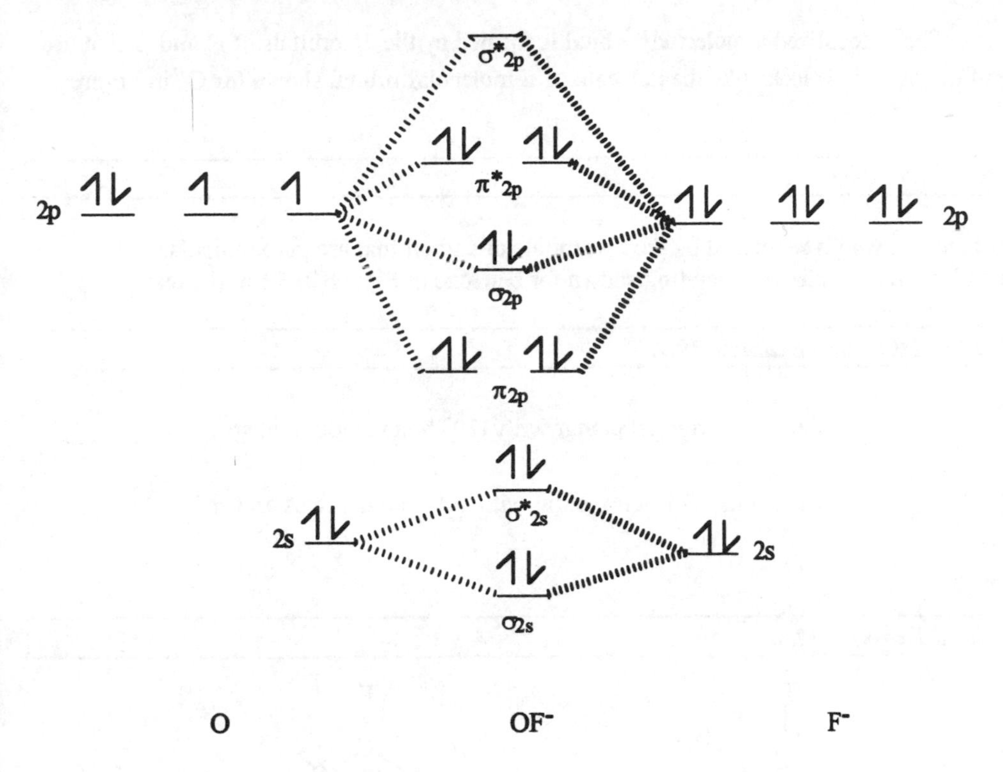

The bond order for OF^- is $\frac{1}{2}[8-6]=1$.

10.102 ***See Section 10.5 and Figure 10.48.***

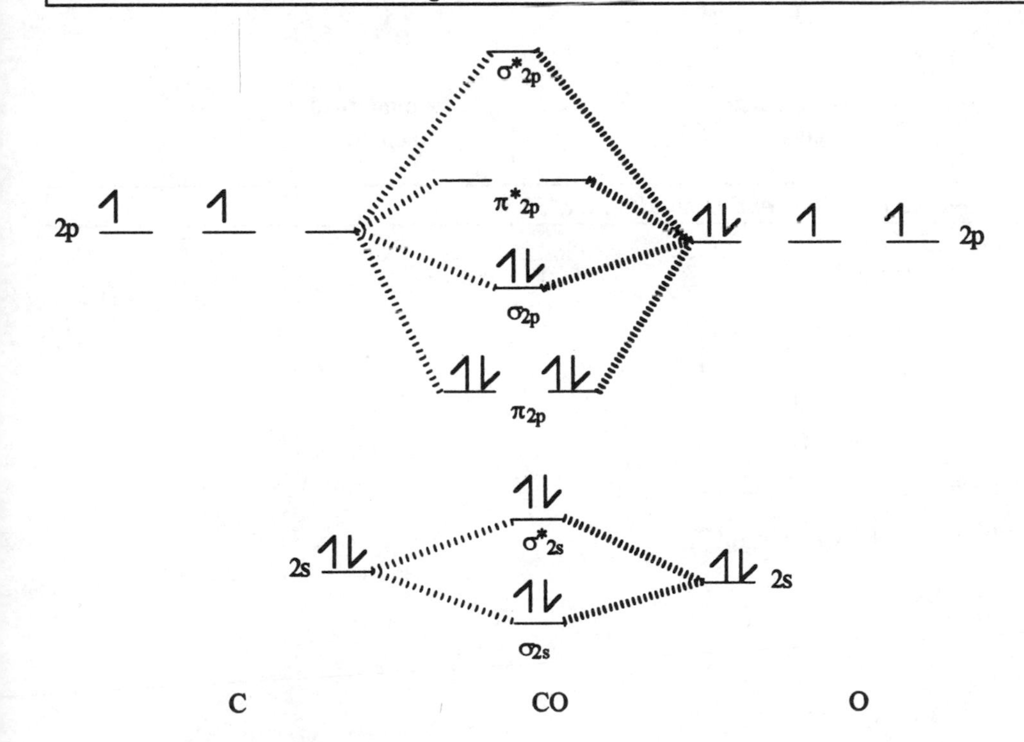

The bond order for CO is $\frac{1}{2}$[8-2]=3.0.

10.103 *See Section 10.6.*

NO_2^- is isoelectronic with O_3. The delocalized π molecular orbital is formed by the 2p orbitals of N and O that are perpendicular to the plane of the atoms. It looks like the delocalized π molecular orbital shown for O_3 in Figure 10.49 in the text.

10.104 *See Section 10.6.*

The π molecular orbital for benzene would be formed by the 2p orbitals of carbon that are perpendicular to the plane of the atoms. It looks like the delocalized pi bonding shown for benzene in Figure 10.32 in the text.

10.105 *See Sections 10.1,10.3, 10.4, and Example 10.7.*

N_2O_5

Each N is trigonal planar with 120° bond angles and sp^2.

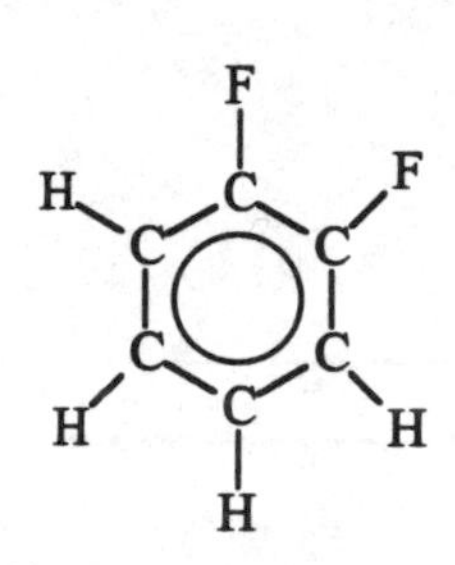

The central O is tetrahedral with 109° bond angles and sp^3.

10.106 *See Section 10.2 and Example 10.4.*

unsymmetrical
polar

unsymmetrical
polar

symmetrical
nonpolar

10.107 *See Section 10.1, Figures 10.1, 10.6, and Examples 10.2, 10.3.*

ClF_2^-

ClF_2^+

$\angle$ F - Cl - F = 180°

$\angle$ F - Cl - F = 109°

10.108 ***See Sections 10.3, 10.4, and Examples 10.8, 10.9.***

(a) There are twenty one sigma bonds and five pi bonds as shown in the Lewis structure. The pi bonding in the six-membered carbon ring is actually delocalized pi bonding.

(b) The SN for the top carbon atom is three, the electron-pair arrangement is trigonal planar and the hybridization for that carbon atom is sp^2.

(c) The SN for the ring carbon atom bonded to the oxygen atom is three, the electron-pair arrangement is trigonal planar and the hybridization for that carbon atom is sp^2. The SN for the oxygen atom attached to the ring carbon atom is four, the electron-pair arrangement is tetrahedral and the hybridization for that oxygen atom is sp^3.

10.109 ***See Sections 10.3, 10.4, and Examples 10.8, 10.9.***

(a) There are thirty nine sigma bonds and six pi bonds shown in the Lewis structure. The pi bonding in the six-membered carbon ring is actually delocalized pi bonding.

(b) The SN for each carbon atom forming a double bond with oxygen is three, the electron-pair arrangement is trigonal planar and the hybridization for each of these carbon atoms is sp^2.

(c) The SN for each N atom is four, the electron-pair arrangement is tetrahedral and the hybridization for each N atom is sp^3.

10.110 ***See Sections 10.1,10.3.***

A bond angle of 109° is associated with a tetrahedral electron-pair arrangement and sp^3 hybrids. A bond angle of 120° is associated with a trigonal planar arrangement and sp^2 hybrids. The lone pair of electrons with the sp^2 hybrids would be located in the unhybridized atomic p orbital.

10.111 *See Sections 10.1, 10.2, 10.3, 10.4, and Examples 10.9, 10.10.*

$COCl_2$

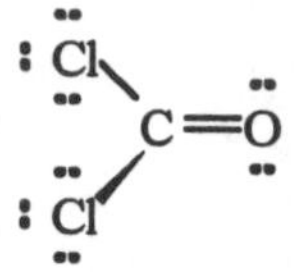

The C has a SN of 3, a trigonal planar electron-pair arrangement and uses sp^2 hybrids. The molecule is unsymmetrical and therefore polar.

10.112 *See Sections 10.1, 10.2, 10.3, 10.4, and Examples 10.9, 10.10.*

NCN^{2-}

$N - C - N^{2-}$

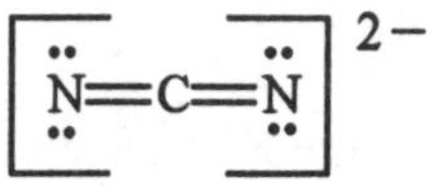

$[N{=}C{=}N]^{2-}$

Total valence electrons = $[2 \times 5(N) + 1 \times 4(C) + 2(\text{charge})] = 16$

Twelve electrons remain after assigning two single bonds, and sixteen unshared electrons are needed to give each atom a noble gas configuration (6 for each N and four for C). Hence, four electrons (16-12) must be used to form two additional bonds. The lowest formal charges result when one additional bond is formed between carbon and each nitrogen atom. The SN for carbon is two, the electron-pair arrangement is linear and the hybridization for carbon is sp.

10.113 *See Section 10.1 and Example 10.1.*

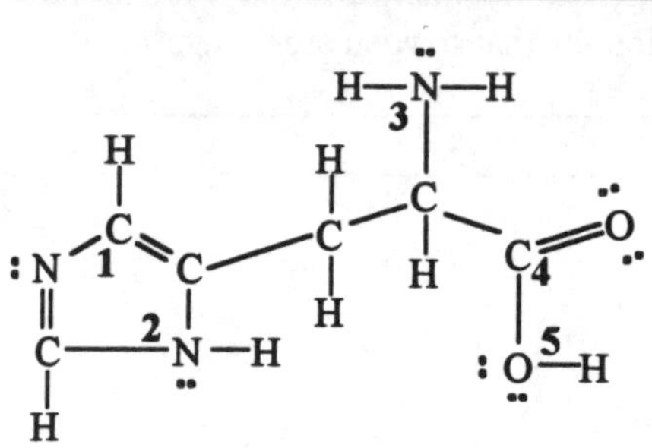

Bond ∠ 1: SN = 3, trigonal planar and therefore 120°.

Bond ∠ 2: SN = 4, tetrahedral and therefore 109°.

Bond ∠ 3: SN = 4, tetrahedral and therefore 109°.

Bond ∠ 4: SN = 3, trigonal planar and therefore 120°.

Bond ∠ 5: SN = 4, tetrahedral and therefore 109°.

10.114 *See Sections 9.5, 10.1.*

HC(O)NH_2

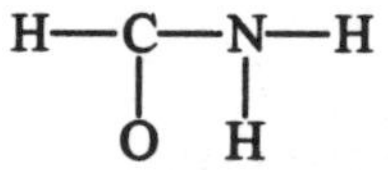

Total valence electrons = $[1 \times 4(C) + 1 \times 5(N) + 1 \times 6(O) + 3 \times 1(H)] = 18$.

Eight electrons remain after assigning five single bonds, and ten unshared electrons are needed to give each atom a noble gas configuration (6 for O, 2 for C, and 2 for N). Hence, two electrons (10-8) must be used to form one additional bond. This gives the following resonance possibilities:

A B

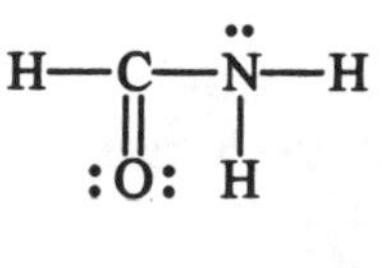

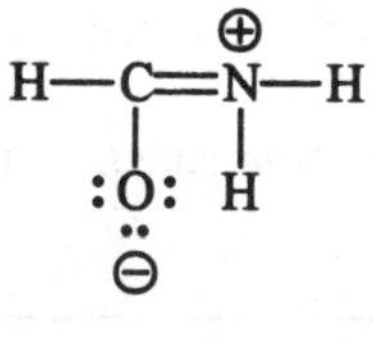

Structure A: For C: SN = 3, trigonal planar, 120° bond angles.
For N: SN = 4, tetrahedral, 109° bond angles.

Structure B: For C: SN = 3, trigonal planar, 120° bond angles.
For N: SN = 3, trigonal planar, 120° bond angles.

Hence, experimental determination of the H-N-H bond angle can be used to determine which resonance form is more important.

10.115 ***See Section 10.5 and Figure 10.48.***

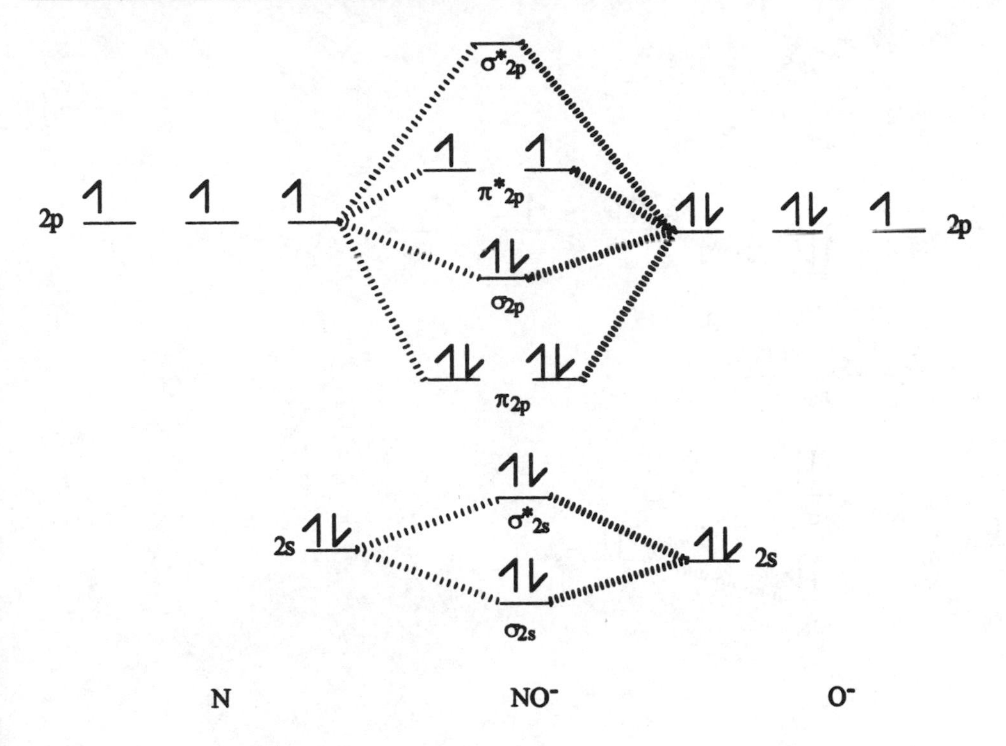

The bond order for $NO^- = \frac{1}{2}[8-4] = 2$.

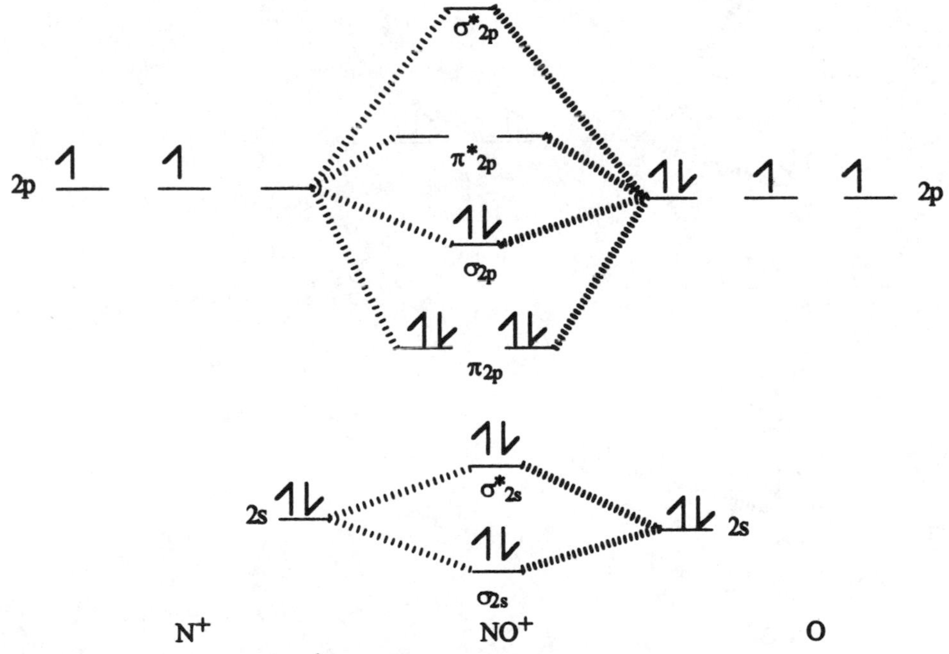

The bond order for $NO^+ = \frac{1}{2}[8-2] = 3$.

Hence, NO^+ has a higher bond order than NO^-.

10.116 ***See Section 10.5 and Figure 10.48.***

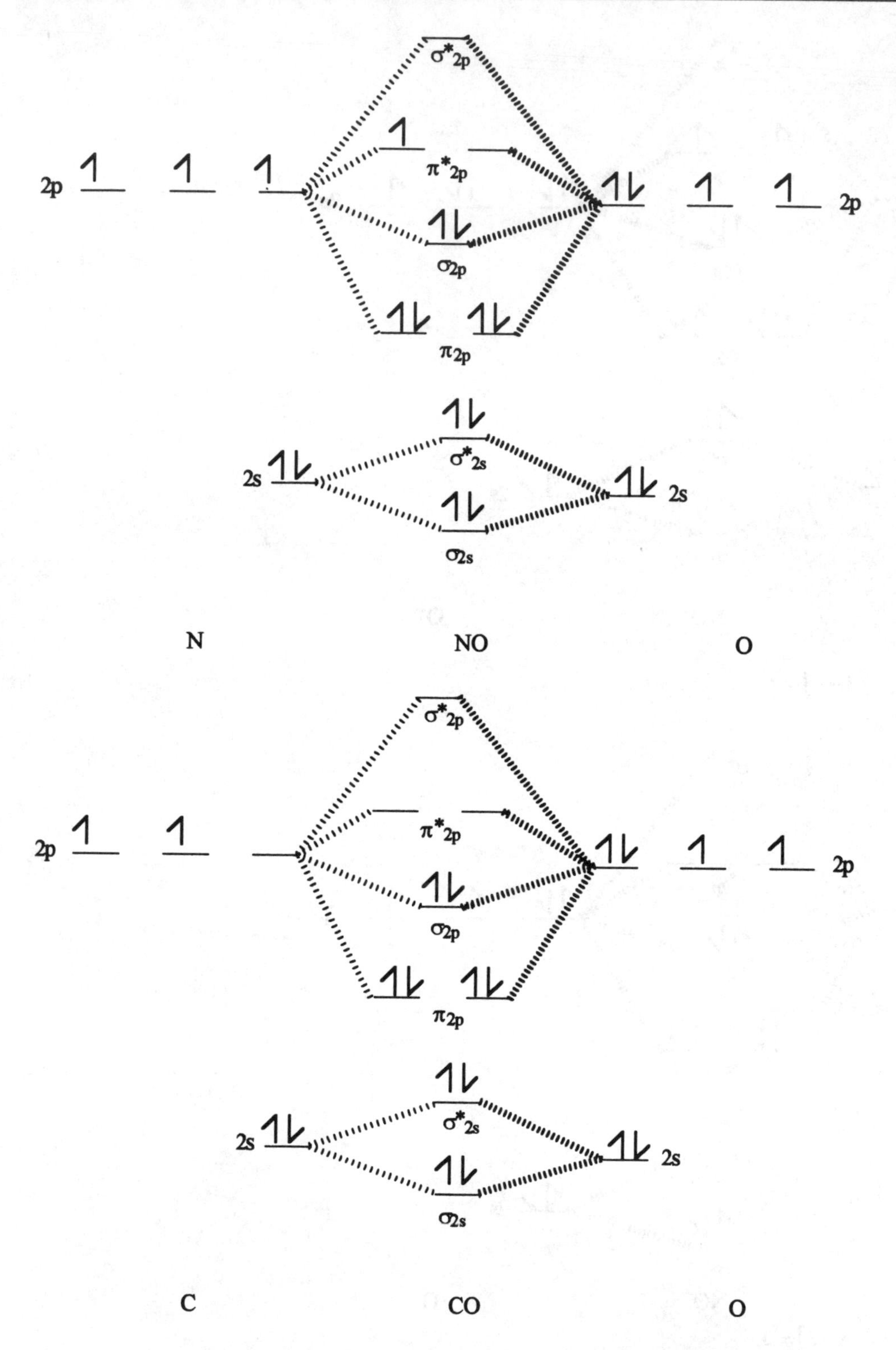

The electron to be ionized from NO is located in a π^*_{2p} orbital, whereas the electron to be ionized from CO is in a lower energy σ_{2p} orbital. Hence, NO is predicted to have a lower ionization energy than CO.

10.117 ***See Sections 10.1, 10.2, 10.3, 10.4.***

(a) NH_2NO_2

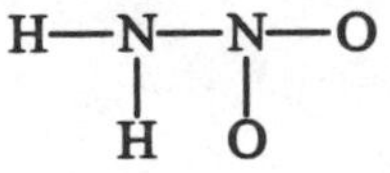

H—N—N—O
 | |
 H O

Total valence electrons = $[2 \times 5(N) + 2 \times 1(H) + 2 \times 6(O)] = 24$.
Fourteen electrons remain after assigning five single bonds, and sixteen unshared electrons are needed to give each atom a noble gas configuration (2 for each N and 6 for each O). Hence , two electrons (16-14) must be used to form one additional bond. Formal charge considerations indicate the additional bond is formed by N and O. SN of 4 for N on the left indicates bond angles of approximately 109°, and SN of 3 for N on right indicates bond angles of 120°. The N on the left uses sp^3 hybrids, and the N on the right uses sp^2 hybrids. The molecule is unsymmetrical and polar.

H—N̈—N⊕=Ö:
 | |
 H :Ö:⊖

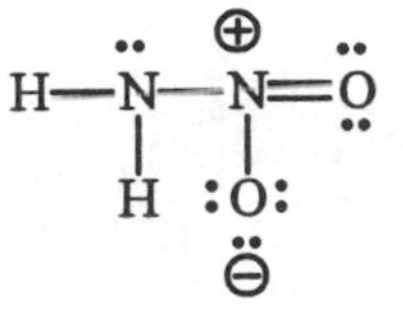

(b) HN_3
H - N - N - N

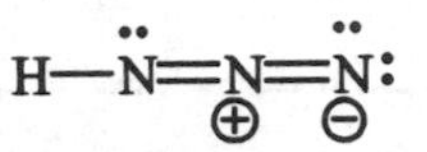

Total valence electrons = $[1 \times 1(H) + 3 \times 5(N)] = 16$
Ten electrons remain after assigning three single bonds, and fourteen unshared electrons are needed to give each atom a noble gas configuration (4 for left N, 4 for center N and 6 for right N). Hence, four electrons (14-10) must be used to form two additional bonds. SN of 3 for N on left indicates bond angles of approximately 120°, and SN of 2 for N in center indicates bond angles of 180°. The N on the left uses sp^2 hybrids and the N in the center uses sp hybrids. The molecule is unsymmetrical and polar.

H—N̈=N⊕=N̈:⊖

Note: H—N̈⊖(:)—N⊕≡N: is plausible with sp^3 for left N and sp for central N.

10.118 ***See Sections 10.1, 10.2, 10.3, 10.4.***

(a) CH_2N_2

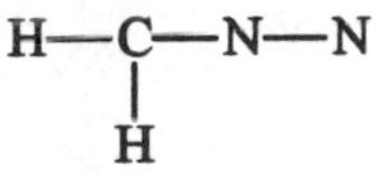

H—C—N—N
 |
 H

Total valence electrons = $[1 \times 4(C) + 2 \times 1(H) + 2 \times 5(N)] = 16$.
Eight electrons remain after assigning four single bonds, and twelve unshared electrons are needed to give each atom a noble gas configuration (2 for C, 4 for left N, and 6 for right N). Hence, four electrons (12-8) must be used to form two additional bonds. This gives the following resonance possibilites:

A: H—C(H)=N⊕=N̈:⊖

B: H—C̈⊖(H)—N⊕≡N:

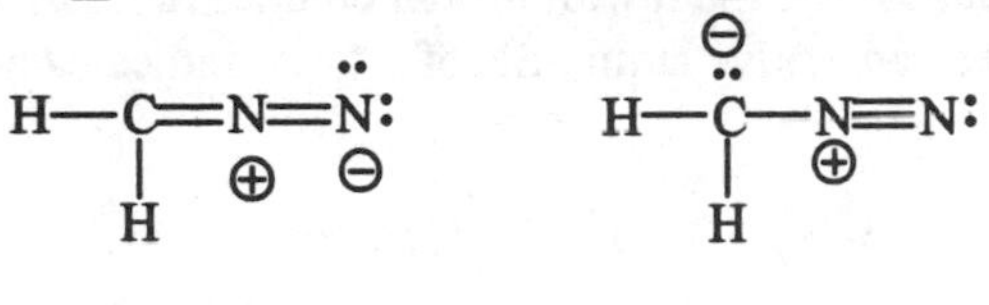

C, N, N Hybrids: sp^2, sp, sp^2 (A); sp^3, sp, sp (B)

Structure A is likely to be more important than Structure B because it places the negative formal charge on N rather than C, on the more electronegative atom. SN of 3 for C indicates 120° bond angles and SN of 2 for N on left indicates bond angles of 120°. The molecule is unsymmetrical and polar.

Note: Since the actual structure is a composite of the resonance forms shown above, it would be best not to try to specify the types of hybrids used by the terminal or end C and N atoms. It is, however, important to note that both C and N must have unhybridized p orbitals available to form multiple bonds.

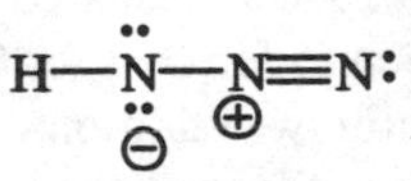

(b) $CH_3C(O)NO_2$

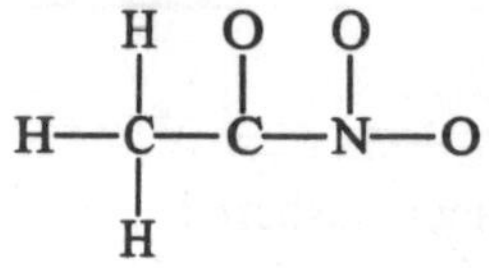

H O O
H—C—C—N—O
H

Total valence electrons = $[2\times4(C)+3\times1(H)+1\times5(N)+3\times6(O)]=34$.
Eighteen electrons remain after assigning eight single bonds, and twenty two unshared electrons are needed to give each atom a noble gas configuration (2 for C, 2 for N, and 6 for each O). Hence, four electrons must be used to form two additional bonds. This gives the following resonance possibilities:

A B

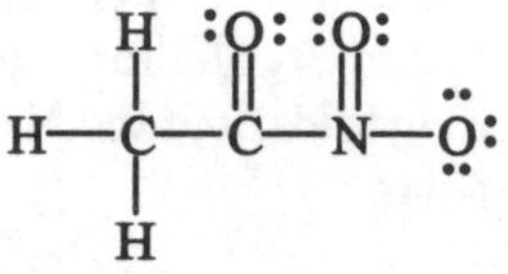

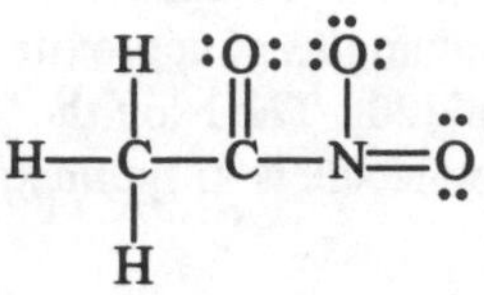

C, C, N hybrids: sp^3, sp^2, sp^2 sp^3, sp^2, sp^2

SN of 4 for C on the left indicates bond angles of 109.5°, SN of 3 for C in center indicates bond angles of 120° and SN of 3 for N indicates bond angles of 120°. The molecule is unsymmetrical and polar.

10.119 ***See Sections 10.1, 10.2, 10.3, 10.4.***

(a) OCN^-

O - C - N$^-$

Total valence electrons = $[1\times6(O)+1\times4(C)+1\times5(N)+1(\text{charge})]=16$.
Twelve electrons remain after assigning two single bonds, and sixteen unshared electrons are needed to give each atom a noble gas configuration (6 for O, 4 for C, and 6 for N). Hence, four electrons (16-12) must be used to form two additional bonds. This gives the following resonance possibilties with lowest formal charges:

A B 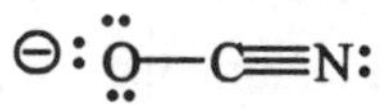 C

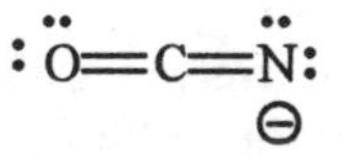

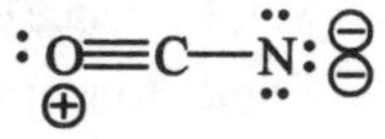

C hybrids: sp sp sp

Structure B is likely to be the most important because it has the lowest formal charges and places the negative formal charge on O rather than N, on the more electronegative atom. SN of 2 for C indicates bond angle of 180°.

(b) NO_3^-

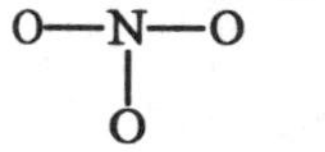

O—N—O $^-$
O

Total valence electrons =$[1\times5(N)+3\times6(O)+1(\text{charge})]=24$.
Eighteen electrons remain after assigning three single bonds, and twenty unshared electrons are needed to give each atom a noble gas configuration. Hence, two electrons (20-18) must be used to form one additional bond. This gives the following resonance possibilities:

A B 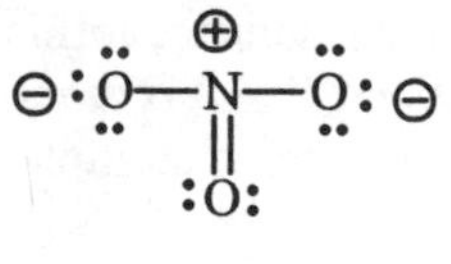 C

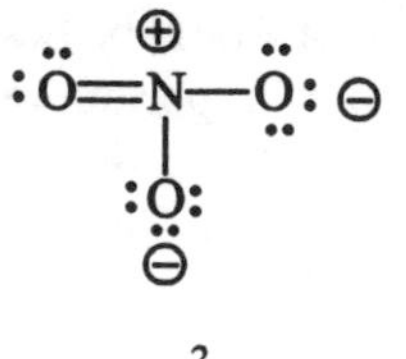

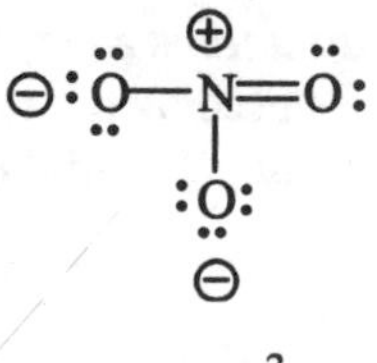

N hybrids: sp^2 sp^2 sp^2

These structures are equivalent and equally important. SN of 3 for N indicates bond angles of 120^0.

10.120 ***See Sections 10.1, 10.2, 10.3, 10.4.***

	ethylene oxide, C_2H_4O	ethylene glycol, $HOCH_2CH_2OH$	acrylonitrile, CH_2CHCN
Hybrids:	C, sp^3; O, sp^3	O, sp^3; C, sp^3; C, sp^3; O, sp^3	C, sp^2; C, sp^2; C, sp
π bonds:	no	no	yes, three

10.121 ***See Sections 10.1, 10.2, 10.3, 10.4.***

Vitamin-A contains ten sp^2 hybridized carbon atoms. These are the ten carbon atoms that are involved in double bonds. Vitamin-A also contains ten sp^3 hybridized carbon atoms. These are the four carbon atoms that are port of the ring and not part of double bonds, the five that are part of CH_3 groups and the one that is bonded to oxygen.

10.122 ***See Sections 10.1, 10.5.***

C_2^{2-}

$C - C^{2-}$

$[:C\equiv C:]^{2-}$

Total valence electrons = $[2 \times 4(C) + 2(\text{charge})] = 10$.

Eight electrons remain after assigning one single bond, and twelve unshared electrons are needed to give each atom a noble gas configuration (6 for each C). Hence, four electrons (12-8) must be used to form two additional bonds.

The molecular orbital diagram for the carbide ion is:

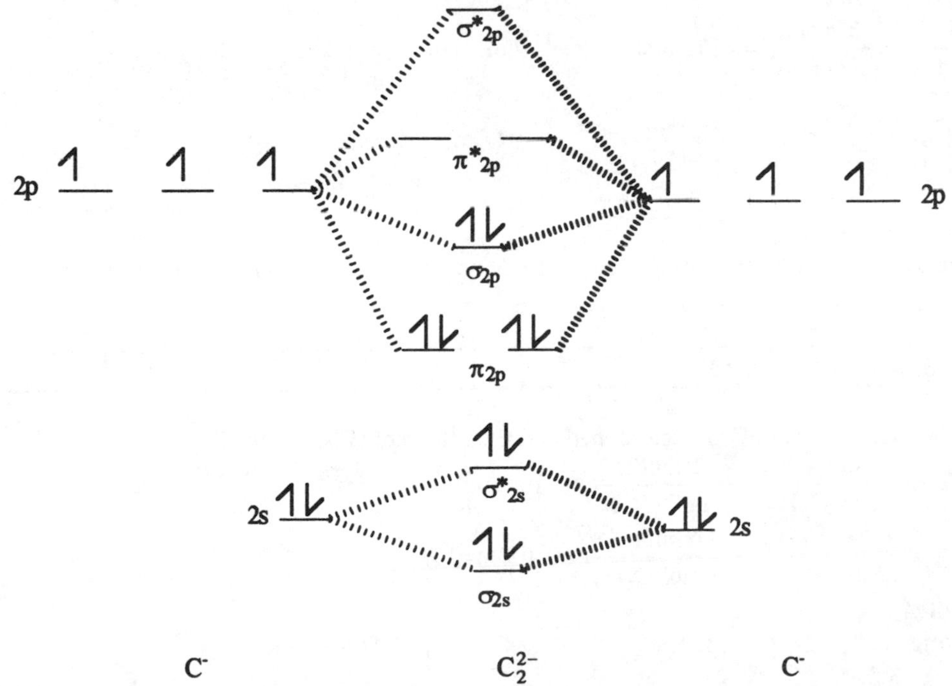

The bond order is $\frac{1}{2}[8-2]=3.0$. There are no unpaired electrons, and this is in agreement with the Lewis structure for C_2^{2-}.

10.123 ***See Sections 3.3, 10.1, 10.3, 10.4, and Example 3.12.***

Assume the sample has a mass of 100.00 g and therefore contains 54.53 g C, 9.15 g H, and 36.32 g O.

$$? \text{ mol C} = 54.53 \text{ g C} \times \frac{1 \text{ mol C}}{12.01 \text{ g C}} = 4.54 \text{ mol C} \qquad \text{relative mol C} = \frac{4.54 \text{ mol C}}{2.27} = 2.00 \text{ mol C}$$

$$? \text{ mol H} = 9.15 \text{ g H} \times \frac{1 \text{ mol H}}{1.008 \text{ g H}} = 9.08 \text{ mol H} \qquad \text{relative mol H} = \frac{9.08 \text{ mol H}}{2.27} = 4.00 \text{ mol H}$$

$$? \text{ mol O} = 36.32 \text{ g O} \times \frac{1 \text{ mol O}}{16.0 \text{ g O}} = 2.27 \text{ mol O} \qquad \text{relative mol O} = \frac{2.27 \text{ mol O}}{2.27} = 1.00 \text{ mol O}$$

The simplest formula is C_2H_4), and the simplest formula molar mass is 44.0 g/mol.

$$n = \frac{\text{molar mass compound}}{\text{molar mass } C_2H_4O} = \frac{44.0 \text{ g / mol}}{44.0 \text{ g / mol}} = 1, \text{ so the molecular formula is also } C_2H_4O.$$

The two possible Lewis structures for C_2H_4O are:

Hybrids: C, sp^3; C, sp^2 C, sp^2; C, sp^2; O, sp^3

10.124 ***See Sections 3.3, 10.1, 10.3, 10.4.***

$$x = \frac{\text{moles F}}{\text{moles S}}$$

$$? \text{ mol F} = 4.76 \text{ g } F_2 \times \frac{1 \text{ mol } F_2}{38.0 \text{ g } F_2} \times \frac{2 \text{ mol F}}{1 \text{ mol } F_2} = 0.250 \text{ mol F}$$

$$? \text{ mol S} = 4.01 \text{ g } S_8 \times \frac{1 \text{ mol } S_8}{256.8 \text{ g } S_8} \times \frac{8 \text{ mol S}}{1 \text{ mol } S_8} = 0.125 \text{ mol S}$$

$$x = \frac{0.250 \text{ mol F}}{0.125 \text{ mol S}} = \frac{2.00 \text{ mol F}}{1.00 \text{ mol S}}$$

The compound is SF_2, and its Lewis structure is: :F̤—S̤—F̤: .
The SN for S is four, the electron-pair arrangement is tetrahedral, the bond angle is 109°, and the hybridization for S is sp^3.

10.125 ***See Sections 3.2, 3.3, 5.4, 10.1, 10.3, 10.4.***

To determine the value of x, the mol H per mol C are determined and then the mol H per 2 mol C.

$$? \text{ mol C in } C_2H_2 = 1.30 \text{ g } C_2H_2 \times \frac{1 \text{ mol } C_2H_2}{26.0 \text{ g } C_2H_2} \times \frac{2 \text{ mol C}}{1 \text{ mol } C_2H_2} = 0.100 \text{ mol C}$$

$$? \text{ mol H in } C_2H_2 = 1.30 \text{ g } C_2H_2 \times \frac{1 \text{ mol } C_2H_2}{26.0 \text{ g } C_2H_2} \times \frac{2 \text{ mol H}}{1 \text{ mol } C_2H_2} = 0.100 \text{ mol H}$$

? mol H in 1.22 L H_2 = 2 x mol H_2.

Known Quantities: P = 1.01 atm   V = 1.22 L   T = 27 + 273 = 300 K

Solving PV = nRT for n gives $n = \frac{PV}{RT}$. $n = \frac{(1.01\ \text{atm})(1.22\ \text{L})}{\left(0.0821\ \frac{\text{L}\cdot\text{atm}}{\text{mol}\cdot\text{K}}\right)(300\ \text{K})} = 0.0500\ \text{mol}\ H_2$? mol H in 1.22 L H_2

$$= 0.0500\ \text{mol}\ H_2 \times \frac{2\ \text{mol H}}{1\ \text{mol}\ H_2} = 0.100\ \text{mol H}$$

$$?\ \frac{\text{mol H}}{\text{mol C}} = \frac{0.100\ \text{mol H} + 0.100\ \text{mol H}}{(0.100\ \text{mol C})} = \frac{2.00\ \text{mol H}}{1.00\ \text{mol C}}$$

Hence, there are 4.00 mol H per 2.00 mol C, and the compound is C_2H_4. The Lewis structure for C_2H_4 is:

```
    H   H
    |   |
H—C=C—H
```

The carbon-carbon double bond is formed by a sp^2-sp^2 σ overlap and a 2p-2p π overlap. See Figures 10.23, 10.24, 10.25 for illustration of these overlaps.

Chapter 11: Liquids and Solids

11.1 *See Chapter Introduction*

In the liquid state SO_2 molecules are close together and occupy a very large fraction of the volume. In the gaseous state SO_2 molecules are much further apart and occupy a very small fraction of the volume. This is why the density of liquid SO_2 (1.43 g/mL) is so much higher than the density of gaseous SO_2 (0.00293 g/mL) at STP.

11.2 *See Section 11.1, Table 11.2, and Example 11.1.*

The physical state of any sample of matter depends on the strength of the intermolecular forces and the average kinetic energy of the molecules and therefore the temperature of the sample. At a low temperature the average kinetic energy of the molecules will be low, and the sample is likely to exist as solid or a liquid. At a high temperature the average kinetic energy of the molecules will be high, and the sample is likely to exist as a liquid.

11.3 *See Section 11.1.*

The molecules are closer together in the solid state than in the liquid state. Hence, the energy of the intermolecular attractions is greater in the solid state than in the liquid state.

11.4 *See Section 11.2.*

The vapor pressure of water at 25° C is lower than that of dimethyl ether at 25° C because the intermolecular forces of attraction between water molecules are stronger than those between dimethyl ether molecules.

The stronger the intermolecular forces of attraction, the lower the vapor pressure and the higher the enthalpy of vaporization. Hence, water has a higher enthalpy of vaporization than dimethyl ether.

11.5 *See Section 11.2.*

The rates of evaporation and condensation are equal at the point at which the equilibrium vapor pressure of a liquid is reached. A state of dynamic equilibrium is reached in which the liquid continues to evaporate, but this is offset by an equal amount of vapor condensing in the same length of time.

11.6 *See Section 11.2.*

The greater cooling is acheived when the wind is blowing because the wind removes the water vapor enabling more water to evaporate in an endothermic process.

11.7 *See Section 11.2.*

There is a high partial pressure of water vapor in the air when the humidity is high. The heat energy that is released by the condensation of water from humid air partially offsets the heat energy that is removed from the body by perspiration and thereby lowers the efficiency of cooling by perspiration.

11.8 *See Section 11.2.*

The apparent difference in cooling is caused by the greater rate of evaporation of diethyl ether compared to water at the same temperature. Since more diethyl ether than water evaporates in an endothermic process in a given amount of time, the evaporation of diethyl ether feels colder.

11.9 *See Section 11.2*

The enthalpy of vaporization and normal boiling point both increase with increasing intermolecular forces. Hence, it is not surprising that the ratio of enthalpy of vaporization to normal boiling point on the Kelvin scale is approximately constant.

11.10 *See Section 11.2.*

The temperature will be the same in the two glasses, since the melting point of a substance is the temperature at which the solid and liquid phases of a substance are in dynamic equilibrium.

11.11 *See Section 11.2.*

For a given substance, $\Delta H_{fus} < \Delta H_{vap} < \Delta H_{sub}$.

11.12 *See Section 11.2.*

Evaporation cools the sweat sufficiently that its temperature drops below body temperature. Heat flows from the body to the sweat to cause further evaporation of the sweat.

11.13 *See Section 11.3.*

An increase in pressure favors the structure having the most mass in the least volume, the structure having the highest density. Hence, an increase in pressure favors the formation of the white form of tin.

11.14 *See Section 11.4.*

In NH_3 there are three hydrogen atoms and one lone pair of electrons about the nitrogen atom. In H_2O there are two hydrogen atoms and two lone pairs of electrons about the oxygen atom. In HF there is one hydrogen atom and three lone pairs of electrons about the fluorine atom. Hence, in water there is an optimum number of hydrogen atoms and lone pairs for hydrogen bonding causing the effect of hydrogen bonding to be greater on the boiling point of water than it is on the boiling points of ammonia and hydrogen fluoride.

11.15 *See Section 11.4 and Figure 11.4.*

The boiling points of the binary hydrides, other than NH_3, H_2O and HF, shown in Figure 11.4 follow the normal patterns of increasing boiling point with increasing size and increasing dispersion forces , in spite of decreasing polarity. Hence, there is no evidence for hydrogen bonding occurring with elements, other than N, O or F, shown by the boiling points for these compounds.

11.16 *See Section 11.5.*

Intermolecular forces of attraction are not the only factor affecting viscosity. Other factors, such as the structure, size and shape of the molecules have an impact on the viscosity of a liquid.

11.17 *See Section 11.5.*

The paint loses its protective coating of wax allowing the water to spread over a greater surface area. The exposed paint also slowly oxidizes and produces polar sites that increase the cohesion of water on the surface.

11.18 *See Sections 11.2, 11.5.*

	Property	Change with Increasing Strength of Intermolecular Forces
(a)	enthalpy of fusion	increase
(b)	melting point	increase
(c)	surface tension	increase
(d)	viscosity	increase
(e)	enthalpy of vaporization	increase
(f)	boiling point	increase

11.19 *See Section 11.5 and Figure 11.16a.*

The adhesive forces between the water in the HCl solution and the silicon dioxide of the glass walls of the buret are greater than the cohesive forces between the molecules in solution causing the surface of the liquid in the buret to be depressed.

11.20 *See Sections 11.4, 11.5.*

The intermolecular forces acting between dimethyl ether (CH_3-O-CH_3) molecules are dipole-dipole and London dispersion forces, and these are weaker than the hydrogen bonding and London dispersion intermolecular forces acting between ethanol (C_2H_5OH) molecules. Hence, ethanol is expected to have the higher surface tension.

11.21 *See Section 11.6.*

An amorphous solid lacks the regular, repeating arrangement of units that is found in a crystalline solid.

11.22 *See Section 11.6.*

Most amorphous solids consist of large molecules that move so slowly that they cannot arrange themselves into the pattern that is present in the crystalline state. Hence, the liquid state becomes very viscous as the temperature is lowered, and the solid state resembles a supercooled liquid.

11.23 *See Section 11.6.*

The slow cooling associated with annealing enables the large molecules of the otherwise amorphous solid to arrange themselves in the pattern of the crystalline state.

11.24 *See Section 11.7 and Figure 11.23.*

In a simple cubic arrangement the atoms touch each other along the edges of the cube. Hence, the length of the unit cell edge, which is taken to be the distance between the centers of the atoms forming the cubic unit cell, is $2r$.

In a body center cubic arrangement the atoms do not touch each other along the edges of the cube but do touch each other along a body diagonal, a diagonal that passes from corner to opposite corner through the center of the cube. Hence, the length of the body diagonal is $1/2(r) + r + r + 1/2(r) = 4r$. The body diagonal is also the hypotenuse of a right triangle formed by an edge of length a and a face diagonal of length $\sqrt{2}a$ (See below.). Hence, $a^2 + (\sqrt{2}a)^2 = d^2$ where d is the length of the body diagonal, and the length of the body diagonal is therefore also given by $d = \sqrt{3}a$. This gives $4r = \sqrt{3}a$ for the the length of the body diagonal and $r = \sqrt{3}r/4 = .43a$.

In a face centered cubic arrangement the atoms do not touch each other along the edges but do touch each other along a face diagonal, a diagonal the passes from corner to opposite corner through a face of the cube. Hence, the length of a face diagonal is $1/2(r) + r + r + 1/2(r) = 4r$. The face diagonal is also the hypotenuse of a right triangle having two edges as legs. Hence, $a^2 + a^2 = d^2$ where d is the length of the face diagonal, and the length of the face diagonal is therefore also given by $d = \sqrt{2}\,a$. This gives $4r = \sqrt{2}\,a$ for the length of the face diagonal and $a = 4r / \sqrt{2} = 2.83r$.

11.25 ***See Section 11.1 and Example 11.1.***

The stronger the intermolecular forces, the higher the melting point and the boiling point. Hence, the order of increasing intermolecular forces of attraction is $H_2S < H_2Te < H_2O$.

11.26 ***See Section 11.1 and Example 11.1.***

The stronger the intermolecular forces, the higher the melting point and the boiling point. Hence, the order of increasing intermolecular forces of attraction is $BCl_3 < BBr_3 < BI_3$.

11.27 ***See Section 11.2.***

(a) The vapor pressure does not change because the fraction of molecules with enough kinetic energy to escape from the liquid does not change.

(b) The vapor pressure increases because the fraction of molecules with enough kinetic energy to escape from the liquid increases.

(c) The vapor pressure does not change because the fraction of molecules with enough kinetic energy to escape from the liquid does not change. The initial increase in vapor pressure vanishes as the system returns to equilibrium.

11.28 ***See Section 11.2.***

(a) The vapor pressure does not change because the fraction of molecules with enough kinetic energy to escape from the liquid does not change.

(b) The vapor pressure decreases because the fraction of molecules with enough kinetic energy to escape from the liquid decreases.

(c) The vapor pressure does not change because the fraction of molecules with enough kinetic energy to escape from the liquid does not change. The initial decrease in vapor pressure vanishes as the system returns to equilibrium.

11.29 ***See Section 11.2.***

(a) The stronger the intermolecular forces the lower the vapor pressure at a given temperature. Hence, the intermolecular forces between isopropanol molecules are stronger than the intermolecular forces between methyl ethyl ether molecules.

(b) The melting point, boiling point, enthalpy of fusion and enthalpy of vaporization are all higher for isopropanol than for methyl ethyl ether.

11.30 ***See Section 11.2.***

(a) The stronger the intermolecular forces the lower the vapor pressure at a given temperature. Hence, the intermolecular forces between n-butanol molecules are stronger than the intermolecular forces between diethyl ether molecules.

(b) The melting point, boiling point, enthalpy of fusion and enthalpy of vaporization are all higher for n-butanol than for diethyl ether.

11.31 *See Section 11.2.*

(a) The stronger the intermolecular forces, the higher the critical temperature. Hence, the intermolecular forces between benzene molecules are stronger than the intermolecular forces between butane molecules.

(b) The vapor pressure at a given temperature would be lower for benzene than for butane. However, the melting point, boiling point, enthalpy of fusion and enthalpy of vaporization would all be higher for benzene than for butane.

11.32 *See Section 11.2.*

(a) The stronger the intermolecular forces, the higher the critical temperature. Hence, the intermolecular forces between nitrogen dioxide (NO_2) molecules are stronger than the intermolecular forces between dintrogen oxide (N_2O) molecules.

(b) The vapor pressure at a given temperature would be lower for nitrogen dioxide than for dintrogen oxide. However, the melting point, boiling point, enthalpy of fusion and enthalpy of vaporization would all be higher for nitrogen dioxide than for dinitrogen oxide.

11.33 *See Section 11.3 and Figure 11.7.*

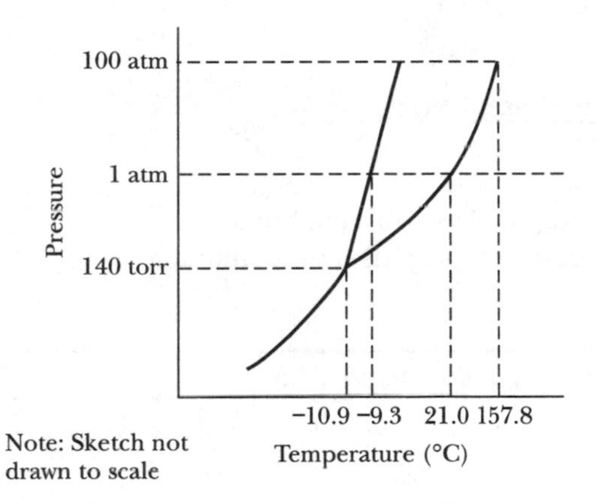

11.34 *See Section 11.3 and Figure 11.7.*

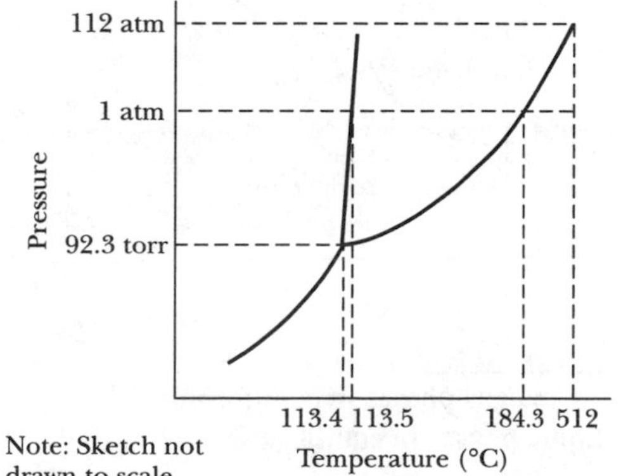

11.35 *See Sections 11.2, 11.3, and Figure 11.6.*

Assuming all three changes occur at the same temperature, $\Delta H_{sub} = \Delta H_{fus} + \Delta H_{vap}$.
Solving for ΔH_{fus} gives $\Delta H_{fus} = H_{sub} - \Delta H_{vap.}$. Hence, ΔH_{fus} = 60.2 kJ/mol - 45.5 kJ/mol = **14.7 kJ/mol.**

11.36 *See Sections 11.2, 11.3, and Figure 11.6.*

Assuming all three changes occur at the same temperature, $\Delta H_{sub} = \Delta H_{fus} + \Delta H_{vap}$.
Solving for ΔH_{vap} gives $\Delta H_{vap} = H_{sub} - \Delta H_{fus}$. Hence, ΔH_{vap} = 68.3 kJ/mol - 17.2 kJ/mol = **51.1 kJ/mol.**

11.37 *See Section 11.3, Figure 11.7, and Example 11.3.*

(a)

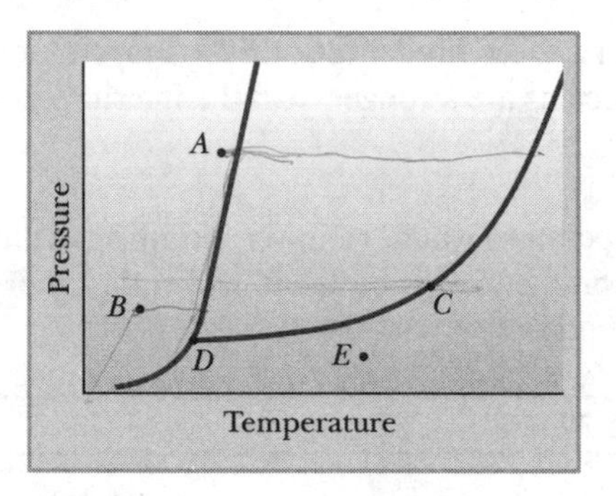

(b)

Point	Phase or Phases Present
A	solid
B	solid
C	liquid and gas phases in equilibrium
D	solid, liquid and gas phases in equilibrium
E	gas

11.38 *See Section 11.3, Figure 11.7, and Example 11.3.*

(a)

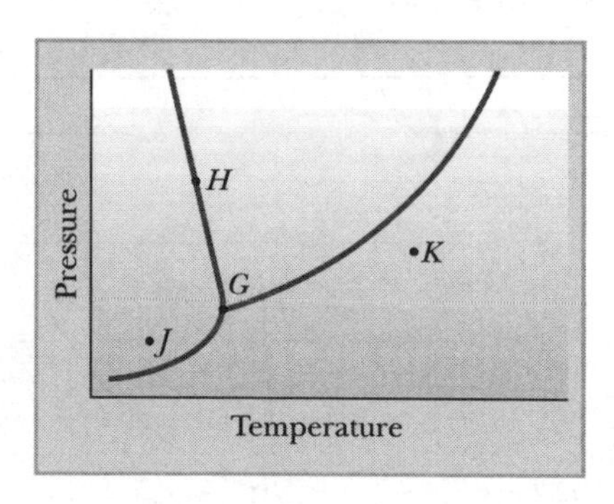

(b)

Point	Phase or Phases Present
G	solid, liquid and gas phases in equilibrium
H	solid and liquid phases in equilibrium
J	solid
K	liquid

11.39 ***See Sections 11.2, 11.3, Example 11.4, and Exercise 11.37.***

(a) The heating curve expected when heat is added to the sample at constant pressure starting with solid at point B of the phase diagram shown in Exercise 11.37 is:

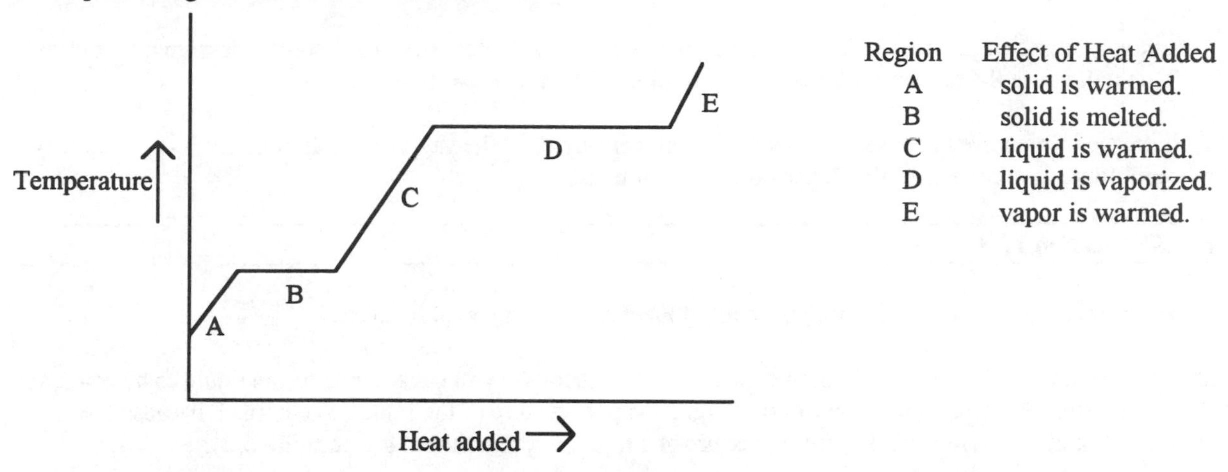

(b) As the pressure is lowered at constant temperature, starting at point A of the phase diagram, solid will be converted to liquid, and eventually liquid will be converted to vapor.

(c) The positive slope of the solid-liquid equilibrium line indicates an increase in pressure at a constant temperature will cause liquid to be converted to solid. This means the solid phase is more dense than the liquid phase of this substance.
(An increase in pressure at constant temperature always favors a decrease in volume, as was noted in Chapter 6.)

11.40 ***See Sections 11.2, 11.3, Example 11.4, and Exercise 11.38.***

(a) The heating curve expected when heat is added to the sample at constant pressure starting with solid at point J of the phase diagram shown in Exercise 11.38 is:

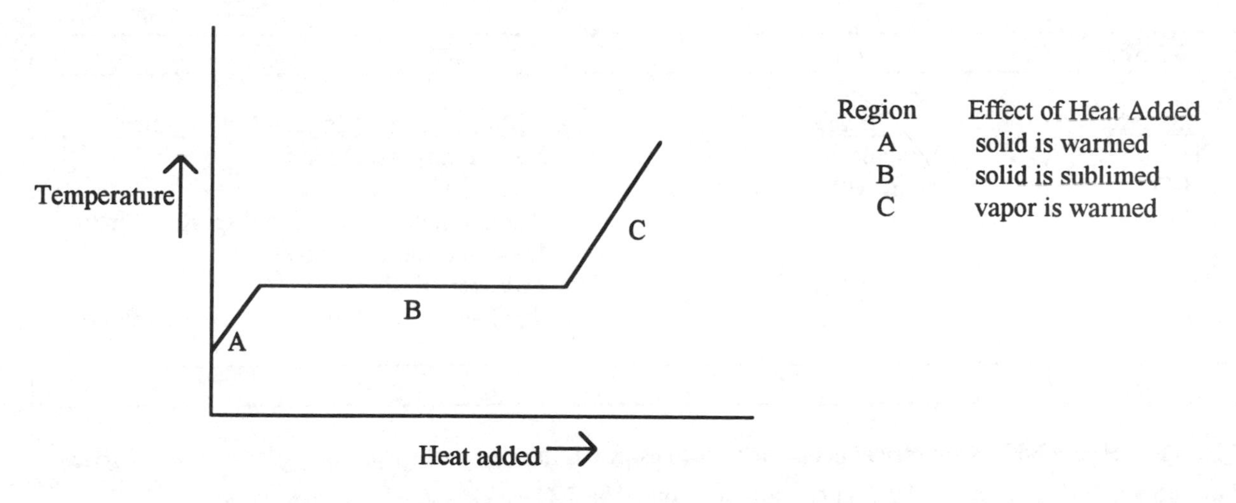

(b) As the pressure is lowered at constant temperature, starting at point K liquid will be converted to solid and eventually solid will be converted to vapor.

(c) The negative slope of the solid-liquid equilibrium line indicates an increase in pressure at a constant temperature will cause solid to be converted to liquid. This means the liquid phase is more dense than the solid phase of this substance.
(An increase in pressure at constant temperature always favors a decrease in volume, as was noted in Chapter 6.)

11.41 *See Section 11.3.*

(a) Addition of heat at constant pressure will convert solid to gas. The equilibrium vapor pressure of a solid increases with increasing temperature.

(b) A sudden increase in pressure will cause some ice to melt, since the density of the solid is less than that of the liquid. This will cause the temperature to decrease until equilibrium is re-established..

(c) An increase in pressure at constant temperature will convert all of the vapor to the liquid, because the density of the liquid phase is greater than the density of the vapor phase.

11.42 *See Section 11.3*

(a) A sudden increase in pressure should cause all of the gas to be converted to solid.

(b) An increase in the volume of the container at constant temperature will cause some of the liquid to be converted to the vapor phase. This will continue until the original vapor pressure of the liquid is obtained, because the temperature remains constant and the vapor pressure of a liquid is a function of temperature only.

(c) Removal of energy as heat will cause gas to be converted to liquid. The equilibrium vapor pressure of a liquid decreases with decreasing temperature.

11.43 *See Section 11.4.*

	Molecule	Comment	Most Important Intermolecular Forces
(a)	C_3H_8	virtually nonpolar	London dispersion forces
(b)	$HO(CH_2)_2OH$	polar with O-H bonds	hydrogen bonds
(c)	C_6H_{12}	virtually nonpolar	London dispersion forces
(d)	PH_3O	polar	dipole-dipole and London dispersion forces
(e)	NO	slightly polar	London dispersion forces
(f)	NH_2OH	polar with N-H and O-H bonds	hydrogen bonds

11.44 *See Section 11.4.*

	Molecule	Comment	Most Important Intermolecular Forces
(a)	CH_4	nonpolar	London dispersion forces
(b)	CH_3OH	polar with O-H bonds	hydrogen bonds
(c)	$CHCl_3$	polar	dipole-dipole and London dispersion forces
(d)	C_6H_6	nonpolar	London dispersion forces
(e)	NH_3	polar with N-H bonds	hydrogen bonds
(f)	SO_2	polar	dipole-dipole and London dispersion forces

11.45 *See Section 11.4.*

(a) C_2H_4 and CH_4 are both symmetrical nonpolar molecules. C_2H_4 has more electrons and a larger size. C_2H_4 is therefore more polarizable and has the higher boiling point (-103.7°C vs. -161°C).

(b) Cl_2 is symmetrical and nonpolar, whereas ClF is unsymmetrical and polar. Cl_2 has more electrons and a larger size. Cl_2 is therefore more polarizable and has the higher boiling point (-35°C vs. -100.8°C).

(c) S_2F_2 and S_2Cl_2 are both unsymmetrical trigonal pyramidal molecules having one sulfur atom as the central atom. S_2F_2 would be expected to be more polar than S_2Cl_2 based on the higher electronegativity of F. However,

S_2Cl_2 has more electrons and a larger size. S_2Cl_2 is therefore more polarizable and has the higher boiling point (135.6°C vs. -38.4°C).

(d) NH_3 is a polar molecule which can hydrogen bond to other NH_3 molecules, whereas PH_3 cannot participate in hydrogen bonding. NH_3 has the higher boiling point (-33°C vs. -88°C).

(e) CH_3I and CHI_3 are both unsymmetrical polar molecules. CHI_3 has more electrons and a larger size. CHI_3 is therefore more polarizable and has the higher boiling point (218°C vs. 42.4°C).

(f) BBr_2 and $BBrI_2$ are both unsymmetrical polar molecules. $BBrI_2$ has more electrons and a larger size. $BBrI_2$ is therefore more polarizable and has the higher boiling point (180°C vs. 125°C).

11.46 ***See Section 11.4.***

(a) C_3H_8 is virtually nonpolar due to the small electronegativity difference between C and H, and CH_4 is symmetrical and nonpolar. C_3H_8 has more electrons and a larger size. C_3H_8 is therefore more polarizable and has the higher boiling point (-42.1°C vs. -164°C).

(b) I_2 is symmetrical and nonpolar, whereas ICl is unsymmetrical and polar. I_2 has more electrons and a larger size. I_2 is therefore more polarizable and has the higher boiling point (184.4°C vs. 97.4°C).

(c) H_2S and H_2Te are both unsymmetrical and polar. H_2Te has more electrons and is larger. H_2Te is therefore more polarizable and has the higher boiling point (-2°C vs. -60.7°C).

(d) H_2Se and H_2O are both unsymmetrical and polar. However, H_2O has polar O-H bonds and hydrogen bonding. Therefore, H_2O has the higher boiling point (-41.5°C vs. 100.0°C).

(e) CH_2Cl_2 and CH_3Cl are both unsymmetrical and polar. CH_2Cl_2 has more electrons and is larger. CH_2Cl_2 is therefore more polarizable and has the higher boiling point (40°C vs. -24.2°C).

(f) NOF and NOCl are both unysmmetrical and polar. NOCl has more electrons and is larger. NOCl is therefore more polarizable and has the higher boiling point (-5.5°C vs. -56°C).

11.47 ***See Section 11.4.***

	Molecule	Molecular Shape	Comment	Type(s) of Forces
(a)	N_2O	linear (NNO)	unsymmetrical, polar	dipole-dipole and London dispersion
(b)	CH_4	tetrahedral	symmetrical, nonpolar	London dispersion
(c)	NH_3	trigonal pyramidal	unsymmetrical, polar with N-H bonds	hydrogen bonds and London dispersion
(d)	SO_2	v-shaped	unsymmetrical, polar	dipole-dipole and London dispersion

11.48 ***See Section 11.4.***

	Molecule	Molecular Shape	Comment	Type(s) of Forces
(a)	F_2	linear	symmetrical, nonpolar	London dispersion
(b)	BF_3	trigonal planar	symmetrical, nonpolar	London dispersion
(c)	HF	linear	unsymmetrical, polar with H-F bonds	hydrogen bonds and London dispersion
(d)	NO_2	v-shaped	unsymmetrical, polar	dipole-dipole and London dispersion

11.49 ***See Sections 11.2, 11.5.***

In each case the substance having the stronger intermolecular forces of attraction will have the larger enthalpy of vaporization.

	Molecules	Comment	Substance with Larger ΔH_{vap}
(a)	C_3H_8 and CH_4	C_3H_8 is larger and more polarizable.	C_3H_8
(b)	I_2 and ICl	I_2 is larger and more polarizable.	I_2
(c)	S_2Cl_2 and S_2F_2	S_2Cl_2 is larger and more polarizable.	S_2Cl_2
(d)	H_2Se and H_2O	H_2O has hydrogen bonding.	H_2O
(e)	CH_2Cl_2 and CH_3Cl	CH_2Cl_2 is larger and more polarizable.	CH_2Cl_2
(f)	NOF and NOCl	NOCl is larger and more polarizable.	NOCl

11.50 ***See Sections 11.2, 11.5.***

In each case the substance having the stronger intermolecular forces of attraction will have the larger enthalpy of vaporization.

	Molecules	Comment	Substance with Larger ΔH_{vap}
(a)	C_2H_4 and CH_4	C_2H_4 is larger and more polarizable.	C_2H_4
(b)	Cl_2 and ClF	Cl_2 is larger and more polarizable.	Cl_2
(c)	H_2S and H_2Te	H_2Te is larger and more polarizable.	H_2Te
(d)	NH_3 and PH_3	NH_3 has hydrogen bonding.	NH_3
(e)	CHI_3 and CH_3I	CHI_3 is larger and more polarizable.	CHI_3
(f)	BBr_2I and $BBrI_2$	$BBrI_2$ is larger and more polarizable.	$BBrI_2$

11.51 ***See Section 11.5.***

The stronger the intermolecular forces of attraction, the greater the surface tension. Hexane (C_6H_{14}) is virtually nonpolar due to the small electronegativity difference between C and H and therefore has London dispersion forces as the primary intermolecular forces. Ethanol (C_2H_5OH) and hydrogen fluoride (HF) are both polar and can participate in hydrogen bonding. However, ethanol molecules are larger and more polarizable than hydrogen fluoride molecules. Hence, the order of increasing intermolecular forces and, therefore, increasing surface tension is: $C_6H_{14} < HF < C_2H_5OH$.

11.52 ***See Section 11.5.***

The stronger the intermolecular forces of attraction, the greater the viscosity. All of the molecules listed have O-H bonds and can participate in hydrogen bonding. The order of decreasing intermolecular forces and, therefore, decreasing viscosity is one of decreasing numbers of hydrogen bonds per molecule: $CH_2(OH)CH\,(OH)CH_2(OH) > (CH_2OH)_2 > C_2H_5OH$.

11.53 ***See Sections 11.4, 11.6.***

	Substance	Type of Solid	Most Important Forces
(a)	H_2O	molecular, polar	hydrogen bonds
(b)	C_6H_6	molecular, nonpolar	London dispersion
(c)	$CaCl_2$	ionic	cation-anion attractions, ionic bonds
(d)	SiO_2	covalent network	Si-O covalent bonds
(e)	Fe	metallic	metal cations-sea of electrons, metallic bonds

11.54 See Sections 11.4, 11.6.

	Substance	Type of Solid	Most Important Forces
(a)	Kr	"molecular"	London dispersion forces
(b)	HF	molecular, polar	hydrogen bonds
(c)	K_2O	ionic	cation-anion attractions, ionic bonds
(d)	CO_2	molecular, nonpolar	London dispersion forces
(e)	Zn	metallic	metal cations-sea of electrons, metallic bonds
(f)	NH_3	molecular, polar	hydrogen bonds

11.55 See Sections 11.4, 11.6.

increasing boiling point ↓

Substance	Comment
N_2	nonpolar with least number of electrons and smallest size
CO_2	nonpolar having more electrons and larger size
H_2O	polar with hydrogen bonds
KCl	ionic with ions of smaller charges
CaO	ionic with ions of higher charges

11.56 See Sections 11.4, 11.6.

increasing boiling point ↓

Substance	Comment
He	nonpolar with least number of electrons and smallest size
NO_2	polar having more electrons and larger size
NH_3	polar with hydrogen bonds
NaBr	ionic with ions of smaller charges
BaO	ionic with ions of higher charges

11.57 See Sections 11.4, 11.6.

CO_2 is a nonpolar molecular solid which experiences only London dispersion forces of attraction between molecules. SiO_2 is a network covalent solid that is held together by strong Si-O covalent bonds. Hence, SiO_2 has a much higher melting point.

11.58 See Sections 11.4, 11.6.

SiC is a covalent network solid that is held together by strong Si-C covalent bonds. This is why SiC is a very hard, high melting solid.

11.59 See Section 11.7.

$$1\ NH_4^+ \text{ ion at center of unit cell} = \mathbf{1\ NH_4^+\ ion/unit\ cell}$$

$$\left(8\ Cl^- \text{ ions at corners}\right) \times \frac{1}{8} = \mathbf{1\ Cl^-\ ion/unit\ cell}$$

Note: We should expect the ratio of cations to anions in the unit cell to be equal to the overall 1:1 ratio in NH_4Cl.

11.60 See Section 11.7.

$$\left[1\ Ca^{2+} \text{ at center} + \left(12\ Ca^{2+} \text{ at edges}\right) \times \frac{1}{4}\right] = \mathbf{4\ Ca^{2+}\ ions / unit\ cell}$$

$$\left[\left(8\ O^{2-} \text{ ions at corners}\right) \times \frac{1}{8} + \left(6\ O^{2-} \text{ ions in faces}\right) \times \frac{1}{2}\right] = \mathbf{4\ O^{2-}\ ions / unit\ cell}$$

Note: We should expect the ratio of cations to anions in the unit cell to be equal to the overall 1:1 ratio in CaO.

11.61 ***See Section 11.7***

$$\left[1 \text{ W at center} + (8 \text{ W at corners}) \times \frac{1}{8}\right] = \mathbf{2\ W\ atoms\ /\ body\ centered\ unit\ cel}$$

11.62 ***See Section 11.7.***

$$\left[(8 \text{ Ag atoms at corners}) \times \frac{1}{8} + (6 \text{ Ag atoms in faces}) \times \frac{1}{2}\right] = \mathbf{4\ Ag\ atoms/\ unit\ cell}$$

11.63 ***See Section 11.7 and Example 11.7.***

Density is mass per unit volume. Mass per unit cell depends on the number of atoms/unit cell.

$$\left[(8 \text{ Ni atoms at corners}) \times \frac{1}{8} + (6 \text{ Ni atoms in faces}) \times \frac{1}{2}\right] = 4 \text{ Ni atoms/ unit cell}$$

$$? \text{ mass per unit cell} = \frac{4 \text{ Ni atoms}}{1 \text{ unit cell}} \times \frac{1 \text{ mol Ni}}{6.022 \times 10^{23} \text{ Ni atoms}} \times \frac{58.69 \text{ g Ni}}{1 \text{ mol Ni}} = 3.898 \times 10^{-22} \text{ g/ unit cell}$$

$$? \text{ volume of unit cell} = \left(351 \text{ pm} \times \frac{1 \text{ m}}{10^{12} \text{ pm}} \times \frac{10^2 \text{ cm}}{1 \text{ m}}\right)^3 = 4.32 \times 10^{-23} \text{ cm}^3 \text{ / unit cell}$$

$$\text{Density of Ni} = \frac{3.898 \times 10^{-22} \text{ g/ unit cell}}{4.32 \times 10^{-23} \text{ cm}^3 \text{ / unit cell}} = \mathbf{9.02\ g/cm^3}$$

11.64 ***See Section 11.7 and Example 11.7.***

Density is mass per unit volume. Mass per unit cell depends on the number of atoms/unit cell.

$$\left[1 \text{ Eu at center} + (8 \text{ Eu at corners}) \times \frac{1}{8}\right] = 2 \text{ Eu atoms / body - centered unit cell}$$

$$? \text{ mass per unit cell} = \frac{2 \text{ Eu atoms}}{1 \text{ unit cell}} \times \frac{1 \text{ mol Eu}}{6.022 \times 10^{23} \text{ Eu atoms}} \times \frac{152.0 \text{ g Eu}}{1 \text{ mol Eu}} = 5.048 \times 10^{-22} \text{ g / unit cell}$$

$$? \text{ volume of unit cell} = \left(458.3 \text{ pm} \times \frac{1 \text{ m}}{10^{12} \text{ pm}} \times \frac{10^2 \text{ cm}}{1 \text{ m}}\right)^3 = 9.626 \times 10^{-23} \text{ cm}^3 \text{ / unit cell}$$

$$\text{Density of Eu} = \frac{5.048 \times 10^{-22} \text{ g / unit cell}}{9.626 \times 10^{-23} \text{ cm}^3 \text{ / unit cell}} = \mathbf{5.244\ g/cm^3}$$

11.65 ***See Section 11.7 and Examples 11.7, 11.8.***

(a) Volume = (length of edge)^3. Hence, the given data must be used to calculate the volume of the unit cell. Since density = mass/ volume and the density is given, the body-centered cubic array data must be used to calculate the number of atoms per unit cell, and then the mass per unit cell to determine the volume of the unit cell.

$$\left[1 \text{ Fe at center} + (8 \text{ Fe at corners}) \times \frac{1}{8}\right] = 2 \text{ Fe atoms/ body - centered unit cel}$$

$$? \text{ mass of unit cell} = \frac{2 \text{ Fe atoms}}{1 \text{ unit cell}} \times \frac{1 \text{ mol Fe}}{6.022 \times 10^{23} \text{ Fe atoms}} \times \frac{55.85 \text{ g Fe}}{1 \text{ mol Fe}} = 1.855 \times 10^{-22} \text{ g}$$

$$? \text{ volume of unit cell} = 1.855 \times 10^{-22} \text{ g} \times \frac{1 \text{ cm}^3}{7.875 \text{ g}} = 2.356 \times 10^{-23} \text{ cm}^3$$

$$\text{length of edge} = \sqrt[3]{\text{volume}} = \sqrt[3]{2.356 \times 10^{-23} \text{ cm}^3} = \mathbf{2.867 \times 10^{-8}\ cm}$$

(b) The length of a diagonal through the center of a cube is equal to $\sqrt{3}a$, where a is the length of the edge, and is also equal to 4r, where r is the radius of an atom. Hence, $r_{Fe} = \frac{\sqrt{3}a}{4} = \frac{\sqrt{3}(2.867 \text{ x } 10^{-8} \text{ cm})}{4} = \mathbf{1.24 \text{ x } 10^{-8} \text{ cm}}$

11.66 ***See Section 11.7 and Examples 11.7, 11.8.***

(a) Volume = (length of edge)^3. Hence, the given data must be used to calculate the volume of the unit cell. Since density = mass/ volume and the density is given, the face-centered cubic array data must be used to calculate the number of atoms per unit cell, and then the mass per unit cell to determine the volume of the unit cell.

$$\left[(8 \text{ Ar atoms at corners}) \times \frac{1}{8} + (6 \text{ Ar atoms in faces}) \times \frac{1}{2}\right] = 4 \text{ Ar atoms/ unit cell}$$

$$? \text{ mass per unit cell} = \frac{4 \text{ Ar atoms}}{1 \text{ unit cell}} \times \frac{1 \text{ mol Ar}}{6.022 \text{ x } 10^{23} \text{ Ar atoms}} \times \frac{39.95 \text{ g Ar}}{1 \text{ mol Ar}} = 2.654 \text{ x } 10^{-22} \text{ g}$$

$$? \text{ volume of unit cell} = 2.654 \text{ x } 10^{-22} \text{ g} \times \frac{1 \text{ cm}^3}{1.65 \text{ g}} = 1.61 \text{ x } 10^{-22} \text{ cm}^3$$

$$\text{length of edge} = \sqrt[3]{\text{volume}} = \sqrt[3]{1.61 \text{ x } 10^{-22} \text{ cm}^3} = \mathbf{5.44 \text{ x } 10^{-8} \text{ cm}}$$

(b) The length of a face diagonal of a cube is equal to $\sqrt{2}a$, where a is the length of the edge, and is also equal to 4r, where r is the radius of an atom. Hence, $r_{Ar} = \frac{\sqrt{2}a}{4} = \frac{\sqrt{2}(5.44 \text{ x } 10^{-8} \text{ cm})}{4} = \mathbf{1.92 \text{ x } 10^{-8} \text{ cm}}$

11.67 ***See Section 11.7.***

$$\left[1 \text{ Ti}^{4+} \text{ at center} + (8 \text{ Ti}^{4+} \text{ at corners}) \times \frac{1}{8}\right] = 2 \text{ Ti}^{4+} \text{ / unit cell}$$

Since the formula is TiO_2, there must be **4 O^{2-}/unit cell.**

11.68 ***See Section 11.7.***

$$\left[(8 \text{ Ca}^{2+} \text{ ions at corners}) \times \frac{1}{8} + (6 \text{ Ca}^{2+} \text{ ions in faces}) \times \frac{1}{2}\right] = 4 \text{ Ca}^{2+} \text{ ions / unit cell}$$

Since the formula is CaF_2, there must be **8 F^-/unit cell.**

11.69 ***See Section 11.7 and Example 11.7.***

Density is mass per unit volume. Mass per unit cell depends on the number of ions of each type per unit cell.

$$\left[1 \text{ Li}^+ \text{ at center} + (12 \text{ Li}^+ \text{ at edges}) \times \frac{1}{4}\right] = 4 \text{ Li}^+ \text{ ions / unit cell}$$

$$\left[(8 \text{ H}^- \text{ ions at corners}) \times \frac{1}{8} + (6 \text{ H}^- \text{ ions in faces}) \times \frac{1}{2}\right] = 4 \text{ H}^- \text{ ions / unit cell}$$

There are 4 LiH units in each unit cell so the mass of the unit cell is

$$? \text{ g per unit cell} = \left[\frac{4 \text{ LiH}}{\text{unit cell}} \times \frac{1 \text{ mol LiH}}{6.022 \text{ x } 10^{23} \text{ LiH}} \times \frac{7.949 \text{ g LiH}}{1 \text{ mol LiH}}\right] = 5.280 \text{ x } 10^{-23} \text{ g / unit cell}$$

$$? \text{ volume of unit cell} = (4.086 \times 10^{-8})^3 = 6.822 \text{ x } 10^{-23} \text{ cm}^3 \text{ / unit cell}$$

$$\text{Density of LiH} = \frac{5.280 \text{ x } 10^{-23} \text{ g / unit cell}}{6.882 \text{ x } 10^{-23} \text{ cm}^3 \text{ / unit cell}} = \mathbf{0.7740 \text{ g / cm}^3}$$

11.70 ***See Section 11.7 and Example 11.7.***

Density is mass per unit volume. Mass per unit cell depends on the number of ions of each type per unit cell.

1 Cs^+ at center of unit cell = 1 Cs^+ ion/unit cell and $\left(8\ I^- \text{ at corners of unit cell}\right) \times \frac{1}{8}$ = 1 I^- ion/unit cell

There is one CsI in each unit cell, so the mass of the unit cell is

$$? \text{ g per unit cell} = \left[\frac{1 \text{ CsI}}{\text{unit cell}} \times \frac{1 \text{ mol CsI}}{6.022 \times 10^{23} \text{ CsI}} \times \frac{259.81 \text{ g CsI}}{1 \text{ mol CsI}}\right] = 4.314 \times 10^{-22} \text{ g / unit cell}$$

$$? \text{ volume of unit cell} = \left(445 \text{ pm} \times \frac{1 \text{ m}}{10^{12} \text{ pm}} \times \frac{10^2 \text{ cm}}{1\text{m}}\right)^3 = 8.81 \times 10^{-23} \text{ cm}^3 \text{ / unit cell}$$

$$\text{Density of CsI} = \frac{4.314 \times 10^{-22} \text{ g/ unit cell}}{8.81 \times 10^{-23} \text{ cm}^3 \text{ / unit cell}} = \mathbf{4.90 \text{ g/cm}^3}$$

11.71 ***See Section 11.7 and Example 11.7.***

The number of atoms per unit cell can be determined from the mass per unit cell. The mass per unit cell can be determined from the volume of the unit cell and the density.

$$? \text{ volume of unit cell} = \left(389 \text{ pm} \times \frac{1 \text{ m}}{10^{12} \text{ pm}} \times \frac{10^2 \text{ cm}}{1 \text{ m}}\right)^3 = 5.89 \times 10^{-23} \text{ cm}^3 \text{ / unit cell}$$

$$? \text{ g per unit cell} = 5.89 \times 10^{-23} \text{ cm}^3 \times 12.02 \text{ g Pd/cm}^3 = 7.08 \times 10^{-22} \text{ g Pd/ unit cell}$$

$$? \text{ Pd atoms per unit cell} = \frac{7.08 \times 10^{-22} \text{ g Pd}}{\text{unit cell}} \times \frac{1 \text{ mol Pd}}{106.4 \text{ g Pd}} \times \frac{6.022 \times 10^{23} \text{ Pd atoms}}{1 \text{ mol Pd}} = 4.00 \text{ Pd atoms/ unit cell}$$

The cubic unit cell which has four atoms per unit cell is the face-centered cubic unit cell

$$\left(8 \text{ corners} \times \frac{1}{8}\right) + \left(6 \text{ faces} \times \frac{1}{2}\right) = 4.$$

Hence, Pd must exist in the **face-centered cubic** unit cell.

11.72 ***See Section 11.7 and Example 11.7.***

The number of atoms per unit cell can be determined from the mass per unit cell. The mass per unit cell can be determined from the volume of the unit cell and the density.

$$? \text{ volume of unit cell} = \left(288 \text{ pm} \times \frac{1 \text{ m}}{10^{12} \text{ pm}} \times \frac{10^2 \text{ cm}}{1 \text{ m}}\right)^3 = 2.39 \times 10^{-23} \text{ cm}^3 \text{ / unit cell}$$

$$? \text{ g per unit cell} = 2.39 \times 10^{-23} \text{ cm}^3 \times 7.20 \frac{\text{g Cr}}{\text{cm}^3} = 1.72 \times 10^{-22} \text{ g Cr / unit cell}$$

$$? \text{ Cr atoms per unit cell} = \frac{1.72 \times 10^{-22} \text{ g Cr}}{\text{unit cell}} \times \frac{1 \text{ mol Cr}}{52.0 \text{ g Cr}} \times \frac{6.022 \times 10^{23} \text{ Cr atoms}}{1 \text{ mol Cr}} = 1.99 \text{ Cr atoms/ unit cell}$$

The cubic unit cell which has two atoms per unit cell is the body centered cubic unit cell.

$$\left[1 \text{ Cr at center} + (8 \text{ Cr at corners}) \times \frac{1}{8}\right] = 2 \text{ Cr atoms / body - centered unit cell}$$

Hence, Cr must exist in the **body centered cubic** unit cell.

11.73 ***See Section 11.7.***

Only (a) KBr and (d) sugar have the highly ordered structures needed to produce well-defined x-ray diffraction patterns.

11.74 ***See Section 11.7.***

Only (c) CaO and (d) diamond have the highly ordered structures needed to produce well-defined x-ray diffraction patterns.

11.75 ***See Section 11.7.***

Solving $n\lambda = 2d\sin\theta$ for $\sin\theta$ gives $\sin\theta = \frac{n\lambda}{2d}$ $\qquad \sin\theta = \frac{(1)(154\text{ pm})}{(2)(293\text{ pm})} = 0.263$

The angle θ having $\sin\theta = 0.263$ is **15.2°.**

Note: On most scientific calculators this angle can be found by entering 0.263 or by starting with 0.263 as the result of the calculation and using 2nd function "sin."

11.76 ***See Section 11.7.***

Solving $n\lambda = 2d\sin\theta$ for $\sin\theta$ gives $\sin\theta = \frac{n\lambda}{2d}$ $\qquad \sin\theta = \frac{(1)(179\text{ pm})}{(2)(325\text{ pm})} = 0.275$

The angle θ having $\sin\theta = 0.275$ is **16.0°.**

Note: On most scientific calculators this angle can be found by entering 0.275 or by starting with 0.275 as the result of the calculation and using 2nd function "sin."

11.77 ***See Section 11.7 and Example 11.6.***

Solving $n\lambda = 2d\sin\theta$ for λ gives $\lambda = \frac{2d\sin\theta}{n}$ $\qquad \lambda = \frac{2(232\text{ pm})(\sin 13.4°)}{1} = \mathbf{108\text{ pm}}$

11.78 ***See Section 11.7 and Example 11.6.***

Solving $n\lambda = 2d\sin\theta$ for λ gives $\lambda = \frac{2d\sin\theta}{n}$ $\qquad \lambda = \frac{2(315\text{ pm})(\sin 16.5°)}{1} = \mathbf{179\text{ pm}}$

11.79 ***See Section 11.6.***

The term allotropes refers to two or more molecular or crystalline forms of an element in the same physical state that exhibit different chemical and physical properties.

(a) $O_2(g)$ and $O_3(g)$ are allotropes. (b) $O_2(l)$ and $O_2(g)$ are not allotropes.

(c) Silicon dioxide is a compound , not an element. Amorphous silicon dioxide and crystalline silicon dioxide (quartz) are not allotropes.

11.80 ***See Section 11.7 and Figures 11.23, 11.24.***

(a) The coordination number of an atom in a simple cubic arrangement is six. Each atom is surrounded by four others in the same plane , one atom above it and one atom below it, all at the same distance. This can be seen by extending the simple cubic structure shown in Figure 11.23 in three dimensions.

(b) The coordination number of an atom in a body centered cubic arrangement is eight. An atom at the center of the body centered cubic unit cell is surrounded by the atoms at the eight corners of the cube. Similarly, an atom on the corner of the body centered cubic unit cell is surrounded the eight atoms that are located at the centers of the eight unit cells that come together at the corner of the body centered unit cell.

11.81 *See Section 11.7 and Example 11.7.*

The number of atoms per cm^3 can be calculated from the crystallographic data. The volume of one mole of Europium can be calculated from its molar mass and density. These values can be combined to calculate the atoms/mol, which is Avogadro's number.

$$\left[1 \text{ Eu at center} + (8 \text{ Eu at corners}) \times \frac{1}{8}\right] = 2 \text{ Eu atoms / unit cell}$$

$$? \text{ volume of unit cell} = \left(458.27 \text{ pm} \times \frac{1 \text{ m}}{10^{12} \text{ pm}} \times \frac{10^2 \text{ cm}}{1 \text{m}}\right)^3 = 9.624 \text{ x } 10^{-23} \text{ cm}^3 \text{ / unit cell}$$

$$? \text{ atoms / cm}^3 = \frac{2 \text{ atoms / unit cell}}{9.624 \text{ x } 10^{-23} \text{ cm}^3 \text{ / unit cell}} = 2.078 \text{ x } 10^{22} \text{ atoms / cm}^3$$

$$? \text{ cm}^3 \text{ / mol} = 151.96 \frac{\text{g}}{\text{mol}} \text{x} \frac{1\text{cm}^3}{5.243 \text{ g}} = 28.98 \text{ cm}^3 \text{ / mol}$$

$$? \text{ atoms / mol} = 2.078 \text{ x } 10^{22} \frac{\text{atoms}}{\text{cm}^3} \text{ x } 28.98 \frac{\text{cm}^3}{\text{mol}} = \mathbf{6.022 \text{ x } 10^{23} \text{ atoms/ mol}}$$

11.82 *See Section 11.7 and Examples 11.6, 11.7.*

(a) Solving $n\lambda = 2d \sin\theta$ for d gives $d = \frac{n\lambda}{2\sin\theta}$ $\qquad d = \frac{(1)(1.790 \text{ x } 10^{-10} \text{ m})}{2\sin 14.7°} = \mathbf{3.527 \text{ x } 10^{-10} \text{ m}}$

(b) Density is mass per unit volume. Mass per unit cell depends on the number of ions of each type per unit cell.

$$\left[1 \text{ K}^+ \text{ at center} + (12 \text{ K}^+ \text{ at edges}) \times \frac{1}{4}\right] = 4 \text{ K}^+ \text{ ions / unit cell}$$

$$\left[(8 \text{ Br}^- \text{ ions at corners}) \times \frac{1}{8} + (6 \text{ Br}^- \text{ ions in faces}) \times \frac{1}{2}\right] = 4 \text{ Br}^- \text{ ions / unit cell}$$

There are 4 KBr units in each unit cell so the mass of the unit cell is

$$? \text{ g per unit cell} = \left[\frac{4 \text{ KBr}}{\text{unit cell}} \times \frac{1 \text{ mol KBr}}{6.022 \text{ x } 10^{23} \text{ KBr}} \times \frac{166.00 \text{ g KBr}}{1 \text{ mol KBr}}\right] = 1.103 \text{ x } 10^{-21} \text{ g / unit cell}$$

$$? \text{ volume of unit cell} = \left[2\left(3.527 \text{ x } 10^{-10} \text{ m x } \frac{10^2 \text{ cm}}{1 \text{ m}}\right)\right]^3 = 3.510 \text{ x } 10^{-22} \text{ cm}^3 \text{ / unit cell}$$

$$\text{Density of KI} = \frac{1.103 \text{ x } 10^{-21} \text{ g / unit cell}}{3.510 \text{ x } 10^{-22} \text{ cm}^3 \text{ / unit cell}} = \mathbf{3.142 \text{ g / cm}^3}$$

11.83 *See Section 11.7 and Example 11.8.*

Solving $4r = \sqrt{2}\,a$ for r gives $r = \sqrt{2}\,a/4$ $\qquad r_{Cu} = \frac{\sqrt{2}(362 \text{ pm})}{4} = \mathbf{128 \text{ pm}}$

11.84 *See Sections 11.4, 11.6.*

(a) N_2O, molecular solid, -90.8°C (b) GeO_2, network solid, 1115°C
(c) $CaCl_2$, ionic solid, 782°C Note: $CaCl_2$ gives ions on melting whereas N_2O and GeO_2 give molecules.

11.85 *See Sections 11.2, 11.4.*

(a) He, London dispersion, 0.083 kJ/mol (b) CH_3OCH_3, London dispersion and dipole-dipole, 19.4 kJ/mol
(c) C_2ClF_5, London dispersion and dipole-dipole, 38.6 kJ/mol Note: C_2ClF_5 molecules are more polarizable than CH_3OCH_3 molecules. Hence, C_2ClF_5 has stronger dispersion forces and a higher ΔH_v than CH_3OCH_3.

11.84 ***See Sections 6.4, 11.2.***

(a) Assuming all 1.50 g of methanol are vaporized,

Known Quantities: $n = 1.50 \text{ g } CH_3OH \times \dfrac{1 \text{ mol } CH_3OH}{32.04 \text{ g } CH_3OH} = 0.0468 \text{ mol}$ $V = 1.00 \text{ L}$

$T = 30 + 273 = 303 \text{ K}$

Solving PV = nRT for P gives $P = \dfrac{nRT}{V}$ $P = \dfrac{(0.0469 \text{ mol})\left(0.0821 \dfrac{L \cdot atm}{mol \cdot K}\right)(303 \text{ K})}{1.00 \text{ L}} = \mathbf{1.17 \text{ atm}}$

or $1.17 \text{ atm} \times \dfrac{760 \text{ torr}}{atm} = \mathbf{885 \text{ torr}}$

(b) Since the answer to part (a) is greater than the vapor pressure, the actual pressure is 158 torr. Using data for quantity of CH_3OH actually vaporized,

Known Quantities: $P = 158 \text{ torr x } \dfrac{1 \text{ atm}}{760 \text{ torr}} = 0.208 \text{ atm}$ $V = 1.00 \text{ L}$ $T = 30 + 273 = 303 \text{ K}$

Solving PV = nRT for n gives $n = \dfrac{PV}{RT}$ $n = \dfrac{(0.208 \text{ atm})(1.00 \text{ L})}{\left(0.0821 \dfrac{L \cdot atm}{mol \cdot K}\right)(303 \text{ K})} = 0.00836 \text{ mol } CH_3OH \text{ vaporized}$

$?\text{g } CH_3OH \text{ vaporized} = 0.00836 \text{ mol } CH_3OH \text{ x } \dfrac{32.0 \text{ g } CH_3OH}{1 \text{ mol } CH_3OH} = \mathbf{0.267 \text{ g } CH_3OH}$

Since only 0.267 g out of the 1.50 g CH_3OH is vaporized, liquid is in equilibrium with vapor in the vessel. This is why the vapor pressure is only 158 torr, compared to the 889 torr it would be if all of the CH_3OH vaporized. Note: See the alternative method used in working 11.87c.

11.87 ***See Sections 6.4, 11.2.***

(a) Assuming all 1.50 g of water are vaporized,

Known Quantities: $n = 1.50 \text{ g } H_2O \times \dfrac{1 \text{ mol } H_2O}{18.0 \text{ g } H_2O} = 0.0833 \text{ mol}$ $V = 1.00 \text{ L}$ $T = 30 + 273 = 303 \text{ K}$

Solving PV = nRT for P gives $P = \dfrac{nRT}{V}$ $P = \dfrac{(0.0833 \text{ mol})\left(0.0821 \dfrac{L \cdot atm}{mol \cdot K}\right)(303 \text{ K})}{1.00 \text{ L}} = 2.07 \text{ atm}$

or $2.07 \text{ atm} \times \dfrac{760 \text{ torr}}{atm} = \mathbf{1.57 \times 10^3 \text{ torr}}$.

(b) At 30°C, the vapor pressure of water is **31.8 torr**.

(c) Since P is directly proportional to n and n is directly proportional to mass,

$? \text{ g } H_2O \text{ actually evaporated at } 30^\circ C = 31.8 \text{ torr} \times \dfrac{1.50 \text{ g } H_2O}{1.57 \times 10^3 \text{ torr}} = \mathbf{0.0304 \text{ g } H_2O}$.

Since only 0.0304 g out of the 1.50 g H_2O is vaporized, liquid is in equilibrium with vapor in the vessel.

11.88 ***See Sections 5.2, 11.2.***

(a) q for lowering temp. of 10.0 g Freon-11 from 23.8°C to 0°C:

$q = mC_s\Delta T$ $q = 10.0 \text{ g} \times \dfrac{0.870 \text{ J}}{g\ ^\circ C} \times [0^\circ C - 23.8^\circ C] = -207 \text{ J}$

Hence, **207 J** must be removed.

(b) Solving q = mΔH for m gives $m = \dfrac{q}{\Delta H}$ $m = \dfrac{207 \text{ J}}{180 \text{ J} \cdot g^{-1}} = \mathbf{1.15 \text{ g}}$

11.89 *See Sections 5.2, 11.2.*

q for raising temp. of 10.0 g of ice from -10°C to 0°C:

$$q_1 = mC_s\Delta T = 10.0\ g \times \frac{2.07\ J}{g\ ^\circ C} \times \left[0^\circ C - \left(-10^\circ C\right)\right] = 207\ J$$

q for melting 10.0 g of ice: $q_2 = n\Delta H = 10.0\ g \times \frac{1\ mol}{18.0\ g} \times 6.01\ x\ 10^3\ \frac{J}{mol} = 3.34\ x\ 10^3\ J$

q for raising temp. of 10.0 g of liquid from 0°C to 95°C:

$$q_3 = mC_s\Delta T = 10.0\ g \times \frac{4.18\ J}{g\ ^\circ C} \times \left[95^\circ C - 0^\circ C\right] = 3.97\ x\ 10^3\ J$$

$$q_{total} = q_1 + q_2 + q_3 = 207\ J + 3.34\ x\ 10^3\ J + 3.97\ x\ 10^3\ J = \mathbf{7.52\ x\ 10^3\ J}$$

11.90 *See Sections 5.2, 11.2.*

q for raising temp. of 15.0 g of liquid from 10°C to 100°C:

$$q_1 = mC_s\Delta T = 15.0\ g \times \frac{4.18\ J}{g\ ^\circ C} \times \left[100^\circ C\ -\ 10^\circ C\right] = 5.64x10^3\ J$$

q for vaporizing 15.0 g of liquid: $q_2 = n\Delta H = 15.0\ g \times \frac{1\ mol}{18.0\ g} \times 40.6 \times 10^3\ \frac{J}{mol} = 3.38 \times 10^4\ J$

q for raising temp. of 15.0 g of steam from 100°C to 105°C:

$$q_3 = mC_s\Delta T = 15.0\ g \times \frac{2.01\ J}{g\ ^\circ C} \times \left[105^\circ C - 100^\circ C\right] = 151\ J$$

$$q_{total} = q_1 + q_2 + q_3 = 5.64x10^3\ J + 3.38\ x\ 10^4\ J + 151\ J = \mathbf{3.96\ x\ 10^4\ J}$$

11.91 *See Section 11.4.*

```
    :F:           :F: :Cl:
     |             |    |
:F—C—F:            C==C
     |             |    |
    :F:           :F: :Cl:
```

CF_4 is tetrahedral, and the C is sp^3 hybridized. CF_4 molecules are totally symmetrical and nonpolar. The intermolecular forces are London dispersion forces. CF_2CCl_2 is triangular planar about each C, and each C is sp^2 hybridized. CF_2CCl_2 molecules are unsymmetrical and polar. The intermolecular forces are London dispersion forces and dipole-dipole forces. CF_2CCl_2 should have the higher normal boiling point because its molecules are larger and more polarizable than CF_4 molecules and there are also dipole-dipole forces with CF_2Cl_2.

11.92 *See Section 11.6.*

$MCl + AgNO_3 \rightarrow MNO_3 + AgCl$

$$?\ mol\ MCl = 4.09\ g\ AgCl \times \frac{1\ mol\ AgCl}{143.32\ g\ AgCl} \times \frac{1\ mol\ MCl}{1\ mol\ AgCl} = 0.0286\ mol\ MCl$$

$$?\ molar\ mass\ MCl = \frac{2.13\ g\ MCl}{0.0286\ mol\ MCl} = 74.5\ g\,/\,mol$$

? molar mass M = molar mass MCl - molar mass Cl = 74.5 g /mol - 35.5 g/mol =39.0 g/mol

Since M and Cl combine in a 1:1 ratio, M must be **potassium** and the compound must be **ionic**. This is consistent with the white crystalline nature of the MCl and its high melting point.

11.93 *See Section 11.4.*

$(CH_3)_3N$ has no H atoms attached to N and no possibility for hydrogen bonding. It has London dispersion forces and dipole-dipole forces. $CH_3CH_2NH_2$ has London dispersion forces and hydrogen bonding and therefore has the higher normal boiling point.

Chapter 12: Solutions

12.1 *See Section 12.1.*

Molarity gives the moles of solute per liter of solution. Mole fraction gives the moles of solute per total moles of solution. Mass percent gives the grams of solute per one hundred grams of solution.

12.2 *See Section 12.1.*

Molality can be used to calculate the moles of solute in a given mass of solvent or vice versa. Mass percent can be used to calculate the grams of solute in a given mass of solution. Mole fraction can be used to calculate the moles of solute in a given total moles of solution or vice versa.

12.3 *See Section 12.1.*

The volume of a dilute solution is very nearly the volume of the solvent. When the density of the solvent is nearly one gram per milliliter, like it is with water, the volume of one kilogram of solvent is nearly one liter. Hence, the number of moles of solute per kilogram of solvent (the molality) is nearly equal to the number of moles of solute per liter of solution (the molarity) when the density of the solvent is nearly one gram per milliliter, but different when it is not.

12.4 *See Section 12.4.*

The general dissolving process can be thought to occur via three steps:

Step	Process	Enthalpy Change
1	solvent is separated	endothermic
2	solute is separated	endothermic
3	solvent and solute interact	exothermic

If the overall dissolving process is exothermic, the energy that is released by solvent-solute interactions must exceed the energy required to separate the solvent and the energy required to separate the solute.

12.5 *See Section 12.2 and Figure 12.3.*

Cooling to room temperature gives a supersaturated solution of sodium acetate. The addition of a small crystal of pure sodium acetate will cause a rapid precipitation of additional sodium acetate.

12.6 *See Section 12.2.*

The dissolving of sulfuric acid in water is a highly exothermic process. If water were added to concentrated sulfuric acid, the dissolving process would cause the water to boil and spatter the solution of sulfuric acid. When concentrated sulfuric acid is added to water, the heat energy is dissipated over a much larger volume of water and spattering can be avoided. However, caution must still be used, and it is wise to place the beaker of water in a pan of ice to chill the water before and during the addition of the sulfuric acid.

12.7 *See Section 12.2.*

As n increases, the overall polarity of the straight chain alcohols ($CH_3(CH_2)_n OH$) decreases. Hence, the solubility in water decreases as n increases.

12.8 *See Section 12.3.*

The solubility of carbon dioxide decreases with increasing temperature, like the solubility of most gases. Hence, opening a warm carbonated beverage results in much more frothing than would be observed if the beverage had been refrigerated.

12.9 *See Section 12.3.*

The solubility of gases decreases with decreasing pressure. When the carbonated beverage is stored in an open container, the carbon dioxide continues to escape from the beverage and it goes flat. When the carbonated beverage is stored in a tightly sealed container, an equilibrium is established between the carbon dioxide in the vapor above the beverage and the carbon dioxide in the beverage, and it does not go flat.

12.10 *See Section 12.3.*

When a solid dissolves in a liquid there is very little difference between the volume of the solution and the sums of the volumes of the solid solute and liquid solvent. Hence, in contrast to the dissolving of gases in liquids, the dissolving of solids in liquids is not appreciably affected by changes in pressure.

12.11 *See Section 12.4 and Examples 12.11, 12.13.*

$$\text{freezing point depression} \xrightarrow[\text{depression constant(kf)}]{\text{divided by}} \text{molality of solution} \xrightarrow[\text{mass of solvent (kg)}]{\text{multiplied by}} \text{moles of solute in sample}$$

$$\text{mass of solute in sample} \xrightarrow[\text{moles of solute in sample}]{\text{divided by}} \text{molar mass of solute}$$

12.12 *See Section 12.4.*

The identical freezing points for the aqueous solutions of the nonelectrolytes sucrose and urea tell us that the solutions have identical concentrations. Hence, these solutions also should have the same vapor pressure, boiling point and osmotic pressure.

12.13 *See Section 12. 5 and Example 12.15.*

$CaCl_2$ and NaCl yield three moles of ions and two moles of ions per mole of compound, respectively, when dissolved in water. Hence, the 0.1 m $CaCl_2$ aqueous solution has a lower freezing point than the 0.1 m NaCl aqueous solution.

12.14 *See Section 12.6 and Figure 12.19.*

In a simple distillation process the more volatile component is vaporized, passed through a condenser and collected in a separate container. In a fractional distillation process two or more components are vaporized and condensed repeatedly as the vapor moves up the column. With each successive evaporation, the vapor becomes richer in the most volatile (lowest boiling) component, which is passed through a condenser and collected in a separate container. Simple distillation can be used when working with substances having widely differing volatilities, whereas fractional distillation must be used when working with substances having comparable volatilities.

12.15 *See Section 12.1 and Examples 12.1, 12.2.*

(a) $\text{mass percent } C_6H_5CO_2H = \frac{1.20 \text{ g } C_6H_5CO_2H}{751.2 \text{ g soln}} \times 100\% = \mathbf{0.160\%\ C_6H_5CO_2H}$

(b) $? \text{ mol } C_6H_5CO_2H = 1.20 \text{ g } C_6H_5CO_2H \times \frac{1 \text{ mol } C_6H_5CO_2H}{122.0 \text{ g } C_6H_5CO_2H} = 0.00984 \text{ mol } C_6H_5CO_2H$

$? \text{ mol } H_2O = 750.0 \text{ g } H_2O \times \frac{1 \text{ mol } H_2O}{18.0 \text{ g } H_2O} = 41.70 \text{ mol } H_2O$

$? \text{ mol total of soln} = 0.00984 \text{ mol} + 41.70 \text{ mol} = 41.71 \text{ mol total of soln}$

$\chi_{C_6H_5CO_2H} = \frac{\text{mol } C_6H_5CO_2H}{\text{mol total of soln}} \qquad \chi_{C_6H_5CO_2H} = \frac{0.00984 \text{ mol } C_6H_5CO_2H}{41.71 \text{ mol total of soln}} = \mathbf{2.36 \times 10^{-4}}$

(c) $\text{molality} = \frac{\text{mol } C_6H_5CO_2H}{\text{kg } H_2O} \qquad \text{molality} = \frac{0.00984 \text{ mol } C_6H_5CO_2H}{0.750 \text{ kg } H_2O} = \mathbf{0.0131\ m\ C_6H_5CO_2H}$

12.16 *See Section 12.1 and Examples 12.1, 12.2.*

(a) $\text{mass percent } BaCl_2 = \frac{25.0 \text{ g } BaCl_2}{525.0 \text{ g soln}} \times 100\% = \mathbf{4.76\%\ BaCl_2}$

(b) $? \text{ mol } BaCl_2 = 25.0 \text{ g } BaCl_2 \times \frac{1 \text{ mol } BaCl_2}{208.2 \text{ g } BaCl_2} = 0.120 \text{ mol } BaCl_2$

$? \text{ mol } H_2O = 500. \text{ g } H_2O \times \frac{1 \text{ mol } H_2O}{18.0 \text{ g } H_2O} = 27.8 \text{ mol } H_2O$

$? \text{ mol total of soln} = 0.120 \text{ mol} + 27.8 \text{ mol} = 27.9 \text{ mol total of soln}$

$\chi_{BaCl_2} = \frac{\text{mol } BaCl_2}{\text{mol total of soln}} \qquad \chi_{BaCl_2} = \frac{0.120 \text{ mol } BaCl_2}{27.9 \text{ mol total of soln}} = \mathbf{4.30 \times 10^{-3}}$

(c) $\text{molality} = \frac{\text{mol } BaCl_2}{\text{kg } H_2O} \qquad \text{molality} = \frac{0.120 \text{ mol } BaCl_2}{0.500 \text{ kg } H_2O} = \mathbf{0.240\ m\ BaCl_2}$

12.17 *See Section 12.1 and Examples 12.1, 12.2.*

(a) $\text{mass percent } Ca(NO_3)_2 = \frac{4.50 \text{ g } Ca(NO_3)_2}{434.5 \text{ g soln}} \times 100\% = \mathbf{1.04\%\ Ca(NO_3)_2}$

(b) $? \text{ mol } Ca(NO_3)_2 = 4.50 \text{ g } Ca(NO_3)_2 \times \frac{1 \text{ mol } Ca(NO_3)_2}{164.1 \text{g } Ca(NO_3)_2} = 0.0274 \text{ mol } Ca(NO_3)_2$

$? \text{ mol } H_2O = 430.0 \text{ g } H_2O \times \frac{1 \text{ mol } H_2O}{18.02 \text{ g } H_2O} = 23.86 \text{ mol } H_2O$

$? \text{ mol total of soln} = 0.0274 \text{ mol} + 23.86 \text{ mol} = 23.89 \text{ mol total of soln}$

$\chi_{Ca(NO_3)_2} = \frac{\text{mol } Ca(NO_3)_2}{\text{mol total of soln}} \qquad \chi_{Ca(NO_3)_2} = \frac{0.0274 \text{ mol } Ca(NO_3)_2}{23.89 \text{ mol total of soln}} = \mathbf{1.15 \times 10^{-3}}$

(c) $\text{molality} = \frac{\text{mol } Ca(NO_3)_2}{\text{kg } H_2O} \qquad \text{molality} = \frac{0.0274 \text{ mol } Ca(NO_3)_2}{0.430 \text{ kg } H_2O} = \mathbf{0.0637\ m\ Ca(NO_3)_2}$

12.18 ***See Section 12.1 and Examples 12.1, 12.2.***

(a) $\text{mass percent CO(NH}_2)_2 = \dfrac{3.80 \text{ g CO(NH}_2)_2}{128.8 \text{ g soln}} \times 100\% = \mathbf{2.95\%\ CO(NH_2)_2}$

(b) $? \text{ mol CO(NH}_2)_2 = 3.80 \text{ g CO(NH}_2)_2 \times \dfrac{1 \text{ mol CO(NH}_2)_2}{60.0 \text{g CO(NH}_2)_2} = 0.0633 \text{ mol CO(NH}_2)_2$

$? \text{ mol H}_2\text{O} = 125.0 \text{ g H}_2\text{O} \times \dfrac{1 \text{ mol H}_2\text{O}}{18.02 \text{ g H}_2\text{O}} = 6.94 \text{ mol H}_2\text{O}$

? mol total of soln = 0.0633 mol + 6.94 mol = 7.00 mol total of soln

$\chi_{\text{CO(NH}_2)_2} = \dfrac{\text{mol CO(NH}_2)_2}{\text{mol total of soln}} \qquad \chi_{\text{CO(NH}_2)_2} = \dfrac{0.0633 \text{mol CO(NH}_2)_2}{7.00 \text{ mol total of soln}} = \mathbf{9.04 \times 10^{-3}}$

(c) $\text{molality} = \dfrac{\text{mol CO(NH}_2)_2}{\text{kg H}_2\text{O}} \qquad \text{molality} = \dfrac{0.0633 \text{ mol CO(NH}_2)_2}{0.125 \text{ kg H}_2\text{O}} = \mathbf{0.506\ m\ CO(NH_2)_2}$

12.19 ***See Section 12.1 and Example 12.1.***

$? \text{ mol H}_2\text{O}_2 = 25.0 \text{ g soln} \times \dfrac{3.0 \text{ g H}_2\text{O}_2}{100 \text{ g soln}} \times \dfrac{1 \text{ mol H}_2\text{O}_2}{34.0 \text{ g H}_2\text{O}_2} = \mathbf{0.022\ mol\ H_2O_2}$

12.20 ***See Section 12.1 and Example 12.1.***

$? \text{ mol NaCl} = 35.0 \text{ g soln} \times \dfrac{3.5 \text{ g NaCl}}{100 \text{ g soln}} = \mathbf{1.2\ g\ NaCl}$

12.21 ***See Section 12.1 and Examples 12.1, 12.2.***

(a) $? \text{ g H}_3\text{BO}_3 \text{ in 15.5 g soln} = 15.5 \text{ g soln} \times \dfrac{1.00 \text{ g H}_3\text{BO}_3}{100.00 \text{ g soln}} = 0.155 \text{ g H}_3\text{BO}_3$

Dissolve 0.155 g H_3BO_3 in sufficient water to give 15.5 g soln, 15.345 g H_2O.

(b) $? \text{ mol H}_3\text{BO}_3 \text{ in 15.5 g soln} = 0.155 \text{ g H}_3\text{BO}_3 \times \dfrac{1 \text{ mol H}_3\text{BO}_3}{61.8 \text{ g H}_3\text{BO}_3} = 0.00251 \text{ mol H}_3\text{BO}_3$

$\text{molality} = \dfrac{\text{mol H}_3\text{BO}_3}{\text{kg H}_2\text{O}} \qquad \text{molality} = \dfrac{0.00251 \text{ mol H}_3\text{BO}_3}{0.015345 \text{ kg H}_2\text{O}} = \mathbf{0.164\ m\ H_3BO_3}$

12.22 ***See Section 12.1 and Examples 12.1, 12.2.***

(a) $? \text{ g KI in 30.0 g soln} = 30.0 \text{ g soln} \times \dfrac{2.50 \text{ g KI}}{100.00 \text{ g soln}} = 0.750 \text{ g KI}$

Dissolve 0.750 g KI in sufficient water to give 30.0 g soln, 29.25 g H_2O.

(b) $? \text{ mol KI in 300. g soln} = 0.750 \text{ g KI} \times \dfrac{1 \text{ mol KI}}{166.0 \text{ g KI}} = 0.00452 \text{ mol KI}$

$\text{molality} = \dfrac{\text{mol KI}}{\text{kg H}_2\text{O}} \qquad \text{molality} = \dfrac{0.00452 \text{ mol KI}}{0.02925 \text{kg H}_2\text{O}} = \mathbf{0.155\ m\ KI}$

12.23 *See Section 12.1 and Example 12.4.*

$$? \text{ g AgNO}_3 \text{ in } 100.00 \text{ g soln} = 100.00 \text{ g soln} \times \frac{0.10 \text{ g AgNO}_3}{100.00 \text{ g soln}} = 0.10 \text{ g AgNO}_3$$

$$? \text{ mol AgNO}_3 \text{ in } 100.00 \text{ g soln} = 0.10 \text{ g AgNO}_3 \times \frac{1 \text{ mol AgNO}_3}{169.9 \text{ g AgNO}_3} = 5.9 \text{ x } 10^{-4} \text{ mol AgNO}_3$$

$$? \text{ g H}_2\text{O in } 100.00 \text{ g soln} = 100.00 \text{ g soln} - 0.10 \text{ g AgNO}_3 = 99.90 \text{ g H}_2\text{O}$$

$$\text{molality} = \frac{\text{mol AgNO}_3}{\text{kg H}_2\text{O}} \qquad \text{molality} = \frac{5.9 \text{ x } 10^{-4} \text{ mol AgNO}_3}{0.09990 \text{ kg H}_2\text{O}} = \mathbf{5.9 \text{ x } 10^{-3} \text{ m AgNO}_3}$$

12.24 *See Section 12.1 and Example 12.4.*

$$? \text{ g CuBr}_2 \text{ in } 100.00 \text{ g soln} = 100.00 \text{ g soln} \times \frac{0.50 \text{ g CuBr}_2}{100.00 \text{ g soln}} = 0.50 \text{ g CuBr}_2$$

$$? \text{ mol CuBr}_2 \text{ in } 100.00 \text{ g soln} = 0.50 \text{ g CuBr}_2 \times \frac{1 \text{ mol CuBr}_2}{223.3 \text{ g CuBr}_2} = 2.2 \text{ x } 10^{-3} \text{ mol CuBr}_2$$

$$? \text{ g H}_2\text{O in } 100.00 \text{ g soln} = 100.00 \text{ g soln} - 0.50 \text{ g CuBr}_2 = 99.50 \text{ g H}_2\text{O}$$

$$\text{molality} = \frac{\text{mol CuBr}_2}{\text{kg H}_2\text{O}} \qquad \text{molality} = \frac{2.2 \text{ x } 10^{-3} \text{ mol CuBr}_2}{0.09950 \text{ kg H}_2\text{O}} = \mathbf{0.022 \text{ m CuBr}_2}$$

12.25 *See Section 12.1.*

$$? \text{ mol C}_6\text{H}_6 = 98.0 \text{ g C}_6\text{H}_6 \times \frac{1 \text{ mol C}_6\text{H}_6}{78.1 \text{ g C}_6\text{H}_6} = 1.25 \text{ mol C}_6\text{H}_6$$

$$? \text{ mol C}_6\text{H}_{14} = 12.0 \text{ g C}_6\text{H}_{14} \times \frac{1 \text{ mol C}_6\text{H}_{14}}{86.2 \text{ g C}_6\text{H}_{14}} = 0.139 \text{ mol C}_6\text{H}_{14}$$

$$? \text{ mol C}_8\text{H}_{18} = 20.0 \text{ g C}_8\text{H}_{18} \times \frac{1 \text{ mol C}_8\text{H}_{18}}{114.2 \text{ g C}_8\text{H}_{18}} = 0.175 \text{ mol C}_8\text{H}_{18}$$

$$? \text{ mol total of soln} = 1.25 \text{ mol C}_6\text{H}_6 + 0.139 \text{ mol C}_6\text{H}_{14} + 0.175 \text{ mol C}_8\text{H}_{18} = 1.56 \text{ mol total of soln}$$

$$\chi_{\text{C}_6\text{H}_6} = \frac{\text{mol C}_6\text{H}_6}{\text{mol total of soln}} \qquad \chi_{\text{C}_6\text{H}_6} = \frac{1.25 \text{ mol C}_6\text{H}_6}{1.56 \text{ mol total of soln}} = \mathbf{0.801}$$

12.26 *See Section 12.1.*

$$? \text{ mol C}_2\text{H}_5\text{OH} = 10.0 \text{ g C}_2\text{H}_5\text{OH} \times \frac{1 \text{ mol C}_2\text{H}_5\text{OH}}{46.0 \text{ g C}_2\text{H}_5\text{OH}} = 0.217 \text{ mol C}_2\text{H}_5\text{OH}$$

$$? \text{ mol C}_2\text{H}_4(\text{OH})_2 = 20.0 \text{ g C}_2\text{H}_4(\text{OH})_2 \times \frac{1 \text{ mol C}_2\text{H}_4(\text{OH})_2}{62.0 \text{ g C}_2\text{H}_4(\text{OH})_2} = 0.323 \text{ mol C}_2\text{H}_4(\text{OH})_2$$

$$? \text{ mol H}_2\text{O} = 90.0 \text{ g H}_2\text{O} \times \frac{1 \text{ mol H}_2\text{O}}{18.0 \text{ g H}_2\text{O}} = 5.00 \text{ mol H}_2\text{O}$$

$$? \text{ mol total of soln} = 0.217 \text{ mol C}_2\text{H}_5\text{OH} + 0.323 \text{ mol C}_2\text{H}_4(\text{OH})_2 + 5.00 \text{ mol H}_2\text{O} = 5.54 \text{ mol total of soln}$$

$$\chi_{\text{H}_2\text{O}} = \frac{\text{mol H}_2\text{O}}{\text{mol total of soln}} \qquad \chi_{\text{H}_2\text{O}} = \frac{5.00 \text{ mol H}_2\text{O}}{5.54 \text{ mol total of soln}} = \mathbf{0.903}$$

12.27 *See Section 12.1 and Example 12.4.*

A 0.020 m N_2O solution contains 0.020 mol N_2O per kg of water.

$$? \text{ mol } H_2O \text{ in one kg } H_2O = 1{,}000 \text{ g } H_2O \times \frac{1 \text{ mol } H_2O}{18.02 \text{ g } H_2O} = 55.49 \text{ mol } H_2O$$

$$? \text{ mol total of solution} = 0.020 \text{ mol } N_2O + 55.49 \text{ mol } H_2O = 55.51 \text{ mol total of soln}$$

$$\chi_{N_2O} = \frac{0.020 \text{ mol } N_2O}{55.51 \text{ mol total of soln}} = \mathbf{3.6 \times 10^{-4}}$$

12.28 ***See Section 12.1 and Example 12.4.***

A 0.10 m Br_2 solution contains 0.10 mol Br_2 per kg of water.

$$? \text{ mol } H_2O \text{ in one kg } H_2O = 1{,}000 \text{ g } H_2O \times \frac{1 \text{ mol } H_2O}{18.02 \text{ g } H_2O} = 55.49 \text{ mol } H_2O$$

$$? \text{ mol total of solution} = 0.10 \text{ mol } Br_2 + 55.49 \text{ mol } H_2O = 55.59 \text{ mol total of soln}$$

$$\chi_{Br_2} = \frac{0.10 \text{ mol } Br_2}{55.59 \text{ mol total of soln}} = \mathbf{1.8 \times 10^{-3}}$$

12.29 ***See Section 12.1 and Example 12.4.***

A 0.75 m NaOCl solution contains 0.75 mol NaOCl per kg of water.

$$? \text{ mol } H_2O \text{ in one kg } H_2O = 1{,}000 \text{ g } H_2O \times \frac{1 \text{ mol } H_2O}{18.02 \text{ g } H_2O} = 55.49 \text{ mol } H_2O$$

$$? \text{ mol total of solution} = 0.75 \text{ mol NaOCl} + 55.49 \text{ mol } H_2O = 56.24 \text{ mol total of soln}$$

$$\chi_{NaOCl} = \frac{0.75 \text{ mol NaOCl}}{56.24 \text{ mol total of soln}} = \mathbf{0.013}$$

12.30 ***See Section 12.1 and Example 12.4.***

A 65% C_3H_7OH solution contains 65 g C_3H_7OH per 100 g solution and therefore per 35 g of water.

$$? \text{ mol } C_3H_7OH \text{ in 100 g soln} = 65 \text{ g } C_3H_7OH \times \frac{1 \text{ mol } C_3H_7OH}{60.1 \text{ g } C_3H_7OH} = 1.1 \text{ mol } C_3H_7OH$$

$$? \text{ mol } H_2O \text{ in 100 g soln} = 35 \text{ g } H_2O \times \frac{1 \text{ mol } H_2O}{18.0 \text{ g } H_2O} = 1.9 \text{ mol } H_2O$$

$$? \text{ mol total of solution} = 1.1 \text{ mol } C_3H_7OH + 1.9 \text{ mol } H_2O = 3.0 \text{ mol total of soln}$$

$$\chi_{C_3H_7OH} = \frac{1.1 \text{ mol } C_3H_7OH}{3.0 \text{ mol total of soln}} = \mathbf{0.37}$$

12.31 ***See Section 12.1 and Examples 12.4, 12.5.***

(a) $$? \text{ g } C_{12}H_{22}O_{11} \text{ in 100.0 g soln} = 100.0 \text{ g soln} \times \frac{10.0 \text{ g } C_{12}H_{22}O_{11}}{100.0 \text{ g soln}} = 10.0 \text{ g } C_{12}H_{22}O_{11}$$

$$? \text{ mol } C_{12}H_{22}O_{11} \text{ in 100.0 g soln} = 10.0 \text{ g } C_{12}H_{22}O_{11} \times \frac{1 \text{ mol } C_{12}H_{22}O_{11}}{342.0 \text{ g } C_{12}H_{22}O_{11}} = 0.0292 \text{ mol } C_{12}H_{22}O_{11}$$

$$? \text{ g } H_2O \text{ in 100.0 g soln} = 100.0 \text{ g soln} - 10.0 \text{ g } C_{12}H_{22}O_{11} = 90.0 \text{ } H_2O$$

$$\text{molality} = \frac{\text{mol } C_{12}H_{22}O_{11}}{\text{kg } H_2O} \qquad \text{molality} = \frac{0.0292 \text{ mol } C_{12}H_{22}O_{11}}{0.090 \text{ kg } H_2O} = \mathbf{0.324 \text{ m } C_{12}H_{22}O_{11}}$$

(b) $$? \text{ L soln having mass of 100.0 g} = 100.0 \text{ g soln} \times \frac{1 \text{ mL}}{1.038 \text{ g soln}} \times \frac{1 \text{ L}}{10^3 \text{ mL}} = 9.634 \times 10^{-2} \text{ L}$$

$$molarity = \frac{mol\ C_{12}H_{22}O_{11}}{L\ soln} \qquad molarity = \frac{\dfrac{0.0292\ mol\ C_{12}H_{22}O_{11}}{100.0\ g\ soln}}{\dfrac{9.634\ x\ 10^{-2}\ L}{100.0\ g\ soln}} = \mathbf{0.303}\ \boldsymbol{M}\ \mathbf{C_{12}H_{22}O_{11}}$$

(c) $? \ mol\ H_2O\ in\ 100.0\ g\ soln = 90.0\ g\ H_2O \times \frac{1\ mol\ H_2O}{18.0\ g\ H_2O} = 5.00\ mol\ H_2O$

$? \ mol\ total\ of\ soln\ for\ 100.0\ g\ soln = 0.0292\ mol\ C_{12}H_{22}O_{11} + 5.00\ mol\ H_2O = 5.03\ mol\ total\ of\ soln$

$$\chi_{C_{12}H_{22}O_{11}} = \frac{mol\ C_{12}H_{22}O_{11}}{mol\ total\ of\ soln} \qquad \chi_{C_{12}H_{22}O_{11}} = \frac{0.0292\ mol\ C_{12}H_{22}O_{11}}{5.03\ mol\ total\ of\ soln} = \mathbf{5.81\ x\ 10^{-3}}$$

12.32 ***See Section 12.1 and Examples 12.4, 12.5.***

(a) $? \ g\ CH_3CO_2H\ \ in\ 100.0\ g\ soln = 100.0\ g\ soln \times \frac{5.0\ g\ CH_3CO_2H}{100.0\ g\ soln} = 5.0\ g\ CH_3CO_2H$

$? \ mol\ CH_3CO_2H\ \ in\ 100.0\ g\ soln = 5.0\ g\ CH_3CO_2H\ \ x\ \frac{1\ mol\ CH_3CO_2H}{60.0\ g\ CH_3CO_2H} = 0.083\ mol\ CH_3CO_2H$

$? \ g\ H_2O\ in\ 100.0\ g\ soln = 100.0\ g\ soln - 5.0\ g\ CH_3CO_2H\ = 95.0\ H_2O$

$$molality = \frac{mol\ CH_3CO_2H}{kg\ H_2O} \qquad molality = \frac{0.083\ mol\ CH_3CO_2OH}{0.0950\ kg\ H_2O} = \mathbf{0.87\ m\ CH_3CO_2OH}$$

(b) $? \ L\ soln\ having\ mass\ of\ 100.0\ g = 100.0\ g\ soln \times \frac{1\ mL}{1.0055\ g\ soln} \times \frac{1\ L}{10^3\ mL} = 9.945\ x\ 10^{-2}\ L$

$$molarity = \frac{mol\ CH_3CO_2H}{L\ soln} \qquad molarity = \frac{\dfrac{0.083\ mol\ CH_3CO_2H}{100.0\ g\ soln}}{\dfrac{9.945\ x\ 10^{-2}\ L}{100.0\ g\ soln}} - \mathbf{0.83}\ \boldsymbol{M}\ \mathbf{CH_3CO_2H}$$

(c) $? \ mol\ H_2O\ in\ 100.0\ g\ soln = 95.0\ g\ H_2O \times \frac{1\ mol\ H_2O}{18.0\ g\ H_2O} = 5.28\ mol\ H_2O$

$? \ mol\ total\ of\ soln\ for\ 100.0\ g\ soln = 0.083\ mol\ CH_3CO_2H + 5.28\ mol\ H_2O = 5.36\ mol\ total\ of\ soln$

$$\chi_{CH_3CO_2H} = \frac{mol\ CH_3CO_2H}{mol\ total\ of\ soln} \qquad \chi_{CH_3CO_2H} = \frac{0.083\ mol\ CH_3CO_2H}{5.36\ mol\ total\ of\ soln} = \mathbf{0.015}$$

12.33 ***See Section 12.1 and Examples 12.4, 12.5.***

(a) $? \ g\ for\ exactly\ 1\ L\ soln = 1\ L\ soln \times \frac{10^3\ mL\ soln}{1\ L\ soln} \times \frac{1.031\ g\ soln}{1\ mL\ soln} = 1,031\ g\ soln$

$? \ g\ H_3PO_4\ in\ 1\ L\ soln = 1\ L\ soln \times \frac{0.631\ mol\ H_3PO_4}{1\ L\ soln} \times \frac{98.0\ g\ H_3PO_4}{1\ mol\ H_3PO_4} = 61.8\ g\ H_3PO_4$

$$mass\ percent\ H_3PO_4 = \frac{61.8\ g\ H_3PO_4}{1,031\ g\ soln} \times 100\% = \mathbf{5.99\%\ H_3PO_4}$$

(b) $? \ g\ H_2O\ in\ 1\ L\ soln = 1,031\ g\ soln - 61.8\ g\ H_3PO_4 = 969\ g\ H_2O$

$? \ mol\ H_2O\ in\ 1\ L\ soln = 969\ g\ H_2O \times \frac{1\ mol\ H_2O}{18.0\ g\ H_2O} = 53.8\ mol\ H_2O$

$? \ mol\ total\ of\ soln\ in\ 1\ L\ soln = 0.631\ mol\ H_3PO_4 + 53.8\ mol\ H_2O = 54.4\ mol\ total\ of\ soln$

$$\chi_{H_3PO_4} = \frac{\text{mol } H_3PO_4}{\text{mol total of soln}} \qquad \chi_{H_3PO_4} = \frac{0.631 \text{ mol } H_3PO_4}{54.4 \text{ mol total of soln}} = \mathbf{0.0116}$$

(c) $$\text{molality} = \frac{\text{mol } H_3PO_4}{\text{kg } H_2O} \qquad \text{molality} = \frac{0.631 \text{ mol } H_3PO_4}{0.969 \text{ kg } H_2O} = \mathbf{0.651 \text{ m } H_3PO_4}$$

12.34 *See Section 12.1 and Examples 12.4 and 12.5.*

(a) $$? \text{ g for exactly 1 L soln} = 1 \text{ L soln} \times \frac{10^3 \text{ mL soln}}{1 \text{ L soln}} \times \frac{1.109 \text{ g soln}}{1 \text{ mL soln}} = 1{,}109 \text{ g soln}$$

$$? \text{ g NaOH in 1 L soln} = 1 \text{ L soln} \times \frac{2.77 \text{ mol NaOH}}{1 \text{ L soln}} \times \frac{40.0 \text{ g NaOH}}{1 \text{ mol NaOH}} = 111 \text{ g NaOH}$$

$$\text{mass percent NaOH} = \frac{111 \text{ g NaOH}}{1{,}109 \text{ g soln}} \times 100\% = \mathbf{10.0\% \text{ NaOH}}$$

(b) $$? \text{ g } H_2O \text{ in 1 L soln} = 1{,}109 \text{ g soln} - 111 \text{ g NaOH} = 998 \text{ g } H_2O$$

$$? \text{ mol } H_2O \text{ in 1 L soln} = 998 \text{ g } H_2O \times \frac{1 \text{ mol } H_2O}{18.0 \text{ g } H_2O} = 55.4 \text{ mol } H_2O$$

$$? \text{ mol total of soln in 1 L soln} = 2.77 \text{ mol NaOH} + 55.4 \text{ mol } H_2O = 58.2 \text{ mol total of soln}$$

$$\chi_{NaOH} = \frac{\text{mol NaOH}}{\text{mol total of soln}} \qquad \chi_{NaOH} = \frac{2.77 \text{ mol NaOH}}{58.2 \text{ mol total of soln}} = \mathbf{0.0476}$$

(c) $$\text{molality} = \frac{\text{mol NaOH}}{\text{kg } H_2O} \qquad \text{molality} = \frac{2.77 \text{ mol NaOH}}{0.998 \text{ kg } H_2O} = \mathbf{2.78 \text{ m NaOH}}$$

12.35 *See Section 12.1 and Examples 12.4, 12. 5.*

(a) $$? \text{ g of exactly 1 L soln} = 1 \text{ L soln} \times \frac{10^3 \text{ mL soln}}{1 \text{ L soln}} \times \frac{0.973 \text{ g soln}}{1 \text{ mL soln}} = 973 \text{ g soln}$$

$$? \text{ g } NH_3 \text{ in 1 L soln} = 973 \text{ g soln} \times \frac{6.00 \text{ g } NH_3}{100.00 \text{ g soln}} = 58.4 \text{ g } NH_3$$

$$? \text{ mol } NH_3 \text{ in 1 L soln} = M_{NH_3} = \frac{58.4 \text{ g } NH_3}{1 \text{ L soln}} \times \frac{1 \text{ mol } NH_3}{17.0 \text{ g } NH_3} = \mathbf{3.44 \; \mathit{M} \; NH_3}$$

$$? \text{ g } H_2O \text{ in 1 L soln} = 973 \text{ g soln} - 58.4 \text{ g } NH_3 = 915 \text{ g } H_2O$$

$$? \text{ mol } H_2O \text{ in 1 L soln} = 915 \text{ g } H_2O \times \frac{1 \text{ mol } H_2O}{18.0 \text{ g } H_2O} = 50.8 \text{ mol } H_2O$$

$$? \text{ mol total of soln} = 3.44 \text{ mol } NH_3 + 50.8 \text{ mol } H_2O = 54.2 \text{ mol total of soln}$$

$$\text{molality} = \frac{\text{mol } NH_3}{\text{kg } H_2O} \qquad \text{molality} = \frac{3.44 \text{ mol } NH_3}{0.915 \text{ kg } H_2O} = \mathbf{3.76 \text{ m } NH_3}$$

$$\chi_{NH_3} = \frac{\text{mol } NH_3}{\text{mol total of soln}} \qquad \chi_{NH_3} = \frac{3.44 \text{ mol } NH_3}{54.2 \text{ mol total of soln}} = \mathbf{0.0635}$$

(b) $$? \text{ g of exactly 1 L soln} = 1 \text{ L soln} \times \frac{10^3 \text{ mL soln}}{1 \text{ L soln}} \times \frac{0.936 \text{ g soln}}{1 \text{ mL soln}} = 936 \text{ g soln}$$

$$? \text{ g } NH_3 \text{ in 1 L soln} = 1 \text{ L soln} \times \frac{8.80 \text{ mol}}{1 \text{ L soln}} \times \frac{17.0 \text{ g } NH_3}{1 \text{ mol } NH_3} = 150 \text{ g } NH_3$$

$$? \text{ g } H_2O \text{ in 1 L soln} = 936 \text{ g soln} - 150 \text{ g } NH_3 = 786 \text{ g } H_2O$$

$$? \text{ mol } H_2O \text{ in 1 L soln} = 786 \text{ g } H_2O \times \frac{1 \text{ mol } H_2O}{18.0 \text{ g } H_2O} = 43.7 \text{ mol } H_2O$$

? mol total of soln = 8.80 mol NH_3 + 43.7 mol H_2O = 52.5 mol total of soln

$$\text{molality} = \frac{\text{mol } NH_3}{\text{kg } H_2O} \qquad \text{molality} = \frac{8.80 \text{ mol } NH_3}{0.786 \text{ kg } H_2O} = \mathbf{11.2 \text{ m } NH_3}$$

$$\text{mass percent } NH_3 = \frac{\text{g } NH_3}{\text{g soln}} \times 100\% \qquad \text{mass percent} = \frac{150 \text{ g } NH_3}{936 \text{ g soln}} \times 100\% = \mathbf{16.0\% \ NH_3}$$

$$\chi_{NH_3} = \frac{\text{mol } NH_3}{\text{mol total of soln}} \qquad \chi_{NH_3} = \frac{8.80 \text{ mol } NH_3}{52.52 \text{ mol total of soln}} = \mathbf{0.168}$$

(c) $$? \text{ g soln containing exactly 1 kg } H_2O = 1{,}000 \text{ g } H_2O + (8.02 \text{ mol } NH_3)\left(\frac{17.0 \text{ g } NH_3}{1 \text{ mol } NH_3}\right) = 1{,}136.3 \text{ g soln}$$

$$? \text{ L soln containing exactly 1 kg } H_2O = 1{,}136.3 \text{ g soln} \times \frac{1 \text{ mL soln}}{0.950 \text{ g soln}} \times \frac{1 \text{ L soln}}{10^3 \text{ mL soln}} = 1.20 \text{ L soln}$$

$$? \text{ mol } H_2O \text{ in exactly 1 kg } H_2O = 1 \text{ kg } H_2O \times \frac{10^3 \text{ g } H_2O}{1 \text{ kg } H_2O} \times \frac{1 \text{ mol } H_2O}{18.02 \text{ g } H_2O} = 55.49 \text{ mol } H_2O$$

? mol total of soln when exactly 1 kg H_2O = 8.02 mol NH_3 + 55.49 mol H_2O = 63.51 mol total of soln

$$\text{molarity} = \frac{\text{mol } NH_3}{\text{L soln}} \qquad \text{molarity} = \frac{8.02 \text{ mol } NH_3}{1.20 \text{ L soln}} = \mathbf{6.68 \ \mathit{M} \ NH_3}$$

$$\text{mass percent } NH_3 = \frac{\text{g } NH_3}{\text{g soln}} \times 100\% \qquad \text{mass percent } NH_3 = \frac{136.3 \text{ g } NH_3}{1{,}136.3 \text{ g soln}} \times 100\% = \mathbf{12.0\% \ NH_3}$$

$$\chi_{NH_3} = \frac{\text{mol } NH_3}{\text{mol total of soln}} \qquad \chi_{NH_3} = \frac{8.02 \text{ mol } NH_3}{63.51 \text{ mol total of soln}} = \mathbf{0.126}$$

(d) The quantity chosen for total mol of soln is arbitrary; 10.00 mol was chosen since this quantity is easy to work with.

$$? \text{ mol } NH_3 \text{ in } 10.00 \text{ total mol of soln} = 10.00 \text{ total mol of soln} \times \frac{0.0738 \text{ mol } NH_3}{1 \text{ mol total of soln}} = 0.738 \text{ mol } NH_3$$

? mol H_2O in 10.00 total mol of soln = 10.00 mol total of soln − 0.738 mol NH_3 = 9.26 mol H_2O

$$? \text{ g of } 10.00 \text{ total mol of soln} = (0.738 \text{ mol } NH_3)\left(\frac{17.0 \text{ g } NH_3}{1 \text{ mol } NH_3}\right) + (9.26 \text{ mol } H_2O)\left(\frac{18.0 \text{ g } H_2O}{1 \text{ mol } H_2O}\right)$$

$$= 12.6 \text{ g } NH_3 + 167 \text{ g } H_2O = 180 \text{ g soln}$$

$$? \text{ L of } 10.00 \text{ total mol of soln} = 180 \text{ g soln} \times \frac{1 \text{ mL soln}}{0.969 \text{ g soln}} \times \frac{1 \text{ L soln}}{10^3 \text{ mL soln}} = 0.186 \text{ L soln}$$

$$\text{molality} = \frac{\text{mol } NH_3}{\text{kg } H_2O} \qquad \text{molality} = \frac{0.738 \text{ mol } NH_3}{0.167 \text{ kg } H_2O} = \mathbf{4.42 \text{ m } NH_3}$$

$$\text{molarity} = \frac{\text{mol } NH_3}{\text{L soln}} \qquad \text{molarity} = \frac{0.738 \text{ mol } NH_3}{0.186 \text{ L soln}} = \mathbf{3.97 \ \mathit{M} \ NH_3}$$

$$\text{mass percent} = \frac{\text{g } NH_3}{\text{g soln}} \times 100\% \qquad \text{mass percent} = \frac{12.6 \text{ g } NH_3}{180 \text{ g soln}} \times 100\% = \mathbf{7.00\% \ NH_3}$$

Summary Table:

	Density(g/cm^3)	Molality	Molarity	Mass % NH_3	Mole Fraction
(a)	0.973	**3.76**	**3.44**	6.00	**0.0635**
(b)	0.936	**11.2**	8.80	**16.0**	**0.168**
(c)	0.950	8.02	**6.68**	**12.0**	**0.126**
(d)	0.969	**4.42**	**3.97**	**7.00**	0.0738

12.36 *See Section 12.1 and Examples 12.4, 12. 5.*

(a) $? \text{ g of exactly 1 L soln} = 1 \text{ L soln} \times \frac{10^3 \text{ mL soln}}{1 \text{ L soln}} \times \frac{1.060 \text{ g soln}}{1 \text{ mL soln}} = 1{,}060 \text{ g soln}$

$? \text{ g } HClO_4 \text{ in 1 L soln} = 1{,}060 \text{ g soln} \times \frac{10.0 \text{ g } HClO_4}{100.00 \text{ g soln}} = 106 \text{ g } HClO_4$

$? \text{ mol } HClO_4 \text{ in 1 L soln} = M_{HClO_4} = \frac{106 \text{ g } HClO_4}{1 \text{ L soln}} \times \frac{1 \text{ mol } HClO_4}{100.4 \text{ g } HClO_4} = \mathbf{1.06}\ \boldsymbol{M}\ HClO_4$

$? \text{ g } H_2O \text{ in 1 L soln} = 1{,}060 \text{ g soln} - 106 \text{ g } HClO_4 = 954 \text{ g } H_2O$

$? \text{ mol } H_2O \text{ in 1 L soln} = 954 \text{ g } H_2O \times \frac{1 \text{ mol } H_2O}{18.02 \text{ g } H_2O} = 52.9 \text{ mol } H_2O$

$? \text{ mol total of soln} = 1.06 \text{ mol } HClO_4 + 52.9 \text{ mol } H_2O = 54.0 \text{ mol total of soln}$

$\text{molality} = \frac{\text{mol } HClO_4}{\text{kg } H_2O}$ $\quad$ $\text{molality} = \frac{1.06 \text{ mol } HClO_4}{0.954 \text{ kg } H_2O} = \mathbf{1.11 \text{ m } HClO_4}$

$\chi_{HClO_4} = \frac{\text{mol } HClO_4}{\text{mol total of soln}}$ $\quad$ $\chi_{HClO_4} = \frac{1.06 \text{ mol } HClO_4}{54.0 \text{ mol total of soln}} = \mathbf{0.0196}$

(b) $? \text{ g of exactly 1 L soln} = 1 \text{ L soln} \times \frac{10^3 \text{ mL soln}}{1 \text{ L soln}} \times \frac{1.011 \text{ g soln}}{1 \text{ mL soln}} = 1{,}011 \text{ g soln}$

$? \text{ g } HClO_4 \text{ in 1 L soln} = 1 \text{ L soln} \times \frac{0.2012 \text{ mol}}{1 \text{ L soln}} \times \frac{100.4 \text{ g } HClO_4}{1 \text{ mol } HClO_4} = 20.2 \text{ g } HClO_4$

$? \text{ g } H_2O \text{ in 1 L soln} = 1{,}011 \text{ g soln} - 20.2 \text{ g } HClO_4 = 991 \text{ g } H_2O$

$? \text{ mol } H_2O \text{ in 1 L soln} = 991 \text{ g } H_2O \times \frac{1 \text{ mol } H_2O}{18.0 \text{ g } H_2O} = 55.1 \text{ mol } H_2O$

$? \text{ mol total of soln} = 0.2012 \text{ mol } HClO_4 + 55.1 \text{ mol } H_2O = 55.3 \text{ mol total of soln}$

$\text{molality} = \frac{\text{mol } HClO_4}{\text{kg } H_2O}$ $\quad$ $\text{molality} = \frac{0.2012 \text{ mol } HClO_4}{0.991 \text{ kg } H_2O} = \mathbf{.203 \text{ m } HClO_4}$

$\text{mass percent } HClO_4 = \frac{\text{g } HClO_4}{\text{g soln}} \times 100\%$ $\quad$ $\text{mass percent} = \frac{20.2 \text{ g } HClO_4}{1{,}011 \text{ g soln}} \times 100\% = \mathbf{2.00\% \ HClO_4}$

$\chi_{HClO_4} = \frac{\text{mol } HClO_4}{\text{mol total of soln}}$ $\quad$ $\chi_{HClO_4} = \frac{0.2012 \text{ mol } HClO_4}{55.3 \text{ mol total of soln}} = \mathbf{0.00364}$

(c)

$? \text{ g soln containing exactly 1 kg } H_2O = 1{,}000 \text{ g } H_2O + (2.807 \text{ mol } HClO_4)\left(\frac{100.4 \text{ g } HClO_4}{1 \text{ mol } HClO_4}\right) = 1{,}281.8 \text{ g soln}$

$? \text{ L soln containing exactly 1 kg } H_2O = 1{,}281.8 \text{ g soln} \times \frac{1 \text{ mL soln}}{1.143 \text{ g soln}} \times \frac{1 \text{ L soln}}{10^3 \text{ mL soln}} = 1.121 \text{ L soln}$

$? \text{ mol } H_2O \text{ in exactly 1 kg } H_2O = 1 \text{ kg } H_2O \times \frac{10^3 \text{ g } H_2O}{1 \text{ kg } H_2O} \times \frac{1 \text{ mol } H_2O}{18.0 \text{ g } H_2O} = 55.6 \text{ mol } H_2O$

$? \text{ mol total of soln when exactly 1 kg } H_2O = 2.807 \text{ mol } HClO_4 + 55.6 \text{ mol } H_2O = 58.4 \text{ mol total of soln}$

$\text{molarity} = \frac{\text{mol } HClO_4}{\text{L soln}}$ $\quad$ $\text{molarity} = \frac{2.807 \text{ mol } HClO_4}{1.121 \text{ L soln}} = \mathbf{2.504}\ \boldsymbol{M}\ \mathbf{HClO_4}$

$\text{mass percent } HClO_4 = \frac{\text{g } HClO_4}{\text{g soln}} \times 100\%$ $\quad$ $\text{mass percent } HClO_4 = \frac{281.8 \text{ g } HClO_4}{1{,}281.8 \text{ g soln}} \times 100\% = \mathbf{21.98\% \ HClO_4}$

$\chi_{HClO_4} = \frac{\text{mol } HClO_4}{\text{mol total of soln}}$ $\quad$ $\chi_{HClO_4} = \frac{2.807 \text{ mol } HClO_4}{58.4 \text{ mol total of soln}} = \mathbf{0.0481}$

(d) The quantity chosen for total mol of soln is arbitrary; 10.00 mol was chosen since this quantity is easy to work with.

$? \text{ mol } HClO_4 \text{ in 10.00 total mol of soln} = 10.00 \text{ total mol of soln} \times \frac{0.0284 \text{ mol } HClO_4}{1 \text{ mol total of soln}} = 0.284 \text{ mol } HClO_4$

$? \text{ mol } H_2O \text{ in } 10.00 \text{ total mol of soln} = 10.00 \text{ mol total of soln} - 0.284 \text{ mol } HClO_4 = 9.72 \text{ mol } H_2O$

$$? \text{ g of } 10.00 \text{ total mol of soln} = (0.284 \text{ mol } HClO_4)\left(\frac{100.4 \text{ g } HClO_4}{1 \text{ mol } HClO_4}\right) + (9.72 \text{ mol } H_2O)\left(\frac{18.0 \text{ g } H_2O}{1 \text{ mol } H_2O}\right)$$

$$= 12.6 \text{ g } HClO_4 + 175 \text{ g } H_2O = 204 \text{ g soln}$$

$$? \text{ L of } 10.00 \text{ total mol of soln} = 204 \text{ g soln} \times \frac{1 \text{ mL soln}}{1.086 \text{ g soln}} \times \frac{1 \text{ L soln}}{10^3 \text{ mL soln}} = 0.188 \text{ L soln}$$

$$\text{molality} = \frac{\text{mol } HClO_4}{\text{kg } H_2O} \qquad \text{molality} = \frac{0.284 \text{ mol } HClO_4}{0.175 \text{ kg } H_2O} = \mathbf{1.62 \text{ m } NH_3}$$

$$\text{molarity} = \frac{\text{mol } HClO_4}{\text{L soln}} \qquad \text{molarity} = \frac{0.284 \text{ mol } HClO_4}{0.188 \text{ L soln}} = \mathbf{1.51\ \mathit{M}\ HClO_4}$$

$$\text{mass percent} = \frac{\text{g } HClO_4}{\text{g soln}} \times 100\% \qquad \text{mass percent} = \frac{28.5 \text{ g } HClO_4}{204 \text{ g soln}} \times 100\% = \mathbf{14.0\% \ HClO_4}$$

Summary Table:

	Density(g/cm^3)	Molality	Molarity	Mass % NH_3	Mole Fraction
(a)	1.060	~~**1.05**~~ 1.11	**1.06**	10.0	**0.0196**
(b)	1.011	**0.203**	0.2012	**2.00**	**0.00364**
(c)	1.143	2.807	**2.504**	**21.98**	**0.0481**
(d)	1.086	**1.62**	**1.51**	**14.0**	0.0284

12.37 ***See Section 12.1 and Examples 12.4, 12. 5.***

$$? \text{ g of exactly 1 L soln} = 1 \text{ L soln} \times \frac{10^3 \text{ mL soln}}{1 \text{ L soln}} \times \frac{1.225 \text{ g soln}}{1 \text{ mL soln}} = 1{,}225 \text{ g soln}$$

$$? \text{ g } H_2SO_4 \text{ in 1 L soln} = 1 \text{ L soln} \times \frac{3.75 \text{ mol } H_2SO_4}{1 \text{ L soln}} \times \frac{98.1 \text{ g } H_2SO_4}{1 \text{ mol } H_2SO_4} = 368 \text{ g } H_2SO_4$$

$? \text{ g } H_2O \text{ in 1 L soln} = 1{,}225 \text{ g soln} - 368 \text{ g } H_2SO_4 = 857 \text{ g } H_2O$

$$? \text{ mol } H_2O \text{ in 1 L soln} = 857 \text{ g } H_2O \times \frac{1 \text{ mol } H_2O}{18.0 \text{ g } H_2O} = 47.6 \text{ mol } H_2O$$

$? \text{ mol total of soln} = 3.75 \text{ mol } H_2SO_4 + 47.6 \text{ mol } H_2O = 51.4 \text{ mol total of soln}$

$$\text{molality} = \frac{\text{mol } H_2SO_4}{\text{kg } H_2O} \qquad \text{molality} = \frac{3.75 \text{ mol } H_2SO_4}{0.857 \text{ kg } H_2O} = \mathbf{4.38 \text{ m } H_2SO_4}$$

$$\text{mass percent } H_2SO_4 = \frac{\text{g } H_2SO_4}{\text{g soln}} \times 100 \qquad \text{mass percent} = \frac{368 \text{ g } H_2SO_4}{1{,}225 \text{ g soln}} \times 100\% = \mathbf{30.0\% \ H_2SO_4}$$

$$\chi_{H_2SO_4} = \frac{\text{mol } H_2SO_4}{\text{mol total of soln}} \qquad \chi_{H_2SO_4} = \frac{3.75 \text{ mol } H_2SO_4}{51.4 \text{ mol total of soln}} = \mathbf{0.0730}$$

12.38 ***See Section 12.1 and Examples 12.4, 12. 5.***

$$? \text{ g of exactly 1 L soln} = 1 \text{ L soln} \times \frac{10^3 \text{ mL soln}}{1 \text{ L soln}} \times \frac{0.973 \text{ g soln}}{1 \text{ mL soln}} = 973 \text{ g soln}$$

$$? \text{ g } NH_3 \text{ in 1 L soln} = 973 \text{ g soln} \times \frac{6.00 \text{ g } NH_3}{100.00 \text{ g soln}} = 58.4 \text{ g } NH_3$$

$$? \text{ mol } NH_3 \text{ in 1 L soln} = M_{NH_3} = \frac{58.4 \text{ g } NH_3}{1 \text{ L soln}} \times \frac{1 \text{ mol } NH_3}{17.04 \text{ g } NH_3} = \mathbf{3.43\ \mathit{M}\ NH_3}$$

$? \text{ g } H_2O \text{ in 1 L soln} = 973 \text{ g soln} - 58.4 \text{ g } NH_3 = 915 \text{ g } H_2O$

$$? \text{ mol } H_2O \text{ in 1 L soln} = 915 \text{ g } H_2O \times \frac{1 \text{ mol } H_2O}{18.02 \text{ g } H_2O} = 50.8 \text{ mol } H_2O$$

$? \text{ mol total of soln} = 3.43 \text{ mol } NH_3 + 50.8 \text{ mol } H_2O = 54.2 \text{ mol total of soln}$

$$\text{molality} = \frac{\text{mol } NH_3}{\text{kg } H_2O} \qquad \text{molality} = \frac{3.43 \text{ mol } NH_3}{0.915 \text{ kg } H_2O} = \mathbf{3.75 \text{ m } NH_3}$$

$$\chi_{NH_3} = \frac{\text{mol } NH_3}{\text{mol total of soln}} \qquad \chi_{NH_3} = \frac{3.43 \text{ mol } NH_3}{54.2 \text{ mol total of soln}} = \mathbf{0.0633}$$

12.39 ***See Section 12.2.***

Substances that have similar intermolecular forces have strong solute-solvent interactions and therefore tend to form solutions. H_2O has polar O-H bonds and hydrogen bonding; C_2H_5OH has polar O-H bonds and is larger and more polarizable than water; C_6H_{14} is virtually nonpolar and the solvent C_6H_6 is nonpolar. Hence, we should obtain the following order of increasing solubility in benzene (C_6H_6): $H_2O < C_2H_5OH < C_6H_{14}$.

12.40 ***See Section 12.2.***

Substances that have similar intermolecular forces have strong solute-solvent interactions and therefore tend to form solutions. H_2O has polar O-H bonds and hydrogen bonding; C_2H_5OH has polar O-H bonds and is larger and is more polarizable than water; $CHCl_3$ is polar and has appreciable London dispersion forces and the solvent CCl_4 is nonpolar. Hence, we should obtain the following order of decreasing solubility in carbon tetrachloride (CCl_4): $CHCl_3 > C_2H_5OH > H_2O$.

12.41 ***See Section 12.2 and Example 12.6.***

(a) Br_2 is nonpolar and therefore more soluble in nonpolar carbon tetrachloride than polar water.

(b) $CaCl_2$ is ionic and therefore more soluble in polar water than nonpolar benzene.

(c) $CHCl_3$ is slightly polar and therefore more soluble in the less polar diethyl ether than hydrogen-bonded water.

(d) $C_2H_4(OH)_2$ has O-H bonds and therefore more soluble in water than nonpolar benzene.

12.42 ***See Section 12.2 and Example 12.6.***

(a) NaCl is ionic and therefore more soluble in polar water than nonpolar carbon tetrachloride.

(b) I_2 is nonpolar and therefore more soluble in nonpolar benzene than polar water.

(c) Ethanol has O-H bonds and therefore more soluble in water than virtually nonpolar hexane (C_6H_{14}).

(d) Ethylene glycol has O-H bonds and therefore more soluble in water than nonpolar benzene (C_6H_6).

12.43 ***See Section 12.2.***

(a) $(CH_3)_2CO$ in water, hydrogen bonding

(b) IBr in $CHCl_3$, dipole-dipole and London dispersion

(c) $CaCl_2$ in water, ion-dipole

(d) Kr in CH_3OH, London dispersion

12.44 ***See Section 12.2.***

(a) CH_3OH in water, hydrogen bonding

(b) IBr in CH_3CN, dipole-dipole and London dispersion

(c) KBr in water, ion-dipole

(d) Ar in H_2O, London dispersion

12.45 ***See Section 12.2.***

(a) $CH_3(CH_2)_{10}OH$ is less polar than $CH_3(CH_2)_2OH$ due to its longer CH_2 chain and therefore more soluble than $CH_3(CH_2)_2OH$ in virtually nonpolar C_6H_{14}.

(b) CCl_4 is nonpolar and therefore more soluble than ionic $BaCl_2$ in virtually nonpolar C_6H_{14}.

(c) $Fe(C_5H_5)_2$ is molecular and nonpolar and therefore more soluble than ionic $FeCl_2$ in virtually nonpolar C_6H_{14}.

12.46 ***See Section 12.2.***

(a) Na_2O is ionic and therefore more soluble than nonpolar CO_2 in polar water.

(b) $TiCl_3$ is ionic and therefore more soluble than slightly polar $CHCl_3$ in polar water.

(c) C_3H_7OH has O-H bonds and therefore more soluble than virtually nonpolar C_3H_8 in polar water.

12.47 ***See Section 12.3 and Example 12.8.***

The solubility of the solute (KCl) increases with increasing temperature indicating the enthalpy of solution is positive and the process is endothermic.

12.48 ***See Section 12.3 and Example 12.8.***

The solubility of the solute ($PbBr_2$) increases with increasing temperature indicating the enthalpy of solution is positive and the process is endothermic.

12.49 ***See Section 12.3 and Example 12.8.***

The solubility of the solute ($Ca(OH)_2$) decreases with increasing temperature indicating the enthalpy of solution is negative and the process is exothermic.

12.50 ***See Section 12.3, Figure 12.11, and Example 12.8.***

The solubility of KNO_3 increases more rapidly with increasing temperature indicating it has a more positive and therefore more endothermic enthalpy of solution than NaCl.

12.51 ***See Section 12.3, Figure 12.11, and Example 12.8.***

The solubility of NH_4Cl increases more rapidly with increasing temperature indicating it has a more positive and therefore more endothermic enthalpy of solution than NaCl.

12.52 ***See Section 12.3.***

The solubility of most gases decreases with increasing temperature indicating the enthalpy of solution for such processes is usually negative and the processes are usually exothermic.

12.53 ***See Section 12.3.***

The solubility of nitrogen should decrease with increasing temperature since the dissolving process is exothermic. The solubility of nitrogen should, however, increase with increasing pressure, as does the solubility of other gases.

Hence,
(a) lower temperature and higher pressure yields higher solubility.
(b) same temperature and lower pressure yields lower solubility.
(c) lower temperature and same pressure yields higher solubility.
(d) higher temperature and same pressure yields lower solubility.
(e) higher temperature and lower pressure yields lower solubility.

12.54 ***See Section 12.3.***

The solubility of neon should decrease with increasing temperature since the dissolving process is exothermic. The solubility of neon should, however, increase with increasing pressure, as does the solubility of other gases. Hence,
(a) lower temperature and higher pressure yields higher solubility.
(b) same temperature and lower pressure yields lower solubility.
(c) lower temperature and same pressure yields higher solubility.
(d) higher temperature and same pressure yields lower solubility.
(e) higher temperature and lower pressure yields lower solubility.

12.55 ***See Section 12.3.***

The solubility of nitrous oxide should decrease with increasing temperature since the dissolving process is exothermic. The solubility of nitrous oxide should, however, increase with increasing pressure, as does the solubility of other gases. Hence,
(a) same pressure and lower temperature yields higher solubility.
(b) same pressure and higher temperature yields lower solubility.
(c) lower pressure and same temperature yields lower solubility.
(d) lower pressure and higher temperature yields lower solubility.

12.56 ***See Section 12.3.***

The solubility of ozone should decrease with increasing temperature since the dissolving process is exothermic. The solubility of ozone should, however, increase with increasing pressure, as does the solubility of other gases. Hence,
(a) same pressure and lower temperature yields higher solubility.
(b) same pressure and higher temperature yields lower solubility.
(c) higher pressure and same temperature yields higher solubility.
(d) lower pressure and higher temperature yields lower solubility.

12.57 ***See Section 12.3 and Example 12.7.***

(a) Solving C = kP for k gives $k = \frac{C}{P}$ $\quad k = \frac{9.38 \times 10^{-3}\ \text{molal}}{0.200\ \text{atm} \times 760\ \text{torr/atm}} = \mathbf{6.17 \times 10^{-5}\ molal/torr}$

(b) Substituting in C = kP gives $C = (6.17 \times 10^{-5}\ \text{molal/torr})(300\ \text{torr}) = 1.85 \times 10^{-2}\ \text{molal}$

Since 1.85×10^{-2} molal corresponds to 1.85×10^{-2} mol C_2H_2/kg H_2O,

$$? \text{ g } C_2H_2 = 1.85 \times 10^{-2} \text{ mol } C_2H_2 \times \frac{26.0 \text{ g } C_2H_2}{1 \text{ mol } C_2H_2} = \mathbf{0.481\ g\ C_2H_2}$$

12.58 ***See Section 12.3 and Example 12.7.***

(a) Solving C = kP for k gives $k = \frac{C}{P}$ $\quad k = \frac{1.27 \times 10^{-4}\ \text{molal}}{0.300\ \text{atm} \times 760\ \text{torr/atm}} = \mathbf{5.57 \times 10^{-7}\ molal/torr}$

(b) Substituting in C = kP gives $C = (5.57 \text{ x } 10^{-7} \text{ molal / torr})(500 \text{ torr}) = 2.79 \text{ x } 10^{-4} \text{ molal}$

Since $2.79 \text{ x } 10^{-4}$ molal corresponds to $2.79 \text{ x } 10^{-4}$ mol C_2H_4/kg H_2O,

$$? \text{ g } C_2H_4 = 2.79 \text{ x } 10^{-4} \text{ mol } C_2H_4 \times \frac{28.0 \text{ g } C_2H_4}{1 \text{ mol } C_2H_4} = \mathbf{0.00781 \text{ g } C_2H_4}$$

12.59 ***See Section 12.3 and Example 12.7.***

(a) $\text{molality} = \dfrac{\text{mol } O_3}{\text{kg } H_2O}$ $\quad$ $\text{molality} = \dfrac{0.105 \text{ g } O_3}{0.100 \text{ kg } H_2O} \times \dfrac{1 \text{ mole } O_3}{48.0 \text{ g } O_3} = 0.0219 \text{ m } O_3$

Solving C = kP for k gives $k = \dfrac{C}{P}$ $\quad$ $k = \dfrac{0.0219 \text{ molal}}{1.0 \text{ atm} \times 760 \text{ torr / atm}} = \mathbf{2.88 \text{ x } 10^{-5} \text{ molal / torr}}$

(b) The solubility of ozone increases as the temperature decreases from 20°C to 10°C because the dissolving process is exothermic. A higher solubility at the same pressure leads to a larger Henry's law constant because k = C/P. Hence, k for ozone will be larger at 10°C than at 20°C.

(c) Substituting in C = kP gives $C = (2.88 \text{ x } 10^{-8}$ molal/torr)(0.500 atm x 760 torr/atm) = **0.0109 molal**

12.60 ***See Section 12.3 and Example 12.7.***

(a) $\text{molality} = \dfrac{\text{mol } N_2O}{\text{kg } H_2O}$ $\quad$ $\text{molality} = \dfrac{0.121 \text{ g } N_2O}{0.100 \text{ kg } H_2O} \times \dfrac{1 \text{ mole } N_2O}{44.0 \text{ g } N_2O} = 0.0275 \text{ m } N_2O$

Solving C = kP for k gives $k = \dfrac{C}{P}$ $\quad$ $k = \dfrac{0.0275 \text{ molal}}{1.0 \text{ atm} \times 760 \text{ torr / atm}} = \mathbf{3.62 \text{ x } 10^{-5} \text{ molal / torr}}$

(b) The solubility of nitrous oxide increases as the temperature decreases from 20°C to 10°C because the dissolving process is exothermic. A higher solubility at the same pressure leads to a larger Henry's law constant because k = C/P. Hence, k for nitrous oxidewill be larger at 10°C than at 20°C.

(c)) Substituting in C = kP gives $C = (3.62 \text{ x } 10^{-5} \text{ molal / torr})(0.500 \text{ atm} \times 760 \text{ torr / atm}) = \mathbf{0.0138 \text{ molal}}$

12.61 ***See Section 12.4 and Example 12.9.***

$\Delta P = \chi_{solute} P^o_{solv}$ and $P_{solv} = \chi_{solv} P^o_{solv}$

$$? \text{ mol } C_6H_5OH = 10.0 \text{ g } C_6H_5OH \times \frac{1 \text{ mol } C_6H_5OH}{94.1 \text{ g } C_6H_5OH} = 0.106 \text{ mol } C_6H_5OH$$

$$? \text{ mol } CHCl_3 = 95.0 \text{ g } CHCl_3 \times \frac{1 \text{ mol } CHCl_3}{119.4 \text{ g } CHCl_3} = 0.796 \text{ mol } CHCl_3$$

$$? \text{ mol total of soln} = 0.106 \text{ mol } C_6H_5OH + 0.796 \text{ mol } CHCl_3 = 0.902 \text{ mol total of soln}$$

$\chi_{solute} = \chi_{C_6H_5OH} = \dfrac{0.106 \text{ mol } C_6H_5OH}{0.902 \text{ mol total of soln}} = 0.118$ $\quad$ $\chi_{solv} = 1 - \chi_{solute}$ $\quad$ $\chi_{solv} = 1 - 0.118 = 0.882$

$\Delta P = \chi_{C_6H_5OH} P^o_{CHCl_3}$ $\quad$ $\Delta P = (0.118)(360 \text{ torr}) = \mathbf{42.5 \text{ torr}}$

$P_{solv} = \chi_{CHCl_3} P^o_{CHCl_3}$ $\quad$ $P_{solv} = (0.882)(360 \text{ torr}) = \mathbf{318 \text{ torr}}$ or $P_{solv} = P^o_{CHCl_3} - \Delta P = \mathbf{318 \text{ torr}}$

12.62 *See Section 12.4 and Example 12.9.*

$\Delta P = \chi_{solute} P^o_{solv}$ and $P_{solv} = \chi_{solv} P^o_{solv}$

$? \text{ mol } C_{10}H_8 = 14.0 \text{ g } C_{10}H_8 \times \dfrac{1 \text{ mol } C_{10}H_8}{128.18 \text{ g } C_{10}H_8} = 0.109 \text{ mol } C_{10}H_8$

$? \text{ mol } C_6H_{12} = 50 \text{ g } C_6H_{12} \times \dfrac{1 \text{ mol } C_6H_{12}}{84.18 \text{ g } C_6H_{12}} = 0.59 \text{ mol } C_6H_{12}$

$? \text{ mol total of soln} = 0.109 \text{ mol } C_{10}H_8 + 0.59 \text{ mol } C_6H_{12} = 0.70 \text{ mol total of soln}$

$\chi_{solute} = \chi_{C_{10}H_8} = \dfrac{0.109 \text{ mol } C_{10}H_8}{0.70 \text{ mol total of soln}} = 0.16$

$\Delta P = \chi_{C_{10}H_8} P^o_{C_6H_{12}}$ $\quad\quad \Delta P = (0.16)(99.0 \text{ torr}) = \mathbf{16 \text{ torr}}$

$P_{solv} = P^o_{CHCl_3} - \Delta P = 99.0 \text{ torr} - 16 \text{ torr} = \mathbf{83 \text{ torr}}$

12.63 *See Section 12.4, Table 12.3, and Examples 12.10, 12.12.*

$\Delta T_f = mk_f$ and $\Delta T_b = mk_b$ $\quad\quad ? \text{ mol urea} = 2.00 \text{ g urea} \times \dfrac{1 \text{ mol urea}}{60.1 \text{ g urea}} = 0.0333 \text{ mol urea}$

$\text{molality} = \dfrac{\text{mol urea}}{\text{kg water}}$ $\quad\quad \text{molality} = \dfrac{0.0333 \text{ mol urea}}{0.0250 \text{ kg water}} = 1.33 \text{ m urea}$

$\Delta T_f = mk_f$ $\quad\quad \Delta T_f = (1.33 \text{ m})(1.86^\circ\text{C/m}) = 2.47^\circ\text{C},$ $\quad\quad T_f = 0.00^\circ\text{C} - 2.47^\circ\text{C} = \mathbf{-2.47^\circ C}$

$\Delta T_b = mk_b$ $\quad\quad \Delta T_b = (1.33 \text{ m})(0.512^\circ\text{C/m}) = 0.681^\circ\text{C},$ $\quad\quad T_b = 100.000^\circ\text{C} + 0.681^\circ\text{C} = \mathbf{100.681^\circ C}$

12.64 *See Section 12.4, Table 12.3, and Examples 12.10, 12.12.*

$\Delta T_f = mk_f$ and $\Delta T_b = mk_b$ $\quad\quad ? \text{ mol sucrose} = 1.00 \text{ g sucrose} \times \dfrac{1 \text{ mol sucrose}}{342.3 \text{ g sucrose}} = 0.00292 \text{ mol sucrose}$

$\text{molality} = \dfrac{\text{mol sucrose}}{\text{kg water}}$ $\quad\quad \text{molality} = \dfrac{0.00292 \text{ mol sucrose}}{0.0100 \text{ kg water}} = 0.292 \text{ m sucrose}$

$\Delta T_f = mk_f$ $\quad\quad \Delta T_f = (0.292 \text{ m})(1.86^\circ\text{C/m}) = 0.543^\circ\text{C},$ $\quad\quad T_f = 0.000^\circ\text{C} - 0.543^\circ\text{C} = \mathbf{-0.543^\circ C}$

$\Delta T_b = mk_b$ $\quad\quad \Delta T_b = (0.292 \text{ m})(0.512^\circ\text{C/m}) = 0.150^\circ\text{C},$ $\quad\quad T_b = 100.000^\circ\text{C} + 0.150^\circ\text{C} = \mathbf{100.150^\circ C}$

12.65 *See Section 12.4, Table 12.3, and Examples 12.10, 12.12.*

$? \text{ mol } C_6H_5OH = 0.500 \text{ g } C_6H_5OH \times \dfrac{1 \text{ mol } C_6H_5OH}{94.1 \text{ g } C_6H_5OH} = 0.00531 \text{ mol } C_6H_5OH$

$\text{molality} = \dfrac{\text{mol } C_6H_5OH}{\text{kg } C_6H_{12}}$ $\quad\quad \text{molality} = \dfrac{0.00531 \text{ mol } C_6H_5OH}{0.0120 \text{ kg } C_6H_{12}} = 0.442 \text{ m } C_6H_5OH$

Solving $\Delta T_f = mk_f$ for k_f gives $k_f = \dfrac{\Delta T_f}{m}$ $\quad\quad k_f = \dfrac{6.50^\circ \text{C} - (-2.44)^\circ \text{C}}{0.442 \text{ m}} = \mathbf{20.2^\circ \text{C/m}}$

12.66 *See Section 12.4, Table 12.3 and Examples 12.10 and 12.12.*

$? \text{ mol } C_6H_5OH = 0.500 \text{ g } C_6H_5OH \times \dfrac{1 \text{ mol } C_6H_5OH}{94.1 \text{ g } C_6H_5OH} = 0.00531 \text{ mol } C_6H_5OH$

$\text{molality} = \dfrac{\text{mol } C_6H_5OH}{\text{kg } C_6H_{12}}$ $\quad\quad \text{molality} = \dfrac{0.00531 \text{ mol } C_6H_5OH}{0.0120 \text{ kg } C_6H_{12}} = 0.442 \text{ m } C_6H_5OH$

Solving $\Delta T_b = mk_b$ for k_b gives $k_b = \frac{\Delta T_b}{m}$ $\qquad k_b = \frac{81.94^\circ C - 80.72^\circ C}{0.442\ m} = \mathbf{2.76^\circ C/m}$

12.67 *See Section 12.4 and Example 12.13.*

Solving $\Delta T_f = mk_f$ for m gives $m = \frac{\Delta T_f}{k_f}$ $\qquad m = \frac{6.50^\circ C - 0.83^\circ C}{20.2^\circ C/\text{molal}} = 0.281\ m$

$$? \text{ mol solute} = 12.0 \text{ g solvent} \times \frac{1 \text{ kg solvent}}{10^3 \text{ g solvent}} \times \frac{0.281 \text{ mol solute}}{1 \text{ kg solvent}} = 3.37 \times 10^{-3} \text{ mol solute}$$

$\text{molar mass} = \frac{\text{g solute}}{\text{mol solute}}$ $\qquad \text{molar mass} = \frac{0.350 \text{ g}}{3.37 \times 10^{-3} \text{ mol}} = \mathbf{104\ g/mol}$

12.68 *See Section 12.4 and Example 12.13.*

Solving $\Delta T_f = mk_f$ for m gives $m = \frac{\Delta T_f}{k_f}$ $\qquad m = \frac{5.51^\circ C - 5.03^\circ C}{4.90^\circ C/\text{molal}} = 0.0980\ m$

$$? \text{ mol solute} = 15.0 \text{ g solvent} \times \frac{1 \text{ kg solvent}}{10^3 \text{ g solvent}} \times \frac{0.0980 \text{ mol solute}}{1 \text{ kg solvent}} = 1.47 \times 10^{-3} \text{ mol solute}$$

$\text{molar mass} = \frac{\text{g solute}}{\text{mol solute}}$ $\qquad \text{molar mass} = \frac{0.500 \text{ g}}{1.47 \times 10^{-3} \text{ mol}} = \mathbf{340\ g/mol}$

12.69 *See Section 12.4 and Example 12.14.*

Known Quantities:

$$? \text{ g protein in } 1.00 \text{ L soln} = 1.00 \text{ L soln} \times \frac{10^3 \text{ mL soln}}{1.00 \text{ L soln}} \times \frac{1.00 \text{ g protein}}{20.0 \text{ mL soln}} = 50.0 \text{ g protein}$$

$\Pi = 35.2 \text{ torr} \times \frac{1 \text{ atm}}{760 \text{ torr}} = 0.0463 \text{ atm}$ $\qquad V = 1.00 \text{ L}$ $\qquad T = 298 \text{ K}$

Solving $\Pi = \frac{nRT}{V}$ for n gives $n = \frac{\Pi V}{RT}$ $\qquad n = \frac{(0.0463 \text{ atm})(1 \text{ L})}{\left(0.0821 \frac{\text{L}\cdot\text{atm}}{\text{mol}\cdot\text{K}}\right)(298 \text{ K})} = 1.89 \times 10^{-3} \text{ mol}$

$\text{molar mass} = \frac{\text{g solute}}{\text{mol solute}}$ $\qquad \text{molar mass} = \frac{50.0 \text{ g protein}}{1.89 \times 10^{-3} \text{ mol protein}} = \mathbf{2.65 \times 10^4\ g/mol}$

12.70 *See Section 12.4 and Example 12.14.*

Known Quantities:

$$? \text{ g enzyme in } 1.00 \text{ L soln} = 1.00 \text{ L soln} \times \frac{10^3 \text{ mL soln}}{1 \text{ L soln}} \times 1.00 \frac{\text{g soln}}{\text{mL soln}} \times \frac{10.0 \text{ g enzyme}}{100.00 \text{ g soln}} = 100. \text{ g enzyme}$$

$\Pi = 13.3 \text{ torr} \times \frac{1 \text{ atm}}{760 \text{ torr}} = 1.75 \times 10^{-2} \text{ atm}$ $\qquad V = 1.00 \text{ L}$ $\qquad T = 298 \text{ K}$

Solving $\Pi = \frac{nRT}{V}$ for n gives $n = \frac{\Pi V}{RT}$ $\qquad n = \frac{(1.75 \times 10^{-2} \text{ atm})(1 \text{ L})}{\left(0.0821 \frac{\text{L}\cdot\text{atm}}{\text{mol}\cdot\text{K}}\right)(298 \text{ K})} = 7.15 \times 10^{-4} \text{ mol}$

$\text{molar mass} = \frac{\text{g solute}}{\text{mol solute}}$ $\qquad \text{molar mass} = \frac{100. \text{ g enzyme}}{7.15 \times 10^{-4} \text{ mol enzyme}} = \mathbf{1.40 \times 10^5\ g/mol}$

12.71 ***See Section 12.5 and Example 12.15.***

The boiling point of a solution depends on the total molal concentration of solute particles. This concentration is given by the product of the van't Hoff factor times the molal concentration of the solute.

Compound	Present As	i	m	$i \times$ m
LiBr	Li^+ & Br^-	2	0.02	0.04
sucrose	molecules	1	0.03	0.03
$MgSO_4$	Mg^{2+} & SO_4^{2-}	2	0.03	0.06
$CaCl_2$	Ca^{2+} & $2Cl^-$	3	0.03	0.09

Hence, the boiling points of these solutions should increase in the order:

0.03 m sucrose < 0.02 m LiBr < 0.03 m $MgSO_4$ < 0.03 m $CaCl_2$

12.72 ***See Section 12.5 and Example 12.15.***

The osmotic pressure of a solution depends on the total molar concentration of solute particles. This concentration is given by the product of the van't Hoff factor times the molar concentration of the solute.

Compound	Present As	i	M	$i \times M$
urea	molecules	1	0.10	0.10
NaCl	Na^+ & Cl^-	2	0.06	0.12
$Ba(NO_3)_2$	Ba^{2+} & $2NO_3^-$	3	0.05	0.15
sucrose	molecules	1	0.06	0.06

Hence, the osmotic pressures of these solutions should decrease in the order:

0.05 m $Ba(NO_3)_2$ > 0.06 m NaCl > 0.10 m urea > 0.06 m sucrose

Note: For dilute aqueous solutions, such as these, $M \cong$ m.

12.73 ***See Sections 12.4, 12.5 and Example 12.15.***

Solving $\Delta T_f = imk_f$ for im, the total molality of solute particles, gives $im = \frac{\Delta T_f}{k_f}$ $\quad im = \frac{2.01°C}{1.86°C/m} = \mathbf{1.08\ m}$

Assuming an ideal i value of 2.0 for NaCl gives $m = \frac{1.08\ m}{2.0} = 0.540\ m\ NaCl$

This corresponds to 0.540 mol NaCl/kg H_2O and

$$?\ g\ NaCl\ per\ kg\ H_2O = \frac{0.540\ mol\ NaCl}{kg\ H_2O} \times \frac{58.5\ g\ NaCl}{1\ mol\ NaCl} = \mathbf{31.6\ g\ NaCl/\ kg\ H_2O}$$

12.74 ***See Sections 12.4, 12.5 and Example 12.15.***

Solving $\Delta T_f = imk_f$ for im, the total molality of solute particles, gives $im = \frac{\Delta T_f}{k_f}$ $\quad im = \frac{1.61°C}{1.86°C/m} = \mathbf{0.866\ m}$

Assuming an ideal i value of 2.0 0 for NaBr gives $m = \frac{0.866\ m}{2.00} = 0.433\ m\ NaBr$

This corresponds to 0.433 mol NaBr/kg H_2O and

$$?\ g\ NaBr\ per\ kg\ H_2O = \frac{0.433\ mol\ NaBr}{kg\ H_2O} \times \frac{102.9\ g\ NaBr}{1\ mol\ NaBr} = \mathbf{44.6\ g\ NaBr/kg\ H_2O}$$

12.75 ***See Sections 12.4, 12.5, Table 12.3, and Examples 12.10, 12.15.***

$$?\ mol\ CaCl_2 = 6.3\ g\ CaCl_2 \times \frac{1\ mol\ CaCl_2}{111.0\ g\ CaCl_2} = 0.057\ mol\ CaCl_2$$

$$molality = \frac{mol\ CaCl_2}{kg\ water} \qquad molality = \frac{0.057\ mol\ CaCl_2}{1.20\ kg\ water} = 0.048\ m\ CaCl_2$$

Assuming an ideal i value of 3 for $CaCl_2$, $\Delta T_b = imk_b$ = (3)(0.048 m)(0.512°C/m) = 0.074°C and
T_b = 100.000°C + 0.074°C = **100.074°C**

12.76 ***See Sections 12.4, 12.5, Table 12.3, and Examples 12.12, 12.15.***

$$? \text{ mol NaCl} = 8.5 \text{ g NaCl} \times \frac{1 \text{ mol NaCl}}{58.5 \text{ g NaCl}} = 0.15 \text{ mol NaCl}$$

$$\text{molality} = \frac{\text{mol NaCl}}{\text{kg water}} \qquad \text{molality} = \frac{0.15 \text{ mol NaCl}}{1.00 \text{ kg water}} = 0.15 \text{ m NaCl}$$

Assuming an ideal i value of 2 for NaCl, $\Delta T_f = imk_f$ = (2)(0.15 m)(1.86°C/m) = 0.56°C and
T_f = 0.00°C - 0.56°C = **-0.56°C**

12.77 ***See Sections 12.4, 12.5 and Example 12.15.***

$$? \text{ mol } CaCl_2 = 3.4 \text{ g } CaCl_2 \times \frac{1 \text{ mol } CaCl_2}{111.0 \text{ g } CaCl_2} = 0.031 \text{ mol } CaCl_2$$

$$\text{molarity} = \frac{\text{mol } CaCl_2}{\text{L soln}} \qquad \text{molarity} = \frac{0.031 \text{ mol } CaCl_2}{0.500 \text{ L soln}} = 0.062\ M\ CaCl_2$$

Assuming an ideal i value of 3 for $CaCl_2$, $\Pi = iMRT = (3)\left(0.062 \frac{\text{mol}}{\text{L}}\right)\left(0.0821 \frac{\text{L} \cdot \text{atm}}{\text{mol} \cdot \text{K}}\right)(298 \text{ K}) = \mathbf{4.6 \text{ atm}}$

12.78 ***See Sections 12.4, 12.5 and Example 12.15.***

$$? \text{ mol NaCl} = 8.5 \text{ g NaCl} \times \frac{1 \text{ mol NaCl}}{58.5 \text{ g NaCl}} = 0.15 \text{ mol NaCl}$$

$$\text{molarity} = \frac{\text{mol NaCl}}{\text{L soln}} \qquad \text{molarity} = \frac{0.15 \text{ mol NaCl}}{1 \text{ L soln}} = 0.15\ M \text{ NaCl}$$

Assuming an ideal i value of 2 for NaCl, $\Pi = iMRT = (2)\left(0.15 \frac{\text{mol}}{\text{L}}\right)\left(0.0821 \frac{\text{L} \cdot \text{atm}}{\text{mol} \cdot \text{K}}\right)(298 \text{ K}) = \mathbf{7.3 \text{ atm}}$

12.79 ***See Sections 12.4, 12.5 and the Insight into Chemistry, entitled Reverse Osmosis.***

The direction of net transport of water molecules depends on whether the pressure that is exerted on the membrane is greater on the pure water side or on the NaCl solution side. The pressure that is exerted on the pure water side will be equal to the pressure that is exerted on the NaCl side when there is no outside force acting on the system and the system is allowed to reach equilibrium. This pressure is given by Π. Hence, Π must be calculated and compared to the applied pressure to determine the direction of net transport of water molecules.

Known Quantities:
Assuming ideal behavior, total molarity of NaCl solute particles = $2 \times 0.010\ M = 0.020\ M$ T = 298 K
Substituting in $\Pi = MRT$ gives $\Pi = \left(0.020 \frac{\text{mol}}{\text{L}}\right)\left(0.0821 \frac{\text{L} \cdot \text{atm}}{\text{mol} \cdot \text{K}}\right)(298 \text{ K}) = 0.49 \text{ atm}$ or $\mathbf{3.7 \times 10^2 \text{ torr}}$
Since Π is less than the 500 torr that is applied to the salt solution side, net transport of water molecules to the pure water side will occur. Reverse osmosis will occur.

12.80 ***See Sections 12.4, 12.5 and the Insight into Chemistry, entitled Reverse Osmosis.***

The direction of net transport of water molecules depends on whether the pressure that is exerted on the membrane is greater on the pure water side or on the $CaCl_2$ solution side. The pressure that is exerted on the pure water side will be equal to the pressure that is exerted on the $CaCl_2$ side when there is no outside force acting on the system and the system is allowed to reach equilibrium. This pressure is given by Π. Hence, Π must be calculated and compared to the applied pressure to determine the direction of net transport of water molecules.

Known Quantities:
Assuming ideal behavior, total molarity of $CaCl_2$ solute particles = $3 \times 0.010\ M = 0.030\ M$ T = 298 K

Substituting in $\Pi = MRT$ gives $\Pi = \left(0.030\frac{\text{mol}}{\text{L}}\right)\left(0.0821\frac{\text{L}\cdot\text{atm}}{\text{mol}\cdot\text{K}}\right)(298\ \text{K}) = 0.734$ atm or **558 torr**

Since Π is greater than the 500 torr that is applied to the calcium chloride solution side, net transport of water molecules to the calcium chloride solution will occur. Reverse osmosis will not occur.

12.81 ***See Sections 12.4, 125.***

(a) Assuming no dissociation of K_2SO_4 occurs, the total expected molarity of K_2SO_4 solute particles is 0.029 M.

Substituting in $\Pi = MRT$ gives $\Pi = \left(0.029\frac{\text{mol}}{\text{L}}\right)\left(0.0821\frac{\text{L}\cdot\text{atm}}{\text{mol}\cdot\text{K}}\right)(298\ \text{K}) = 0.71$ atm

$$i = \frac{\text{measured value}}{\text{expected value}} \qquad i = \frac{1.79\ \text{atm}}{0.71\ \text{atm}} = 2.5$$

(b) The i value of 2.5 is less than the ideal i value of 3 for K_2SO_4 due to ion association. At higher concentrations i will be smaller since more extensive ion association will occur.

12.82 ***See Sections 12.4 and 125.***

(a) Assuming no dissociation of $CuSO_4$ occurs, the total expected molality of $CuSO_4$ solute particles is 0.031 m. Hence, the expected ΔT_f is given by $\Delta T_f = mk_f$ = (0.031 molal)(1.86°C/molal) = 0.056°C.

$$i = \frac{\text{measured value}}{\text{expected value}} \qquad i = \frac{0.075^\circ\ \text{C}}{0.056^\circ\ \text{C}} = 1.3$$

(b) The i value of 1.3 is less than the ideal i value of 2 for $CuSO_4$ due to ion association. At higher concentrations i will be smaller since more extensive ion association will occur.

12.83 ***See Section 12.6 and Example 12.16.***

(a) $?\ \text{mol}\ C_6H_{14} = 15.0\ \text{g}\ C_6H_{14} \times \frac{1\ \text{mol}\ C_6H_{14}}{86.0\ \text{g}\ C_6H_{14}} = 0.174\ \text{mol}\ C_6H_{14}$

$?\ \text{mol}\ C_7H_{16} = 20.0\ \text{g}\ C_7H_{16} \times \frac{1\ \text{mol}\ C_7H_{16}}{100.0\ \text{g}\ C_7H_{16}} = 0.200\ \text{mol}\ C_7H_{16}$

? mol total of soln = $0.174\ \text{mol}\ C_6H_{14} + 0.200\ \text{mol}\ C_7H_{16} = 0.374$ mol total of soln

$$\chi_{C_6H_{14}} = \frac{\text{mol}\ C_6H_{14}}{\text{mol total of soln}} \qquad \chi_{C_6H_{14}} = \frac{0.174\ \text{mol}\ C_6H_{14}}{0.374\ \text{mol total of soln}} = \mathbf{0.465}$$

$$\chi_{C_7H_{16}} = 1 - \chi_{C_6H_{14}} \qquad \chi_{C_7H_{16}} = 1 - 0.465 = \mathbf{0.535}$$

(b) $P_{C_6H_{14}} = \chi_{C_6H_{14}} P^\circ_{C_6H_{14}}$ $\qquad P_{C_6H_{14}} = (0.465)(278\ \text{torr}) = \mathbf{129\ torr}$

$P_{C_7H_{16}} = \chi_{C_7H_{16}} P^\circ_{C_7H_{16}}$ $\qquad P_{C_7H_{16}} = (0.535)(92.3\ \text{torr}) = \mathbf{49.4\ torr}$

(c) $\chi_{C_6H_{14}(g)} = \frac{129\ \text{torr}}{129\ \text{torr} + 49.4\ \text{torr}} = \mathbf{0.725}$ $\qquad \chi_{C_7H_{16}(g)} = 1 - \chi_{C_6H_{14}(g)}$ $\qquad \chi_{C_7H_{16}(g)} = 1 - 0.725 = \mathbf{0.275}$

12.84 *See Section 12.6 and Example 12.6.*

(a) $? \text{ mol } C_6H_{12} = 25.0 \text{ g } C_6H_{12} \times \frac{1 \text{ mol } C_6H_{12}}{84.0 \text{ g } C_6H_{12}} = 0.298 \text{ mol } C_6H_{12}$

$? \text{ mol } C_6H_{14} = 44.0 \text{ g } C_6H_{14} \times \frac{1 \text{ mol } C_6H_{14}}{86.0 \text{ g } C_6H_{14}} = 0.512 \text{ mol } C_6H_{14}$

? mol total of soln = $0.298 \text{ mol } C_6H_{12} + 0.512 \text{ mol } C_6H_{14} = 0.810$ mol total of soln

$\chi_{C_6H_{12}} = \frac{\text{mol } C_6H_{12}}{\text{mol total of soln}}$ $\qquad \chi_{C_6H_{12}} = \frac{0.298 \text{ mol } C_6H_{12}}{0.810 \text{ mol total of soln}} = \mathbf{0.368}$

$\chi_{C_6H_{14}} = 1 - \chi_{C_6H_{12}}$ $\qquad \chi_{C_6H_{14}} = 1 - 0.368 = \mathbf{0.632}$

(b) $P_{C_6H_{12}} = \chi_{C_6H_{12}} P^{o}_{C_6H_{12}}$ $\qquad P_{C_6H_{12}} = (0.368)(150 \text{ torr}) = \mathbf{55.2 \text{ torr}}$

$P_{C_6H_{14}} = \chi_{C_6H_{14}} P^{o}_{C_6H_{14}}$ $\qquad P_{C_6H_{14}} = (0.632)(313 \text{ torr}) = \mathbf{198 \text{ torr}}$

(c) $\chi_{C_6H_{12}(g)} = \frac{55.2 \text{ torr}}{55.2 \text{ torr} + 198 \text{ torr}} = \mathbf{0.218}$ $\qquad \chi_{C_6H_{14}(g)} = 1 - \chi_{C_6H_{12}(g)}$ $\qquad \chi_{C_6H_{14}(g)} = 1 - 0.218 = \mathbf{0.782}$

12.85 *See Section 12.6 and Example 12.16.*

(a) $P_{CH_3CO_2H} = \chi_{CH_3CO_2H} P^{o}_{CH_3CO_2H}$, $\qquad P_{C_2H_4Br_2} = \chi_{C_2H_4Br_2} P^{o}_{C_2H_4Br_2}$, $\qquad P_T = P_{CH_3CO_2H} + P_{C_2H_4Br_2}$

Arbitrarily assuming a 100 g mixture of CH_3CO_2H and $C_2H_4Br_2$,

$? \text{ mol } CH_3CO_2H = 100 \text{ g soln} \times \frac{25 \text{ g } CH_3CO_2H}{100 \text{ g soln}} \times \frac{1 \text{ mol } CH_3CO_2H}{60.0 \text{ g } CH_3CO_2H} = 0.42 \text{ mol } CH_3CO_2H$

$? \text{ mol } C_2H_4Br_2 = 100 \text{ g soln} \times \frac{75 \text{ g } C_2H_4Br_2}{100 \text{ g soln}} \times \frac{1 \text{ mol } C_2H_4Br_2}{189.7 \text{ g } C_2H_4Br_2} = 0.40 \text{ mol } C_2H_4Br_2$

? mol total of soln = 0.42 mol CH_3CO_2H + 0.40 $C_2H_4Br_2$ = 0.82 mol total of soln

$\chi_{CH_3CO_2H} = \frac{0.42 \text{ mol } CH_3CO_2H}{0.82 \text{ mol total of soln}} = 0.512$ $\qquad \chi_{C_2H_4Br_2} = 1 - 0.512 = 0.488$

$P_{CH_3CO_2H} = \chi_{CH_3CO_2H} P^{o}_{CH_3CO_2H}$ $\qquad P_{CH_3CO_2H} = (0.512)(471 \text{ torr}) = \mathbf{241 \text{ torr}}$

$P_{C_2H_4Br_2} = \chi_{C_2H_4Br_2} P^{o}_{C_2H_4Br_2}$ $\qquad P_{C_2H_4Br_2} = (0.488)(631 \text{ torr}) = \mathbf{311 \text{ torr}}$

$P_T = P_{CH_3CO_2H} + P_{C_2H_4Br_2}$ $\qquad P_T = 241 \text{ torr} + 311 \text{ torr} = \mathbf{552 \text{ torr}}$

(b) The actual vapor pressure of the azeotropic mixture at 103.7°C is 760 torr, the prevailing atmospheric pressure. Since the actual vapor pressure is greater than that calculated (760 torr vs. 552 torr), the deviation from Raoult's Law is positive.

(c) Positive deviation implies the intermolecular forces between CH_3CO_2H and $C_2H_4Br_2$ are weaker than the average of those in the pure substances.

12.86 *See Section 12.6 and Example 12.16.*

(a) $P_{HCO_2H} = \chi_{HCO_2H} P^{o}_{HCO_2H}$, $\qquad P_{H_2O} = \chi_{H_2O} P^{o}_{H_2O}$, $\qquad P_T = P_{HCO_2H} + P_{H_2O}$

Arbitrarily assuming a 100.0 g mixture of HCO_2H and H_2O,

$? \text{ mol } HCO_2H = 100.0 \text{ g soln} \times \frac{77.5 \text{ g } HCO_2H}{100.0 \text{ g soln}} \times \frac{1 \text{ mol } HCO_2H}{46.0 \text{ g } HCO_2H} = 1.68 \text{ mol } HCO_2H$

$? \text{ mol } H_2O = 100.0 \text{ g soln} \times \frac{22.5 \text{ g } H_2O}{100.0 \text{ g soln}} \times \frac{1 \text{ mol } H_2O}{18.0 \text{ g } H_2O} = 1.25 \text{ mol } H_2O$

? mol total of soln = 1.68 mol HCO_2H + 1.25 mol H_2O = 2.93 mol total of soln

$$\chi_{HCO_2H} = \frac{1.68 \text{ mol } HCO_2H}{2.93 \text{ mol total of soln}} = 0.573 \qquad \chi_{H_2O} = 1 - 0.573 = 0.427$$

$$P_{HCO_2H} = \chi_{HCO_2H} P^{\circ}_{HCO_2H} \qquad P_{HCO_2H} = (0.573)(917 \text{ torr}) = \mathbf{525\ torr}$$

$$P_{H_2O} = \chi_{H_2O} P^{\circ}_{H_2O} \qquad P_{H_2O} = (0.427)(974 \text{ torr}) = \mathbf{416\ torr}$$

$$P_T = P_{HCO_2H} + P_{H_2O} \qquad P_T = 525 \text{ torr} + 416 \text{ torr} = \mathbf{941\ torr}$$

(b) The actual vapor pressure of the azeotropic mixture at 107.1°C is 760 torr, the prevailing atmospheric pressure. Since the actual vapor pressure is less than that calculated (760 torr vs. 941 torr), the deviation from Raoult's Law is negative.

(c) Negative deviation implies the intermolecular forces between HCO_2H and H_2O are stronger than the average of those in the pure substances.

12.87 *See Section 12.1 and Examples 12.4, 12. 5.*

(a) $? \text{ g of exactly 1 L soln} = 1 \text{ L soln} \times \frac{10^3 \text{ mL soln}}{1 \text{ L soln}} \times \frac{0.9877 \text{ g soln}}{1 \text{ mL soln}} = 987.7 \text{ g soln}$

$? \text{ g } C_2H_5OH \text{ in 1 L soln} = 987.7 \text{ g soln} \times \frac{7.23 \text{ g } C_2H_5OH}{100.00 \text{ g soln}} = 71.4 \text{ g } C_2H_5OH$

$? \text{ g } H_2O \text{ in 1 L soln} = 987.7 \text{ g soln} - 71.4 \text{ g } C_2H_5OH = 916.3 \text{ g } H_2O$

$$\text{molality} = \frac{\text{mol } C_2H_5OH}{\text{kg } H_2O} \qquad \text{molality} = \frac{71.4 \text{ g } C_2H_5OH}{0.9163 \text{ kg } H_2O} \times \frac{1 \text{ mol } C_2H_5OH}{46.0 \text{ g } C_2H_5OH} = \mathbf{1.69\ m\ C_2H_5OH}$$

(b) $? \text{ mol } H_2O \text{ in 1 L soln} = 916.3 \text{ g } H_2O \times \frac{1 \text{ mol } H_2O}{18.02 \text{ g } H_2O} = 50.85 \text{ mol } H_2O$

$? \text{ mol total of soln} = 1.55 \text{ mol } C_2H_5OH + 50.85 \text{ mol } H_2O = 52.40 \text{ mol total of soln}$

$$\chi_{C_2H_5OH} = \frac{\text{mol } C_2H_5OH}{\text{mol total of soln}} \qquad \chi_{C_2H_5OH} = \frac{1.55 \text{ mol } C_2H_5OH}{52.40 \text{ mol total of soln}} = \mathbf{0.0296}$$

(c) $\text{molarity} = \frac{\text{mol } C_2H_5OH}{\text{L soln}} \qquad M_{C_2H_5OH} = \frac{71.4 \text{ g } C_2H_5OH}{1 \text{ L soln}} \times \frac{1 \text{ mol } C_2H_5OH}{46.0 \text{ g } C_2H_5OH} = \mathbf{1.55\ \mathit{M}\ C_2H_5OH}$

(d) $? \text{ g } C_2H_5OH \text{ per 100 mL wine} = 100 \text{ mL wine} \times 0.9877 \frac{\text{g wine}}{\text{mL wine}} \times \frac{7.23 \text{ g } C_2H_5OH}{100.00 \text{ g wine}} = \mathbf{7.14\ g\ C_2H_5OH}$

12.88 *See Sections 12.1, 12.4 , 12.5 and Examples 12.10, 12.15.*

A 0.0520 m aqueous Na_2CO_3 solution contains 0.0520 mol Na_2CO_3 per kilogram of water, 5.51 g Na_2CO_3 per kilogram of water, and can be made by adding 5.51 g $NaCO_3$ to 1.00 kg water.

Assuming an ideal *i* value of 3 for Na_2CO_3 gives

$$\Delta T_f = imk_f = 3 \times 0.0520 \text{ molal} \times 1.86 \frac{^\circ C}{\text{molal}} = 0.290^\circ C \text{ and}$$

$$T_f = 0.000^\circ C - 0.290^\circ C = \mathbf{-0.290^\circ C}$$

12.89 *See Section 12.2 and Example 12.6.*

(a) KI is ionic and therefore more soluble in highly polar water than slightly polar methylenechloride, CH_2Cl_2.

(b) $C_6H_5CH_3$ is slightly polar and therefore more soluble in nonpolar benzene, C_6H_6, than highly polar water.

(c) $C_2H_4(OH)_2$ is polar with O-H bonds for hydrogen bonding and therefore more soluble in ethanol, C_2H_5OH, than virtually nonpolar hexane, C_6H_{14}.

12.90 ***See Section 12.3.***

The enthalpy of solution is positive indicating the dissolving process for $KClO_4$ in water is endothermic. Hence, the solubility of $KClO_4$ in water will be greater at 92°C than it is at 25°C.

12.91 ***See Section 12.3 and Example 12.7.***

(a) Solving C = kP for k gives $k = \frac{C}{P}$. Substituting the given data yields

$$k = \frac{\frac{0.240 \text{ g } CO_2}{100 \text{ mL}}}{1.00 \text{ atm}} = 2.40 \times 10^{-3} \text{ g } CO_2 \cdot mL^{-1} \cdot atm^{-1}$$

Using this value of k to calculate C when P = 4.00 atm yields

$$C = \left(2.40 \times 10^{-3} \text{ g } CO_2 \cdot mL^{-1} \cdot atm^{-1}\right)(4.00 \text{ atm}) = 9.60 \times 10^{-3} \text{ g } CO_2 \cdot mL^{-1}$$

$$\text{Hence, ? g } CO_2 \text{ in 12 oz can} = 12 \text{ oz} \times \frac{28.35 \text{ mL}}{1 \text{ oz}} \times 9.60 \times 10^{-3} \frac{\text{g } CO_2}{\text{mL}} = \mathbf{3.27 \text{ g } CO_2}$$

(b) Neglecting the equilibrium pressure of CO_2 in the atmosphere as the pressure of CO_2 remaining, we can assume that all of the CO_2 is expelled. This gives

$$\text{? L } CO_2 \text{ expelled when measured at STP} = 3.27 \text{ g } CO_2 \times \frac{1 \text{ mol } CO_2}{44.0 \text{ g } CO_2} \times \frac{22.4 \text{ L } CO_2}{1 \text{ mol } CO_2} = \mathbf{1.66 \text{ L } CO_2}$$

12.92 ***See Section 12.4 and Example 12.9.***

$$\Delta P = \chi_{solute} P^{o}_{solv} \quad \text{and} \quad P_{solv} = \chi_{solv} P^{o}_{solv}$$

$$\text{? mol } Fe(C_5H_5)_2 = 2.00 \text{ g } Fe(C_5H_5)_2 \times \frac{1 \text{ mol } Fe(C_5H_5)_2}{185.8 \text{ g } Fe(C_5H_5)_2} = 0.0108 \text{ mol } Fe(C_5H_5)_2$$

$$\text{? mol } C_2H_3Cl_3 = 25.0 \text{ g } C_2H_3Cl_3 \times \frac{1 \text{ mol } C_2H_3Cl_3}{133.4 \text{ g } C_2H_3Cl_3} = 0.187 \text{ mol } C_2H_3Cl_3$$

$$\text{? mol total of soln} = 0.0108 \text{ mol } Fe(C_5H_5)_2 + 0.187 \text{ mol } C_2H_3Cl_3 = 0.198 \text{ mol total of soln}$$

$$\chi_{solv} = \chi_{C_2H_3Cl_3} = \frac{0.187 \text{ mol } C_2H_3Cl_3}{0.198 \text{ mol total of soln}} = 0.944$$

$$P_{solv} = \chi_{C_2H_3Cl_3} P^{o}_{C_2H_3Cl_3} \quad P_{solv} = (0.944)(100 \text{ torr}) = \mathbf{94.4 \text{ torr}}$$

12.93 ***See Sections 2.4, 2.5, 12.4 and Examples 2.7, 12.13.***

(a) Solving $\Delta T_f = mk_f$ for m gives $m = \frac{\Delta T_f}{k_f}$ $\quad m = \frac{9.80°\text{ C} - 7.97°\text{ C}}{11.8 \text{ °C / molal}} = 0.155 \text{ m}$

$$\text{? mol solute} = 12.2 \text{ g solvent} \times \frac{1 \text{ kg solvent}}{10^3 \text{ g solvent}} \times \frac{0.155 \text{ mol solute}}{1 \text{ kg solvent}} = 1.89 \times 10^{-3} \text{ mol solute}$$

$$\text{molar mass} = \frac{\text{g solute}}{\text{mol solute}} \qquad \text{molar mass} = \frac{0.315 \text{ g}}{1.89 \times 10^{-3} \text{ mol}} = \mathbf{172 \text{ g / mol}}$$

$$\text{(b) ? Fe atoms per molecule} = \frac{172\ \mu \text{ compound}}{\text{molecule}} \times \frac{32.5\ \mu \text{ Fe}}{100.0\ \mu \text{ compound}} \times \frac{1 \text{ Fe atom}}{55.85\ \mu} = \mathbf{1 \text{ Fe atom/ molecule}}$$

12.94 *See Sections 12.4, 12.5 and the Insight into Chemistry, entitled Reverse Osmosis.*

The direction of net transport of water molecules depends on whether the pressure that is exerted on the membrane is greater on the pure water side or on the NaCl solution side. The pressure that is exerted on the pure water side will be equal to the pressure that is exerted on the NaCl side when there is no outside force acting on the system and the system is allowed to reach equilibrium. This pressure is given by Π. Hence, Π must be calculated to determine the minimum pressure that must be applied to the NaCl side to cause reverse osmosis to occur.

Known Quantities:

$\text{molarity} = \frac{\text{mol NaCl}}{\text{L soln}}$ $\quad\quad \text{molarity} = \frac{0.500 \text{ g NaCl}}{\text{L soln}} \times \frac{1 \text{ mol NaCl}}{58.44 \text{ g NaCl}} = 0.00856 \text{ M NaCl}$

Assuming ideal behavior, total molarity of NaCl solute particles $= 2 \times 0.00856\ M = 0.0172\ M$

T = 298 K

Substituting in $\Pi = MRT$ gives $\Pi = \left(0.0172 \frac{\text{mol}}{\text{L}}\right)\left(0.0821 \frac{\text{L} \cdot \text{atm}}{\text{mol} \cdot \text{K}}\right)(298 \text{ K}) = 0.421 \text{ atm}$ or **320 torr**

12.95 *See Section 12.6 and Figures 12.21, 12.22, 12.23.*

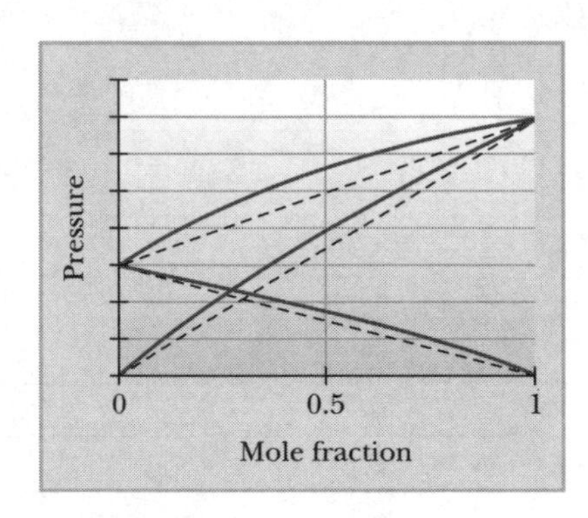

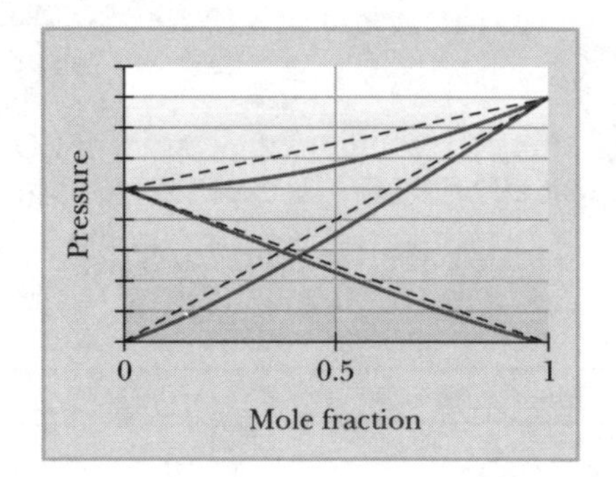

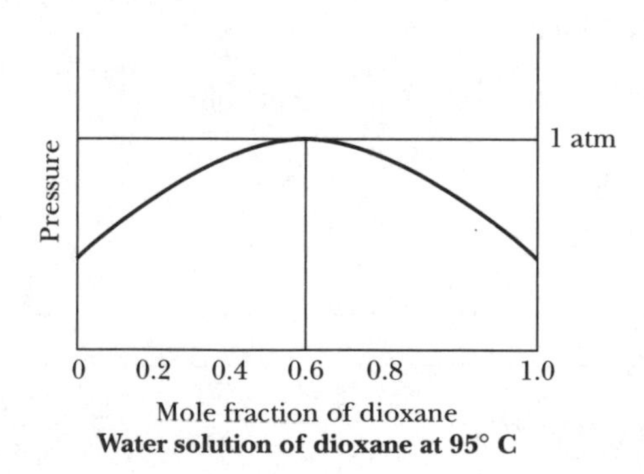

Water solution of dioxane at 95° C

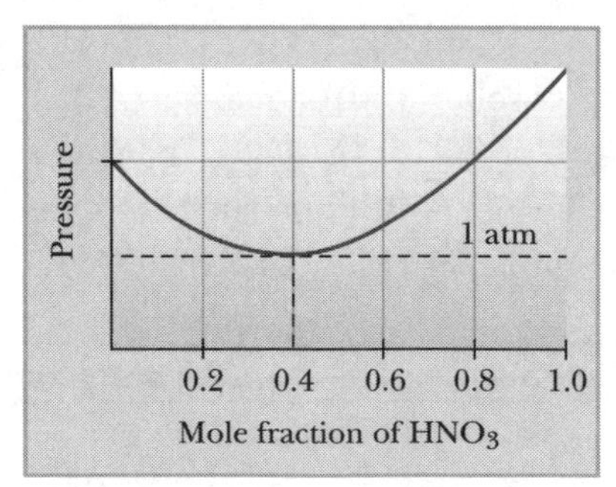

Water solution of nitric acid at 120.5°C

12.96 *See Sections 3.1, 12.2, 12.4, 12.5.*

(a) Solving $\Delta T_f = mk_f$ for m gives $m = \frac{\Delta T_f}{k_f}$ $\quad\quad m = \frac{5.51°\text{C} - 4.04°\text{C}}{4.90\ °\text{C / molal}} = \mathbf{0.300\ m}$

(b) Solving $\Delta T_f = mk_f$ for m gives $m = \frac{\Delta T_f}{k_f}$ $\quad\quad m = \frac{0.00°\text{C} - (-1.07)°\text{C}}{1.86\ °\text{C / molal}} = \mathbf{0.575\ m}$

(c) The total molality of solute particles from HCl in water is almost twice that in benzene. HCl ionizes in polar water to produce primarily H^+ (H_3O^+) and Cl^- ions, whereas HCl does not ionize appreciably in nonpolar benzene.

12.97 See Sections 6.4, 12.4 and Examples 6.6, 12.13.

(a) Known Quantities: m = 0.262 g $P = 745 \text{ torr} \times \frac{1 \text{ atm}}{760 \text{ torr}} = 0.980 \text{ atm}$

$V = 51.0 \text{ mL} \times \frac{1 \text{ L}}{10^3 \text{ mL}} = 0.0510 \text{ L}$ T = 25 + 273 = 298 K

Solving PV = nRT for n gives $n = \frac{PV}{RT}$ $n = \frac{(0.980 \text{ atm})(0.0510 \text{ L})}{\left(0.0821 \frac{\text{L}\cdot\text{atm}}{\text{mol}\cdot\text{K}}\right)(298 \text{ K})} = 2.04 \times 10^{-3} \text{ mol}$

and using $\mathcal{M} = \frac{m}{n}$ yields $\mathcal{M} = \frac{0.262 \text{ g}}{2.04 \times 10^{-3} \text{ mol}} = \mathbf{128 \text{ g/mol}}$

(b) Solving $\Delta T_f = mk_f$ for m gives $m = \frac{\Delta T_f}{k_f}$ $m = \frac{0°\text{C} - (-0.61°\text{C})}{1.86° \text{ C/molal}} = 0.33 \text{ m}$

$? \text{ mol solute} = 12.0 \text{ g solvent} \times \frac{1 \text{ kg solvent}}{10^3 \text{ g solvent}} \times \frac{0.33 \text{ mol solute}}{1 \text{ kg solvent}} = 4.0 \times 10^{-3} \text{ mol solute}$

$\text{molar mass} = \frac{\text{g solute}}{\text{mol solute}}$ $\mathcal{M} = \frac{0.262 \text{ g}}{4.0 \times 10^{-3} \text{ mol}} = \mathbf{66 \text{ g/mol}}$

(c) The key to understanding this difference is to note that the number of moles of solute particles that are obtained in solution is almost twice the number of moles of gas. The gaseous substance must therefore act as an electrolyte in water and have an *i* value of approximately 2. This could occur if the gaseous compound acts as an acid in water: $HX(g) + H_2O \rightarrow H_3O^+ + X^-$. (See Section 3.1).

12.98 See Section 12.4 and Example 12.14.

Known Quantities:

$\Pi = 5.5 \text{ torr} \times \frac{1 \text{ atm}}{760 \text{ torr}} = 0.0072 \text{ atm}$ V = 0.010 L T = 298 K

Solving $\Pi = \frac{nRT}{V}$ for n gives $n = \frac{\Pi V}{RT}$ $n = \frac{(0.0072 \text{ atm})(0.010 \text{ L})}{\left(0.0821 \frac{\text{L}\cdot\text{atm}}{\text{mol}\cdot\text{K}}\right)(298 \text{ K})} = 2.94 \times 10^{-6} \text{ mol}$

$\text{molar mass} = \frac{\text{g solute}}{\text{mol solute}}$ $\text{molar mass} = \frac{0.200 \text{ g hemoglobin}}{2.94 \times 10^{-6} \text{ mol hemoglobin}} = 6.80 \times 10^4 \text{ g/mol}$

$? \text{ mol Fe} = \frac{6.80 \times 10^4 \text{ g hemoglobin}}{\text{mol hemoglobin}} \times \frac{0.33 \text{ g Fe}}{100.00 \text{ g hemoglobin}} \times \frac{1 \text{ mol Fe}}{55.85 \text{ g Fe}} = \mathbf{4 \text{ mol Fe atoms/ mol hemoglobin}}$

12.99 See Sections 12.4, 12.5 and Example 12.4.

(a) Solving $\Delta T_f = mk_f$ for m gives $m = \frac{\Delta T_f}{k_f}$

For C_6H_6: $m = \frac{0.205°\text{C}}{4.90°\text{C/m}} = \mathbf{0.0418 \text{ m}}$ For H_2O: $m = \frac{0.159°\text{C}}{1.86°\text{C/m}} = \mathbf{0.0855 \text{ m}}$

(b) Assuming 100.0 g soln for each solution and recognizing that 0.50 g is solute and 99.50 g is solvent gives

$? \text{ mol } CH_3COOH = 0.50 \text{ g } CH_3COOH \times \frac{1 \text{ mol } CH_3COOH}{60.0 \text{ g } CH_3COOH} = 8.3 \times 10^{-3} \text{ mol } CH_3COOH$

$$\text{molality} = \frac{\text{mol } CH_3COOH}{\text{kg solvent}} \qquad \text{molality} = \frac{8.3 \times 10^{-3} \text{ mol } CH_3COOH}{0.0995 \text{ kg solvent}} = 8.3 \times 10^{-2} \text{ m}$$

The experimental molality that is obtained with benzene is approximately one-half that calculated and thus corresponds to an i value of one-half:

$$im = 0.0418 \text{ molal and m} = 8.3 \times 10^{-2} \text{ molal yields } \boldsymbol{i \cong \tfrac{1}{2}}$$

The experimental molality that is obtained with water is approximately equal to that calculated and thus corresponds to *an* i value of one:

$$im = 0.0855 \text{ molal and m} = 8.3 \times 10^{-2} \text{ molal yields } \boldsymbol{i \cong 1}$$

(c) CH_3COOH forms a hydrogen-bonded dimer in nonpolar benzene (C_6H_6) causing it to give one-half as many solute particles as might be expected.

Hence, the effective molality of CH_3COOH is approximately one-half that in water.

```
          O''''H—O
         //        \
   H3C—C            C—CH3
         \        //
          O—H''''O
```

Note: The observation that the experimental molality for CH_3COOH in water is slightly higher than calculated can be attributed to the observation that CH_3COOH remains monomeric and acts as a weak acid in water.

12.100 ***See Sections 4.1, 12.1 and Examples 4.12, 12.1.***

(a) Balanced: $NH_4Br(aq) + AgNO_3(aq) \rightarrow NH_4NO_3(aq) + AgBr(s)$

Strategy: L $AgNO_3$ soln $\rightarrow$ mol $AgNO_3$ $\rightarrow$ mol NH_4Br $\rightarrow$ M NH_4Br

$$? \text{ mol } NH_4Br = 0.02341 \text{ L } AgNO_3 \text{ soln} \times \frac{1.200 \text{ mol } AgNO_3}{1 \text{ L } AgNO_3 \text{ soln}} \times \frac{1 \text{ mol } NH_4Br}{1 \text{ mol } AgNO_3} = 0.02809 \text{ mol } NH_4Br$$

$$\text{molarity } NH_4Br \text{ soln} = \frac{\text{mol } NH_4Br}{\text{L } NH_4Br \text{ soln}} \qquad \text{molarity } NH_4Br \text{ soln} = \frac{0.02809 \text{ mol } NH_4Br}{0.01000 \text{ L } NH_4Br \text{ soln}} = \mathbf{2.809}\ \boldsymbol{M}\ \mathbf{NH_4Br}$$

$$\text{(b) } ? \text{ g } NH_4Br = 1.000 \text{ L soln} \times \frac{2.809 \text{ mol } NH_4Br}{1 \text{ L soln}} \times \frac{97.9 \text{ g } NH_4Br}{1 \text{ mol } NH_4Br} = \mathbf{275 \text{ g } NH_4Br / L \text{ soln}}$$

$$\text{(c) } ? \text{ g } NH_4Br \text{ soln} = \frac{275 \text{ g } NH_4Br}{1 \text{ L soln}} \times \frac{100.00 \text{ g } NH_4Br \text{ soln}}{24.00 \text{ g } NH_4Br} = \mathbf{1{,}146 \text{ g } NH_4Br \text{ soln/ Lsoln}}$$

$$\text{(d) } ? \text{ g } NH_4Br \text{ soln / mL soln} = \frac{1{,}146 \text{ g soln}}{1 \text{ L soln}} \times \frac{1 \text{ L soln}}{10^3 \text{ mL soln}} = \mathbf{1.146 \text{ g / mL}}$$

12.101 ***See Sections 3.3, 12.4, 12.5 and Examples 3.11, 12.13, 12.15.***

(a) Assume the sample has a mass of 100.0 g and therefore contains 33.0 g Na, 17.7 g C and 47.0 g O.

$$? \text{ mol Na} = 33.8 \text{ g Na} \times \frac{1 \text{ mol Na}}{23.0 \text{ g Na}} = 1.47 \text{ mol Na} \qquad \text{relative mol Na} = \frac{1.47 \text{ mol Na}}{1.47} = 1.00 \text{ mol Na}$$

$$? \text{ mol C} = 17.7 \text{ g C} \times \frac{1 \text{ mol C}}{12.0 \text{ g C}} = 1.48 \text{ mol C} \qquad \text{relative mol C} = \frac{1.48 \text{ mol Na}}{1.47} = 1.01 \text{ mol C}$$

$$? \text{ mol O} = 47.0 \text{ g O} \times \frac{1 \text{ mol O}}{16.0 \text{ g O}} = 2.94 \text{ mol O} \qquad \text{relative mol O} = \frac{2.94 \text{ mol O}}{1.47} = 2.00 \text{ mol O}$$

The empirical formula is $\mathbf{NaCO_2}$. The empirical formula molar mass is 67.0 g/mol.

(b) Solution One: The crystalline compound is composed of a metal and two nonmetals and is most likely an ionic compound that dissociates into ions in aqueous solution to give the solution that conducts an electric current. The empirical formula can be used to write the equation for this dissociation as $Na_nC_nO_{2n}(s) \rightarrow nNa^+(aq) + C_nO_{2n}^{n-}(aq)$. The total moles of solute particles in solution can then be determined using the freezing point data.

Solving $\Delta T_f = imk_f$ for im gives $im = \frac{\Delta T_f}{k_f}$. $\quad im = \frac{0.00^\circ C - (-1.01)^\circ C}{1.86^\circ C / \text{molal}} = 0.543 \text{ molal}$

Hence, the number of moles of ions from the 0.500 g of sample is

$$? \text{ mol ions} = 20.0 \text{ g solvent} \times \frac{1 \text{ kg solvent}}{10^3 \text{ g solvent}} \times \frac{0.543 \text{ mol ions}}{\text{kg solvent}} = 1.09 \times 10^{-2} \text{ mol ions}$$

the number of moles of empirical formula units from the 0.500 g of sample is

$$? \text{ mol formula units} = 0.500 \text{ g } NaCO_2 \times \frac{1 \text{ mol } NaCO_2}{67.0 \text{ g } NaCO_2} = 7.46 \times 10^{-3} \text{ mol } NaCO_2$$

and the number of moles of ions per mole of formula units is

$$? \frac{\text{mol ions}}{\text{mol formula units}} = \frac{1.09 \times 10^{-2} \text{ mol ions}}{7.46\ 10^{-3} \text{ mol } NaCO_2} = 1.46 \text{ mol ions/ mol } NaCO_2 .$$

Since the number of ions per mole of compound must be a whole number, we can conclude that there must be 3 mol ions/ 2mol $NaCO_2$ or 3 mol ions/ 1 mol $Na_2C_2O_4$ and therefore conclude that the compound is **$Na_2C_2O_4$**.

The equation for the dissociation of $Na_2C_2O_4$ (sodium oxalate) in water is:

$$Na_2C_2O_4(aq) \rightarrow 2Na^+(aq) + C_2O_4^{2-}(aq)$$

and the difference between the experimental value of 2.92 mol ions/ mol compound and the theoretical value of 3 mol ions/ mol compound is due to nonideal behavior, such as ion-pairing.

Solution Two: The crystalline compound is composed of a metal and two nonmetals and is most likely an ionic compound that dissociates into ions in aqueous solution to give the solution that conducts an electric current. The molar mass of the ionic compound must be a whole-number multiple of the molar mass of the empirical formula. We can determine the value of the multiple by assuming ideal i values of 2, 3, etc. and comparing the calculated molar mass values to the mass of the empirical formula.

Solving $\Delta T_f = imk_f$ for m gives $m = \frac{\Delta T_f}{ik_f}$.

Assuming i=2, $m = \frac{\Delta T_f}{ik_f} = \frac{0.00^\circ C - (-1.01)^\circ C}{2 \times 1.86^\circ C / \text{molal}} = 0.272 \text{ m}$

$$? \text{ mol solute} = 20.0 \text{ g solvent} \times \frac{1 \text{ kg solvent}}{10^3 \text{ g solvent}} \times \frac{0.272 \text{ mol solute}}{1 \text{ kg solvent}} = 5.44 \times 10^{-3} \text{ mol solute}$$

$\text{molar mass} = \frac{\text{g solute}}{\text{mol solute}} \qquad \text{molar mass} = \frac{0.500 \text{ g}}{5.44 \times 10^{-3} \text{ mol}} = \mathbf{91.9\ g / mol}$

However, 91.9 g/mol is not equal to a whole-number multiple of the molar mass of the empirical formula (67.0 g/mol). Hence, i cannot be two.

Assuming i=3, $m = \frac{\Delta T_f}{ik_f} = \frac{0.00^\circ C - (-1.01)^\circ C}{3 \times 1.86^\circ C / \text{molal}} = 0.181 \text{ m}$

$$? \text{ mol solute} = 20.0 \text{ g solvent} \times \frac{1 \text{ kg solvent}}{10^3 \text{ g solvent}} \times \frac{0.181 \text{ mol solute}}{1 \text{ kg solvent}} = 3.62 \times 10^{-3} \text{ mol solute}$$

$\text{molar mass} = \frac{\text{g solute}}{\text{mol solute}} \qquad \text{molar mass} = \frac{0.500 \text{ g}}{3.62 \times 10^{-3} \text{ mol}} = \mathbf{138\ g / mol}$

Within experimental error, 138 g/mol is equal to two times the molar mass of the empirical formula (2 x 67.0 g/mol = 134 g/mol). Hence, we can conclude that the compound is $Na_2C_2O_4$ with an ideal i value of 3.

The equation for the dissociation of $Na_2C_2O_4$ (sodium oxalate) in water is:

$$Na_2C_2O_4(aq) \rightarrow 2Na^+(aq) + C_2O_4^{2-}(aq)$$

Chapter 13: Kinetics

13.1 ***See Chapter Introduction and Sections 13.2, 13.4, 13.5.***

The factors which affect rate of reaction are 1) nature of the reactants (i.e. what is reacting), 2) concentrations of the reactants (and in some cases, products), 3) temperature of the reactants and 4) presence or absence of a catalyst. If the reaction involves a solid, the surface area/particle size can also affect rate of reaction.

13.2 ***See Section 13.2 and Example 13.2.***

The rate of the reaction $CH_3Br + OH^- \rightarrow CH_3OH + Br^-$ can be determined by following the change in a property that is related to the molar concentration of a reactant or product with time. With the method of initial rates, only the rate of reaction within the first few percent of reaction is determined.

Several experiments would be conducted in which the molar concentration of CH_3Br is varied and the molar concentration of OH^- is held constant. These experiments would be used to determine the order "x" with respect to CH_3Br in the rate law: rate = $k[CH_3Br]^x[OH^-]^y$. This would be accomplished by comparing the relative concentrations of CH_3Br to the relative rates for these reactions. Then several experiments would be conducted in which the molar concentration of CH_3Br is held constant and the molar concentration of OH^- is varied. These experiments would be used to determine the order "y" with respect to OH^- by comparing the relative concentrations of OH^- to the relative rates for these reactions. All of these experiments would need to be conducted at the same temperature.

Lastly, the value of k would be determined by dividing the initial rate for each reaction by the concentrations of CH_3Br and OH^- for each reaction raised to the appropriate powers, raised to the appropriate orders. An average value of k would be calculated using the value of k for each reaction.

13.3 ***See Section 13.3.***

A differential form of the rate law relates changes in concentration and time to the concentrations of the reactants, as in rate = $-\Delta[A]/\Delta t = k[A]$ for a first order reaction. An integrated form of the rate law relates actual concentrations to time, rather than changes in concentration to changes in time, as in $\ln[A] = -kt + \ln[A]_o$ for a first order reaction.

13.4 ***See Section 13.3.***

Half-lives for first order reactions do not change with changes in the starting concentrations of the reactants, whereas half-lives for the other order reactions do change with changes in the starting concentrations. This means the half-lives for other orders depend on the starting concentrations and even vary from one half-live to another for a given reaction because the starting concentration for each successive half-live is one-half of what it was for the just completed half-live.

13.5 ***See Section 13.3.***

At $t = t_{½}$, $[A] = ½[A]_o$. Hence, substituting in $[A]^{½} = [A]_o^{½} - ½kt$, yields

$(½[A]_o)^{½} = [A]_o^{½} - ½kt_{½}$, $(½)^{½}[A]_o^{½} = [A]_o^{½} - ½kt_{½}$, $0.707[A]_o^{½} = [A]_o^{½} - ½kt_{½}$ and $t_{½} = \dfrac{0.586[A]_o^{½}}{k}$.

13.6 *See Section 13.3.*

At $t = t_{1/2}$, $[A] = \frac{1}{2}[A]_o$. Hence, substituting in $\frac{1}{[A]^2} - \frac{1}{[A]_o^2} = 2kt$ yields

$$\frac{1}{(\frac{1}{2}[A]_o)^2} - \frac{1}{[A]_o^2} = 2kt_{1/2},\ \frac{4}{[A]_o^2} - \frac{1}{[A]_o^2} = 2kt_{1/2} \text{ and } t_{1/2} = \frac{3}{2k[A]_o^2} = \frac{1.5}{k[A]_o^2}.$$

13.7 *See Section 13.4.*

The activation energy, E_a, is the minimum collision energy required for a reaction to occur. The higher the activation energy, the slower the reaction at a given temperature because there are fewer collisions with energies that exceed the activation energy.

13.8 *See Section 13.4.*

Only collisions with sufficient energy to rearrange bonds can result in the formation of products.

13.9 *See Section 13.4.*

Rate of reaction is generally much lower than the rate of collisions. Only collisions with energies greater than the minimum energy, the activation energy E_a, and proper geometric orientation can lead to products.

13.10 *See Section 13.4.*

Doubling the concentration of a reactant doubles the number of collisions per time, doubles the collision frequency.

13.11 *See Sections 13.4, 13.5, and Solution for Excercise 13.13.*

Increasing the temperature increases the rate of the forward reaction of an endothermic reaction more than the rate of the reverse reaction because the forward reaction of an endothermic process has the higher activation energy.

The activation energies for the forward and reverse reactions of a catalyzed endothermic reaction are much smaller than those for the corresponding uncatalyzed reaction. Hence, the rates of the forward and reverse reactions of a catalyzed endothermic reaction increase less with increasing temperature than those of the uncatalyzed reaction.

13.12 *See Sections 13.4, 13.5, Figure 13.13, and Solution for Excercise 13.14.*

Increasing the temperature increases the rate of the reverse reaction of an exothermic reaction more than the rate of the forward reaction because the reverse reaction of an exothermic process has the higher activation energy.

The activation energies for the forward and reverse reactions of a catalyzed exothermic reaction are much smaller than those for the corresponding uncatalyzed reaction. Hence, the rates of the forward and reverse reactions of a catalyzed exothermic reaction increase less with increasing temperature than those of the uncatalyzed reaction.

13.13 *See Section 13.5 and Figure 13.13.*

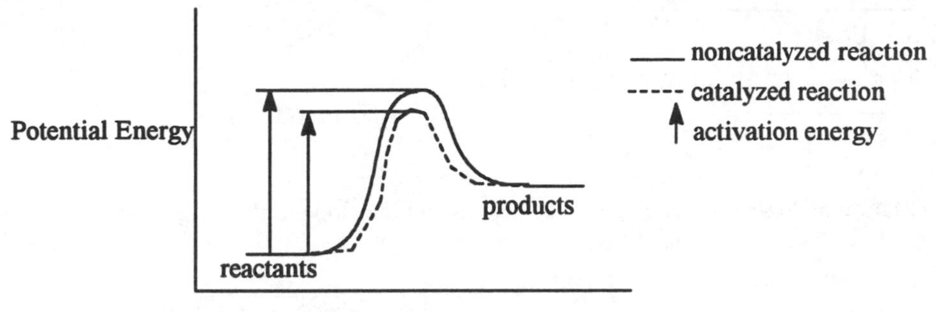

13.14 *See Section 13.5 and Figure 13.13.*

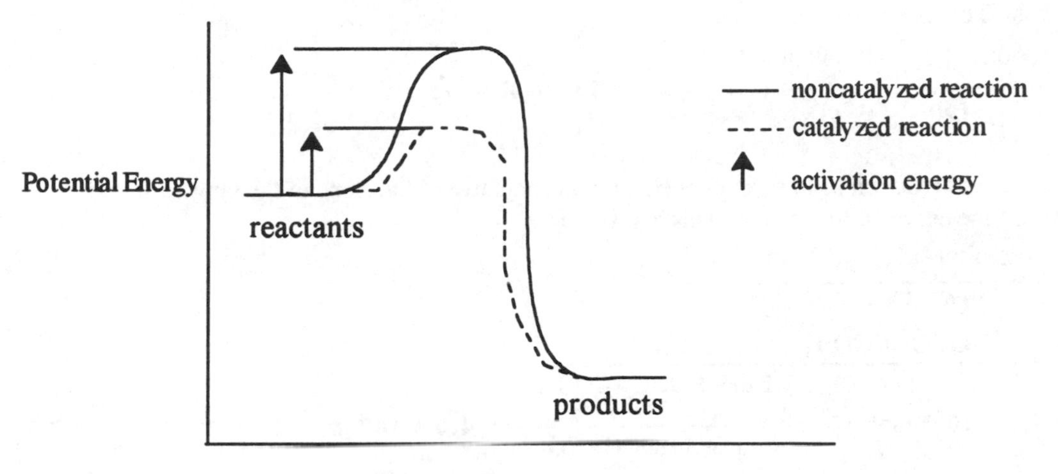

13.15 *See Section 13.6.*

An elementary step is an equation that describes an actual molecular-level event.

13.16 *See Section 13.6.*

An elementary step is an equation that describes an actual molecular-level event. Hence, the orders of reaction for an elementary step are equal to the stoichiometric coefficients of an elementary step.

An overall equation is equal to the sum of the elementary steps of the reaction. Some of these elementary steps can occur after the rate determining step of the reaction. Such steps affect the overall stoichiometry of the reaction but do not affect the rate of reaction. Hence, there is no necessary relation between the orders of reaction in the rate law for an overall reaction and the stoichiometric coefficients for the overall equation.

13.17 *See Section 13.1 and Example 13.1.*

(a) For chemical reactions, rate is usually expressed in terms of change in concentration per unit time. For the reaction HOOC-COOH(g) $\rightarrow$ HCOOH(g) + CO_2(g), the rate of reaction can be expressed as

$$\text{rate} = \frac{-\Delta[\text{HOOC-COOH}]}{\Delta t} = \frac{\Delta[\text{HCOOH}]}{\Delta t} = \frac{\Delta[\text{CO}_2]}{\Delta t}$$

(b) The average rate of reaction between 10 and 30 seconds is given by

$$\text{ave rate} = \frac{-([\text{HOOC-COOH}]_{30} - [\text{HOOC-COOH}]_{10})}{(30\text{ s} - 10\text{ s})}$$

$$\text{ave rate} = \frac{-(0.016\text{ mol}\cdot\text{L}^{-1} - 0.034\text{ mol}\cdot\text{L}^{-1})}{(30\text{ s} - 10\text{ s})} = \frac{-(-0.018\text{ mol}\cdot\text{L}^{-1})}{20\text{ s}} = \mathbf{9.0 \times 10^{-4}\ mol\cdot L^{-1}\cdot s^{-1}}$$

(c) The instantaneous rate of reaction at 20 seconds is equal to the negative of the slope of the tangent to the [HOOC-COOH] versus time curve at 20 seconds. This is given by

$$\text{rate} = -\text{slope} = \frac{-(0.000\text{ mol}\cdot\text{L}^{-1} - 0.040\text{ mol}\cdot\text{L}^{-1})}{(50\text{ s} - 0.\text{ s})} = \frac{-(-0.040\text{ mol}\cdot\text{L}^{-1})}{50\text{ s}} = \mathbf{8.0 \times 10^{-4}\ mol\cdot L^{-1}\cdot s^{-1}}$$

(d) The initial rate of reaction is equal to the negative of the slope of the tangent to the [HOOC-COOH] versus time curve at 0 seconds. This is given by

$$\text{rate} = -\text{slope} = \frac{-(0.000\text{ mol}\cdot\text{L}^{-1} - 0.050\text{ mol}\cdot\text{L}^{-1})}{(20\text{ s} - 0.\text{ s})} = \mathbf{2.5 \times 10^{-3}\ mol\cdot L^{-1}\cdot s^{-1}}$$

(e) The instantaneous rate of reaction at 40 seconds is equal to the negative of the slope of the tangent to the [HOOC-COOH] versus time curve at 40 seconds. This is given by

$$\text{rate} = -\text{slope} = \frac{-(0.000\text{ mol}\cdot\text{L}^{-1} - 0.027\text{ mol}\cdot\text{L}^{-1})}{(67.5\text{ s} - 0.\text{ s})} = 4.0 \times 10^{-4}\text{ mol}\cdot\text{L}^{-1}\cdot\text{s}^{-1}$$

Hence, $$\frac{\Delta[CO_2]}{\Delta t} = \frac{-\Delta[\text{HOOC-COOH}]}{\Delta t} \times \frac{1\text{ mol } CO_2}{1\text{ mol HOOC-COOH}}$$

$$= 4.0 \times 10^{-4}\text{ mol}\cdot\text{L}^{-1}\cdot\text{s}^{-1} \times \frac{1\text{ mol } CO_2}{1\text{ mol HOOC-COOH}} = \mathbf{4.0 \times 10^{-4}\ mol\cdot L^{-1}\cdot s^{-1}}$$

13.18 ***See Section 13.1 and Example 13.1.***

(a) For chemical reactions, rate is usually expressed in terms of change in concentration per unit time. For the reaction $C_4H_8(g) \rightarrow 2C_2H_4(g)$, the rate of reaction can be expressed as

$$\text{rate} = \frac{-\Delta[C_4H_8]}{\Delta t} = \frac{\Delta[C_2H_4]}{2\Delta t}$$

(b) The average rate of reaction between 10 and 30 seconds is given by

$$\text{ave rate} = \frac{-([C_4H_8]_{30} - [C_4H_8]_{10})}{(30\text{ s} - 10\text{ s})}$$

$$\text{ave rate} = \frac{-(0.085\text{mol}\cdot\text{L}^{-1} - 0.170\text{ mol}\cdot\text{L}^{-1})}{(30\text{ s} - 10\text{ s})} = \frac{-(-0.085\text{ mol}\cdot\text{L}^{-1})}{20\text{ s}} = \mathbf{0.0043\ mol\cdot L^{-1}\cdot s^{-1}}$$

(c) The instantaneous rate of reaction at 20 seconds is equal to the negative of the slope of the tangent to the $[C_4H_8]$ versus time curve at 20 seconds. This is given by

$$\text{rate} = -\text{slope} = \frac{-(0.070\text{ mol}\cdot\text{L}^{-1} - 0.16\text{ mol}\cdot\text{L}^{-1})}{(30\text{ s} - 10\text{ s})} = \frac{-(-0.090\text{ mol}\cdot\text{L}^{-1})}{20\text{ s}} = \mathbf{0.0045}\text{ mol}\cdot\text{L}^{-1}\cdot\text{s}^{-1}$$

(d) The initial rate of reaction is equal to the negative of the slope of the tangent to the $[C_4H_8]$ versus time curve at 0 seconds. This is given by

$$\text{rate} = -\text{slope} = \frac{-(0.25\ \text{mol}\cdot\text{L}^{-1} - 0.21\ \text{mol}\cdot\text{L}^{-1})}{(5\ \text{s} - 0\ \text{s})} = \mathbf{0.008\ mol\cdot L^{-1}\cdot s^{-1}}$$

(e) The instantaneous rate of reaction at 40 seconds is equal to the negative of the slope of the tangent to the $[C_4H_8]$ versus time curve at 40 seconds. This is given by

$$\text{rate} = -\text{slope} = \frac{-(0.14\ \text{mol}\cdot\text{L}^{-1} - 0.00\ \text{mol}\cdot\text{L}^{-1})}{(69\ \text{s} - 0\ \text{s})} = 0.0020\ \text{mol}\cdot\text{L}^{-1}\cdot\text{s}^{-1}$$

Hence, $\frac{\Delta[C_2H_4]}{\Delta t} = \frac{-\Delta[C_4H_8]}{\Delta t} \times \frac{2\ \text{mol}\ C_2H_4}{1\ \text{mol}\ C_4H_8} = 0.0020\ \text{mol}\cdot\text{L}^{-1}\cdot\text{s}^{-1} \times \frac{2\ \text{mol}\ C_2H_4}{1\ \text{mol}\ C_4H_8} = \mathbf{0.0040\ mol\cdot L^{-1}\cdot s^{-1}}$

13.19 ***See Section 13.1 and Example 13.1.***

(a) For chemical reactions, rate is usually expressed in terms of change in concentration per unit time. For the reaction $NO(g) + \frac{1}{2}Cl_2(g) \rightarrow NOCl(g)$, rate of reaction can be expressed as

$$\text{rate} = \frac{-\Delta[NO(g)]}{\Delta t} = \frac{-2\Delta[Cl_2]}{\Delta t} = \frac{\Delta[NOCl]}{\Delta t}.$$

(b) The average rate of reaction between 40 and 120 seconds is given by

$$\text{ave rate} = \frac{([NOCl]_{120} - [NOCl]_{40})}{(120\ \text{s} - 40\ \text{s})}$$

$$\text{ave rate} = \frac{(0.575\ \text{mol}\cdot\text{L}^{-1} - 0.450\ \text{mol}\cdot\text{L}^{-1})}{(120\ \text{s} - 40\ \text{s})} = \frac{(0.125\ \text{mol}\cdot\text{L}^{-1})}{80\ \text{s}} = \mathbf{1.6 \times 10^{-3}\ mol\cdot L^{-1}\cdot s^{-1}}$$

(c) The instantaneous rate of reaction at 80 seconds is equal to the slope of the tangent to the [NOCl] versus time curve at 80 seconds. This is given by

$$\text{rate} = \text{slope} = \frac{(0.700\ \text{mol}\cdot\text{L}^{-1} - 0.433\ \text{mol}\cdot\text{L}^{-1})}{(200\ \text{s} - 0.\ \text{s})} = \frac{(0.267\ \text{mol}\cdot\text{L}^{-1})}{200\ \text{s}} = \mathbf{1.3 \times 10^{-3}\ mol\cdot L^{-1}\cdot s^{-1}}$$

(d) The instantaneous rate of reaction at 60 seconds is equal to the slope of the tangent to the [NOCl] versus time curve at 60 seconds. This is given by

$$\text{rate} = \text{slope} = \frac{(0.70\ \text{mol}\cdot\text{L}^{-1} - 0.38\ \text{mol}\cdot\text{L}^{-1})}{(160\ \text{s} - 0.\ \text{s})} = \frac{(0.32\ \text{mol}\cdot\text{L}^{-1})}{160\ \text{s}} = 2.0 \times 10^{-3}\ \text{mol}\cdot\text{L}^{-1}\cdot\text{s}^{-1}$$

Hence, $\frac{-\Delta[Cl_2]}{\Delta t} = \frac{\Delta[NOCl]}{\Delta t} \times \frac{\frac{1}{2}\ \text{mol}\ Cl_2}{1\ \text{mol NOCl}} = 2.0 \times 10^{-3}\ \text{mol}\cdot\text{L}^{-1}\cdot\text{s}^{-1} \times \frac{\frac{1}{2}\ \text{mol}\ Cl_2}{1\ \text{mol NOCl}} = \mathbf{1.0 \times 10^{-3}\ mol\cdot L^{-1}\cdot s^{-1}}$

13.20 ***See Section 13.1 and Example 13.1.***

(a) For chemical reactions, rate is usually expressed in terms of change in concentration per unit time. For the reaction $H_2(g) + I_2(g) \rightarrow 2HI(g)$, rate of reaction can be expressed as

$$\text{rate} = \frac{-\Delta[H_2(g)]}{\Delta t} = \frac{-\Delta[I_2(g)]}{\Delta t} = \frac{\Delta[HI]}{2\Delta t}.$$

(b) The average rate of reaction between 20 and 60 seconds is given by

$$\text{ave rate} = \frac{([\text{HI}]_{60} - [\text{HI}]_{20})}{2(60\text{ s} - 20\text{ s})}$$

$$\text{ave rate} = \frac{(0.60\text{ mol}\cdot\text{L}^{-1} - 0.35\text{ mol}\cdot\text{L}^{-1})}{2(60\text{ s} - 20\text{ s})} = \frac{(0.25\text{ mol}\cdot\text{L}^{-1})}{80\text{ s}} = \mathbf{0.0031}\text{ mol}\cdot\text{L}^{-1}\cdot\text{s}^{-1}$$

(c) The instantaneous rate of reaction at 40 seconds is equal to one-half the slope of the tangent to the [HI] versus time curve at 40 seconds. This is given by

$$\text{rate} = \text{slope} = \frac{(0.70\text{ mol}\cdot\text{L}^{-1} - 0.30\text{ mol}\cdot\text{L}^{-1})}{2(70\text{ s} - 0\text{ s})} = \frac{(0.40\text{ mol}\cdot\text{L}^{-1})}{140\text{ s}} = \mathbf{0.0029}\text{ mol}\cdot\text{L}^{-1}\cdot\text{s}^{-1}$$

(d) The initial rate of reaction is equal to one-half the slope of the tangent to the [HI] versus time curve at 0 seconds. This is given by

$$\text{rate} = \text{slope} = \frac{(0.20\text{ mol}\cdot\text{L}^{-1} - 0.0\text{ mol}\cdot\text{L}^{-1})}{2(10\text{ s} - 0.\text{ s})} = \frac{(0.20\text{ mol}\cdot\text{L}^{-1})}{20\text{ s}} = \mathbf{0.010}\text{ mol}\cdot\text{L}^{-1}\cdot\text{s}^{-1}$$

(e) The instantaneous rate of reaction at 60 seconds is equal to one-half the slope of the tangent to the [HI] versus time curve at 60 seconds. This is given by

$$\text{rate} = \text{slope} = \frac{(0.76\text{ mol}\cdot\text{L}^{-1} - 0.38\text{ mol}\cdot\text{L}^{-1})}{2(100\text{ s} - 0\text{ s})} = \frac{(0.38\text{ mol}\cdot\text{L}^{-1})}{200\text{ s}} = 0.0019\text{ mol}\cdot\text{L}^{-1}\cdot\text{s}^{-1}$$

Hence, $\frac{-\Delta[\text{H}_2]}{\Delta t} = \frac{\Delta[\text{HI}]}{2\Delta t} = \mathbf{0.0019\ mol\cdot L^{-1}\cdot s^{-1}}$.

13.21 ***See Section 13.1 and Example 13.2.***

(a) The stoichiometry of the reaction $N_2O_4(g) \rightarrow 2NO_2(g)$ tells us $[NO_2]$ increases twice as fast as $[N_2O_4]$ decreases. Hence,

Time, μs	$[N_2O_4], M$	$[NO_2], M$
0.000	0.050	0.000
20.00	0.033	**0.034**
40.00	**0.025**	0.050
60.00	0.020	**0.060**

(b) The instantaneous rate of reaction at 30 μs can be determined from the negative of the slope of the tangent to the $[N_2O_4]$ versus time curve at 30 μs.

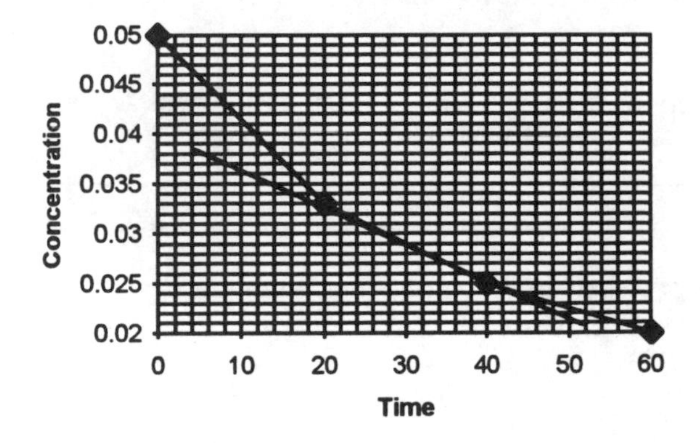

This gives:

$$\text{rate} = \frac{-(0.020\ M - 0.040\ M)}{(55\text{ s} - 0.\text{ s})} = 3.6 \times 10^{-4}\text{ mol}\cdot\text{L}^{-1}\cdot\text{s}^{-1}$$

13.22 *See Section 13.1 and Example 13.2.*

The stoichiometry of the reaction $2C_6H_{12} \rightarrow C_{12}H_{10} + 7H_2$ tells us $[C_{12}H_{10}]$ increases one-half as fast as $[C_6H_{12}]$ decreases and $[H_2]$ increases $\frac{7}{2}$ times as fast as $[C_6H_{12}]$ decreases. Hence,

Time, s	$[C_6H_{12}]$, *M*	$[C_{12}H_{10}]$, *M*	$[H_2]$, *M*
0.0	0.200	0.000	0.000
1.00	0.159	0.021	**0.144**
2.00	0.132	**0.034**	**0.238**
3.00	**0.112**	0.044	**0.308**

(b) The instantaneous rate of reaction at 1.55 s can be determined from one-half the negative of the slope of the tangent to the $[C_6H_{12}]$ versus time cuve at 1.55 s. In this case, given the specified time is at the midpoint of a very short time interval, the rate will be very nearly the average rate of reaction between 1 s and 2 s, which is given by

$$\frac{-\Delta[C_6H_{12}]}{2\Delta t} = \frac{-(0.132\ mol \cdot L^{-1} \cdot s^{-1} - 0.159\ mol \cdot L^{-1} \cdot s^{-1})}{2(2\ s - 1\ s)} = \mathbf{0.014\ mol \cdot L^{-1} \cdot s^{-1}}.$$

13.23 *See Section 13.1 and Example 13.2.*

(a) The stoichiometry of the reaction $2NO_2(g) + O_3(g) \rightarrow N_2O_5(g) + O_2(g)$ tells us that

$$\frac{-\Delta[NO_2]}{\Delta t} = \frac{\Delta[N_2O_5]}{\Delta t} \times \frac{2\ mol\ NO_2}{1\ mol\ N_2O_5} = 0.0055\ mol \cdot L^{-1} \cdot s^{-1} \times \frac{2\ mol\ NO_2}{1\ mol\ N_2O_5} = \mathbf{0.11\ mol \cdot L^{-1} \cdot s^{-1}}$$

(b) rate of reaction $= \frac{-\Delta[NO_2]}{2\Delta t} = \frac{-\Delta[O_3]}{\Delta t} = \frac{\Delta[N_2O_5]}{\Delta t} = \frac{\Delta[O_2]}{\Delta t} = \mathbf{0.0055\ mol \cdot L^{-1} \cdot s^{-1}}$

13.24 *See Section 13.1 and Example 13.2.*

(a) The stoichiometry of the reaction $NO(g) + \frac{1}{2}O_2(g) \rightarrow NO_2(g)$ tells us

$$\frac{-\Delta[O_2]}{\Delta t} = \frac{\Delta[NO_2]}{\Delta t} \times \frac{\frac{1}{2}\ mol\ O_2}{1\ mol\ NO_2} = 0.026\ mol \cdot L^{-1} \cdot s^{-1} \times \frac{\frac{1}{2}\ mol\ O_2}{1\ mol\ NO_2} = \mathbf{0.013\ mol \cdot L^{-1} \cdot s^{-1}}$$

(b) rate of reaction $= \frac{-\Delta[NO]}{\Delta t} = \frac{-2\Delta[O_2]}{\Delta t} = \frac{\Delta[NO_2]}{\Delta t} = \mathbf{0.026\ mol \cdot L^{-1} \cdot s^{-1}}$

13.25 *See Section 13.1 and Example 13.2.*

(a) The stoichiometry of the reaction $2NO(g) + Cl_2(g) \rightarrow 2NOCl(g)$ tells us

$$\frac{-\Delta[Cl_2]}{\Delta t} = \frac{\Delta[NOCl]}{\Delta t} \times \frac{1\ mol\ Cl_2}{2\ mol\ NOCl} = 0.030\ mol \cdot L^{-1} \cdot s^{-1} \times \frac{1\ mol\ Cl_2}{2\ mol\ NOCl} = \mathbf{0.015\ mol \cdot L^{-1} \cdot s^{-1}}$$

(b) rate of reaction $= \frac{-\Delta[NO]}{2\Delta t} = \frac{-\Delta[Cl_2]}{\Delta t} = \frac{\Delta[NOCl]}{2\Delta t} = \mathbf{0.015\ mol \cdot L^{-1} \cdot s^{-1}}$

13.26 ***See Section 13.1 and Example 13.2.***

The stoichiometry of the reaction $3N_2O(g) + C_2H_2(g) \rightarrow 3N_2(g) + 2CO(g) + H_2O(g)$ tells us

$$\frac{-\Delta[N_2O]}{\Delta t} = 0.10\ \text{mol}\cdot L^{-1}\cdot s^{-1} \times \frac{3\ \text{mol}\ N_2O}{1\ \text{mol}\ H_2O} = \mathbf{0.30\ mol\cdot L^{-1}\cdot s^{-1}}$$

$$\frac{-\Delta[C_2H_2]}{\Delta t} = 0.10\ \text{mol}\cdot L^{-1}\cdot s^{-1} \times \frac{1\ \text{mol}\ C_2H_2}{1\ \text{mol}\ H_2O} = \mathbf{0.10\ mol\cdot L^{-1}\cdot s^{-1}}$$

$$\frac{\Delta[N_2]}{\Delta t} = 0.10\ \text{mol}\cdot L^{-1}\cdot s^{-1} \times \frac{3\ \text{mol}\ N_2}{1\ \text{mol}\ H_2O} = \mathbf{0.30\ mol\cdot L^{-1}\cdot s^{-1}}$$

$$\frac{\Delta[CO]}{\Delta t} = 0.10\ \text{mol}\cdot L^{-1}\cdot s^{-1} \times \frac{2\ \text{mol}\ CO}{1\ \text{mol}\ H_2O} = \mathbf{0.20\ mol\cdot L^{-1}\cdot s^{-1}}$$

(b) rate of reaction $= \dfrac{-\Delta[N_2O]}{3\Delta t} = \dfrac{-\Delta[C_2H_2]}{\Delta t} = \dfrac{\Delta[N_2]}{3\Delta t} = \dfrac{\Delta[CO]}{2\Delta t} = \dfrac{\Delta[H_2O]}{\Delta t} = \mathbf{0.10\ mol\cdot L^{-1}\cdot s^{-1}}$

13.27 ***See Section 13.1 and Example 13.2.***

(a) The stoichiometry of the reaction $2CrO_2^- + 3H_2O_2 + 2OH^- \rightarrow 2CrO_4^{2-} + 4H_2O$ tells us

$$\frac{-\Delta[CrO_2^-]}{\Delta t} = 0.0050\ \text{mol}\cdot L^{-1}\cdot s^{-1} \times \frac{2\ \text{mol}\ CrO_2^-}{2\ \text{mol}\ CrO_4^{2-}} = \mathbf{0.0050\ mol\cdot L^{-1}\cdot s^{-1}}$$

$$\frac{-\Delta[H_2O_2]}{\Delta t} = 0.0050\ \text{mol}\cdot L^{-1}\cdot s^{-1} \times \frac{3\ \text{mol}\ H_2O_2}{2\ \text{mol}\ CrO_4^{2-}} = \mathbf{0.0075\ mol\cdot L^{-1}\cdot s^{-1}}$$

$$\frac{-\Delta[OH^-]}{\Delta t} = 0.0050\ \text{mol}\cdot L^{-1}\cdot s^{-1} \times \frac{2\ \text{mol}\ OH^-}{2\ \text{mol}\ CrO_4^{2-}} = \mathbf{0.0050\ mol\cdot L^{-1}\cdot s^{-1}}$$

$$\frac{\Delta[H_2O]}{\Delta t} = 0.0050\ \text{mol}\cdot L^{-1}\cdot s^{-1} \times \frac{4\ \text{mol}\ H_2O}{2\ \text{mol}\ CrO_4^{2-}} = \mathbf{0.0100\ mol\cdot L^{-1}\cdot s^{-1}}$$

(b) rate of reaction $= \dfrac{-\Delta[CrO_2^-]}{2\Delta t} = \dfrac{-\Delta[H_2O_2]}{3\Delta t} = \dfrac{-\Delta[OH^-]}{2\Delta t} = \dfrac{\Delta[CrO_4^{2-}]}{2\Delta t} = \dfrac{\Delta[H_2O]}{4\Delta t} = \mathbf{0.0025\ mol\cdot L^{-1}\cdot s^{-1}}$

13.28 ***See Section 13.1 and Example 13.2.***

(a) The stoichiometry of the reaction $2MnO_4^- + 5HNO_2 + H^+ \rightarrow 2Mn^{2+} + 5NO_3^- + 3H_2O$ tells us

$$\frac{-\Delta[HNO_2]}{\Delta t} = 0.012\ \text{mol}\cdot L^{-1}\cdot s^{-1} \times \frac{5\ \text{mol}\ HNO_2}{2\ \text{mol}\ MnO_4^-} = \mathbf{0.030\ mol\cdot L^{-1}\cdot s^{-1}}$$

$$\frac{-\Delta[H^+]}{\Delta t} = 0.012\ \text{mol}\cdot L^{-1}\cdot s^{-1} \times \frac{1\ \text{mol}\ H^+}{2\ \text{mol}\ MnO_4^-} = \mathbf{0.0060\ mol\cdot L^{-1}\cdot s^{-1}}$$

$$\frac{\Delta[Mn^{2+}]}{\Delta t} = 0.012\ \text{mol}\cdot L^{-1}\cdot s^{-1} \times \frac{2\ \text{mol}\ Mn^{2+}}{2\ \text{mol}\ MnO_4^-} = \mathbf{0.012\ mol\cdot L^{-1}\cdot s^{-1}}$$

$$\frac{\Delta[NO_3^-]}{\Delta t} = 0.012\ \text{mol}\cdot L^{-1}\cdot s^{-1} \times \frac{5\ \text{mol}\ NO_3^-}{2\ \text{mol}\ MnO_4^-} = \mathbf{0.030\ mol\cdot L^{-1}\cdot s^{-1}}$$

$$\frac{\Delta[H_2O]}{\Delta t} = 0.012\ \text{mol}\cdot L^{-1}\cdot s^{-1} \times \frac{3\ \text{mol}\ H_2O}{2\ \text{mol}\ MnO_4^-} = \mathbf{0.018\ mol\cdot L^{-1}\cdot s^{-1}}$$

(b) $\text{rate} = \frac{-\Delta[MnO_4^-]}{2\Delta t} = \frac{-\Delta[HNO_2]}{5\Delta t} = \frac{-\Delta[H^+]}{\Delta t} = \frac{\Delta[Mn^{2+}]}{2\Delta t} = \frac{\Delta[NO_3^-]}{5\Delta t} = \frac{\Delta[H_2O]}{3\Delta t} = \mathbf{0.0060\ mol \cdot L^{-1} \cdot s^{-1}}$

13.29 ***See Section 13.2 and Example 13.3.***

$2NO(g) + 2H_2(g) \rightarrow N_2(g) + 2H_2O(g)$ rate = $k[NO]^x[H_2]^y$

Experiments 1, 4 and 5 have the same concentration of H_2 and can be used to determine the order with respect to NO.

Expt.	Initial $[NO]$, M	Initial Rate, $mol \cdot L^{-1} \cdot s^{-1}$	Relative Concentration*	Relative Rate**
1	0.10	3.8	1.0	1.0
4	0.20	15.3	2.0	4.0
5	0.30	34.2	3.0	9.0

* Obtained by dividing each concentration by the smallest concentration.
** Obtained by dividing each initial rate by the smallest initial rate.

As the relative concentration of NO increases from 1.0 to 2.0 to 3.0, the relative rate of reaction increases from 1.0 to 4.0 to 9.0. This indicates that the rate of reaction is proportional to the concentration of NO raised to the second power. The reaction is therefore second order with respect to NO.

Experiments 1, 2 and 3 have the same concentration of NO and can be used to determine the order with respect to H_2.

Expt.	Initial $[H_2]$, M	Initial Rate, $mol \cdot L^{-1} \cdot s^{-1}$	Relative Concentration	Relative Rate
1	0.10	3.8	1.0	1.0
2	0.20	7.7	2.0	2.0
3	0.30	11.4	3.0	3.0

As the relative concentration of H_2 increases from 1 to 2 to 3, the relative rate of reaction increases from 1.0 to 2.0 to 3.0. This indicates that the rate of reaction is proportional to the concentration of H_2 raised to the first power. The reaction is therefore first order with respect to H_2.

The rate law is rate= $k[NO]^2[H_2]$, and using data for Expt. 1 in the original table yields

$$k = \frac{\text{rate}}{[NO]^2[H_2]} \qquad k = \frac{3.8\ mol \cdot L^{-1} \cdot s^{-1}}{(0.10\ mol \cdot L^{-1})^2(0.10\ mol \cdot L^{-1})} = 3.8 \times 10^3\ L^2 \cdot mol^{-2} \cdot s^{-1}$$

This gives **rate= $\mathbf{3.8 \times 10^3\ L^2 \cdot mol^{-2} \cdot s^{-1}[NO]^2[H_2]}$** as the rate law.

13.30 ***See Section 13.2 and Example 13.3.***

$2NO(g) + Cl_2(g) \rightarrow 2NOCl(g)$ rate = $k[NO]^x[Cl_2]^y$

Experiments 1 and 3 have the same concentration of Cl_2 and can be used to determine the order with respect to NO.

Expt.	Initial $[NO]$, M	Initial Rate, $mol \cdot L^{-1} \cdot s^{-1}$	Relative Concentration*	Relative Rate**
1	0.010	2.5	1.0	1.0
3	0.020	5.2	2.0	2.1

* Obtained by dividing each concentration by the smallest concentration.
** Obtained by dividing each initial rate by the smallest initial rate.

As the relative concentration of NO increases from 1.0 to 2.0, the relative rate of reaction increases from 1.0 to 2.1. This indicates that the rate of reaction is proportional to the concentration of NO raised to the first power. The reaction is therefore first order with respect to NO.

Experiments 1 and 2 have the same concentration of NO and can be used to determine the order with respect to Cl_2.

Expt.	Initial $[Cl_2]$, M	Initial Rate, mol·L^{-1}·s^{-1}	Relative Concentration	Relative Rate
1	0.010	2.5	1.0	1.0
2	0.020	5.1	2.0	2.0

As the relative concentration of Cl_2 increases from 1 to 2, the relative rate of reaction increases from 1.0 to 2.0. This indicates that the rate of reaction is proportional to the concentration of Cl_2 raised to the first power. The reaction is therefore first order with respect to Cl_2.

The rate law is rate= $k[NO][Cl_2]$, and using data for Expt. 1 in the original table yields

$$k = \frac{\text{rate}}{[NO][Cl_2]} \qquad k = \frac{2.5\ \text{mol}\cdot\text{L}^{-1}\cdot\text{s}^{-1}}{(0.010\ \text{mol}\cdot\text{L}^{-1})(0.010\ \text{mol}\cdot\text{L}^{-1})} = 2.5 \times 10^4\ \text{L}\cdot\text{mol}^{-1}\cdot\text{s}^{-1}$$

This gives **rate=** $\mathbf{2.5 \times 10^4\ L \cdot mol^{-1} \cdot s^{-1}[NO][Cl_2]}$ as the rate law.

13.31 ***See Section 13.2 and Example 13.3.***

$N_2O(g) + H_2O(g) \rightarrow 2NO(g) + H_2(g)$ rate = $k[N_2O]^x[H_2O]^y$

Experiments 1 and 2 have the same concentration of N_2O and can be used to determine the order with respect to H_2.

Expt.	Initial $[H_2]$, M	Initial Rate, mol·L^{-1}·s^{-1}	Relative Concentration*	Relative Rate**
1	0.10	0.051	1.0	1.0
2	0.20	0.100	2.0	2.0

* Obtained by dividing each concentration by the smallest concentration.

** Obtained by dividing each initial rate by the smallest initial rate.

As the relative concentration of H_2 increases from 1.0 to 2.0 , the relative rate of reaction increases from 1.0 to 2.0. This indicates that the rate of reaction is proportional to the concentration of H_2 raised to the first power. The reaction is therefore first order with respect to H_2.

There are no experiments which have the same concentration of H_2 and can be easily used to determine the order with respect to N_2O. Instead, Experiments 2 and 3 having changing $[N_2O]$ and $[H_2]$ are used.

Expt.	Initial $[N_2O]$, M	Initial $[H_2]$, M	Initial Rate mol·L^{-1}·s^{-1}	Relative $[N_2O]$	Relative $[H_2]$	Relative Rate	Relative Rate / Relative $[H_2]$
2	0.12	0.20	0.100	1.0	1.0	1.0	1.0
3	0.25	0.30	0.313	2.1	1.5	3.13	2.1

The change in Relative Rate/ Relative $[H_2]$ gives the change in Relative Rate that is due to change in Relative $[N_2O]$, since the reaction has been shown to be first order with respect to H_2. As the relative concentration of N_2O changes from 1.0 to 2.1 there is an accompanying change in Relative Rate/ Relative $[H_2]$ from 1.0 to 2.1. This indicates the rate of reaction is proportional to the concentration of N_2O raised to the first power. The reaction is therefore first order with respect to N_2O.

The rate law is rate= $k[N_2O][H_2]$, and using data for Expt. 1 in the original table yields

$$k = \frac{\text{rate}}{[N_2O][H_2]} \qquad k = \frac{0.0051\ \text{mol}\cdot\text{L}^{-1}\cdot\text{s}^{-1}}{(0.12\ \text{mol}\cdot\text{L}^{-1})(0.10\ \text{mol}\cdot\text{L}^{-1})} = 4.2\ \text{L}\cdot\text{mol}^{-1}\cdot\text{s}^{-1}$$

This gives **rate=** $\mathbf{4.2\ L \cdot mol^{-1} \cdot s^{-1}[N_2O][H_2]}$ as the rate law.

13.32 ***See Section 13.2 and Example 13.3.***

$HI(g) + C_2H_5(g) \rightarrow C_2H_6(g) + I_2(g)$ rate = $k[HI]^x[C_2H_5I]^y$

Experiments 1 and 2 have the same concentration of C_2H_5I and can be used to determine the order with respect to HI.

Expt.	Initial $[HI], M$	Initial Rate, $mol \cdot L^{-1} \cdot s^{-1}$	Relative Concentration*	Relative Rate**
1	0.053	3.7×10^{-5}	1.0	1.0
2	0.106	7.4×10^{-5}	2.0	2.0

* Obtained by dividing each concentration by the smallest concentration.
** Obtained by dividing each initial rate by the smallest initial rate.

As the relative concentration of HI increases from 1.0 to 2.0 , the relative rate of reaction increases from 1.0 to 2.0. This indicates that the rate of reaction is proportional to the concentration of HI raised to the first power. The reaction is therefore first order with respect to HI.

Experiments 2 and 3 have the same concentration of HI and can be used to determine the order with respect to C_2H_5I.

Expt.	Initial $[C_2H_5I], M$	Initial Rate, $mol \cdot L^{-1} \cdot s^{-1}$	Relative Concentration	Relative Rate
2	0.23	7.4×10^{-5}	1.0	1.0
3	0.46	14.8×10^{-5}	2.0	2.0

As the relative concentration of C_2H_5I increases from 1 to 2, the relative rate of reaction increases from 1.0 to 2.0. This indicates that the rate of reaction is proportional to the concentration of C_2H_5I raised to the first power. The reaction is therefore first order with respect to C_2H_5I.

The rate law is rate= $k[C_2H_5I][HI]$, and using data for Expt. 1 in the original table yields

$$k = \frac{\text{rate}}{[C_2H_5I][HI]} \qquad k = \frac{3.7 \times 10^{-5}\ mol \cdot L^{-1} \cdot s^{-1}}{(0.053\ mol \cdot L^{-1})(0.23\ mol \cdot L^{-1})} = 3.0 \times 10^{-3}\ L \cdot mol^{-1} \cdot s^{-1}$$

This gives **rate=** $\mathbf{3.0 \times 10^{-3}\ L \cdot mol^{-1} \cdot s^{-1}[C_2H_5I][HI]}$ as the rate law.

13.33 ***See Section 13.2 and Example 13.3.***

$NO_2(g) + O_3(g) \rightarrow NO_3(g) + O_2(g)$ rate = $k[NO_2]^x[O_3]^y$

Experiments 1 and 2 have the same concentration of O_3 and can be used to determine the order with respect to NO_2.

Expt.	Initial $[NO_2], M$	Initial Rate, $mol \cdot L^{-1} \cdot s^{-1}$	Relative Concentration*	Relative Rate**
1	2.0×10^{-6}	2.1×10^{-7}	1.0	1.0
2	3.0×10^{-6}	3.1×10^{-7}	1.5	1.5

* Obtained by dividing each concentration by the smallest concentration.
** Obtained by dividing each initial rate by the smallest initial rate.

As the relative concentration of NO_2 increases from 1.0 to 1.5 , the relative rate of reaction increases from 1.0 to 1.5. This indicates that the rate of reaction is proportional to the concentration of NO_2 raised to the first power. The reaction is therefore first order with respect to NO_2.

Experiments 3 and 4 have the same concentration of NO_2 and can be used to determine the order with respect to O_3.

Expt.	Initial $[O_3]$, M	Initial Rate, mol·L^{-1}·s^{-1}	Relative Concentration	Relative Rate
3	3.0×10^{-6}	6.2×10^{-7}	1.0	1.0
4	4.0×10^{-6}	8.3×10^{-7}	1.3	1.3

As the relative concentration of O_3 increases from 1.0 to 1.3, the relative rate of reaction increases from 1.0 to 1.3. This indicates that the rate of reaction is proportional to the concentration of O_3 raised to the first power. The reaction is therefore first order with respect to O_3.

The rate law is rate= $k[NO_2][O_3]$, and using data for Expt. 1 in the original table yields

$$k = \frac{\text{rate}}{[NO_2][O_3]} \qquad k = \frac{2.1 \times 10^{-7}\ \text{mol} \cdot \text{L}^{-1} \cdot \text{s}^{-1}}{(2.0 \times 10^{-6}\ \text{mol} \cdot \text{L}^{-1})(2.0 \times 10^{-6}\ \text{mol} \cdot \text{L}^{-1})} = 5.2 \times 10^{4}\ \text{L} \cdot \text{mol}^{-1} \cdot \text{s}^{-1}$$

This gives **rate=** $\mathbf{5.2 \times 10^{4}\ L \cdot mol^{-1} \cdot s^{-1}[NO_2][O_3]}$ as the rate law.

13.34 ***See Section 13.2 and Example 13.3.***

$PH_3(g) + B_2H_6(g) \rightarrow PH_3BH_3(g) + BH_3(g)$ rate $= k(P_{PH_3})^x(P_{B_2H_6})^y$

Experiments 1 and 2 have the same pressure of B_2H_6 and can be used to determine the order with respect to PH_3.

Expt.	Initial P_{PH_3}, torr	Initial Rate, torr·s^{-1}	Relative Pressure*	Relative Rate**
1	1.2	1.9×10^{-3}	1.0	1.0
2	3.0	4.7×10^{-3}	2.5	2.5

* Obtained by dividing each pressure by the smallest pressure.
** Obtained by dividing each initial rate by the smallest initial rate.

As the relative pressure of PH_3 increases from 1.0 to 2.5 , the relative rate of reaction increases from 1.0 to 2.5. This indicates that the rate of reaction is proportional to the pressure of PH_3 raised to the first power. The reaction is therefore first order with respect to PH_3.

Experiments 3 and 4 have the same pressure of PH_3 and can be used to determine the order with respect to B_2H_6.

Expt.	Initial $P_{B_2H_6}$, torr	Initial Rate, torr·s^{-1}	Relative Pressure	Relative Rate
3	1.2	6.4×10^{-3}	1.0	1.0
4	3.0	1.6×10^{-2}	2.5	2.5

As the relative pressure of B_2H_6 increases from 1.0 to 2.5, the relative rate of reaction increases from 1.0 to 2.5. This indicates that the rate of reaction is proportional to the pressure of B_2H_6 raised to the first power. The reaction is therefore first order with respect to B_2H_6.

The rate law is rate= $k(P_{PH_3})(P_{B_2H_6})$, and using data for Expt. 1 in the original table yields

$$k = \frac{\text{rate}}{(P_{PH_3})(P_{B_2H_6})} \qquad k = \frac{1.9 \times 10^{-3}\ \text{torr} \cdot \text{s}^{-1}}{(1.2\ \text{torr})(1.2\ \text{torr})} = 1.3 \times 10^{-3}\ \text{torr}^{-1} \cdot \text{s}^{-1}$$

This gives **rate=** $\mathbf{1.3 \times 10^{-3}\ torr^{-1} \cdot s^{-1}(P_{PH_3})(P_{B_2H_6})}$ as the rate law.

13.35 *See Section 13.3, Table 13.4, and Examples 13.4,13.9.*

Time,s	[Reactant], M	ln[Reactant]	$\frac{1}{[\text{Reactant}]}$, M^{-1}
0	0.250	-1.386	4.00
1	0.216	-1.532	4.63
2	0.182	-1.704	5.49
3	0.148	-1.911	6.76
4	0.114	-2.172	8.77
5	0.080	-2.53	12

For a zero order reaction, a plot of [Reactant] vs. time gives a straight line.
For a first order reaction, a plot of ln[Reactant] vs. time gives a straight line.
For a second order reaction ,a plot of $\frac{1}{[\text{Reactant}]}$ vs. time gives a straight line.

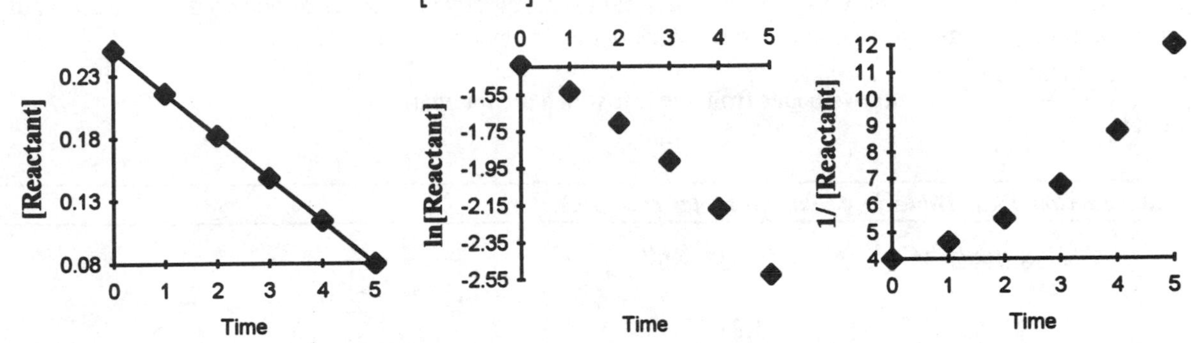

The data give a straight line for a **zero order** reaction. The rate consant for a zero order reaction is equal to the negative of the slope of the line. A linear regression analysis gives a slope of -0.034, and a correlation coefficient of 1.000 indicating excellent agreement between the data and the equation. Hence, **k = 0.034 mol·L^{-1}·s^{-1}**.

Note: If your calculator does not have the ability to perform linear regression analyses or you do not have access to a computer graphing program, you can caculate the slope of the line by using

$$\frac{([\text{Reactant}]_2 - [\text{Reactant}]_1)}{(t_2 - t_1)}$$ with two points from the line which are far apart.

13.36 *See Section 13.3, Table 13.4, and Examples 16.4, 16.9.*

Time,s	[Reactant], M	ln[Reactant]	$\frac{1}{[\text{Reactant}]}$, M^1
0	0.0451	-3.099	22.2
2	0.0421	-3.168	23.8
5	0.0376	-3.281	26.6
9	0.0316	-3.455	31.6
15	0.0226	-3.790	44.2

For a zero order reaction, a plot of [Reactant] vs. time gives a straight line.
For a first order reaction, a plot of ln[Reactant] vs. time gives a straight line.
For a second order reaction ,a plot of $\frac{1}{[\text{Reactant}]}$ vs. time gives a straight line.

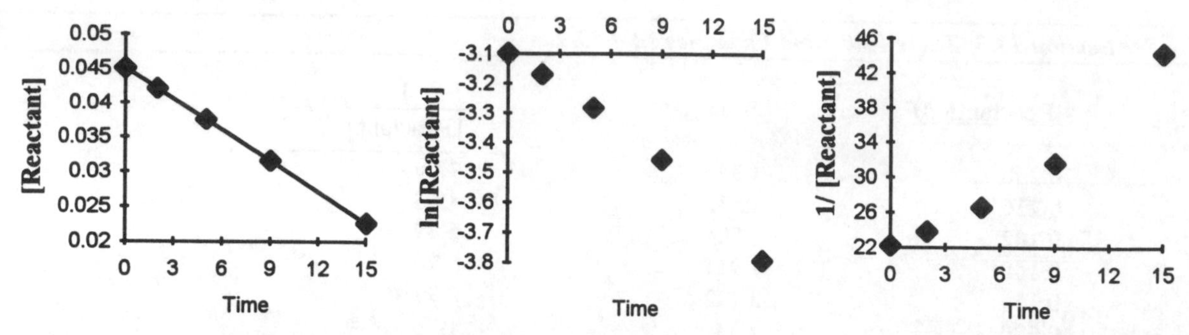

The data give a straight line for a **zero order** reaction. The rate consant for a zero order reaction is equal to the negative of the slope of the line. A linear regression analysis gives a slope of -0.00150, and a correlation coefficient of 1.000 indicating excellent agreement between the data and the equation. Hence, **k = 0.00150 mol·L^{-1}·s^{-1}**.

Note: If your calculator does not have the ability to perform linear regression analyses or you do not have access to a computer graphing program, you can caculate the slope of the line by using $\dfrac{([\text{Reactant}]_2 - [\text{Reactant}]_1)}{(t_2 - t_1)}$ with two points from the line which are far apart.

13.37 ***See Section 13.3, Table 13.4, and Examples 16.4, 16.9.***

Time,s	[Reactant], M	ln[Reactant]	$\dfrac{1}{[\text{Reactant}]}$, M^{1}
0.001	0.220	-1.51	4.5
0.002	0.140	-1.97	7.1
0.003	0.080	-2.53	12
0.004	0.050	-3.00	20
0.005	0.030	-3.51	33

For a zero order reaction, a plot of [Reactant] vs time gives a straight line.
For a first order reaction, a plot of ln[Reactant] vs. time gives a straight line.
For a second order reaction ,a plot of $\dfrac{1}{[\text{Reactant}]}$ vs. time gives a straight line.

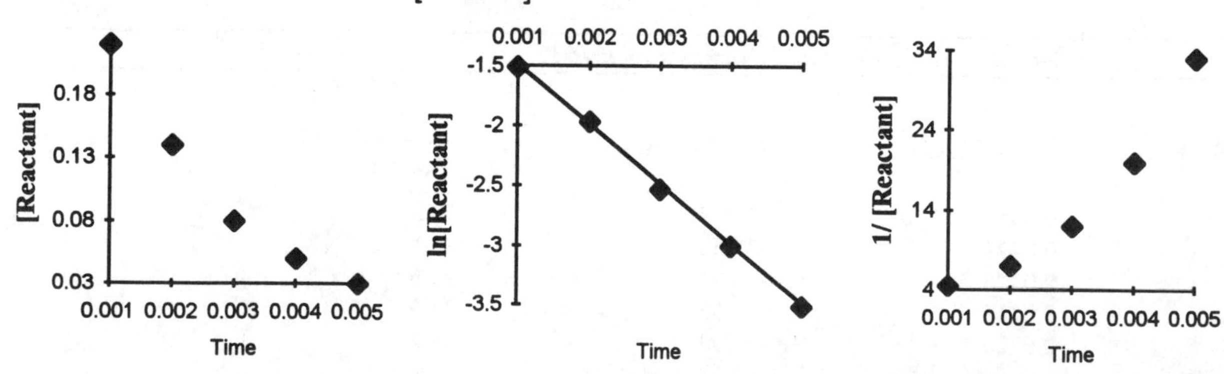

The data give a straight line plot for a **first order** reaction. The rate constant for a first order reaction is equal to the negative of the slope of the line when ln[conc] is plotted vs. time. A linear regression analysis gives a slope of -503, and a correlation coefficient of 0.9996 indicates excellent agreement between the data and equation. Hence, **k = 503 s^{-1}**.

Note: If your calculator does not have the ability to perform linear regression analyses or you do not have access to a computer graphing program, you can calculate the slope of the line by using

$$\frac{(\ln[\text{Reactant}]_2 - \ln[\text{Reactant}]_1)}{(t_2 - t_1)}$$ with two points from the line which are far apart.

13.38 ***See Section 13.3, Table 13.4, and Examples 16.4, 16.9.***

Time,s	[Reactant], M	ln[Reactant]	$\frac{1}{[\text{Reactant}]}$, M^{-1}
0	0.0350	-3.352	28.6
10	0.0223	-3.803	44.8
20	0.0142	-4.255	70.4
50	0.0037	-5.60	270
70	0.0015	-6.50	667

For a zero order reaction, a plot of [Reactant] vs time gives a straight line.
For a first order reaction, a plot of ln[Reactant] vs. time gives a straight line.
For a second order reaction ,a plot of $\frac{1}{[\text{Reactant}]}$ vs. time gives a straight line.

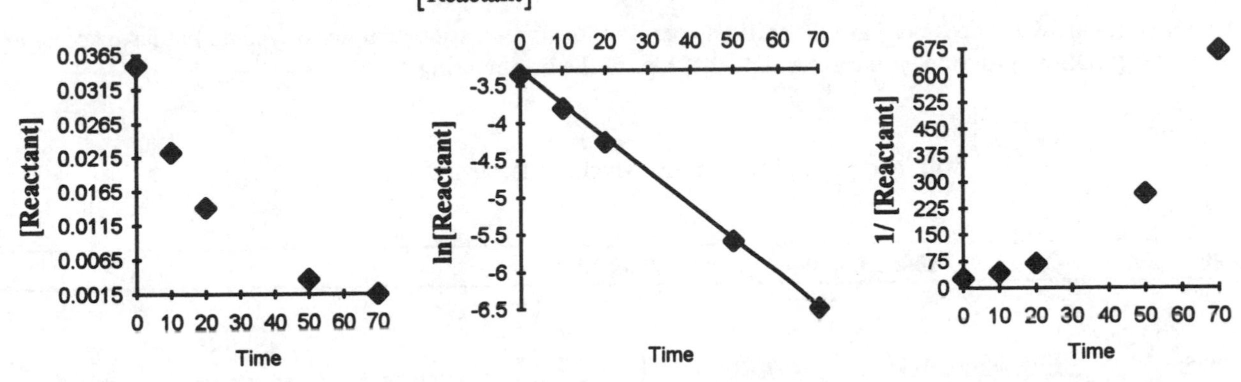

The data give a straight line plot for a **first order** reaction. The rate constant for a first order reaction is equal to the negative of the slope of the line when ln[conc] is plotted vs. time. A linear regression analysis gives a slope of -0.0450, and a correlation coefficient of 0.999999 indicates excellent agreement between the data and equation. Hence, $\mathbf{k = 0.0450\ s^{-1}}$.

Note: If your calculator does not have the ability to perform linear regression analyses or you do not have access to a computer graphing program, you can caculate the slope of the line by using

$$\frac{(\ln[\text{Reactant}]_2 - \ln[\text{Reactant}]_1)}{(t_2 - t_1)}$$ with two points from the line which are far apart.

13.39 ***See Section 13.3, Table 13.4, and Examples 13.4, 13.9.***

Time,s	[NOCl], M	ln[NOCl]	$\frac{1}{[\text{NOCl}]}$, M^{-1}
0	0.100	-2.303	10.0
30	0.064	-2.75	15.6
60	0.047	-3.06	21.3
100	0.035	-3.35	28.6
200	0.021	-3.86	47.6
300	0.015	-4.20	66.7
400	0.012	-4.42	83.3

For a zero order reaction, a plot of [NOCl] vs time gives a straight line.
For a first order reaction, a plot of ln[NOCl] vs. time gives a straight line.
For a second order reaction ,a plot of $\frac{1}{[NOCl]}$ vs. time gives a straight line.

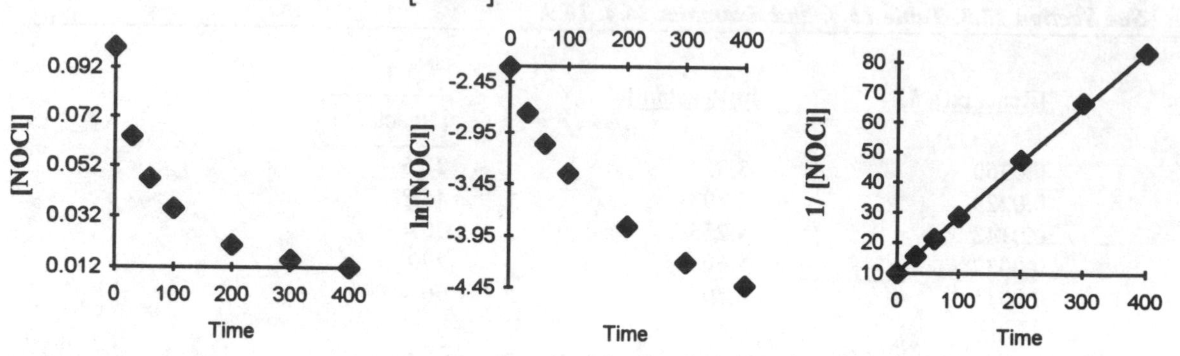

The data give a straight line plot for a **second order** reaction. The rate constant for a second order reaction is equal to the slope of the line. A linear regression analysis gives a slope of 0.185, and a correlation coefficient of 0.99998 indicates excellent agreement between the data and equation. Hence, **k = 0.185 L·mol^{-1}·s^{-1}**.

Note: If your calculator does not have the ability to perform linear regression analyses or you do not have access to a computer graphing program, you can caculate the slope of the line by using

$$\frac{\left(\frac{1}{[NOCl]_2} - \frac{1}{[NOCl]_1}\right)}{(t_2 - t_1)}$$ with two points from the line which are far apart.

13.40 ***See Section 13.3, Table 13.4, and Examples 13.4, 13.9.***

Time,s	[Reactant], *M*	ln[Reactant]	$\frac{1}{[Reactant]}$, M^{-1}
0	0.52	-0.65	1.9
20	0.27	-1.31	3.7
40	0.18	-1.71	5.6
60	0.15	-1.90	6.7
80	0.11	-2.21	9.1
100	0.09	-2.4	11

For a zero order reaction, a plot of [Reactant] vs time gives a straight line.
For a first order reaction, a plot of ln[Reactant] vs. time gives a straight line.
For a second order reaction ,a plot of $\frac{1}{[Reactant]}$ vs. time gives a straight line.

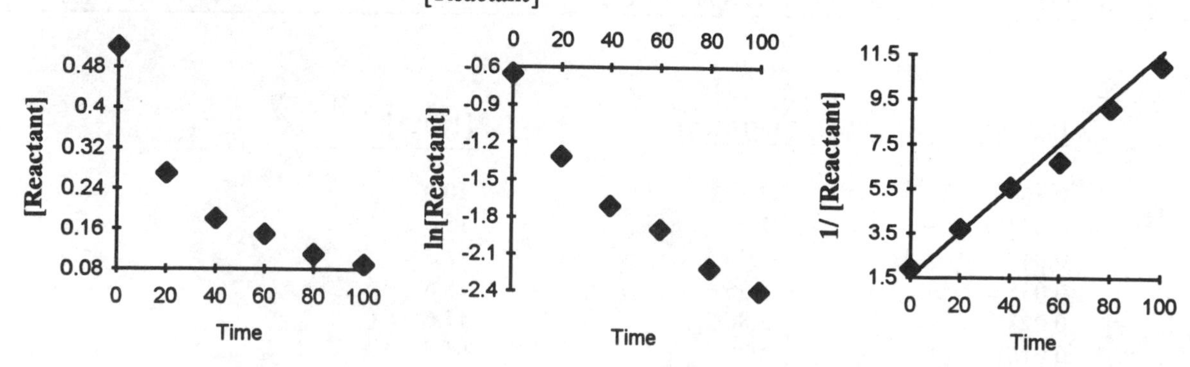

The data give a straight line plot for a **second order** reaction. The rate constant for a second order reaction is equal to the slope of the line. A linear regression analysis gives a slope of 0.090, and a correlation coefficient of 0.996884 indicates very good agreement between the data and equation. Hence, $\mathbf{k = 0.090\ L{\cdot}mol^{-1}{\cdot}s^{-1}}$.

Note: If your calculator does not have the ability to perform linear regression analyses or you do not have access to a computer graphing program, you can caculate the slope of the line by using

$$\frac{\left(\dfrac{1}{[\text{Reactant}]_2} - \dfrac{1}{[\text{Reactant}]_1}\right)}{(t_2 - t_1)}$$ with two points from the line which are far apart.

13.41 ***See Section 13.3 and Example 13.6..***

(a) Since the reaction is first order, $t_{1/2} = \dfrac{0.693}{k}$. Solving for k and substituting gives

$$k = \frac{0.693}{t_{1/2}} \qquad k = \frac{0.693}{24.5\ \text{min}} = \mathbf{2.83 \times 10^{-2}\ min^{-1}}$$

(b) Solving $\ln[A] = -kt + \ln[A]_o$ for t and substituting gives

$$t = \frac{\ln[A]_o - \ln[A]}{k} \qquad t = \frac{\ln(0.15) - \ln(0.015)}{2.83 \times 10^{-2}\ \text{min}^{-1}} = \mathbf{81.4\ min} \text{ or } \mathbf{4.88 \times 10^3\ s}$$

13.42 ***See Section 13.3 and Example 13.6..***

(a) Since the reaction is first order, $t_{1/2} = \dfrac{0.693}{k} = \dfrac{0.693}{3.05 \times 10^{-2}\ \text{s}^{-1}} = \mathbf{22.7\ s}$.

(b) Solving $\ln[A] = -kt + \ln[A]_o$ for t and substituting gives

$$t = \frac{\ln[A]_o - \ln[A]}{k} \qquad t = \frac{\ln(760) - \ln(1)}{3.05 \times 10^{-2}\ \text{s}^{-1}} = \mathbf{217\ s}$$

13.43 ***See Section 13.3.***

Since the reaction is first order, $t_{1/2} = \dfrac{0.693}{k}$. Solving $\ln[A] = -kt + \ln[A]_o$ for k and substituting gives

$$k = \frac{\ln[A]_o - \ln[A]}{t} \qquad k = \frac{\ln(0.012) - \ln(0.0082)}{66.2\ \text{s}} = 5.75 \times 10^{-3}\ \text{s}^{-1}$$

Hence, $t_{1/2} = \dfrac{0.693}{5.75 \times 10^{-3}\ \text{s}^{-1}} = \mathbf{120\ s}$.

13.44 ***See Section 13.3.***

Since the reaction is first order, $t_{1/2} = \dfrac{0.693}{k}$. Solving $\ln[A] = -kt + \ln[A]_o$ for k and substituting gives

$$k = \frac{\ln[A]_o - \ln[A]}{t} \qquad k = \frac{\ln(0.155) - \ln(0.104)}{(15.0\ \text{s} - 2.1\text{s})} = 0.0309\ \text{s}^{-1}$$

Hence, $t_{1/2} = \dfrac{0.693}{0.0309\ \text{s}^{-1}} = \mathbf{22.4\ s}$.

13.45 *See Section 13.3.*

(a) Since the reaction is first order, $t_{1/2} = \frac{0.693}{k}$. Solving $\ln[A] = -kt + \ln[A]_o$ for k and substituting gives

$$k = \frac{\ln[A]_o - \ln[A]}{t} \qquad k = \frac{\ln(1.02 \times 10^{-3}) - \ln(7.4 \times 10^{-4})}{116.7\ s} = 2.7 \times 10^{-3}\ s^{-1}$$

Hence, $t_{1/2} = \frac{0.693}{2.7 \times 10^{-3}\ s^{-1}} = \mathbf{2.6 \times 10^{2}\ s}$.

(b) Solving $\ln[A] = -kt + \ln[A]_o$ for t and substituting gives

$$t = \frac{\ln[A]_o - \ln[A]}{k} \qquad t = \frac{\ln(7.4 \times 10^{-4}) - \ln(2.0 \times 10^{-4})}{2.7 \times 10^{-3}\ min^{-1}} = \mathbf{4.8 \times 10^{2}\ s}$$

13.46 *See Section 13.3 and Example 13.6..*

(a) Since the reaction is first order, $t_{1/2} = \frac{0.693}{k}$. Solving $\ln[A] = -kt + \ln[A]_o$ for k and substituting gives

$$k = \frac{\ln[A]_o - \ln[A]}{t} \qquad k = \frac{\ln(0.0451) - \ln(0.0321)}{(45.0\ s - 30.5\ s)} = 0.0235\ s^{-1}$$

Hence, $t_{1/2} = \frac{0.693}{0.0235\ s^{-1}} = \mathbf{29.4\ s}$.

(b) Solving $\ln[A] = -kt + \ln[A]_o$ for t and substituting gives

$$t = \frac{\ln[A]_o - \ln[A]}{k} \qquad t = \frac{\ln(0.0321) - \ln(0.0100)}{0.0235\ s^{-1}} = 49.6\ s$$

Hence, it will take 49.6 s for the reactant concentration to go from 0.0321 *M* to 0.0100 *M* and will therefore reach 0.0100 *M* at **94.6 s** (45.0 + 49.6) after the reaction starts.

13.47 *See Section 16.3.*

Solving $\frac{1}{[A]} = \frac{1}{[A]_o} + kt$ for t and substituting gives

$$t = \frac{\frac{1}{[A]} - \frac{1}{[A]_o}}{k} \qquad t = \frac{\left(\frac{1}{1.0\ torr \times \frac{1\ atm}{760\ torr}}\right) - \left(\frac{1}{21.0\ torr \times \frac{1\ atm}{760\ torr}}\right)}{30.6\ atm^{-1} \cdot s^{-1}} = \mathbf{24\ s}$$

Note: The conversion factors for converting reactant pressures to *M* and the units of k to *M* would be identical and cancel, since one of these conversion factors would be in the numerator of the expression for t and the other in the denominator of the same expression.

13.48 *See Section 16.3.*

Solving $\frac{1}{[A]} = \frac{1}{[A]_o} + kt$ for t and substituting gives

$$t = \frac{\frac{1}{[A]} - \frac{1}{[A]_o}}{k} \qquad t = \frac{\left(\frac{1}{22 \text{ torr} \times \frac{1 \text{ atm}}{760 \text{ torr}}}\right) - \left(\frac{1}{44 \text{ torr} \times \frac{1 \text{ atm}}{760 \text{ torr}}}\right)}{15.4 \text{ atm}^{-1} \cdot \text{s}^{-1}} = \mathbf{1.1\ s}$$

Note: The conversion factors for converting reactant pressures to *M* and the units of k to *M* would be identical and cancel, since one of these conversion factors would be in the numerator of the expression for t and the other in the denominator of the same expression.

(b) To obtain k in $\text{mol} \cdot \text{L}^{-1} \cdot \text{s}^{-1}$, multipy k in $\text{atm}^{-1} \cdot \text{s}^{-1}$ by RT.

$$k = \left(15.4 \text{ atm}^{-1} \cdot \text{s}^{-1}\right)\left(0.0821 \frac{\text{L} \cdot \text{atm}}{\text{mol} \cdot \text{K}}\right)(450 \text{ K}) = 569 \text{ mol} \cdot \text{L}^{-1} \cdot \text{s}^{-1}$$

Hence,

$$t = \frac{\frac{1}{[A]} - \frac{1}{[A]_o}}{k} \qquad t = \frac{\left(\frac{1}{0.0022 \text{ mol} \cdot \text{L}^{-1}}\right) - \left(\frac{1}{0.0044 \text{ mol} \cdot \text{L}^{-1}}\right)}{569 \text{ mol} \cdot \text{L}^{-1} \cdot \text{s}^{-1}} = \mathbf{0.40\ s}$$

13.49 *See Section 13.4.*

The Arrhenius equation can be written as $\ln k = \frac{-E_a}{R}\left(\frac{1}{T}\right) + \ln A$. A plot of ln k vs. 1/ T will have a slope of $-E_a/R$ and an intercept of ln A. Determine the value of the slope of the line, and use E_a = -(slope)(R) to obtain E_a.

$k, \text{L} \cdot \text{mol}^{-1} \cdot \text{s}^{-1}$	ln k	T, K	$1/T, \text{K}^{-1}$
0.36×10^6	12.79	500	0.00200
3.7×10^6	15.12	550	0.00182
27×10^6	17.11	600	0.00167

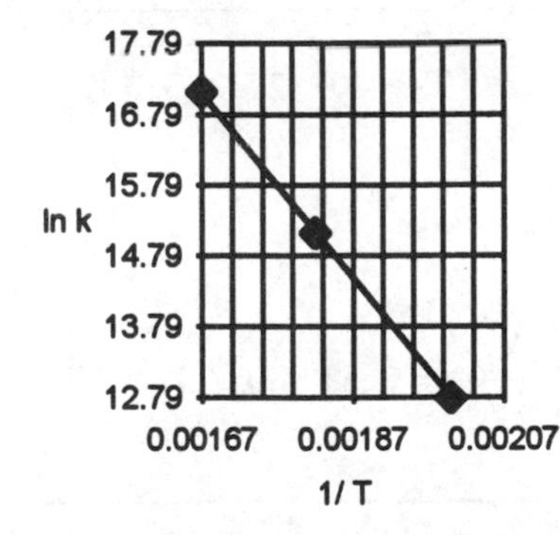

A linear regression analysis gives a slope of -1.31×10^4 and an intercept of 38.95 with a correlation coefficient of 0.99998 indicating excellent agreement between the data and equation. Hence,

$$E_a = -\left(-1.31 \times 10^4 \text{ K}^{-1}\right)\left(8.314 \text{ J} \cdot \text{mol}^{-1} \cdot \text{K}^{-1}\right) = \mathbf{1.09 \times 10^5\ J \cdot mol^{-1}}$$

and

$$A = \text{inv} \ln 38.95 = \mathbf{7.83 \times 10^{16}}.$$

Note: If your calculator does not have the ability to perform linear regression analyses or you do not have access to a computer graphing program, you can caculate the slope of the line by using

$\frac{(\ln k_2 - \ln k_1)}{(t_2 - t_1)}$ with two points from the line which are far apart.

The slope of the line can then be used to calculate the value of E_a, and this value can be used with the given data to solve the equation for lnA and therefore A. It isn't feasible to obtain lnA from the graph, since the y-intercept occurs when x = 0 and this would require a graph of unreasonable size.

13.50 *See Section 13.4.*

The Arrhenius equation can be written as $\ln k = \frac{-E_a}{R}\left(\frac{1}{T}\right) + \ln A$. A plot of ln k vs. 1/ T will have a slope of $-E_a/R$ and an intercept of ln A. Determine the value of the slope of the line, and use E_a = -(slope)(R) to obtain E_a.

k, s^{-1}	ln k	T, K	$1/T, K^{-1}$
0.00027	-8.22	800	0.00125
0.00049	-7.62	825	0.00121
0.00086	-7.06	850	0.00118
0.00143	-6.550	875	0.00114
0.00234	-6.058	900	0.00111
0.00372	-5.594	925	0.00108

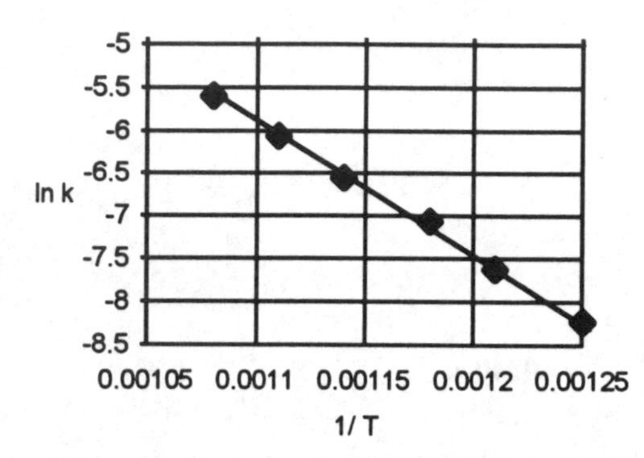

A linear regression analysis gives a slope of -1.54×10^4 and an intercept of 11.02 with a correlation coefficient of 0.999258 indicating excellent agreement between the data and equation. Hence,

$$\mathbf{E_a} = -\left(-1.54 \times 10^4\ K^{-1}\right)\left(8.314\ J\cdot mol^{-1}\cdot K^{-1}\right) = \mathbf{1.28 \times 10^5\ J\cdot m}$$

and

$$\mathbf{A} = \text{inv } \ln 11.02 = \mathbf{6.1 \times 10^4}\ .$$

Note: If your calculator does not have the ability to perform linear regression analyses or you do not have access to a computer graphing program, you can caculate the slope of the line by using

$$\frac{(\ln k_2 - \ln k_1)}{(t_2 - t_1)}$$ with two points from the line which are far apart.

The slope of the line can then be used to calculate the value of E_a, and this value can be used with the given data to solve the equation for lnA and therefore A. It isn't feasible to obtain lnA from the graph, since the y-intercept occurs when x = 0 and this would require a graph of unreasonable size.

13.51 *See Section 13.4 and Example 13.11.*

Known Quantities: $k_1 = k_1$ $k_2 = 2k_1$ $T_1 = 25 + 273 = 298$ K $T_2 = 40 + 273 = 313$ K

Solving $\ln\left(\frac{k_1}{k_2}\right) = \frac{-E_a}{R}\left(\frac{1}{T_1} - \frac{1}{T_2}\right) = \frac{-E_a}{R}\left(\frac{T_2 - T_1}{T_1T_2}\right)$ for E_a and substituting gives

$$E_a = \frac{-RT_1T_2}{(T_2 - T_1)}\ln\left(\frac{k_1}{k_2}\right) \qquad E_a = \frac{-\left(8.314\ J\cdot mol^{-1}\cdot K^{-1}\right)(298\ K)(313\ K)}{(313\ K - 298\ K)}\ln\left(\frac{k_1}{2k_1}\right) = \mathbf{3.58 \times 10^4\ J\cdot mol^{-1}}$$

13.52 *See Section 13.4 and Example 13.11.*

Known Quantities: $k_1 = k_1$ $k_2 = 3k_1$ $T_1 = 15 + 273 = 288$ K $T_2 = 35 + 273 = 308$ K

Solving $\ln\left(\frac{k_1}{k_2}\right) = \frac{-E_a}{R}\left(\frac{1}{T_1} - \frac{1}{T_2}\right) = \frac{-E_a}{R}\left(\frac{T_2 - T_1}{T_1T_2}\right)$ for E_a and substituting gives

$$E_a = \frac{-RT_1T_2}{(T_2 - T_1)}\ln\left(\frac{k_1}{k_2}\right) \qquad E_a = \frac{-\left(8.314\ J\cdot mol^{-1}\cdot K^{-1}\right)(288\ K)(308\ K)}{(308\ K - 288\ K)}\ln\left(\frac{k_1}{3k_1}\right) = \mathbf{4.05 \times 10^4\ J\cdot mol^{-1}}$$

13.53 ***See Section 13.4.***

Known Quantities: $E_a = 274\ kJ{\cdot}mol^{-1}$ $T_1 = 500 + 273 = 773\ K$ $T_2 = 550 + 273 = 823\ K$

Substituting in $\ln\left(\frac{k_1}{k_2}\right) = \frac{-E_a}{R}\left(\frac{1}{T_1} - \frac{1}{T_2}\right) = \frac{-E_a}{R}\left(\frac{T_2 - T_1}{T_1 T_2}\right)$ and solving for $\frac{k_1}{k_2}$ gives,

$$\ln\left(\frac{k_1}{k_2}\right) = \frac{-274 \times 10^3\ J \cdot mol^{-1}}{8.314\ J \cdot mol^{-1} \cdot K^{-1}}\left(\frac{823\ K - 773\ K}{(773\ K)(823\ K)}\right) = -2.59 \qquad \frac{k_1}{k_2} = e^{-2.59} = \text{invln}(-2.59) = 7.5 \times 10^{-2}$$

So, $\frac{k_2}{k_1} = \frac{1}{7.5 \times 10^{-2}} = 13$, meaning the rate of reaction increases by a factor of 13 when the temperature increases from 500°C to 550°C.

13.54 ***See Section 13.4.***

Known Quantities: $E_a = 274\ kJ{\cdot}mol^{-1}$ $T_1 = 250 + 273 = 523\ K$ $T_2 = 280 + 273 = 553\ K$

Substituting in $\ln\left(\frac{k_1}{k_2}\right) = \frac{-E_a}{R}\left(\frac{1}{T_1} - \frac{1}{T_2}\right) = \frac{-E_a}{R}\left(\frac{T_2 - T_1}{T_1 T_2}\right)$ and solving for $\frac{k_1}{k_2}$ gives,

$$\ln\left(\frac{k_1}{k_2}\right) = \frac{-274 \times 10^3\ J \cdot mol^{-1}}{8.314\ J \cdot mol^{-1} \cdot K^{-1}}\left(\frac{553\ K - 523\ K}{(523\ K)(553\ K)}\right) = -3.42 \qquad \frac{k_1}{k_2} = e^{-3.42} = \text{invln}(-3.42) = 3.3 \times 10^{-2}$$

So, $\frac{k_2}{k_1} = \frac{1}{3.3 \times 10^{-2}} = 30$, meaning the rate of reaction increases by a factor of 30 when the temperature increases from 250°C to 280°C.

Note: This increase can be compared to the factor of 13 increase that occurs for the same reaction from 523 K to 553 K; see the solution for Exercise 13.53 above.

13.55 ***See Section 13.4.***

Known Quantities: $E_a = 29 \times 10^3\ J{\cdot}mol^{-1}$ $T_1 = 13 + 273 = 286\ K$ $T_2 = 29 + 273 = 302\ K$

Substituting in $\ln\left(\frac{k_1}{k_2}\right) = \frac{-E_a}{R}\left(\frac{1}{T_1} - \frac{1}{T_2}\right) = \frac{-E_a}{R}\left(\frac{T_2 - T_1}{T_1 T_2}\right)$ and solving for $\frac{k_1}{k_2}$ gives,

$$\ln\left(\frac{k_1}{k_2}\right) = \frac{-29 \times 10^3\ J \cdot mol^{-1}}{8.314\ J \cdot mol^{-1} \cdot K^{-1}}\left(\frac{302\ K - 286\ K}{(286\ K)(302\ K)}\right) = -0.65 \qquad \frac{k_1}{k_2} = e^{-.065} = \text{invln}(-0.65) = 0.52$$

So, $\frac{k_1}{k_2} = \frac{1}{0.52} = 1.9$, meaning the rate of reaction increases by a factor of 1.9 when the temperature increases from 13°C to 29°C.

13.56 ***See Section 13.4.***

Known Quantities: $E_a = 261\ kJ{\cdot}mol^{-1}$ $T_1 = 500 + 273 = 773\ K$ $T_2 = 600 + 273 = 873\ K$

Substituting in $\ln\left(\frac{k_1}{k_2}\right) = \frac{-E_a}{R}\left(\frac{1}{T_1} - \frac{1}{T_2}\right) = \frac{-E_a}{R}\left(\frac{T_2 - T_1}{T_1 T_2}\right)$ and solving for $\frac{k_1}{k_2}$ gives,

$$\ln\left(\frac{k_1}{k_2}\right) = \frac{-261 \times 10^3\ J\cdot mol^{-1}}{8.314\ J\cdot mol^{-1}\cdot K^{-1}}\left(\frac{873\ K - 773\ K}{(773\ K)(873\ K)}\right) = -4.65 \qquad \frac{k_1}{k_2} = e^{-4.65} = \text{invln}(-4.65) = 9.6 \times 10^{-3}$$

So, $\frac{k_2}{k_1} = \frac{1}{9.6 \times 10^{-3}} = 104$, meaning the rate of reaction increases by a factor of 104 when the temperature increases from 500°C to 600°C. Hence, $\frac{\Delta[C_2H_4]}{\Delta t} = 104(0.043\ g/s) = \mathbf{4.45\ g/s}$.

13.57 ***See Section 13.4.***

For the catalyzed path, $k_{cat} = Ae^{\frac{-E_a}{RT}}$ $\qquad k_{cat} = Ae^{\frac{-28 \times 10^3\ J\cdot mol^{-1}}{(8.314\ J\cdot mol^{-1}\cdot K^{-1})(298\ K)}}$

For the uncatalyzed path, $k = Ae^{\frac{-E_a}{RT}}$ $\qquad k = Ae^{\frac{-72 \times 10^3\ J\cdot mol^{-1}}{(8.314\ J\cdot mol^{-1}\cdot K^{-1})(298\ K)}}$

Hence, $\frac{k_{cat}}{k} = \frac{Ae^{\frac{-E_a}{RT}}}{Ae^{\frac{-E_a}{RT}}}$ $\qquad \frac{k_{cat}}{k} = \frac{e^{\frac{-28 \times 10^3\ J\cdot mol^{-1}}{(8.314\ J\cdot mol^{-1}\cdot K^{-1})(298\ K)}}}{e^{\frac{-72 \times 10^3\ J\cdot mol^{-1}}{(8.314\ J\cdot mol^{-1}\cdot K^{-1})(298\ K)}}} = \frac{1.2 \times 10^{-5}}{2.4 \times 10^{-13}} = 5.0 \times 10^7$

This means the catalyzed reaction will go 5.0×10^7 times faster than the uncatalyzed reaction at 298 K.

13.58 ***See Section 13.4.***

For the catalyzed path, $k_{cat} = Ae^{\frac{-E_a}{RT}}$ $\qquad k_{cat} = Ae^{\frac{-25 \times 10^3\ J\cdot mol^{-1}}{(8.314\ J\cdot mol^{-1}\cdot K^{-1})(288\ K)}}$

For the uncatalyzed path, $k = Ae^{\frac{-E_a}{RT}}$ $\qquad k = Ae^{\frac{-50 \times 10^3\ J\cdot mol^{-1}}{(8.314\ J\cdot mol^{-1}\cdot K^{-1})(288\ K)}}$

Hence, $\frac{k_{cat}}{k} = \frac{Ae^{\frac{-E_a}{RT}}}{Ae^{\frac{-E_a}{RT}}}$ $\qquad \frac{k_{cat}}{k} = \frac{e^{\frac{-25 \times 10^3\ J\cdot mol^{-1}}{(8.314\ J\cdot mol^{-1}\cdot K^{-1})(288\ K)}}}{e^{\frac{-50 \times 10^3\ J\cdot mol^{-1}}{(8.314\ J\cdot mol^{-1}\cdot K^{-1})(288\ K)}}} = \frac{2.9 \times 10^{-5}}{8.5 \times 10^{-10}} = 3.4 \times 10^4$

This means the catalyzed reaction will go 1.7×10^4 times faster than the uncatalyzed reaction at 288 K.

13.59 ***See Section 13.5 and Figure 13.13.***

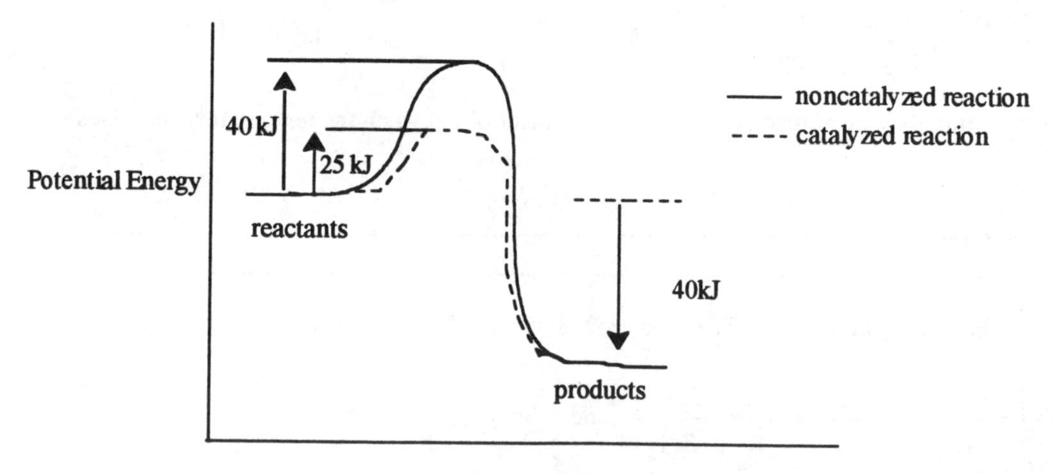

13.60 *See Section 13.5 and Figure 13.13.*

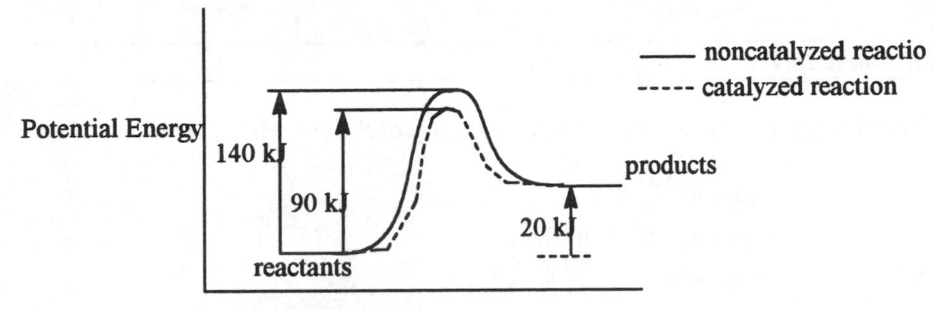

13.61 *See Section 13.6 and Example 13.12.*

$$NO_2(g) + NO_2(g) \rightarrow NO_3(g) + NO(g)$$
$$NO_3(g) + CO(g) \rightarrow NO_2(g) + CO_2(g)$$

$$NO_2(g) + CO(g) \rightarrow NO(g) + CO_2(g)$$

13.62 *See Section 13.6 and Example 13.12.*

$$Cl_2 \rightarrow 2Cl$$
$$Cl + O \rightarrow COCl$$
$$COCl + Cl \rightarrow COCl_2$$

$$Cl_2 + CO \rightarrow COCl_2$$

13.63 *See Section 13.6 and Example 13.12.*

$$NO \rightarrow N + O$$
$$O_3 + O \rightarrow 2O_2$$
$$O_2 + N \rightarrow NO_2$$

$$NO + O_3 \rightarrow O_2 + NO_2$$

13.64 *See Section 13.6 and Example 13.12.*

$$Cl_2 \rightarrow Cl^+ + Cl^-$$
$$Cl^- + H_2O \rightarrow HCl + OH^-$$
$$Cl^+ + OH^- \rightarrow HCl + O$$

$$Cl_2 + H_2O \rightarrow 2HCl + O$$

Note: Reactions yielding highly reactive oxygen atoms as final products are not likely. This is why the reaction was classified as being hypothetical

13.65 *See Section 13.6 and Example 13.13.*

The rate law for an elementary reaction can be deduced from the stoichiometry of the elementary reaction. This gives:

(a) $HCl \rightarrow H + Cl$ **rate= k[HCl]**

(b) $H_2 + Cl \rightarrow HCl + H$ rate= $k[H_2][Cl]$

(c) $2NO_2 \rightarrow N_2O_4$ rate= $k[NO_2]^2$

13.66 ***See Section 13.6 and Example 13.13.***

The rate law for an elementary reaction can be deduced from the stoichiometry of the elementary reaction. This gives:

(a) $H + Br \rightarrow HBr$ rate= $k[H][Br]$

(b) $CH_3Br + OH^- \rightarrow CH_3OH + Br^-$ rate= $k[CH_3Br][OH^-]$

(c) $NO + NO_2 + O_2 \rightarrow NO_2 + NO_3$ rate= $k[NO][NO_2][O_2]$

13.67 ***See Section 13.6 and Example 13.13.***

The rate law for an elementary reaction can be deduced from the stoichiometry of the elementary reaction. This gives:

(a) $C_2H_5Cl \rightarrow C_2H_4 + HCl$ rate= $k[C_2H_5Cl]$

(b) $NO + O_3 \rightarrow NO_2 + O_2$ rate= $k[NO][O_3]$

(c) $HI + C_2H_5I \rightarrow C_2H_6 + I_2$ rate= $k[HI][C_2H_5I]$

13.68 ***See Section 13.6 and Example 13.13.***

The rate law for an elementary reaction can be deduced from the stoichiometry of the elementary reaction. This gives:

(a) $NO + NO_2Cl \rightarrow NO_2 + NOCl$ rate= $k[NO][NO_2Cl]$

(b) $NO_2 + SO_2 \rightarrow NO + SO_3$ rate= $k[NO_2][SO_2]$

(c) $N_2O_4 \rightarrow 2NO_2$ rate= $k[N_2O_4]$

13.69 ***See Section 13.6 and Examples 13.13, 13.14.***

Since the first step in the mechanism

$$NO_2Cl \rightarrow NO_2 + Cl$$
$$NO_2Cl + Cl \rightarrow NO_2 + Cl_2$$
$$\overline{2NO_2Cl \rightarrow 2NO_2 + Cl_2}$$

is a unimolecular elementary reaction and the reaction is a first-order reaction , **the first step is the rate-limiting step.**

13.70 ***See Section 13.6 and Examples 13.13, 13.14.***

Since the first step in the mechanism

$$NO_2 + O_3 \rightarrow NO_3 + O_2$$
$$NO_3 + NO_2 \rightarrow N_2O_5$$
$$\overline{2NO_2 + O_3 \rightarrow N_2O_5 + O_2}$$

is a bimolecular elementary reaction involving one NO_2 and one O_3 and the reaction is first-order in each, **the first step is the rate-limiting step.**

13.71 ***See Section 13.6 and Examples 13.13, 13.14.***

The rate law for Mechanism I $NO_2(g) + CO(g) \rightarrow NO(g) + CO_2(g)$
would be: **rate=** $\mathbf{k[NO_2][CO]}$.
The rate law for Mechanism II

$NO_2(g) + NO_2(g) \rightarrow NO_3(g) + NO(g)$ slow
$NO_3(g) + CO(g) \rightarrow NO_2(g) + CO_2(g)$ fast

would be: **rate=** $\mathbf{k[NO_2][NO_2] = k[NO_2]^2}$.

13.72 ***See Section 13.6 and Examples 13.13, 13.14.***

The rate law for Mechanism I $NO(g) + O_3(g) \rightarrow NO_2(g) + O_2(g)$
would be: **rate=** $\mathbf{k[NO][O_3]}$.
The rate law for Mechanism II

$O_3(g) \rightarrow O_2(g) + O(g)$ slow
$NO(g) + O(g) \rightarrow NO_2(g)$ fast

would be: **rate=** $\mathbf{k[O_3]}$.

13.73 ***See Section 13.6.***

$2NO(g) \rightleftarrows N_2O_2$ fast, reversible
$N_2O_2(g) + Cl_2(g) \rightarrow 2NOCl$ slow step

The rate of reaction is limited by the rate of the slowest step, Step 2. Hence, rate= $k_2[N_2O_2][Cl_2]$. However, **N_2O_2 is an intermediate**, and it is customary to express the concentrations of intermediates in terms of concentrations of reactants and when appropriate products. Setting the rates of the forward and reverse reactions of Step 1 equal to each other gives $k_1[NO]^2 = k_{-1}[N_2O_2]$. Solving this expression for $[N_2O_2]$ and substituting for $[N_2O_2]$ in the rate law for Step 2 gives

$$\text{rate} = k_2[N_2O_2][Cl_2] = k_2\left(\frac{k_1[NO]^2}{k_{-1}}\right)[Cl_2] = k[NO]^2[Cl_2]$$

Hence, this mechanism is **consistent** with the experimental rate law.

13.74 ***See Section 13.6.***

(a) $2NO_2 \rightleftarrows N_2O_4$ fast, reversible
$N_2O_4 + O_3 \rightarrow NO_3 + O_2$ slow step

The rate of reaction is limited by the rate of the slowest step, Step 2. Hence, rate= $k_2[N_2O_4][O_3]$. However, **N_2O_4 is an intermediate**, and it is customary to express the concentrations of intermediates in terms of concentrations of reactants and when appropriate products. Setting the rates of the forward and reverse reactions of Step 1 equal to each other gives $k_1[NO_2]^2 = k_{-1}[N_2O_4]$. Solving this expression for $[N_2O_4]$ and substituting for $[N_2O_2]$ in the rate law for Step 2 gives

$$\text{rate} = k_2[N_2O_4][O_3] = k_2\left(\frac{k_1[NO_2]^2}{k_{-1}}\right)[O_3] = k[NO]^2[O_3]$$

Hence, this mechanism is **not consistent** with the experimental rate law of rate = $k[NO_2][O_3]$.

(b) $NO_2 + O_3 \rightleftarrows NO_3 + O_2$ slow
$NO_3 + NO_2 \rightarrow N_2O_5$ fast

Since the first step in the mechanism is the rate limiting step, rate = $k[NO_2][O_3]$. Hence, this mechanism is **consistent** with the experimental rate law of rate = $k[NO_2][O_3]$.

13.75 *See Section 13.1.*

For aA + bB → cC + dD, rate of reaction= $\frac{-\Delta[A]}{a\Delta t} = \frac{-\Delta[B]}{b\Delta t} = \frac{\Delta[C]}{c\Delta t} = \frac{\Delta[D]}{d\Delta t}$

13.76 *See Section 13.3, Table 13.4, and Examples 13.4, 13.9.*

Time,s	[pTSA], M	ln[pTSA]	$\frac{1}{[pTSA]}$, M^{-1}
0	0.100	-2.303	10.0
2000	0.072	-2.63	13.9
3000	0.060	-2.81	16.7
5000	0.055	-2.90	18.2
10000	0.030	-3.51	33.3
18000	0.020	-3.91	50.0

For a zero order reaction, a plot of [pTSA] vs time gives a straight line.
For a first order reaction, a plot of ln[pTSA] vs. time gives a straight line.
For a second order reaction ,a plot of $\frac{1}{[pTSA]}$ vs. time gives a straight line.

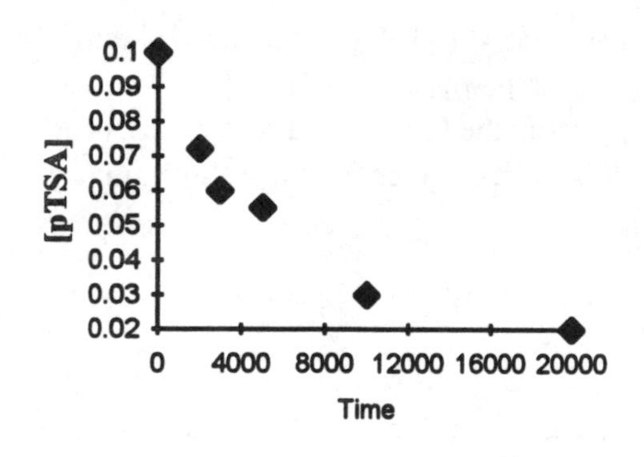

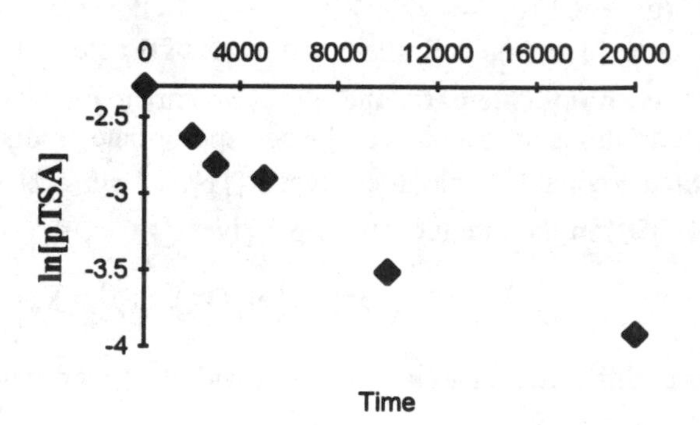

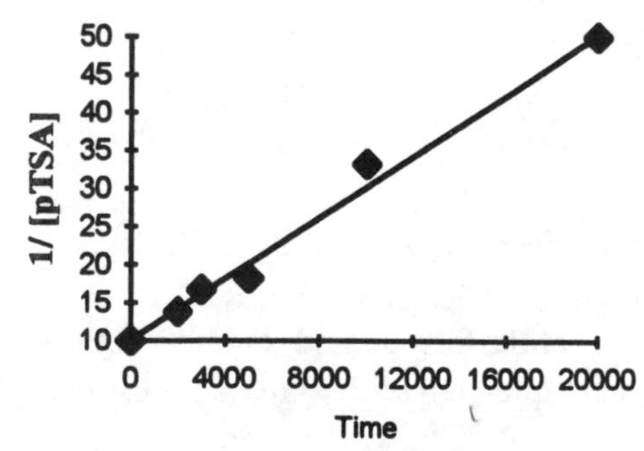

The data give a straight line plot for a **second order** reaction. The rate constant for a second order reaction is equal to the slope of the line. A linear regression analysis gives a slope of 0.0023, and a correlation coefficient of 0.9962 indicates reasonably good agreement between the data and equation. Hence, **k = 0.0023 L·mol^{-1}·s^{-1}**.

Note: If your calculator does not have the ability to perform linear regression analyses or you do not have access to a computer graphing program, you can caculate the slope of the line by using

$$\frac{\left(\frac{1}{[pTSA]_2} - \frac{1}{[pTSA]_1}\right)}{(t_2 - t_1)}$$ with two points from the line which are far apart.

13.77 *See Section 13.3, Table 13.4 and Examples 13.4, 13.9.*

Time,s	$[H_2O_2]$, M	$\ln[H_2O_2]$	$\frac{1}{[H_2O_2]}$, M^{-1}
0	0.0334	-3.399	29.9
10	0.0300	-3.507	33.3
20	0.0283	-3.565	35.3
30	0.0249	-3.692	40.2
40	0.0198	-3.922	50.5
50	0.0164	-4.110	61.0
60	0.0130	-4.343	76.9

For a zero order reaction, a plot of $[H_2O_2]$ vs time gives a straight line.
For a first order reaction, a plot of $\ln[H_2O_2]$ vs. time gives a straight line.
For a second order reaction ,a plot of $\frac{1}{[H_2O_2]}$ vs. time gives a straight line.

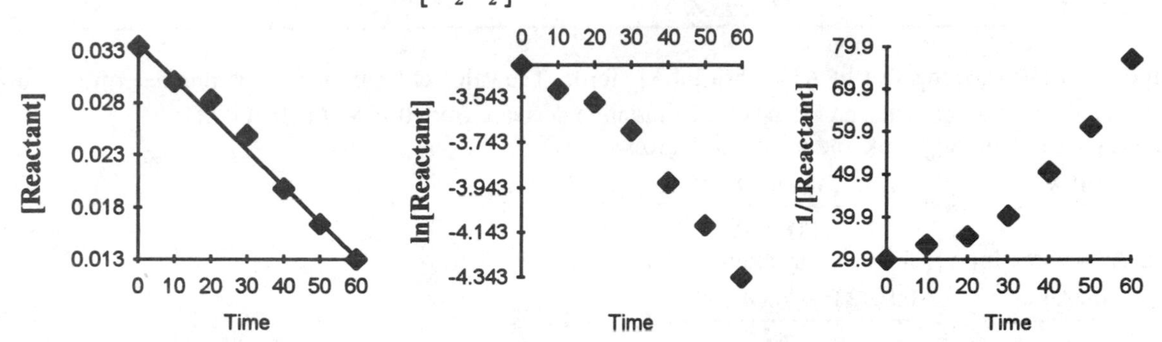

The data give a straight line for a **zero order** reaction. The rate constant for a zero order reaction is equal to the negative of the slope of the line. A linear regression analysis gives a slope of -0.00035, and a correlation coefficient of .9941 indicating reasonably good agreement between the data and the equation.
Hence, $\mathbf{k = 3.5 \times 10^{-4}\ mol \cdot L^{-1} \cdot s^{-1}}$.

Note: If your calculator does not have the ability to perform linear regression analyses or you do not have access to a computer graphing program, you can caculate the slope of the line by using $\frac{([H_2O_2]_2 - [H_2O_2]_1)}{(t_2 - t_1)}$ with two points from the line which are far apart.

13.78 *See Section 13.2 and Example 13.3.*

$CO(g) + Cl_2(g) \rightarrow COCl_2(g)$ rate = $k[CO]^x[Cl_2]^y$

Experiments 1 and 2 have the same concentration of Cl_2 and can be used to determine the order with respect to CO.

Expt.	Initial $[CO]$, M	Initial Rate, $mol \cdot L^{-1} \cdot s^{-1}$	Relative Concentration*	Relative Rate**
1	0.053	3.7×10^{-5}	1.0	1.0
2	0.106	7.4×10^{-5}	2.0	2.0

* Obtained by dividing each concentration by the smallest concentration.
** Obtained by dividing each initial rate by the smallest initial rate.

As the relative concentration of CO increases from 1.0 to 2.0, the relative rate of reaction increases from 1.0 to 2.0. This indicates that the rate of reaction is proportional to the concentration of CO raised to the first power. The reaction is therefore first order with respect to CO.

Experiments 2 and 3 have the same concentration of CO and can be used to determine the order with respect to Cl_2.

Expt.	Initial $[Cl_2]$, M	Initial Rate, mol·L^{-1}·s^{-1}	Relative Concentration	Relative Rate
2	0.23	7.4×10^{-5}	1.0	1.0
3	0.46	10.4×10^{-5}	2.0	1.4

As the relative concentration of Cl_2 increases from 1.0 to 2.0, the relative rate of reaction increases from 1.0 to 1.4. This indicates that the rate of reaction is proportional to the concentration of Cl_2 raised to the one-half power ($2^{\frac{1}{2}} = \sqrt{2} = 1.414$). The reaction is therefore one-half order with respect to Cl_2.

The rate law is rate= $k[CO][Cl_2]^{\frac{1}{2}}$, and using data for Expt. 1 in the original table yields

$$k = \frac{\text{rate}}{[CO][Cl_2]^{\frac{1}{2}}} \qquad k = \frac{3.7 \times 10^{-5}\ \text{mol} \cdot \text{L}^{-1} \cdot \text{s}^{-1}}{(0.53\ \text{mol} \cdot \text{L}^{-1})(0.230\ \text{mol} \cdot \text{L}^{-1})^{\frac{1}{2}}} = 1.5 \times 10^{-4}\ \text{L}^{\frac{1}{2}} \cdot \text{mol}^{-\frac{1}{2}} \cdot \text{s}^{-1}$$

This gives **rate= $1.5 \times 10^{-4}\ \text{L}^{\frac{1}{2}} \cdot \text{mol}^{-\frac{1}{2}} \cdot \text{s}^{-1}[CO][Cl_2]^{\frac{1}{2}}$** as the rate law.

13.79 ***See Section 13.3 and Examples 13.5, 13.6.***

The given data can be used to solve $\ln[A] = -kt + \ln[A]_o$ for k. The value of k can then be used to determine the amount of time that would be required for the concentration to decrease from 0.235 M to 0.100 M.

Solving $\ln[A] = -kt + \ln[A]_o$ for k and substituting gives

$$k = \frac{\ln[A]_o - \ln[A]}{t} \qquad k = \frac{\ln(0.451) - \ln(0.235)}{131\ \text{s}} = 4.98 \times 10^{-3}\ \text{s}^{-1}$$

Solving $\ln[A] = -kt + \ln[A]_o$ for t and substituting gives

$$t = \frac{\ln[A]_o - \ln[A]}{k} \qquad t = \frac{\ln(0.235) - \ln(0.100)}{4.98 \times 10^{-3}\ \text{s}^{-1}} = \mathbf{1.72 \times 10^2\ s}$$

13.80 ***See Section 13.3, Table 13.4, and Examples 13.4, 13.9.***

For $2NO_2(g) \rightarrow 2NO(g) + O_2(g)$ at 310K,

Time,s	$[P_{NO_2}]$, torr	$\ln(P_{NO_2})$	$\frac{1}{[P_{NO_2}]}$, torr^{-1}
0	24.0	3.178	0.0417
1	18.1	2.896	0.0552
2	13.7	2.617	0.0730
3	10.3	2.332	0.0971
4	7.8	2.05	0.13
5	5.9	1.78	0.17
6	4.5	1.50	0.22
7	3.4	1.22	0.29
8	2.6	0.96	0.38
9	1.9	0.64	0.53
10	1.5	0.41	0.67

For a zero order reaction, a plot of (Pressure) vs time gives a straight line.
For a first order reaction, a plot of ln(Pressure) vs. time gives a straight line.
For a second order reaction, a plot of $\frac{1}{(\text{Pressure})}$ vs. time gives a straight line.

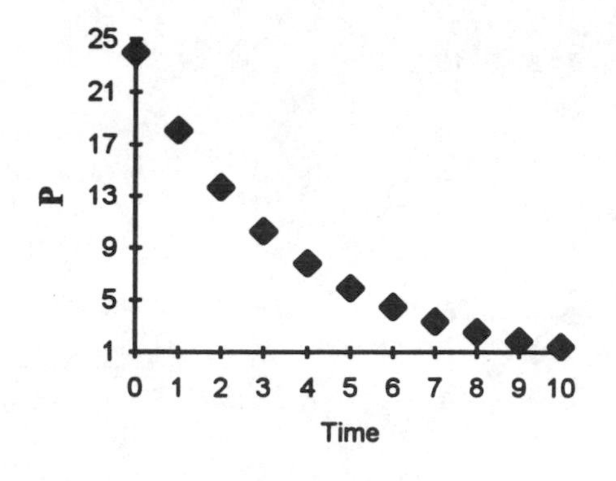

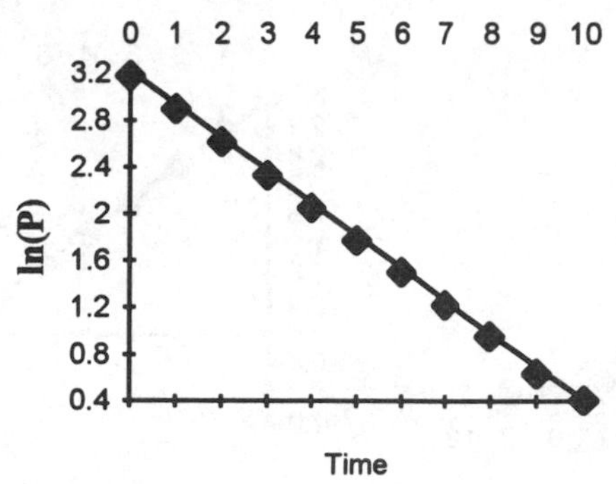

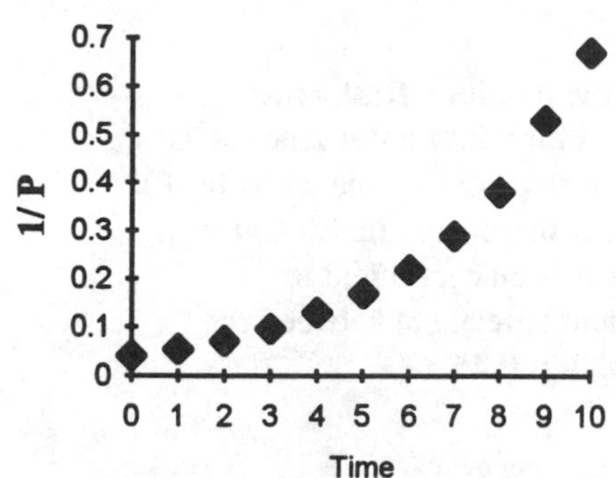

The data give a straight line plot for a **first order** reaction. The rate constant for a first order reaction is equal to the negative of the slope of the line when ln (P) is plotted vs. time. A linear regression analysis gives a slope of -0.28, and a correlation coefficient of 0.999933 indicates excellent agreement between the data and equation. Hence, $\mathbf{k = 0.28\ s^{-1}}$.

Note: If your calculator does not have the ability to perform linear regression analyses or you do not have access to a computer graphing program, you can caculate the slope of the line by using

$$\frac{(\ln(\text{Pressure})_2 - \ln(\text{Pressure})_1)}{(t_2 - t_1)}$$ with two points from the line which are far apart.

For $2NO_2(g) \rightarrow 2NO(g) + O_2(g)$ at 315K,

Time,s	$[P_{NO_2}]$, torr	$\ln(P_{NO_2})$	$\frac{1}{[P_{NO_2}]}$, $torr^{-1}$
0	24.0	3.178	0.0417
1	15.2	2.721	0.0658
2	9.7	2.27	0.10
3	6.1	1.81	0.16
4	3.9	1.36	0.26
5	2.5	0.92	0.40
6	1.6	0.47	0.63
7	1.0	0.00	1.0
8	0.6	-0.51	1.7
9	0.4	-0.92	2.5
10	0.3	-1.2	3.3

For a zero order reaction, a plot of (Pressure) vs time gives a straight line.
For a first order reaction, a plot of ln(Pressure) vs. time gives a straight line.
For a second order reaction ,a plot of $\frac{1}{(\text{Pressure})}$ vs. time gives a straight line.

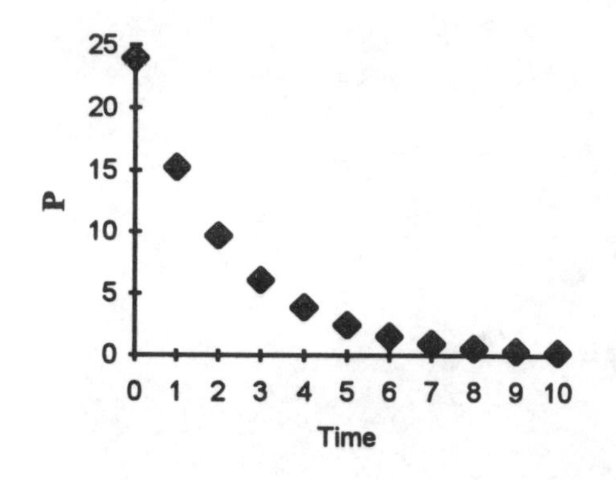

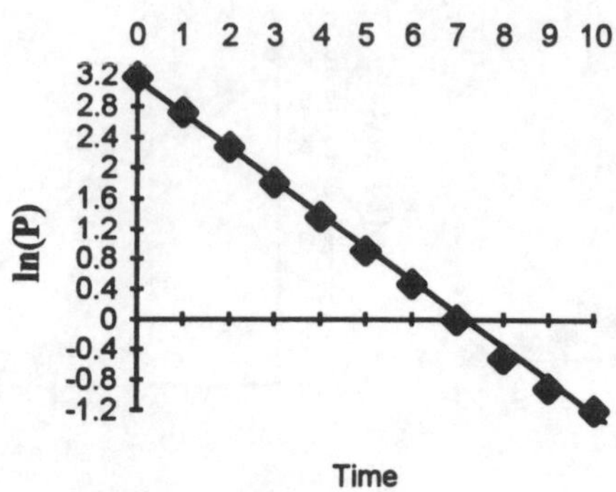

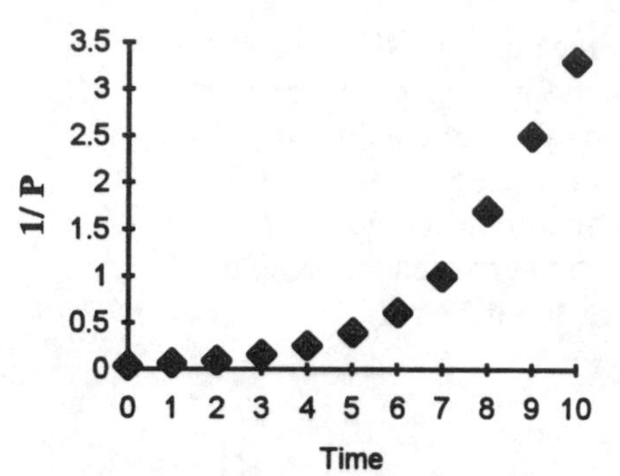

The data give a straight line plot for a **first order** reaction. The rate constant for a first order reaction is equal to the negative of the slope of the line when ln (P) is plotted vs. time. A linear regression analysis gives a slope of -0.45, and a correlation coefficient of 0.999456 indicates excellent agreement between the data and equation. Hence, $\mathbf{k = 0.45\ s^{-1}}$.

Note: If your calculator does not have the ability to perform linear regression analyses or you do not have access to a computer graphing program, you can caculate the slope of the line by using $\dfrac{\left(\ln(\text{Pressure})_2 - \ln(\text{Pressure})_1\right)}{(t_2 - t_1)}$ with two points from the line which are far apart.

(a) The reaction is a first order reaction at both temperatures.
(b) The rate constant at 310 K is 0.28 s^{-1}, and the rate constant at 315 K is 0.45 s^{-1}.
(c) Known Quantities: $k_1 = 0.28\ s^{-1}$ $k_2 = 0.45\ s^{-1}$ $T_1 = 310\ K$ $T_2 = 315\ K$

Solving $\ln\left(\dfrac{k_1}{k_2}\right) = \dfrac{-E_a}{R}\left(\dfrac{1}{T_1} - \dfrac{1}{T_2}\right) = \dfrac{-E_a}{R}\left(\dfrac{T_2 - T_1}{T_1 T_2}\right)$ for E_a and substituting gives

$$E_a = \frac{-RT_1T_2}{(T_2 - T_1)}\ln\left(\frac{k_1}{k_2}\right) = \frac{-(8.314\ J\cdot mol^{-1}\cdot K^{-1})(310\ K)(315\ K)}{(315\ K - 310\ K)}\ln\left(\frac{0.28}{0.45}\right) = 7.7 \times 10^4\ J\cdot mol^{-1} \text{ or } \mathbf{77\ kJ}$$

13.81 ***See Section 13.3, Table 13.4, and Examples 13.4, 13.9.***

Time,s	[Reactant], M	ln[Reactant]	$\dfrac{1}{[\text{Reactant}]}$, M^{-1}
0.00	0.7014	-0.3547	1.426
10.00	0.4534	-0.7910	2.206
20.00	0.3304	-1.107	3.027
40.00	0.2181	-1.523	4.585
60.00	0.1638	-1.809	6.105
80.00	0.1297	-2.043	7.710
100.00	0.1084	-2.222	9.225

For a zero order reaction, a plot of [Reactant] vs time gives a straight line.
For a first order reaction, a plot of ln[Reactant] vs. time gives a straight line.

For a second order reaction ,a plot of $\dfrac{1}{[\text{Reactant}]}$ vs. time gives a straight line.

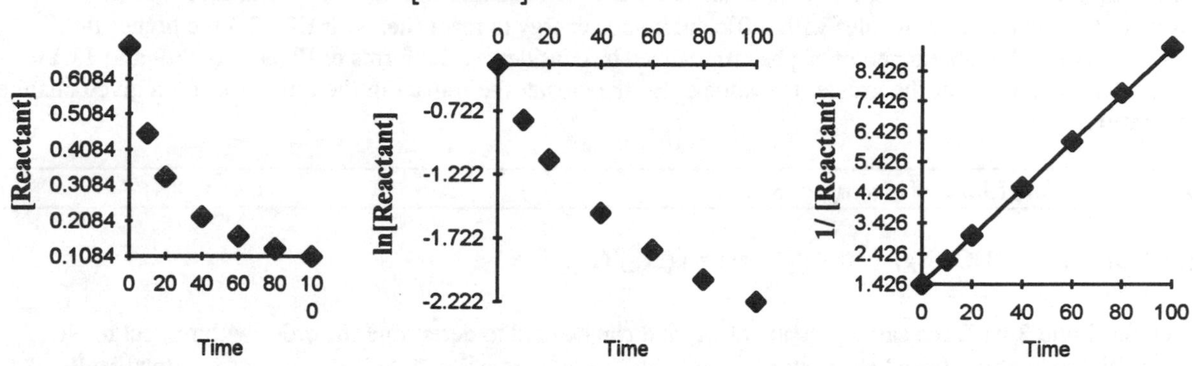

The data give a straight line plot for a **second order** reaction. The rate constant for a second order reaction is equal to the slope of the line. A linear regression analysis gives a slope of 0.078, and a correlation coefficient of 0.99998 indicates excellent agreement between the data and equation. Hence, **k = 0.078 L·mol^{-1}·s^{-1}**.

Note: If your calculator does not have the ability to perform linear regression analyses or you do not have acess to a computer graphing program, you can caculate the slope of the line by using

$$\frac{\left(\dfrac{1}{[\text{Reactant}]_2} - \dfrac{1}{[\text{Reactant}]_1}\right)}{(t_2 - t_1)}$$ with two points from the line which are far apart.

13.82 *See Section 13.4.*

Assuming the ratio of the rates is given by the ratio of the rate constants,

$$R_{25} = \frac{k_A(25)}{k_B(25)},\ R_{35} = \frac{k_A(35)}{k_B(35)} \text{ and } \frac{R_{25}}{R_{35}} = \frac{\dfrac{k_A(25)}{k_B(25)}}{\dfrac{k_A(35)}{k_B(35)}} = \frac{k_A(25)}{k_A(35)} \times \frac{k_B(35)}{k_B(25)}.$$

Known Quantities for Reaction A:

$E_a = 30\ \text{kJ·mol}^{-1}$ $\quad T_1 = 35 + 273 = 308\ \text{K}$ $\quad T_2 = 25 + 273 = 298\ \text{K}$

Substituting in $\ln\left(\dfrac{k_1}{k_2}\right) = \dfrac{-E_a}{R}\left(\dfrac{1}{T_1} - \dfrac{1}{T_2}\right) = \dfrac{-E_a}{R}\left(\dfrac{T_2 - T_1}{T_1T_2}\right)$ and solving for $\dfrac{k_1}{k_2}$ gives,

$$\ln\left(\frac{k_{25}}{k_{35}}\right) = \frac{-30 \times 10^3\ \text{J}\cdot\text{mol}^{-1}}{8.314\ \text{J}\cdot\text{mol}^{-1}\cdot\text{K}^{-1}}\left(\frac{298\ \text{K} - 308\ \text{K}}{(308\ \text{K})(298\ \text{K})}\right) = -0.39 \qquad \frac{k_{25}}{k_{35}} = e^{-0.39} = \text{invln}(-0.39) = 0.68$$

Known Quantities for Reaction B:

$E_a = 40\ \text{kJ·mol}^{-1}$ $\quad T_1 = 25 + 273 = 298\ \text{K}$ $\quad T_2 = 35 + 273 = 308\ \text{K}$

Substituting in $\ln\left(\dfrac{k_1}{k_2}\right) = \dfrac{-E_a}{R}\left(\dfrac{1}{T_1} - \dfrac{1}{T_2}\right) = \dfrac{-E_a}{R}\left(\dfrac{T_2 - T_1}{T_1T_2}\right)$ and solving for $\dfrac{k_1}{k_2}$ gives,

$$\ln\left(\frac{k_{35}}{k_{25}}\right) = \frac{-40 \times 10^3\ \text{J}\cdot\text{mol}^{-1}}{8.314\ \text{J}\cdot\text{mol}^{-1}\cdot\text{K}^{-1}}\left(\frac{308\ \text{K} - 298\ \text{K}}{(298\ \text{K})(308\ \text{K})}\right) = 0.52 \qquad \frac{k_{35}}{k_{25}} = e^{0.52} = \text{invln}(0.52) = 1.7$$

Hence, $\dfrac{R_{25}}{R_{35}} = 0.68 \times 1.7 = 1.2.$

13.83 *See Section 13.4.*

Rates of reaction with high activation energies are influenced more by temperature changes than are rates of reaction with low activation energies. This is due to the fact that equal temperature changes cause greater increases in the fraction of molecules with sufficient kinetic energy to react (i.e. with $KE \geq E_a$) the higher the activation energy. This phenomenon can be rationalized by considering the forms of Equations 13.9 and 13.11. The latter clearly shows that the greater the value of E_a, the greater the impact in the ratio k_1/k_2 for a given change in temperature.

13.84 *See Section 13.2 and Example 13.3.*

$2NO(g) + Br_2(g) \rightarrow 2NOBr(g)$ rate = $k(P_{NO})^x(P_{Br_2})^y$

Experiments 1 and 3 have the same pressure of Br_2 and can be used to determine the order with respect to NO.

Expt.	Initial P_{NO}, atm	Initial Rate, $atm \cdot s^{-1}$	Relative P_{NO} *	Relative Rate**
1	1.1×10^{-5}	0.37	1.0	1.0
2	3.5×10^{-5}	1.2	3.2	3.2

* Obtained by dividing each pressure by the smallest pressure.

** Obtained by dividing each initial rate by the smallest initial rate.

As the relative pressure of NO increases from 1.0 to 3.2 , the relative rate of reaction increases from 1.0 to 3.2. This indicates that the rate of reaction is proportional to the pressure of NO raised to the first power. The reaction is therefore first order with respect to NO.

There are no experiments which have the same pressure of NO and can be easily used to determine the order with respect to Br_2. Instead, Experiments 1 and 4 having changing P_{NO} and P_{Br_2} are used.

Expt.	Initial P_{NO}, atm	Initial P_{Br_2}, atm	Initial Rate $atm \cdot s^{-1}$	Relative P_{NO}	Relative P_{Br_2}	Relative Rate	Relative Rate / Relative P_{NO}
1	1.1×10^{-5}	1.2×10^{-5}	0.37	1.0	1.0	1.0	1.0
4	4.0×10^{-5}	3.0×10^{-5}	3.4	3.6	2.5	9.2	2.5

The change in Relative Rate/ Relative P_{NO} gives the change in Relative Rate that is due to change in Relative P_{Br_2}, since the reaction has been shown to be first order with respect to NO. As the relative pressure of Br_2 changes from 1.0 to 2.5 there is an accompanying change in Relative Rate/ Relative P_{NO} from 1.0 to 2.5. This indicates the rate of reaction is proportional to the pressure of Br_2 raised to the first power. The reaction is therefore first order with respect to Br_2.

The rate law is rate = $k(P_{NO})(P_{Br_2})$, and using data for Expt. 1 in the original table yields

$$k = \frac{\text{rate}}{P_{NO}\,P_{Br_2}} \qquad k = \frac{0.37\ atm \cdot s^{-1}}{(1.1 \times 10^{-5}\ atm)(1.2 \times 10^{-5}\ atm)} = 2.8 \times 10^{9}\ atm^{-1} \cdot s^{-1}$$

This gives **rate= $2.8 \times 10^9\ atm^{-1} \cdot s^{-1}\ P_{NO}\ P_{Br_2}$** as the rate law.

13.85 ***See Section 13.3, Table 13.4, and Examples 13.4, 13.9.***

Time,s	P, torr	P_f - P, torr	ln(P_f -P)	$\frac{1}{(P_f - P)}$, $torr^{-1}$
0.00	17.54	10.00	2.303	0.100
0.50	18.98	8.56	2.15	0.117
1.00	20.09	7.45	2.01	0.134
2.00	21.65	5.89	1.77	0.170
4.00	23.46	4.08	1.41	0.245
6.00	24.48	3.06	1.12	0.327
8.00	25.14	2.40	0.875	0.417
10.00	25.59	1.95	0.668	0.513

For a zero order reaction, a plot of $(P_f - P)$ vs time gives a straight line.

For a first order reaction, a plot of $\ln(P_f - P)$ vs. time gives a straight line.

For a second order reaction ,a plot of $\frac{1}{(P_f - P)}$ vs. time gives a straight line.

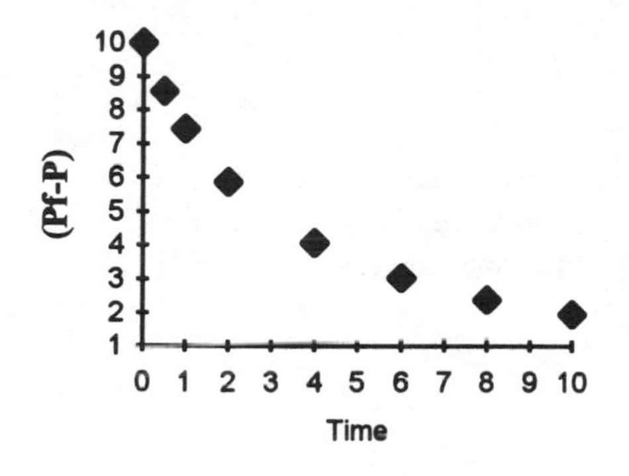

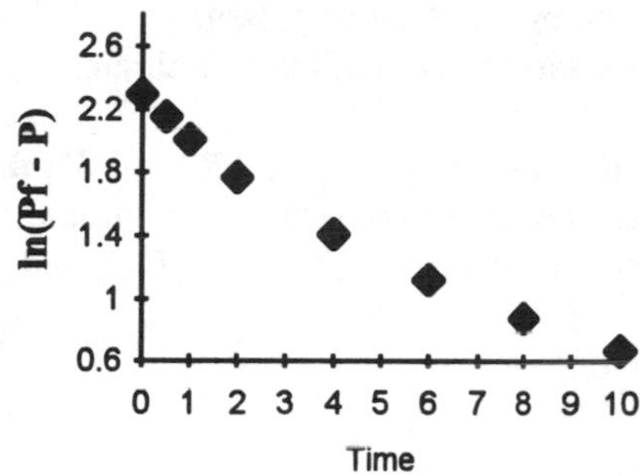

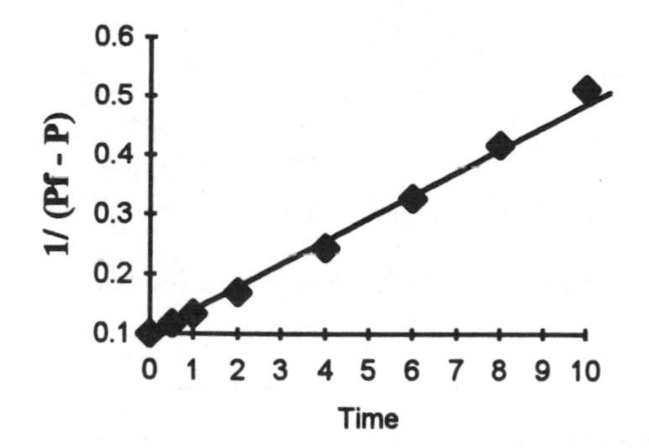

The data give a straight line plot for a **second order** reaction. The rate constant for a second order reaction is equal to the slope of the line. A linear regression analysis gives a slope of 0.0409, and a correlation coefficient of 0.9985 indicates very good agreement between the data and equation.

The rate law for the reaction is **rate = $k[H_2NNO_2]^2$**. The value of k expressed in units of torr and time in seconds is **0.0409 $torr^{-1}s^{-1}$**.

13.86 ***See Section 13.6 and Examples 13.13, 13.14.***

The rate law for Mechanism I

$$CH_3Br \xrightarrow{slow} CH_3^+ + Br^-$$
$$CH_3^+ + OH^- \xrightarrow{fast} CH_3OH$$

would be: **rate= $k[CH_3Br]$**.

The rate law for Mechanism II $CH_3Br + OH^- \rightarrow Br^- + CH_3OH$

would be: **rate= $k[CH_3Br][OH^-]$**.

13.87 *See Section 13.2 and Example 13.3.*

(a) $C_2H_5Cl(g) \rightarrow C_2H_4(g) + Hcl(g)$ rate = $k[C_2H_5Cl]^x$

$$\text{Rate 1} = \frac{\frac{0.0321 \text{ g NaOH}}{4.0 \text{ L}} \times \frac{1 \text{ mol NaOH}}{40.0 \text{ g NaOH}} \times \frac{1 \text{ mol HCl}}{1 \text{ mol NaOH}}}{11.4 \text{ s}} = 1.8 \times 10^{-5} \text{ mol} \cdot \text{L}^{-1} \cdot \text{s}^{-1}$$

$$\text{Rate 2} = \frac{\frac{0.0399 \text{ g NaOH}}{4.0 \text{ L}} \times \frac{1 \text{ mol NaOH}}{40.0 \text{ g NaOH}} \times \frac{1 \text{ mol HCl}}{1 \text{ mol NaOH}}}{11.4 \text{ s}} = 2.2 \times 10^{-5} \text{ mol} \cdot \text{L}^{-1} \cdot \text{s}^{-1}$$

$$\text{Rate 3} = \frac{\frac{0.0336 \text{ g NaOH}}{4.0 \text{ L}} \times \frac{1 \text{ mol NaOH}}{40.0 \text{ g NaOH}} \times \frac{1 \text{ mol HCl}}{1 \text{ mol NaOH}}}{11.4 \text{ s}} = 2.3 \times 10^{-5} \text{ mol} \cdot \text{L}^{-1} \cdot \text{s}^{-1}$$

Expt.	Initial $P_{C_2H_5Cl}$, torr	Initial Rate, $mol \cdot L^{-1} \cdot s^{-1}$	Relative Pressure*	Relative Rate**
1	131	1.8×10^{-5}	1.00	1.0
2	160	2.2×10^{-5}	1.22	1.2
3	172	2.3×10^{-5}	1.31	1.3

* Obtained by dividing each pressure by the smallest pressure.

** Obtained by dividing each initial rate by the smallest initial rate.

As the relative pressure of C_2H_5Cl increases from 1.00 to 1.22 to 1.31, the relative rate of reaction increases from 1.0 to 1.2 to 1.3. This indicates that the rate of reaction is proportional to the pressure of C_2H_5Cl raised to the first power. The reaction is therefore first order with respect to C_2H_5Cl, and the rate law is: **rate = $k[C_2H_5Cl]$**.

(b) Using data for the first experiment,

$$M\,C_2H_5Cl = \frac{n}{V} = \frac{P}{RT} = \frac{131 \text{ torr} \times \frac{1 \text{ atm}}{760 \text{ torr}}}{\left(0.0821 \frac{\text{L} \cdot \text{atm}}{\text{mol} \cdot \text{K}}\right)(437 \text{ K})} = 4.8 \times 10^{-3} \text{ mol} \cdot \text{L}^{-1}$$

$$k = \frac{\text{rate 1}}{[C_2H_5Cl]} = \frac{1.8 \times 10^{-5} \text{ mol} \cdot \text{L}^{-1} \cdot \text{s}^{-1}}{4.8 \times 10^{-3} \text{ mol} \cdot \text{L}^{-1}} = \mathbf{3.8 \times 10^{-3}\ s^{-1}}$$

Note: Pressures, rather than molar concentrations, can be used to determine the rate law because the solution for Part b indicates the pressure of a gas is directly proportional to its molarity.

13.88 *See Section 13.3 and Example 13.4.*

$HCOOH(g) \rightarrow CO_2(g) + H_2(g)$

Since the stoichiometric coefficients are all one, we can represent the decrease in the pressure of HCOOH(g) during each time interval by -y and the increase in the pressure of $CO_2(g)$ and $H_2(g)$ each by +y during the same time interval. This gives $P_T = P_{HCOOH} + P_{CO_2} + P_{H_2}$ and $P_T = (220 - y) + y + y = 220 + y$ at each point during the reaction. Solving for y and using (220-y) for P_{HCOOH} gives

Time,s	P_{HCOOH}	$\ln(P_{HCOOH})$
0	220	5.394
50	96	4.56
100	61	4.11
150	32	3.47
200	17	2.83
250	9	2.2
300	5	1.6

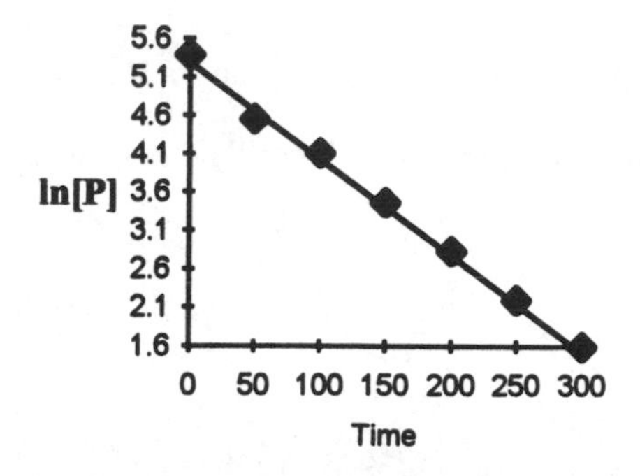

The rate constant for a first order reaciton is equal to the slope of the line when lnP is plotted versus time. A linear regression analysis gives a slope of -0.0124, and a correlation coefficient of 0.9988 indicates very good agreement between the data and first-order reaction. Hence, **k = 0.0124 s^{-1}**.

Since $kt_{\frac{1}{2}} = 0.693$ for a first-order process,

$$t_{\frac{1}{2}} = \frac{0.693}{k} \qquad t_{\frac{1}{2}} = \frac{0.693}{0.0124\ s^{-1}} = \mathbf{55.9\ s}$$

Note: If your calculator does not have the ability to perform linear regression analyses or you do not have access to a computer graphing program, you can caculate the slope of the line by using $\frac{(\ln[P]_2 - \ln[P]_1)}{(t_2 - t_1)}$ with two points from the line which are far apart.

13.89 ***See Section 13.2 and Example 13.3.***

(a) $2NO(g) + 2H_2(g) \rightarrow N_2(g) + 2H_2O(g)$ $\qquad$ rate = $k[NO]^x[H_2]^y$

Experiments 1, 2 and 3 have the same pressure of NO and can be used to determine the order with respect H_2.

Expt.	Initial P_{H_2}, torr	Initial Rate, torr·s^{-1}	Relative Pressure*	Relative Rate**
1	289	0.160	1.97	2.02
2	205	0.110	1.39	1.39
3	147	0.179	1.00	1.00

* Obtained by dividing each pressure by the smallest pressure.
** Obtained by dividing each initial rate by the smallest initial rate.

As the relative pressure of H_2 increases from 1.00 to 1.39 to 1.97, the relative rate of reaction increases from 1.00 to 1.39 to 2.02. This indicates that the rate of reaction is proportional to the pressure of H_2 raised to the first power. The reaction is therefore first order with respect to H_2.

Experiments 4, 5 and 6 have the same pressure of H_2 and can be used to determine the order with respect to NO.

Expt.	Initial P_{NO}, torr	Initial Rate, torr·s^{-1}	Relative Pressure	Relative Rate
4	359	0.150	2.36	6.0
5	300	0.103	1.97	4.1
6	152	0.025	1.00	1.0

As the relative pressure of NO increases from 1.00 to 1.97 to 2.36, the relative rate of reaction increases from 1.0 to 4.1 to 6.0. This indicates that the rate of reaction is proportional to the pressure of NO raised to the second power. The reaction is therefore second order with respect to NO.

Hence, the rate law is **rate = $k[H_2][NO]^2$**.

(b) Using data from experiment one,

$$k = \frac{\text{rate 1}}{(P_{H_2})_1 (P_{NO})_1^2} = \frac{0.160 \text{ torr} \cdot s^{-1}}{(289 \text{ torr})(400 \text{ torr})^2} = \mathbf{3.46 \times 10^{-9} \text{ torr}^{-2} \cdot s^{-1}}$$

(c) The Arrhenius equation can be written as $\ln k = \frac{-E_a}{R}\left(\frac{1}{T}\right) + \ln A$. A plot of ln k vs. 1/ T will have a slope of $-E_a/R$ and an intercept of ln A. Determine the value of the slope of the line, and use E_a = -(slope)(R) to obtain E_a.

k, relative	ln k	T, K	1/ T, K^{-1}
1.00	0.000	956	0.00105
2.34	0.850	984	0.00102
5.15	1.64	1024	0.000977
10.9	2.39	1061	0.000943
18.8	2.93	1099	0.000910

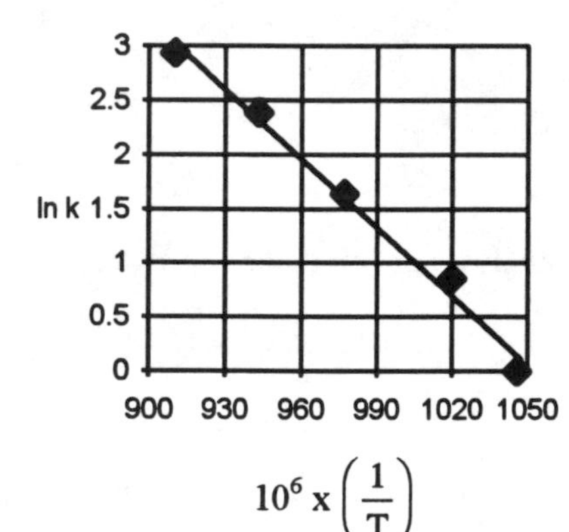

A linear regression analysis gives a slope of -2.11×10^4 and an intercept of 22.23 with a correlation coefficient of 0.9946 indicating reasonably good agreement between the data and equation. Hence,

$$E_a = -(-2.11 \times 10^4 \text{ K}^{-1})(8.314 \text{ J} \cdot \text{mol}^{-1} \cdot \text{K}^{-1}) = \mathbf{1.75 \times 10^5 \text{ J} \cdot \text{mol}^{-1}}$$

Note: If your calculator does not have the ability to perform linear regression analyses or you do not have access to a computer graphing program, you can caculate the slope of the line by using

$\frac{(\ln k_2 - \ln k_1)}{(1/T_2 - 1/T_1)}$ with two points from the line which are far apart.

The slope of the line can then be used to calculate the value of E_a.

13.90, 13.91 ***Solutions for these Spreadsheet Exercises are on the Reger, Goode and Mercer Spreadsheet Solutions Diskette available from Saunders College Publishing.***

Chapter 14: Chemical Equilibrium

14.1 *See Section 14.1.*

By following some physical property, such as the absorption of light or pressure of the system, it is possible to determine that no further change in concentration or pressure is occurring, and the system is at equilibrium.

14.2 *See Section 14.1.*

If cars enter and leave a parking lot at the same rate, there is no net increase in the number of cars in the parking lot and the system is in a state of dynamic equilibrium. If the rate of cars entering the parking lot increased, the rate of cars leaving the parking lot would need to increase to maintain the state of equilibrium. Conversely, if the rate of cars leaving the parking lot increased, the rate of cars entering the parking lot would need to increase to maintain the state of equilibrium.

14.3 *See Section 14.1.*

If the rate of cars entering a parking lot exceeds the rate of cars leaving the parking lot, there is a net change in the number of cars in the parking lot and the system is not at equilibrium.

14.4 *See Section 14.1.*

The production of ozone by sunlight striking the upper atmosphere cannot in and of itself yield a state of equilibrium. A state of equilibrium can only be attained if the ozone were destroyed at an equal rate by sunlight striking the ozone or by other means giving a constant concentration of ozone in the upper atmosphere.

14.5 *See Section 14.1.*

A battery which produces an electrical current as a result of a spontaneous chemical reaction is not a system that is at equilibrium. When the chemical reaction that produces the electrical current reaches equilibrium, it proceeds in opposite directions at equal rates. When this occurs, no electrical current can be produced, and we say the battery is "dead." Electrochemistry is described in greater detail in Chapter 18.

14.6 *See Section 14.1.*

K_{eq} is the value of Q that is obtained at equilibrium. If Q is greater than K_{eq}, spontaneous reaction occurs to the left to reach equilibrium. If Q is less than K_{eq}, spontaneous reaction occurs to the right to reach equilibrium. A system always reacts spontaneously to make $Q = K_{eq}$. Hence, Q and K_{eq} are the same when a system is at equilibrium and are different in value when the system is not at equilibrium.

14.7 *See Section 14.2.*

K_p and K_c are equal when there is no change in the moles of gas during the chemical reaction. They are unequal whenever there is an increase or decrease in moles of gas during the chemical reaction, whenever the moles of gaseous reactants is not equal to the moles of gaseous products during the chemical reaction.

14.8 See Section 14.3.

The principle of LeChatelier is applicable to any system at equilibrium. As a person squeezes a partially filled balloon there is an increase in pressure inside the balloon that seeks to restore the volume of the gas in the balloon to its original volume.

14.9 See Section 14.3.

A decrease in the volume of a gaseous system favors a decrease in the moles of gas present, whereas an increase in the volume of a gaseous system favors an increase in the moles of gas present. However, when the number of moles of gaseous reactants is equal to the number of moles of gaseous products, there can be no change in the number of moles of gases present with changes in the volume of the system. Under these conditions, there is no change in the equilibrium concentrations of the reactants and products with changes in the volume of the ystem.

14.10 See Sections 14.2, 14.3.

Heating a chemical reaction system causes reaction to occur in the endothermic direction , the direction consuming heat. Heating a reaction that is exothermic in the forward direction therefore causes the formation of more reactants and decreases the value of K_{eq}. Heating a reaction that is endothermic in the forward direction therefore causes the formation of more products and increases the value of K_{eq}. Changes in the value of Q with temperature follow the changes in the value of K_{eq}, because K_{eq} is just a special value of Q pertaining to the state of equilibrium.

14.11 See Section 14.5.

The concentration of a pure solid or liquid is related to its density and molar mass and does not change during the course of a reaction because it does not depend on the amount of solid or liquid present. Hence, concentration terms for pure solids and liquids are not included in equilibrium constant expressions.

14.12 See Section 14.6.

Heating a chemical reaction system causes the reaction to occur in the endothermic direction , the direction consuming heat. Hence, the solubility of substances having endothermic dissolving processes increases with increasing temperature, whereas the solubility of substances having exothermic dissolving processes decreases with increasing temperature.

14.13 See Section 14.1 and Example 14.1.

(a) $PCl_5(g) \rightleftarrows PCl_3(g) + Cl_2(g)$

$$K_c = \frac{[PCl_3][Cl_2]}{[PCl_5]}$$

(b) $2NO_2(g) \rightleftarrows 2NO(g) + O_2(g)$

$$K_c = \frac{[NO]^2[O_2]}{[NO_2]^2}$$

(c) $2SO_3(g) \rightleftarrows 2SO_2(g) + O_2(g)$

$$K_c = \frac{[SO_2]^2[O_2]}{[SO_3]^2}$$

(d) $H_2(g) + I_2(g) \rightleftarrows 2HI(g)$

$$K_c = \frac{[HI]^2}{[H_2][I_2]}$$

14.14 See Section 14.1 and Example 14.1.

(a) $2H_2O(g) \rightleftarrows 2H_2(g) + O_2(g)$

$$K_c = \frac{[H_2]^2[O_2]}{[H_2O]^2}$$

(b) $2HCl(g) \rightleftarrows H_2(g) + Cl_2(g)$

$$K_c = \frac{[H_2][Cl_2]}{[HCl]^2}$$

(c) $CO(g) + Cl_2(g) \rightleftarrows COCl_2(g)$

$$K_c = \frac{[COCl_2]}{[CO][Cl_2]}$$

(d) $2CO(g) + O_2(g) \rightleftarrows 2CO_2(g)$

$$K_c = \frac{[CO_2]^2}{[CO]^2[O_2]}$$

14.15 ***See Section 14.1 and Example 14.1.***

(a) $HCl(g) + \frac{1}{4}O_2(g) \rightleftarrows \frac{1}{2}Cl_2(g) + \frac{1}{2}H_2O(g)$

$$K_p = \frac{(P_{Cl_2})^{\frac{1}{2}}(P_{H_2O})^{\frac{1}{2}}}{(P_{HCl})(P_{O_2})^{\frac{1}{4}}}$$

(b) $\frac{1}{2}N_2O_4(g) \rightleftarrows NO_2(g)$

$$K_p = \frac{(P_{NO_2})}{(P_{N_2O_4})^{\frac{1}{2}}}$$

(c) $N_2O_4(g) \rightleftarrows N_2(g) + 2O_2(g)$

$$K_p = \frac{(P_{N_2})(P_{O_2})^2}{(P_{N_2O_4})}$$

(d) $\frac{1}{2}O_2(g) + SO_2(g) \rightleftarrows SO_3(g)$

$$K_p = \frac{(P_{SO_3})}{(P_{O_2})^{\frac{1}{2}}(P_{SO_2})}$$

14.16 ***See Section 14.1 and Example 14.1.***

(a) $Cl_2(g) + H_2O(g) \rightleftarrows 2HCl(g) + \frac{1}{2}O_2(g)$

$$K_p = \frac{(P_{HCl})^2(P_{O_2})^{\frac{1}{2}}}{(P_{Cl_2})(P_{H_2O})}$$

(b) $2NO_2(g) \rightleftarrows N_2O_4(g)$

$$K_p = \frac{(P_{N_2O_4})}{(P_{NO_2})^2}$$

(c) $3O_2(g) \rightleftarrows 2O_3(g)$

$$K_p = \frac{(P_{O_3})^2}{(P_{O_2})^3}$$

(d) $CO_2(g) \rightleftarrows CO(g) + \frac{1}{2}O_2(g)$

$$K_p = \frac{(P_{CO})(P_{O_2})^{\frac{1}{2}}}{(P_{CO_2})}$$

14.17 ***See Section 14.1 and Example 14.2.***

$2H_2(g) + O_2(g) \rightleftarrows 2H_2O(g)$

$$K_1 = \frac{[H_2O]^2}{[H_2]^2[O_2]} = 1.6 \times 10^{10}$$

$H_2(g) + \frac{1}{2}O_2(g) \rightleftarrows H_2O(g)$

$$K_2 = \frac{[H_2O]}{[H_2][O_2]^{\frac{1}{2}}}$$

Examination shows $K_2 = (K_1)^{\frac{1}{2}}$. Hence, $K_2 = \sqrt{1.6 \times 10^{10}}$ **$= 1.3 \times 10^5$**.

14.18 ***See Section 14.1 and Example 14.2.***

$H_2(g) + I_2(g) \rightleftarrows 2HI(g)$

$$K_1 = \frac{[HI]^2}{[H_2][I_2]} = 155$$

$\frac{1}{2}H_2(g) + \frac{1}{2}I_2(g) \rightleftarrows HI(g)$ $\qquad K_2 = \frac{[HI]}{[H_2]^{\frac{1}{2}}[I_2]^{\frac{1}{2}}}$

Examination shows $K_2 = (K_1)^{\frac{1}{2}}$. Hence, $K_2 = \sqrt{155} = \mathbf{12.4}$.

14.19 *See Section 14.1 and Example 14.2.*

$PCl_3(g) + Cl_2(g) \rightleftarrows PCl_5(g)$ $\qquad K_1 = \frac{[PCl_5]}{[PCl_3][Cl_2]} = 1.7 \times 10^4$

$PCl_5(g) \rightleftarrows PCl_3(g) + Cl_2(g)$ $\qquad K_2 = \frac{[PCl_3][Cl_2]}{[PCl_5]}$

Examination shows $K_2 = \frac{1}{K_1}$. Hence, $K_2 = \frac{1}{1.7 \times 10^4} = \mathbf{5.9 \times 10^{-5}}$.

14.20 *See Section 14.1 and Example 14.2.*

$2SO_2(g) + O_2(g) \rightleftarrows 2SO_3(g)$ $\qquad K_1 = \frac{[SO_3]^2}{[SO_2]^2[O_2]} = 6.81$

$2SO_3(g) \rightleftarrows 2SO_2(g) + O_2(g)$ $\qquad K_2 = \frac{[SO_2]^2[O_2]}{[SO_3]^2}$

Examination shows $K_2 = \frac{1}{K_1}$. Hence, $K_2 = \frac{1}{6.81} = \mathbf{0.147}$.

14.21 *See Section 14.1 and Example 14.2.*

$NO(g) + \frac{1}{2}Br_2(g) \rightleftarrows NOBr(g)$ $\qquad K_1 = \frac{[NOBr]}{[NO][Br_2]^{\frac{1}{2}}} = 4.42$

$2NOBr(g) \rightleftarrows 2NO(g) + Br_2(g)$ $\qquad K_2 = \frac{[NO]^2[Br_2]}{[NOBr]^2}$

Examination shows $K_2 = \frac{1}{(K_1)^2}$. Hence, $K_2 = \frac{1}{(4.42)^2} = \mathbf{0.0512}$.

14.22 *See Section 14.1 and Example 14.2.*

$N_2(g) + O_2(g) \rightleftarrows 2NO(g)$ $\qquad K_1 = \frac{[NO]^2}{[N_2][O_2]} = 3.6 \times 10^{-6}$

$2NO(g) \rightleftarrows N_2(g) + O_2(g)$ $\qquad K_2 = \frac{[N_2][O_2]}{[NO]^2}$

Examination shows $K_2 = \frac{1}{K_1}$. Hence, $K_2 = \frac{1}{3.6 \times 10^{-6}} = \mathbf{2.8 \times 10^5}$.

14.23 ***See Section 14.2 and Example 14.4.***

$CO(g) + \frac{1}{2}O_2 \rightleftarrows CO_2(g)$

$[CO] = \frac{0.050 \text{ mol}}{5.0 \text{ L}} = 0.010\ M$

$[O_2] = \frac{0.050 \text{ mol}}{5.0 \text{ L}} = 0.010\ M$

$[CO_2] = \frac{4.95 \text{ mol}}{5.0 \text{ L}} = 0.99\ M$

$Q_c = \frac{[CO_2]}{[CO][O_2]^{\frac{1}{2}}}$ $Q_c = \frac{(0.99)}{(0.010)(0.010)^{\frac{1}{2}}} = 9.9 \times 10^2$

Number line: ----K-----------Q----

Since Q_c is greater than K_c (1.05×10^{-5}), the reaction proceeds to the left forming more CO.

14.24 ***See Section 14.2 and Example 14.4.***

$2SO_2(g) + O_2 \rightleftarrows 2SO_3(g)$

$[SO_2] = 0.015\ M$

$[O_2] = 0.012\ M$

$[SO_3] = 1.45\ M$

$Q_c = \frac{[SO_3]^2}{[SO_2]^2[O_2]}$ $Q_c = \frac{(1.45)^2}{(0.015)^2(0.012)} = 7.8 \times 10^5$

Number line: -------Q--K-----------

Since Q_c is less than K_c (5.0×10^6), the reaction proceeds to the right

14.25 ***See Section 14.2 and Example 14.4.***

$PCl_5(g) \rightleftarrows PCl_3(g) + Cl_2(g)$

$[PCl_5] = \frac{1.0 \text{ mol}}{20.0 \text{ L}} = 0.050\ M$

$[PCl_3] = \frac{2.0 \text{ mol}}{20.0 \text{ L}} = 0.10\ M$

$[Cl_2] = \frac{2.0 \text{ mol}}{20.0 \text{ L}} = 0.10\ M$

$Q_c = \frac{[PCl_3][Cl_2]}{[PCl_5]}$ $Q_c = \frac{(0.10)(0.10)}{(0.050)} = 0.20$

Number line: ----Q-----------K----

Since Q_c is less than K_c (4.0), the reaction proceeds to the right

14.26 ***See Section 14.2 and Example 14.4.***

$2NO_2(g) \rightleftarrows N_2O_4(g)$

$[NO_2] = \frac{2.0 \text{ mol}}{2.5 \text{ L}} = 0.80\ M$

$[N_2O_4] = \frac{0. \text{ mol}}{2.5 \text{ L}} = 0.00\ M$

$Q_c = \frac{[N_2O_4]}{[NO_2]^2}$ $Q_c = \frac{(0.00)}{(0.80)^2} = 0.00$

Number line: ----Q-----------K----

Since Q_c is less than K_c (17.2), the reaction proceeds to the right

14.27 ***See Section 14.2 and Example 14.4.***

$2HI(g) \rightleftarrows H_2(g) + I_2(g)$ $K_c = 0.010$ $Q_c = \frac{[H_2][I_2]}{[HI]^2}$

Converting the initial amounts to initial molar concentrations and calculating Q_c gives:

Part	[HI]	$[H_2]$	$[I_2]$	Q_c	Q_c vs. K_c	Direction of Spontaneous Reaction
(a)	5×10^{-6}	2.0×10^{-5}	1.0×10^{-5}	8	$Q_c > K_c$	←
(b)	2.0×10^{-5}	2.0×10^{-4}	2.0×10^{-4}	1×10^2	$Q_c > K_c$	←
(c)	1.05×10^{-3}	1.0×10^{-4}	9.0×10^{-5}	8×10^{-3}	$Q_c < K_c$	→
(d)	2.00×10^{-3}	5.0×10^{-5}	8.0×10^{-5}	1×10^{-3}	$Q_c < K_c$	→

14.28 ***See Section 14.2 and Example 14.4.***

$CO_2(g) + H_2O(g) \rightleftarrows CO_2(g) + H_2(g)$ $\quad K_c = 5.0 \quad Q_c = \dfrac{[CO_2][H_2]}{[CO][H_2O]}$

Converting the initial amounts to initial molar concentrations and calculating Q_c gives:

Part	[CO]	$[H_2O]$	$[CO_2]$	$[H_2]$	Q_c	Q_c vs. K_c	Direction of Spontaneous Reaction
(a)	0.50	0.40	0.80	0.90	3.6	$Q_c < K_c$	→
(b)	0.01	0.02	0.03	0.04	6	$Q_c > K_c$	←
(c)	1.22	1.22	2.78	2.78	5.19	$Q_c > K_c$	←
(d)	0.61	1.22	1.39	2.39	4.46	$Q_c < K_c$	→

14.29 ***See Section 14.1 and Example 14.3.***

$NO(g) + \frac{1}{2}Br_2(g) \rightleftarrows NOBr_2(g)$ $\quad \Delta n_g = \left[1-\left(1+\frac{1}{2}\right)\right] = -\frac{1}{2}$

Substituting in $K_c = K_p(RT)^{-\Delta n}$ gives $K_c = 116(0.0821 \times 298)^{\frac{1}{2}} = \mathbf{574}$.

14.30 ***See Section 14.1 and Example 14.3.***

$CO_2(g) \rightleftarrows CO(g) + \frac{1}{2}O_2(g)$ $\quad \Delta n_g = \left[\left(1+\frac{1}{2}\right)-1\right] = \frac{1}{2}$

Substituting in $K_c = K_p(RT)^{-\Delta n}$ gives $K_c = 2.48(0.0821 \times 3{,}000)^{-\frac{1}{2}} = \mathbf{0.158}$.

14.31 ***See Section 14.1 and Example 14.3.***

$SO_2(g) + Cl_2(g) \rightleftarrows SO_2Cl_2(g)$ $\quad \Delta n_g = [1-(1+1)] = -1$

Substituting in $K_p = K_c(RT)^{\Delta n}$ gives $K_p = 20.9(0.0821 \times 500)^{-1} = \mathbf{0.509}$.

14.32 ***See Section 14.1 and Example 14.3.***

$CO(g) + Cl_2(g) \rightleftarrows COCl_2(g)$ $\quad \Delta n_g = [1-(1+1)] = -1$

Substituting in $K_p = K_c(RT)^{\Delta n}$ gives $K_p = 15.8(0.0821 \times 871)^{-1} = \mathbf{0.211}$.

14.33 ***See Section 14.3 and Examples 14.5, 14.6, 14.7.***

(a) Adding SO_3, a reactant, causes $Q < K$ and spontaneous reaction to the right.
(b) Removing O_2, a product, causes $Q < K$ and spontaneous reaction to the right.

(c) Increasing the volume of the container favors a net increase in the number of mole of gas. In this case, this causes spontaneous reaction to the right (1.5 mol gas for products vs. 1 mol gas for reactants).
(d) An increase in temperature favors the endothermic direction. In this case, this causes spontaneous reaction to the right.
(e) The pressure of materials other than reactants or products has no effect on the equilibrium. Hence, adding argon has no effect on this equilibrium.

14.34 *See Section 14.3 and Examples 14.5, 14.6, 14.7.*

(a) Adding NH_3, a reactant, causes Q < K and spontaneous reaction to the right.
(b) Adding H_2, a product, causes Q > K and spontaneous reaction to the left.
(c) Decreasing the volume of the container favors a net decrease in the number of mole of gas. In this case, this causes spontaneous reaction to the left (2 mol gas for reactants vs. 4 mol gas for products).
(d) A decrease in temperature favors the exothermic direction. In this case, this causes spontaneous reaction to the left.
(e) The pressure of materials other than reactants or products has no effect on the equilibrium. Hence, adding argon has no effect on this equilibrium.

14.35 *See Sections 14.1, 14.4, and Example 14.8.*

(a) $2NH_3(g) \rightleftarrows N_2(g) + 3H_2(g)$ The change in the concentration of NH_3 can be determined from the initial and equilibrium concentrations given for NH_3. The change in the concentrations of N_2 and H_2 can be determined from the change for NH_3 and the coefficients in the balanced equation giving

Conc, M	NH_3	N_2	H_2
initial	0.0240	0.0000	0.0000
Change	−0.0200	+0.0100	+0.0300
equilibrium	0.0040	0.0100	0.0300

(b) $K_c = \dfrac{[N_2][H_2]^3}{[NH_3]^2}$ $K_c = \dfrac{(0.0100)(0.0300)^3}{(0.0040)^2} = \mathbf{0.017}$

14.36 *See Sections 14.1, 14.4 and Example 14.8.*

(a) $2SO_2(g) + O_2(g) \rightleftarrows 2SO_3(g)$ The change in the concentration of SO_2 can be determined from the initial and equilibrium concentrations given for SO_2. The change in the concentrations of O_2 and SO_3 can be determined from the change for SO_2 and the coefficients in the balanced equation giving

Conc, *M*	SO_2	O_2	SO_3
initial	0.0076	0.0036	0.0000
Change	−0.0044	-0.0022	+0.0044
equilibrium	0.0032	0.0014	0.0044

(b) $K_c = \dfrac{[SO_3]^2}{[SO_2]^2[O_2]}$ $K_c = \dfrac{(0.0044)^2}{(0.0032)^2(0.0014)} = \mathbf{1.4 \times 10^3}$

14.37 ***See Sections 14.1, 14.4 and Example 14.8.***

(a) $2HI(g) \rightleftarrows H_2(g) + I_2(g)$ The change in the pressure of HI can be determined from the initial and equilibrium pressures given for HI. The change in the pressures of H_2 and I_2 can be determined from the change for HI and the coefficients in the balanced equation giving

P, atm	HI	H_2	I_2
initial	1.25	0.00	0.00
Change	−0.20	+0.10	+0.10
equilibrium	1.05	0.10	0.10

(b) $K_p = \frac{(P_{H_2})(P_{I_2})}{(P_{HI})^2}$ $K_p = \frac{(0.10)(0.10)}{(1.05)^2} = \mathbf{9.1 \times 10^{-3}}$

14.38 ***See Sections 14.1, 14.4 and Example 14.8.***

(a) $CO(g) + Cl_2(g) \rightleftarrows COCl_2(g)$

P, atm	CO	Cl_2	$COCl_2$
initial	0.96	1.02	0.00
Change	-y	-y	+y
equilibrium	0.96 - y	1.02 - y	y

At equilibrium, $P_{CO} + P_{Cl_2} + P_{COCl_2} = P_T$. Hence, $(0.96 - y) + (1.02 - y) + y = 1.22$, $1.98 - y = 1.22$ and $y = 0.76$ atm, giving

P, atm	CO	Cl_2	$COCl_2$
initial	0.96	1.02	0.00
Change	-0.76	-0.76	+0.76
equilibrium	0.20	0.26	0.76

(b) $K_p = \frac{(P_{COCl_2})}{(P_{CO})(P_{Cl_2})}$ $K_p = \frac{(0.76)}{(0.20)(0.26)} = \mathbf{15}$

14.39 ***See Sections 14.1, 14.4 and Example 14.8.***

(a) $2SO_3(g) \rightleftarrows 2SO_2(g) + O_2(g)$ The change in the concentration of SO_3 can be determined from the initial and equilibrium concentrations given for SO_3. The change in the concentrations of SO_2 and O_2 can be determined from the change for SO_3 and the coefficients in the balanced equation giving

Conc, *M*	SO_3	SO_2	O_2
initial	1.00	0.00	0.00
Change	−0.50	+0.50	+0.25
equilibrium	0.50	0.50	0.25

(b) $K_c = \frac{[SO_2]^2[O_2]}{[SO_3]^2}$ $K_c = \frac{(0.50)^2(0.25)}{(0.50)^2} = \mathbf{0.25}$

14.40 *See Sections 14.1, 14.4 and Example 14.8.*

(a) $SO_2Cl_2\,(g) \rightleftarrows SO_2\,(g) + Cl_2\,(g)$ The change in the concentration of SO_2Cl_2 can be determined from the initial and equilibrium concentrations given for SO_2Cl_2. The change in the concentrations of SO_2 and Cl_2 can be determined from the change for SO_2Cl_2 and the coefficients in the balanced equation giving

Conc, *M*	SO_2Cl_2	SO_2	Cl_2
initial	0.0060	0.0000	0.0000
Change	−0.0058	+0.0058	+0.0058
equilibrium	0.00020	0.0058	0.0058

(b) $K_c = \dfrac{[SO_2][Cl_2]}{[SO_2Cl_2]}$ $K_c = \dfrac{(0.0058)(0.0058)}{(0.00020)} = \mathbf{0.17}$

14.41 *See Sections 14.1, 14.4.*

$2NO_2\,(g) \rightleftarrows N_2O_4\,(g)$

(a) $?\ \text{mol } NO_2 = \dfrac{0.010\ \text{mol } NO_2}{1\ \text{L}} \times 20.0\ \text{L} = \mathbf{0.20\ mol\ NO_2}$

(b) $?\ \text{mol } N_2O_4 \text{ formed} = \text{mol } NO_2 \text{ reacted} \times \dfrac{1\ \text{mol } N_2O_4}{2\ \text{mol } NO_2 \text{ reacted}}$

$?\ \text{mol } N_2O_4 \text{ formed} = (2.0 - 0.20)\ \text{mol } NO_2 \text{ reacted} \times \dfrac{1\ \text{mol } N_2O_4}{2\ \text{mol } NO_2 \text{ reacted}} = \mathbf{0.90\ mol\ N_2O_4}$

$[N_2O_4] = \dfrac{0.90\ \text{mol } N_2O_4}{20.0\ \text{L}} = \mathbf{0.045}\ \boldsymbol{M}\ \mathbf{N_2O_4}$

(c) $K_c = \dfrac{[N_2O_4]}{[NO_2]^2}$ $K_c = \dfrac{(0.045)}{(0.010)^2} = \mathbf{4.5 \times 10^2}$

14.42 *See Sections 14.1, 14.4 and Example 14.8.*

(a) $NOBr(g) \rightleftarrows NO(g) + \frac{1}{2}Br_2\,(g)$ The change in the concentration of Br_2 can be determined from the initial and equilibrium concentrations given for Br_2. The change in the concentrations of NOBr and NO can be determined from the change for Br_2 and the coefficients in the balanced equation giving

Conc, M	NOBr	NO	Br_2
initial	0.0050	0.0000	0.0000
Change	−0.0040	+0.0040	+0.0020
equilibrium	0.0010	0.0040	0.0020

(b) $K_c = \dfrac{[NO][Br_2]^{\frac{1}{2}}}{[NOBr]}$ $K_c = \dfrac{(0.0040)(0.0020)^{\frac{1}{2}}}{(0.0010)} = \mathbf{0.18}$

14.43 *See Section 14.4 and Example 14.9.*

(a) $H_2(g) + I_2(g) \rightleftarrows 2HI(g)$

Conc, M	H_2	I_2	HI
initial	0.50	0.50	0.
Change	$-\frac{1}{2}y$	$-\frac{1}{2}y$	+y
equilibrium	$0.50 - \frac{1}{2}y$	$0.50 - \frac{1}{2}y$	y

(b) $K_c = \dfrac{[HI]^2}{[H_2][I_2]}$ (c) $K_c = \dfrac{(y)^2}{(0.50 - \frac{1}{2}y)(0.50 - \frac{1}{2}y)} = 4.6$

14.44 *See Section 14.4 and Example 14.9.*

(a) $SO_2(g) + NO_2(g) \rightleftarrows SO_3(g) + NO(g)$

Conc, *M*	SO_2	NO_2	SO_3	NO
initial	1.0	1.0	0.0	0.0
Change	-y	-y	+y	+y
equilibrium	1.0 - y	1.0 - y	y	y

(b) $K_c = \dfrac{[SO_3][NO]}{[SO_2][NO_2]}$ (c) $K_c = \dfrac{(y)(y)}{(1.0 - y)(1.0 - y)} = 1.98$

14.45 *See Section 14.4 and Example 14.9.*

(a) $SO_3(g) + NO(g) \rightleftarrows SO_2(g) + NO_2(g)$

Conc, *M*	SO_3	NO	SO_2	NO_2
initial	0.025	0.025	0.00	0.00
Change	-y	-y	+y	+y
equilibrium	0.025-y	0.025-y	y	y

(b) $K_c = \dfrac{[SO_2][NO_2]}{[SO_3][NO]}$ (c) $K_c = \dfrac{(y)(y)}{(0.025-y)(0.025-y)} = 0.50$

14.46 *See Section 14.4 and Example 14.9.*

(a) $CO(g) + H_2O(g) \rightleftarrows CO_2(g) + H_2(g)$

Conc, *M*	CO_3	H_2O	CO_2	H_2
initial	0.50	0.50	0.00	0.00
Change	-y	-y	+y	+y
equilibrium	0.50 - y	0.50 - y	y	y

(b) $K_c = \dfrac{[CO_2][H_2]}{[CO][H_2O]}$ (c) $K_c = \dfrac{(y)(y)}{(0.50 - y)(0.50 - y)} = 0.55$

14.47 *See Section 14.4 and Example 14.10.*

$COCl_2 \rightleftarrows CO + Cl_2$

Conc, *M*	$COCl_2$	CO	Cl_2
initial	0.250	0.00	0.00
Change	-y	+y	+y
equilibrium	0.250-y	y	y

$$K_c = \frac{[CO][Cl_2]}{[COCl_2]} \qquad K_c = \frac{(y)(y)}{(0.250-y)} = 4.93 \times 10^{-3}$$

Rearranging yields $y^2 + 4.93 \times 10^{-3}\, y - 1.23 \times 10^{-3} = 0$.

Substituting into the quadratic formula gives $y = \dfrac{-4.93 \times 10^{-3} \pm \sqrt{(4.93 \times 10^{-3})^2 - 4(1)(-1.23 \times 10^{-3})}}{2(1)}$

and $y = \dfrac{-4.93 \times 10^{-3} \pm 7.03 \times 10^{-2}}{2} = 3.27 \times 10^{-2}$ or -3.76×10^{-2}.

The only reasonable value for y is 3.27×10^{-2}, since -3.76×10^{-2} predicts negative concentrations for CO and Cl_2. Hence, $[\mathbf{COCl_2}] = 0.250 - 0.0327 = \mathbf{0.217}$ ***M*** and $[\mathbf{CO}] = [\mathbf{Cl_2}] = \mathbf{0.0327}$ ***M***.

Inserting these equilibrium concentrations into the equilibrium expression gives

$$K_c = \frac{[CO][Cl_2]}{[COCl_2]} \qquad K_c = \frac{(0.0327)(0.0327)}{(0.217)} = 4.93 \times 10^{-3}$$

and this is in good agreement with the given value of 4.93×10^{-3}.

14.48 *See Section 14.4 and Example 14.10.*

$PCl_5 \rightleftarrows PCl_3 + Cl_2$

Conc, *M*	PCl_5	PCl_3	Cl_2
initial	0.100	0.000	0.000
Change	-y	+y	+y
equilibrium	0.100 - y	y	y

$$K_c = \frac{[PCl_3][Cl_2]}{[PCl_5]} \qquad K_c = \frac{(y)(y)}{(0.100-y)} = 0.60$$

Rearranging yields $y^2 + 0.60y - 0.060 = 0$.

Substituting into the quadratic formula gives $y = \dfrac{-0.60 \pm \sqrt{(0.60)^2 - 4(1)(-0.060)}}{2(1)}$

and $y = \dfrac{-0.60 \pm 0.77}{2} = 0.085$ or -0.68.

The only reasonable value for y is 0.085, since -0.68 predicts negative concentrations for PCl_3 and Cl_2. Hence, $[\mathbf{PCl_5}] = 0.100 - 0.085 = \mathbf{0.015}$ ***M*** and $[\mathbf{PCl_3}] = [\mathbf{Cl_2}] = \mathbf{0.085}$ ***M***.

Inserting these equilibrium concentrations into the equilibrium expression gives

$$K_c = \frac{[PCl_3][Cl_2]}{[PCl_5]} \qquad K_c = \frac{(0.085)(0.085)}{(0.015)} = 0.48$$

and this is in reasonably good agreement with the given value of 0.60 considering the limitations of significant figures.

14.49 *See Section 14.4 and Example 14.10.*

$CO(g) + Cl_2(g) \rightleftarrows COCl_2(g)$

Conc, *M*	CO	Cl_2	$COCl_2$
initial	0.200	0.200	0.0
Change	-y	-y	+y
equilibrium	0.200-y	0.200-y	y

$$K_c = \frac{[COCl_2]}{[CO][Cl_2]} \qquad K_c = \frac{y}{(0.200-y)(0.200-y)} = 7.52$$

Rearranging and combining terms yields $7.52y^2 - 4.01y + 0.301 = 0$.

Substituting into the quadratic formula gives $y = \dfrac{4.01 \pm \sqrt{(-4.01)^2 - 4(7.52)(0.301)}}{2(7.52)}$

and $y = \dfrac{4.01 \pm 2.65}{15.04} = 0.443$ or 0.0904.

The only reasonable value for y is 0.0904 , since 0.443 predicts negative concentrations for CO and CL_2.
Hence, $[\mathbf{CO}] = [\mathbf{Cl_2}] = 0.200 - 0.0904 = \mathbf{0.110}$ ***M*** and $[\mathbf{COCl_2}] = \mathbf{0.0904}$ ***M***.

Inserting these equilibrium concentrations into the equilibrium expression gives

$$K_c = \frac{[COCl_2]}{[CO][Cl_2]} \qquad K_c = \frac{0.0904}{(0.110)(0.110)} = 7.47$$

and this is in good agreement with the given value of 7.52.

14.50 *See Section 14.4 and Example 14.10.*

$PCl_3(g) + Cl_2(g) \rightleftarrows PCl_5(g)$

Conc, *M*	PCl_3	Cl_2	PCl_5
initial	0.0100	0.0100	0.0000
Change	-y	-y	+y
equilibrium	0.0100 - y	0.0100 - y	y

$$K_c = \frac{[PCl_5]}{[PCl_3][Cl_2]} \qquad K_c = \frac{y}{(0.0100 - y)(0.0100 - y)} = 8.18$$

Rearranging and combining terms yields $8.18y^2 - 1.164y + 0.000818 = 0$.

Substituting into the quadratic formula gives $y = \dfrac{1.164 \pm \sqrt{(-1.164)^2 - 4(8.18)(0.000818)}}{2(8.18)}$

and $y = \dfrac{1.164 \pm 1.152}{16.36} = 0.142$ or 7.335×10^{-4}.

The only reasonable value for y is 7.335×10^{-4} , since 0.142 predicts negative concentrations for PCl_3 and Cl_2.
Hence, $[\mathbf{PCl_3}] = [\mathbf{Cl_2}] = 0.0100 - 7.335 \times 10^{-4} = \mathbf{0.009}$ ***M*** and $[\mathbf{PCl_5}] = \mathbf{7.335 \times 10^{-4}}$ ***M***.

Inserting these equilibrium concentrations into the equilibrium expression gives

$$K_c = \frac{[PCl_5]}{[PCl_3][Cl_2]} \qquad K_c = \frac{7.335 \times 10^{-4}}{(0.0009)(0.0009)} = 9$$

and this is in good agreement with the given value of 8.18.

14.51 *See Section 14.4 and Example 14.10.*

$SO_2(g) + Cl_2(g) \rightleftarrows SO_2Cl_2(g)$

Conc, M	SO_2	Cl_2	SO_2Cl_2
initial	0.0100	0.0200	0.
Change	-y	-y	+y
equilibrium	0.0100-y	0.0200-y	y

$$K_c = \frac{[SO_2Cl_2]}{[SO_2][Cl_2]} \qquad K_c = \frac{y}{(0.0100-y)(0.0200-y)} = 89.3$$

Rearranging and combining terms yields $89.3y^2 - 3.68y + 0.0179 = 0$.

Substituting into the quadratic formula gives $y = \frac{3.68 \pm \sqrt{(-3.68)^2 - 4(89.3)(0.0179)}}{2(89.3)}$

and $y = \frac{3.68 \pm 2.67}{178.6} = 3.56 \times 10^{-2}$ or 5.66×10^{-3}.

The only reasonable answer is 5.66×10^{-3}, since 3.56×10^{-2} predicts negative concentrations for SO_2 and Cl_2. Hence, $[SO_2] = 0.0100 - 5.66 \times 10^{-3} = \mathbf{4.34 \times 10^{-3}}$ ***M***, $[Cl_2] = 0.0200 - 5.66 \times 10^{-3} = \mathbf{1.43 \times 10^{-2}}$ ***M*** and $[SO_2Cl_2] = \mathbf{5.66 \times 10^{-3}}$ ***M***.

Inserting these equilibrium expressions into the equilibrium expression gives

$$K_c = \frac{[SO_2Cl_2]}{[SO_2][Cl_2]} \qquad K_c = \frac{(5.66 \times 10^{-3})}{(4.34 \times 10^{-3})(1.43 \times 10^{-2})} = 91.1$$

and this value is in good agreement with the given value of 89.3.

14.52 ***See Section 14.4 and Example 14.10.***

$CO_2(g) + Cl_2(g) \rightleftarrows COCl_2(g)$

Conc, M	CO_2	Cl_2	$COCl_2$
initial	0.333	0.667	0.000
Change	-y	-y	+y
equilibrium	0.333 - y	0.667 - y	y

$$K_c = \frac{[COCl_2]}{[CO][Cl_2]} \qquad K_c = \frac{y}{(0.333-y)(0.667-y)} = 2.5$$

Rearranging and combining terms yields $2.5y^2 - 3.5y + 0.56 = 0$.

Substituting into the quadratic formula gives $y = \frac{3.5 \pm \sqrt{(-3.5)^2 - 4(2.5)(0.56)}}{2(2.5)}$

and $y = \frac{3.5 \pm 2.6}{5.0} = 1.2$ or 0.18.

The only reasonable answer is 0.18, since 1.2 predicts negative concentrations for CO and Cl_2. Hence, $[CO] = 0.333 - 0.18 = \mathbf{0.15}$ ***M***, $[Cl_2] = 0.667 - 0.18 = \mathbf{0.49}$ ***M*** and $[COCl_2] = \mathbf{0.18}$ ***M***.

Inserting these equilibrium expressions into the equilibrium expression gives

$$K_c = \frac{[COCl_2]}{[CO][Cl_2]} \qquad K_c = \frac{(0.18)}{(0.15)(0.49)} = 2.4$$

and this value is in good agreement with the given value of 2.5.

14.53 *See Section 14.4 and Example 14.10.*

$CO(g) + H_2(g) \rightleftarrows CH_2O(g)$

P, atm	CO	H_2	CH_2O
initial	2.00	3.00	0.0
Change	-y	-y	+y
equilibrium	2.00-y	3.00-y	+y

$$K_p = \frac{P_{CH_2O}}{P_{CO}P_{H_2}} \qquad K_p = \frac{y}{(2.00-y)(3.00-y)} = 5.60$$

Rearranging and combining terms yields $5.60y^2 - 29.0y + 33.6 = 0$.

Substituting into the quadratic formula gives $y = \dfrac{29.0 \pm \sqrt{(-29.0)^2 - 4(5.60)(33.6)}}{2(5.60)}$

and $y = \dfrac{29.0 \pm 9.4.0}{11.2} = 3.43$ or 1.75.

The only reasonable value for y is 1.75, since 3.43 predicts negative pressures for CO and H_2.
Hence, $\mathbf{P_{CO}} = 2.00 - 1.75 = \mathbf{0.25\ atm}$, $\mathbf{P_{H_2}} = 3.00 - 1.75 = \mathbf{1.25\ at}$ and $\mathbf{P_{CH_2O}} = \mathbf{1.75\ atm}$.

Inserting these equilibrium pressures into the equilibrium expression gives

$$K_p = \frac{P_{CH_2O}}{P_{CO}P_{H_2}} \qquad K_p = \frac{1.75}{(0.25)(1.25)} = 5.6$$

and this is in agreement with the given value of 5.6.

14.54 *See Section 14.4 and Example 14.10.*

$PCl_3(g) + Cl_2(g) \rightleftarrows PCl_5(g)$ $\qquad Q_p = \dfrac{(P_{PCl_5})}{(P_{PCl_3})(P_{Cl_2})}$

Known Quantities: $n_{PCl_3} = 0.15$ mol $\quad n_{Cl_2} = 0.20$ mol $\quad n_{PCl_5} = 0.25$ mol

V = 10.0 L $\quad$ T = 332 + 273 = 605 K

Solving PV = nRT for P gives $P = \dfrac{nRT}{V}$.

$$P_{PCl_3} = \frac{(0.15\ mol)\left(0.08212\frac{L\cdot atm}{mol\cdot K}\right)(605\ K)}{10.0\ L} = 0.75\ atm$$

$$P_{Cl_2} = \frac{(0.20\ mol)\left(0.08212\frac{L\cdot atm}{mol\cdot K}\right)(605\ K)}{10.0\ L} = 0.99\ atm$$

$$P_{PCl_5} = \frac{(0.25\ mol)\left(0.08212\frac{L\cdot atm}{mol\cdot K}\right)(605\ K)}{10.0\ L} = 1.2\ atm$$

$$Q_P = \frac{(P_{PCl_5})}{(P_{PCl_3})(P_{Cl_2})} \qquad Q_P = \frac{(1.2)}{(0.75)(0.99)} = 1.6$$

Since $Q_p < K_p$ (2.5), the reaction proceeds to the right to attain equilibrium, giving

P, atm	PCl_3	Cl_2	PCl_5
initial	0.75	0.99	1.2
Change	-y	-y	+y
equilibrium	0.75 - y	0.99 - y	1.2 + y

$$K_p = \frac{P_{PCl_5}}{P_{PCl_3}P_{Cl_2}} \qquad K_p = \frac{(1.2+y)}{(0.75-y)(0.99-y)} = 2.5$$

Rearranging and combining terms yields $2.5y^2 - 5.35y + 0.66 = 0$.

Substituting into the quadratic formula gives $y = \dfrac{-5.35 \pm \sqrt{(-5.35)^2 - 4(2.5)(+0.66)}}{2(2.5)}$

and $y = \dfrac{5.35 \pm 4.69}{5.0} = 2.0$ or 0.13.

The only reasonable value for y is 0.13, since 2.0 predicts a negative pressure for PCl_3 and Cl_2.
Hence, $\mathbf{P_{PCl_3}} = 0.75 - 0.13 = \mathbf{0.62\ atm}$, $\mathbf{P_{Cl_2}} = 0.99 - 0.13 = \mathbf{0.86\ atm}$ and $\mathbf{P_{PCl_5}} = 1.2 + 0.13 = \mathbf{0.13\ atm}$.

Inserting these equilibrium pressures into the equilibrium expression gives

$K_p = \dfrac{P_{PCl_5}}{P_{PCl_3}P_{Cl_2}}$ $\quad K_p = \dfrac{0.17}{(0.23)(0.28)} = 2.6$ and this is in agreement with the given value of 2.5.

14.55 ***See Sections 14.1, 14.4 and Examples 14.3, 14.10.***

$2SO_3(g) \rightleftarrows 2SO_2(g) + O_2(g)$ $\qquad \Delta n_g = [(2+1) - 2] = 1$

Substituting in $K_c = K_p(RT)^{-\Delta n}$ gives $K_c = 1150(0.0821 \times 1529)^{-1} = 9.16$ and

Conc, *M*	SO_3	SO_2	O_2
initial	0.80	0.	0.0
Change	-2y	+2y	+y
equilibrium	0.80 - 2y	2y	y

$K_c = \dfrac{[SO_2]^2[O_2]}{[SO_3]^2}$ $\quad K_c = \dfrac{(2y)^2(y)}{(0.80 - 2y)^2} = 9.16$.

However, this set-up leads to a cubic equation and ignores the fact that the reaction goes essentially to completion.

Assuming the reaction goes to completion gives $\dfrac{(0.80)^2(0.40)}{[SO_3]^2} \cong 9.16$ and $[SO_3] \cong 0.17$, which means $0.80 - 2y = 0.17$ and $y \cong 0.32$.

Reiterating yields $\dfrac{(0.64)^2(0.32)}{[SO_3]^2} \cong 9.16$ and $[SO_3] \cong 0.12$, which means $0.80 - 2y = 0.12$ and $y \cong 0.34$.

Another reiteration yields $\dfrac{(0.68)^2(0.34)}{[SO_3]^2} \cong 9.16$ and $[SO_3] \cong 0.13$, which is in good agreement with the result of the previous calculation. Hence, $\mathbf{[SO_3] = 0.13}$ ***M***, $\mathbf{[SO_2] = 0.68}$ ***M*** and $\mathbf{[O_2] = 0.34}$ ***M***.

14.56 ***See Sections 4.2, 14.1, 14.4, and Examples 4.8, 14.3, 14.10.***

$H_2(g) + F_2(g) \rightleftarrows 2HF(g)$ $\qquad \Delta n_g = [2 - (1+1)] = 0$

Substituting in $K_c = K_p(RT)^{-\Delta n}$ gives $K_c = 7.8 \times 10^{14}(0.0821 \times T)^0 = 7.8 \times 10^{14}$.

Solving $M(\text{con}) \times V(\text{con}) = M(\text{dil}) \times V(\text{dil})$ for $M(\text{dil})$ gives $M(\text{dil}) = \dfrac{M(\text{con}) \times V(\text{con})}{V(\text{dil})}$.

$$M(\text{dil})_{H_2} = \frac{0.010\ M \times 2.0\ \text{L}}{7.0\ \text{L}} = 0.0029\ M \qquad M(\text{dil})_{F_2} = \frac{0.020\ M \times 5.0\ \text{L}}{7.0\ \text{L}} = 0.014\ M$$

Conc, M	H_2	F_2	HF
initial	0.0029	0.014	0.0
Change	-y	-y	+2y
equilibrium	0.0029 - y	0.014 - y	2y

$$K_c = \frac{[HF]^2}{[H_2][F_2]} \qquad K_c = \frac{(2y)^2}{(0.0029 - y)(0.014 - y)} = 7.8 \times 10^{14}$$

Rearranging and combining terms yields $7.8 \times 10^{14} y^2 - 1.3 \times 10^{13} y + 3.2 \times 10^{10} = 0$.

Substituting into the quadratic formula gives $y = \dfrac{1.3 \times 10^{13} \pm \sqrt{(1.3 \times 10^{13})^2 - 4(7.8 \times 10^{14})(3.2 \times 10^{10})}}{2(7.8 \times 10^{14})}$

and $y = \dfrac{1.3 \times 10^{13} \pm 8.3 \times 10^{12}}{2(7.8 \times 10^{14})} = 0.014$ or 0.0030.

The only reasonable answer is 0.0030, since 0.014 predicts a negative concentration for H_2.
Hence, $[\mathbf{H_2}] = 0.0029 - 0.0030 \cong \mathbf{0.00}\ \boldsymbol{M}$, $[\mathbf{F_2}] = 0.014 - 0.0030 = \mathbf{0.011}\ \boldsymbol{M}$ and $[\mathbf{HF}] = 2(0.0030) = \mathbf{0.0060}\ \boldsymbol{M}$.

Note that these results are consistent with the observation that the very large value of K_c indicates the reaction goes virtually to completion and that H_2 is the limiting reactant. In fact, assuming the reaction goes to completion and noting that H_2 is limiting reactant gives [HF] = 2y = 2(0.0029) = 0.0058 M, $[H_2]$ = 0.00 M and $[F_2]$ = 0.014 - .0029 = 0.011 M.

14.57 ***See Section 14.2 and Example 14.4.***

For $CaCO_3(s) \rightleftarrows CaO(s) + CO_2(g)$, $K_p = \mathbf{P_{CO_2}} = \mathbf{0.12\ atm}$ at 1000K.

14.58 ***See Section 14.2 and Example 14.4.***

For $NaHCO_3(s) \rightleftarrows NaOH(s) + CO_2(g)$, $K_p = \mathbf{P_{CO_2}} = \mathbf{0.25\ atm}$ at 700K.

14.59 ***See Section 14.2 and Example 14.4.***

For $CaSO_4(s) \rightleftarrows CaO(s) + SO_3(g)$, $K_p = \mathbf{P_{SO_3}} = \mathbf{0.74\ atm}$ at 2100K.

14.60 ***See Section 14.2 and Example 14.4.***

For $Na_2CO_3(s) \rightleftarrows Na_2O(s) + CO_2(g)$, $K_p = \mathbf{P_{CO_2}} = \mathbf{1.25\ atm}$ at 1500K.

14.61 ***See Section 14.6 and Example 14.13.***

(a) $Mg_2F(s) \rightleftarrows Mg^{2+}(aq) + 2F^-(aq)$ $K_{sp} = [Mg^{2+}][F^-]^2$

(b) $Ca_3(PO_4)_2(s) \rightleftarrows 3Ca^{2+}(aq) + 2PO_4^{3-}(aq)$ $K_{sp} = [Ca^{2+}]^3[PO_4^{3-}]^2$

(c) $Al_2(CO_3)_3(s) \rightleftarrows 2Al^{3+}(aq) + 3CO_3^{2-}(aq)$ $K_{sp} = [Al^{3+}]^2[CO_3^{2-}]^3$

(d) $LaF_3(s) \rightleftarrows La^{3+}(aq) + 3F^-(aq)$ $K_{sp} = [La^{3+}][F^-]^3$

14.62 ***See Section 14.6 and Example 14.13.***

(a) $BaSO_4(s) \rightleftarrows Ba^{2+}(aq) + SO_4^{2-}(aq)$ $K_{sp} = [Ba^{2+}][SO_4^{2-}]$

(b) $AgCH_3CO_2(s) \rightleftarrows Ag^+(aq) + CH_3CO_2^-(aq)$ $K_{sp} = [Ag^+][CH_3CO_2^-]$

(c) $Cu_2CO_3(s) \rightleftarrows 2Cu^+(aq) + CO_3^{2-}(aq)$ $K_{sp} = [Cu^+]^2[CO_3^{2-}]$

(d) $AuCl_3(s) \rightleftarrows Au^{3+}(aq) + 3Cl^-(aq)$ $K_{sp} = [Au^{3+}][Cl^-]^3$

14.63 ***See Section 14.6 and Examples 14.14, 14.15.***

$AgI(s) \rightleftarrows Ag^+(aq) + I^-(aq)$

Conc, *M*	AgI	Ag^+	I^-
initial	excess	0.	0.
Change	-9×10^{-9}	$+9 \times 10^{-9}$	$+9 \times 10^{-9}$
equilibrium	excess	9×10^{-9}	9×10^{-9}

$K_{sp} = [Ag^+][I^-]$ $K_{sp} = (9 \times 10^{-9})(9 \times 10^{-9}) = \mathbf{8 \times 10^{-17}}$

14.64 ***See Section 14.6 and Examples 14.14, 14.15.***

$AgIO_3(s) \rightleftarrows Ag^+(aq) + IO_3^-(aq)$

Conc, M	$AgIO_3$	Ag^+	IO_3^-
initial	excess	0.	0.
Change	-1.8×10^{-4}	$+1.8 \times 10^{-4}$	$+1.8 \times 10^{-4}$
equilibrium	excess	1.8×10^{-4}	1.8×10^{-4}

$K_{sp} = [Ag^+][IO_3^-]$ $K_{sp} = (1.8 \times 10^{-4})(1.8 \times 10^{-4}) = \mathbf{3.2 \times 10^{-8}}$

14.65 ***See Section 14.6, Examples 14.14, 14.15, and the Solution for Excercise 14.63.***

(a) $BaCrO_4(s) \rightleftarrows Ba^{2+}(aq) + CrO_4^{2-}(aq)$

$K_{sp} = [Ba^{2+}][CrO_4^{2-}]$ $K_{sp} = (1.1 \times 10^{-5})(1.1 \times 10^{-5}) = \mathbf{1.2 \times 10^{-10}}$

(b) $CsMnO_4(s) \rightleftarrows Cs^+(aq) + MnO_4^-(aq)$

$$? \text{ s } CsMnO_4 = \frac{0.22 \text{ g } CsMnO_4}{100 \text{ mL}} \times \frac{10^3 \text{ mL}}{1 \text{ L}} \times \frac{1 \text{ mol } CsMnO_4}{251.8 \text{ g } CsMnO_4} = 0.0087\ M\ CsMnO_4$$

$K_{sp} = [Cs^+][MnO_4^-]$ $\qquad$ $K_{sp} = (0.0087)(0.0087) = \mathbf{7.6 \times 10^{-5}}$

(c) $Ag_3PO_4(s) \rightleftarrows 3Ag^+(aq) + PO_4^{3-}(aq)$

$K_{sp} = [Ag^+]^3[PO_4^{3-}]$ $\qquad$ $K_{sp} = (3 \times 4.4 \times 10^{-5})^3(4.4 \times 10^{-5}) = \mathbf{1.0 \times 10^{-16}}$

14.66 ***See Section 14.6, Examples 14.14, 14.15, and the Solution for Excercise 14.63.***

(a) $Ag_2SO_4(s) \rightleftarrows 2Ag^+(aq) + SO_4^{2-}(aq)$

$K_{sp} = [Ag^+]^2[SO_4^{2-}]$ $\qquad$ $K_{sp} = (2 \times 1.0 \times 10^{-2})^2(1.0 \times 10^{-2}) = \mathbf{4.0 \times 10^{-6}}$

(b) $KIO_3(s) \rightleftarrows K^+(aq) + IO_3^-(aq)$

$$? \text{ s } KIO_3 = \frac{43 \text{ g } KIO_3}{1 \text{ L}} \times \frac{1 \text{ mol } KIO_3}{214.0 \text{ g } KIO_3} = 0.20\ M\ KIO_3$$

$K_{sp} = [K^+][IO_3^-]$ $\qquad$ $K_{sp} = (0.20)(0.20) = \mathbf{0.040}$

(c) $CdF_2(s) \rightleftarrows Cd^{2+}(aq) + 2F^-(aq)$

$K_{sp} = [Cd^{2+}][F^-]^2$ $\qquad$ $K_{sp} = [0.12][2 \times 0.12]^2 = \mathbf{6.9 \times 10^{-3}}$

14.67 ***See Section 14.7, Table 14.6, and Example 14.16.***

$Ag_2WO_4(s) \rightleftarrows 2Ag^+(aq) + WO_4^{2-}(aq)$

Conc, *M*	Ag_2WO_4	Ag^+	WO_4^{2-}
initial	excess	0.	0.
Change	-s	+2s	+s
equilibrium	excess	2s	s

$K_{sp} = [Ag^+]^2[WO_4^{2-}]$ $\qquad$ $K_{sp} = (2s)^2(s) = 5.5 \times 10^{-12}$

Hence, $4s^3 = 5.5 \times 10^{-12}$ and $\mathbf{s = 1.1 \times 10^{-4}}$ ***M***.

14.68 ***See Section 14.7, Tables 14.4, 14.6, and Example 14.16.***

$Cu(IO_3)_2(s) \rightleftarrows Cu^{2+}(aq) + 2IO_3^-(aq)$

Conc, *M*	$Cu(IO_3)_2$	Cu^{2+}	IO_3^-
initial	excess	0.	0.
Change	-s	+s	+2s
equilibrium	excess	s	2s

$K_{sp} = [Cu^{2+}][IO_3^-]^2$ $\qquad$ $K_{sp} = (s)(2s)^2 = 7.4 \times 10^{-8}$

Hence, $4s^3 = 7.4 \times 10^{-8}$ and $\mathbf{s = 2.6 \times 10^{-3}}$ ***M***.

14.69 *See Section 14.7, Tables 14.4, 14.6 and Example 14.16.*

$BaSO_4\,(s) \rightleftarrows Ba^{2+}\,(aq) + SO_4^{2-}\,(aq)$

Conc, M	$BaSO_4$	Ba^{2+}	SO_4^{2-}
initial	excess	0.	0.
Change	-s	+s	+s
equilibrium	excess	s	s

$K_{sp} = [Ba^{+}][SO_4^{2-}]$ $\quad K_{sp} = (s)(s) = 8.7 \times 10^{-11}$

Hence, $s^2 = 8.7 \times 10^{-11}$ and $s = 9.3 \times 10^{-6}\,M$.

$$?\frac{g\ BaSO_4}{L} = \frac{9.3 \times 10^{-6}\ mol\ BaSO_4}{1\ L\ soln} \times \frac{233.4\ g\ BaSO_4}{1\ mol\ BaSO_4} = \mathbf{2.2 \times 10^{-3}\ \frac{g\ BaSO_4}{L}}$$

14.70 *See Section 14.7, Appendix F, and Example 14.16.*

$PbCO_3\,(s) \rightleftarrows Pb^{2+}\,(aq) + CO_3^{2-}\,(aq)$

Conc, M	$PbCO_3$	Pb^{2+}	CO_3^{2-}
initial	excess	0.	0.
Change	-s	+s	+s
equilibrium	excess	s	s

$K_{sp} = [Pb^{2+}][CO_3^{2-}]$ $\quad K_{sp} = (s)(s) = 1.1 \times 10^{-13}$

Hence, $s^2 = 1.1 \times 10^{-13}$ and $s = 3.2 \times 10^{-7}\,M$.

$$?\frac{g\ Pb}{day} = \frac{1\ L}{day} \times \frac{365\ days}{yr} \times \frac{3.2 \times 10^{-7}\ mol\ Pb}{L} \times \frac{207.2\ g\ Pb}{1\ mol\ Pb} = \mathbf{0.024\ g\ Pb / yr}$$

14.71 *See Section 14.7 and Example 14.17.*

$BaSO_4\,(s) \rightleftarrows Ba^{2+}\,(aq) + SO_4^{2-}\,(aq)$

(a)

Conc, M	$BaSO_4$	Ba^{2+}	SO_4^{2-}
initial	excess	0.	0.
Change	-s	+s	+s
equilibrium	excess	s	s

$K_{sp} = [Ba^{2+}][SO_4^{2-}]$ $\quad K_{sp} = (s)(s) = 8.7 \times 10^{-11}$

Hence, $s^2 = 8.7 \times 10^{-11}$ and $\mathbf{s = 9.3 \times 10^{-6}\,M}$.

(b) This problem is solved in the same manner as part (a), but the initial concentration of Ba^{2+} is 0.10 M from 0.10 *M BaCl₂*,

Conc, M	$BaSO_4$	Ba^{2+}	SO_4^{2-}
initial	excess	0.10	0.
Change	-s	+s	+s
equilibrium	excess	0.10 + s	s

$K_{sp} = [Ba^{2+}][SO_4^{2-}]$ $\quad K_{sp} = (0.10 + s)(s) = 8.7 \times 10^{-11}$

Assuming s is small compared to 0.10 leads to $(0.10)(s) = 8.7 \times 10^{-11}$ and $\mathbf{s = 8.7 \times 10^{-10}}$.
Since s is indeed small compared to 0.10, our assumption is valid.

Note: We had good reason to believe s would be small compared to 0.10. The value of s that was obtained in the absence of $BaCl_2$ in part (a) was $9.3 \times 10^{-6}\,M$, and application of the Principle of Le Chatelier leads us to predict it will be even smaller in the presence of 0.10 M $BaCl_2$.

14.72 ***See Section 14.7 and Example 14.17.***

$Cu(IO_3)_2(s) \rightleftarrows Cu^{2+}(aq) + 2IO_3^-(aq)$

(a)

Conc, *M*	$Cu(IO_3)_2$	Cu^{2+}	IO_3^-
initial	excess	0.	0.
Change	-s	+s	+2s
equilibrium	excess	s	2s

$K_{sp} = [Cu^{2+}][IO_3^-]^2 \qquad K_{sp} = (s)(2s)^2 = 7.4 \times 10^{-8}$

Hence, $4s^3 = 7.4 \times 10^{-8}$ and **s = 2.6 x 10⁻³ *M*.**

(b) This problem is solved in the same manner as part (a), but the initial concentration of Cu^{2+} is 0.10 *M* from 0.10 *M* $Cu(NO_3)_2$,

Conc, *M*	$Cu(IO_3)_2$	Cu^{2+}	IO_3^-
initial	excess	0.10	0.0
Change	-s	+s	+2s
equilibrium	excess	0.10 + s	2s

$K_{sp} = [Cu^{2+}][IO_3^-]^2 \qquad K_{sp} = (0.10 + s)(2s)^2 = 7.4 \times 10^{-8}$

Assuming s is small compared to 0.10 leads to $(0.10)(2s)^2 = 7.4 \times 10^{-8}$ and $\mathbf{s = 4.3 \times 10^{-4}}$.
Since s is indeed small compared to 0.10, our assumption is valid.

Note: We had good reason to believe s would be small compared to 0.10. The value of s that was obtained in the absence of $NaIO_3$ in part (a) was 2.6×10^{-3} *M*, and application of the Principle of Le Chatelier leads us to predict it will be even smaller in the presence of 0.10 *M* $NaIO_3$.

14.73 ***See Section 14.7 and Example 14.17.***

$PbF_2(s) \rightleftarrows Pb^{2+}(aq) + 2F^-(aq)$

(a)

Conc, *M*	PbF_2	Pb^{2+}	F^-
initial	excess	0.	0.
Change	-s	+s	+2s
equilibrium	excess	s	2s

$K_{sp} = [Pb^{2+}][F^-]^2 \qquad K_{sp} = (s)(2s)^2 = 7.1 \times 10^{-7}$

Hence, $4s^3 = 7.1 \times 10^{-7}$ and $\mathbf{s = 5.6 \times 10^{-3}}$ ***M*.**

(b) This problem is solved in the same manner as part (a), but the initial concentration of F^- is 0.050 *M* from 0.050 *M* KF.

Conc, *M*	PbF_2	Pb^{2+}	F^-
initial	excess	0.	0.050
Change	-s	+s	+2s
equilibrium	excess	s	0.050 + 2s

$K_{sp} = [Pb^{2+}][F^-]^2 \qquad K_{sp} = (s)(0.050 + 2s)^2 = 7.1 \times 10^{-7}$

Assuming 2s is small compared to 0.050 leads to $(s)(0.050)^2 = 7.1 \times 10^{-7}$ and $\mathbf{s = 2.8 \times 10^{-4}}$.
Since 2s is indeed small compared to 0.050, our assumption is valid.

Note: We had good reason to believe 2s would be small compared to 0.050. The value of s that was obtained in the absence of KF in part (a) was 5.6×10^{-3} *M*, and application of the Principle of Le Chatelier leads us to predict it will be even smaller in the presence of 0.050 *M* KF.

14.74 *See Section 14.7 and Example 14.18.*

$CuI(s) \rightleftarrows Cu^{+}(aq) + I^{-}(aq)$

(a)

Conc, *M*	CuI	Cu^{+}	I^{-}
initial	excess	0.	0.
Change	-s	+s	+s
equilibrium	excess	s	s

$K_{sp} = [Cu^{+}][I^{-}]$ $\qquad K_{sp} = (s)(s) = 1.1 \times 10^{-12}$

Hence, $s^2 = 1.1 \times 10^{-12}$ and **$s = 1.0 \times 10^{-6}$ *M*.**

(b) This problem is solved in the same manner as part (a), but the initial concentration of I^{-} is 0.050 *M* from 0.050 *M* NaI,

Conc, *M*	CuI	Cu^{+}	I^{-}
initial	excess	0.	0.050
Change	-s	+s	+s
equilibrium	excess	s	0.050 + s

$K_{sp} = [Cu^{+}][I^{-}]$ $\qquad K_{sp} = (s)(0.050 + s) = 1.1 \times 10^{-12}$

Assuming s is small compared to 0.050 leads to (s)(0.050) = 1.1×10^{-12} and **$s = 2.2 \times 10^{-11}$**.
Since s is indeed small compared to 0.050, our assumption is valid.

Note: We had good reason to believe s would be small compared to 0.050. The value of s that was obtained in the absence of NaI in part (a) was 1.0×10^{-6} *M*, and application of the Principle of Le Chatelier leads us to predict it will be even smaller in the presence of 0.050 *M* NaI.

14.75 *See Section 14.7 and Example 14.19.*

We need to calculate Q_{sp} and compare it to K_{sp} for $Ag_2SO_4(s) \rightleftarrows 2Ag^{+}(aq) + SO_4^{2-}(aq)$ to determine whether Ag_2SO_4 will precipitate. Hence, we need to calculate the concentrations of Ag^{+} and SO_4^{2-} in the mixed soultion.
Solving $M(con) \times V(con) = M(dil) \times V(dil)$ for M(dil) gives $M(dil) = \frac{M(con) \times V(con)}{V(dil)}$.

Hence, $[Ag^{+}]$ in mixed soln $= \frac{0.0010\ M \times 10\ mL}{20\ mL} = 0.00050\ M$

and $[SO_4^{2-}]$ in mixed soln $= \frac{0.0010\ M \times 10\ mL}{20\ mL} = 0.00050\ M$

$Q_{sp} = [Ag^{+}]^2[SO_4^{2-}]$ $\qquad Q_{sp} = (5.0 \times 10^{-4})^2(5.0 \times 10^{-4}) = 1.2 \times 10^{-10}$

Since Q_{sp} is less than K_{sp} (1.2×10^{-5}), **no precipitate will form**.

14.76 *See Section 14.7, Example 14.19, and the Solution for Exercise 14.69.*

We need to calculate Q_{sp} and compare it to K_{sp} for $MgF_2(s) \rightleftarrows Mg^{2+}(aq) + 2F^{-}(aq)$ to determine whether MgF_2 will precipitate. Hence, we need to calculate the concentrations of Mg^{2+} and F^{-} in the mixed solution, using the same procedure that was used in working Exercise 14.69.

$[Mg^{2+}] = \frac{1.0 \times 10^{-6}\ M \times 20\ mL}{100\ mL} = 2.0 \times 10^{-7}\ M$ $\qquad [F^{-}] = \frac{(1.0 \times 10^{-6}\ M) \times 80\ mL}{100\ mL} = 8.0 \times 10^{-7}\ M$

$Q_{sp} = [Mg^{2+}][F^-]^2$ $\qquad Q_{sp} = (2.0 \times 10^{-7})(2.0 \times 10^{-7})^2 = 1.3 \times 10^{-19}$

Since Q_{sp} is less than K_{sp} (7.4×10^{-11}), no precipitate will form.

14.77 ***See Section 14.7, Example 14.19, and the Solution for Exercise 14.69.***

We need to calculate Q_{sp} and compare it to K_{sp} for $Fe(OH)_2\,(s) \rightleftarrows Fe^{2+}\,(aq) + 2OH^-\,(aq)$ to determine whether $Fe(OH)_2$ will precipitate. Hence, we need to calculate the concentrations of Fe^{2+} and OH^- in the mixed solution, using the same procedure that was used in working Exercise 14.69.

$$[Fe^{2+}] = \frac{1.0 \times 10^{-6}\ M \times 10\ \text{mL}}{30\ \text{mL}} = 3.3 \times 10^{-7}\ M \qquad [OH^-] = \frac{2(3.0 \times 10^{-4}\ M) \times 20\ \text{mL}}{30\ \text{mL}} = 4.0 \times 10^{-4}\ M$$

$$Q_{sp} = [Fe^{2+}][OH^-]^2 \qquad Q_{sp} = (3.3 \times 10^{-7})(4.0 \times 10^{-4})^2 = 5.3 \times 10^{-14}$$

Since Q_{sp} is greater than K_{sp} (1.6×10^{-14}), **a precipitate will form.**

14.78 ***See Section 14.7, Example 14.19, and the Solution for Exercise 14.69.***

We need to calculate Q_{sp} and compare it to K_{sp} for $PbCl_2\,(s) \rightleftarrows Pb^{2+}\,(aq) + 2Cl^-\,(aq)$ to determine whether $PbCl_2$ will precipitate. Hence, we need to calculate the concentrations of Pb^{2+} and Cl^- in the mixed solution, using the same procedure that was used in working Exercise 14.69.

$$[Pb^{2+}] = \frac{0.10\ M \times 5.0\ \text{mL}}{10.0\ \text{mL}} = 0.050\ M \qquad [Cl^-] = \frac{(0.020\ M) \times 5.0\ \text{mL}}{10.0\ \text{mL}} = 0.010\ M$$

$$Q_{sp} = [Pb^{2+}][Cl^-]^2 \qquad Q_{sp} = (0.050)(0.010)^2 = 5.6 \times 10^{-6}$$

Since Q_{sp} is less than K_{sp} (1.6×10^{-5}), **no precipitate will form.**

14.79 ***See Section 14.7, Example 14.19, and Insights into Chemistry: Selective Precipitation.***

(a) The barium compound requiring the lowest concentration of Ba^{2+} for precipitation will precipitate first. Hence, we need to calculate the concentrations of Ba^{2+} required to precipitate $BaSO_4$ and $Ba_3(PO_4)_2$.

Since Q_{sp} must equal K_{sp} for precipitation to occur and K_{sp} for $BaSO_4 = [Ba^{2+}][SO_4^{2-}] = 8.7 \times 10^{-11}$ and K_{sp} for $Ba_3(PO_4)_2 = [Ba^{2+}]^3[PO_4^{3-}]^2 = 6.0 \times 10^{-39}$,

$$[Ba^{2+}] \text{ for } BaSO_4 \text{ to ppt} = \frac{K_{sp}\,BaSO_4}{[SO_4^{2-}]} = \frac{8.7 \times 10^{-11}}{0.050} = 1.7 \times 10^{-9}\ M$$

$$[Ba^{2+}] \text{ for } Ba_3(PO_4)_2 \text{ to ppt} = \sqrt[3]{\frac{K_{sp}\ Ba_3(PO_4)_2}{[PO_4^{3-}]^2}} = \sqrt[3]{\frac{6.0 \times 10^{-39}}{(0.020)^2}} = 2.5 \times 10^{-12}$$

$Ba_3(PO_4)_2$ requires the lower $[Ba^{2+}]$ for precipitation and will therefore precipitate first.

Note: $\sqrt[3]{y} = (y)^{1/3} = (y)^{0.33333}$ using the y^x key on a calculator.

(b) The concentration of Ba^{2+} is $1.7 \times 10^{-9}\ M$ when $BaSO_4$ begins to precipitate. Hence,

$$[PO_4^{3-}] = \sqrt[3]{\frac{K_{sp}\ Ba_3(PO_4)_2}{[Ba^{2+}]^3}} = \sqrt[3]{\frac{6.0 \times 10^{-39}}{(1.7 \times 10^{-9})^3}} = 1.1 \times 10^{-6}$$

Note: The $[PO_4^{3-}]$ when $Ba_3(PO_4)_2$ just begins to precipitate is 0.020 *M*. However, $Ba_3(PO_4)_2$ continues to precipitate as $BaCl_2$ is added. The $[PO_4^{3-}]$ is thus reduced from 0.020 *M* to 1.0 x 10^{-6} *M* as the $[Ba^{2+}]$ increases from 2.5 x 10^{-12} *M* to 1.7 x 10^{-9} *M*, the $[Ba^{2+}]$ at which the more soluble $BaSO_4$ just begins to precipitate.

14.80 ***See Section 14.7, Example 14.19, and Insights into Chemistry: Selective Precipitation.***

(a) The phosphate requiring the lowest concentration of PO_4^{3-} for precipitation will precipitate first. Hence, we need to calculate the concentrations of PO_4^{3-} required to precipitate $AlPO_4$ and $Ca_3(PO_4)_2$ in the solution. Since Q must equal K_{sp} for precipitation to occur and K_{sp} for $AlPO_4$ is $K_{sp} = [Al^{3+}][PO_4^{3-}] = 9.8 \times 10^{-21}$ and K_{sp} for $Ca_3(PO_4)_2$ is $K_{sp} = [Ca^{2+}]^3[PO_4^{3-}]^2 = 1.3 \times 10^{-32}$,

$$[PO_4^{3-}] \text{ for } AlPO_4 \text{ to ppt} = \frac{K_{sp}\,AlPO_4}{[Al^{3+}]} = \frac{9.8 \times 10^{-21}}{0.0020} = 4.9 \times 10^{-18}\ M$$

$$[PO_4^{3-}] \text{ for } Ca_3(PO_4)_2 \text{ to ppt} = \sqrt{\frac{K_{sp}\,Ca_3(PO_4)_2}{[Ca^{2+}]^3}} = \sqrt{\frac{1.3 \times 10^{-32}}{(0.0040)^3}} = 4.5 \times 10^{-13}$$

$AlPO_4$ requires the lower $[PO_4^{3-}]$ for precipitation and will therefore precipitate first.

(b) The concentration of PO_4^{3-} is 4.5 x 10^{-13} *M* when $Ca_3(PO_4)_2$ begins to precipitate. Hence,

$$[Al^{3+}] = \frac{K_{sp}\,AlPO_4}{[PO_4^{3-}]} \qquad [Al^{3+}] = \frac{9.8 \times 10^{-21}}{4.5 \times 10^{-13}} = 2.2 \times 10^{-8}\ M$$

Note: The $[Al^{3+}]$ when $AlPO_4$ just begins to precipitate is 0.0020 *M*. However, $AlPO_4$ continues to precipitate as Na_3PO_4 is added. The $[Al^{3+}]$ is thus reduced from 0.0020 *M* to 2.2 x 10^{-6} *M* as the $[PO_4^{3-}]$ increases from 4.9 x 10^{-18} *M* to 4.5 x 10^{-13} *M*, the $[PO_4^{3-}]$ at which the more soluble $Ca_3(PO_4)_2$ just begins to precipitate.

14.81 ***See Sections 14.1, 14.4, and Example 14.8.***

$$PCl_5(g) \rightleftarrows PCl_3(g) + Cl_2(g)$$

$$[PCl_5] = \frac{0.10\text{ mol}}{20.0\text{ L}} = 0.0050\ M$$

$$[PCl_3] = \frac{0.20\text{ mol}}{20.0\text{ L}} = 0.010\ M$$

$$[Cl_2] = \frac{0.20\text{ mol}}{20.0\text{ L}} = 0.010\ M$$

$$K_c = \frac{[PCl_3][Cl_2]}{[PCl_5]} \qquad K_c = \frac{(0.010)(0.010)}{(0.0050)} = \mathbf{0.020}$$

14.82 ***See Sections 14.1, 14.4, and Example 14.8.***

$$2NO_2(g) \rightleftarrows N_2O_4(g)$$

$$[NO_2] = \frac{0.15\text{ mol}}{2.5\text{ L}} = 0.060\ M$$

$$[N_2O_4] = \frac{0.025\text{ mol}}{2.5\text{ L}} = 0.010\ M$$

$$K_c = \frac{[N_2O_4]}{[NO_2]^2} \qquad K_c = \frac{(0.010)}{(0.060)^2} = \mathbf{2.8}$$

14.83 *See Sections 14.1, 14.4, and Example 14.8.*

(a) $2SO_3(g) \rightleftarrows 2SO_2(g) + O_2(g)$ $\qquad\qquad$ $\Delta n_g = [(2+1)-2] = 1$

The change in the concentration of SO_3 can be determined from the initial and equilibrium concentrations given for SO_3. The change in the concentrations of SO_2 and O_2 can be determined from the change for SO_3 and the coefficients in the balanced equation giving

Conc, *M*	SO_3	SO_2	O_2
initial	0.0050	0.0000	0.0000
Change	–0.0040	+0.0040	+0.0020
equilibrium	0.0010	0.0040	0.0020

(b) $K_c = \dfrac{[SO_2]^2[O_2]}{[SO_3]^2}$ $\qquad$ $K_c = \dfrac{(0.0040)^2(0.0020)}{(0.0010)^2} = \mathbf{0.032}$

(c) Substituting into $K_p = K_c(RT)^{\Delta n}$ gives $K_p = 0.032(0.0821 \times 1105)^1 = \mathbf{2.9}$.

14.84 *See Sections 14.1, 14.4, and Example 14.8.*

(a) $PCl_5(g) \rightleftarrows PCl_3(g) + Cl_2(g)$ $\qquad\qquad$ $\Delta n_g = [(1+1)-1] = 1$

The change in the concentration of PCl_5 can be determined from the initial and equilibrium concentrations given for PCl_5. The change in the concentrations of PCl_3 and Cl_2 can be determined from the change for PCl_5 and the coefficients in the balanced equation giving

Conc, *M*	PCl_5	PCl_3	Cl_2
initial	0.060	0.000	0.000
Change	–0.040	+0.040	+0.040
equilibrium	0.020	0.040	0.040

(b) $K_c = \dfrac{[PCl_3][Cl_2]}{[PCl_5]}$ $\qquad$ $K_c = \dfrac{(0.040)(0.040)}{(0.020)} = \mathbf{0.080}$

(c) Substituting into $K_p = K_c(RT)^{\Delta n}$ gives $K_p = 0.080(0.0821 \times 811)^1 = \mathbf{5.3}$.

14.85 *See Section 14.4 and Example 14.10.*

$2HI(g) \rightleftarrows H_2(g) + I_2(g)$ $\qquad\qquad$ $\Delta n_g = [(1+1)-2] = 0$

Conc, *M*	HI	H_2	I_2
initial	0.0020	0.	0.
Change	-2y	+y	+y
equilibrium	0.0020-2y	y	y

$K_c = K_p(RT)^{-\Delta n}$ $\quad$ $K_c = K_p(RT)^{-0}$ $\quad$ $K_c = K_p$ $\quad$ $K_c = \dfrac{[H_2][I_2]}{[HI]^2}$ $\quad$ $K_c = \dfrac{(y)(y)}{(0.0020-2y)^2} = 0.050$

Rearranging yields $0.80y^2 + 4.0 \times 10^{-4} y - 2.0 \times 10^{-7} = 0$.

Substituting into the quadratic formula gives $y = \dfrac{-4.0 \times 10^{-4} \pm \sqrt{(4.0 \times 10^{-4})^2 - 4(0.80)(-2.0 \times 10^{-7})}}{2(0.80)}$

and $y = \dfrac{-4.0 \times 10^{-4} \pm 8.9 \times 10^{-4}}{1.6} = 3.1 \times 10^{-4}$ or -8.1×10^{-4}.

The only reasonable value of y is 3.1×10^{-4}, since -8.1×10^{-4} predicts negative concentrations for H_2 and I_2.
Hence, $[\mathbf{HI}] = 0.0020 - (2)(3.1 \times 10^{-4}) = \mathbf{1.4 \times 10^{-3}}$ ***M*** and $[\mathbf{H_2}] = [\mathbf{I_2}] = \mathbf{3.1 \times 10^{-4}}$ ***M***.

Inserting these equilibrium concentrations into the equilibrium expression gives

$K_c = \dfrac{[H_2][I_2]}{[HI]^2}$ $\qquad K_c = \dfrac{(3.1 \times 10^{-4})(3.1 \times 10^{-4})}{(1.4 \times 10^{-3})^2} 0.049$ $\qquad$ and this is in good agreement with the given value of 0.050.

14.86 ***See Section 14.6 and Example 14.16.***

$AgCl(s) \rightleftarrows Ag^+(aq) + Cl^-(aq)$

Since Q_{sp} must equal K_{sp} for precipitation to occur, and K_{sp} for AgCl is $K_{sp} = [Ag^+][Cl^-] = 1.8 \times 10^{-10}$,

$[Ag^+]$ for AgCl to ppt $= \dfrac{K_{sp} AgCl}{[Cl^-]}$ $\qquad [Ag^+]$ for AgCl to ppt $= \dfrac{1.8 \times 10^{-10}}{7.4 \times 10^{-4}} = \mathbf{2.4 \times 10^{-7}}$ ***M***

14.87 ***See Sections 3.5, 4.1, 4.2, 4.3, 14.6.***

(a) Overall: $AgNO_3(aq) + KCl(aq) \rightarrow AgCl(s) + KNO_3(aq)$

Complete ionic: $Ag^+(aq) + NO_3^-(aq) + K^+(aq) + Cl^-(aq) \rightarrow AgCl(s) + K^+(aq) + Cl^-(aq)$

Net ionic: $Ag^+(aq) + Cl^-(aq) \rightarrow AgCl(s)$

(b) Overall: $AgNO_3(aq) + KCl(aq) \rightarrow AgCl(s) + KNO_3(aq)$

Strategy: L $AgNO_3$ soln → mol $AgNO_3$ → mol AgCl → g AgCl

$$? \text{ g AgCl based on } AgNO_3 = 0.0200 \text{ L } AgNO_3 \text{ soln} \times \frac{0.0259 \text{ mol } AgNO_3 \text{ soln}}{1 \text{ L } AgNO_3 \text{ soln}} \times \frac{1 \text{ mol AgCl}}{1 \text{ mol } AgNO_3} \times \frac{143.3 \text{ g AgCl}}{1 \text{ mol AgCl}} = 0.0742 \text{ g AgCl}$$

Strategy: L KCl soln → mol KCl → mol AgCl → g AgCl

$$? \text{ g AgCl based on KCl soln} = 0.0100 \text{ L KCl soln} \times \frac{0.0502 \text{ mol KCl}}{1 \text{ L KCl soln}} \times \frac{1 \text{ mol AgCl}}{1 \text{ mol KCl}} \times \frac{143.3 \text{ g AgCl}}{1 \text{ mol AgCl}} = 0.0719 \text{ g AgCl}$$

KCl is the limiting reactant because it produces less AgCl. The maximum amount of AgCl which can be produced from 20.0 mL of 0.0259 *M* $AgNO_3$ and 10.0 mL of 0.0502 *M* KCl is **0.0719 g AgCl**.

(c) All of the Cl^- from the KCl is consumed in reaction with Ag^+ to form AgCl. The only Cl^- remaining in solution is the amount that can exist in equilibrium with the excess Ag^+ via $AgCl(s) \rightleftarrows Ag^+(aq) + Cl^-(aq)$.

? mol Ag^+ excess = mol Ag^+ original - mol Ag^+ reacting with Cl^-

$$? \text{ mol } Ag^+ \text{ original} = 0.0200 \text{ L } AgNO_3 \text{ soln} \times \frac{0.0259 \text{ mol } AgNO_3}{1 \text{ L } AgNO_3 \text{ soln}} \times \frac{1 \text{ mol } Ag^+}{1 \text{ mol } AgNO_3} = 5.18 \times 10^{-4} \text{ mol } Ag^+$$

$$? \text{ mol } Ag^+ \text{ reacting} = 0.0719 \text{ g AgCl} \times \frac{1 \text{ mol AgCl}}{143.3 \text{ g AgCl}} \times \frac{1 \text{ mol } Ag^+}{1 \text{ mol AgCl}} = 5.02 \times 10^{-4} \text{ mol } Ag^+$$

$$? \text{ mol } Ag^+ \text{ excess} = 5.18 \times 10^{-4} - 5.02 \times 10^{-4} \text{ mol } Ag^+ = 0.16 \times 10^{-4} \text{ mol } Ag^+ = 1.6 \times 10^{-5} \text{ mol } Ag^+$$

$$? M\ Ag^+ \text{ excess} = \frac{1.6 \times 10^{-5} \text{ mol } Ag^+}{0.0200 \text{ L} + 0.0100 \text{ L}} = 5.3 \times 10^{-4}\ M$$

$$\left[Cl^-\right] = \frac{K_{sp}\ AgCl}{\left[Ag^+\right]} = \frac{1.8 \times 10^{-10}}{5.3 \times 10^{-4}} = \mathbf{3.4 \times 10^{-7}}\ \boldsymbol{M}$$

All of the NO_3^- from the $AgNO_3$ remains in solution. The molarity of the NO_3^- in the mixed solution can be calculated using the formula for dilution calculations.

Solving $M(\text{con}) \times V(\text{con}) = M(\text{dil}) \times V(\text{dil})$ for M(dil) gives $M(\text{dil}) = \dfrac{M(\text{con}) \times V(\text{con})}{V(\text{dil})}$.

Hence, $\left[NO_3^-\right]$ in mixed solution $= \dfrac{0.0259\ M \times 20.0 \text{ mL}}{30.0 \text{ mL}} = \mathbf{0.0173}\ \boldsymbol{M}$

14.88 ***See Section 14.1 and Examples 14.2, 14.3.***

$CO_2(g) \rightleftarrows CO(g) + \frac{1}{2}O_2(g)$ $\qquad K_1 = \dfrac{[CO][O_2]^{\frac{1}{2}}}{[CO_2]} = ?$

$2CO(g) + O_2(g) \rightleftarrows 2CO_2(g)$ $\qquad K_2 = \dfrac{[CO_2]^2}{[CO]^2[O_2]}$

Examination shows $K_2 = \dfrac{1}{(K_1)^2}$ and Δn_g for reaction one $= \left[\left(1 + \frac{1}{2}\right) - 1\right] = \frac{1}{2}$.

Substituting into $K_c = K_p(RT)^{-\Delta n}$ gives $K_1 = 2.48(0.0821 \times 3000)^{\frac{1}{2}} = 0.158$.

Hence, $K_2 = \dfrac{1}{(K_1)^2} = \dfrac{1}{(0.158)^2} = \mathbf{40.1}$.

14.89 ***See Sections 14.1, 14.4, and Examples 14.8, 14.10.***

$2SO_2(g) + O_2(g) \rightleftarrows 2SO_3(g)$

(a) Since 2.0 mmol SO_3 are formed from 5.0 mmol SO_2 and 5.0 mmol O_2, it might seem reasonable to estimate that twice as much SO_3 (4.0 mmol SO_3) would be formed from twice as much SO_2 and O_2. However, this isn't likely to give a ratio of $[SO_3]^2/[SO_2]^2[O_2]$ equal to K_c.

(b) For the first set of conditions, the change in the concentration of SO_3 can be determined from the initial and equilibrium concentrations given for SO_3. The change in the concentrations of SO_2 and O_2 can then be determined from the change for SO_3 and the coefficients in the balanced equation giving

Conc, M	SO_2	O_2	SO_3
initial	0.0050	0.0050	0.0000
Change	-0.0020	-0.0010	+0.0020
equilibrium	0.0030	0.0040	0.0020

$$K_c = \frac{[SO_3]^2}{[SO_2]^2[O_2]} \qquad K_c = \frac{(0.0020)^2}{(0.0030)^2(0.0040)} = 1.1 \times 10^2$$

(c) For the second set of conditions,

Conc, M	SO_2	O_2	SO_3
initial	0.0100	0.0100	0.0000
Change	-2y	-y	+2y
equilibrium	0.0100 - 2y	0.010 - y	2y

$$K_c = \frac{[SO_3]^2}{[SO_2]^2[O_2]}$$

Error! Objects cannot be created from editing field codes.

14.90 ***See Sections 14.1, 14.2, 14.4, and Examples 14.3, 14.4, and 14.10.***

$SO_2(g) + Cl_2(g) \rightleftarrows SO_2Cl_2(g)$ $\qquad \Delta n_g = [1-(1+1)] = -1$

Substituting in $K_c = K_p(RT)^{-\Delta n}$ gives $K_c = 0.051(0.0821 \times 500)^1 = 2.1.$

$$? \text{ initial } M\ SO_2 = \frac{1.00 \text{ g } SO_2}{1.00 \text{ L}} \times \frac{1 \text{ mol } SO_2}{64.1 \text{ g } SO_2} = 0.0156\ M\ SO_2$$

$$? \text{ initial } M\ Cl_2 = \frac{1.00 \text{ g } Cl_2}{1.00 \text{ L}} \times \frac{1 \text{ mol } Cl_2}{70.9 \text{ g } Cl_2} = 0.0141\ M\ Cl_2$$

$$? \text{ M } SO_2Cl_2 \text{ after 15 min} = \frac{45.5\ \mu\text{g } SO_2Cl_2}{1.00 \text{ L}} \times \frac{1 \text{ g } SO_2Cl_2}{10^6\ \mu\text{g } SO_2Cl_2} \times \frac{1 \text{ mol } SO_2Cl_2}{135.0 \text{ g } SO_2Cl_2} = 3.37 \times 10^{-7}\ M\ SO_2Cl_2$$

After 15 min,

Conc, M	SO_2	Cl_2	SO_2Cl_2
initial	0.0156	0.0141	0.
Change	-3.37×10^{-7}	-3.37×10^{-7}	$+3.37 \times 10^{-7}$
equilibrium	0.0156	0.0141	3.37×10^{-7}

After 15 min, $Q_c = \frac{[SO_2Cl_2]}{[SO_2][Cl_2]} = \frac{(3.37 \times 10^{-7})}{(0.0156)(0.0141)} = 1.53 \times 10^{-3}$

Since Q_c is less than K_c (2.1), the reaction is not at equilibrium and will continue to proceed to the right to attain equilibrium.

At equilibrium,

Conc, M	SO_2	Cl_2	SO_2Cl_2
initial	0.0156	0.0141	0.0000
Change	-y	-y	+y
equilibrium	0.0156 - y	0.0141 - y	y

$$K_c = \frac{[SO_2Cl_2]}{[SO_2][Cl_2]} \qquad K_c = \frac{y}{(0.0156 - y)(0.0141 - y)} = 2.1$$

Rearranging and combining terms yields $2.1y^2 - 1.062y + 0.00046 = 0$.

Substituting into the quadratic formula gives $y = \dfrac{1.062 \pm \sqrt{(-1.062)^2 - 4(2.1)(0.00046)}}{2(2.1)}$

and $y = \dfrac{1.062 \pm 1.060}{2(2.1)} = 0.5 \text{ or } 5 \times 10^{-4}$.

The only reasonable answer is 5×10^{-4}, since 0.5 predicts negative concentrations for SO_2 and Cl_2.
Hence, $[SO_2] = 0.0156 - 5 \times 10^{-4} = \mathbf{0.0151}\ M$, $[Cl_2] = 0.01410 - 5 \times 10^{-4} = \mathbf{0.0136}\ M$ and $[SO_2Cl_2] = \mathbf{5 \times 10^{-4}}\ M$.

Inserting these equilibrium expressions into the equilibrium expression gives

$$K_c = \frac{[SO_2Cl_2]}{[SO_2][Cl_2]} \qquad K_c = \frac{(5 \times 10^{-4})}{(0.0151)(0.0136)} = 2.4$$

and this value is in good agreement with the value of 2.1 considering the limitations of significant figures.

$$\text{Hence, ? g } SO_2Cl_2 = 1.00 \text{ L} \times \frac{5 \times 10^{-4} \text{ mol } SO_2Cl_2}{1.0 \text{ L}} \times \frac{135.0 \text{ g } SO_2Cl_2}{1 \text{ mol } SO_2Cl_2} = \mathbf{0.07 \text{ g } SO_2Cl_2}$$

14.91 ***See Sections 14.1, 14.3, 14.4, and Examples 14.6, 14.11.***

$N_2(g) + 3H_2(g) \rightleftarrows 2NH_3(g)$

$$\text{(a) } K_c = \frac{[NH_3]^2}{[N_2][H_2]^3} \qquad K_c = \frac{(0.0016)^2}{(0.00485)(0.022)^3} = \mathbf{50}$$

(b) A decrease in volume favors a decrease in moles of gas. In this case, the decrease in volume favors the forward reaction, the left-to-right reaction. Hence, the concentration of nitrogen would decrease as the volume is halved, and the change in the concentration of the nitrogen would be **negative**.

(c)

Conc, M	N_2	H_2	NH_3
initial	0.00970	0.044	0.0032
Change	−y	−3y	+2y
equilibirum	0.00970-y	0.044-3y	0.0032+2y

$$K_c = \frac{[NH_3]^2}{[N_2][H_2]^3} \qquad K_c = \frac{(0.0032 + 2y)^2}{(0.00970 - y)(0.044 - 3y)^3} = 50$$

14.92 ***See Section 14.6, Examples 14.14, 14.15, and the Solution for Exercise 14.63.***

$BaSO_4(s) \rightleftarrows Ba^{2+}(aq) + SO_4^{2-}(aq)$

$$[Ba^{2+}] = [SO_4^{2-}] = = \frac{1.2 \mu g\ Ba^{2+}}{1 \text{ mL}} \times \frac{10^3 \text{ mL}}{1 \text{ L}} \times \frac{1 \text{ g } Ba^{2+}}{10^6\ \mu g\ Ba^{2+}} \times \frac{1 \text{ mol } Ba^{2+}}{137.3 \text{ g } Ba^{2+}} = 8.7 \times 10^{-6}\ M$$

$$K_{sp} = [Ba^{2+}][SO_4^{2-}] \qquad K_{sp} = (8.7 \times 10^{-6})(8.7 \times 10^{-6}) = \mathbf{7.6 \times 10^{-11}}$$

14.93 See Section 14.7 and Example 14.19.

$PbCl_2(s) \rightleftarrows Pb^{2+}(aq) + 2Cl^-(aq)$ $\quad K_{sp} = [Pb^{2+}][Cl^-]^2$

$$[Pb^{2+}]_{max} = \frac{5.0\,mg\,Pb^{2+}}{1.0\,L} \times \frac{1\,g\,Pb^{2+}}{10^3\,mg\,Pb^{2+}} \times \frac{1\,mol\,Pb^{2+}}{207.2\,g\,Pb^{2+}} = 2.4 \times 10^{-5}\,mol/L$$

$$[Cl^-]\text{ for this } M\,Pb^{2+} = \sqrt{\frac{K_{sp}\text{ for }PbCl_2}{[Pb^{2+}]}} = \sqrt{\frac{1.6 \times 10^{-5}}{2.4 \times 10^{-5}}} = 0.82\,mol/L$$

$$?\,kg\,NaCl = 500\,L \times \frac{0.82\,mol\,Cl^-}{1\,L} \times \frac{1\,mol\,NaCl}{1\,mol\,Cl^-} \times \frac{58.5\,g\,NaCl}{1\,mol\,NaCl} \times \frac{1\,kg\,NaCl}{10^3\,g\,NaCl} = \mathbf{24\,kg\,NaCl}$$

14.94 See Sections 3.5, 4.1, 4.2, 4.3, 14.6.

To determine whether $BaSO_4$ will precipitate, we need to calculate the value of Q_{sp} at the instant of mixing and compare it to the value of the K_{sp} for $BaSO_4$. $Q_{sp} = [Ba^{2+}][SO_4^{2-}]$

Strategy: mL Ba^{2+} → g Ba^{2+} → mol Ba^{2+} → M Ba^{2+}

$$?\,mol\,Ba^{2+} = 20.0\,mL\,Ba^{2+}\,soln \times \frac{10\,\mu g\,Ba^{2+}}{1\,mL\,Ba^{2+}\,soln} \times \frac{1\,g\,Ba^{2+}}{10^6\,\mu g\,Ba^{2+}} \times \frac{1\,mol\,Ba^{2+}}{137.3\,g\,Ba^{2+}} = 1.5 \times 10^{-6}\,mol\,Ba^{2+}$$

$$M\,Ba^{2+}\text{ on mixing} = \frac{mol\,Ba^{2+}}{total\,L\,soln} \qquad M\,Ba^{2+}\text{ on mixing} = \frac{1.5 \times 10^{-6}\,mol\,Ba^{2+}}{0.0450\,L\,soln} = 3.3 \times 10^{-5}\,M$$

Strategy: L K_2SO_4 soln → mol K_2SO_4 → mol SO_4^{2-} → M SO_4^{2-}

$$?\,mol\,SO_4^{2-} = 0.0250\,L\,K_2SO_4\,soln \times \frac{0.050\,mol\,K_2SO_4}{1\,L\,K_2SO_4\,soln} \times \frac{1\,mol\,SO_4^{2-}}{1\,mol\,K_2SO_4} = 1.2 \times 10^{-3}\,mol\,SO_4^{2-}$$

$$M\,SO_4^{2-}\text{ on mixing} = \frac{mol\,SO_4^{2-}}{total\,L\,soln} \qquad M\,SO_4^{2-}\text{ on mixing} = \frac{1.2 \times 10^{-3}\,mol\,SO_4^{2-}}{0.0450\,L} = 2.7 \times 10^{-2}\,M$$

$Q_{sp} = [Ba^{2+}][SO_4^{2-}] = (3.3 \times 10^{-5})(2.7 \times 10^{-2}) = 8.9 \times 10^{-7}$
Since Q_{sp} is greater than K_{sp} (8.7×10^{-11}), $BaSO_4$ will precipitate.

(a) Net ionic: $Ba^{2+}(aq) + SO_4^{2-}(aq) \rightarrow BaSO_4(s)$

(b) *Strategy: mL Ba^{2+} → g Ba^{2+} → mol Ba^{2+} → mol $BaSO_4$ → g $BaSO_4$*

$$?\,g\,BaSO_4\text{ based on }Ba^{2+} = 20.0\,mL\,Ba^{2+}\,soln \times \frac{10\,\mu g\,Ba^{2+}}{1\,mL\,Ba^{2+}\,soln} \times \frac{1\,g\,Ba^{2+}}{10^6\,\mu g\,Ba^{2+}} \times \frac{1\,mol\,Ba^{2+}}{137.3\,g\,Ba^{2+}} \times \frac{1\,mol\,BaSO_4}{1\,mol\,Ba^{2+}} \times \frac{233.4\,g\,BaSO_4}{1\,mol\,BaSO_4} = 3.4 \times 10^{-4}\,g\,BaSO_4$$

Strategy: L K_2SO_4 soln → mol K_2SO_4 → mol SO_4^{2-} → mol $BaSO_4$ → g $BaSO_4$

$$? \text{ g } BaSO_4 \text{ based on } K_2SO_4 = 0.0250 \text{ L } K_2SO_4 \text{ soln} \times \frac{0.050 \text{ mol } K_2SO_4}{1 \text{ L } K_2SO_4 \text{ soln}} \times \frac{1 \text{ mol } SO_4^{2-}}{1 \text{ mol } K_2SO_4}$$
$$\times \frac{1 \text{ mol } BaSO_4}{1 \text{ mol } SO_4^{2-}} \times \frac{233.4 \text{ g } BaSO_4}{1 \text{ mol } BaSO_4} = 0.29 \text{ g } BaSO_4$$

Ba^{2+} is the limiting reactant because it produces less $BaSO_4$. The maximum amount of $BaSO_4$ that can be produced from 20.0 mL of 10 μg/mL Ba^{2+} and 25.0 mL of 0.050 M K_2SO_4 is **3.4×10^{-4} g**.

(c) All of the K^+ from the K_2SO_4 remains in solution. The molarity of the K^+ in the mixed solution can be calculated using the formula for dilution calculations.

Solving $M(\text{con}) \times V(\text{con}) = M(\text{dil}) \times V(\text{dil})$ for M(dil) gives $M(\text{dil}) = \dfrac{M(\text{con}) \times V(\text{con})}{V(\text{dil})}$.

Hence, $\left[K^+\right]$ in mixed solution $= \dfrac{(2 \times 0.050)\ M \times 25.0 \text{ mL}}{45.0 \text{ mL}} = \mathbf{0.056}\ \boldsymbol{M}$.

The calculations in part b above indicate that SO_4^{2-} is present in large excess compared to Ba^{2+} (could form $0.29/3.4 \times 10^{-4} = 850$ times as much $BaSO_4$ from SO_4^{2-} compared to Ba^{2+}). Hence, very little SO_4^{2-} is consumed in reaction with Ba^{2+}, and $[SO_4^{2-}]$ in mixed solution $= \dfrac{0.050\ M \times 25.0 \text{ mL}}{45.0 \text{ mL}} = \mathbf{0.028}\ \boldsymbol{M}$.

All of the Ba^{2+} is consumed in reaction with SO_4^{2-} to form $BaSO_4$. The only Ba^{2+} remaining in solution is the amount that can exist in equilibrium with the excess SO_4^{2-} according to $BaSO_4(s) \rightleftarrows Ba^{2+}(aq) + SO_4^{2-}(aq)$.

$$\left[Ba^{2+}\right] = \frac{K_{sp}\ BaSO_4}{\left[SO_4^{2-}\right]} = \frac{8.7 \times 10^{-11}}{0.028} = \mathbf{3.1 \times 10^{-9}}\ \boldsymbol{M}.$$

14.95-14.97 ***Solutions for these Spreadsheet Exercises are on the Reger, Goode and Mercer Spreadsheet Solutions Diskette available from Saunders College Publishing.***

Chapter 15: Solutions of Acids and Bases

15.1 ***See Section 15.1.***

In the Arrhenius model an acid is a substance that increases the concentration of hydrogen ions when dissolved in water. In the Bronsted-Lowry model an acid is a substance that is a proton donor. The Bronsted-Lowry model is an extension of the Arrhenius model. All Arrhenius acids are also Bronsted acids. However, the Bronsted model includes substances that act as proton donors in gas phase reactions and are not classified as Arrhenius acids.

15.2 ***See Section 15.1.***

The Arrhenius model of a base is a substance that increases the concentration of hydroxide ions when dissolved in water. The Bronsted-Lowry model of a base is a substance that is a proton acceptor. The Bronsted-Lowry model is an extension of the Arrhenius model. All Arrhenius bases are also Bronsted bases, so it is not possible for a compound to be an Arrhenius base and not be a Bronsted-Lowry base.

15.3 ***See Section 15.2.***

Autoionization involves the transfer of a proton from one molecule of the solvent to another molecule of the solvent. Hence, the equation for the autoionization of acetic acid is: $CH_3COOH + CH_3COOH \rightleftarrows CH_3COOH_2^+ + CH_3COO^-$.

15.4 ***See Section 15.3.***

A strong acid ionizes completely in solution, whereas a weak acid does not ionize completely in solution. For example, nitric acid, HNO_3, is a strong acid. Experiments show that the equilibrium concentrations of H_3O^+ and NO_3^- formed by the ionization of nitric acid are equal to the starting concentration of HNO_3. On the other hand, nitrous acid, HNO_2, is a weak acid. Experiments show that the equilibrium concentrations of H_3O^+ and NO_2^- are very, very small compared to the starting concentration of HNO_2. Most of the HNO_2 remains in the molecular form in aqueous solution.

15.5 ***See Sections 15.3, 15.4, 15.6.***

The conjugate base anions of strong acids are usually very weak bases that have very little tendency to remove a proton from water. Hence, these anions act as spectator ions in solution. However, the conjugate base of the strong diprotic acid H_2SO_4 is HSO_4^- and is an important exception. It acts as a weak acid in solution.

15.6 ***See Sections 15.3, 15.6.***

A weak acid is one that does not ionize completely when dissolved in water.

15.7 ***See Section 15.4, 15.6.***

The conjugate acid cations of weak bases act as weak acids. The weaker the base, the stronger its conjugate acid; the relation between K_a and K_b for members of conjugate acid-base pairs being $K_aK_b = K_w$.

15.8 *See Section 15.5.*

The fraction of an acid that is ionized varies with the starting concentration of the acid. For two solutions of the same acid, the fraction ionized is larger in the solution in which the concentration is lower. Hence, it would be necessary to tabulate the fraction ionized for every possible starting concentration of an acid, and this would not be practical.

15.9 *See Section 15.5.*

A base that is positively charged is not likely to accept a positively charged proton, H^+, because there would be considerable repulsion between the positive charges of the base and the proton.

15.10 *See Section 15.6.*

(a) Using the relation $K_aK_b = K_w$ for members of conjugate acid-base pairs gives

$$K_a\ NH_4^+ = \frac{K_w}{K_b\ NH_3} = \frac{1.0 \times 10^{-14}}{1.8 \times 10^{-5}} = 5.6 \times 10^{-10}$$

(b) K_a and K_b for the NH_3/NH_2^- pair cannot be calculated from the preceding data and the value of K_w. The K_a for NH_3 would need to be calculated from the preceding data to calculate the value of K_b for NH_2^-. However, NH_3 is not the conjugate acid of either member of the NH_4^+/NH_3 pair.

15.11 *See Section 15.8.*

The acid strengths of the binary hydrides increase going "southeast" from carbon, that is from CH_4 to PH_3 to H_2Se to HI. The primary reason for this is the decrease in the strength of the H-X bond with the increase in the size of the X atom on descending the periods of the periodic table. This effect dominates the decrease in electronegativity that occurs on descending the periods of the periodic table, just as it does when considering the binary hydrides of any given group of nonmetals and descending that group.

The same trend is observed going from SiH_4 to AsH_3 to H_2Te because the reasoning is the same as it is for going from CH_4 to HI. The acid strength of HAt would be expected greater than that of H_2Te, but At is radioactive and it is doubtful that the chemistry of HAt has been studied extensiviely.

15.12 *See Section 15.6.*

Determining the molecular geometry of NH_4^+ is one approach that could be used to determine whether all four N-H bonds are equivalent or whether the coordinate covalent N-H^+ bond is different. If the NH_4^+ ion is a symmetrical tetrahedron, we can conclude that all four N-H bonds are equivalent. If it is not a symmetrical tetrahedron, we can conclude that the coordinate covalent bond is different. Experiment shows NH_4^+ is a symmetrical tetrahedron indicating all four N-H bonds are equivalent.

15.13 *See Section 15.8.*

HAt is expected to be a stronger acid than HI. The dominant factor in determining the strengths of the hydrohalic acids is the strength of the H-X bond, and this decreases with the increasing size of X going down the period.

15.14 *See Sections 15.3, 15.8.*

The concentration of H_3O^+ can be determined relatively easily using a pH meter and should be used. The concentration of Cl^- can also be determined relatively easily (See note 2 below.), but there is no way to directly measure the concentration of HCl molecules in solution.

Notes:

1) HCl is a strong acid that is expected to be completely ionized in solution, so that no molecules of HCl are expected to remain in solution. This furthermore means that attempting to directly measure the concentration of HCl would not be feasible. However, we only know this as the result of experiments and should not assume this in designing the experiment to determine the extent of ionization.

2) The concentration of Cl^- can be determined relatively easily using an ion-selective electrode that acts on the same electrochemical principle as the pH electrode; these principles are discussed in greater detail in Chapter 18. Hence, we also can determine the concentration of Cl^- and can calculate the extent of ionization from the concentration of Cl^- in solution, but you wouldn't necessarily know this at this point in your study of chemistry.

3) The vapor pressure lowering or the freezing point depression of a solution of HCl can also be used to determine the extent of ionization of HCl in solution.

15.15 *See Section 15.1, Tables 15.1, 15.6, 15.8, and Example 15.1.*

To obtain the formula of a conjugate acid, add H^+. This gives:

Part	Base	Base Fromula	Conj. Acid Formula	Conj. Acid Name
(a)	hydrogen sulfate	HSO_4^-	H_2SO_4	sulfuric acid
(b)	water	H_2O	H_3O^+	hydronium ion
(c)	ammonia	NH_3	NH_4^+	ammonium ion
(d)	pyridine	C_5H_5N	$C_5H_5NH^+$	pyridinium ion

15.16 *See Section 15.1, Tables 15.1, 15.6, 15.8, and Example 15.1.*

To obtain the formula of a conjugate base, remove H^+. This gives:

Part	Acid	Acid Formula	Conj. Base Formula	Conj. Base Name
(a)	nitric acid	HNO_3	NO_3^-	nitrate ion
(b)	hydrogen sulfate ion	HSO_4^-	SO_4^{2-}	sulfate ion
(c)	water	H_2O	OH^-	hydroxide ion
(d)	hydrochloric acid	HCl	Cl^-	chloride ion

15.17 *See Section 15.1, Tables 15.1, 15.6, 15.8, and Example 15.1.*

To obtain the formula of a conjugate base, remove H^+. This gives:

Part	Acid Formula	Conj. Base Formula	Conj. Base Name
(a)	HCN	CN^-	cyanide ion
(b)	HSO_4^-	SO_4^{2-}	sulfate ion
(c)	$H_2PO_3^-$	HPO_3^{2-}	hydrogen phosphite ion
(d)	HCO_3^-	CO_3^{2-}	carbonate ion

15.18 *See Section 15.1, Tables 15.1, 15.6, 15.8, and Example 15.1.*

To obtain the formula of a conjugate acid, add H^+. This gives:

Part	Base Formula	Conj. Acid Formula	Conj. Acid Name
(a)	N_2H_4	$N_2H_5^+$	hydrazinium ion
(b)	NO_2^-	HNO_2	nitrous acid
(c)	ClO_4^-	$HClO_4$	perchloric acid
(d)	I^-	HI	hydroiodic acid

15.19 *See Section 15.1, Tables 15.1, 15.2, 15.6, 15.8, and Example 15.1.*

Part	Reaction	Acid/Conj. Base	Base/Conj. Acid
(a)	$NH_3 + CH_3COOH \rightarrow NH_4CH_3COO$	CH_3COOH / CH_3COO^-	NH_3 / NH_4^+
(b)	$N_2H_5^+ + CO_3^{2-} \rightarrow N_2H_4 + HCO_3^-$	$N_2H_5^+ / N_2H_4$	CO_3^{2-} / HCO_3^-
(c)	$H_3O^+ + OH^- \rightarrow 2H_2O$	H_3O^+ / H_2O	OH^- / H_2O
(d)	$HSO_4^- + HCOO^- \rightarrow SO_4^{2-} + HCOOH$	HSO_4^- / SO_4^{2-}	$HCOO^- / HCOOH$

15.20 *See Section 15.1, Tables 15.1, 15.2, 15.6, 15.8, and Example 15.1.*

Part	Reaction	Acid/Conj. Base	Base/Conj. Acid
(a)	$NH_3 + HCl \rightarrow NH_4Cl$	HCl / Cl^-	NH_3 / NH_4^+
(b)	$HCO_3^- + HNO_3 \rightarrow H_2CO_3 + NO_3^-$	HNO_3 / NO_3^-	HCO_3^- / H_2CO_3
(c)	$HCOOH + CN^- \rightarrow HCN + HCOO^-$	$HCOOH / HCOO^-$	CN^- / HCN
(d)	$CH_3COO^- + H_2O \rightarrow CH_3COOH + OH^-$	H_2O / OH^-	CH_3COOH / CH_3COO^-

15.21 *See Section 15.2 and Examples 15.3, 15.4.*

Using $pH = -\log[H_3O^+]$ and $[H_3O^+] = 10^{-pH} = \text{inv}\log(-pH)$ yields:

Part	pH	$[H_3O^+]$, M	Solution
(a)	2.34	**4.6×10^{-3}**	**acidic**
(b)	**12.98**	1.04×10^{-13}	**basic**
(c)	-1.09	**12.**	**acidic**
(d)	**10.67**	2.12×10^{-11}	**basic**
(e)	**1.13**	7.40×10^{-2}	**acidic**
(f)	13.41	**3.9×10^{-14}**	**basic**
(g)	**4.15**	7.07×10^{-5}	**acidic**
(h)	9.80	**1.6×10^{-10}**	**basic**
(i)	**0.30**	0.505	**acidic**

15.22 *See Section 15.2 and Examples 15.3, 15.4.*

Using $pH = -\log[H_3O^+]$ and $[H_3O^+] = 10^{-pH} = \text{inv}\log(-pH)$ yields:

Part	pH	$[H_3O^+]$, M	Solution
(a)	2.00	**1.0×10^{-2}**	**acidic**
(b)	**2.98**	1.04×10^{-3}	**acidic**
(c)	9.84	**1.4×10^{-10}**	**basic**

Part	pH	$[H_3O^+]$, *M*	Solution
(d)	**0.70**	2.00×10^{-1}	**acidic**
(e)	**8.03**	9.40×10^{-9}	**basic**
(f)	11.34	$\mathbf{4.6 \times 10^{-12}}$	**basic**
(g)	**3.34**	4.57×10^{-4}	**acidic**
(h)	4.51	$\mathbf{3.1 \times 10^{-5}}$	**acidic**
(I)	**14.18**	6.65×10^{-15}	**basic**

15.23 *See Section 15.2 and Examples 15.2, 15.3, 15.4.*

Using $pH = -\log[H_3O^+]$ $\qquad [H_3O^+] = 10^{-pH} = \text{inv}\log(-pH)$

$pOH = -\log[OH^-]$ $\qquad [OH^-] = 10^{-pOH} = \text{inv}\log(-pOH)$

$K_w = [H_3O^+][OH^-] = 1.0 \times 10^{-14}$ $\qquad pH + pOH = 14.00$ yields

Part	pH	$[H_3O^+]$, *M*	pOH	$[OH^-]$, *M*	Solution
(a)	-1.04	**11.**	**15.04**	$\mathbf{9.1 \times 10^{-16}}$	**acidic**
(b)	**13.66**	$\mathbf{2.2 \times 10^{-14}}$	0.34	**0.46**	**basic**
(c)	**6.70**	1.98×10^{-7}	**7.30**	$\mathbf{5.05 \times 10^{-8}}$	**acidic (very weak)**
(d)	**12.65**	$\mathbf{2.2 \times 10^{-13}}$	**1.35**	4.4×10^{-2}	**basic**

15.24 *See Section 15.2 and Examples 15.2, 15.3, 15.4.*

Using $pH = -\log[H_3O^+]$ $\qquad [H_3O^+] = 10^{-pH} = \text{inv}\log(-pH)$

$pOH = -\log[OH^-]$ $\qquad [OH^-] = 10^{-pOH} = \text{inv}\log(-pOH)$

$K_w = [H_3O^+][OH^-] = 1.0 \times 10^{-14}$ $\qquad pH + pOH = 14.00$ yields

Part	pH	$[H_3O^+]$, *M*	pOH	$[OH^-]$, *M*	Solution
(a)	10.34	$\mathbf{4.6 \times 10^{-11}}$	**3.66**	$\mathbf{2.2 \times 10^{-4}}$	**basic**
(b)	**3.66**	$\mathbf{2.2 \times 10^{-4}}$	10.34	$\mathbf{4.6 \times 10^{-11}}$	**acidic**
(c)	**0.38**	0.412	**13.62**	$\mathbf{2.43 \times 10^{-14}}$	**acidic**
(d)	**3.05**	$\mathbf{8.93 \times 10^{-4}}$	**10.95**	11.2×10^{-12}	**acidic**

15.25 *See Section 15.2, Table 15.4, and Examples 15.2, 15.4, 15.5.*

Using $[H_3O^+] = 10^{-pH} = \text{inv}\log(-pH)$ and $[OH^-] = \frac{K_w}{[H_3O^+]} = \frac{1.0 \times 10^{-14}}{[H_3O^+]}$ yields

Part	Substance	pH	$[H_3O^+]$, *M*	$[OH^-]$, *M*
(a)	vinegar	2.9	$\mathbf{1 \times 10^{-3}}$	$\mathbf{1 \times 10^{-11}}$
(b)	stomach acid	1.7	$\mathbf{2 \times 10^{-2}}$	$\mathbf{5 \times 10^{-13}}$
(c)	coffee	5.0	$\mathbf{1 \times 10^{-5}}$	$\mathbf{1 \times 10^{-9}}$
(d)	milk	6.9	$\mathbf{1 \times 10^{-7}}$	$\mathbf{1 \times 10^{-7}}$

15.26 *See Section 15.2, Table 15.4, and Examples 15.2, 15.4, 15.5.*

Using $\left[H_3O^+\right] = 10^{-pH} = \text{inv}\log(-pH)$ and $\left[OH^-\right] = \frac{K_w}{\left[H_3O^+\right]} = \frac{1.0 \times 10^{-14}}{\left[H_3O^+\right]}$ yields

Part	Substance	pH	$[H_3O^+]$, *M*	$[OH^-]$, *M*
(a)	lemon juice	2.2	**6×10^{-3}**	**2×10^{-12}**
(b)	wine	3.5	**3×10^{-4}**	**3×10^{-11}**
(c)	blood	7.4	**4×10^{-8}**	**2×10^{-7}**
(d)	household ammonia	11.9	**1×10^{-12}**	**1×10^{-2}**

15.27 *See Sections 15.1, 15.3, and Examples 15.5, 15.6.*

(a) $HCl + H_2O \rightarrow H_3O^+ + Cl^-$
pH = -log[H_3O^+] = -log (0.050) = **1.30**

0.050 *M* HCl → 0.050 *M* H_3O^+
pOH = 14.00 - pH = **12.70**

(b) $KOH \rightarrow K^+ + OH^-$
pOH = -log [OH^-] = -log (0.024) = **1.62**

0.024 *M* KOH → 0.024 *M* OH^-
pH = 14.00 - pOH = **12.38**

(c) $HClO_4 + H_2O \rightarrow H_3O^+ + ClO_4^-$
pH = -log[H_3O^+] = -log (0.014) = **1.85**

0.014 *M* $HClO_4$ → 0.014 *M* H_3O^+
pOH = 14.00 - pH = **12.15**

(d) $NaOH \rightarrow Na^+ + OH^-$
pOH = -log [OH^-] = -log (1.05) = **-0.02**

1.05 *M* NaOH → 1.05 *M* OH^-
pH = 14.00 - pOH = **14.02**

15.28 *See Sections 15.1, 15.3, and Examples 15.5, 15.6.*

(a) $CsOH \rightarrow Cs^+ + OH^-$
pOH = -log [OH^-] = -log (0.51) = **0.29**

0.51 *M* CsOH → 0.51 *M* OH^-
pH = 14.00 - pOH = **13.71**

(b) $HI + H_2O \rightarrow H_3O^+ + I^-$
pH = -log[H_3O^+] = -log (0.0040) = **2.40**

0.0040 *M* HI → 0.0040 *M* H_3O^+
pOH = 14.00 - pH = **11.60**

(c) $LiOH \rightarrow Li^+ + OH^-$
pOH = -log [OH^-] = -log (0.13) = **0.89**

0.13 *M* LiOH → 0.13 *M* OH^-
pH = 14.00 - pOH = **13.11**

(d) $HClO_4 + H_2O \rightarrow H_3O^+ + ClO_4^-$
pH = -log[H_3O^+] = -log (0.66) = **0.18**

0.66 *M* $HClO_4$ → 0.66 *M* H_3O^+
pOH = 14.00 - pH = **13.82**

15.29 *See Sections 15.1, 15.3, and Examples 15.5, 15.6.*

(a) $HBr + H_2O \rightarrow H_3O^+ + Br^-$
pH = -log[H_3O^+] = -log (0.94) = **0.03**

0.94 *M* HBr → 0.94 *M* H_3O^+
pOH = 14.00 - pH = **13.97**

(b) $Sr(OH)_2 \rightarrow Sr^{2+} + 2OH^-$
pOH = -log [OH^-] = -log (0.084) = **1.08**

0.042 *M* $Sr(OH)_2$ → 0.084 *M* OH^-
pH = 14.00 - pOH = **12.92**

(c) $HCl + H_2O \rightarrow H_3O^+ + Cl^-$
pH = -log[H_3O^+] = -log (0.00033) = **3.48**

0.00033 *M* HCl → 0.00033 *M* H_3O^+
pOH = 14.00 - pH = **10.52**

(d) $RbOH \rightarrow Rb^{+} + OH^{-}$

0.88 *m* RbOH → 0.88 *M* OH^{-}

pOH = -log [OH^{-}] = -log (0.88) = **0.06**

pH = 14.00 - pOH = **13.94**

15.30 ***See Sections 15.2, 15.3, and Examples 15.5, 15.6.***

(a) $Ba(OH)_2 \rightarrow Ba^{2+} + 2OH^{-}$

0.0045 *M* $Ba(OH)_2 \rightarrow$ 0.0090 *M* OH^{-}

$\mathbf{pOH} = -\log\left[OH^{-}\right] = -\log(0.0090) = \mathbf{2.05}$

$\mathbf{pH} = 14.00 - pOH = \mathbf{11.95}$

(b) $HI + H_2O \rightarrow H_3O^{+} + I^{-}$

0.080 *M* HI → 0.080 *M* H_3O^{+}

$\mathbf{pH} = -\log\left[H_3O^{+}\right] = -\log(0.080) = \mathbf{1.10}$

$\mathbf{pOH} = 14.00 - pH = \mathbf{12.90}$

(c) $Sr(OH)_2 \rightarrow Sr^{2+} + 2OH^{-}$

0.030 *M* $Sr(OH)_2 \rightarrow$ 0.060 *M* OH^{-}

$\mathbf{pOH} = -\log\left[OH^{-}\right] = -\log(0.060) = \mathbf{1.22}$

$\mathbf{pH} = 14.00 - pOH = \mathbf{12.78}$

(d) $HNO_3 + H_2O \rightarrow H_3O^{+} + NO_3^{-}$

12.3 *M* $HNO_3 \rightarrow$ 12.3 *M* H_3O^{+}

$\mathbf{pH} = -\log\left[H_3O^{+}\right] = -\log(12.3) = \mathbf{-1.09}$

$\mathbf{pOH} = 14.00 - pH = \mathbf{15.09}$

15.31 ***See Section 15.5 and Example 15.7.***

The percent ionization can be used to calculate the equilibrium concentrations of H_3O^{+} and CN^{-} formed via $HCN + H_2O \rightleftarrows H_3O^{+} + CN^{-}$. Since 0.0050 *M* HCN is 0.038% ionized, $[H_3O^{+}] = [CN^{-}]$ = (0.00038)(0.0050 *M*) = 1.9 x 10^{-6} *M*, and [HCN] = 0.0050 *M* - 1.9 x 10^{-6} M = 0.0050 *M*. Using an iCe table this would be shown as:

Conc, *M*	HCN	H_3O^{+}	CN^{-}
initial	0.0050	0.	0.
Change	−(0.00038)(0.0050)	+(0.00038)(0.0050)	+(0.00038)(0.0050)
equilibrium	0.0050	1.9×10^{-6}	1.9 x 10^{-6}

Hence, $K_a = \dfrac{[H_3O^{+}][CN^{-}]}{[HCN]}$ $\quad K_a = \dfrac{(1.9 \times 10^{-6})(1.9 \times 10^{-6})}{0.0050}$ = **7.2 x 10^{-10}**

15.32 ***See Section 15.5 and Example 15.8.***

$$? M\ C_5H_3N_4O_3H = \frac{0.121\ g\ C_5H_3N_4O_3H}{0.0100\ L} \times \frac{1\ mol\ C_5H_3N_4O_3H}{168\ g\ C_5H_3N_4O_3H} = 0.0720\ M$$

The percent ionization can be used to calculate the equilibrium concentrations of H_3O^{+} and $C_5H_3N_4O_3^{-}$ formed via $C_5H_3N_4O_3H + H_2O \rightleftarrows H_3O^{+} + C_5H_3N_4O_3^{-}$. Since 0.0720 *M* $C_5H_3N_4O_3H$ is 4.2% ionized, $[H_3O^{+}] = [C_5H_3N_4O_3^{-}]$ = (0.042)(0.0720 *M*) = 3.0 x 10^{-3} *M*, and $[C_5H_3N_4O_3H]$ = 0.0720 *M* - 0.0030 *M* = 0.0069 *M*.

Using an iCe table this would be shown as:

Conc, M	$C_5H_3N_4O_3H$	H_3O^+	$C_5H_3N_4O_3^-$
initial	0.0720	0.	0.
Change	−(0.042)(0.0720)	+(0.042)(0.0720)	+(0.042)(0.0720)
equilibrium	0.0069	0.0030	0.0030

Hence, $K_a = \dfrac{[H_3O^+][C_5H_3N_4O_3^-]}{[C_5H_3N_4O_3H]}$ $K_a = \dfrac{(3.0 \times 10^{-3})(3.0 \times 10^{-3})}{0.0069} = \mathbf{1.3 \times 10^{-4}}$

15.33 ***See Section 15.5 and Example 15.8.***

$$C_3H_7COOH + H_2O \rightleftarrows H_3O^+ + C_3H_7COO^-$$

$$? \text{ initial} M\ C_3H_7COOH = \frac{50.0 \text{ mg } C_3H_7COOH}{1.00 \text{ mL soln}} \times \frac{1 \text{ mol } C_3H_7COOH}{88.0 \times 10^3 \text{ mg } C_3H_7COOH} \times \frac{10^3 \text{ mL soln}}{1 \text{ L soln}}$$
$$= 0.568\ M\ C_3H_7COOH$$

? equilibrium M $H_3O^+ = 10^{-pH}$ = inv log(-pH) = inv log(-2.52) = 3.0×10^{-3}

Conc, M	C_3H_7COOH	H_3O^+	$C_3H_7COO^-$
initial	0.568	0.	0.
Change	-3.0×10^{-3}	$+3.0 \times 10^{-3}$	$+3.0 \times 10^{-3}$
equilibrium	0.565	3.0×10^{-3}	3.0×10^{-3}

$K_a = \dfrac{[H_3O^+][C_3H_7COO^-]}{[C_3H_7COOH]}$ $K_a = \dfrac{(3.0 \times 10^{-3})(3.0 \times 10^{-3})}{0.565} = 1.6 \times 10^{-5}$

$pK_a = -\log K_a$ $pK_a = -\log 1.6 \times 10^{-5} = \mathbf{4.80}$

15.34 ***See Section 15.5 and Example 15.8.***

$HVit^+ + H_2O \rightleftarrows H_3O^+ + Vit$

$$? \text{ initial } M \text{ HVitCl} = \frac{1.00 \text{ g HVitCl}}{0.01000 \text{ L}} \times \frac{1 \text{ mol HVitCl}}{337.28 \text{ g HVitCl}} = 0.296\ M$$

? equilibrium M $H_3O^+ = 10^{-pH}$ = inv log(-pH) = inv log(-4.50) = $3.2 \times 10^{-5}\ M$

Conc, M	HVit +	H_3O^+	Vit
initial	0.296	0.	0.
Change	-3.2×10^{-5}	$+3.2 \times 10^{-5}$	$+3.2 \times 10^{-5}$
equilibrium	0.296	3.2×10^{-5}	3.2×10^{-5}

$K_a = \dfrac{[H_3O^+][Vit]}{[HVit^+]}$ $K_a = \dfrac{(3.2 \times 10^{-5})(3.2 \times 10^{-5})}{(0.296)} = \mathbf{3.5 \times 10^{-9}}$

15.35 *See Section 15.5 and Example 15.8.*

$C_5H_{11}COOH + H_2O \rightleftarrows H_3O^+ + C_5H_{11}COO^-$

$[H_3O^+] = 10^{-pH} = \text{inv}\log(-pH)$ $[H_3O^+] = \text{inv}\log(-3.43) = 3.7 \times 10^{-4}$ *M*

Conc, *M*	$C_5H_{11}COOH$	H_3O^+	$C_5H_{11}COO^-$
initial	0.0100	0.	0.
Change	-3.7×10^{-4}	$+3.7 \times 10^{-4}$	$+3.7 \times 10^{-4}$
equilibrium	9.6×10^{-3}	3.7×10^{-4}	3.7×10^{-4}

$$K_a = \frac{[H_3O^+][C_5H_{11}COO^-]}{[C_5H_{11}COOH]} \qquad K_a = \frac{(3.7 \times 10^{-4})(3.7 \times 10^{-4})}{9.6 \times 10^{-3}} = \mathbf{1.4 \times 10^{-5}}$$

$pK_a = -\log K_a$ $pK_a = -\log 1.4 \times 10^{-5} = \mathbf{4.85}$

15.36 *See Section 15.5 and Example 15.8.*

$C_2H_5OCOOH + H_2O \rightleftarrows H_3O^+ + C_2H_5OCOO^-$

$[H_3O^+] = 10^{-pH} = \text{inv}\log(-pH)$ $[H_3O^+] = \text{inv}\log(-2.66) = 2.2 \times 10^{-3}$ *M*

Conc, *M*	C_2H_5OCOOH	H_3O^+	$C_2H_5OCOO^-$
initial	0.0376	0.	0.
Change	−0.0022	+0.0022	+0.0022
equilibrium	0.0354	0.0022	0.0022

$$K_a = \frac{[H_3O^+][C_2H_5OCOO^-]}{[C_2H_5OCOOH]} \qquad K_a = \frac{(0.0022)(0.0022)}{(0.0354)} = \mathbf{1.4 \times 10^{-4}}$$

$pK_a = -\log K_a$ $pK_a = -\log(1.4 \times 10^{-4}) = \mathbf{3.85}$

15.37 *See Section 15.5 and Table 15.6*

(a) $C_6H_5COOH + H_2O \rightleftarrows H_3O^+ + C_6H_5COO^-$ $K_a = 6.3 \times 10^{-5}$

Conc, *M*	C_6H_5COOH	H_3O^+	$C_6H_5COO^-$
initial	0.20	0.	0.
Change	−y	+y	+y
equilibrium	0.20 − y	y	y

$$K_a = \frac{[H_3O^+][C_6H_5COO^-]}{[C_6H_5COOH]} \qquad K_a = \frac{(y)(y)}{(0.20 - y)} \cong \frac{y^2}{0.20} = 6.3 \times 10^{-5}$$

(b) $HCOOH + H_2O \rightleftarrows H_3O^+ + HCOO^-$ $K_a = 1.8 \times 10^{-4}$

Conc, *M*	HCOOH	H_3O^+	$HCOO^-$
initial	1.50	0.	0.
Change	−y	+y	+y
equilibrium	1.50 − y	y	y

$$K_a = \frac{[H_3O^+][HCOO^-]}{[HCOOH]} \qquad K_a = \frac{(y)(y)}{(1.50-y)} \cong \frac{y^2}{1.50} = 1.8 \times 10^{-4}$$

(c) $HCN + H_2O \rightleftarrows H_3O^+ + CN^-$ $K_a = 7.2 \times 10^{-10}$

Conc, *M*	HCN	H_3O^+	CN^-
initial	0.0055	0.	0.
Change	−y	+y	+y
equilibrium	0.0055 − y	y	y

$$K_a = \frac{[H_3O^+][CN^-]}{[HCN]} \qquad K_a = \frac{(y)(y)}{(0.0055-y)} \cong \frac{y^2}{0.0055} = 7.2 \times 10^{-10}$$

(d) $HNO_2 + H_2O \rightleftarrows H_3O^+ + NO_2^-$ $K_a = 4.6 \times 10^{-4}$

Conc, *M*	HNO_2	H_3O^+	NO_2^-
initial	0.075	0.	0.
Change	−y	+y	+y
equilibrium	0.075 − y	y	y

$$K_a = \frac{[H_3O^+][NO_2^-]}{[HNO_2]} \qquad K_a = \frac{(y)(y)}{(0.075-y)} \cong \frac{y^2}{0.075} = 4.6 \times 10^{-4}$$

15.38 ***See Section 15.5 and Table 15.6.***

(a) $HOCl + H_2O \rightleftarrows H_3O^+ + OCl^-$ $K_a = 3.0 \times 10^{-8}$

Conc, *M*	HOCl	H_3O^+	OCl^-
initial	1.25	0.	0.
Change	−y	+y	+y
equilibrium	1.25 − y	y	y

$$K_a = \frac{[H_3O^+][OCl^-]}{[HOCl]} \qquad K_a = \frac{(y)(y)}{(1.25-y)} \cong \frac{y^2}{1.25} = 3.0 \times 10^{-8}$$

(b) $HF + H_2O \rightleftarrows H_3O^+ + F^-$ $K_a = 3.5 \times 10^{-4}$

Conc, M	HF	H_3O^+	F^-
initial	0.80	0.	0.
Change	−y	+y	+y
equilibrium	0.80 − y	y	y

$$K_a = \frac{[H_3O^+][F^-]}{[HF]} \qquad K_a = \frac{(y)(y)}{(0.80-y)} \cong \frac{y^2}{0.80} = 3.5 \times 10^{-4}$$

(c) $CH_3COOH + H_2O \rightleftarrows H_3O^+ + CH_3COO^-$ $K_a = 1.8 \times 10^{-5}$

Conc, M	CH_3COOH	H_3O^+	CH_3COO^-
initial	0.14	0.	0.
Change	−y	+y	+y
equilibrium	0.14 − y	y	y

$$K_a = \frac{[H_3O^+][CH_3COO^-]}{[CH_3COOH]} \qquad K_a = \frac{(y)(y)}{(0.14-y)} \cong \frac{y^2}{0.14} = 1.8 \times 10^{-5}$$

(d) $HCOOH + H_2O \rightleftarrows H_3O^+ + HCOO^-$

Conc, M	HCOOH	H_3O^+	$HCOO^-$
initial	0.25	0.	0.
Change	−y	+y	+y
equilibrium	0.25 − y	y	y

$$K_a = \frac{[H_3O^+][HCOO^-]}{[HCOOH]} \qquad K_a = \frac{(y)(y)}{(0.25-y)} \cong \frac{y^2}{0.25} = 1.8 \times 10^{-4}$$

15.39 ***See Section 15.5, Table 15.6, and Example 15.9.***

(a) $HNO_2 + H_2O \rightleftarrows H_3O^+ + NO_2^-$ $K_a = 4.6 \times 10^{-4}$

Conc, M	HNO_2	H_3O^+	NO_2^-
initial	0.33	0.	0.
Change	−y	+y	+y
equilibrium	0.33−y	y	y

$$K_a = \frac{\left[H_3O^+\right]\left[NO_2^-\right]}{\left[HNO_2\right]} \qquad K_a = \frac{(y)(y)}{(0.33-y)} \cong \frac{y^2}{0.33} = 4.6 \times 10^{-4}$$

$\left[H_3O^+\right] = y = 1.2 \times 10^{-2}$ $pH = -\log\left(1.2 \times 10^{-2}\right) = \mathbf{1.92}$

Approximation check: $\left(\frac{1.2 \times 10^{-2}}{0.33}\right) \times 100\% = 3.6\%$, so the approximation is valid.

(b) $C_6H_5OH + H_2O \rightleftarrows H_3O^+ + C_6H_5O^-$ $\qquad K_a = 1.3 \times 10^{-10}$

Conc, *M*	C_6H_5OH	H_3O^+	$C_6H_5O^-$
initial	0.016	0.	0.
Change	−y	+y	+y
equilibrium	0.016−y	y	y

$$K_a = \frac{[H_3O^+][C_6H_5O^-]}{[C_6H_5OH]} \qquad K_a = \frac{(y)(y)}{(0.016-y)} \cong \frac{y^2}{0.016} = 1.3 \times 10^{-10}$$

$$[H_3O^+] = y = 1.4 \times 10^{-6} \qquad pH = -\log(1.4 \times 10^{-6}) = \mathbf{5.85}$$

Approximation check: $\left(\frac{1.4 \times 10^{-6}}{0.016}\right) \times 100\% = 0.0088\%$, so the approximation is valid.

(c) $HF + H_2O \rightleftarrows H_3O^+ + F^-$ $\qquad K_a = 3.5 \times 10^{-4}$

Conc, *M*	HF	H_3O^+	F^-
initial	0.25	0.	0.
Change	−y	+y	+y
equilibrium	0.25 − y	y	y

$$K_a = \frac{[H_3O^+][F^-]}{[HF]} \qquad K_a = \frac{(y)(y)}{(0.25-y)} \cong \frac{y^2}{0.25} = 3.5 \times 10^{-4}$$

$$[H_3O^+] = y = 9.4 \times 10^{-3} \qquad pH = -\log(9.4 \times 10^{-3}) = \mathbf{2.03}$$

Approximation check: $\left(\frac{7.2 \times 10^{-3}}{0.15}\right) \times 100\% = 4.8\%$, so the approximation is valid.

(d) $HCOOH + H_2O \rightleftarrows H_3O^+ + HCOO^-$ $\qquad K_a = 1.8 \times 10^{-4}$

Conc, *M*	HCOOH	H_3O^+	$HCOO^-$
initial	0.010	0.	0.
Change	−y	+y	+y
equilibrium	0.010−y	y	y

$$K_a = \frac{[H_3O^+][HCOO^-]}{[HCOOH]} \qquad K_a = \frac{(y)(y)}{(0.010-y)} \cong \frac{y^2}{0.010} = 1.8 \times 10^{-4}$$

$$[H_3O^+] = y = 1.3 \times 10^{-3} \qquad pH = -\log(1.3 \times 10^{-3}) = 2.89$$

Approximation check: $\left(\frac{1.3 \times 10^{-3}}{0.010}\right) \times 100\% = 13\%$, so the approximation is not valid. The quadaratic formula or successive approximation method must be used to solve the equilibrium expression.

$$K_a = \frac{(y)(y)}{(0.010 - y)} = 1.8 \times 10^{-4} \text{ yields } y^2 + 1.8 \times 10^{-4} y - 1.8 \times 10^{-6} = 0$$

Substituting into the quadratic formula gives $y = \frac{-1.8 \times 10^{-4} \pm \sqrt{(1.8 \times 10^{-4})^2 - 4(1)(-1.8 \times 10^{-6})}}{2(1)}$

and $y = \frac{-1.8 \times 10^{-4} \pm 2.7 \times 10^{-3}}{2} = 1.3 \times 10^{-3}$ or -1.4×10^{-3}.

The only reasonable value of y is 1.3×10^{-3}, since -1.4×10^{-3} predicts negative concentrations for H_3O^+ and $HCOO^-$. Hence, $[H_3O^+] = y = 1.3 \times 10^{-3}$ and $pH = -\log(1.3 \times 10^{-3}) = \mathbf{2.89}$

15.40 ***See Sections 15.3, 15.5, Tables 15.5, 15.6, and Examples 15.5, 15.9.***

(a) $HI + H_2O \rightarrow H_3O^+ + I^-$ $\qquad pH = -\log[H_3O^+] = -\log(0.050) = \mathbf{1.30}$

(b) $HF + H_2O \rightleftarrows H_3O^+ + F^-$ $\qquad K_a = 3.5 \times 10^{-4}$

Conc, *M*	HF	H_3O^+	F^-
initial	0.85	0.	0.
Change	−y	+y	+y
equilibrium	0.85 − y	y	y

$$K_a = \frac{[H_3O^+][F^-]}{[HF]} \qquad K_a = \frac{(y)(y)}{(0.85 - y)} \cong \frac{y^2}{0.85} = 3.5 \times 10^{-4}$$

$[H_3O^+] = y = 0.017\ M$ $\qquad pH = -\log(0.017) = \mathbf{1.77}$

Approximation check: $\left(\frac{0.017}{0.85}\right) \times 100\% = 2.0\%$, so the approximation is valid.

(c) $CH_3COOH + H_2O \rightleftarrows H_3O^+ + CH_3COO^-$ $\qquad K_a = 1.8 \times 10^{-5}$

Conc, *M*	CH_3COOH	H_3O^+	CH_3COO^-
initial	0.15	0.	0.
Change	−y	+y	+y
equilibrium	0.15 − y	y	y

$$K_a = \frac{[H_3O^+][CH_3COO^-]}{[CH_3COOH]} \qquad K_a = \frac{(y)(y)}{(0.15 - y)} \cong \frac{y^2}{0.15} = 1.8 \times 10^{-5}$$

$[H_3O^+] = y = 1.6 \times 10^{-3}$ $\qquad pH = -\log(1.6 \times 10^{-3}) = \mathbf{2.80}$

Approximation check: $\left(\dfrac{1.6 \times 10^{-4}}{0.15}\right) \times 100\% = 0.11\%$, so the approximation is valid.

(d) $C_6H_5COOH + H_2O \rightleftarrows H_3O^+ + C_6H_5COO^-$ $K_a = 6.3 \times 10^{-5}$

Conc, *M*	C_6H_5COOH	H_3O^+	$C_6H_5COO^-$
initial	0.017	0.	0.
Change	−y	+y	+y
equilibrium	0.017 − y	+y	+y

$$K_a = \frac{[H_3O^+][C_6H_5COO^-]}{[C_6H_5COOH]} \qquad K_a = \frac{(y)(y)}{(0.017 - y)} \cong \frac{y^2}{0.017} = 6.3 \times 10^{-5}$$

$[H_3O^+] = y = 1.0 \times 10^{-3}$ pH = $-\log(1.0 \times 10^{-3}) = 3.00$

Approximation check: $\left(\dfrac{1.0 \times 10^{-3}}{0.017}\right) \times 100\% = 5.9\%$, so the approximation is not valid. The quadratic formula or successive approximation method must be used to solve the equilibrium expression.

$$K_a = \frac{(y)(y)}{(0.017 - y)} = 6.3 \times 10^{-5} \text{ yields } y^2 + 6.3 \times 10^{-5}\, y - 1.07 \times 10^{-6} = 0.$$

Substituting into the quadratic formula gives $y = \dfrac{-6.3 \times 10^{-5} \pm \sqrt{(6.3 \times 10^{-5})^2 - 4(1)(-1.07 \times 10^{-6})}}{2(1)}$

and $y = \dfrac{-6.3 \times 10^{-5} \pm 2.1 \times 10^{-3}}{2} = 1.0 \times 10^{-3}$ or -1.1×10^{-3}.

The only reasonable value of y is 1.0×10^{-3}, since -1.1×10^{-3} predicts negative concentrations for H_3O^+ and $C_6H_5COO^-$. Hence, $[H_3O^+] = 1.0 \times 10^{-3}$ and pH = $-\log(1.0 \times 10^{-3})$ = **3.00**.

15.41 ***See Sections 15.3, 15.5, and Figure 15.4.***

An increase in the concentration of a strong acid produces a directly proportional increase in the concentrations of ions and thus in conductivity whereas an increase in the concentration of a weak acid does not. This is due to the observation that strong acids ionize completely whereas weak acids do not.

We can determine whether conductivity is directly proportional to concentration by graphing conductivity versus concentration for each acid.

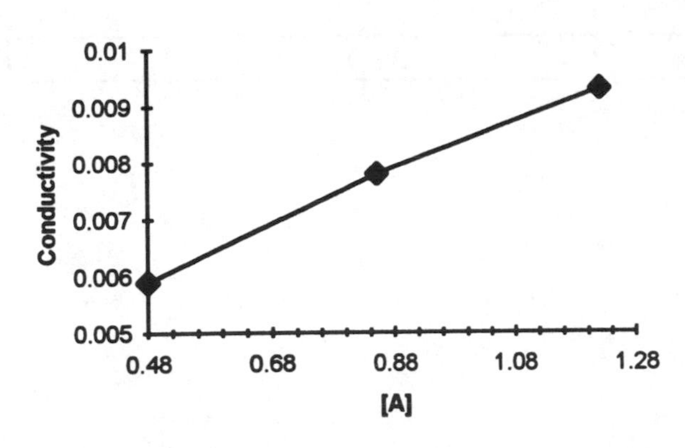

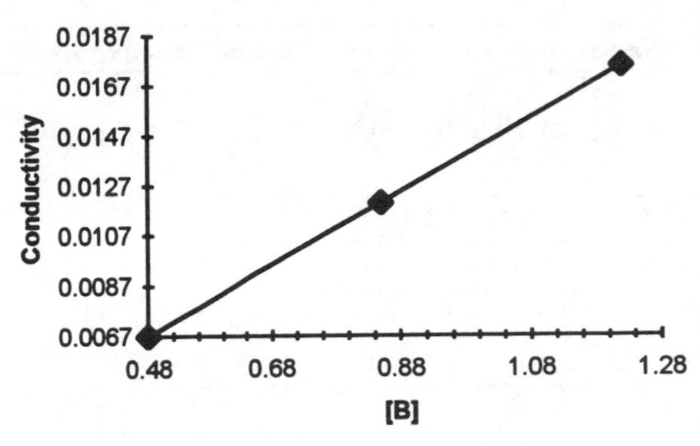

Since there is no directly proportional increase in conductivity with increase in concentration for Acid A, we must conclude Acid A is a weak acid. Since there is a directly proportional increase in conductivity with increase in concentration for Acid B, (within the limits of experimental error), we must conclude Acid B is a strong acid.

15.42 ***See Sections 15.3, 15.5, andFigure 15.4.***

An increase in the concentration of a strong acid produces a directly proportional increase in the concentrations of ions and thus in conductivity whereas an increase in the concentration of a weak acid does not. This is due to the observation that strong acids ionize completely whereas weak acids do not. Hence, we obtain

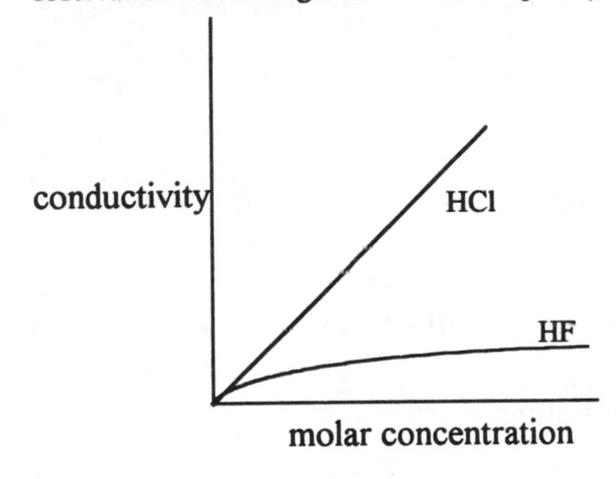

15.43 ***See Section 15.6, Table 15.8, and Example 15.10.***

$N_2H_4 + H_2O \rightleftarrows N_2H_5^+ + OH^-$ $\qquad K_b = 1.7 \times 10^{-6}$

Conc, *M*	N_2H_4	$N_2H_5^+$	OH^-
initial	0.10	0.	0.
Change	−y	+y	+y
equilibrium	0.10 − y	y	y

$$K_b = \frac{[N_2H_5^+][OH^-]}{[N_2H_4]} \qquad K_b = \frac{(y)(y)}{(0.10 - y)} \cong \frac{y^2}{0.10} = 1.7 \times 10^{-6}$$

15.44 *See Section 15.6, Table 15.8, and Example 15.10.*

$NH_2OH + H_2O \rightleftarrows NH_3OH^+ + OH^-$ $\quad K_b = 1.1 \times 10^{-8}$

Conc, *M*	NH_2OH	NH_2OH^+	OH^-
initial	0.10	0.	0.
Change	−y	+y	+y
equilibrium	0.10 − y	y	y

$$K_b = \frac{[NH_2OH^+][OH^-]}{[NH_2OH]} \qquad K_b = \frac{(y)(y)}{(0.10 - y)} \cong \frac{y^2}{0.10} = 1.1 \times 10^{-8}$$

15.45 *See Section 15.6 and Example 15.10.*

$C_8H_{17}N + H_2O \rightleftarrows C_8H_{17}NH^+ + OH^-$

Conc, *M*	$C_8H_{17}N$	$C_8H_{17}NH^+$	OH^-
initial	0.50	0.	0.
Change	−y	+y	+y
equilibrium	0.50 − y	y	y

$pK_b = 3.1 \qquad K_b = 10^{-pK_b} = \text{inv}\log(-pK_b) = 7.9 \times 10^{-4}$

$$K_b = \frac{[C_8H_{17}NH^+][OH^-]}{[C_8H_{17}N]} \qquad K_b = \frac{(y)(y)}{(0.50 - y)} \cong \frac{y^2}{0.50} = 7.9 \times 10^{-4}$$

$[OH^-] = y = 2.0 \times 10^{-2} \qquad pOH = -\log(2.0 \times 10^{-2}) = 1.70 \qquad pH = 14.00 - 1.70 = \mathbf{12.30}$

Approximation check: $\left(\frac{2.0 \times 10^{-2}}{0.50}\right) \times 100\% = 4\%$, so the approximation is valid.

15.46 *See Section 15.6 and Example 15.10.*

$C_{17}H_{19}O_3N + H_2O \rightleftarrows C_{17}H_{19}O_3NH^+ + OH^-$ $\quad K_b = 1.6 \times 10^{-6}$

Conc, *M*	$C_{17}H_{19}O_3N$	$C_{17}H_{19}O_3NH^+$	OH^-
initial	0.0010	0.	0.
Change	−y	+y	+y
equilibrium	0.0010 − y	y	y

$$K_b = \frac{[C_{17}H_{19}O_3NH^+][OH^-]}{[C_{17}H_{19}O_3N]} \qquad K_b = \frac{(y)(y)}{(0.0010 - y)} \cong \frac{y^2}{0.0010} = 1.6 \times 10^{-6}$$

$[OH^-] = y = 4.0 \times 10^{-5} \qquad pOH = -\log(4.0 \times 10^{-5}) = \mathbf{4.40} \qquad pH = 14.00 - 4.40 = \mathbf{9.60}$

15.47 ***See Section 15.6, Tables 15.6, 15.8, and Example 15.12.***

(a) $HCOO^- + H_2O \rightleftarrows HCOOH + OH^-$ $\quad K_b\ HCOO^- = \dfrac{K_w}{K_a\ HCOOH} = \dfrac{1.0 \times 10^{-14}}{1.8 \times 10^{-4}} = \mathbf{5.6 \times 10^{-11}}$

(b) $NO_2^- + H_2O \rightleftarrows HNO_2 + OH^-$ $\quad K_b\ NO_2^- = \dfrac{K_w}{K_a\ HNO_2} = \dfrac{1.0 \times 10^{-14}}{4.6 \times 10^{-4}} = \mathbf{2.2 \times 10^{-11}}$

15.48 ***See Section 15.6, Tables 15.6, 15.8, and Example 15.12.***

(a) $ClO_2^- + H_2O \rightleftarrows HClO_2 + OH^-$ $\quad K_b\ ClO_2^- = \dfrac{K_w}{K_a\ HClO_2} = \dfrac{1.0 \times 10^{-14}}{1.0 \times 10^{-2}} = \mathbf{1.0 \times 10^{-12}}$

(b) $F^- + H_2O \rightleftarrows HF + OH^-$ $\quad K_b\ F^- = \dfrac{K_w}{K_a\ HF} = \dfrac{1.0 \times 10^{-14}}{3.50 \times 10^{-4}} = \mathbf{2.9 \times 10^{-11}}$

15.49 ***See Section 15.6, Tables 15.6, 15.8, and Example 15.12.***

(a) $NH_2OH_2^+ + H_2O \rightleftarrows H_3O^+ + NH_2OH$ $\quad K_a\ NH_2OH_2^+ = \dfrac{K_w}{K_b\ NH_2OH} = \dfrac{1.0 \times 10^{-14}}{1.1 \times 10^{-8}} = \mathbf{9.1 \times 10^{-7}}$

(b) $NH_4^+ + H_2O \rightleftarrows H_3O^+ + OH^-$ $\quad K_a\ NH_4^+ = \dfrac{K_w}{K_b NH_3} = \dfrac{1.0 \times 10^{-14}}{1.8 \times 10^{-5}} = \mathbf{5.6 \times 10^{-10}}$

15.50 ***See Section 15.6, Tables 15.6, 15.8, and Example 15.12.***

(a) $C_5H_5NH^+ + H_2O \rightleftarrows H_3O^+ + C_5H_5NH$ $\quad K_a\ C_5H_5NH^+ = \dfrac{K_w}{K_b C_5H_5NH} = \dfrac{1.0 \times 10^{-14}}{1.8 \times 10^{-9}} = \mathbf{5.6 \times 10^{-6}}$

(b) $N_2H_5^+ + H_2O \rightleftarrows H_3O^+ + N_2H_4$ $\quad K_a = N_2H_5^+ = \dfrac{K_w}{K_b N_2H_4} = \dfrac{1.0 \times 10^{-14}}{1.7 \times 10^{-6}} = \mathbf{5.9 \times 10^{-9}}$

15.51 ***See Section 15.6, Tables 15.6, 15.8, and Example 15.12.***

According to Table 15.6, $H_2O < HF < HSO_4^-$. According to Table 15.8, NH_3 is a weaker base than OH^-. Hence, NH_4^+ is a stronger acid than H_2O. To determine where NH_4^+ fits in this series, we must calculate $K_a\ NH_4^+$ using

$K_a\ NH_4^+ = \dfrac{K_w}{K_b\ NH_3}$ $\quad K_a\ NH_4^+ = \dfrac{1.0 \times 10^{-14}}{1.8 \times 10^{-5}} = 5.6 \times 10^{-10}$

This gives $H_2O < NH_4^+ < HF < HSO_4^-$.

15.52 ***See Section 15.6, Tables 15.5, 15.6, 15.8, and Example 15.12.***

According to Tables 15.5 and 15.6, $HF < HCl$. To determine where NH_4^+ fits in this series, we must calculate K_a NH_4^+ using $K_a\ NH_4^+ = \dfrac{K_w}{K_b\ NH_3}$ $\quad K_a\ NH_4^+ = \dfrac{1.0 \times 10^{-14}}{1.8 \times 10^{-5}} = 5.6 \times 10^{-10}$

Knowing NH_3 acts as a base gives $NH_3 < NH_4^+ < HF < HCl$.

15.53 See Section 15.6 and Table 15.6.

According to Table 15.6, $H_2O < CH_3COOH < HCOOH$. F^- cannot act as Bronsted acid, since it doesn't contain partially positive (i.e. protonic) hydrogen atoms.

15.54 See Section 15.6, Tables 15.5, 15.6, 15.8, and Example 15.12.

According to Tables 15.5 and 15.6, HF < HCl. Knowing that NH_3 acts a base and Na^+ is an aqueous cation derived from a strong base and does not appreciably react with water gives $NH_3 < Na^+ < HF < HCl$.

15.55 See Section 15.6.

(a) 0.050 *M* NaF is slightly basic. Na^+ comes from the strong base NaOH and does not appreciably react with water. F^- is the conjugate base of the weak acid HF and therefore acts as a weak base. The pH might be in the range 7-9.

(b) 0.100 *M* KCl is neutral. K^+ comes from the strong base KOH and does not appreciably react with water. Cl^- comes from the strong acid HCl and does not appreciably react with water. The pH should be 7, provided the water used to make the solution is neutral.

(c) 0.080 *M* NH_4Br is slightly acidic. NH_4^+ comes from the weak base NH_3 and acts as a weak acid. Br^- comes from the strong acid HBr and does not appreciably react with water. The pH might be in the range 5-7.

15.56 See Section 15.6.

(a) 0.150 *M* $NaHSO_4$ is slightly acidic. Na^+ comes from the strong base NaOH and does not appreciably react with water. HSO_4^- is a weak acid. The pH might be in the range 3-5.

(b) 0.050 *M* Na_3PO_4 is quite basic. Na^+ comes from the strong base NaOH and does not appreciably react with water. PO_4^{3-} comes from the weak acid H_3PO_4 and is a relatively strong base. The pH might be in the range 10-12.

(c) 0.100 *M* KBr is neutral. K^+ comes from the strong base KOH and does not appreciably react with water. Br^- comes from the strong acid HBr and does not appreciably react with water. The pH should be 7, provided the water used to make the solution is neutral.

15.57 See Section 15.6, Tables 15.8, 15.9, and Example 15.13.

The pyridium ion, $C_5H_5NH^+$, is the conjugate acid of the weak base pyridine, C_5H_5N, and therefore acts as a weak acid. The iodide ion, I^-, is the conjugate base of the strong acid HI and therefore does not react appreciably with water. The equilibrium of interest is $C_5H_5NH^+ + H_2O \rightleftarrows H_3O^+ + C_5H_5N$.

Conc, *M*	$C_5H_5NH^+$	H_3O^+	C_5H_5N
initial	0.060	0.	0.
Change	−y	+y	+y
equilibrium	0.060 − y	y	y

$$K_a\ C_5H_5NH^+ = \frac{K_w}{K_b\ C_5H_5N} = \frac{[H_3O^+][C_5H_5N]}{[C_5H_5NH^+]}$$

$$K_a\ C_5H_5NH^+ = \frac{(y)(y)}{0.060 - y} \cong \frac{y^2}{0.060} = \frac{1.0 \times 10^{-14}}{1.8 \times 10^{-9}} = 5.6 \times 10^{-6}$$

15.58 ***See Section 15.6, Tables 15.8, 15.9, and Example 15.13.***

The ammonium ion, NH_4^+, is the conjugate acid of the weak base ammonia, NH_3, and therefore acts as a weak acid. The chloride ion, Cl^-, is the conjugate base of the strong acid HCl and therefore does not react appreciably with water. The equilibrium of interest is $NH_4^+ + H_2O \rightleftarrows H_3O^+ + NH_3$.

Conc, *M*	NH_4^+	H_3O^+	NH_3
initial	1.5	0.	0.
Change	−y	+y	+y
equilibrium	1.5 − y	y	y

$$K_a\ NH_4^+ = \frac{K_w}{K_b\ NH_3} = \frac{[H_3O^+][NH_3]}{[NH_4^+]}$$

$$K_a = \frac{(y)(y)}{1.5 - y} \cong \frac{y^2}{1.5} = \frac{1.0 \times 10^{-14}}{1.8 \times 10^{-5}} = 5.6 \times 10^{-10}$$

15.59 ***See Section 15.6, Tables 15.6, 15.9, and Example 15.13.***

The sodium ion, Na^+, comes from the strong base NaOH and does not react appreciably with water. The acetate ion, CH_3COO^-, is the conjugate base of the weak acid acetic acid, CH_3COOH, and therefore acts as a weak base. The equilibrium of interest is : $CH_3COO^- + H_2O \rightleftarrows CH_3COOH + OH^-$.

Conc, *M*	CH_3COO^-	CH_3COOH	OH^-
initial	0.010	0.	0.
Change	−y	+y	+y
equilibrium	0.010 − y	y	y

$$K_b\ CH_3COO^- = \frac{K_w}{K_a\ CH_3COOH} = \frac{[CH_3COO^-][OH^-]}{[CH_3COOH]}$$

$$K_b\ CH_3COO^- = \frac{(y)(y)}{0.010 - y} \cong \frac{y^2}{0.010} = \frac{1.0 \times 10^{-14}}{1.8 \times 10^{-5}} = 5.6 \times 10^{-10}$$

$[OH^-] = y = 2.4 \times 10^{-6}$, **pOH** $= -\log(2.4 \times 10^{-6}) = 5.62$ **and pH** $= 14.00 - 5.62 =$ **8.38**

Approximation check: $\left(\frac{2.4 \times 10^{-6}}{0.010}\right) \times 100\% = 0.024\%$, **so the approximtion is valid.**

15.60 *See Section 15.6, Tables 15.8, 15.9, and Example 15.13.*

The potassium ion , K^+ , comes from the strong base KOH and does not react appreciably with water. The nitrite ion, NO_2^- , is the conjugate base of the weak acid HNO_2 and therefore acts as a weak base. The equilibrium of interest is $NO_2^- + H_2O \rightleftarrows HNO_2 + OH^-$.

Conc, *M*	NO_2^-	HNO_2	OH^-
initial	0.25	0.	0.
Change	−y	+y	+y
equilibrium	0.25 − y	y	y

$$K_b\, NO_2^- = \frac{K_w}{K_a HNO_2} = \frac{[HNO_2][OH^-]}{[NO_2^-]} \qquad K_b\, NO_2^- = \frac{(y)(y)}{0.25-y} \cong \frac{y^2}{0.25} = \frac{1.0 \times 10^{-14}}{4.6 \times 10^{-4}} = 2.2 \times 10^{-11}$$

$[OH^-] = y = 2.3 \times 10^{-6}$, $pOH = -\log(2.3 \times 10^{-6}) = 5.64$ and **pH** = 14.00 - 5.64 = **8.36** .

Approximation check: $\left(\frac{2.3 \times 10^{-6}}{0.25}\right) \times 100\% = 0.00092\%$, so the approximation is valid.

15.61 *See Section 15.7, Tables 15.5, 15.6, and Example 15.15.*

HI is a strong acid, whereas CH_3COOH is a weak acid. The pH will be determined by the strong acid HI: $HI + H_2O \rightarrow H_3O^+ + I^-$. Hence, $[H_3O^+] = C_{HI} = 0.050\, M$ and $pH = -\log(0.050) = \mathbf{1.30}$.

15.62 *See Section 15.7, Tables 15.5, 15.6, and Example 15.15.*

HCl is a strong acid, whereas HF is a weak acid. The pH will be determined by the strong acid HCl: $HCl + H_2O \rightarrow H_3O^+ + Cl^-$. Hence, $[H_3O^+] = C_{HCl} = 0.050\, M$, and pH = -log(0.050) = **1.30**.

15.63 *See Section 15.7, Table 15.6, and Example 15.15.*

Acetic acid, CH_3COOH, with $K_a = 1.8 \times 10^{-5}$ is a much stronger acid than hydrocyanic acid, HCN, with $K_a = 7.2 \times 10^{-10}$. We can therefore consider this to be just a solution of acetic acid: $CH_3COOH + H_2O \rightleftarrows H_3O^+ + CH_3COO^-$.

Conc, *M*	CH_3COOH	H_3O^+	CH_3COO^-
initial	0.10	0.	0.
Change	−y	+y	+y
equilibrium	0.10 − y	y	y

$$K_a = \frac{[H_3O^+][CH_3COO^-]}{[CH_3COOH]} \qquad K_a = \frac{(y)(y)}{0.10-y} \cong \frac{y^2}{0.10} = 1.8 \times 10^{-5}$$

$[H_3O^+] = y = 1.3 \times 10^{-3}$ and **pH** $= -\log(1.3 \times 10^{-3}) = \mathbf{2.89}$.

Approximation check: $\left(\frac{1.3 \times 10^{-3}}{0.10}\right) \times 100\% = 1.3\%$, so the approximation is valid.

15.64 *See Section 15.7, Table 15.6, and Example 15.15.*

Formic acid, HCOOH, with $K_a = 1.8 \times 10^{-4}$ is a much stronger acid than phenol, C_6H_5OH, with $K_a = 1.3 \times 10^{-10}$. We can therefore consider this to be just a solution of formic acid.

$HCOOH + H_2O \rightleftarrows H_3O^+ + HCOO^-$

Conc, M	HCOOH	H_3O^+	$HCOO^-$
initial	0.050	0.	0.
Change	−y	+y	+y
equilibrium	0.050 − y	y	y

$$K_a = \frac{[H_3O^+][HCOO^-]}{[HCOOH]} \qquad K_a = \frac{(y)(y)}{0.050 - y} \cong \frac{y^2}{0.050} = 1.8 \times 10^{-4}$$

$[H_3O^+] = y = 3.0 \times 10^{-3}$ and $pH = -\log(3.0 \times 10^{-3}) = 2.52$..

Approximation check: $\left(\frac{3.0 \times 10^{-3}}{0.050}\right) \times 100\% = 6.0\%$, so the approximation is not valid. The quadratic formula or successive approximation method must be used to solve the equilibrium expression.

$K_a = \frac{(y)(y)}{(0.050 - y)} = 1.8 \times 10^{-4}$ yields $y^2 + 1.8 \times 10^{-4}y - 9.0 \times 10^{-6} = 0$.

Substituting into the quadratic formula gives $y = \frac{-1.8 \times 10^{-4} \pm \sqrt{(1.8 \times 10^{-4})^2 - 4(1)(-9.0 \times 10^{-6})}}{2(1)}$

and $y = \frac{1.8 \times 10^{-4} \pm 6.0 \times 10^{-3}}{2} = 2.9 \times 10^{-3}$ or -3.1×10^{-3}.

The only reasonable value of y is 2.9×10^{-3}, since -3.1×10^{-3} predicts negative concentrations for H_3O^+ and $HCOO^-$. Hence, $[H_3O^+] = y = 2.9 \times 10^{-3}$, and **pH** $= -\log(2.9 \times 10^{-3}) =$ **2.54**.

15.65 *See Section 15.8.*

(a) AsH_3 is a stronger acid than GeH_4. Within any row of the periodic table the acidities of the binary hydrides increase as the central atom becomes more electronegative and can more easily accommodate negative charge. Note: This increase in acidity occurs in spite of increasing A-H bond strengths moving from left-to-right across a row or period; see text.

(b) HNO_3 is a stronger acid than HNO_2. The additional oxygen atom withdraws electron density from the N, which in turn withdraws additional electron density from the O-H bond, making it weaker and easier to ionize in water. The additional oxygen atom also helps stabilize the anion that is formed when the acid ionizes, so NO_3^- is more stable than NO_2^-.

15.66 *See Section 15.8.*

(a) H_3AsO_4 is a stronger acid than H_3AsO_3. The additional oxygen atom withdraws electron density from the As, which in turn withdraws additional electron density from the O-H bond, making it weaker and easier to ionize in water. The additional oxygen atom also helps stabilize the anion that is formed when the acid ionizes, so $H_2AsO_4^-$ is more stable than $H_2AsO_3^-$.

(b) H_2S is a stronger acid than PH_3. Within any row of the periodic table the acidities of the binary hydrides increase as the central atom becomes more electronegative and can more easily accommodate negative charge. Note: This increase in acidity occurs in spite of increasing A-H bond strengths moving from left-to-right across a row or period; see text.

15.67 ***See Section 15.8.***

(a) $HClO_4$ is a stronger acid than $HClO_3$. The additional oxygen atom withdraws electron density from the Cl, which in turn withdraws additional electron density from the O-H bond, making it weaker and easier to ionize in water. The additional oxygen atom also helps stabilize the anion that is formed when the acid ionizes, so ClO_4^- is more stable than ClO_3^-.

(b) H_2Se is a stronger acid than H_2S. Within any group of nonmetals the acidities of the binary hydrides increase as the central atom becomes larger and less electronegative giving weaker A-H bonds from top to bottom within the group.

15.68 ***See Section 15.8.***

(a) $HClO_2$ is a stronger acid than HClO. The additional oxygen atom withdraws electron density from the Cl, which in turn withdraws additional electron density from the O-H bond, making it weaker and easier to ionize in water. The additional oxygen atom also helps stabilize the anion that is formed when the acid ionizes, so ClO_2^- is more stable than ClO^-.

(b) H_2S is a stronger acid than H_2O. Within any group of nonmetals the acidities of the binary hydrides increase as the central atom becomes larger and less electronegative giving weaker A-H bonds from top to bottom within the group.

15.69 ***See Sections 3.1, 15.1, 15.9.***

System	Acid	Base
Arrhenius	Any substance that provides H^+ in water.	Any substance that provides OH^- in water.
Bronsted-Lowry	A proton donor.	A proton acceptor.
Lewis	An electron pair acceptor.	An electron pair donor.

(a) $HCl(aq) + NH_3(aq) \rightarrow NH_4Cl(aq)$ — Arrhenius, Bronsted-Lowry and Lewis reaction.

(b) $SO_2(g) + NaOH(s) \rightarrow NaHSO_3(s)$ — Lewis reaction.

Note: The acid-base theories become more general moving from Arrhenius to Bronsted-Lowry to Lewis. Hence, any Arrhenius reaction is also a Bronsted-Lowry reaction and a Lewis reaction. Any Bronsted-Lowry reaction is also a Lewis reaction. However, some Lewis reactions are neither Arrhenius nor Bronsted-Lowry reactions, since the Lewis theory is the most general and therefore includes additional reactions.

15.70 ***See Sections 3.1, 15.1, 15.9, and the Solution for Exercise 15.69.***

(a) $HCl(aq) + H_2O(\ell) \rightarrow H_3O^+(aq) + Cl^-(aq)$ — Arrhenius, Bronsted-Lowry and Lewis reaction.

(b) $Zn(OH)_3^-(aq) + OH^-(aq) \rightarrow Zn(OH)_4^{2-}(aq)$ — Lewis reaction.

15.71 ***See Sections 3.1, 15.1, 15.9, and the Solution for Exercise 15.69.***

(a) $LiH(s) + H_2O(\ell) \rightarrow LiOH(aq) + H_2(g)$ — Bronsted-Lowry and Lewis reaction.

(b) $HSO_4^-(aq) + F^-(aq) \rightleftarrows HF(aq) + SO_4^{2-}(aq)$ Arrhenius, Bronsted-Lowry and Lewis reaction.

15.72 *See Sections 3.1, 15.1, 15.9, and the Solution for Exercise 15.69.*

(a) $CO_2(g) + LiOH(s) \rightarrow LiHCO_3(s)$ Lewis reaction.

(b) $SO_2(g) + H_2O(g) \rightarrow H_2SO_3(g)$ Lewis reaction.

15.73 *See Section 15.6.*

(a) 0.250 *M* HBr is strongly acidic. HBr is a strong acid. The pH is in the range 0-1.

(b) 0.50 *M* HF is weakly acidic. HF is, however, a relatively strong weak acid. The pH might be in the range 2-3.

(c) 0.020 *M* $Ba(OH)_2$ is strongly basic. $Ba(OH)_2$ is a strong base. The pH might be in the range 12-13.

(d) 0.44 *M* NH_3 is weakly basic. NH_3 is a weak base. The pH might be in the range 9-10.

15.74 *See Section 15.6.*

(a) 0.30 *M* NH_4Cl is slightly acidic. NH_4^+ comes from the weak base NH_3 and acts as a weak acid. Cl^- comes from the strong acid HCl and does not appreciably react with water. The pH might be in the range 5-7.

(b) 0.25 *M* Na_3PO_4 is quite basic. Na^+ comes from the strong base NaOH and does not appreciably react with water. PO_4^{3-} comes from the weak acid H_3PO_4 and is a relatively strong base. The pH might be in the range 10-12.

(c) 0.080 *M* HI is strongly acidic. HI is a strong acid. The pH is in the range 1-2.

(d) 0.12 *M* LiI is neutral. Li^+ comes from the strong base LiOH and does not appreciably react with water. I^- comes from the strong acid HI and does not appreciably react with water. The pH is 7.

15.75 *See Section 15.6.*

(a) 0.45 *M* NaCl is neutral. Na^+ comes from the strong base NaOH and does not appreciably react with water. Cl^- comes from the strong acid HCl and does not appreciably react with water. The pH should be 7.

(b) 0.18 *M* BaF_2 is weakly basic. Ba^{2+} comes from the strong base $Ba(OH)_2$ and does not appreciably react with water. F^- comes from the weak acid HF and acts as a weak base. The pH might be in the range 8-10.

(c) 0.25 *M* $KHSO_4$ is weakly acidic. K^+ comes from the strong base KOH and does not appreciably react with water. HSO_4^- comes from the strong acid H_2SO_4 and is a weak acid. The pH might be in the range 4-6.

(d) 0.33*M* $NaNO_2$ is weakly basic. Na^+ comes from the strong base NaOH and does not appreciably react with water. NO_2^- comes from the weak acid HNO_2 and acts as a weak base. The pH might be in the range 8-10.

15.76 *See Section 15.6.*

(a) 0.30 *M* NH_4Cl is slightly acidic. NH_4^+ comes from the weak base NH_3 and acts as a weak acid. Cl^- comes from the strong acid HCl and does not appreciably react with water. The pH might be in the range 5-7.

(b) 0.15 M N_2H_5Cl is slightly acidic. $N_2H_5^+$ comes from the weak base N_2H_4 and acts as a weak acid. Cl^- comes from the strong acid HCl and does not appreciably react with water. The pH might be in the range 4-6.

(c) 0.50 M KNO_3 is neutral. K^+ comes from the strong base KOH and does not appreciably react with water. NO_3^- comes from the strong acid HNO_3 and does not appreciably react with water. The pH is 7.

(d) 0.50 M NaHCOO is weakly basic. Na^+ comes from the strong base NaOH and does not appreciably react with water. $HCOO^-$ comes from the weak acid HCOOH and acts as a weak base. The pH might be in the rage 8-10.

15.77 ***See Section 15.5, Table 15.6, and Example 15.9.***

(a) $C_6H_5COOH + H_2O \rightleftarrows H_3O^+ + C_6H_5COO^-$ $\qquad K_a = 6.3 \times 10^{-5}$

Conc, M	C_6H_5COOH	H_3O^+	$C_6H_5COO^-$
initial	0.010	0.	0.
Change	−y	+y	+y
equilibrium	0.010 − y	y	y

$$K_a = \frac{[H_3O^+][C_6H_5COO^-]}{[C_6H_5COOH]} \qquad K_a = \frac{(y)(y)}{(0.010-y)} \cong \frac{y^2}{0.010} = 6.3 \times 10^{-5}$$

Hence, $y = 7.9 \times 10^{-4}$ and $\left(\frac{7.9 \times 10^{-4}}{0.010}\right) \times 100\% = 7.9\%$, so the approximation is not valid. The quadratic formula or successive approximation method must be used to solve the equilibrium expression.

$K_a = \frac{(y)(y)}{(0.010-y)} = 6.3 \times 10^{-5}$ yields $y^2 + 6.3 \times 10^{-5}\, y - 6.3 \times 10^{-7} = 0$.

Substituting into the quadratic formula gives $y = \frac{-6.3 \times 10^{-5} \pm \sqrt{(6.3 \times 10^{-5})^2 - 4(1)(-6.3 \times 10^{-7})}}{2(1)}$

and $y = \frac{-6.3 \times 10^{-5} \pm 1.6 \times 10^{-3}}{2} = 7.7 \times 10^{-4}$ or -8.3×10^{-4}.

The only reasonable value of y is 7.7×10^{-4}, since -8.3×10^{-4} predicts negative concentrations for H_3O^+ and $C_6H_5COO^-$. Hence, fraction ionized $= \frac{[C_6H_5COO^-]}{C_{C_6H_5COOH}} \times 100\% = \frac{7.7 \times 10^{-4}}{0.010} \times 100\% = 7.7\%$.

(b) $C_6H_5COOH + H_2O \rightleftarrows H_3O^+ + C_6H_5COO^-$ $\quad K_a = 6.3 \times 10^{-5}$

Conc, M	C_6H_5COOH	H_3O^+	$C_6H_5COO^-$
initial	0.0010	0.	0.
Change	−y	+y	+y
equilibrium	0.0010−y	y	y

$$K_a = \frac{[H_3O^+][C_6H_5COO^-]}{[C_6H_5COOH]} \qquad K_a = \frac{(y)(y)}{(0.0010-y)} \cong \frac{y^2}{0.0010} = 6.3 \times 10^{-5}$$

Hence, $y = 2.5 \times 10^{-4}$ and $\left(\dfrac{2.5 \times 10^{-4}}{0.010}\right) \times 100\% = 25\%$, so the approximation is not valid. The quadratic formula or successive approximation method must be used to solve the equilibrium expression.

$$K_a = \frac{(y)(y)}{(0.0010 - y)} = 6.3 \times 10^{-5} \text{ yields } y^2 + 6.3 \times 10^{-5}y - 6.3 \times 10^{-8} = 0$$

Substituting into the quadratic formula gives $y = \dfrac{-6.3 \times 10^{-5} \pm \sqrt{(6.3 \times 10^{-5})^2 - 4(1)(-6.3 \times 10^{-8})}}{2(1)}$

and $y = \dfrac{-6.3 \times 10^{-5} \pm 5.1 \times 10^{-4}}{2} = 2.2 \times 10^{-4}$ or -2.9×10^{-4}.

The only reasonable value of y is 2.2×10^{-4}, since -2.9×10^{-4} predicts negative concentrations for H_3O^+ and $C_6H_5COO^-$. Hence, fraction ionized $= \dfrac{[C_6H_5COO^-]}{C_{C_6H_5COOH}} \times 100\% = \dfrac{2.2 \times 10^{-4}}{0.0010} \times 100\% = 22\%$.

15.78 ***See Section 15.5, Table 15.6, and Example 15.9.***

(a) $C_6H_5OH = H_2O \rightleftarrows H_3O^+ + C_6H_5O^-$

$pK_a = 9.89$ $\quad K_a = \text{inv log}(-pK_a) = \text{inv log}(-9.89) = 1.3 \times 10^{-10}$

Conc, *M*	C_6H_5OH	H_3O^+	$C_6H_5O^-$
initial	0.010	0.	0.
Change	−y	+y	+y
equilibrium	0.010 − y	y	y

$$K_a = \frac{[H_3O^+][C_6H_5O^-]}{[C_6H_5OH]} \qquad K_a = \frac{(y)(y)}{(0.010 - y)} \cong \frac{y^2}{0.010} = 1.3 \times 10^{-10}$$

Hence, $y = 1.1 \times 10^{-6}$ and $\left(\dfrac{1.1 \times 10^{-6}}{0.010}\right) \times 100\% = 0.011\%$, so the approximation is valid and the fraction ionized is **0.011%**.

(b)

Conc, *M*	C_6H_5OH	H_3O^+	$C_6H_5O^-$
initial	0.0010	0.	0.
Change	−y	+y	+y
equilibrium	0.0010 − y	y	y

$$K_a = \frac{[H_3O^+][C_6H_5O^-]}{[C_6H_5OH]} \qquad K_a = \frac{(y)(y)}{(0.0010 - y)} \cong \frac{y^2}{0.0010} = 1.3 \times 10^{-10}$$

Hence, $y = 3.6 \times 10^{-7}$ and $\left(\dfrac{3.6 \times 10^{-7}}{0.0010}\right) \times 100\% = 0.036\%$, so the approximation is valid and the fraction ionized is **0.036%**.

15.79 *See Section 15.5, Table 15.6, and Example 15.8.*

$HNO_2 + H_2O \rightleftarrows H_3O^+ + NO_2^-$ $\qquad K_a = 4.6 \times 10^{-4}$

$[H_3O^+] = 10^{-pH} = \text{inv}\log(-pH)$ $\qquad [H_3O^+] = \text{inv}\log(-4.75) = 1.8 \times 10^{-5}\ M$

Conc, M	HNO_2	H_3O^+	NO_2^-
initial	y	0.	0.
Change	-1.8×10^{-5}	$+1.8 \times 10^{-5}$	$+1.8 \times 10^{-5}$
equilibrium	$y - 1.8 \times 10^{-5}$	1.8×10^{-5}	1.8×10^{-5}

$$K_a = \frac{[H_3O^+][NO_2^-]}{[HNO_2]} \qquad K_a = \frac{(1.8 \times 10^{-5})(1.8 \times 10^{-5})}{(y - 1.8 \times 10^{-5})} = 4.6 \times 10^{-4}$$

Rearranging and solving for y gives
$3.2 \times 10^{-10} = 4.6 \times 10^{-4}\,y - 8.3 \times 10^{-9}$ and $[HNO_2]$ initial $= y = 1.9 \times 10^{-5}\ M$.

Solving $M(\text{con}) \times V(\text{con}) = M(\text{dil}) \times V(\text{dil})$ for (con) gives $(\text{con}) = \frac{M(\text{dil}) \times V(\text{dil})}{M(\text{con})}$

$$(\text{con}) = \frac{1.9 \times 10^{-5}\ M \times 1.00\ \text{L}}{0.083\ M} = 2.3 \times 10^{-4}\ \text{L or } \mathbf{0.23\ mL}.$$

15.80 *See Section 15.5, Table 15.6, and Example 15.8.*

$CH_3COOH + H_2O \rightleftarrows H_3O^+ + CH_3COO^-$ $\qquad K_a = 1.8 \times 10^{-5}$

$[H_3O^+] = 10^{-pH} = \text{inv}\log(-pH)$ $\qquad [H_3O^+] = \text{inv}\log(-4.00) = 1.0 \times 10^{-4}\ M$

Conc, M	CH_3COOH	H_3O^+	CH_3COO^-
initial	y	0.	0.
Change	-1.0×10^{-4}	$+1.0 \times 10^{-4}$	$+1.0 \times 10^{-4}$
equilibrium	$y - 1.0 \times 10^{-4}$	1.0×10^{-4}	1.0×10^{-4}

$$K_a = \frac{[H_3O^+][CH_3COO^-]}{[CH_3COOH]} \qquad K_a = \frac{(1.0 \times 10^{-4})(1.0 \times 10^{-4})}{(y - 1.0 \times 10^{-4})} = 1.8 \times 10^{-5}$$

Rearranging and solving for y gives $1.0 \times 10^{-8} = 1.0 \times 10^{-5}\,y - 1.8 \times 10^{-9}$ and $[CH_3COOH]$ initial $= y = 6.6 \times 10^{-4}\ M$.

Solving M (con) x V (con) = M (dil) x V (dil) for V(con) gives

$$V(\text{con}) = \frac{M(\text{dil}) \times V(\text{dil})}{M(\text{con})} \qquad V(\text{con}) = \frac{6.6 \times 10^{-4}\ M \times 5.0\ \text{L}}{0.10\ M} = 0.033\ \text{L or } \mathbf{33\ mL}.$$

15.81 *See Section 15.5, Table 15.6, and Example 15.8.*

$C_6H_5COOH + H_2O \rightleftarrows H_3O^+ + C_6H_5COO^-$ $K_a = 6.3 \times 10^{-5}$

$[H_3O^+] = 10^{-pH} = \text{inv}\log(-pH)$ $[H_3O^+] = \text{inv}\log(-3.50) = 3.2 \times 10^{-4}\ M$

Conc, M	C_6H_5COOH	H_3O^+	$C_6H_5COO^-$
initial	y	0.	0.
Change	-3.2×10^{-4}	$+3.2 \times 10^{-4}$	$+3.2 \times 10^{-4}$
equilibrium	$y - 3.2 \times 10^{-4}$	3.2×10^{-4}	3.2×10^{-4}

$$K_a = \frac{[H_3O^+][C_6H_5COO^-]}{[C_6H_5COOH]} \qquad K_a = \frac{(3.2 \times 10^{-4})(3.2 \times 10^{-4})}{(y - 3.2 \times 10^{-4})} = 6.3 \times 10^{-5}$$

Rearranging and solving for y gives

$1.0 \times 10^{-7} = 6.3 \times 10^{-5}\,y - 2.0 \times 10^{-8}$, and $[C_6H_5COOH]\ \text{initial} = y = 1.9 \times 10^{-3}\ M$.

$$\text{Hence, ? g } C_6H_5COOH = 1.00\text{ L } C_6H_5COOH\text{ soln} \times \frac{1.9 \times 10^{-3}\text{ mol } C_6H_5COOH}{1\text{ L } C_6H_5COOH\text{ soln}} \times \frac{122.0\text{ g } C_6H_5COOH}{1\text{ mol } C_6H_5COOH}$$

$$= \mathbf{0.23\ g\ C_6H_5COO}$$

15.82 *See Sectiond 15.2, 15.3.*

$HCl + H_2O \rightarrow H_3O^+ + Cl^-$

$C_{HCl} - [H_3O^+] = 10^{-pH} = \text{inv}\log(-pH)$ $[H_3O^+] = \text{inv}\log(-3.50) = 3.2 \times 10^{-4}$

Solving M(con) x V(con) = M(dil) x V(dil) for V(con) gives

$$V(\text{con}) = \frac{M(\text{dil}) \times V(\text{dil})}{M(\text{con})} \qquad V(\text{con}) = \frac{3.2 \times 10^{-4}\ M \times 100.0\text{ L}}{14.3\ M} = 0.0022\text{ L or } \mathbf{2.2\ mL}.$$

15.83 *See Section 15.6 and Example 15.10.*

$NH_3 + H_2O \rightleftarrows NH_4^+ + OH^-$ $K_b = 1.8 \times 10^{-5}$

$$?\ M\,NH_3 = \frac{10.0\text{ mL conc } NH_3\text{ soln}}{1.00\text{ L}} \times \frac{0.90\text{ g conc } NH_3\text{ soln}}{1\text{ mL conc } NH_3\text{ soln}} \times \frac{28\text{ g } NH_3}{100\text{ g conc } NH_3\text{ soln}} \times \frac{1\text{ mol } NH_3}{17.0\text{ g } NH_3} = 0.15\ M\,NH_3$$

Conc, M	NH_3	NH_4^+	OH^-
initial	0.15	0.	0.
Change	$-y$	$+y$	$+y$
equilibrium	$0.15 - y$	y	y

$$K_b = \frac{[NH_4^+][OH^-]}{[NH_3]} \qquad K_b = \frac{(y)(y)}{0.15 - y} \cong \frac{y^2}{0.15} = 1.8 \times 10^{-5}$$

$[OH^-] = y = 1.6 \times 10^{-3}$, pOH $= -\log(1.6 \times 10^{-3}) = 2.80$ and pH $= 14.00 - 2.80 = \mathbf{11.20}$

Approximation check: $\left(\dfrac{1.6 \times 10^{-3}}{0.15}\right) \times 100\% = 1.1\%$, so the approximation is valid.

15.84 *See Section 15.3.*

$HCl + H_2O \rightarrow H_3O^+ + Cl^-$

$$[H_3O^+] = C_{HCl} = \frac{25.0\text{ mL conc HCl soln}}{0.500\text{ L}} \times \frac{1.19\text{ g conc HCl soln}}{1\text{ mL conc HCl soln}} \times \frac{37\text{ g HCl}}{100\text{ g conc HCl soln}} \times \frac{1\text{ mol HCl}}{36.5\text{ g HCl}} = 0.60\ M$$

$pH = -\log[H_3O^+]$ $\quad$ $pH = -\log(0.60) = \mathbf{0.22}$

15.85 *See Sections 15.1, 15. 3, 15.4.*

$2HF \rightleftarrows H_2F^+ + F^-$ $\quad$ H_2F^+ is the conjugate acid of HF and F^- is the conjugate base of HF.

(a) KF is a source of F^- and therefore acts as a base.
(b) $HClO_4 + HF \rightarrow H_2F^+ + ClO_4^-$
(c) $NH_3 + HF \rightarrow NH_4^+ + F^-$
(d) $H_2F^+ + NH_3 \rightarrow HF + NH_4^+$

15.86 *See Sections 15.1, 15.3, 15.4.*

(a) $NH_3 + NH_3 \rightleftarrows NH_4^+ + NH_2^-$
(b) NH_4^+ is the conjugate acid of NH_3, and NH_2^- is the conjugate base of NH_3.
(c) $NaNH_2$ is a source of NH_2^- and therefore acts as a base.
(d) NH_4Br is a source of NH_4^+ and therefore acts as an acid.

15.87 *See Sections 15.5, 15.6, and Examples 15.9, 15.13.*

(a) $C_6H_5COOH + H_2O \rightleftarrows H_3O^+ + C_6H_5COO$ $\quad$ $K_a = 6.3 \times 10^{-5}$

Conc, M	C_6H_5COOH	H_3O^+	$C_6H_5COO^-$
initial	0.050	0.	0.
Change	−y	+y	+y
equilibrium	0.050 − y	y	−y

$$K_a = \frac{[H_3O^+][C_6H_5COO^-]}{[C_6H_5COOH]} \qquad K_a = \frac{(y)(y)}{0.050 - y} \cong \frac{y^2}{0.050} = 6.3 \times 10^{-5}$$

$[H_3O^+] = y = 1.8 \times 10^{-3}$ $\quad$ $pH = -\log(1.8 \times 10^{-3}) = \mathbf{2.74}$

Approximation check: $\left(\frac{1.8 \times 10^{-3}}{0.050}\right) \times 100\% = 0.0064\%$, so the approximation is valid.

(b) Na^+ comes from the strong base NaOH and does not appreciably react with water. $C_6H_5COO^-$ is the conjugate base of the weak acid C_6H_5COOH and acts as a weak base. The equilibrium of interest is $C_6H_5COO^- + H_2O \rightleftarrows C_6H_5COOH + OH^-$.

Conc, M	$C_6H_5COO^-$	C_6H_5COOH	OH^-
initial	0.050	0.	0.
Change	−y	+y	+y
equilibrium	0.050 − y	y	y

$$K_b\ C_6H_5COO^- = \frac{K_w}{K_a\ C_6H_5COOH} = \frac{[C_6H_5COOH][OH^-]}{[C_6H_5COO^-]}$$

$$K_b = \frac{(y)(y)}{(0.050 - y)} \cong \frac{y^2}{0.050} = \frac{1.0 \times 10^{-14}}{6.3 \times 10^{-5}} = 1.6 \times 10^{-10}$$

$[OH^-] = y = 2.8 \times 10^{-6}$

$pOH = -\log(2.8 \times 10^{-6}) = 5.55$

$pH = 14.00 - pOH$

$pH = 14.00 - 5.55 = \mathbf{8.45}$

Approximation check: $\left(\frac{2.8 \times 10^{-6}}{0.050}\right)$ x 100% = 0.0056%, so the approximation is valid.

15.88 ***See Sections 15.5, 15.6, and Examples 15.9, 15.13.***

(a) $HF + H_2O \rightleftarrows H_3O^+ + F^-$ $K_a = 3.5 \times 10^{-4}$

Conc, *M*	HF	H_3O^+	F^-
initial	0.25	0.	0.
Change	−y	+y	+y
equilibrium	0.25 − y	y	y

$$K_a = \frac{[H_3O^+][F^-]}{HF}$$

$$K_a = \frac{(y)(y)}{0.25 - y} \cong \frac{y^2}{0.25} = 3.5 \times 10^{-4}$$

$[H_3O^+] = y = 9.4 \times 10^{-3}$

$pH = -\log(9.4 \times 10^{-3}) = \mathbf{2.03}$

Approximation check: $\left(\frac{9.4 \times 10^{-3}}{0.25}\right)$ x 100% = 3.8%, so the approximation is valid.

(b) K^+ comes from the strong base KOH and does not appreciably react with water. F^- is the conjugate base of the weak acid HF and acts as a weak base. The equilibrium of interest is $F^- + H_2O \rightleftarrows HF + OH^-$.

Conc, *M*	F^-	HF	OH^-
initial	0.25	0.	0.
Change	−y	+y	+y
equilibrium	0.25 − y	y	y

$$K_b\ F^- = \frac{K_w}{K_a HF} = \frac{[HF][OH^-]}{[F^-]}$$

$$K_b = \frac{(y)(y)}{0.25 - y} \cong \frac{y^2}{0.25} = \frac{1.0 \times 10^{-14}}{3.5 \times 10^{-4}} = 2.9 \times 10^{-11}$$

$[OH^-] = y = 2.7 \times 10^{-6}$

$pOH = -\log(2.7 \times 10^{-6}) = 5.57$

$pH = 14.00 - pOH$

$pH = 14.00 - 5.57 = \mathbf{8.43}$

Approximation check: $\left(\frac{2.7 \times 10^{-6}}{0.25}\right)$ x 100% = 0.0011%, so the approximation is valid.

15.89 ***See Section 15.4.***

The equilibrium will favor the formation of the weaker acid and the weaker base.

(a) $HCl(aq) + NH_3(aq) \rightleftarrows NH_4Cl(aq)$: The reaction favors **products**, since HCl(aq) is a stronger acid than $NH_4^+(aq)$ and $NH_3(aq)$ is a stronger base than $Cl^-(aq)$.

(b) $HNO_3(aq) + NaOH(aq) \rightleftarrows NaNO_3(aq) + H_2O(l)$: The reaction favors **products**, since HNO_3 is a strong acid and NaOH is a strong base.

(c) $2KCl(aq) + Ba(OH)_2(aq) \rightleftarrows BaCl_2(aq) + 2KOH(aq)$: **No reaction** occurs.

(d) $HSO_4^-(aq) + NH_3(aq) \rightleftarrows NH_4^+(aq) + SO_4^{2-}(aq)$: The reaction favors **products**, since HSO_4^- is a stronger acid than NH_4^+ and NH_3 is a stronger base than SO_4^{2-}.

15.90 ***See Section 15.4.***

The equilibrium will favor the formation of the weaker acid and the weaker base.

(a) $NaOH(aq) + KCl(aq) \rightleftarrows NaCl(aq) + KOH(aq)$: **Neither direction** is favored.

(b) $HF(aq) + NaOH(aq) \rightleftarrows NaF(aq) + H_2O(\ell)$: The reaction favors **products**, since HF is a stronger acid than H_2O and OH^- is a stronger base than F^-.

(c) $NaCH_3COO(aq) + H_2O(\ell) \rightleftarrows CH_3COOH(aq) + NaOH(aq)$: The reaction favors **reactants** since CH_3COOH is a stronger acid than H_2O and OH^- is a stronger base than CH_3COO^-.

(d) $NH_4^+(aq) + H_2O(\ell) \rightleftarrows NH_3(aq) + H_3O^+(aq)$: The reaction favors **reactants** since H_3O^+ is a stronger acid than NH_4^+ and NH_3 is a stronger base than H_2O.

15.91 ***See Sections 15.2, 15.3, 15.4, 15.5, 15.6, 15.7.***

$HCOOH(aq) + H_2O(\ell) \rightleftarrows H_3O^+(aq) + HCOO^-(aq)$ $\qquad K_a = 1.8 \times 10^{-4}$

(a) HCl is a strong acid. Hence, pH will decrease.

(b) HSO_4^- is a weak acid but is one that is stronger than formic acid. Hence, pH will decrease.

(c) CH_3COONa contains the weak base CH_3COO^- which will react with HCOOH causing the $HCOOH/HCOO^-$ equilibrium to shift to the left. Hence, pH will increase.

(d) KBr contains ions which do not appreciably react with water or anything in the $HCOOH/HCOO^-$ equilibrium reaction. Hence, pH will not change.

(e) H_2O will not change the $HCOOH/HCOO^-$ equilibrium. Hence, pH will not change.

15.92 ***See Sections 15.2, 15.3, and and Example 15.6.***

(a) $NaOH(s) \xrightarrow{H_2O} Na^+(aq) + OH^-(aq)$

$$? \, M \, OH^- = \frac{15.0 \text{ g NaOH}}{0.500 \text{ L soln}} \times \frac{1 \text{ mol NaOH}}{40.1 \text{ g NaOH}} \times \frac{1 \text{ mol OH}^-}{1 \text{ mol NaOH}} = 0.748 \, M \, OH^-$$

$\mathbf{pOH} = -\log[OH^-] = -\log(0.748) = \mathbf{0.13}$ $\qquad$ $\mathbf{pH} = 14.00 - pOH = \mathbf{13.87}$

(b) The relations $[H_3O^+][OH^-] = 1.0 \times 10^{-14}$ and pH + pOH = 14.00 are only applicable to room temperature, are actually only applicable to 25°C.

15.93 *See Section 15.7, Table 15.6, and Example 15.6.*

The benzoate ion, $C_6H_5COO^-$, is the conjugate base of the weak acid benzoic acid, C_6H_5 COOH, and therefore acts as a weak base. KOH is a strong base, and the pH of the solution is determined by the KOH.

$\left[OH^-\right] = C_{OH^-} = 0.10\ M$, $pOH = -\log(0.10) = 1.00$ and $pH = 14.00 - 1.00 = \mathbf{13.00}$

15.94 *See Sections 15.2, 15.3, and Example 15.6.*

Solving M(con) x V(con) = M(dil) x V(dil) for M(dil) gives $M(\text{dil}) = \dfrac{M(\text{con}) \times V(\text{con})}{V(\text{dil})}$.

$$M(\text{dil}) = \frac{14.8\ M \times 10.0\ \text{mL}}{250\ \text{mL}} = 0.592\ M\ \text{KOH}$$

$KOH \rightarrow K^+ + OH^-$ $\quad$ $0.592\ M$ KOH $\rightarrow 0.592\ M\ OH^-$

pOH = -log $[OH^-]$ = -log (0.592) = 0.23 $\quad$ pH = 14.00 - pOH = **13.77**

15.95 *See Sections 15.2, 15.3, and Example 15.6.*

NaOH is a strong base, whereas NH_3 is a weak base. The pH will be determined by the strong base NaOH.

Solving M(con) x V(con) = M(dil) x V(dil) for M(dil) gives $M(\text{dil}) = \dfrac{M(\text{con}) \times V(\text{con})}{V(\text{dil})}$.

$$M(\text{dil}) = \frac{1.0\ M \times 10.0\ \text{mL}}{110\ \text{mL}} = 0.091\ M\ \text{NaOH}$$

$NaOH \rightarrow Na^+ + OH^-$ $\quad$ $0.091\ M$ NaOH $\rightarrow 0.091\ M\ OH^-$

pOH = -log $[OH^-]$ = -log (0.091) = 1.04 $\quad$ pH = 14.00 - pOH = **12.96**

15.96-15.99 *Solutions for these Spreadsheet Exercises are on the Reger, Goode and Mercer Spreadsheet Solutions Diskette available from Saunders College Publishing.*

Chapter 16: Acid/Base Reactions

16.1 ***See Section 16.1.***

A buffer solution usually contains a weak acid and its conjugate base or a weak base and its conjugate acid.
(a) NH_3 is a weak base, and NH_4^+ is its conjugate acid. Hence, this is a buffer system.
(b) NH_3 is a weak base that reacts with the CH_3CO_2H that is present in equimolar amounts to give an $NH_4CH_3CO_2$ solution which is a buffer system because NH_4^+ can react with added base and $CH_3CO_2^-$ can react with added acid.
(c) CH_3CO_2H and NH_4NO_3 is not a buffer because neither the conjugate base of CH_3CO_2H or NH_4^+ is present.

16.2 ***See Section 16.1.***

The Henderson-Haselbach equation assumes the equilibrium concentrations of the weak acid and its conjugate base are equal to their analytical concentration, their initial concentrations, and are not appreciably changed as the system reacts to achieve equilibrium. This assumption is only valid when the concentrations of both the weak acid and its conjugate base are at least 100 times K_a and are not applicable to more dilute solutions.

16.3 ***See Section 16.1.***

The first test to be conducted would be to prepare the buffer solution by adding one package to 500 mL of distilled water and determining whether the resulting pH was as advertised. The second test to be conducted would be to determine the buffer capacity of the resulting buffer solution by determining the amount of strong acid or base that is needed to change the pH of one liter of the buffer solution by 1 unit. If the buffer capacity is too low, add more buffer mix. If it is too high, dilute the buffer solution by adding water. These changes will not change the pH.

16.4 ***See Sections 4.4, 16.2.***

Titration is a procedure that is used to determine the quantity of one substance by adding a measured amount of a second substance. The substance that is being determined is called the analyte, and the substance being added is called the titrant. The titrant is generally chosen so its reaction with the analyte goes to completion.

16.5 ***See Section 16.2.***

Approximately 25 mL of the approximately 0.2 *M* HCl would react with 50 mL of the 0.100 *M* base, given that the base contains one mole of hydroxide ions per mole of base. Hence, it would be desirable to use less than 25 mL HCl to avoid refilling the buret with base and losing accuracy. The student should pipet 20 mL of the approximately 0.2 *M* HCl acid into a 125 mL Erlenmyer flask containing about 25 mL of distilled or deionized water, add 2-3 drops of phenolphthalein, and titrate with the 0.100 *M* NaOH until a faint pink color persists for at least 30 seconds. Take care to slow the rate of addition of NaOH when the pink color of the basic form of phenolphthalein begins to linger and the titration approaches the endpoint.

16.6 ***See Sections 16.3, 16.4, 16.5, and Figures 16.2, 16.5, 16.7, 16.10.***

A titration curve is a plot of pH versus mL of titrant added. If an acid is being titrated with a strong base, the titration curve begins at a low pH, increases slowly up to the equivalence point region, rises rapidly near the equivalence and levels off beyond the equivalence point. In the case of titration of a weak acid with the strong base there is an initial jump in pH followed by a buffer region. If a base is being titrated with a strong acid, the acid by base curves are inverted. In the case of titration of weak base by a strong acid there is an initial drop in pH followed by a buffer region.

16.7 *See Sections 16.2, 16.3, and Figure 16.5.*

The titration curve for titration of 0.100 *M* KOH by 0.100 *M* HCl is:

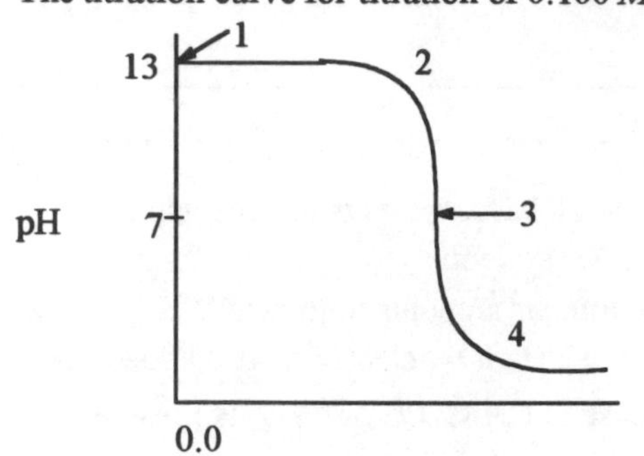

Region 1: pH is due to initial concentration of hydroxide from KOH.
Region 2: pH is due to concentration of OH^- from excess KOH.
Region 3: pH is due to autoionization of H_2O.
Region 4: pH is due to concentration of H_3O^+ from excess HCl.

16.8 *See Sections 16.4, 16.5, Tible 16.3, and Figure 16.7.*

The titration curve for titration of 0.100 *M* CH_3COOH by 0.100 *M* NaOH is:

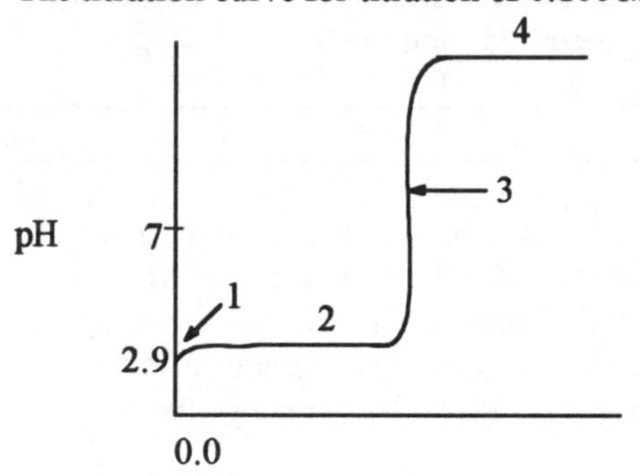

Region 1: pH due to H_3O^+ from initial concentration of CH_3COOH.
Region 2: pH due to CH_3COOH (excess) / CH_3COO^- buffer system.
Region 3: pH at equivalence point due to CH_3COO^- base hydrolysis.
Region 4: pH due to concentration of OH^- from excess NaOH.

16.9 *See Section 16.6.*

Most common indicators are weak acids or bases that would exist in primarily molecular form in pure water and therefore be only slightly soluble in pure water. They are more soluble in the less polar alcohol. The amount of indicator, alcohol and water are important because they affect the solubility and concentration of the indicator and the extent to which the indicator ionizes.

16.10 *See Section 16.6.*

Bromothymol blue is a weak acid. When the student added the bromothymol blue to the beaker of water, the water became acidified and pH dropped from 6.50 to 4.90. However, most active aquariums act as dilute buffer systems that will not change pH when bromothymol blue is added. Hence, bromothymol can be used as an indicator in fish tanks.

16.11 *See Section 16.7.*

Phosphoric acid contributes to the flavor of soft drinks because of its tartness. It also stabilizes the carbon dioxide / / carbonic acid system, $CO_2(aq) + H_2O(\ell) \rightleftarrows H_2CO_3(aq) \rightleftarrows H^+(aq) + HCO_3^-(aq)$, by acting as a source of H^+ / H_3O^+. Any weak acid that has a K_a near K_{a2} of H_3PO_4 and is not poisonous and does not taste bad could be used in place of phosphoric acid. Hence, the polyprotic acid nature of phosphoric acid is not important in this application.

16.12 *See Section 16.8.*

Substances more soluble in acidic solutions:*
1. $Pb(OH)_2$
2. BaF_2
3. $CaCO_3$
4. $Ca_3(PO_4)_2$
5. $Cu_2C_2O_4$

Substances with solubility not affected by acidity:
1. KNO_3
2. AgBr
3. $CaSO_4$
4. $BaCl_2$
5. PbI_2

*The solubility of any salt containing an anion that is a base is increased by increasing the acidity.

16.13 *See Section 16.1, Tables 15.6, 15.8, Example 16.1, and Appendix F.*

(a) $pH = pK_a + \log\left(\frac{C_b}{C_a}\right)$ The K_a for HIO_3 is 1.7×10^{-1}, and its pK_a is $-\log(1.7 \times 10^{-1}) = 0.77$.

$$? C_b = \frac{12.5 \text{ g NaIO}_3}{1.00 \text{ L soln}} \times \frac{1 \text{ mol NaIO}_3}{197.89 \text{ g NaIO}_3} \times \frac{1 \text{ mol IO}_3^-}{1 \text{ mol NaIO}_3} = 0.0632\ M\ \text{IO}_3^-$$

$$? C_a = \frac{6.0 \text{ g HIO}_3}{1.00 \text{ L soln}} \times \frac{1 \text{ mol HIO}_3}{175.91 \text{ g HIO}_3} = 0.034\ M\ \text{HIO}_3$$

$$pH = 0.77 + \log\left(\frac{0.0632}{0.034}\right) = \mathbf{1.04}$$

(b) $pH = pK_a + \log\left(\frac{C_b}{C_a}\right)$ The K_a for HCO_2H is 1.8×10^{-4}, and its pK_a is $-\log(1.8 \times 10^{-4}) = 3.74$..

$$? C_b = \frac{20.0 \text{ g NaHCO}_2}{1.00 \text{ L soln}} \times \frac{1 \text{ mol NaHCO}_2}{68.01 \text{ g NaHCO}_2} \times \frac{1 \text{ mol HCO}_2^-}{1 \text{ mol NaHCO}_2} = 0.294\ M\ \text{HCO}_2^-$$

$$?C_a = \frac{45.0 \text{ g HCO}_2\text{H}}{1.00 \text{ L soln}} \times \frac{1 \text{ mol HCO}_2\text{H}}{46.03 \text{ g HCO}_2\text{H}} = 0.978\ M\ \text{HCO}_2\text{H}$$

$$pH = 3.74 + \log\left(\frac{0.294}{0.978}\right) = \mathbf{3.22}$$

16.14 *See Section 16.1, Tables 15.6, 15.8, and Example 16.1.*

(a) $pH = pK_a + \log\left(\frac{C_b}{C_a}\right)$ The K_a for $C_6H_5CO_2H$ is 6.3×10^{-5}, and its pK_a is $-\log(6.3 \times 10^{-5}) = 4.20$.

$$? C_b = \frac{10.0 \text{ g NaC}_6\text{H}_5\text{CO}_2}{1.00 \text{ L soln}} \times \frac{1 \text{ mol NaC}_6\text{H}_5\text{CO}_2}{144.1 \text{ g NaC}_6\text{H}_5\text{CO}_2} \times \frac{1 \text{ mol C}_6\text{H}_5\text{CO}_2^-}{1 \text{ mol NaC}_6\text{H}_5\text{CO}_2} = 0.0694\ M\ \text{C}_6\text{H}_5\text{CO}_2^-$$

$$? C_a = \frac{3.00 \text{ g C}_6\text{H}_5\text{CO}_2\text{H}}{1.00 \text{ L soln}} \times \frac{1 \text{ mol C}_6\text{H}_5\text{CO}_2\text{H}}{122.1 \text{ g C}_6\text{H}_5\text{CO}_2\text{H}} = 0.0246\ M\ \text{C}_6\text{H}_5\text{CO}_2\text{H}$$

$$pH = 4.20 + \log\left(\frac{0.0694}{0.0246}\right) = \mathbf{4.65}$$

(b) $pH = pK_a + \log\left(\frac{C_b}{C_a}\right)$ The K_a for CH_3CO_2H is 1.8×10^{-5}, and its pK_a is $-\log(1.8 \times 10^{-5}) = 4.74$.

$$? C_b = \frac{25.0 \text{ g NaCH}_3\text{CO}_2}{0.500 \text{ L soln}} \times \frac{1 \text{ mol NaCH}_3\text{CO}_2}{82.05 \text{ g NaCH}_3\text{CO}_2} \times \frac{1 \text{ mol CH}_3\text{CO}_2^-}{1 \text{ mol NaCH}_3\text{CO}_2} = 0.609\ M\ \text{CH}_3\text{CO}_2^-$$

$$? C_a = \frac{9.0 \text{ g CH}_3\text{CO}_2\text{H}}{0.500 \text{ L soln}} \times \frac{1 \text{ mol CH}_3\text{CO}_2\text{H}}{60.1 \text{ g CH}_3\text{CO}_2\text{H}} = 0.300\ M\ \text{CH}_3\text{CO}_2\text{H}$$

$$pH = 4.74 + \log\left(\frac{0.609}{0.300}\right) = \mathbf{5.05}$$

16.15 ***See Section 16.1, Tables 15.6, 15.8, and Example 16.1.***

(a) $pH = pK_a + \log\left(\frac{C_b}{C_a}\right)$ The K_a for HF is 3.5×10^{-4}, and its pK_a is $-\log(3.5 \times 10^{-4}) = 3.46$.

$$?\,C_b = \frac{15.45 \text{ g KF}}{0.1000 \text{ L soln}} \times \frac{1 \text{ mol KF}}{58.10 \text{ g KF}} \times \frac{1 \text{ mol F}^-}{1 \text{ mol KF}} = 2.66\ M \qquad pH = 3.46 + \log\left(\frac{2.66}{0.850}\right) = \mathbf{3.96}$$

(b) $pH = pK_a + \log\left(\frac{C_b}{C_a}\right)$ The K_b for NH_3 is 1.8×10^{-5}.

$$\text{Hence, } K_a NH_4^+ = \frac{K_w}{K_b NH_3} \qquad K_a NH_4^+ = \frac{1.0 \times 10^{-14}}{1.8 \times 10^{-5}} = 5.6 \times 10^{-10} \qquad pK_a NH_4^+ = -\log(5.6 \times 10^{-10}) = 9.26$$

$$?\,C_a = \frac{45.00 \text{ g } NH_4Cl}{0.2500 \text{ L soln}} \times \frac{1 \text{ mol } NH_4Cl}{53.49 \text{ g } NH_4Cl} \times \frac{1 \text{ mol } NH_4^+}{1 \text{ mol } NH_4Cl} = 3.36\ M \qquad pH = 9.26 + \log\left(\frac{0.455}{3.37}\right) = \mathbf{8.39}$$

16.16 ***See Section 16.1, Tables 15.6, 15.8, and Example 16.1.***

(a) $pH = pK_a + \log\left(\frac{C_b}{C_a}\right)$ The K_a for HCO_2H is 1.8×10^{-4}, and its pK_a is $-\log(1.8 \times 10^{-4}) = 3.74$.

$$?\,C_b = \frac{30.0 \text{ g } NaHCO_2}{0.300 \text{ L soln}} \times \frac{1 \text{ mol } NaHCO_2}{68.01 \text{ g } NaHCO_2} \times \frac{1 \text{ mol } HCO_2^-}{1 \text{ mol } NaHCO_2} = 1.47\ M\ NaHCO_2^-$$

$$pH = 3.74 + \log\left(\frac{1.47}{0.30}\right) = \mathbf{4.43}$$

(b) $pH = pK_a + \log\left(\frac{C_b}{C_a}\right)$ The K_a for CH_3CO_2H is 1.8×10^{-5}, and its pK_a is $-\log(1.8 \times 10^{-5}) = 4.74$.

$$?\,C_b = \frac{30.0 \text{ g } NaCH_3CO_2}{0.300 \text{ L soln}} \times \frac{1 \text{ mol } NaCH_3CO_2}{82.05 \text{ g } NaCH_3CO_2} \times \frac{1 \text{ mol } CH_3CO_2^-}{1 \text{ mol } NaCH_3CO_2} = 1.22\ M\ CH_3CO_2^-$$

$$pH = 4.74 + \log\left(\frac{1.22}{0.30}\right) = \mathbf{5.35}$$

16.17 ***See Section 16.1, Tables 15.6, 15.8, and Example 16.1.***

(a) $pH = pK_a + \log\left(\frac{C_b}{C_a}\right)$ The K_a for HCOOH is 1.8×10^{-4}, and its pK_a is $-\log(1.8 \times 10^{-4}) = 3.74$.

V_T = 100.0 mL + 200.0 mL = 300.0 mL

Solving $M(\text{con}) \times V(\text{con}) = M(\text{dil}) \times V(\text{dil})$ for M(dil) gives $M(\text{dil}) = M(\text{con}) \times \frac{V(\text{con})}{V(\text{dil})}$. Hence,

$$C_b = 0.100\ M \times \frac{200.0 \text{ mL}}{300.0 \text{ mL}} = 0.0667\ M \text{ and } C_a = 0.800\ M \times \frac{100.0 \text{ mL}}{300.0 \text{ mL}} = 0.267\ M$$

$$pH = 3.74 + \log\left(\frac{0.0667}{0.267}\right) = \mathbf{3.14}$$

(b) $pH = pK_a + \log\left(\frac{C_b}{C_a}\right)$ The K_b for NH_3 is 1.8×10^{-5}.

Hence, $K_a NH_4^+ = \frac{K_w}{K_b\ NH_3}$ $K_a NH_4^+ = \frac{1.0 \times 10^{-14}}{1.8 \times 10^{-5}} = 5.6 \times 10^{-10}$ $pK_a NH_4^+ = -\log(5.6 \times 10^{-10}) = 9.25$

$V_T = 300.0 \text{ mL} + 200.0 \text{ mL} = 500.0 \text{ mL}$

Solving $M(\text{con}) \times V(\text{con}) = M(\text{dil}) \times V(\text{dil})$ for M(dil) gives $M(\text{dil}) = M(\text{con}) \times \frac{V(\text{con})}{V(\text{dil})}$. Hence,

$C_b = 0.350\ M \times \frac{300.0 \text{ mL}}{500.0 \text{ mL}} = 0.210\ M$ and $C_a = 0.150\ M \times \frac{200.0 \text{ mL}}{500.0 \text{ mL}} = 0.0600\ M$

$pH = 9.25 + \log\left(\frac{0.210}{0.0600}\right) = \mathbf{9.79}$

16.18 ***See Section 16.1, Tables 15.6, 15.8, and Example 16.1.***

(a) $pH = pKa + \log\left(\frac{C_b}{C_a}\right)$ The K_a for CH_3CO_2H is 1.8×10^{-5}, and its pK_a is $-\log(1.8 \times 10^{-5}) = 4.74$.

$V_T = 10.00 \text{ mL} + 20.00 \text{ mL} = 30.00 \text{ mL}$

Solving $M(\text{con}) \times V(\text{con}) = M(\text{dil}) \times V(\text{dil})$ for M(dil) gives $M(\text{dil}) = M(\text{con}) \times \frac{V(\text{con})}{V(\text{dil})}$. Hence,

$C_b = 0.500\ M \times \frac{10.00 \text{ mL}}{30.00 \text{ mL}} = 0.167\ M$ and $C_a = 0.350\ M \times \frac{20.0 \text{ mL}}{30.0 \text{ mL}} = 0.233\ M$

$pH = 4.74 + \log\left(\frac{0.167}{0.233}\right) = \mathbf{4.60}$

(b) $pH = pK_a + \log\left(\frac{C_b}{C_a}\right)$ The K_b for C_5H_5N is 1.8×10^{-9}.

Hence, $K_a\ C_5H_5NH^+ = \frac{K_w}{K_b C_5H_5N} = \frac{1.0 \times 10^{-14}}{1.8 \times 10^{-9}} = 5.6 \times 10^{-6}$ $pK_a\ C_5H_5NH^+ = -\log(5.6 \times 10^{-6}) = 5.25$

$V_T = 350.0 \text{ mL} + 650.0 \text{ mL} = 1000.0 \text{ mL}$

$C_b = 0.450\ M \times \frac{650.0 \text{ mL}}{1000.0 \text{ mL}} = 0.292\ M$ $C_a = 0.150\ M \times \frac{350.0 \text{ mL}}{1000.0 \text{ mL}} = 0.0525\ M$

$pH = 5.25 + \log\left(\frac{0.292}{0.0525}\right) = \mathbf{6.00}$

16.19 ***See Section 16.1, Table 15.6, and Example 16.2.***

$pH = pK_a + \log\left(\frac{n_b}{n_a}\right)$ The K_a for HF is 3.5×10^{-4}, and its pK_a is $-\log(3.5 \times 10^{-4}) = 3.46$.

$?\ n_b = 0.750 \text{ L NaF soln} \times \frac{0.200 \text{ mol NaF}}{1 \text{ L NaF soln}} = 0.150 \text{ mol NaF}$

Substituting into the Henderson-Hasselbach equation gives $3.95 = 3.46 + \log\left(\frac{0.150}{n_a}\right)$

$\log\left(\frac{0.150}{n_a}\right) = 3.95 - 3.46 = 0.49$ $\frac{0.150}{n_a} = 10^{0.49} = \text{invlog}(0.49) = 3.09$

$n_a = \frac{0.150}{3.09} = 0.0485 \text{ mol}$ $?\ \text{L HF soln} = 0.0485 \text{ mol HF} \times \frac{1 \text{ L HF soln}}{0.500 \text{ mol HF}} = \mathbf{0.0970\ L\ HF\ sol}$

16.20 *See Section 16.1, Table 15.6, and Example 16.2.*

(a) $pH = pK_b + \log\left(\frac{C_b}{C_a}\right)$ The K_a for CH_3CO_2H is 1.8×10^{-5}, and its pK_a is $-\log(1.8 \times 10^{-5}) = 4.74$.

Substituting into the Henderson-Hasselbach equation gives $4.35 = 4.74 + \log\left(\frac{C_b}{0.500}\right)$

$\log\left(\frac{C_b}{0.500}\right) = 4.35 - 4.74 = -0.39$ $\frac{C_b}{0.500} = 10^{-0.39} = \text{inv}\log(-0.39) = 0.407$ $C_b = (0.500\ M)(0.407) = 0.204\ M$

$$\text{g } NaCH_3CO_2 = 0.400 \text{ L soln} \times \frac{0.204 \text{ mol } NaCH_3CO_2}{1 \text{ L soln}} \times \frac{82.05 \text{ g } NaCH_3CO_2}{1 \text{ mol } NaCH_3CO_2} = \mathbf{6.70 \text{ g } NaCH_3CO_2}$$

16.21 *See Section 16.1, Table 15.6, and Example 16.2.*

(a) $pH = pK_a + \log\left(\frac{C_b}{C_a}\right)$ The K_a for CH_3CO_2H is 1.8×10^{-5}, and its pK_a is $-\log(1.8 \times 10^{-5}) = 4.74$.

$$?\ n_b = 0.500 \text{ L } CH_3CO_2H \text{ soln} \times \frac{0.250 \text{ mol } CH_3CO_2H}{1 \text{ L } CH_3CO_2H \text{ soln}} = 0.125 \text{ mol } CH_3CO_2H$$

Substituting into the Henderson-Hasselbach equation gives $5.00 = 4.74 + \log\frac{n_b}{0.125}$

$\log\left(\frac{n_b}{0.125}\right) = 5.00 - 4.74 = 0.26$ $\frac{n_b}{0.125} = 10^{0.26} = \text{inv}\log(0.26) = 1.82$

$n_b = (0.125)(1.82) = 0.228 \text{ mol}$

$$?\text{ L } NaCH_3CO_2 \text{ soln} = 0.228 \text{ mol } NaCH_3CO_2 \times \frac{1 \text{ L } NaCH_3CO_2 \text{ soln}}{0.250 \text{ mol } NaCH_3CO_2} = \mathbf{0.912 \text{ L } NaCH_3CO_2 \text{ soln}}$$

16.22 *See Section 16.1, Table 15.6, and Example 16.2.*

$pH = pK_a + \log\left(\frac{C_b}{C_a}\right)$ The K_b for NH_3 is 1.8×10^{-5}.

Hence, $K_a NH_4^+ = \frac{K_w}{K_b NH_3}$ $K_a NH_4^+ = \frac{1.0 \times 10^{-14}}{1.8 \times 10^{-5}} = 5.6 \times 10^{-10}$ $pK_a NH_4^+ = -\log(5.6 \times 10^{-10}) = 9.26$

Substituting into the Henderson-Hasselbach equation gives $9.80 = 9.25 + \log\frac{0.137}{C_a}$

$\log\left(\frac{0.137}{C_a}\right) = 9.80 - 9.25 = 0.55$ $\frac{0.137}{C_a} = 10^{0.55} = \text{inv}\log(0.55) = 3.55$ $C_a = \frac{0.137}{3.55} = 0.0386\ M$

$$\text{g } NH_4Cl = 0.500 \text{ L} \times \frac{0.0386 \text{ mol}}{1 \text{ L soln}} \times \frac{53.50 \text{ g } NH_4Cl}{1 \text{ mol } NH_4Cl} = \mathbf{1.03 \text{ g } NH_4Cl}$$

16.23 *See Section 16.1, Table 15.6, and Examples 16.1, 16.2.*

The objective is to calculate the ratio n_b/ n_a that is needed to give a pH of 5.00 using the CH_3COOH/ CH_3COO^- buffer system and then use these numbers of moles to calculate the concentrations of acetic acid and sodium acetate in solution. The K_a for CH_3COOH is 1.8×10^{-5}, and its pK_a is 4.74. Substituting into the Henderson-Hasselbach equation yields

$$5.00 = 4.74 + \log\left(\frac{n_b}{n_a}\right), \quad \log\left(\frac{n_b}{n_a}\right) = 5.00 - 4.74 \text{ and } \frac{n_b}{n_a} = 10^{0.26} = \text{inv}\log(0.26) = 1.8$$

This means the initial n_b/ n_a ratio of 0.0500 mol/ 0.0500 mol or 1.00 must be adjusted to 1.8 to give a pH of 5.00.

Hence, $\frac{0.0500+y}{0.0500-y} = 1.8$, $2.8y = 0.0400$ and $y = 0.014$.

This gives $[CH_3COO^-] = \frac{(0.0500+0.014)\text{ mol}}{1.00\text{ L}} = \mathbf{0.064}\ \boldsymbol{M}$

and $[CH_3COOH] = \frac{(0.0500-0.014)\text{ mol}}{1.00\text{ L}} = \mathbf{0.036}\ \boldsymbol{M}$.

16.24 ***See Section 16.1 and Example 16.2.***

$pH = pK_a + \log\left(\frac{C_b}{C_a}\right)$ Initial $C_a = \frac{3.0\text{ mg } C_7H_5NSO_3}{350\text{ mL}} \times \frac{1\text{ mmol } C_7H_5NSO_3}{183.19\text{ mg } C_7H_5NSO_3} = 4.7 \times 10^{-5} M$

Letting C_b equal molar concentration of saccharin converted to saccharide by ionization gives

$4.50 = 11.68 + \log\left(\frac{C_b}{4.7\times10^{-5} - C_b}\right)$ $\log\left(\frac{C_b}{4.7\times10^{-5} - C_b}\right) = 4.50 - 11.68 = -7.18$

$\frac{C_b}{4.7\times10^{-5} - C_b} = 10^{-7.18} = \text{inv}\log(-7.18) = 6.6 \times 10^{-8}$

$C_b = (4.7 \times 10^{-5} - C_b)(6.6 \times 10^{-8})$, $C_b + 6.6 \times 10^{-8}\, C_b = 3.1 \times 10^{-12}$, $C_b = 3.1 \times 10^{-12}$

Hence, **[saccharin]** $= 4.7 \times 10^{-5} - 3.1 \times 10^{-12} = \mathbf{4.7 \times 10^{-5}}\ \boldsymbol{M}$, and **[saccharide]** $= \mathbf{3.1 \times 10^{-12}}\ \boldsymbol{M}$ when pH = 4.50.

16.25 ***See Section 16.1, Table 15.6, and Examples 16.2, 16.3, 16.5.***

(a) $pH = pK_a + \log\left(\frac{C_b}{C_a}\right) = pK_a + \log\left(\frac{n_b}{n_a}\right)$

The K_a for CH_3COOH is 1.8×10^{-5}, and its pK_a is $-\log(1.8 \times 10^{-5}) = 4.74$.

Initial $pH = 4.74 + \log\left(\frac{1.00}{0.500}\right) = \mathbf{5.04}$

When NaOH is added, OH^- reacts with CH_3COOH: $OH^- + CH_3COOH \rightarrow H_2O + CH_3COO^-$

? mmol OH^- added $= 1.00\text{ mL} \times \frac{1.00\text{ mmol}}{1\text{ mL}} = 1.00\text{ mmol}$

? starting mmol $CH_3COOH = 100.0\text{ mL} \times \frac{0.500\text{ mmol}}{1\text{ mL}} = 50.0\text{ mmol}$

? starting mmol $CH_3COO^- = 100.0\text{ mL} \times \frac{1.00\text{ mmol}}{1\text{ mL}} = 100.0\text{ mmol}$

$V_T = 100.0\text{ mL} + 1.00\text{ mL} = 101.0\text{ mL}$

	OH^-	CH_3COOH	CH_3COO^-	H_2O
s, mmol *	1.00	50.00	100.0	excess
R, mmol	-1.00	-1.00	+1.00	+1.00
f, mmol	0.	49.00	101.0	excess

* s = starting mmol, R = reacting mmol, and f = final mmol

Final $pH = 4.74 + \log\left(\frac{101.0}{49.00}\right) = \mathbf{5.05}$

Change in pH = 5.05 - 5.04 = **0.01**

(b) Initial pH = **7.00**, assuming we are using pure water at 25°C.

$$\text{Final}\left[OH^-\right] = \frac{1.00\text{ mmol}}{101.00\text{ mL}} = 9.90 \times 10^{-3}\ M$$

pOH = -log(9.90 x 10^{-3}) = 2.00 and pH = 14.00 - 2.00 = **12.00**

Change in pH = 12.00 - 7.00 = **5.00**

16.26 ***See Section 16.1, Table 15.6, and Examples 16.2, 16.3, 16.5.***

(a) $$pH = pK_a + \log\left(\frac{C_b}{C_a}\right) = pK_a + \log\left(\frac{n_b}{n_a}\right)$$

The K_a for CH_3COOH is 1.8 x 10^{-5}, and its pK_a is -log(1.8 x 10^{-5}) = 4.74.

$$\text{Initial pH} = 4.74 + \log\left(\frac{0.100}{0.200}\right) = \mathbf{4.44}$$

When HCl is added, H_3O^+ reacts with CH_3COO^-: $H_3O^+ + CH_3COO^- \rightarrow H_2O + CH_3CO_2H$

$$?\text{ mmol } H_3O^+ \text{ added} = 1.00\text{ mL HCl soln} \times \frac{0.100\text{ mmol } H_3O^+}{1\text{ mL HCl soln}} = 0.100\text{ mmol } H_3O^+$$

$$?\text{ starting mmol } CH_3CO_2H = 100.0\text{ mL buffer} \times \frac{0.200\text{ mmol } CH_3CO_2H}{1\text{ mL buffer}} = 20.0\text{ mmol } CH_3CO_2H$$

$$?\text{ starting mmol } CH_3COO^- = 100.0\text{ mL buffer} \times \frac{0.100\text{ mmol } CH_3COO^-}{1\text{ mL buffer}} = 10.0\text{ mmol } CH_3COO^-$$

V_T = 100.0 mL + 1.00 mL = 101.0 mL

	H_3O^+	CH_3COO^-	CH_3COOH	H_2O
s, mmol *	0.100	10.00	20.0	excess
R, mmol	-0.100	-0.100	+0.100	+0.100
f, mmol	0.	9.9	20.1	excess

$$\text{Final pH} = 4.74 + \log\left(\frac{9.9}{20.1}\right) = \mathbf{4.43}$$

Change in pH = 4.43 - 4.44 = **-0.01**

* s = starting mmol, R = reacting mmol, and f = final mmol

(b) Initial pH = **7.00**, assuming we are using pure water at 25°C.

$$\text{Final}\left[H_3O^+\right] = \frac{0.100\text{ mmol}}{101.00\text{ mL}} = 9.90 \times 10^{-4}\ M$$

pH = -log(9.90 x 10^{-4}) = 3.00 and change in pH = 3.00 - 7.00 = **-4.00**

16.27 ***See Section 16.1, Table 15.7, and Examples 16.2, 16.3.***

$$pH = pK_a + \log\left(\frac{C_b}{C_a}\right) = pK_a + \log\left(\frac{n_b}{n_a}\right)$$

The K_a for HCOOH is 1.8 x 10^{-4} , and its pK_a is -log(1.8 x 10^{-4}) = 3.74.

$$?\text{ mmol strong acid or strong base to be added} = 1.00\text{ mL} \times \frac{0.100\text{ mmol}}{1\text{ mL}} = 0.100\text{ mmol}$$

For pH = 3.80: $$3.80 = 3.74 + \log\left(\frac{n_b}{n_a}\right)$$

$$\log\left(\frac{n_b}{n_a}\right) = 3.80 - 3.74 = 0.06 \qquad \frac{n_b}{n_a} = 10^{0.06} = \text{inv}\log(0.06) = 1.15$$

To achieve pH = 3.90, n_b must increase by 0.100 mmol and n_a must decrease by 0.100 mmol due to the addition of strong base and its reaction with the acid component of the buffer. This gives:

$$3.90 = 3.74 + \log\left(\frac{n_b + 0.100}{n_a - 0.100}\right)$$

$$\log\left(\frac{n_b + 0.100}{n_a - 0.100}\right) = 3.90 - 3.74 = 0.16 \qquad \frac{n_b + 0.100}{n_a - 0.100} = 10^{0.16} = \text{inv}\log(0.16) = 1.45$$

Hence, we have $\frac{n_b}{n_a} = 1.15$ and $\frac{n_b + 0.100 \text{ mmol}}{n_a - 0.100 \text{ mmol}} = 1.45.$

Solving the first expression for n_b and substituting into the second expression gives

$$\frac{1.15n_a + 0.100 \text{ mmol}}{n_a - 0.100 \text{ mmol}} = 1.45 \qquad 1.15n_a + 0.100 \text{ mmol} = 1.45n_a - 0.145 \text{ mmol}$$

$$0.30n_a = 0.245 \text{ mmol} \qquad n_a = 0.82 \text{ mmol}$$

Hence, $n_b = 1.15n_a = (1.15)(0.82) = .94$ mmol

The minimum concentration of formic acid is therefore $\frac{0.82 \times 10^{-3} \text{ mol}}{0.5010 \text{ L}} = \mathbf{1.64 \times 10^{-3}}$ ***M*** and that of sodium formate is therefore $\frac{0.94 \times 10^{-3} \text{ mol}}{0.5010 \text{ L}} = \mathbf{1.88 \times 10^{-3}}$ ***M***.

16.28 ***See Section 16.1, Table 15.7, and Examples 16.2, 16.3.***

$$\text{pH} = \text{p}K_a + \log\left(\frac{C_b}{C_a}\right) = \text{p}K_a + \log\left(\frac{n_b}{n_a}\right)$$

The K_a for CH_3CO_2H is 1.8×10^{-5}, and its $\text{p}K_a$ is $-\log(1.8 \times 10^{-5}) = 4.74$.

? mmol strong acid or strong base to be added = $1.00 \text{ mL} \times \frac{0.100 \text{ mmol}}{1 \text{ mL}} = 0.100$ mmol

For pH = 4.50: $4.50 = 4.74 + \log\left(\frac{n_b}{n_a}\right)$

$$\log\left(\frac{n_b}{n_a}\right) = 4.50 - 4.74 = -0.24 \qquad \frac{n_b}{n_a} = 10^{-0.24} = \text{inv}\log(-0.24) = 0.58$$

To achieve pH = 4.60, n_b nust increase by 0.100 mmol and n_a must decrease by 0.100 mmol due to the addition of strong base and its reaction with the acid component of the buffer. This gives:

$$4.35 = 4.74 + \log\left(\frac{n_b + 0.100}{n_a - 0.100}\right) = 4.55 - 4.74 = -0.19 \qquad \frac{n_b + 0.100}{n_a - 0.100} = 10^{-0.19} = \text{inv}\log(-0.19) = 0.65$$

Hence, we have $\frac{n_b}{n_a} = 0.58$ and $\frac{n_b + 0.100}{n_a - 0.100} = 0.65.$

Solving the first expression for n_b and substituting into the second expression gives

$$\frac{0.58\ n_a + 0.100 \text{ mmol}}{n_a - 0.100 \text{ mmol}} = 0.66 \qquad 0.58\ n_a + 0.100 \text{ mmol} = 0.65\ n_a - 0.065 \text{ mmol}$$

$0.07\ n_a = 0.58\ n_a = (0.58)(2.4 \text{ mmol}) = 1.4$ mmol

The minimum concentration of acetic acid is therefore $\frac{2.4 \text{ mmol}}{100 \text{ mL}} = \mathbf{0.024}$ ***M*** and that of sodium acetate is therefore $\frac{1.4 \text{ mmol}}{100 \text{ mL}} = \mathbf{0.014}$ ***M***.

16.29 ***See Section 16.2 and Example 16.4.***

(a) $HCl(aq) + KOH(aq) \rightarrow KCl(aq) + H_2O(\ell)$

Strategy: L KOH soln → mol KOH → mol HCl → L HCl soln → mL HCl soln

$$? \text{ mL HCl soln} = 0.0100 \text{ L KOH soln} \times \frac{0.150 \text{ mol KOH}}{1 \text{ L KOH soln}} \times \frac{1 \text{ mol HCl}}{1 \text{ mol KOH}} \times \frac{1 \text{ L HCl soln}}{0.100 \text{ mol HCl}} \times \frac{10^3 \text{ mL HCl soln}}{1 \text{ L HCl soln}}$$
$$= \mathbf{15.0 \text{ mL HCl soln}}$$

(b) $2HCl(aq) + Ba(OH)_2(aq) \rightarrow BaCl_2(aq) + 2H_2O(\ell)$

Strategy: L $Ba(OH)_2$ soln $\rightarrow$ mol $Ba(OH)_2$ $\rightarrow$ mol HCl $\rightarrow$ L HCl soln $\rightarrow$ mL HCl soln

$$? \text{ mL HCl soln} = 0.2500 \text{ L Ba(OH)}_2 \text{ soln} \times \frac{0.00520 \text{ mol Ba(OH)}_2}{1 \text{ L Ba(OH)}_2 \text{ soln}} \times \frac{2 \text{ mol HCl}}{1 \text{ mol Ba(OH)}_2} \times \frac{1 \text{ L HCl soln}}{0.100 \text{ mol HCl}}$$
$$\times \frac{10^3 \text{ mL HCl soln}}{1 \text{ L HCl soln}} = \mathbf{26.0 \text{ mL HCl soln}}$$

(c) $HCl(aq) + NH_3(aq) \rightarrow NH_4Cl(aq)$

Strategy: L NH_3 soln $\rightarrow$ mol NH_3 $\rightarrow$ mol HCl $\rightarrow$ L HCl soln $\rightarrow$ mL HCl soln

$$? \text{ mL HCl soln} = 0.1000 \text{ L NH}_3 \text{ soln} \times \frac{0.100 \text{ mol NH}_3}{1 \text{ L NH}_3 \text{ soln}} \times \frac{1 \text{ mol HCl}}{1 \text{ mol NH}_3} \times \frac{1 \text{ L HCl soln}}{0.100 \text{ mol HCl}} \times \frac{10^3 \text{ mL HCl soln}}{1 \text{ L HCl soln}}$$
$$= \mathbf{100 \text{ mL HCl soln}}$$

16.30 ***See Section 16.2 and Example 16.4.***

(a) $HCl(aq) + NaOH(aq) \rightarrow NaCl(aq) + H_2O(\ell)$

Strategy: L HCl soln $\rightarrow$ mol HCl $\rightarrow$ mol NaOH $\rightarrow$ L NaOH soln $\rightarrow$ mL NaOH soln

$$? \text{ mL NaOH soln} = 0.04500 \text{ L HCl soln} \times \frac{0.0500 \text{ mol HCl}}{1 \text{ L HCl soln}} \times \frac{1 \text{ mol HaOH}}{1 \text{ mol HCl}}$$
$$\times \frac{1 \text{ L NaOH soln}}{0.100 \text{ mol NaOH}} \times \frac{10^3 \text{ mL NaOH soln}}{1 \text{ L NaOH soln}} = \mathbf{22.5 \text{ mL NaOH soln}}$$

(b) $H_2SO_4(aq) + 2NaOH(aq) \rightarrow Na_2SO_4(aq) + 2\, H_2O(\ell)$

Stategy: L H_2SO_4 soln $\rightarrow$ mol H_2SO_4 $\rightarrow$ mol NaOH $\rightarrow$ L NaOH soln $\rightarrow$ mL NaOH soln

$$? \text{ mL NaOH soln} = 0.00500 \text{ L H}_2\text{SO}_4 \text{ soln} \times \frac{0.350 \text{ mol H}_2\text{SO}_4}{1 \text{ L H}_2\text{SO}_4 \text{ soln}} \times \frac{2 \text{ mol NaOH}}{1 \text{ mol H}_2\text{SO}_4}$$
$$\times \frac{1 \text{ L NaOH soln}}{0.100 \text{ mol NaOH}} \times \frac{10^3 \text{ mL NaOH soln}}{1 \text{ L NaOH soln}} = \mathbf{35.0 \text{ mL NaOH soln}}$$

(c) $CH_3CO_2H(aq) + NaOH(aq) \rightarrow NaCH_3CO_2(aq) + H_2O(\ell)$

Strategy: L CH_3CO_2H soln $\rightarrow$ mol CH_3CO_2H $\rightarrow$ mol NaOH $\rightarrow$ L NaOH soln $\rightarrow$ mL NaOH soln

$$? \text{ mL NaOH soln} = 0.01000 \text{ L CH}_3\text{CO}_2\text{H soln} \times \frac{0.100 \text{ mol CH}_3\text{CO}_2\text{H}}{1 \text{ L CH}_3\text{CO}_2\text{H soln}} \times \frac{1 \text{ mol NaOH}}{1 \text{ mol CH}_3\text{CO}_2\text{H}}$$
$$\times \frac{1 \text{ L NaOH soln}}{0.100 \text{ mol NaOH}} \times \frac{10^3 \text{ mL NaOH soln}}{1 \text{ L NaOH soln}} = \mathbf{10.0 \text{ mL NaOH soln}}$$

16.31 *See Section 16.2.*

$HCl(aq) + NaOH(aq) \rightarrow NaCl(aq) + H_2O(\ell)$

Strategy: mL NaOH → mmol NaOH → mmol HCl

$$? \text{ mmol HCl} = 50.0 \text{ mL NaOH soln} \frac{0.0233 \text{ mmol NaOH}}{1 \text{ mL NaOH soln}} \times \frac{1 \text{ mmol HCl}}{1 \text{ mmol NaOH}} = \mathbf{1.16 \text{ mmol HCl}}$$

16.32 *See Section 16.2.*

$HNO_3(aq) + KOH(aq) \rightarrow KNO_3(aq) + H_2O(\ell)$

Strategy: mL HNO_3 → mmol HNO_3 → mmol KOH

$$? \text{ mmol } HNO_3 = 35.1 \text{ mL } HNO_3 \text{ soln} \times \frac{0.101 \text{ mmol } HNO_3}{1 \text{ mL } HNO_3 \text{ soln}} \times \frac{1 \text{ mmol KOH}}{1 \text{ mmol } HNO_3} = \mathbf{3.55 \text{ mmol KOH}}$$

16.33 *See Section 16.2.*

$H_2SO_4(aq) + 2LiOH(aq) \rightarrow Li_2SO_4(aq) + 2H_2O(\ell)$

Strategy: mL LiOH → mmol $Li(OH)_2$ → mmol H_2SO_4

$$? \text{ mmol } H_2SO_4 = 25.0 \text{ mL LiOH soln} \times \frac{0.244 \text{ mmol LiOH}}{1 \text{ mL LiOH soln}} \times \frac{1 \text{ mmol } H_2SO_4}{2 \text{ mmol LiOH}} = \mathbf{3.05 \text{ mmol } H_2SO_4}$$

16.34 *See Section 16.2.*

$3H_2SO_4(aq) + 2La(OH)_3(aq) \rightarrow La_2(SO_4)_3(s) + 6H_2O(\ell)$

Strategy: mL $La(OH)_3$ → mmol $La(OH)_3$ → mmol H_2SO_4

$$? \text{ mmol } H_2SO_4 = 100.0 \text{ mL } La(OH)_3 \text{ soln} \times \frac{0.045 \text{ mmol } La(OH)_3}{1 \text{ mL } La(OH)_3 \text{ soln}} \times \frac{3 \text{ mmol } H_2SO_4}{2 \text{ mmol } La(OH)_3} = \mathbf{6.8 \text{ mmol } H_2SO_4}$$

16.35 *See Section 16.3 and Example 16.5.*

Overall: $HNO_3(aq) + KOH(aq) \rightarrow KNO_3(aq) + H_2O(\ell)$
Net ionic: $H_3O^+(aq) + OH^-(aq) \rightarrow 2H_2O(\ell)$

1. **0. mL KOH soln added**:

$$? \text{ mmol } H_3O^+ = 50.00 \text{ mL } HNO_3 \text{ soln} \times \frac{0.250 \text{ mmol } H_3O^+}{1 \text{ ml } HNO_3 \text{ soln}} = 12.5 \text{ mmol } H_3O^+$$

$$[H_3O^+] = \frac{12.5 \text{ mmol } H_3O^+}{50.00 \text{ mL}} = 0.250\ M, \text{ and pH} = -\log(0.250) = \mathbf{0.60}.$$

2. **12.50 mL KOH soln added**:

$$? \text{ mmol } OH^- \text{ added} = 12.50 \text{ mL KOH soln} \times \frac{0.500 \text{ mmol } OH^-}{1 \text{ mL KOH soln}} = 6.25 \text{ mmol } OH^-$$

$V_T = 50.00 \text{ mL} + 12.50 \text{ mL} = 62.50 \text{ mL}$

	H_3O^+	OH^-	H_2O
s, mmol	12.5	6.25	excess
R, mmol	-6.25	−6.25	+12.5
f, mmol	6.25	0.	excess
c, *M*	0.100	0.	excess

pH = -log(0.100) = **1.00**

3. **25.00 mL KOH soln added**:

$$? \text{ mmol OH}^- \text{ added} = 25.00 \text{ mL KOH soln} \times \frac{0.500 \text{ mmol OH}^-}{1 \text{ mL KOH soln}} = 12.5 \text{ mmol OH}^-$$

V_T = 50.00 mL + 25.00 mL = 75.00 mL

	H_3O^+	OH^-	H_2O
s, mmol	12.5	12.5	excess
R, mmol	-12.5	−12.5	+25.0
f, mmol	0.	0.	excess
c, *M*	1.0×10^{-7} *	1.0×10^{-7} *	excess

* Concentrations due to autoionization of water at 25° C

pH = -log(1.0×10^{-7}) = **7.00**

4. **40.00 mL KOH soln added**:

$$? \text{ mmol OH}^- \text{ added} = 40.00 \text{ mL KOH soln} \times \frac{0.500 \text{ mmol OH}^-}{1 \text{ mL KOH soln}} = 20.0 \text{ mmol OH}^-$$

V_T = 50.00 mL + 40.00 mL = 90.00 mL

	H_3O^+	OH^-	H_2O
s, mmol	12.5	20.0	excess
R, mmol	-12.5	−12.5	+25.0
f, mmol	0.0	7.5	excess
c, *M*	0.	0.083	excess

pOH = -log(0.083) = 1.08 pH = 14.00 - 1.08 = **12.92**

Summary:

mL KOH soln	pH
0.	0.60
12.50	1.00
25.00	7.00
40.00	12.92

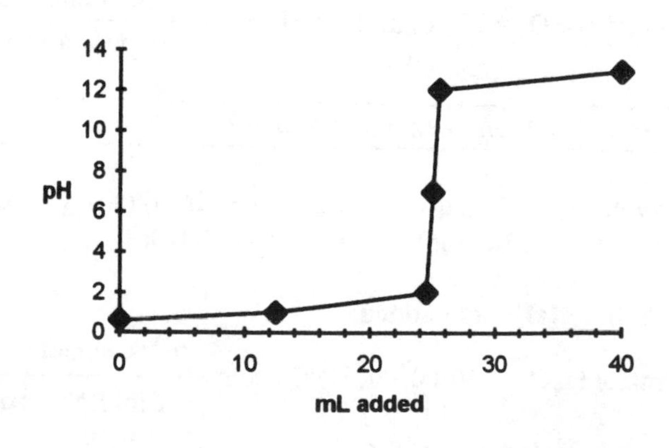

16.36 *See Section 16.3 and Example 16.5.*

Overall: HCl(aq) + NaOH(aq) → NaCl(aq) + $H_2O(\ell)$

Net ionic: $H_3O^+(aq) + OH^- \rightarrow 2H_2O(\ell)$

1. **0. mL NaOH added:**

$$? \text{ mmol } H_3O^+ = 100.0 \text{ mL HCl soln} \times \frac{0.200 \text{ mmol } H_3O^+}{1 \text{ mL HCl soln}} = 20.0 \text{ mmol } H_3O^+$$

$$[H_3O^+] = \frac{20.0 \text{ mmol } H_3O^+}{100.0 \text{ mL}} = 0.200\ M, \text{ and pH} = -\log(0.200) = \mathbf{0.70}$$

2. **25.00 mL NaOH added:**

$$? \text{ mmol } OH^- \text{ added} = 25.00 \text{ mL NaOH soln} \times \frac{0.400 \text{ mmol } OH^-}{1 \text{ mL NaOH soln}} = 10.0 \text{ mmol } OH^-$$

V_T = 100.0 mL + 25.00 mL = 125.0 mL

	H_3O^+	OH^-	H_2O
s, mmol	20.0	10.0	excess
R, mmol	-10.0	−10.0	+10.0
f, mmol	10.0	0.	excess
c, *M*	0.0800	0.	excess

pH = -log(0.0800) = **1.10**

3. **50.0 mL NaOH added:**

$$? \text{ mmol } OH^- \text{ added} = 50.0 \text{ mL NaOH soln} \times \frac{0.400 \text{ mmol } OH^-}{1 \text{ mL NaOH soln}} = 20.0 \text{ mmol } OH^-$$

V_T = 100.0 mL + 50.0 mL = 150.0 mL

	H_3O^+	OH^-	H_2O
s, mmol	20.0	20.0	excess
R, mmol	-20.0	−20.0	+20.0
f, mmol	0.	0.	excess
c, *M*	$1.0\text{x}10^{-7}$ *	$1.0\text{x}10^{-7}$ *	excess

*Concentrations due to autoionization of water at 25°C.

pH = -log(1.0 x 10^{-7}) = **7.00**

4. **75.0 mL NaOH added:**

$$? \text{ mmol } OH^- \text{ added} = 75.0 \text{ mL NaOH solution} \times \frac{0.400 \text{ mmol } OH^-}{1 \text{ mL NaOH soln}} = \mathbf{30.0 \text{ mmol } OH^-}$$

V_T = 100.0 mL + 75.0 mL = 175.0 mL

	H_3O^+	OH^-	H_2O
s, mmol	20.0	30.0	excess
R, mmol	-20.0	−20.0	+20.0
f, mmol	0.	10.0	excess
c, *M*	0.	0.0571	excess

pH = -log(0.0571) = **1.24**   pH = 14.00 - 1.24 = **12.76**

Summary:

mL NaOH soln	pH
0.0	0.70
25.0	1.10
50.0	7.00
75.0	12.76

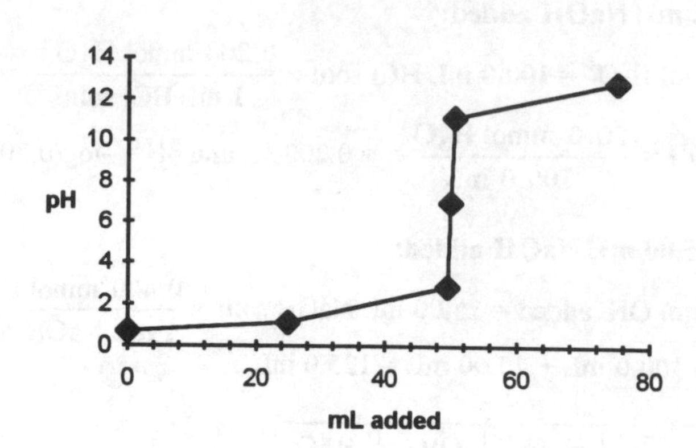

16.37 ***See Section 16.3 and Example 16.5.***

Overall: $HNO_3(aq) + LiOH(aq) \rightarrow LiNO_3(aq) + H_2O(\ell)$
Net ionic: $H_3O^+(aq) + OH^-(aq) \rightarrow 2H_2O(\ell)$

1. **0. mL HNO_3 soln added**:

$$? \text{ mmol } OH^- = 1.00 \text{ mL LiOH soln} \times \frac{0.240 \text{ mmol } OH^-}{1 \text{ mL LiOH soln}} = 0.240 \text{ mmol } OH^-$$

$$[OH^-] = \frac{0.240 \text{ mmol } OH^-}{1.00 \text{ mL}} = 0.240\ M$$

pOH = -log(0.240) = 0.620 pH = 14.00 - 0.620 = **13.38**

2. **0.25 mL HNO_3 soln added**:

$$? \text{ mmol } H_3O^+ \text{ added} = 0.25 \text{ mL } HNO_3 \text{ soln} \times \frac{0.200 \text{ mmol } H_3O^+}{1 \text{ mL } HNO_3 \text{ soln}} = 5.0 \times 10^{-2} \text{ mmol } H_3O^+$$

V_T = 1.00 mL + 0.25 mL = 1.25 mL

	H_3O^+	OH^-	H_2O
s, mmol	0.050	0.240	excess
R, mmol	-0.050	-0.050	+0.100
f, mmol	0.	0.190	excess
c, *M*	0.	0.152	excess

pOH = -log(0.152) = 0.818 pH = 14.00 - 0.818 = **13.18**

3. **0.50 mL HNO_3 soln added**:

$$? \text{ mmol } H_3O^+ \text{ added} = 0.50 \text{ mL } HNO_3 \text{ soln} \times \frac{0.200 \text{ mmol } H_3O^+}{1 \text{ mL } HNO_3 \text{ soln}} = 0.10 \text{ mmol } H_3O^+$$

V_T = 1.00 mL + 0.50 mL = 1.50 mL

	H_3O^+	OH^-	H_2O
s, mmol	0.10	0.240	excess
R, mmol	-0.10	-0.10	+0.20
f, mmol	0.	0.14	excess
C, *M*	0.	0.093	excess

pOH = -log(0.093) = 1.03 pH = 14.00 - 1.03 = **12.97**

4. **1.20 mL HNO_3 soln added:**

$$? \text{ mmol } H_3O^+ \text{ added} = 1.20 \text{ mL } HNO_3 \text{ soln} \times \frac{0.200 \text{ mmol } H_3O^+}{1 \text{ mL } HNO_3 \text{ soln}} = 0.240 \text{ mmol } H_3O^+$$

$V_T = 1.00$ mL + 1.20 mL = 2.20 mL

	H_3O^+	OH^-	H_2O
s, mmol	0.240	0.240	excess
R, mmol	-0.240	-0.240	+0.480
f, mmol	0.	0.	excess
c, *M*	1.0×10^{-7} *	1.0×10^{-7} *	excess

* Concentrations due to autoionization of water at 25° C.

pH = -log(1.0×10^{-7}) = **7.00**

5. **1.50 mL HNO_3 soln added:**

$$? \text{ mmol } H_3O^+ \text{ added} = 1.50 \text{ mL } HNO_3 \text{ soln} \times \frac{0.200 \text{ mmol } H_3O^+}{1 \text{ mL } HNO_3 \text{ soln}} = 0.300 \text{ mmol } H_3O^+$$

$V_T = 1.00$ mL + 1.50 mL = 2.50 mL

	H_3O^+	OH^-	H_2O
s, mmol	0.300	0.240	excess
R, mmol	-0.240	-0.240	+0.480
f, mmol	0.060	0.	excess
c, *M*	0.024	0.	excess

pH = -log(0.024) = **1.62**

Summary:

mL HNO_3 soln	pH
0.	13.38
0.25	13.18
0.50	12.97
1.20	7.00
1.50	1.62

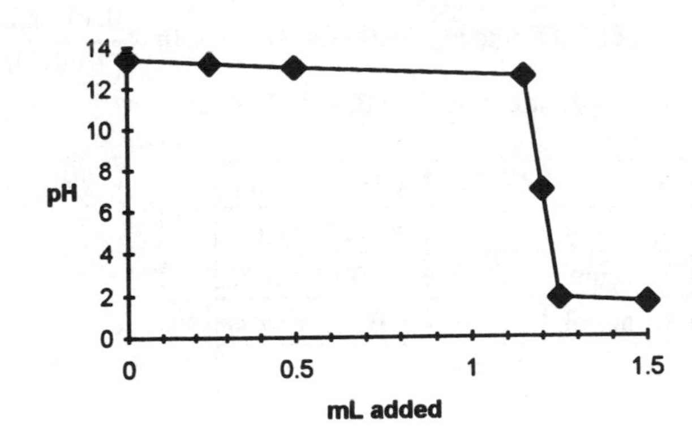

16.38 ***See Section 16.3 and Example 16.5.***

Overall: $HNO_3(aq) + NaOH(aq) \rightarrow NaNO_3(aq) + H_2O(\ell)$
Net ionic: $H_3O^+(aq) + OH^-(aq) \rightarrow 2H_2O(\ell)$

1. **0. mL HNO_3 added:**

$$? \text{ mmol } OH^- = 50.0 \text{ mL NaOH soln} \times \frac{0.100 \text{ mmol } OH^-}{1 \text{ mL NaOH soln}} = 5.00 \text{ mmol } OH^-$$

$[OH^-] = \frac{5.00 \text{ mmol } OH^-}{50.0 \text{ mL}} = 0.100\,M$ pOH = -log(0.100) = 1.00 pH = 14.00 - 1.00 = **13.00**

2. **25.00 mL HNO_3 added:**

$$? \text{ mmol } H_3O^+ \text{ added} = 25.00 \text{ mL } HNO_3 \text{ soln} \times \frac{0.100 \text{ mmol } H_3O^+}{1 \text{ mL } HNO_3 \text{ soln}} = 2.50 \text{ mmol } H_3O^+$$

$V_T = 50.00 \text{ mL} + 25.00 \text{ mL} = 75.00 \text{ mL}$

	H_3O^+	OH^-	H_2O
s, mmol	2.50	5.00	excess
R, mmol	-2.50	−2.50	+2.50
f, mmol	0.	2.50	excess
c, *M*	0.	0.0333	excess

pOH = -log(0.0333) = 1.48 pH = 14.00 - 1.48 = **12.52**

3. **50.00 mL HNO_3 added:**

$$? \text{ mmol } H_3O^+ \text{ added} = 50.00 \text{ mL } HNO_3 \text{ soln} \times \frac{0.100 \text{ mmol } H_3O^+}{1 \text{ mL } HNO_3 \text{ soln}} = 5.00 \text{ mmol } H_3O^+$$

$V_T = 50.00 \text{ mL} + 50.00 \text{ mL} = 100.00 \text{ mL}$

	H_3O^+	OH^-	H_2O
s, mmol	5.5	5.00	excess
R, mmol	-5.00	−5.00	+5.00
f, mmol	0.	0.	excess
c, *M*	1.0×10^{-7} *	1.0×10^{-7} *	excess

***Concentrations due to autoionization of water at 25°C.**

pH = -log(1.0×10^{-7}) = **7.00**

4. **75.00 mL HNO_3 added:**

$$? \text{ mmol } H_3O^+ \text{ added: } 75.00 \text{ mL } HNO_3 \text{ soln} \times \frac{0.100 \text{ mmol } H_3O^+}{1 \text{ mL } HNO_3 \text{ soln}} = 7.50 \text{ mmol } H_3O^+$$

$V_T = 50.00 \text{ mL} + 75.00 \text{ mL} = 125.00 \text{ mL}$

	H_3O^+	OH^-	H_2O
s, mmol	7.50	5.00	excess
R, mmol	-5.00	−5.00	+5.00
f, mmol	2.50	0.	excess
c, *M*	0.0200	0.	excess

pH = -log(0.0200) = **1.70**

Summary:

mL HNO_3 soln	pH
0.	13.00
25.00	12.52
50.00	7.00
75.00	1.70

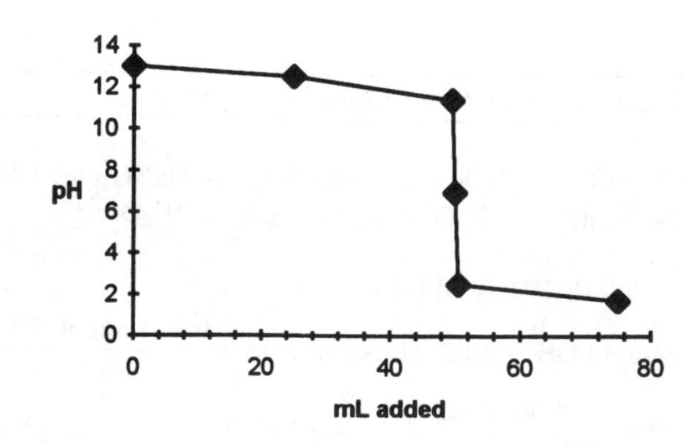

16.39 ***SeeSections 16.4, 16.5.***

Weak acid - Strong base Curve:

mL 0.20 *M* strong base	pH
0.0	2.50
20.0	3.82
40.0	4.60
49.5	6.00
50.0	8.45
50.5	10.82
70.0	12.37

Strong acid - Strong base Curve:

mL 0.20 *M* strong base	pH
0.0	1.00
20.0	1.30
40.0	1.85
49.5	3.17
50.0	7.00
50.5	10.82
70.0	12.37

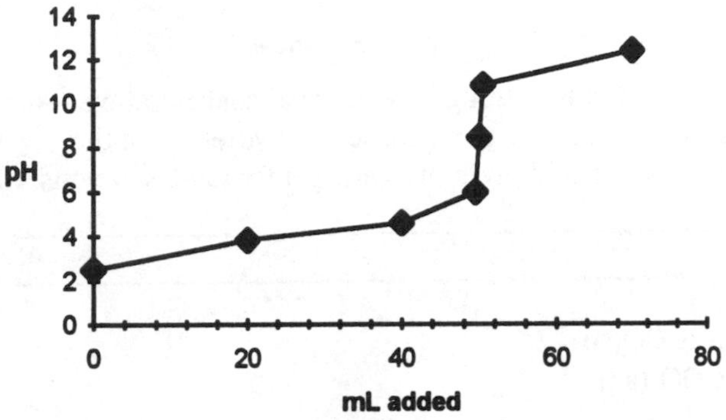

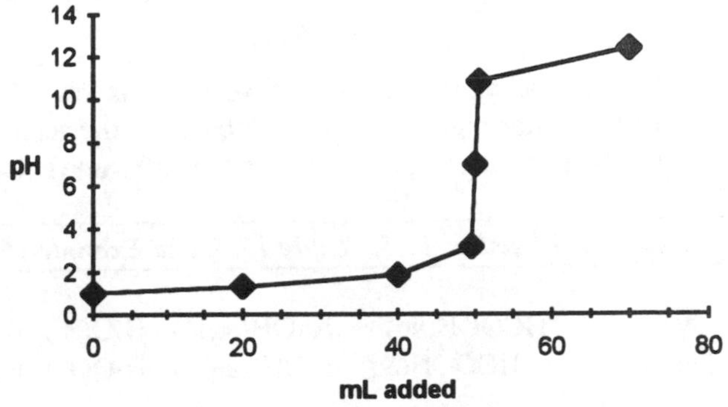

Notice how the starting pH for the weak acid is higher than that of the strong acid of the same concentration due to less ionization. Also notice the buffer region for the weak acid and the basicity at the equivalence point caused by hydrolysis of the conjugate base anion. Lastly, notice how the same amount of excess base gives the same pH for these solutions.

16.40 ***See Sections 16.4, 16.5.***

Weak base - Strong acid Case:

mL strong acid	pH
0.0	11.00
10.0	9.18
20.0	8.40
24.5	7.31
25.0	5.09
25.5	2.88
40.0	1.48

Strong base - Strong Acid Case:

mL strong acid	pH
0.0	13.00
10.0	12.70
20.0	12.15
24.5	11.13
25.0	7.00
25.5	2.88
40.0	1.48

Weak base - Strong acid Case: Strong base - Strong Acid Case:

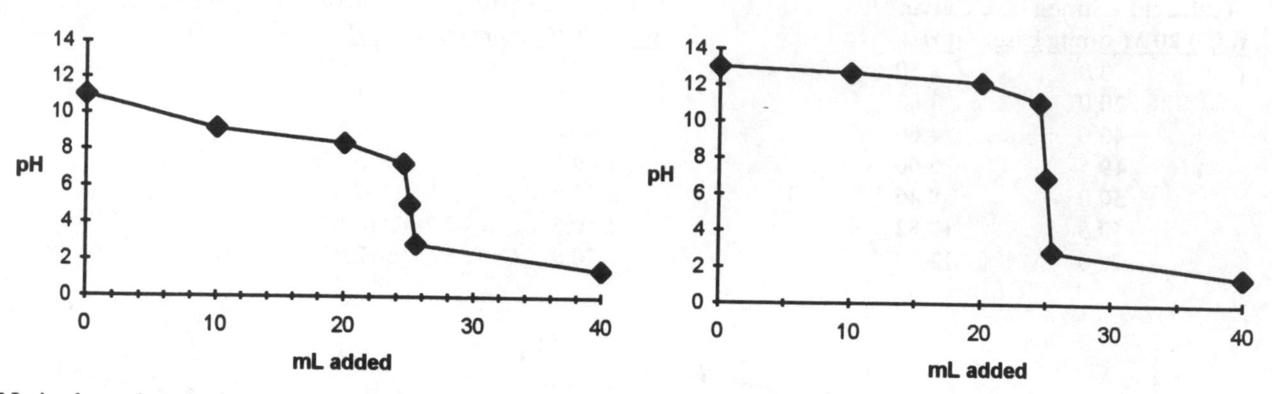

Notice how the starting pH for the weak base is lower for than that of the strong base of equal concentration. Also notice the buffer region for the weak base and the acidity at the equivalence point caused by hydrolysis of the conjugate acid cation. Lastly, notice how the same amount of excess acid gives the same pH for these solutions.

16.41 ***See Section 16.5, Table 15.6, and Exanple 16.5.***

Overall: $HCOOH(aq) + NaOH(aq) \rightarrow H_2O(\ell) + NaHCOO(aq)$
Net ionic: $HCOOH(aq) + OH^-(aq) \rightarrow H_2O(\ell) + HCOO^-(aq)$

1. 0. **NaOH soln added**:

$$HCOOH(aq) + H_2O(\ell) \rightleftarrows H_3O^+(aq) + HCOO^-(aq)$$

The K_a for HCOOH is 1.8×10^{-4}, and its pK_a is $-\log(1.8 \times 10^{-4}) = 3.74$

Conc, M	HCOOH	H_3O^+	$HCOO^-$
initial	0.250	0.	0.
Change	$-y$	$+y$	$+y$
equilibrium	$0.250-y$	y	y

$$K_a = \frac{[H_3O^+][HCOO^-]}{[HCOOH]} \qquad K_a = \frac{(y)(y)}{(0.250-y)} \cong \frac{(y)(y)}{0.250} = 1.8 \times 10^{-4}$$

$[H_3O^+] = y = 6.7 \times 10^{-3}$ and $pH = -\log(6.7 \times 10^{-3}) = \mathbf{2.17}$.

Approximation check: $\left(\frac{6.7 \times 10^{-3}}{0.250}\right) \times 100\% = 2.68\%$, so the approximation is valid.

2. **5.00 mL NaOH soln added**:

$$? \text{ starting mmol HCOOH} = 20.00 \text{ mL HCOOH soln} \times \frac{0.250 \text{ mmol HCOOH}}{1 \text{ mL HCOOH soln}} = 5.00 \text{ mmol HCOOH}$$

$$? \text{ mmol OH}^- \text{ added} = 5.00 \text{ mL NaOH soln} \times \frac{0.500 \text{ mmol OH}^-}{1 \text{ mL NaOH soln}} = 2.50 \text{ mmol OH}^-$$

$V_T = 20.00 \text{ mL} + 5.00 \text{ mL} = 25.00 \text{ mL}$

	HCOOH	OH^-	$HCOO^-$
s, mmol	5.00	2.50	0.
R, mmol	-2.50	−2.50	+2.50
f, mmol	2.50	0.	2.50
C, *M*	0.100	0.	0.100

$$pH = pK_a + \log\left(\frac{C_b}{C_a}\right) \qquad pH = 3.74 + \log\left(\frac{0.100}{0.100}\right) = \mathbf{3.74}$$

3. **10.00 mL NaOH added**:

$$? \text{ mmol OH}^- \text{ added} = 10.00 \text{ mL NaOH soln} \times \frac{0.500 \text{ mmol OH}^-}{1 \text{ mL NaOH soln}} = 5.00 \text{ mmol OH}^-$$

$V_T = 20.00 \text{ mL} + 10.00 \text{ mL} = 30.00 \text{ mL}$

	HCOOH	OH^-	$HCOO^-$
s, mmol	5.00	5.00	0.
R, mmol	−5.00	−5.00	+5.00
f, mmol	0.	0.	5.00
C, *M*	0.	0.	0.167

The pH at the equivalence point is due to the reaction of $HCOO^-$ with H_2O:

$$HCOO^- + H_2O \rightleftarrows HCOOH + OH^-$$

$$K_b\ HCOO^- = \frac{K_w}{K_a\ HCOOH} \qquad K_b\ HCOO^- = \frac{1.0 \times 10^{-14}}{1.8 \times 10^{-4}} = 5.6 \times 10^{-11}$$

Conc, *M*	$HCOO^-$	HCOOH	OH^-
initial	0.167	0.	0.
Change	−y	+y	+y
equilibrium	0.167 − y	y	y

$$K_b\ HCOO^- = \frac{[HCOOH][OH^-]}{[HCOO^-]} \qquad K_b\ HCOO^- = \frac{(y)(y)}{(0.167-y)} \cong \frac{(y)(y)}{0.167} = 5.6 \times 10^{-11}$$

$[OH^-] = y = 3.1 \times 10^{-6}$, $pOH = -\log(3.1 \times 10^{-6}) = 5.51$ and $pH = 14.00 - 5.51 = \mathbf{8.49}$

4. **15.00 mL NaOH added**:

$$? \text{ mmol OH}^- \text{ added} = 15.00 \text{ mL NaOH soln} \times \frac{0.500 \text{ mmol OH}^-}{1 \text{ mL NaOH soln}} = 7.50 \text{ mmol OH}^-$$

$V_T = 20.00 \text{ mL} + 15.00 \text{ mL} = 35.00 \text{ mL}$

	HCOOH	OH^-	$HCOO^-$
s, mmol	5.00	7.50	0.
R, mmol	−5.00	−5.00	+5.00
f, mmol	0.	2.50	5.00
C, *M*	0.	0.071	0.143

The excess of strong base determines the pH.

$[OH^-] = 0.071\ M$, pOH = -log(0.071) = 1.15 and pH = 14.00 - 1.15 = **12.85**

Summary:

mL NaOH soln	pH
0.	2.17
5.00	3.74
10.00	8.49
15.00	12.85

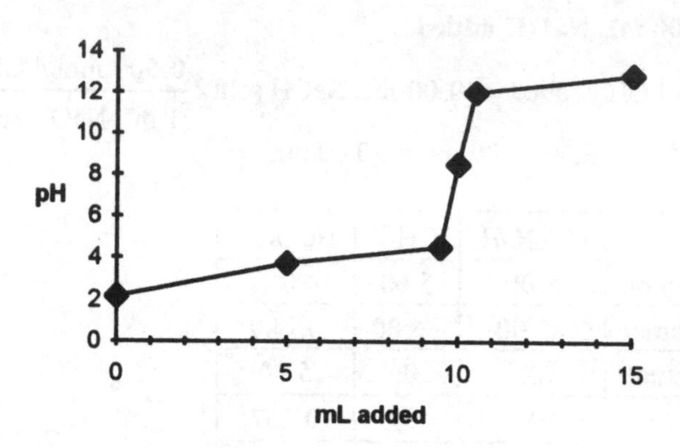

Four regions of importance:
1. Initial point: HCOOH weak acid equilibrium
2. Any point between initial point and equivalence point: HCOOH / $HCOO^-$ buffer equilibrium
3. Equivalence point: $HCOO^-$ base hydrolysis equilibrium
4. Beyond equivalence point: Excess NaOH strong base

16.42 ***See Section 16.5, Table 15.6, and Example 16.5.***

Overall: $CH_3CO_2H(aq) + NaOH(aq) \rightarrow NaCH_3CO_2(aq) + H_2O(\ell)$

Net ionic: $CH_3CO_2H(aq) + OH^-(aq) \rightarrow CH_3CO_2^-(aq) + H_2O(\ell)$

1. **0. mL NaOH soln added:**

$$CH_3CO_2H(aq) + H_2O(\ell) \rightleftarrows H_3O^+(aq) + CH_3CO_2^-(aq)$$

The K_a for CH_3CO_2H is 1.8×10^{-5}, and its pK_a is $-\log(1.8 \times 10^{-5}) = 4.74$.

Conc, *M*	CH_3CO_2H	H_3O^+	$CH_3OO_2^-$
initial	0.400	0.	0.
Change	−y	+y	+y
equilibrium	0.400 − y	y	y

$$K_a = \frac{[H_3O^+][CH_3CO_2^-]}{[CH_3CO_2H]} \qquad K_a = \frac{(y)(y)}{0.400 - y} \cong \frac{(y)(y)}{0.400} = 1.8 \times 10^{-5}$$

$[H_3O^+] = y = 2.7 \times 10^{-3}$ and pH = $-\log(2.7 \times 10^{-3})$ = **2.57**

2. **10.00 mL NaOH soln added:**

? starting mmol CH_3CO_2H = 25.00 mL CH_3CO_2H soln $\times \frac{0.400 \text{ mmol } CH_3CO_2H}{1 \text{ mL } CH_3CO_2H \text{ soln}}$ = 10.0 mmol CH_3CO_2H

$? \text{ mmol OH}^- \text{ added} = 10.00 \text{ mL NaOH soln} \times \dfrac{0.500 \text{ mmol OH}^-}{1 \text{ mL NaOH soln}} = 5.00 \text{ mmol OH}^-$

$V_T = 25.00 \text{ mL} + 10.00 \text{ mL} = 35.00 \text{ mL}$

	CH_3CO_2H	OH^-	$CH_3CO_2^-$
s, mmol	10.00	5.00	0.
R, mmol	–5.00	–5.00	+5.00
f, mmol	5.0	0.	5.00
C, *M*	0.14	0.	0.14

$pH = pK_a + \log\left(\dfrac{C_b}{C_a}\right)$ $\quad pH = 4.74 + \log\left(\dfrac{0.14}{0.14}\right) = \mathbf{4.74}$

3. **20.00 mL NaOH soln added:**

$? \text{ mmol OH}^- \text{ added} = 20.00 \text{ mL NaOH soln} \times \dfrac{0.500 \text{ mol OH}^-}{1 \text{ mL NaOH soln}} = 10.0 \text{ mmol OH}^-$

$V_T = 25.00 \text{ mL} + 20.00 \text{ mL} = 45.00 \text{ mL}$

	CH_3CO_2H	OH^-	$CH_3CO_2^-$
s, mmol	10.0	10.0	0.
R, mmol	–10.0	–10.0	+10.0
f, mmol	0.	0.	10.0
C, *M*	0.	0.	0.222

The pH at the equivalence point is due to the reaction of $CH_3CO_2^-$ with H_2O:

$$CH_3CO_2^- + H_2O \rightleftarrows CH_3CO_2H + OH^-$$

$K_b\ CH_3CO_2^- = \dfrac{K_w}{K_a\ CH_3CO_2H}$ $\quad K_b\ CH_3CO_2^- = \dfrac{1.00 \times 10^{-14}}{1.8 \times 10^{-5}} = 5.6 \times 10^{-10}$

Conc, *M*	$CH_3CO_2^-$	CH_3CO_2H	OH^-
initial	0.222	0.	0.
Change	–y	+y	+y
equilibrium	0.222 – y	y	y

$K_b = \dfrac{[CH_3CO_2H][OH^-]}{[CH_3CO_2^-]}$ $\quad K_b\ CH_3CO_2^- = \dfrac{(y)(y)}{(0.222 - y)} \cong \dfrac{(y)(y)}{0.222} = 5.6 \times 10^{-10}$

$[OH^-] = y = 1.1 \times 10^{-5}$, $pOH = -\log(1.1 \times 10^{-5}) = 4.96$ and $pH = 14.00 - 4.96 = \mathbf{9.04}$

4. **25.00 mL NaOH soln added:**

$? \text{ mmol OH}^- \text{ added} = 25.00 \text{ mL NaOH soln} \times \dfrac{0.500 \text{ mmol OH}^-}{1 \text{ mL NaOH soln}} = 12.5 \text{ mmol}$

$V_T = 25.00 \text{ mL} + 25.00 \text{ mL} = 50.00 \text{ mL}$

	CH_3CO_2H	OH^-	$CH_3CO_2^-$
s, mmol	10.0	12.5	0.
R, mmol	–10.0	–10.0	+10.0
f, mmol	0.	2.5	10.0
C, *M*	0.	0.050	0.200

The excess of strong base determines the pH.
$[OH^-] = 0.050$, pOH = -log (0.050) = 1.30 and pH = 14.00 - 1.30 = 12.70

Summary:

mL NaOH soln	pH
0.	2.57
10.00	4.74
20.00	9.04
25.00	12.70

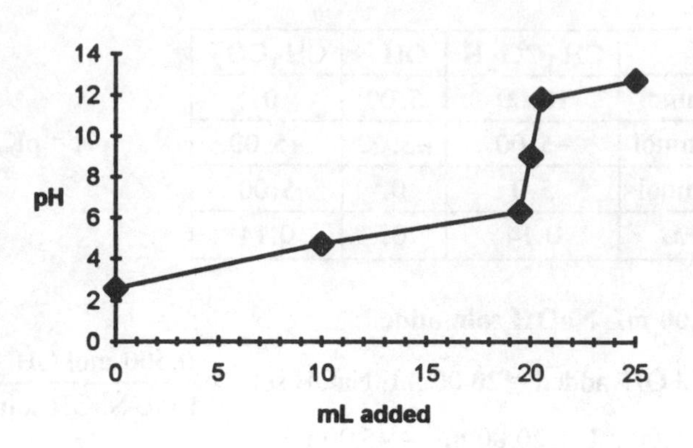

Four regions of importance:
1. Initial point: CH_3CO_2H weak acid equilibrium
2. Any point between intial point and equivalence point: CH_3CO_2H / $CH_3CO_2^-$ buffer equilibrium
3. Equivalence point: $CH_3CO_2^-$ base hydrolysis equilibrium
4. Beyond equivalence point: Excess NaOH strong base

16.43 ***See Section 16.5, Table 15.6, and Example 16.6.***

Overall: $HF(aq) + NaOH(aq) \rightarrow H_2O(\ell) + NaF(aq)$
Net ionic: $HF(aq) + OH^-(aq) \rightarrow H_2O(\ell) + F^-(aq)$

1. **0. mL NaOH soln added**:

$$HF(aq) + H_2O(\ell) \rightleftarrows H_3O^+(aq) + F^-(aq)$$

The K_a for HF is 3.5×10^{-4}, and its pK_a is $-\log(3.5 \times 10^{-4}) = 3.46$

Conc, M	HF	H_3O^+	F^-
initial	0.230	0.	0.
Change	$-y$	$+y$	$+y$
equilibrium	$0.230-y$	y	y

$$K_a = \frac{[H_3O^+][F^-]}{[HF]} \qquad K_a = \frac{(y)(y)}{(0.230-y)} \cong \frac{(y)(y)}{0.230} = 3.5 \times 10^{-4}$$

$[H_3O^+] = y = 9.0 \times 10^{-3}$ pH = **2.05**

Approximation check: $\left(\frac{9.0 \times 10^{-3}}{0.230}\right) \times 100\% = 3.9\%$, so the approximation is valid.

2. **50% of amount of base needed to reach the equivalence point**:

$$? \text{ starting mmol HF} = 100.0 \text{ mL HF soln} \times \frac{0.230 \text{ mmol HF}}{1 \text{ mL HF soln}} = 23.0 \text{ mmol HF}$$

$$? \text{ mmol HF at 50\%} = \text{mmol F}^- = (0.50)(23.0 \text{ mmol}) = 11.5 \text{ mmol}$$

$$pH = pK_a + \log\left(\frac{\text{mmol b}}{\text{mmol a}}\right) \qquad pH = 3.46 + \log\left(\frac{11.5}{11.5}\right) = \mathbf{3.46}$$

3. **95% of amount of base needed to reach the equivalence point**:

$? \text{ mmol HF} = (0.05)(23.0 \text{ mmol}) = 1.15 \text{ mmol HF}$

$? \text{ mmol F}^- = (0.95)(23.0 \text{ mmol}) = 21.9 \text{ mmol F}^-$

$$pH = pK_a + \log\left(\frac{\text{mmol b}}{\text{mmol a}}\right) \qquad pH = 3.46 + \log\left(\frac{21.9}{1.15}\right) = \mathbf{4.74}$$

4. **100% of amount of base needed to reach the equivalence point**:

$? \text{ mmol HF} = (0.00)(23.0 \text{ mmol}) = 0. \text{ mmol HF}$

$? \text{ mmol F}^- = (1.00)(23.0 \text{ mmol}) = 23.0 \text{ mmol F}^-$

The pH at the equivalence point is due to the reaction of F^- with H_2O:

$$F^-(aq) + H_2O(l) \rightleftarrows HF(aq) + OH^-(aq)$$

$$K_b\ F^- = \frac{K_w}{K_a\ HF} \qquad K_b\ F^- = \frac{1.0 \times 10^{-14}}{3.5 \times 10^{-4}} = 2.9 \times 10^{-11}$$

We need to know $\left[F^-\right]$ at the equivalence point and therefore need to know the total volume of the solution at the equivalence point. Hence, we need to know the volume of NaOH containing 23.0 mmol OH^-, since HF and OH^- react in a 1:1 ratio.

$$? \text{ L NaOH soln} = 23.0 \text{ mmol OH}^- \times \frac{1 \text{ mmol NaOH}}{1 \text{ mmol OH}^-} \times \frac{1 \text{ mL NaOH soln}}{0.500 \text{ mmol NaOH}} = 46.0 \text{ mL NaOH soln}$$

$V_T = 100.0 \text{ mL} + 46.0 \text{ mL} = 146.0 \text{ mL}$

$$\left[F^-\right] \text{ at the equivalence point} = \frac{23.0 \text{ mmol F}^-}{146.0 \text{ mL soln}} = 0.158\ M$$

Conc, *M*	F^-	HF	OH^-
initial	0.158	0.	0.
Change	−y	+y	+y
equilibrium	0.158 − y	y	y

$$K_b = \frac{[HF]\left[OH^-\right]}{\left[F^-\right]} \qquad K_b = \frac{(y)(y)}{(0.158 - y)} \cong \frac{(y)(y)}{0.158} = 2.9 \times 10^{-11}$$

$\left[OH^-\right] = y = 2.1 \times 10^{-6}$, $pOH = -\log(2.1 \times 10^{-6}) = 5.68$ and $pH = 14.00 - 5.68 = \mathbf{8.32}$

5. **105% of amount of base needed to reach equivalence point**:

$? \text{ mmol excess OH}^- = (0.05)(23.0 \text{ mmol}) = 1.15 \text{ mmol}$

$? \text{ mmol total OH}^- = (1.05)(23.0 \text{ mmol}) = 24.2 \text{ mmol}$

$$? \text{ mL NaOH soln added} = 24.2 \text{ mmol OH}^- \times \frac{1 \text{ mmol NaOH}}{1 \text{ mmol OH}^-} \times \frac{1 \text{ mL NaOH soln}}{0.500 \text{ mmol NaOH}} = 48.4 \text{ mL NaOH soln}$$

$V_T = 100.0 \text{ mL} + 48.4 \text{ mL} = 148.4 \text{ mL}$

$$\left[OH^-\right] = \frac{1.15 \text{ mmol}}{148.4 \text{ mL soln}} = 7.75 \times 10^{-3}\ M$$

$pOH = -\log(7.75 \times 10^{-3}) = 2.11$ and $pH = 14.00 - 2.11 = \mathbf{11.89}$

Summary:

%OH added	pH
0.	2.05
50	3.46
95	4.74
100	8.32
105	11.89

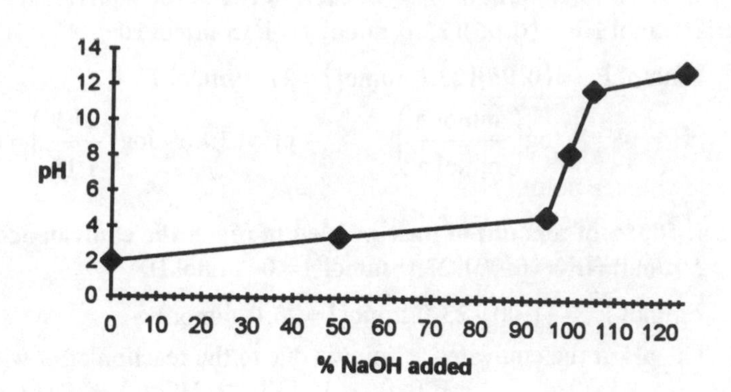

Four regions of importance:

1. Initial point: HF weak acid equilibrium
2. Any point between initial point and equivalence point: HF / F^- buffer equilibrium
3. Equivalence point: F^- base hydrolysis equilibrium
4. Beyond equivalence point: Excess NaOH strong base

16.44 ***See Section 16.5, Table 15.6, and Example 16.6.***

Overall: $HClO(aq) + KOH(aq) \rightarrow KClO(aq) + H_2O(\ell)$

Net ionic: $HClO(aq) + OH^-(aq \rightarrow ClO^-(aq) + H_2O(\ell)$

1. 0. mL KOH soln added:

$$HClO(aq) + H_2O(\ell) \rightleftarrows H_3O^+(aq) + ClO^-(aq)$$

The K_a for HClO is 3.0×10^{-8}, and its pK_a is $-\log(3.0 \times 10^{-8}) = 7.52$.

Conc, M	HClO	H_3O^+	ClO^-
initial	0.400	0.	0.
Change	−y	+y	+y
equilibrium	0.400 − y	y	y

$$K_a = \frac{[H_3O^+][ClO^-]}{[HClO]} \qquad K_a = \frac{(y)(y)}{0.400 - y} \cong \frac{(y)(y)}{0.400} = 3.0 \times 10^{-8}$$

$[H_3O^+] = y = 1.1 \times 10^{-4}$ $\quad$ $pH = -\log(1.1 \times 10^{-4}) = \mathbf{3.96}$

2. 50% of amount of base needed to reach the equivalence point:

$$? \text{ starting mmol HClO} = 10.00 \text{ mL HClO soln} \times \frac{0.400 \text{ mmol HClO}}{1 \text{ mL HClO soln}} = 4.00 \text{ mmol HClO}$$

$? \text{ mmol HClO at } 50\% = \text{mmol } ClO^- = (0.50)(4.00 \text{ mmol}) = 2.00 \text{ mmol}$

$$pH = pK_a + \log\left(\frac{\text{mmol b}}{\text{mmol a}}\right) \qquad pH = 7.52 + \log\left(\frac{2.00}{2.00}\right) = \mathbf{7.52}$$

3. 95% of amount of base needed to reach the equivalence point:

$? \text{ mmol HClO} = (0.05)(4.00 \text{ mmol}) = 0.200 \text{ mmol}$

$? \text{ mmol } ClO^- = (0.95)(4.00 \text{ mmol}) = 3.80 \text{ mmol}$

$$pH = pK_a + \log\left(\frac{\text{mmol b}}{\text{mmol a}}\right) \qquad pH = 7.52 + \log\left(\frac{3.80}{0.200}\right) = \mathbf{8.80}$$

4. 100% of amount of base needed to reach equivalence point:

$? \text{ mmol HClO} = (0.00)(4.00 \text{ mmol}) = 0. \text{ mmol HClO}$

$? \text{ mmol } ClO^- = (1.00)(4.00 \text{ mmol}) = 4.00 \text{ mmol } ClO^-$

The pH at the equivalence point is due to the reaction of ClO^- with H_2O:

$$ClO^-(aq) + H_2O(\ell) \rightleftarrows HClO(aq) + OH^-(aq)$$

$K_b\ ClO^- = \dfrac{K_w}{K_a\ HClO}$ $\qquad K_b\ ClO^- = \dfrac{1.0 \times 10^{-14}}{3.0 \times 10^{-8}} = 3.3 \times 10^{-7}$

We need to know $[ClO^-]$ at the equivalence point and therefore need to know the total volume of the solution at the equivalence point. Hence, we need to know the volume of 0.500 *M* KOH containing 4.00 mmol OH^-, since HClO and KOH react in a 1:1 ratio.

$$? \text{ L KOH} = 4.00 \text{ mmol } OH^- \times \frac{1 \text{ mmol KOH}}{1 \text{ mmol } OH^-} \times \frac{1 \text{ mL KOH soln}}{0.500 \text{ mmol KOH}} = 8.00 \text{ mL KOH soln}$$

$V_T = 10.00 \text{ mL} + 8.00 \text{ mL} = 18.00 \text{ mL}$

$$[ClO^-] \text{ at equivalence point} = \frac{4.00 \text{ mmol } ClO^-}{18.00 \text{ mL soln}} = 0.222\ M$$

Conc, *M*	ClO^-	HClO	OH^-
initial	0.222	0.	0.
Change	−y	+y	+y
equilibrium	0.222 − y	y	y

$K_b\ ClO^- = \dfrac{[HClO][OH^-]}{[ClO^-]}$ $\qquad K_b = \dfrac{(y)(y)}{(0.222 - y)} \cong \dfrac{(y)(y)}{0.222} = 3.3 \times 10^{-7}$

$[OH^-] = y = 2.7 \times 10^{-4}$, pOH = -log ($2.7 \times 10^{-4}$) = 3.57 and pH = 14.00 - 3.57 = **10.43**

5. **105% of amount of base needed to reach equivalence point:**

? mmol excess OH^- = (0.05)(4.00 mmol) = 0.200 mmol

? mmol total OH^- = (1.05)(4.00 mmol) = 4.20 mmol

$$? \text{ total mL NaOH soln added} = 4.20 \text{ mmol } OH^- \times \frac{1 \text{ mmol KOH}}{1 \text{ mmol } OH^-} \times \frac{1 \text{ mL KOH soln}}{0.500 \text{ mmol KOH}} = 8.40 \text{ mL KOH soln}$$

$V_T = 10.00 \text{ mL} \div 8.40 \text{ mL} = 18.40 \text{ mL}$

$$[OH^-] = \frac{0.200 \text{ mmol}}{18.40 \text{ mL}} = 0.0109\ M$$

pOH = -log (0.0109) = 1.96 and pH = 14.00 - 1.96 = **12.04**

Summary:

% KOH added:	pH
0	3.96
50	7.52
95	8.80
100	10.43
105	12.04

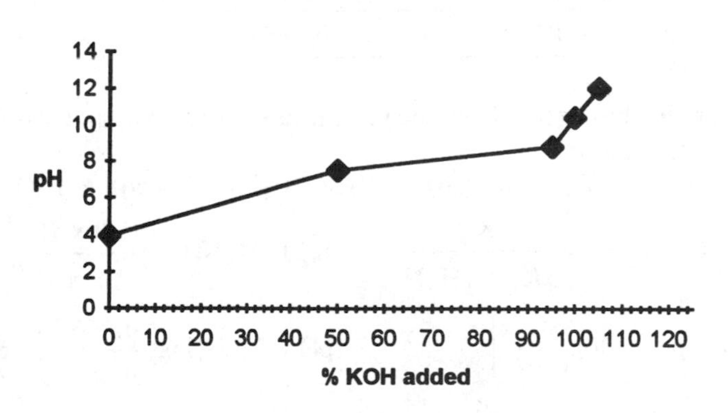

Four regions of importance:

1. Initial point: HClO weak acid equilibrium
2. Any point between inital point and equivalence point: HClO / ClO^- buffer equilibrium
3. Equivalence point: ClO^- base hydrolysis equilibrium
4. Beyond equivalence point: Excess KOH strong base

16.45 ***See Section 16.5 and Table 15.7.***

Overall: $C_5H_5N(aq) + HCl(aq) \rightarrow C_5H_5NH^+(aq) + Cl^-(aq)$

Net ionic: $C_5H_5N(aq) + H_3O^+(aq) \rightarrow C_5H_5NH^+(aq) + H_2O(\ell)$

1. **0. mL HCl added:**

$$C_5H_5N(aq) + H_2O(\ell) \rightleftarrows C_5H_5NH^+(aq) + OH^-(aq) \quad K_b = 1.8 \times 10^{-9}$$

Conc, M	C_5H_5N	$C_5H_5NH^+$	OH^-
initial	0.200	0.	0.
Change	−y	+y	+y
equilibrium	0.200 − y	y	y

$$K_b = \frac{[C_5H_5NH^+][OH^-]}{[C_5H_5N]} \qquad K_b = \frac{(y)(y)}{(0.200-y)} \cong \frac{(y)(y)}{0.200} = 1.8 \times 10^{-9}$$

$[OH^-] = y = 1.9 \times 10^{-5}$, pOH = -log(1.9 x 10^{-5}) = 4.72 and pH = 14.00 - 4.72 = **9.28**

Approximation check: $\left(\frac{1.9 \times 10^{-5}}{0.200}\right) \times 100\% = 9.5 \times 10^{-3}$ %, so the approximation is valid.

2. **15.00 mL HCl added:**

$$? \text{ starting mmol } C_5H_5N = 30.00 \text{ mL } C_5H_5N \text{ soln} \times \frac{0.200 \text{ mmol } C_5H_5N}{1 \text{ mL } C_5H_5N \text{ soln}} = 6.00 \text{ mmol } C_5H_5N$$

$$? \text{ mmol } H_3O^+ \text{ added} = 15.00 \text{ mL HCl soln} \times \frac{0.200 \text{ mmol } H_3O^+}{1 \text{ mL HCl soln}} = 3.00 \text{ mmol } H_3O^+$$

V_T = 30.00 mL + 15.00 mL = 45.00 mL

	C_5H_5N	H_3O^+	$C_5H_5NH^+$
s. mmol	6.00	3.00	0.
R, mmol	−3.00	−3.00	+3.00
f, mmol	3.00	0.	3.00
C, M	0.0667	0.	0.0667

After HCl has been added, the system can be treated as an $C_5H_5NH^+$/ C_5H_5NH buffer system until the equivalence point is reached.

$$C_5H_5NH^+(aq) + H_2O(\ell) \rightleftarrows C_5H_5N(aq) + H_3O^+(aq)$$

$$K_a\ C_5H_5NH^+ = \frac{K_w}{K_b\ C_5H_5N} \qquad K_a\ C_5H_5NH^+ = \frac{1.0 \times 10^{-14}}{1.8 \times 10^{-9}} = 5.6 \times 10^{-6} \qquad pK_a = -\log(5.6 \times 10^{-6}) = 5.25$$

$$pH = pK_a + \log\left(\frac{C_b}{C_a}\right) \qquad pH = 5.25 + \log\left(\frac{0.0667}{0.0667}\right) = \mathbf{5.25}$$

3. **30.00 mL HCl added:**

$$? \text{ mmol } H_3O^+ \text{ added} = 30.00 \text{ mL HCl soln} \times \frac{0.200 \text{ mmol } H_3O^+}{1 \text{ mL HCl soln}} = 6.00 \text{ mmol } H_3O^+$$

V_T = 30.00 mL + 30.00 mL = 60.00 mL

	C_5H_5N	H_3O^+	$C_5H_5NH^+$
s. mmol	6.00	6.00	0.
R, mmol	−6.00	−6.00	+6.00
f, mmol	0.	0.	6.00
C, *M*	0.	0.	0.100

The pH at the equivalence point is due to the reaction of $C_5H_5NH^+$ with H_2O:

$$C_5H_5NH^+(aq) + H_2O(\ell) \rightleftarrows C_5H_5N(aq) + H_3O^+(aq) \qquad K_a = 5.6 \times 10^{-6}$$

Conc, *M*	$C_5H_5NH^+$	C_5H_5N	H_3O^+
initial	0.100	0.	0.
Change	−y	+y	+y
equilibrium	0.100 − y	y	y

$$K_a\ C_5H_5NH^+ = \frac{[C_5H_5N][H_3O^+]}{[C_5H_5NH^+]} \qquad K_a\ C_5H_5NH^+ = \frac{(y)(y)}{(0.100-y)} \cong \frac{(y)(y)}{0.100} = 5.6 \times 10^{-6}$$

$[H_3O^+] = y = 7.5 \times 10^{-4}$ and $pH = -\log(7.5 \times 10^{-4}) = \mathbf{3.12}$

Approximation check: $\left(\frac{7.5 \times 10^{-4}}{0.100}\right) \times 100\% = 0.75\%$, so the approximation is valid.

4. **40.00 mL HCl added**:

$$? \text{ mmol } H_3O^+ \text{ added} = 40.00 \text{ mL HCl soln} \times \frac{0.200 \text{ mol } H_3O^+}{1 \text{ mL HCl soln}} = 8.00 \text{ mmol } H_3O^+$$

$V_T = 30.00 \text{ mL} + 40.00 \text{ mL} = 70.00 \text{ mL}$

	C_5H_5N	H_3O^+	$C_5H_5NH^+$
s. mmol	6.00	6.00	0.
R, mmol	−6.00	−6.00	+6.00
f, mmol	0.	2.00	6.00
C, *M*	0.	0.0286	0.0857

The excess of strong acid determines the pH: $[H_3O^+] = 0.0286$ and $pH = -\log(0.0286) = \mathbf{1.54}$

Summary:

mL HCl soln	pH
0.	9.28
15.	5.25
30.00	3.12
40.00	1.54

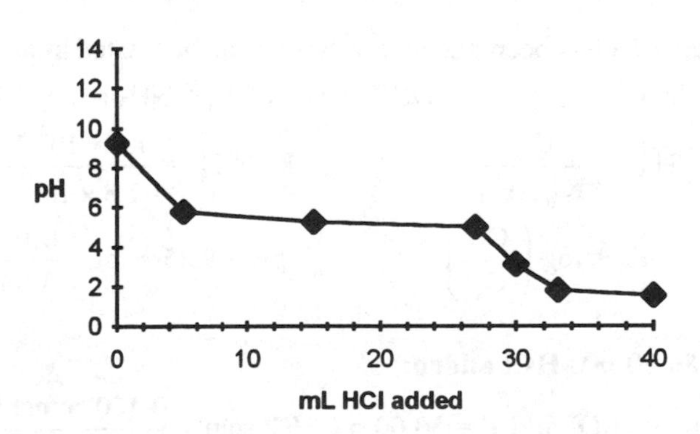

Four regions of importance:
1. Initial point: C_5H_5N weak base equilibrium
2. Any point between initial point and equivalence point: $C_5H_5NH^+$ / C_5H_5N buffer equilibrium
3. Equivalence point: $C_5H_5NH^+$ acid hydrolysis equilibrium
4. Beyond equivalence point: Excess HCl strong acid

Note: Pyridine is too weak a base to show a titration curve with a clear inflection point.

16.46 ***See Section 16.5, Table 15.6, and Example 16.5.***

Overall: $HCl(aq) + NH_3(aq) \rightarrow NH_4Cl(aq)$

Net ionic: $H_3O^+(aq) + NH_3(aq) \rightarrow NH_4^+(aq) + H_2O(\ell)$

1. **0. mL HCl soln added:**

$$NH_3(aq) + H_2O(\ell) \rightleftarrows NH_4^+(aq) + OH^-(aq) \qquad K_b = 1.8 \times 10^{-5}$$

Conc, M	NH_3	NH_4^+	OH^-
initial	0.100	0.	0.
Change	−y	+y	+y
equilibrium	0.100 − y	y	y

$$K_b = \frac{[NH_4^+][OH^-]}{[NH_3]} \qquad K_b = \frac{(y)(y)}{(0.100 - y)} \cong \frac{(y)(y)}{0.100} = 1.8 \times 10^{-5}$$

$[OH^-] = y = 1.3 \times 10^{-3}$, pOH = 2.89 and pH = 14.00 - 2.89 = **11.11**

Approximation check: $\left(\frac{1.3 \times 10^{-3}}{0.100}\right) \times 100\% = 1.3\%$, so the approximation is valid.

2. **25.00 mL HCl soln added:**

$$? \text{ starting mmol } NH_3 = 50.00 \text{ mL } NH_3 \text{ soln} \times \frac{0.100 \text{ mmol } NH_3}{1 \text{ mL } NH_3 \text{ soln}} = 5.00 \text{ mmol } NH_3$$

$$? \text{ mmol } H_3O^+ \text{ added} = 25.00 \text{ mL HCl soln} \times \frac{0.100 \text{ mmol } H_3O^+}{1 \text{ mL HCl soln}} = 2.50 \text{ mmol } H_3O^+$$

$V_T = 50.00 \text{ mL} + 25.00 \text{ mL} = 75.00 \text{ mL}$

	NH_3	H_3O^+	NH_4^+
s. mmol	5.00	2.50	0.
R, mmol	−2.50	−2.50	+2.50
f, mmol	2.50	0.	2.50
C, M	0.0333	0.	0.0333

After HCl has been added, the system can be treated as an NH_4^+ / NH_3 buffer system until the equivalence point is reached. $\quad NH_4^+ + H_2O(\ell) \rightleftarrows NH_3(aq) + H_3O^+(aq)$

$$K_a\,NH_4^+ = \frac{K_w}{K_b NH_3} \qquad K_a\,NH_4^+ = \frac{1.0 \times 10^{-14}}{1.8 \times 10^{-5}} = 5.6 \times 10^{-10} \qquad pK_a = -\log(5.6 \times 10^{-10}) = \mathbf{9.25}$$

$$pH = pK_a + \log\left(\frac{C_b}{C_a}\right) \qquad pH = 9.25 + \log\left(\frac{0.0333}{0.0333}\right) = \mathbf{9.25}$$

3. **50.00 mL HCl added:**

$$? \text{ mmol } H_3O^+ \text{ added} = 50.00 \text{ mL HCl soln} \times \frac{0.100 \text{ mmol } H_3O^+}{1 \text{ mL HCl soln}} = 5.00 \text{ mmol } H_3O^+$$

$V_T = 50.00 \text{ mL} + 50.00 \text{ mL} = 100.00 \text{ mL}$

	NH_3	H_3O^+	NH_4^+
s. mmol	5.00	5.00	0.
R, mmol	−5.00	−5.00	+5.00
f, mmol	0.	0.	5.00
C, *M*	0.	0.	0.0500

The pH at the equivalence point is due to the reaction of NH_4^+ with H_2O:

$$NH_4^+(aq) + H_2O(\ell) \rightleftarrows NH_3(aq) + H_3O^+(aq) \qquad K_a = 5.6 \times 10^{-10}$$

Conc, *M*	NH_4^+	NH_3	H_3O^+
initial	0.0500	0.	0.
Change	−y	+y	+y
equilibrium	0.0500 − y	y	y

$$K_a\,NH_4^+ = \frac{[NH_3][H_3O^+]}{[NH_4^+]} \qquad K_a\,NH_4^+ = \frac{(y)(y)}{(0.0500 - y)} \cong \frac{(y)(y)}{0.0500} = 5.6 \times 10^{-10}$$

$[H_3O^+] = y = 5.3 \times 10^{-6}$ and pH = -log (5.3×10^{-6}) = **5.28**

Approximation check: $\left(\frac{5.3 \times 10^{-6}}{0.0500}\right)$ x 100% = 0.011%, so the approximation is valid.

4. **75.00 mL HCl soln added:**

? mmol H_3O^+ added = 75.00 mL HCl $\times \frac{0.100 \text{ mmol } H_3O^+}{1 \text{ mL HCl soln}}$ = 7.50 mmol H_3O^+

V_T = 50.00 mL + 75.00 mL = 125.00 mL

	NH_3	H_3O^+	NH_4^+
s. mmol	5.00	7.50	0.
R, mmol	−5.00	−5.00	+5.00
f, mmol	0.	2.50	5.00
C, *M*	0.	0.0200	0.0400

The excess of strong acid determines the pH: $[H_3O^+]$ = 0.0200 and pH = -log (0.0200) = **1.70**

Summary:

mL HCl soln added	pH
0.	11.11
25.00	9.25
50.00	5.28
75.00	1.70

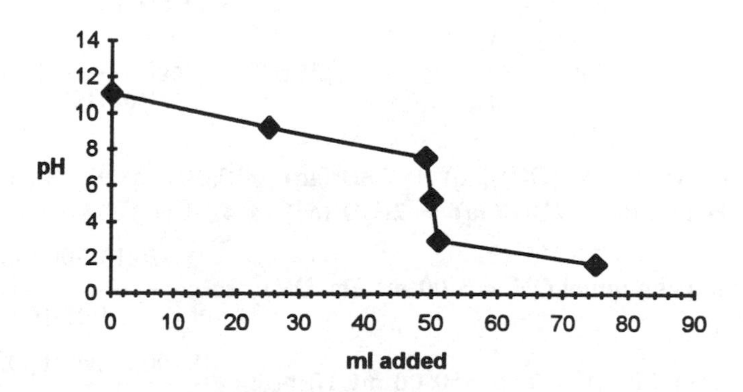

Four regions of importance:

1. Initial point: NH_3 weak base equilibrium
2. Any point between initial point and equivalence point: NH_4^+ / NH_3 buffer equilibrium
3. Equivalence point: NH_4^+ acid hydrolysis equilibrium
4. Beyond equivalence point: Excess HCl strong acid

16.47 *See Sections 16.4, 16.5, and Tables 15.6, 15.8.*

(a)

Overall: $HCOOH(aq) + NaOH(aq) \rightarrow NaHCOO(aq) + H_2O(\ell)$

Net ionic: $HCOOH(aq) + OH^-(aq) \rightarrow HCOO^-(aq) + H_2O(\ell)$

$$? \text{ starting mmol HCOOH} = 1.00 \text{ mL HCOOH soln} \times \frac{0.150 \text{ mmol HCOOH}}{1 \text{ mL HCOOH soln}} = 0.150 \text{ mmol HCOOH}$$

$$? \text{ mmol OH}^- \text{ added} = 2.00 \text{ mL NaOH soln} \times \frac{0.100 \text{ mmol OH}^-}{1 \text{ mL NaOH soln}} = 0.200 \text{ mmol OH}^-$$

$V_T = 1.00 \text{ mL} + 2.00 \text{ mL} = 3.00 \text{ mL}$

An excess of 0.050 mmol of OH^- has been added, and this excess will determine the pH of the solution.

$$\left[OH^-\right] = \frac{0.050 \text{ mmol OH}^-}{3.00 \text{ mL}} = 0.017, \qquad pOH = -\log(0.017) = 1.77 \quad \text{and} \quad pH = 14.00 - 1.77 = \mathbf{12.23}$$

(b)

Overall: $NH_3(aq) + HI(aq) \rightarrow NH_4I(aq)$

Net ionic: $NH_3(aq) + H_3O^+(aq) \rightarrow NH_4^+(aq)$

$$? \text{ starting mmol NH}_3 = 25.00 \text{ mL NH}_3 \text{ soln} \times \frac{0.250 \text{ mmol NH}_3}{1 \text{ mL NH}_3 \text{ soln}} = 6.25 \text{ mmol NH}_3$$

$$? \text{ mmol H}_3\text{O}^+ \text{ added} = 5.00 \text{ mL HI soln} \times \frac{0.100 \text{ mmol H}_3\text{O}^+}{1 \text{ mL HI soln}} = 0.500 \text{ mmol H}_3\text{O}^+$$

$V_T = 25.00 \text{ mL} + 5.00 \text{ mL} = 30.00 \text{ mL}$

	NH_3	H_3O^+	NH_4^+
s, mmol	6.25	0.500	0.
R, mmol	−0.500	−0.500	+0.500
f, mmol	5.75	0.	0.500
C, *M*	0.192	0.	0.0167

An excess of the weak base NH_3 remains, and the system can be treated as a NH_4^+/NH_3 buffer system.

For $NH_4^+(aq) + H_2O(\ell) \rightleftarrows NH_3(aq)$,

$$K_a\ NH_4^+ = \frac{K_w}{K_b\ NH_3} \qquad K_a\ NH_4^+ = \frac{1.0 \times 10^{-14}}{1.8 \times 10^{-5}} = 5.6 \times 10^{-10} \quad \text{and} \quad pK_a = -\log(5.6 \times 10^{-10}) = 9.25$$

$$pH = pK_a + \log\left(\frac{C_b}{C_a}\right) \qquad pH = 9.25 + \log\left(\frac{0.192}{0.0167}\right) = \mathbf{10.31}$$

(c) Overall: $Ba(OH)_2(aq) + 2HBr(aq) \rightarrow BaBr_2(aq) + 2H_2O(\ell)$

Net ionic: $2OH^-(aq) + 2H_3O^+(aq) \rightarrow 4H_2O(\ell)$

$$? \text{ starting mmol OH}^- = 5.00 \text{ mL Ba(OH)}_2 \text{ soln} \times \frac{2 \times (0.200 \text{ mmol OH}^-)}{1 \text{ mL Ba(OH)}_2 \text{ soln}} = 2.00 \text{ mmol OH}^-$$

$$? \text{ mmol H}_3\text{O}^+ \text{ added} = 50.00 \text{ mL HBr soln} \times \frac{0.400 \text{ mmol H}_3\text{O}^+}{1 \text{ mL HBr soln}} = 20.0 \text{ mmol H}_3\text{O}^+$$

$V_T = 5.00 \text{ mL} + 50.00 \text{ mL} = 55.00 \text{ mL}$

An excess of 18.0 mmol of H_3O^+ has been added, and this excess will determine the pH of the solution.

$$\left[H_3O^+\right] = \frac{18.00 \text{ mmol H}_3\text{O}^+}{55.00 \text{ mL}} = 0.327M \quad \text{and} \quad pH = -\log(0.327) = \mathbf{0.48}$$

16.48 ***See Sections 16.4, 16.5, and Tables 15.6, 15.8.***

(a) Overall: $HF(aq) + NaOH(aq) \rightarrow NaF(aq) + N_2O(\ell)$

Net ionic: $HF(aq) + OH^-(aq) \rightarrow F^-(aq) + H_2O(\ell)$

$$? \text{ starting mmol HF} = 10.0 \text{ mL HF soln} \times \frac{0.300 \text{ mmol HF}}{1 \text{ mL HF soln}} = 3.00 \text{ mmol HF}$$

$$? \text{ mmol } OH^- \text{ added} = 30.0 \text{ mL NaOH soln} \times \frac{0.100 \text{ mol } OH^-}{1 \text{ mL NaOH soln}} = 3.00 \text{ mmol } OH^-$$

$V_T = 10.0 \text{ mL} + 30.0 \text{ mL} = 40.0 \text{ mL}$

	HF	OH^-	F^-
s, mmol	3.00	3.00	0.
R, mmol	−3.00	−3.00	+3.00
f, mmol	0.	0.	3.00
C, *M*	0.	0.	0.0750

The pH at the equivalence point is due to the reaction of F^- with H_2O:

$$F^-(aq) + H_2O(\ell) \rightleftarrows HF(aq) + OH^-(aq)$$

Conc, *M*	F^-	HF	OH^-
initial	0.0750	0.	0.
Change	−y	+y	+y
equilibrium	0.0750 − y	y	y

$$K_b F^- = \frac{K_w}{K_a HF} \qquad K_b F^- = \frac{1.0 \times 10^{-14}}{3.5 \times 10^{-4}} = 2.9 \times 10^{-11}$$

$$K_b F^- = \frac{[HF][OH^-]}{[F^-]} \qquad K_b F^- = \frac{(y)(y)}{(0.0750 - y)} \cong \frac{(y)(y)}{0.0750} = 2.9 \times 10^{-11}$$

$[OH^-] = y = 1.5 \times 10^{-6}$, pOH = -log (1.5×10^{-6}) = 5.82 and pH = 14.00 - 5.82 = **8.18**

(b) Overall: $HCl(aq) + NH_3(aq) \rightarrow NH_4Cl(aq)$

Net ionic: $H_3O^+(aq) + NH_3(aq) \rightarrow NH_4^+(aq) + H_2O(\ell)$

$$? \text{ starting mmol } NH_3 = 100.0 \text{ mL } NH_3 \text{ soln} \times \frac{0.250 \text{ mmol } NH_3}{1 \text{ mL } NH_3 \text{ soln}} = 25.0 \text{ mmol } NH_3$$

$$? \text{ mmol } H_3O^+ \text{ added} = 50.0 \text{ mL HCl soln} \times \frac{0.100 \text{ mmol } H_3O^+}{1 \text{ mL HCl soln}} = 5.00 \text{ mmol } H_3O^+$$

$V_T = 100.0 \text{ mL} + 50.0 \text{ mL} = 150.0 \text{ mL}$

	NH_3	H_3O^+	NH_4^+
s, mmol	25.0	5.00	0.
R, mmol	−5.00	−5.00	+5.00
f, mmol	20.0	0.	5.00
C, *M*	0.133	0.	0.0333

An excess of the weak base NH_3 remains, and the system can be treated as a NH_4^+ / NH_3 buffer system.

$$K_a\ NH_4^+ = \frac{K_w}{K_b\ NH_3} \qquad K_a\ NH_4^+ = \frac{1.0 \times 10^{-14}}{1.8 \times 10^{-5}} = 5.6 \times 10^{-10} \quad \text{and} \quad pK_a = -\log(5.6 \times 10^{-10}) = 9.25$$

$$pH = pK_a + \log\left(\frac{C_b}{C_a}\right) \qquad pH = 9.25 + \log\left(\frac{0.133}{0.0333}\right) = \mathbf{9.85}$$

(c) Overall: $H_2SO_4(aq) + 2NaOH(aq) \rightarrow Na_2SO_4(aq) + 2H_2O(\ell)$
Net ionic: $H_3O^+(aq) + OH^-(aq) \rightarrow H_2O(\ell)$

$$?\ \text{mmol } H_3O^+ \text{ available} = 25.0 \text{ mL } H_2SO_4 \text{ soln} \times \frac{(2 \times 0.200 \text{ mmol } H_3O^+}{1 \text{ mL } H_2SO_4 \text{ soln}} = 10.0 \text{ mmol } H_3O^+$$

$$?\ \text{mmol } OH^- \text{ added} = 50.0 \text{ mL naOH soln} \times \frac{0.400 \text{ mmol } OH^-}{1 \text{ mL NaOH soln}} = 20.0 \text{ mmol } OH^-$$

$V_T = 25.0 \text{ mL} + 50.0 \text{ mL} = 75.0 \text{ mL}$

An excess of 10.0 mmol OH^- has been added, and this excess will determine the pH of the solution.

$$[OH^-] = \frac{10.0 \text{ mmol } OH^-}{75.0 \text{ mL}} = 0.133\ M,\ pOH = -\log(0.133) = 0.88 \text{ and } pH = 14.00 - 0.88 = \mathbf{13.12}$$

16.49 ***See Section 16.1 and Example 16.2.***

The NaOH that is added reacts with some of the CH_3CO_2H that is present to form the $CH_3CO_2^-$ ion that is needed for the CH_3CO_2H / $CH_3CO_2^-$ buffer system.

$$NaOH(s) + CH_3CO_2H(aq) \rightarrow NaCH_3CO_2(aq) + H_2O(\ell)$$

The number of moles of OH^- added is equal to the number of moles of CH_3CO_2H reacting and hence the number of moles of $NaCH_3CO_2$ formed. We can call this number of moles y and can solve the Henderson-Hasselbach equation for the value of y.

$$pH = pK_a + \log\left(\frac{n_b}{n_a}\right) \qquad pH = pK_a + \log\left(\frac{y}{\text{initial } n_a - y}\right)$$

The K_a for CH_3CO_2H is 1.8×10^{-5}, and its pK_a is $-\log(1.8 \times 10^{-5}) = 4.74$.

$$?\ \text{initial mol } CH_3CO_2H = 0.500 \text{ L } CH_3CO_2H \text{ soln} \times \frac{0.10 \text{ mol } CH_3CO_2H}{1 \text{ L } CH_3CO_2H \text{ soln}} = 0.050 \text{ mol } CH_3CO_2H$$

$$\text{Hence, } 5.00 = 4.74 + \log\left(\frac{y}{0.050 - y}\right) \text{ and } \log\left(\frac{y}{0.050 - y}\right) = 5.00 - 4.74 = 0.26$$

$$\frac{y}{0.050 - y} = 10^{0.26} = \text{inv log}(0.26) = 1.8,\ y = 0.090 - 1.8\ y, \text{ and } y = 0.032$$

$$?\ \text{g NaOH} = 0.032 \text{ mol NaOH} \times \frac{40.0 \text{ g NaOH}}{1 \text{ mol NaOH}} = \mathbf{1.3\ g\ NaOH}$$

16.50 ***See Section 16.1 and Example 16.2.***

The NaOH that is added reacts with some of the NH_4Cl that is present to form the NH_3 that is needed for the NH_4^+ / NH_3 buffer system.

$$NH_4^+(aq) + OH^-(aq) \rightarrow NH_3(aq) + H_2O(\ell)$$

The number of moles/L of NH_3 formed is equal to the number of moles/L NH_4^+ reacting. We can call this number of moles/L y and can solve the Henderson-Hasselbach equation for the value of y.

$$pH = pK_a + \log\left(\frac{C_b}{C_a}\right) \qquad pH = pK_a + \log\left(\frac{y}{\text{initial } C_a - y}\right)$$

$$K_a\ NH_4^+ = \frac{K_w}{K_b\ NH_3} \qquad K_a\ NH_4^+ = \frac{1.0 \times 10^{-14}}{1.8 \times 10^{-5}} = 5.6 \times 10^{-10} \quad \text{and} \quad pK_a = -\log(5.6 \times 10^{-10}) = 9.25$$

Hence, $9.00 = 9.25 + \log\left(\frac{y}{0.100 - y}\right)$ and $\log\left(\frac{y}{0.100 - y}\right) = 9.00 - 9.25 = -0.25$

$\frac{y}{0.100 - y} = 10^{-0.26} = \text{inv log }(-0.26) = 0.55$, $y = 0.055 - 0.55\,y$, and $y = 0.035$

$[NH_4^+] = 0.100 - y = 0.100 - 0.035 =$ **0.065 *M***

16.51 *See Section 16.6 and Table 16.4.*

(a) methyl red (b) methyl red (c) methyl orange

16.52 *See Section 16.6 and Table 16.4.*

(a) bromthymol blue (b) phenolphthalein (c) phenolphthalein

16.53 *See Sections 16.5, 16.6, Table 16.4, and Example 16.6.*

Overall: $HLac(aq) + NaOH(aq) \rightarrow NaLac(aq) + H_2O(\ell)$

Net ionic: $HLac(aq) + OH^-(aq) \rightarrow Lac^-(aq) + H_2O(\ell)$

$? \text{ starting mmol HLac} = 100 \text{ mL HLac soln} \times \frac{0.100 \text{ mmol HLac}}{1 \text{ mL HLac soln}} = 10.0 \text{ mmol HLac}$

$? \text{ mmol } OH^- \text{ required to reach equivalence point} = 10.0 \text{ mmol HLac} \times \frac{1 \text{ mmol } OH^-}{1 \text{ mmol HLac}} = 10.0 \text{ mmol } OH^-$

$? \text{ mL NaOH required to reach equivalence point} = 10.0 \text{ mmol } OH^- \times \frac{1 \text{ mL NaOH soln}}{0.500 \text{ mmol } OH^-} = 20.0 \text{ mL NaOH soln}$

$V_T = 100 \text{ mL} + 20.0 \text{ mL} = 120 \text{ mL}$

	HLac	OH^-	Lac^-
s, mmol	10.0	10.0	0.
R, mmol	−10.0	−10.0	+10.0
f, mmol	0.	0.	10.0
C, *M*	0.	0.	0.0833

The pH at the equivalence point is due to the reaction of Lac^- with H_2O:

$$Lac^-(aq) + H_2O(\ell) \rightleftarrows HLac(aq) + OH^-(aq)$$

$K_b\ Lac^- = \frac{K_w}{K_a\ HLac}$ $\quad K_b\ Lac^- = \frac{1.0 \times 10^{-14}}{8.4 \times 10^{-4}} = 1.2 \times 10^{-11}$

Conc, *M*	Lac^-	HLac	OH^-
initial	0.0833	0.	0.
Change	−y	+y	+y
equilibrium	0.0833−y	y	y

$K_b\ Lac^- = \frac{[HLac][OH^-]}{[Lac^-]}$ $\quad K_b\ Lac^- = \frac{(y)(y)}{(0.0833-y)} \cong \frac{(y)(y)}{0.0833} = 1.2 \times 10^{-11}$

$[OH^-] = y = 1.0 \times 10^{-6}$, $pOH = -\log(1.0 \times 10^{-6}) = 6.00$ and **pH** $= 14.00 - 6.00 =$ **8.00**

Approximation check: $\left(\dfrac{1.0 \times 10^{-6}}{0.0833}\right) \times 100\% = 1.2 \times 10^{-3}\ \%$, so the approximation is valid.

According to Table 16.4, phenolphthalein has a pK_{In} of 8.7 and pH range of 8.3 - 10.0. Hence, **phenolphthalein** can be used for titrating lactic acid with sodium hydroxide.

16.54 ***See Sections 16.5, 16.6, Table 16.4, and Example 16.6.***

Overall: $ClCH_2CH_2COOH(aq) + KOH(aq) \rightarrow ClCH_2CH_2COOK(aq) + H_2O(\ell)$

Net ionic: $ClCH_2CH_2COOH(aq) + OH^-(aq) \rightarrow ClCH_2CH_2COO^-(aq) + H_2O(\ell)$

? starting mmol $ClCH_2CH_2COOH$(HA) = 10.00 mL HA soln $\times \dfrac{0.100 \text{ mmol HA}}{1 \text{ mL HA soln}}$ = 1.00 mmol HA

? mmol OH^- required to reach equivalence point = 1.00 mmol HA $\times \dfrac{1 \text{ mmol OH}^-}{1 \text{ mmol HA}}$ = 1.00 mmol OH^-

? mL KOH required to reach equivalence point = 1.00 mmol OH^- $\times \dfrac{1 \text{ mL KOH soln}}{0.100 \text{ mmol OH}^-}$ = 10.0 mL KOH soln

V_T = 10.0 mL + 10.0 mL = 20.0 mL

	HA	OH^-	A^-
s, mmol	1.00	1.00	0.
R, mmol	−1.00	−1.00	+1.00
f, mmol	0.	0.	1.00
C, *M*	0.	0.	0.0500

The pH at the equivalence point is due to the reaction of $ClCH_2CH_2COO^-(A^-)$ with H_2O:

$$A^-(aq) + H_2O(\ell) \rightleftarrows HA(aq) + OH^-(aq)$$

$K_b\ A^- = \dfrac{K_w}{K_a HA}$ $\qquad K_b\ A^- = \dfrac{1.00 \times 10^{-14}}{7.94 \times 10^{-5}} = 1.26 \times 10^{-10}$

Conc, *M*	A^-	HA	OH^-
initial	0.0500	0.	0.
Change	−y	+y	+y
equilibrium	0.0500 − y	y	y

$K_b\ A^- = \dfrac{[HA][OH^-]}{[A^-]}$ $\qquad K_b\ A^- = \dfrac{(y)(y)}{0.0500 - y} \cong \dfrac{(y)(y)}{0.0500} = 1.26 \times 10^{-10}$

$[OH^-] = y = 2.5 \times 10^{-6}$, pOH = -log (2.5×10^{-6}) = 5.60 and pH = 14.00 - 5.60 = **8.40**

Approximation check: $\left(\dfrac{2.5 \times 10^{-6}}{0.0500}\right) \times 100\% = 0.0050\%$, so the approximation is valid.

According to Table 16.4, phenolphthalein has a pK_{In} of 8.7 and pH range of 8.3 - 10.0. Hence, **phenolphthalein** cna be used for titrating chloroproplonic acid with potassium hydroxide.

16.55 ***See Sections 16.5, 16.6, Table 16.4, and Example 16.6.***

Overall: $NH_3(aq) + HCl(aq) \rightarrow NH_4Cl(aq)$

Net ionic: $NH_3(aq) + H_3O^+(aq) \rightarrow NH_4^+(aq) + H_2O(\ell)$

? starting mmol NH_3 = 25.0 mL NH_3 soln $\times \dfrac{1.44 \text{ mmol NH}_3}{1 \text{ mL NH}_3 \text{ soln}}$ = 36.0 mmol NH_3

$$? \text{ mmol } H_3O^+ \text{ required to reach equivalence point} = 36.0 \text{ mmol } NH_3 \times \frac{1 \text{ mmol } H_3O^+}{1 \text{ mmol } NH_3} = 36.0 \text{ mmol } H_3O^+$$

$$? \text{ mL HCl required to reach equivalence point} = 36.0 \text{ mmol } H_3O^+ \times \frac{1 \text{ mL HCl soln}}{1.50 \text{ mmol } H_3O^+} = 24.0 \text{ mL HCl soln}$$

$V_T = 25.0 \text{ ml} + 24.0 \text{ ml} = 49.0 \text{ mL}$

	NH_3	H_3O^+	NH_4^+
s, mmol	36.0	36.0	0.
R, mmol	–36.0	–36.0	+36.0
f, mmol	0.	0.	36.0
C, *M*	0.	0.	0.735

The pH at the equivalence point is due to the reaction of NH_4^+ with H_2O:

$$NH_4^+(aq) + H_2O(\ell) \rightleftarrows NH_3(aq) + H_3O^+(aq)$$

$$K_a\ NH_4^+ = \frac{K_w}{K_b\ NH_3} \qquad K_a\ NH_4^+ = \frac{1.0 \times 10^{-14}}{1.8 \times 10^{-5}} = 5.6 \times 10^{-10}$$

Conc, *M*	NH_4^+	NH_3	H_3O^+
initial	0.735	0.	0.
Change	–y	+y	+y
equilibrium	0.735–y	y	y

$$K_a\ NH_4^+ = \frac{[NH_3][H_3O^+]}{[NH_4^+]} \qquad K_a\ NH_4^+ = \frac{(y)(y)}{(0.735-y)} \cong \frac{(y)(y)}{0.735} = 5.6 \times 10^{-10}$$

$[H_3O^+] = y = 2.0 \times 10^{-5}$ and **$pH = -\log(2.0 \times 10^{-5}) = 4.69$**

Approximation check: $\left(\frac{2.0 \times 10^{-5}}{0.735}\right) \times 100\% = 2.7 \times 10^{-3}$ %, so the approximation is valid.

According to Table 16.4, methyl red has $pK_{In} = 5.00$ and a pH range of 4.2 - 6.3. Hence, **methyl red can be used** for titrating NH_3 with HCl.

16.56 ***See Sections 16.5, 16.6, Table 16.4, Example 16.6, and Appendix F.***

Overall: $C_2H_5NH_2(aq) + HNO_3(aq) \rightarrow C_2H_5NH_3NO_3(aq)$
Net ionic: $C_2H_5NH_2(aq) + H_3O^+(aq) \rightarrow C_2H_5NH_3^+(aq) + H_2O(\ell)$

$$? \text{ starting mmol } C_2H_5NH_2(aq) = 50.00 \text{ mL } C_2H_5NH_2(aq) \text{ soln} \times \frac{0.0500 \text{ mmol } C_2H_5NH_2}{1 \text{ mL } C_2H_5NH_2 \text{ soln}} = 2.50 \text{ mmol } C_2H_5NH_2$$

$$? \text{ mmol } H_3O^+ \text{ required to reach equivalence point} = 2.50 \text{ mmol } C_2H_5NH_2 \times \frac{1 \text{ mmol } H_3O^+}{1 \text{ mmol } C_2H_5NH_2}$$
$$= 2.50 \text{ mmol } H_3O^+$$

$$? \text{ mL } HNO_3 \text{ required to reach equivalence point} = 2.50 \text{ mmol } H_3O^+ \times \frac{1 \text{ mL } HNO_3 \text{ soln}}{0.100 \text{ mmol } H_3O^+} = 25.0 \text{ mL } HNO_3 \text{ soln}$$

$V_T = 50.00 \text{ mL} + 25.0 \text{ mL} = 75.0 \text{ mL}$

	$C_2H_5NH_2$	H_3O^+	$C_2H_5NH_3^+$
s, mmol	2.50	2.50	0.
R, mmol	−2.50	−2.50	+2.50
f, mmol	0.	0.	2.50
C, M	0.	0.	0.0333

The pH at the equivalence point is due to the reaction of $C_2H_5NH_3^+$ with H_2O:

$$C_2H_5NH_3^+(aq) + H_2O(\ell) \quad C_2H_5NH_2(aq) + H_3O^+(aq)$$

$$K_a\ C_2H_5NH_3^+ = \frac{K_w}{K_b\ C_2H_5NH_2} \qquad K_a\ C_2H_5NH_3^+ = \frac{1.0\times10^{-14}}{6.5\times10^{-4}} = 1.5 \times 10^{-11}$$

Conc, M	$C_2H_5NH_3^+$	$C_2H_5NH_2$	H_3O^+
initial	0.0333	0.	0.
Change	−y	+y	+y
equilibrium	0.0333 − y	y	y

$$K_a\ C_2H_5NH_3^+ = \frac{[C_2H_5NH_2][H_3O^+]}{[C_2H_5NH_3^+]} \qquad K_a = \frac{(y)(y)}{(0.0333 - y)} = \frac{(y)(y)}{0.0333} = 1.5 \times 10^{-11}$$

$[H_3O^+] = y = 7.1 \times 10^{-7}$ and pH = -log (7.1×10^{-7}) = **6.14**

Approximation check: $\left(\frac{7.1\times10^{-7}}{0.0333}\right)$ x 100% = 0.0021%, so the approximation is valid.

According to Table 16.4, methyl red has pK_{In} = 5.00 and a pH range of 4.2 to 6.3. Hence, **methyl red can be used** for titrating $C_2H_5NH_2$ with HNO_3.

16.57 ***See Section 16.7.***

(a)

$$H_2C_2O_4(aq) + H_2O(\ell) \rightleftarrows H_3O^+(aq) + HC_2O_4^-(aq) \qquad K_{a1} = \frac{[H_3O^+][HC_2O_4^-]}{[H_2C_2O_4]}$$

$$HC_2O_4^-(aq) + H_2O(\ell) \rightleftarrows H_3O^+(aq) + C_2O_4^{2-}(aq) \qquad K_{a2} = \frac{[H_3O^+][C_2O_4^{2-}]}{[HC_2O_4^-]}$$

(b)

$$H_2SO_3(aq) + H_2O(\ell) \rightleftarrows H_3O^+(aq) + HSO_3^-(aq) \qquad K_{a1} = \frac{[H_3O^+][HSO_3^-]}{[H_2SO_3]}$$

$$HSO_3^-(aq) + H_2O(\ell) \rightleftarrows H_3O^+(aq) + SO_3^{2-}(aq) \qquad K_{a2} = \frac{[H_3O^+][SO_3^{2-}]}{[HSO_3^-]}$$

16.58 ***See Section 16.7.***

(a)

$$H_2C_4H_4O_6(aq) + H_2O(\ell) \rightleftarrows H_3O^+(aq) + HC_4H_4O_6^-\ aq) \qquad K_{a1} = \frac{[H_3O^+][HC_4H_4O_6^-]}{[H_2C_4H_4O_6]}$$

$$HC_4H_4O_6^-(aq) + H_2O(\ell) \rightleftarrows H_3O^+(aq) + C_4H_4O_6^{2-}(aq) \qquad K_{a2} = \frac{[H_3O^+][C_4H_4O_6^{2-}]}{[HC_4H_4O_6^-]}$$

(b) $H_2C_4H_4O_5(aq) + H_2O(\ell) \rightleftarrows H_3O^+(aq) + HC_4H_4O_5^-(aq)$ $K_{a1} = \frac{[H_3O^+][HC_4H_4O_5^-]}{[H_2C_4H_4O_5]}$

$HC_4H_4O_5^-(aq) + H_2O(\ell) \rightleftarrows H_3O^+(aq) + C_4H_4O_5^{2-}(aq)$ $K_{a2} = \frac{[H_3O^+][C_4H_4O_5^{2-}]}{[HC_4H_4O_5^-]}$

16.59 ***See Section 16.7 and Example 16.7.***

$H_2C_6H_6O_6(aq) + H_2O(\ell) \rightleftarrows H_3O^+(aq) + HC_6H_6O_6^-(aq)$ $K_{a1} = 8.0 \times 10^{-5}$

Conc, *M*	$H_2C_6H_6O_6$	H_3O^+	$HC_6H_6O_6^-$
initial	0.010	0.	0.
Change	−y	+y	+y
equilibrium	0.0100 − y	y	y

$K_{a1}\ H_2C_6H_6O_6 = \frac{[H_3O^+][HC_6H_6O_6^-]}{[H_2C_6H_6O_6]}$ $K_{a1} = \frac{(y)(y)}{(0.010 - y)} \cong \frac{(y)(y)}{0.010} = 8.0 \times 10^{-5}$

$[H_3O^+] = [HC_6H_6O_6^-] = y = 8.9 \times 10^{-4}$, $[H_2C_6H_6O_6] = 0.010 = y = 0.010 - 8.9 \times 10^{-4} = .009$

Conc, *M*	$HC_6H_6O_6^-$	H_3O^+	$C_6H_6O_6^{2-}$
initial	8.9×10^{-4}	8.9×10^{-4}.	0.
Change	−z	$8.9 \times 10^{-4} + z$	+z
equilibrium	$8.9 \times 10^{-4} - z$	$8.9 \times 10^{-4} + z$	z

$K_{a2} = \frac{[H_3O^+][C_6H_6O_6^{2-}]}{[HC_6H_6O_6^-]}$ $K_{a2} = \frac{(8.9 \times 10^{-4} + z)(z)}{(8.9 \times 10^{-4} - z)} \cong z = 1.6 \times 10^{-12}$

$[C_6H_6O_6^{2-}] = z = 1.6 \times 10^{-12}$

$[H_3O^+] = 8.9 \times 10^{-4} + z = 8.9 \times 10^{-4} + 1.6 \times 10^{-12} = 8.9 \times 10^{-4}$ and pH = $-\log(8.9 \times 10^{-4})$ = **3.05**

Approximation check: $\left(\frac{1.6 \times 10^{-12}}{8.9 \times 10^{-4}}\right) \times 100\% = 1.8 \times 10^{-7}\%$, so the approximation is valid.

16.60 ***See Section 16.7, Example 16.7, and the Solution for Exercise 16.59.***

The pH for a solution of a weak polyprotic acid having K_{a1} and K_{a2} values differing by a factor of 1000 or more is due to the $[H_3O^+]$ produced by the first ionization of the acid. For H_3PO_4,

$H_3PO_4(aq) + H_2O(\ell) \rightleftarrows H_3O^+(aq) + H_2PO_4^-(aq)$ $K_{a1} = 7.5 \times 10^{-3}$

Conc, *M*	H_3PO_4	H_3O^+	$H_2PO_4^-$
initial	0.050	0.	0.
Change	−y	+y	+y
equilibrium	0.050 − y	y	y

$K_{a1} = \frac{[H_3O^+][H_2PO_4^-]}{[H_3PO_4]}$ $K_{a1} = \frac{(y)(y)}{(0.050 - y)} \cong \frac{(y)(y)}{0.050} = 7.5 \times 10^{-3}$

$[H_3O^+(aq)] = y = 1.9 \times 10^{-2}$

Approximation check: $\left(\dfrac{1.9 \times 10^{-2}}{0.050}\right)$ x 100% = 38%, so the approximation is not valid. The quadratic formula or successive approximation method must be used to solve the equilibrium expression.

$K_{a1} = \dfrac{(y)(y)}{(0.050 - y)} = 7.5 \times 10^{-3}$ yields $y^2 + 7.5 \times 10^{-3}y - 3.8 \times 10^{-4} = 0$.

Substituting into the quadratic formula gives: $y = \dfrac{-7.5 \times 10^{-3} \pm \sqrt{(7.5 \times 10^{-3})^2 - 4(1)(-3.8 \times 10^{-4})}}{2(1)}$

and $y = \dfrac{-7.5 \times 10^{-3} \pm 4.0 \times 10^{-2}}{2} = 0.024$ or -0.024

The only reasonable value of y is 0.024, since -0.024 predicts negative concentrations for H_3O^+ and $H_2PO_4^-$.
Hence, $[H_3O^+] = y = 0.024$ and pH = -log (0.024) = **1.62**

16.61 ***See Section 16.7, Tables 16.5, 16.6, 16.7, and Example 16.8.***

(a) The reaction of $HC_2O_4^-$ as an acid is:

$$HC_2O_4^-(aq) + H_2O(\ell) \rightleftarrows C_2O_4^{2-}(aq) + H_3O^+(aq)$$

$K_a = K_{a2}\ H_2C_2O_4 = 6.4 \times 10^{-5}$

The reaction of $HC_2O_4^-$ as a base is:

$$HC_2O_4^-(aq) + H_2O(\ell) \rightleftarrows H_2C_2O_4(aq) + OH^-(aq)$$

$$K_{b2} = \frac{K_w}{K_{a1}\ H_2C_2O_4} = \frac{1.0 \times 10^{-14}}{5.9 \times 10^{-2}} = 1.7 \times 10^{-13}$$

Since $K_a\ HC_2O_4^- > K_b\ HC_2O_4^-$, the solution will be **acidic**.

(b) The reaction of $HC_3H_2O_4^-$ as an acid is:

$$HC_3H_2O_4^-(aq) + H_2O(\ell) \rightleftarrows C_3H_2O_4^{2-}(aq) + H_3O^+(aq)$$

$K_a = K_{a2}\ H_2C_3H_2O_4 = 2.1 \times 10^{-6}$

The reaction of $HC_3H_2O_4^-$ as a base is:

$$HC_3H_2O_4^-(aq) + H_2O(\ell) \rightleftarrows H_2C_3H_2O_4(aq) + OH^-(aq)$$

$$K_{b2} = \frac{K_w}{K_{a1}\ H_2C_3H_2O_4} = \frac{1.0 \times 10^{-14}}{1.6 \times 10^{-2}} = 6.2 \times 10^{-13}$$

Since $K_a\ HC_3H_2O_4^- > K_b\ HC_3H_2O_4^-$, the solution will be **acidic**.

16.62 ***See Section 16.7, Tables 16.5, 16.6, 16.7, and Example 16.8.***

(a) The reaction of $H_2C_6H_5O_7^-$ as an acid is:

$$H_2C_6H_5O_7^-(aq) + H_2O(\ell) \rightleftarrows HC_6H_5O_7^{2-}(aq) + H_3O^+(aq)$$

$K_a = K_{a2}\ H_3C_6H_5O_7 = 1.7 \times 10^{-5}$

The reaction of $H_2C_6H_5O_7^-$ as a base is:

$$H_2C_6H_5O_7^-(aq) + H_2O(\ell) \rightleftarrows H_3C_6H_5O_7(aq) + OH^-(aq)$$

$$k_{b2} = \frac{K_w}{K_{a1} H_3C_6H_5O_7} = \frac{1.0 \times 10^{-14}}{7.4 \times 10^{-4}} = 1.4 \times 10^{-11}$$

Since $K_a\ H_2C_6H_5O_7^- > K_b\ H_2C_6H_5O_7^-$, the solution will be **acidic**.

(b) The reaction of $HC_6H_5O_7^{2-}$ as an acid is:

$$HC_6H_5O_7^{2-}(aq) + H_2(\ell) \rightleftarrows C_6H_5O_7^{3-}(aq) + H_3O^+(aq)$$

$K_a = K_{a3}\ H_3C_6H_5O_7 = 4.0 \times 10^{-7}$

The reaction of $HC_6H_5O_7^{2-}$ as a base is:

$$HC_6H_5O_7^{2-}(aq) + H_2O(\ell) \rightleftarrows H_2C_6H_5O_7^{2-}(aq) + OH^-(aq)$$

$$K_{b2} = \frac{K_w}{K_{a2}H_3C_6H_5O_7} = \frac{1.0 \times 10^{-14}}{1.7 \times 10^{-5}} = 5.9 \times 10^{-10}$$

Since $K_a\ HC_6H_5O_7^{2-} > K_b\ HC_6H_5O_7^{2-}$, the solution will be **acidic**.

16.63 ***See Section 16.7, Tables 16.5, 16.6, 16.7, and Example 16.8.***

(a) The reaction of $H_2PO_4^-$ as an acid is:

$$H_2PO_4^-(aq) + H_2O(\ell) \rightleftarrows HPO_4^{2-}(aq) + H_3O^+(aq)$$

$K_a = K_{a2}\ H_3PO_4 = 6.2 \times 10^{-8}$

The reaction of $H_2PO_4^-$ as a base is:

$$H_2PO_4^-(aq) + H_2O(\ell) \rightleftarrows H_3PO_4(aq) + OH^-(aq)$$

$$K_{b3} = \frac{K_w}{K_{a1}\ H_3PO_4} = \frac{1.0 \times 10^{-14}}{7.5 \times 10^{-3}} = 1.3 \times 10^{-12}$$

Since $K_a\ H_2PO_4^-$ is greater than $K_b\ H_2PO_4^-$, the solution will be **acidic**.

(b) The reaction of HCO_3^- as an acid is:

$$HCO_3^-(aq) + H_2O(\ell) \rightleftarrows CO_3^{2-}(aq) + H_3O^+(aq)$$

$K_a = K_{a2}$ H2CO3 $= 5.6 \times 10^{-11}$

The reaction of HCO_3^- as a base is:

$$HCO_3^-(aq) + H_2O(\ell) \rightleftarrows H_2CO_3(aq) + OH^-(aq)$$

$$K_{b2} = \frac{K_w}{K_{a1}\ H_2CO_3} \qquad K_{b1} = \frac{1.0 \times 10^{-14}}{4.3 \times 10^{-7}} = 2.3 \times 10^{-8}$$

Since $K_b\ HCO_3^- > K_a\ HCO_3^-$, the solution will be **basic**.

16.64 ***See Section 16.7, Tables 16.5, 16.6, 16.7, and Example 16.8.***

(a) The reaction of HPO_4^{2-} as an acid is:

$$HPO_4^{2-}(aq) + H_2O(\ell) \rightleftarrows PO_4^{3-}(aq) + H_3O^+(aq)$$

$K_a = K_{a3}$ H_3PO4 $= 2.2 \times 10^{-13}$

The reaction of HPO_4^{2-} as a base is:

$$HPO_4^{2-}(aq) + H_2O(\ell) \rightleftarrows H_2PO_4^{2-}(aq) + OH^-(aq)$$

$$K_{b2} = \frac{K_w}{K_{a2}H_3PO_4} = \frac{1.0 \times 10^{-14}}{6.2 \times 10^{-8}} = 1.6 \times 10^{-7}$$

Since $K_b\ HPO_4^{2-} > K_a\ HPO_4^{2-}$, the solution will be **basic**.

(b) The reaction of $HC_4H_4O_6^-$ as an acid is:

$$HC_4H_4O_6^-(aq) + H_2O(\ell) \rightleftarrows C_4H_4O_6^{2-}(aq) + H_3O^+(aq)$$

$K_a = K_{a2}\ H_2C_4H_4O_6 = 4.6 \times 10^{-5}$

The reaction of $HC_4H_4O_6^-$ as a base is:

$$HC_4H_4O_6^-(aq) + H_2O(\ell) \rightleftarrows H_2C_4H_4O_6(aq) + OH^-(aq)$$

$$K_{b2} = \frac{K_w}{K_{a1}H_2C_4H_4O_6} = \frac{1.0 \times 10^{-14}}{1.0 \times 10^{-3}} = 1.0 \times 10^{-11}$$

Since K_a $HC_4H_4O_6^-$ > K_b $HC_4H_4O_6^-$, the solution will be **acidic.**

16.65 *See Section 16.8.*

(a) The acetate ion, CH_3COO^-, is the conjugate base of the weak acid acetic acid, CH_3COOH, and therefore acts as a weak base. The acetate ion will therefore react with the H_3O^+ from HCl. The pertinent reactions are:

$$Ca(CH_3COO)_2(s) \rightleftarrows Ca^{2+}(aq) + 2CH_3COO^-(aq)$$

$$2CH_3COO^-(aq) + 2H_3O^+(aq) \rightleftarrows 2CH_3COOH(aq) + 2H_2O(l)$$

$$Ca(CH_3COO)_2(s) + 2H_3O^+(aq) \rightleftarrows Ca^{2+}(aq) + 2CH_3COOH(aq) + 2H_2O(\ell)$$

Hence, HCl **increases** the solubility of calcium acetate.

(b) The fluoride ion, F^-, is the conjugate base of the weak acid hydrofluoric acid, HF, and therfore acts as a weak base. The fluoride ion will therefore react with the H_3O^+ from HCl. The pertinent reactions are:

$$MgF_2(s) \rightleftarrows Mg^{2+}(aq) + 2F^-(aq)$$

$$2F^-(aq) + 2H_3O^+(aq) \rightleftarrows 2HF(aq) + 2H_2O(\ell)$$

$$MgF_2(s) + 2H_3O^+(aq) \rightleftarrows Mg^{2+}(aq) + 2HF(aq) + 2H_2O(\ell)$$

Hence, HCl **increases** the solubility of MgF_2.

16.66 *See Section 16.8.*

(a) NH_3 forms a complex ion with Ag^+, $Ag(NH_3)_2^+$. The pertinent reactions are:

$$AgCl(s) \rightleftarrows Ag^+(aq) + Cl^-(aq)$$

$$Ag^+(aq) + 2NH_3(aq) \rightleftarrows Ag(NH_3)_2^+(aq)$$

$$AgCl(s) + 2NH_3(aq) \rightleftarrows Ag(NH_3)_2^+(aq) + Cl^-(aq)$$

Hence, NH_3 **increases** the solutility of AgCl(s).

(b) The Pb^{2+} from the $Pb(NO_3)_2$ causes the $PbCl_2$ equilibrium to shift to the left in accord with the Principle of LeChatelier. $PbCl_2(s) \rightleftarrows Pb^{2+}(aq) + 2Cl^-(aq)$

Hence, $Pb(NO_3)_2$ **decreases** the solubility of $PbCl_2$.

16.67 *See Section 16.8.*

(a) The oxalate ion is related to the weak acid oxalic acid, $H_2C_2O_4$, and therefore acts as a weak base. The oxalate ion therefore reacts with the H_3O^+ from HCl. The pertinent reactions are:

$$CaC_2CO_4(s) \rightleftarrows Ca^{2+}(aq) + C_2O_4^{2-}(aq)$$

$$C_2O_4^{2-}(aq) + 2H_3O^+(aq) \rightleftarrows H_2C_2O_4(aq) + 2H_2O(\ell)$$

$$CaC_2O_4(s) + 2H_3O^+(aq) \rightleftarrows Ca^{2+}(aq) + H_2C_2O_4(aq) + 2H_2O(\ell)$$

Hence, HCl **increases** the solubility of CaC_2O_4.

(b) NH_3 forms a complex ion with Cu^{2+}, $Cu(NH_3)_4^{2+}$. The pertinent reactions are:

$$CuSO_4(s) \rightleftarrows Cu^{2+}(aq) + SO_4^{2-}(aq)$$

$$Cu^{2+}(aq) + 4NH_3(aq) \rightleftarrows Cu(NH_3)_4^{2+}(aq)$$

$$CuSO_4(s) + 4NH_3(aq) \rightleftarrows Cu(NH_3)_4^{2+}(aq) + SO_4^{2-}(aq)$$

Hence, NH_3 **increases** the solubility of $CuSO_4$.

16.68 ***See Section 16.8.***

(a) The OH^- from the NaOH causes the $Zn(OH)_2$ equilibrium to shift to the left in accord with the Principle of LeChatelier. $Zn(OH)_2(s) \rightleftarrows Zn^{2+}(aq) + 2OH^-(aq)$

Hence, NaOH **decreases** the solubility of $Zn(OH)_2$, provided the concentration of the hydroxide remains too low to form $Zn(OH)_4^{2-}$ as a soluble complex ion. This possibility exists because $Zn(OH)_2$ is amphoteric.

(b) The phosphate ion is related to the weak acid H_3PO_4 and therefore acts as base in a reaction with the H_3O^+ from HNO_3.

$$Mg_3(PO_4)_2(s) \rightleftarrows 3Mg^{2+}(aq) + 2PO_4^{3-}(aq)$$

$$2PO_4^{3-}(aq) + 6H_3O^+(aq) \rightleftarrows 2H_3PO_4(aq) + 6H_2O(\ell)$$

$$Mg_3(PO_4)_2(s) + 6H_3O^+(\ell) \rightleftarrows 3Mg^{2+}(aq) + 2H_3PO_4(aq) + 6H_2O(\ell)$$

Hence, HNO_3 **increases** the solubility of $Mg_3(PO_4)_2(s)$.

16.69 ***See Sections 4.4. 16.2, and Examples 4.12, 16.4.***

$$H_2SO_4(aq) + 2NaOH(aq) \rightarrow Na_2SO_4(aq) + 2H_2O(\ell)$$

Strategy: *L NaOH soln → mol NaOH → mol* H_2SO_4 *→ M* H_2SO_4

$$?\ \text{mol } H_2SO_4 = 0.03177\ \text{L NaOH soln} \times \frac{0.102\ \text{mol NaOH}}{1\ \text{L NaOH soln}} \times \frac{1\ \text{mol } H_2SO_4}{2\ \text{mol NaOH}} = 0.00162\ \text{mol } H_2SO_4$$

$$?\ M\ H_2SO_4 = \frac{0.00162\ \text{mol } H_2SO_4}{0.01000\ \text{L } H_2SO_4\ \text{soln}} = \mathbf{0.162\ M}$$

16.70 ***See Sections 4.4. 16.2, and Examples 4.12, 16.4.***

$$HCl(aq) + NaOH(aq) \rightarrow NaCl(s) + H_2O(\ell)$$

Strategy: L NaOH soln → mol NaOH → mol HCl → M HCl soln

$$?\ \text{mol HCl} = 0.01834\ \text{L NaOH soln} \times \frac{0.0982\ \text{mol NaOH}}{1\ \text{L NaOH soln}} \times \frac{1\ \text{mol HCl}}{1\ \text{mol NaOH}} = 0.00180\ \text{mol HCl}$$

$$?\ M\ \text{HCl} = \frac{0.00180\ \text{mol HCl}}{0.02000\ \text{L HCl soln}} = \mathbf{0.0900\ M\ HCl}$$

16.71 ***See Section 16.1 and Example 16.1.***

(a) $pH = pK_a + \log\left(\frac{C_b}{C_a}\right)$ The K_a for CH_3CO_2H is 1.8×10^{-5}, and its pK_a is $-\log(1.8 \times 10^{-5}) = 4.74$.

The pH of the initial buffer solution will be the pH of the buffered biological system because the concentrations of $CH_3CO_2^-$ and CH_3CO_2H will be changed by the same factor when the buffer is added to the biological system.

Hence, $pH = 4.74 + \log\left(\frac{0.020}{0.010}\right) = \mathbf{5.04}$

(b) The buffer could be prepared using 0.020 *M* potassium acetate in place of 0.0200 *M* sodium acetate since each salt gives one mole of acetate ion per mole of salt.

(c) The Henderson-Hasselbach equation assumes the concentrations of the acid and its conjugate base are equal to their analytical (starting) concentrations and are not changed by ionization of the acid. This assumption is only valid when the analytical concentrations are relatively large and is not valid for dilute buffer solutions. Knowing the total volume of the system would allow constructing an ice table, calculating the degree of ionization, and checking the validity of the assumption that y is small. Measuring the pH of the resulting solution would also give an indication of the applicability of the Henderson-Hasselbach equation to the system of interest.

16.72 ***Solutions for Spreadhseet Exercises are on the Reger, Goode and Mercer Spreadsheet Solutions Diskette available from Saunders College Publishing.***

16.73 ***See Section 16.7 and Table 16.4.***

(a) $SO_4^{2-}(aq) + H_2O(\ell) \rightleftarrows HSO_4^-(aq) + OH^-(aq)$

$$K_b SO_4^{2-} = \frac{[HSO_4^-][OH^-]}{[SO_4^{2-}]} = \frac{K_w}{K_{a2}\ H_2SO_4} \qquad K_b SO_4^{2-} = \frac{1.0 \times 10^{-14}}{1.2 \times 10^{-2}} = \mathbf{8.3 \times 10^{-13}}$$

(b) $C_6H_5O_7^{3-}(aq) + H_2O(\ell) \rightleftarrows HC_6H_5O_7^{2-}(aq) + OH^-(aq)$

$$K_b C_6H_5O_7^{3-} = \frac{[HC_6H_5O_7^{2-}][OH^-]}{[C_6H_5O_7^{3-}]} = \frac{K_w}{K_{a3}\ H_3C_6H_5O_7^{2-}} \qquad K_b C_6H_5O_7^{3-} = \frac{1.0 \times 10^{-14}}{4.0 \times 10^{-7}} = \mathbf{2.5 \times 10^{-8}}$$

16.74 ***See Section 16.7 and Table 16.4.***

(a) $C_3H_2O_4^{2-}(aq) + H_2O(\ell) \rightleftarrows HC_3H_2O_4^-(aq) + OH^-(aq)$

$$K_b\ C_3H_2O_4^{2-} = \frac{K_w}{K_a HC_3H_2O_4^-} = \frac{K_w}{K_{a2}H_2C_3H_2O_4} = \frac{1.0 \times 10^{-14}}{2.1 \times 10^{-6}} = \mathbf{4.8 \times 10^{-9}}$$

(b) $CO_3^{2-}(aq) + H_2O(\ell) \rightleftarrows HCO_3^-(aq) + OH^-(aq)$

$$K_b\ CO_3^{2-} = \frac{K_w}{K_a HCO_3^-} = \frac{K_w}{K_{a2}H_2CO_3} = \frac{1.0 \times 10^{-14}}{5.6 \times 10^{-11}} = \mathbf{1.8 \times 10^{-4}}$$

16.75 ***See Section 16.6.***

For $HIn(aq) + H_2O(\ell) \rightleftarrows H_3O^+(aq) + In^-(aq)$,

$$K_{In} = \frac{[H_3O^+][In^-]}{[HIn]}, \quad -\log K_{In} = -\log[H_3O^+] - \log\left(\frac{[In^-]}{[HIn]}\right), \quad \text{and} \quad pK_{In} = pH - \log\left(\frac{[In^-]}{[HIn]}\right)$$

(a) For pH to be one unit higher than pK_{In} , $\frac{[In^-]}{[HIn]} = \frac{10}{1}$.

$$pK_{In} = pH - \log\left(\frac{[In^-]}{[HIn]}\right) \quad pK_{In} = pH - \log\left(\frac{10}{1}\right) \quad pK_{In} = pH - 1$$

(b) For pH to be one unit lower than pK_{In} , $\frac{[In^-]}{[HIn]} = \frac{1}{10}$.

$$pK_{In} = pH - \log\left(\frac{[In^-]}{[HIn]}\right) \quad pK_{In} = pH - \log\left(\frac{1}{10}\right) \quad pK_{In} = pH + 1$$

16.76 ***See Section 16.6.***

For $HIn(aq) + H_2O(\ell) \rightleftarrows H_3O^+(aq) + In^-(aq)$,

$$K_{In} = \frac{[H_3O^+][In^-]}{[HIn]}, \quad -\log K_{In} = -\log[H_3O^+] - \log\left(\frac{[In^-]}{[HIn]}\right), \quad \text{and} \quad pK_{In} = pH - \log\left(\frac{[In^-]}{[HIn]}\right)$$

For acid color, pH = 8.30 and $8.70 = 8.30 - \log\left(\frac{[In^-]}{[HIn]}\right)$.

$$\log\left(\frac{[In^-]}{[HIn]}\right) = 8.30 - 8.70 = -0.40 \qquad \frac{[In-]}{[HIn]} = 10^{-0.40} = \text{inv log}(-0.40) = 0.40$$

$[In^-] = 0.40\ [HIn]$ when acid color is observed and letting $[HIn] = 1.00$ and $[In^-] = 0.40$, fraction in the acid form = $\frac{1.00}{1.00 + 0.40}$ = **0.71**. This means 71% of the indicator molecules are in the HIn acid form.

For base color, pH = 10.00 and $8.70 = 10.00 - \log\left(\frac{[In^-]}{[HIn]}\right)$

$$\log\left(\frac{[In^-]}{[HIn]}\right) = 10.00 - 8.70 = 1.30 \qquad \frac{[In-]}{[HIn]} = 10^{1.30} = \text{inv log}(1.30) = 20$$

$[In^-] = 20\ [HIn]$ when base color is observed and letting $[HIn] = 1.00$ and $[In^-] = 20.00$, fraction in the base form = $\frac{20.00}{20.00 + 1.00}$ = **0.95**. This means 95% of the indicator molecules are in the base In^- form.

16.77 ***See Section 16.6.***

For $HIn(aq) + H_2O(\ell) \rightleftarrows H_3O^+(aq) + In^-(aq)$,

$$K_{In} = \frac{[H_3O^+][In^-]}{[HIn]}, \quad -\log K_{In} = -\log[H_3O^+] - \log\left(\frac{[In^-]}{[HIn]}\right), \quad \text{and} \quad pK_{In} = pH - \log\left(\frac{[In^-]}{[HIn]}\right)$$

So, $pH = pK_{In} + \log\left(\frac{[In^-]}{[HIn]}\right)$

When 1% of methyl red is in the HIn form: $\textbf{pH} = 5.00 + \log\left(\dfrac{0.99y}{0.01y}\right) = \textbf{7.00}$

When 5% of methyl red is in the HIn form: $\textbf{pH} = 5.00 + \log\left(\dfrac{0.95y}{0.05y}\right) = \textbf{6.28}$

When 95% of methyl red is in the HIn form: $\textbf{pH} = 5.00 + \log\left(\dfrac{0.05y}{0.95y}\right) = \textbf{3.72}$

When 99% of methyl red is in the HIn form: $\textbf{pH} = 5.00 + \log\left(\dfrac{0.01y}{0.99y}\right) = \textbf{3.00}$

16.78 ***See Section 16.6.***

For $HIn(aq) + H_2O(\ell) \rightleftarrows H_3O^+(aq) + In^-(aq)$,

$$K_{In} = \frac{[H_3O^+][In^-]}{[HIn]}, \quad -\log K_{In} = -\log[H_3O^+] - \log\left(\frac{[In^-]}{[HIn]}\right), \quad \text{and} \quad pK_{In} = pH - \log\left(\frac{[In^-]}{[HIn]}\right)$$

For acid color, pH = 4.20 and 5.00 = 4.20 -log $\left(\dfrac{[In^-]}{[HIn]}\right)$.

$$\log\left(\frac{[In^-]}{[HIn]}\right) = 4.20 - 5.00 = 0.80 \qquad \frac{[In^-]}{[HIn]} = 10^{-0.80} = \text{inv log}(-0.80) = 0.16$$

Letting [HIn] = 1.00 and [In⁻] = 0.16, fraction of indicator molecules in the acid form = $\dfrac{1.00}{1.00 + 0.16} = \mathbf{0.86}$. This means 86% of the indicator molecules are in the HIn acid form.

For base color, pH = 6.30 and 5.00 = 6.30 -log $\left(\dfrac{[In^-]}{[HIn]}\right)$.

$$\log\left(\frac{[In^-]}{[HIn]}\right) = 6.30 - 5.00 = 1.30 \qquad \frac{[In^-]}{[HIn]} = 10^{1.30} = \text{inv log}(1.30) = 20.0$$

Letting [HIn] = 1.00 and [In⁻] = 20.0, fraction of molecules in the In⁻ base form = $\dfrac{20.0}{20.0 + 1.00} = \mathbf{0.95}$. This means 95% of the indicator molecules are in the base In⁻ form.

16.79 ***See Sections 16.3, 16.5.***

(a) Equivalent mmol strong acid and strong base giving salt of strong acid and strong base. Therefore, **7** for neutral solution.

(b) Equivalent mmol strong acid and weak base giving salt of strong acid and weak base. Therefore, **3** due to acid hydrolysis of NH_4^+.

(c) Mixture of strong and weak bases. Therefore, **13** for strong base.

(d) Excess mmol weak base. Therefore, **9** for weak base/conjugate acid buffer.

16.80 ***See Sections 16.3, 16.5.***

(a) Excess mmol strong acid. Therefore, **1** for strong acid.

(b) Equivalent mmol weak acid and strong base. Therefore, **11** due to base hydrolysis of CH_3COO^-.

(c) Excess mmol weak acid. Therefore, **5** for a weak acid buffer.

(d) Equivalent mmol of strong diprotic acid and strong dibasic base. Therefore, **7** for neutral solution.

16.81 ***See Sections 16.3, 16.5.***

(a) Excess mmol strong acid. Therefore, **1** for strong acid solution.

(b) Excess mmol H_3O^+ from diprotic strong acid. However, one-half the H_2SO_4 reacts leaving the weak acid HSO_4^-. Therefore, **3** for weak acid solution.

(c) Equivalent mmol weak base and strong acid. Therefore, **3** due to acid hydrolysis of NH_4^+.

16.82 ***See Section 16.7.***

The equivalence point occurs when two moles of NaOH have been added per one mole of acid. The observation that there is no separation between the first and second equivalence points indicates the pK_{a1} and pK_{a2} values are reasonably close together. The acid is **malic acid**.

16.83 ***See Section 16.7.***

The equivalence point occurs when two moles of NaOH have been added per one mole of acid. The observation that there is a slight pH rise after the addition of one equivalent of NaOH and that the pH half-way to the first equivalence point is about equal to pK_{a1} for phthalic acid indicates the acid is **phthalic acid**.

16.84 ***See Section 16.7.***

The shape of the titration curve indicates the acid is a diprotic acid. The separation of the equivalence points and the pH values at the half-way points for the two equivalence points indicate the acid being titrated is **oxalic acid**.

16.85 ***See Section 16.7.***

The equivalence point occurs when three moles of NaOH have been added per one mole of acid. The only triprotic acid in the list is **citric acid**. The pK_a values for the three ionizable hydrogen atoms are too close to give separate equivalence points.

16.86 ***See Sections 13.6, 14.2, and Appendix F.***

$$Fe(OH)_2(s) \rightleftarrows Fe^{2+}(aq) + 2OH^-(aq)$$

Conc, M	$Fe(OH)_2$	Fe^{2+}	OH^-
initial	excess	0.	0.
Change	−s	+s	+2s
equilibrium	excess	s	2s

$K_{sp} = [Fe^{2+}][OH^-]^2 \quad K_{sp} = (s)(2s)^2 = 1.6 \times 10^{-14} \quad$ Hence, $4s^3 = 1.6 \times 10^{-14}$ and $s = 1.6 \times 10^{-5}$.

$[OH^-] = 2s = (2)(1.6 \times 10^{-5}) = 3.2 \times 10^{-5}$, $pOH = -\log(3.2 \times 10^{-5}) = 4.49$, **pH** $= 14.00 - 4.49 =$ **9.51**

16.87 ***See Sections 4.4. 16.2.***

The key to solving this problem is recognizing that $\%N = \dfrac{g\ N}{g\ \text{plant material}} \times 100\%$.

To calculate g N, we need to know moles NH_3 released by the plant material.
Since $NH_3(g) + HCl(aq) \rightarrow NH_4Cl(aq)$, mol NH_3 released = mol HCl reacted with NH_3.
Furthermore, since $HCl(aq) + NaOH(aq) \rightarrow NaCl(aq) + H_2O(\ell)$, 3mol HCl remaining after reaction with NH_3 = mol NaOH required for titration.
Hence, mol NH_3 released = mol HCl reacted with NH_3
= (mol HCl initial - mol HCl remaining)
= (mol HCl initial - mol NaOH required).
This gives

$$? \text{ mol HCl initial} = 0.1000 \text{ L HCl soln} \times \frac{0.121 \text{ mol HCl}}{1 \text{ L HCl soln}} = 0.0121 \text{ mol HCl}$$

$$? \text{ mol NaOH required} = 0.03422 \text{ L NaOH soln} \times \frac{0.118 \text{ mol NaOH}}{1 \text{ L NaOH soln}} = 0.00404 \text{ mol NaOH}$$

$$? \text{ mol HCl reacted with } NH_3 = 0.0121 \text{ mol HCl initial} - 0.00404 \text{ mol HCl remaining} = 0.0081 \text{ mol HCl}$$

$$? \text{ mol } NH_3 \text{ released} = \text{mol HCl reacted with } NH_3 = 0.0081 \text{ mol } NH_3$$

$$? \text{ g } NH_3 = 0.0081 \text{ mol } NH_3 \times \frac{14.0 \text{ g N}}{1 \text{ mol } NH_3} = 0.11 \text{ g N}$$

$$\%N = \frac{g\ N}{g\ \text{sample}} \times 100\% \qquad \%N = \frac{0.11 \text{ g N}}{21.34 \text{ g plant material}} \times 100\% = \mathbf{0.52\ \%N}$$

16.88 ***See Sections 4.4, 16.2.***

$HA(s) + NaOH(aq) \rightarrow NaA(aq) + H_2O(\ell)$
$HCl(aq) + xs\ NaOH(aq) \rightarrow NaCl(aq) + H_2O(\ell)$

Stategy: start mol NaOH - xs mol NaOH → mol NaOH reacted with HA → mol HA → gHA → mass percent HA in sample

L HCl → mol HCl → xs mol NaOH

$$? \text{ mol NaOH start} = 0.1000 \text{ L NaOH soln } \frac{0.1050 \text{ mol NaOH}}{1 \text{ LNaOH soln}} = 0.01050 \text{ mol NaOH}$$

$$? \text{ mol NaOH xs} = 0.00328 \text{ L HCl soln } \frac{0.0970 \text{ mol HCl}}{1 \text{ L HCl soln}} \times \frac{1 \text{ mol NaOH}}{1 \text{ mol HCl}} = 3.18 \times 10^{-4} \text{ mol NaOH}$$

$$? \text{ mol NaOH reacting with HA} = 0.01050 \text{ mol} - 3.18 \times 10^{-4} \text{ mol} = 0.01018 \text{ mol NaOH}$$

$$? \text{ mol HA} = 0.01018 \text{ mol NaOH } \frac{1 \text{ mol HA}}{1 \text{ mol NaOH}} = 0.01018 \text{ mol HA}$$

$$? \text{ g HA} = 0.01018 \text{ mol HA} \times \frac{176.1 \text{ g HA}}{1 \text{ mol HA}} = 1.793 \text{ g HA}$$

$$\%HA = \frac{g\ HA}{g\ \text{sample}} \times 100\% \qquad \%HA = \frac{1.793 \text{ g}}{1.8457 \text{ g}} \times 100\% = \mathbf{97.14\%}$$

16.89 *See Sections 4.4, 16.2, and Examples 4.12, 16.4.*

$H_2A(aq) + 2KOH(aq) \rightarrow K_2A(aq) + 2H_2O(\ell)$

Strategy: g H_2A → mol H_2A → mol KOH → L KOH soln

$$?\text{L KOH soln} = 24.93\text{ g }H_2A \times \frac{1\text{ mol }H_2A}{124.0\text{ g }H_2A} \times \frac{2\text{ mol KOH}}{1\text{ mol }H_2A} \times \frac{1\text{ L KOH soln}}{0.221\text{ mol KOH}} = \mathbf{1.82\text{ L KOH sol}}$$

16.90 *See Sections 4.4, 16.2, and Examples 4.12, 16.4.*

$HA(aq) + NaOH(aq) \rightarrow NaA(aq) + H_2O(\ell)$

Strategy: L NaOH → mol NaOH → mol HA → molar mass HA

$$?\text{ mol HA} = 0.03669\text{ L NaOH soln} \times \frac{0.404\text{ mol NaOH}}{1\text{ L NaOH soln}} \times \frac{1\text{ mol HA}}{1\text{ mol NaOH}} = 0.0148\text{ mol HA}$$

$$?\,M\text{ HA} = \frac{1.2451\text{ g}}{0.0148\text{ mol}} = \mathbf{84.1\text{ g/mol}}$$

If the acid were diprotic, the reaction would be $H_2A(aq) + 2NaOH(aq) \rightarrow 2NaA(aq) + 2H_2O(\ell)$, and the ratio 1 mol H_2A / 2 mol NaOH would yield 0.00741 mol H_2A, so

$$?\,M\,H_2A = \frac{1.2451\text{ g}}{0.00741\text{ mol}} = \mathbf{168\text{ g/mol}}$$

16.91 *See Section 16.6 and Table 16.4.*

The pH at the equivalence point depends on the concentrations of the acid and base involved because these items affect the moles of salt formed, the total volume of the solution and therefore the concentration of the conjugate base of the acid. However, we should expect the pH at the equivalence point to be in the range 7.0 to 8.0 for normal concentrations due to $A^-(aq) + H_2O(\ell) \rightleftarrows HA(aq) + OH^-(aq)$ with $K_b = K_w / K_a = 1.0 \times 10^{-14} / 1.5 \times 10^{-2} = 6.7 \times 10^{-13}$. Hence, **bromthymol blue** would probably be the best choice for the indicator.

16.92 *See Section 16.2*

(a) *Strategy: mL conc HCl soln → g HCl soln → g HCl → mol HCl → M HCl*

$$?\,M\text{ mol HCl} = 83\text{ mL conc HCl soln} \times \frac{1.19\text{ g conc HCl soln}}{1\text{ mL conc HCl soln}} \times \frac{0.38\text{ g HCl}}{1.00\text{ g conc HCl soln}} \times \frac{1\text{ mol HCl}}{36.46\text{ g HCl}}$$

$$= 1.0\text{ mol HCl}$$

$$?\,M\text{ HCl} = \frac{1.0\text{ mol HCl}}{1.00\text{L HCl soln}} = 1.0\,M\text{ HCl}$$

(b) $HCl(aq) + NaOH(aq) \rightarrow NaCl(aq) + H_2O(\ell)$

Strategy: L NaOH soln → mol NaOH → mol HCl → M HCl

$$?\text{ mol HCl} = 0.02388\text{ mL NaOH soln} \times \frac{1.04\text{ mol NaOH}}{1\text{ L NaOH soln}} \times \frac{1\text{ mol HCl}}{1\text{ mol NaOH}} = 0.0248\text{ mol HCl}$$

$$?\,M\text{ HCl} = \frac{0.0248\text{ mol HCl}}{0.02500\text{ L HCl soln}} = \mathbf{0.992\,\mathit{M}\text{ HCl}}$$

16.93 ***See Sections 16.2, 16.3, 16.6, and Table 16.4.***

Overall: $HI(aq) + NaOH(aq) \rightarrow NaI(s) + H_2O(\ell)$

Net ionic: $H_3O^+(aq) + OH^-(aq) \rightarrow H_2O(l)$

$$? \text{ mol } H_3O^+ \text{ present} = 10.0 \text{ mol HI soln} \times \frac{1.70 \text{ g HI soln}}{1 \text{ mL HI soln}} \times \frac{57 \text{ g HI}}{100 \text{ g HI soln}} \times \frac{1 \text{ mol HI}}{127.9 \text{ g HI}} \times \frac{1 \text{ mol } H_3O^+}{1 \text{ mol HI}}$$

$$= 7.6 \times 10^{-2} \text{ mol } H_3O^+$$

$$? \text{ mol } OH^- \text{ needed to reach equivalence point} = 7.6 \times 10^{-2} \text{ mol } H_3O^+ \times \frac{1 \text{ mol } OH^-}{1 \text{ mol } H_3O^+}$$

$$= 7.6 \times 10^{-2} \text{ mol } OH^-$$

$$? \text{ mL NaOH soln required to reach equivalence point} = 7.6 \times 10^{-2} \text{ mol } OH^- \times \frac{1 \text{ mol NaOH}}{1 \text{ mol } OH^-}$$

$$\times \frac{10^3 \text{ mL NaOH soln}}{0.988 \text{ mol NaOH}} = \mathbf{77 \text{ mL NaOH soln}}$$

Since the titration involves titrating a strong acid with a strong base, we should expect the pH to be 7.0 at the equivalence point. **Phenolphthalein** (pK_{In} = 8.7, pH range 8.3 - 10.0) is a suitable indicator, since there is a very rapid change in pH near the equivalence point of a strong acid-strong base titration (Figures 16.2, 16.3, 16.5).

16.94-16.96 ***Solutions for these Spreadsheet Exercises are on the Reger, Goode and Mercer Spreadsheet Solutions Diskette available from Saunders College Publishing.***

Chapter 17: Thermodynamics

17.1 *See Section 17.1.*

Work is positive ($w > 0$) when the surroundings do work on the system, and work is negative ($w < 0$) when the system does work on the surroundings. Hence, a positive sign for work (w) corresponds to an increase in the energy of the system, and a negative sign for work (w) corresponds to a decrease in the energy of the system.

17.2 *See Section 17.1.*

Heat is positive ($q > 0$) when the system absorbs energy as heat, and heat is negative ($q < 0$) when the system releases energy as heat. Hence, a positive sign for heat (q) corresponds to an increase in the energy of the system, and a negative sign for heat (q) corresponds to a decrease in the energy of the system.

17.3 *See Section 17.1*

When a fuel-oxygen mixture burns propelling an automobile, the system does work on the surroundings. Hence, the sign of w is **negative**.

17.4 *See Section 17.1.*

When a refrigerator compresses a gas to a liquid, the surroundings do work on the system. Hence, the sign of w is **negative**.

17.5 *See Section 17.1.*

The burning gases from the rocket do work on the surroundings propelling the rocket into space. Hence, the sign of w is **negative**.

17.6 *See Section 17.1.*

No work is done when a gas expands against an opposing pressure of 0.0 atmospheres because $w = -P\Delta V$ where P is the pressure opposing the system.

17.7 *See Section 17.2.*

The first Law of Thermodynamics states: *Energy can be neither created or destroyed.* This is expressed in equation form as $\Delta E = q + w$ where ΔE is the change in the internal energy of the system, q is the heat absorbed by the system and w is the work done on the system. This equation is based on the observation that heat and work are the ways in which energy enters or leaves a system.

17.8 *See Section 17.2.*

Heat absorbed by the system under conditions of constant pressure when only PV work can be done is equal to the change in enthalpy.

17.9 *See Section 17.2.*

The change in internal energy, ΔE, is the heat absorbed by the system when the volume is constant and no PV work can occur. The change in enthalpy, ΔH, is the heat absorbed by the system when the pressure is constant and PV work can occur. The difference between the enthalpy change and the internal energy change for a chemical reaction is the amount of PV work that occurs when the reaction is conducted at constant pressure. Hence, $\Delta H - \Delta E = P\Delta V$ and $\Delta H = \Delta E + P\Delta V$.

17.10 *See Sections 17.1, 17.2.*

Enthalpy is related to internal energy by $H = E + PV$. Internal energy, E, represents the total energy of the system which involves both kinetic and potential energy. Potential energy is dependent on coulombic forces of attraction and repulsions within substances. Since we can only know the internal energy of ideal monatomic gases experiencing no forces of attraction and repulsion, it is generally impossible to know the absolute internal energies and enthalpies of substances. Hence, we can only measure changes in internal energies and enthalpies.

17.11 *See Section 17.3.*

We can meassure the absolute entropy of a substance at a given temperature because the Third Law of Thermodynamics gives us a reference point. The Third Law of Thermodynamics states: *The entropy of a pure crystalline substance is 0 at a temperature of zero kelvin.* As the temperature rises above 0 K the motion of the particles increases, increasing the disorder and entropy of the substance in ways that can be calculated.

17.12 *See Section 17.3.*

The Third Law of Thermodynamics states: *The entropy of a pure crystalline substance is 0 at a temperature of zero kelvin.*

17.13 *See Sections 1.2, 17.3.*

An alloy is a solid solution that is a homogeneous mixture and therefore has the same composition throughout. However, at the atomic level the positions of the atoms are random, not crystalline, and an alloy can have variable composition from one sample to another. Hence, an alloy does not meet the criterion for being classified as a pure substance and should not be expected to have an entropy of zero at 0 K.

17.14 *See Section 17.4.*

When the enthalpy change for a spontaneous process is unfavorable ($\Delta H = +$ and process is endothermic), it must have a favorable entropy change ($\Delta S = +$ and process leading to greater disorder). In the case of soluble ionic salts dissolving in water in endothermic processes, there is an increase in disorder as the crystal lattice of the solid and the solvent structure of water are broken. There is also an increase in order associated with the orientation of water molecules about the ations and anions. However, this contribution is small when the ions have low charges and large sizes, and the overall entropy change is favorable, is positive.

17.15 *See Section 17.4.*

Whether reactions can occur to give predominantly products under a given set of conditons is a matter of thermodynamics, i.e., negative or positive free energy change. Whether a spontaneous reaction actually occurs to give predominantly products under a given set of conditions is a matter of kinetics. For example, the combustion of methane does not occur to any appreciable extent under ordinary conditions because the activation energies for reactions involving double-bonded oxygen molecules are usually very high. However, the reaction of methane with oxygen is still classified as being spontaneous from a thermodynamic point of view because it can proceed to give predominantly products.

17.16 *See Section 17.4.*

When ice is formed from liquid water at 0°C and 1 atmosphere pressure, there is a decrease in disorder, and the sign of the entropy change for the system is negative ($\Delta S_{sys} = S_{H_2O(s)} - S_{H_2O(\ell)}$ and $S_{H_2O(s)} < S_{H_2O(\ell)}$). Since $\Delta G = -T\Delta S_{univ} = 0$, S_{univ} must be zero. In addition ΔS_{surr} must be positive but equal in size to ΔS_{sys} because $\Delta S_{univ} = \Delta S_{sys} + \Delta S_{surr} = 0$.

17.17 *See Sections 17.3, 17.4.*

The Second Law of Thermodynamics states there is an increase in the entropy of the universe in any spontaneous process, and it can be shown that $\Delta G_{sys} = -T\Delta S_{univ}$. Hence, ΔG_{sys} must be negative for a process to be spontaneous.

17.18 *See Sections 17.4, 17.5.*

(a) $\Delta G = \Delta G^\circ + RT \ln Q$ and since $\Delta G^\circ = -RT \ln K_{eq}$, $\Delta G = -RT \ln K_{eq} + RT \ln Q = RT \ln\left(\frac{Q}{K_{eq}}\right)$.

(b) $\Delta G = w_{max}$

(c) $\Delta G = -T\Delta S_{univ}$.

17.19 *See Section 17.5.*

According to the principle of LeChatelier, an increase in temperature favors the heat absorbing or endothermic direction. Since the value of the equilibrium constant of the reaction decreases with increasing temperature, meaning there are less products and more reactants with increasing temperature, the reverse reaction must be the endothermic direction. The forward reaction is therefore exothermic and sign of ΔH° is negative, whereas nothing can be said about the sign of ΔS° based on the given information.

17.20 *See Section 17.4.*

Since the free energy change for the reaction decreases as the temperature increases and $\Delta G^\circ = \Delta H^\circ - T\Delta S^\circ$, the sign of ΔS° must be positive. Nothing can be said about the sign of ΔH° based on the given information.

17.21 *See Section 17.1.*

(a) $CaCO_3(s) + H_2SO_4(aq) \rightarrow CaSO_4(aq) + H_2O(l) + CO_2(g)$: There is an increase in mole of gas during this reaction causing the system to expand against an opposing pressure exerted by the surroundings. The sign of **w is negative**, since the system transfers energy to the surroundings.

(b) $HCl(g) + NH_3(g) \rightarrow NH_4Cl(s)$: There is a decrease in moles of gas during this reaction causing the system to contract under the influenece of pressure exerted by the surroundings. The sign of **w is positive**, since the surroundings transfers energy to the system as it compresses it.

17.22 *See Section 17.1.*

(a) $Fe_2S_3(s) + 6HNO_3(aq) \rightarrow 2Fe(NO_3)_2(aq) + 3H_2S(g)$: There is an increase in mole of gas during this reaction causing the system to expand against an opposing pressure exerted by the surroundings. The sign of **w is negative**, since the system transfers energy to the surroundings.

(b) $CH_4(g) + 2O_2(g) \rightarrow CO_2(g) + 2H_2O(\ell)$: There is a decrease in moles of gas during this reaction causing the system to contract under the influenece of pressure exerted by the surroundings. The sign of **w is positive**, since the surroundings transfers energy to the system as it compresses it.

17.23 *See Sections 17.1, 17.2, and Example 17.3.*

Since the volumes of solids and liquids are small compared to the volumes of gases, the amount of PV work done during physical changes of state and chemical reactions involving gases can be calculated from changes in the numbers of moles of gases.

The volume of 1 mole of an ideal gas at 298 K can be calculated using $V = \frac{nRT}{P}$ and is 24.5 L.

(a) $CO_2\,(g) + NaOH(s) \rightarrow NaHCO_3\,(s)$ involves a decrease of one mole of gas, so Δn_g = -1..

Hence, $w = -P\Delta V$ $\quad w = -(1\text{ atm})(-24.5\text{ L}) = \mathbf{24.5\ L \cdot atm}$

or $\quad w = -\Delta nRT$ $\quad w = -(-1\text{ mol})\left(8.314\frac{J}{mol \cdot K}\right)(298\text{ K}) = 2.48 \times 10^3$ J or **2.48 kJ.**

(b) $2O_3\,(g) \rightarrow 3O_2\,(g)$ involves an increase of one mole of gas, so Δn = +1..

Hence, $w = -P\Delta V$ $\quad w = -(1\text{ atm})(+24.5\text{ L}) = \mathbf{-24.5\ L \cdot atm}$

or $\quad w = -\Delta nRT$ $\quad w = -(1\text{ mol})\left(8.314\frac{J}{mol \cdot K}\right)(298\text{ K}) = -2.48 \times 10^3$ J or **-2.48 kJ.**

Note: The reaction in (a) involves a decrease in the number of moles of gas, and the sign of w is positive. The reaction in (b) involves an increase in the number of moles of gas, and the sign of w is negative.

17.24 *See Sections 17.1, 17.2, and Example 17.3.*

Since the volumes of solids and liquids are small compared to the volumes of gases, the amount of PV work done during physical changes of state and chemical reactions involving gases can be calculated from changes in the numbers of moles of gases.

The volume of 1 mole of an ideal gas at 298 K can be calculated using $V = \frac{nRT}{P}$ and is 24.5 L.

(a) $2Na(s) + 2H_2O(\ell) \rightarrow 2NaOH(s) + H_2(g)$ involves an increase of one mole gas, so Δn = 1.

Hence, $w = -P\Delta V$ $\quad w = -(1\text{ atm})(24.5\text{ L}) = \mathbf{-24.5\ L \cdot atm}$

or $\quad w = -\Delta nRT$ $\quad w = -(1\text{ mol})\left(8.314\frac{J}{mol \cdot K}\right)(298\text{ K}) = -2.48 \times 10^3$ J or **-2.48 kJ**

(b) $Fe_2O_3 + 3C(s) \rightarrow 4Fe(s) + 3CO_2(g)$ involves an increase of three moles of gas, so Δn = 3..

Hence, $w = -P\Delta V$ $\quad w = -(1\text{ atm})(3 \times 24.5\text{ L}) = \mathbf{-73.5\ L \cdot atm}$

or $\quad w = -\Delta nRT$ $\quad w = -(3\text{ mol})\left(8.314\frac{J}{mol \cdot K}\right)(298\text{ K}) = -7.43 \times 10^3$ J or **-7.43 kJ.**

Note: The reaction in a and b involve an increase in the number of moles of gas, and the sign of w is negative.

17.25 *See Sections 15.1, 15.2, Example 17.3, and the Solution for 17.23.*

(a) $Ni(CO)_4\,(g) \rightarrow Ni(s) + 4CO(g)$ involves an increase of three mole of gas, so Δn = 3..

Hence, $w = -P\Delta V$ $\quad w = -(1\text{ atm})[3 \times 24.5\text{L}] = \mathbf{-73.5\ L \cdot atm}$

or $\quad w = -\Delta nRT$ $\quad w = -(3\text{ mol})\left(8.314\frac{J}{mol \cdot K}\right)(298\text{ K}) = -7.43 \times 10^3$ J or **-7.43 kJ.**

(b) $2NO(g) \rightarrow N_2\,(g) + O_2\,(g)$ does not involve a change in moles of gases.
Hence, no pressure-volume work is done during this reaction.

17.26 *See Sections 15.1, 15.2, Example 17.3, and the Solution for 17.23.*

(a) $Fe_2O_3(s) + 2Al(s) \rightarrow 2Fe(s) + Al_2O_3(s)$ does not involve a change in moles of gases. Hence, no pressure-volume work is done during this reaction.

(b) $2H_2(g) + O_2(g) \rightarrow 2H_2O(\ell)$ involves a decrease of three moles of gas, so $\Delta n = -3$.

Hence, $w = -P\Delta V$ $\quad w = -(1\text{ atm})(3 \times (-24.5\text{ L})) = 73.5\text{ L·atm}$

or $\quad w = -\Delta nRT$ $\quad w = -(3\text{ mol})\left(8.314\dfrac{\text{J}}{\text{mol}\cdot\text{K}}\right)(298\text{ K}) = -7.43 \times 10^3\text{ J}$ or $\mathbf{-7.43\ kJ}$.

Note: The molar volume of a gas at 298 K and 1 atm is 24.5 L. This value can be calculated using Charles' Law expression and the knowledge that the standard molar volume of a gas is 22.4 L.

17.27 *See Section 17.1 and Example 17.1.*

Initial conditions: $V_1 = 220\text{ L}$ $\quad P_1 = 150\text{ atm}$

Final conditions: $V_2 = ?$, unknown $\quad P_2 = P_{ext} = 1.0\text{ atm}$

Solving $P_1V_1 = P_2V_2$ for V_2 gives $V_2 = V_1 \times \dfrac{P_1}{P_2}$ $\quad V_2 = 220\text{ L} \times \dfrac{150\text{ atm}}{1.0\text{ atm}} = 3.3 \times 10^4\text{ L}$

Hence, $w = -P\Delta V = -(1.0\text{ atm})(3.3 \times 10^4\text{ L} - 220\text{ L}) = -3.3 \times 10^4\text{ L·atm}$.

Converting L·atm to kJ gives $w = -3.3 \times 10^4\text{ L·atm} \times 101.3\dfrac{\text{J}}{\text{L·atm}} \times \dfrac{1\text{ kJ}}{10^3\text{ J}} = \mathbf{-3.3 \times 10^3\ kJ}$

17.28 *See Section 17.1 and Example 17.1.*

Initial conditions: $V_1 = 2.0\text{ L}$ $\quad P_1 = 1.10\text{ atm}$

Final conditions: $V_2 = ?$, unknown $\quad P_2 = P_{ext} = 754\text{ torr} \times \dfrac{1\text{ atm}}{760\text{ torr}} = 0.992\text{ atm}$

Solving $P_1V_1 = P_2V_2$ for V_2 gives $V_2 = V_1 \times \dfrac{P_1}{P_2}$ $\quad V_2 = 2.0\text{ L} \times \dfrac{1.10\text{ atm}}{0.992\text{ atm}} = 2.2\text{ L}$

Hence, $w = -P\Delta V = -(0.992\text{ atm})(2.2\text{ L} - 2.0\text{ L}) = -0.2\text{ L·atm}$.

Converting L·atm to kJ gives $w = -0.2\text{ L}\cdot\text{atm} \times 101.3\dfrac{\text{J}}{\text{L}\cdot\text{atm}} \times \dfrac{1\text{ kJ}}{10^3\text{ J}} = \mathbf{-0.02\ kJ}$

17.29 *See Sections 17.1, 17.2, and Example 17.1.*

According to the First Law of Thermodynamics, $\Delta E = q + w$. In this case, $q = +4.35\text{ kJ}$ and $w = -P\Delta V = -(0.94\text{ atm})(4.5\text{ L}) = -4.2\text{ L·atm}$.

Converting L·atm to kJ gives $w = -4.2\text{ L·atm} \times 101.3\dfrac{\text{J}}{\text{L·atm}} \times \dfrac{1\text{ kJ}}{10^3\text{ J}} = -0.43\text{ kJ}$

and $\Delta E = 4.35\text{ kJ} + (-0.43\text{ kJ}) = \mathbf{3.92\ kJ}$.

17.30 *See Sections 17.1, 17.2, and Example 17.1.*

According to the First Law of Thermodynamics, $\Delta E = q + w$. In this case, $q = -2.71\text{ kJ}$ and $w = -P\Delta V = -(1.02\text{ atm})(3.5\text{ L}) = -3.6\text{ L·atm}$.

Converting L·atm to kJ gives $w = -3.6\text{ L}\cdot\text{atm} \times 101.3\dfrac{\text{J}}{\text{L}\cdot\text{atm}} \times \dfrac{1\text{ kJ}}{10^3\text{ J}} = -0.36\text{ kJ}$

and $\Delta E = -2.71\text{ kJ} + (-0.36\text{ kJ}) = \mathbf{-3.07\ kJ}$.

17.31 *See Sections 17.1, 17.2, and Example 17.1.*

(a) Initial conditions: $V_1 = 1.00$ L $P_1 = 9.00$ atm
Final conditions: $V_2 = ?$, unknown $P_2 = P_{ext} = 1.00$ atm

Solving $P_1V_1 = P_2V_2$ for V_2 gives $V_2 = V_1 \times \frac{P_1}{P_2}$ $V_2 = 1.00\text{ L} \times \frac{9.00\text{ atm}}{1.00\text{ atm}} = 9.00\text{ L}$

So, $w = -P\Delta V = -(1.0\text{ atm})(9.00\text{ L} - 1.00\text{ L}) = -8.00$ L·atm.

Converting L·atm to kJ gives $w = -8.00\text{ L}\cdot\text{atm} \times 101.3\frac{\text{J}}{\text{L}\cdot\text{atm}} \times \frac{1\text{ kJ}}{10^3\text{ J}} = \mathbf{-0.810\text{ kJ}}$

Since $\Delta E = 0$ for an isothermal expansion, and $\Delta E = q + w$, $\mathbf{q = +0.810\text{ kJ}}$.

(b) Following the plan outlined in the solution for (a) gives:
(1) V_2 at the end of expansion against 3.00 atm = 3.00 L,
$w_1 = -P\Delta V = -(3.00\text{ atm})(3.00\text{ L} - 1.00\text{ L}) = -6.00$ L·atm or -0.608 kJ and $q_1 = +0.608$ kJ.
(2) V_2' at the end of expansion from (3.00 L, 3.00 atm) to (? L, 1.00 atm) = 9.00 L,
$w_2 = -(1.0\text{ atm})(9.00\text{ L} - 3.00\text{ L}) = -6.00$ L·atm or -0.608 kJ and $q_2 = +0.608$ kJ.

Hence, $\mathbf{w = w_1 + w_2 = -0.608\text{ kJ} + -0.608\text{ kJ} = -1.216\text{ kJ}}$
and $\mathbf{q = q_1 + q_2 = 0.608\text{ kJ} + 0.608\text{ kJ} = 1.216\text{ kJ}}$.

(c) Following the plan outlined in the solution for (a) gives:
1) V_2 at the end of expansion against 3.00 atm = 3.00 L,
$w_1 = -P\Delta V = -(3.00\text{ atm})(3.00\text{ L} - 1.00\text{ L}) = -6.00$ L·atm or -0.608 kJ and $q_1 = +0.608$ kJ.
2) V_2' at the end of expansion from (3.00 L, 3.00 atm) to (? L, 2.00 atm) = 4.50 L,
$w_2 = -(2.00\text{ atm})(4.50\text{ L} - 3.00\text{ L}) = -3.00$ L·atm or -0.304 kJ and $q_2 = +0.304$ kJ.
3) V_2'' at the end of expansion from (4.50 L, 2.00 atm) to (? L, 1.00 atm) = 9.00 L,
$w_3 = -(1.00\text{ atm})(9.00\text{ L} - 4.50\text{ L}) = -4.50$ L·atm or -0.456 kJ and $q_3 = +0.456$ kJ.

Hence, $\mathbf{w = w_1 + w_2 + w_3} = -0.608\text{ kJ} - 0.304\text{ kJ} - 0.456\text{ kJ} = \mathbf{-1.368\text{ kJ}}$,
and $\mathbf{q = q_1 + q_2 + q_3} = 0.608\text{ kJ} + 0.304\text{ kJ} + 0.456\text{ kJ} = \mathbf{1.368\text{ kJ}}$.

The inital and final states for the gas are the same in a, b and c. However, expansion occurs in a one-step process in a, a two-step process in b, and in a three-step process in 15.18:
a. (1.00 L, 9.00 atm) → (9.00 L, 1.00 atm)
b. (1.00 L, 9.00 atm) → (3.00 L, 3.00 atm) → (9.00 L, 1 atm)
c. (1.00 L, 9.00 atm) → (3.00 L, 3.00 atm) → (4.50 L, 2.00 atm) → (9.00 L, 1.00 atm)

Since the inital and final states are the same in all three processes and all involve isothermal expansions, $\Delta E = 0$ for all three. However,

$w_a = -0.810$ kJ $q_a = -0.810$ kJ
$w_b = -1.216$ kJ $q_b = +1.216$ kJ
$w_c = -1.368$ kJ $q_c = +1.368$ kJ

These calculations illustrate that heat and work (q and w) are path dependent and therefore not state functions whereas ΔE is independent of the pathway and therefore internal energy (E) is a state function.

(d) The calculations for a, b, and c indicate that the amount of work done on the surroundings increases as the number of increments of expansion increases. Hence, the maximum amount of work would be done on the surroundings when the gas expands against infinestimally small decreases in pressure.

17.32 *See Sections 17.1, 17.2, and Example 17.1.*

(a) Initial conditions: $V_1 = 9.00$ L $P_1 = 1.00$ atm
Final conditions: $V_2 = ?$, unknown $P_2 = P_{ext} = 9.00$ atm

Solving $P_1V_1 = P_2V_2$ for V_2 gives $V_2 = V_1 \times \frac{P_1}{P_2}$ $V_2 = 9.00\text{ L} \times \frac{1.00\text{ atm}}{9.00\text{ atm}} = 1.00\text{ L}$

So, $w = -P\Delta V = -(9.00\text{ atm})(1.00\text{ L} - 9.00\text{ L}) = 72.0$ L·atm.

Converting L·atm to kJ gives $w = 72.0\ L \cdot atm \times 101.3 \frac{J}{L \cdot atm} \times \frac{1\ kJ}{10^3\ J} = \mathbf{7.29\ kJ}$

Since $\Delta E = 0$ for an isothermal compression, and $\Delta E = q + w$, **q = -7.29 kJ.**

(b) Following the plan outlined in the solution for (a) gives:

(1) V_2 at the end of compression against 3.00 atm = 3.00 L,
$w_1 = -P\Delta V = -(3.00\ atm)(3.00\ L - 9.00\ L) = 18.0$ L·atm or 1.82 kJ and $q_1 = -1.82$ kJ.

(2) V_2' at the end of compression from (4.50 L, 3.00 atm) to (? L, 9.00 atm) = 1.00 L,
$w_2 = -(9.00\ atm)(1.00\ L - 3.00\ L) = 18.0$ L·atm or 1.82 kJ and $q_2 = -1.82$ kJ.

Hence, $\mathbf{w = w_1 + w_2 = 1.82\ kJ + 1.82\ kJ = 3.64\ kJ}$

and $\mathbf{q = q_1 + q_2 = -1.82\ kJ + (-1.82)\ kJ = -3.64\ kJ}$.

(c) Following the plan outlined in the solution for (a) gives:

1) V_2 at the end of compression against 2.00 atm = 4.50 L,
$w_1 = -P\Delta V = -(2.00\ atm)(4.50\ L - 9.00\ L) = 9.00$ L·atm or 0.912 kJ and $q_1 = -0.912$ kJ.

2) V_2' at the end of compression from (4.50 L, 2.00 atm) to (? L, 3.00 atm) = 3.00 L,
$w_2 = -(3.00\ atm)(3.00\ L - 4.50\ L) = 4.50$ L·atm or 0.456 kJ and $q_2 = -0.456$ kJ.

3) V_2'' at the end of compression from (3.00 L, 3.00 atm) to (? L, 9.00 atm) = 1.00 L,
$w_3 = -(9.00\ atm)(1.00\ L - 3.00\ L) = 18.0$ L·atm or 1.82 kJ and $q_3 = -1.82$ kJ.

Hence, $\mathbf{w = w_1 + w_2 + w_3} = 0.912\ kJ + 0.456\ kJ + 1.82\ kJ = \mathbf{3.19\ kJ}$,

and $\mathbf{q = q_1 + q_2 + q_3} = -0.912\ kJ + (-0.456)\ kJ + -1.82\ kJ = \mathbf{-3.19\ kJ.}$

The inital and final states for the gas are the same in a, b and c. However, compression occurs in a one-step process in a, a two-step process in b, and in a three-step process in c:

a. (9.00 L, 1.00 atm) → (1.00 L, 9.00 atm)
b. (9.00 L, 1.00 atm) → (3.00 L, 3.00 atm) → (1.00 L, 9 atm)
c. (9.00 L, 1.00 atm) → (4.50 L, 2.00 atm) → (3.00 L, 3.00 atm) → (1.00 L, 9.00 atm)

Since the inital and final states are the same in all three processes and all involve isothermal compressions, $\Delta E = 0$ for all three. However,

$w_a = +7.29$ kJ	$q_a = -7.29$ kJ
$w_b = +3.64$ kJ	$q_b = -3.64$ kJ
$w_c = +3.19$ kJ	$q_c = -3.19$ kJ

These calculations illustrate that heat and work (q and w) are path dependent and therefore not state functions whereas ΔE is independent of the pathway and therefore internal energy (E) is a state function.

(d) The calculations for a, b, and c indicate that the amount of work done on the system decreases as the number of increments of compression increases. Hence, the minimum amount of work would be done on the system when the gas contracts against infinestimally small increases in pressure.

17.33 ***See Section 17.2 and Example 17.2.***

$q_{surr} = C\Delta t \qquad q_{surr} = 5.374 \frac{kJ}{^oC} \times 4.375°C = 23.51\ kJ$

$q_{sys} = -q_{surr} \qquad q_{sys} = -23.51\ kJ$

$\Delta E = \frac{-23.51\ kJ}{1.241\ g\ C_2H_4(OH)_2} \times \frac{62.07\ g\ C_2H_4(OH)_2}{1\ mol\ C_2H_4(OH)_2} = \mathbf{-1.176 \times 10^3\ kJ}$

17.34 ***See Section 17.2 and Example 17.2.***

$q_{surr} = C\Delta t \qquad q_{surr} = 5.694 \frac{kJ}{^oC} \times 6.021°C = 34.28\ kJ$

$q_{sys} = -q_{surr} \qquad q_{sys} = -34.28\ kJ$

$\Delta E = \frac{-34.28\ kJ}{1.022\ g\ C_3H_7OH} \times \frac{60.10\ g\ C_3H_7OH}{1\ mol\ C_3H_7OH} = \mathbf{-2.016 \times 10^3\ kJ}$

17.35 *See Section 17.2 and Example 17.2.*

$q_{surr} = C\Delta t$ $\quad q_{surr} = 4.736\ \frac{kJ}{^{o}C} \times 2.548^{o}C = 12.07\ kJ$

$q_{sys} = -q_{surr}$ $\quad q_{sys} = -12.07\ kJ$

$$\Delta E = \frac{-12.07\ kJ}{2.273\ g\ Zn} \times \frac{65.39\ g\ Zn}{1\ mol\ Zn} = \mathbf{-3.472 \times 10^2\ kJ}$$

17.36 *See Section 17.2 and Example 17.2.*

$q_{surr} = C\Delta t$ $\quad q_{surr} = 4.597 \frac{kJ}{^{o}C} \times 5.019^{o}C = 23.07\ kJ$

$q_{sys} = -q_{surr}$ $\quad q_{sys} = -23.07\ kJ$

$$\Delta E = \frac{-23.07\ kJ}{1.056\ g\ Cr} \times \frac{51.996\ g\ Cr}{1\ mol\ Cr} = \mathbf{-1.136 \times 10^3\ kJ}$$

17.37 *See Section 17.2, Example 17.3, and the Solution for Exercise 17.33.*

$C_2H_4(OH)_2(\ell) + \frac{5}{2}O_2(g) \rightarrow 2CO_2(g) + 3H_2O(\ell)$ involves a decrease of $\frac{1}{2}$ mol gas, so $\Delta n = -\frac{1}{2}$ mol.

$\Delta H = \Delta E + P\Delta V = \Delta E + \Delta nRT$

$$\Delta H = -1.176 \times 10^3\ kJ + \left(-\tfrac{1}{2}\ mol\right)\left(8.314 \times 10^{-3}\ \frac{kJ}{mol \cdot K}\right)(298\ K) = \mathbf{-1.177 \times 10^3\ kJ}$$

17.38 *See Section 17.2, Example 17.3, and the Solution for Exercise 17.34.*

$C_3H_7OH(\ell) + \frac{9}{2}O_2(g) \rightarrow 3CO_2(g) + 4H_2O(\ell)$ involves a decrease of $\frac{3}{2}$ mol gas, so $\Delta n = -\frac{3}{2}$ mol.

$\Delta H = \Delta E + P\Delta V = \Delta E + \Delta nRT$

$$\Delta H = -2.016 \times 10^3 kJ + \left(-\tfrac{3}{2}\ mol\right)\left(8.314 \times 10^{-3}\ \frac{kJ}{mol \cdot K}\right)(298\ K) = \mathbf{-2.020 \times 10^3\ kJ}$$

17.39 *See Section 17.2, Example 17.3, and the Solution for Exercise 17.33.*

$Zn(s) + \frac{1}{2}O_2(g) \rightarrow ZnO(s)$ involves a decrease of $\frac{1}{2}$ mol gas, so $\Delta n = -\frac{1}{2}$ mol.

$\Delta H = \Delta E + P\Delta V = \Delta E + \Delta nRT$

$$\Delta H = \Delta H^o_f\ ZnO(s) = -3.472 \times 10^2\ kJ + \left(-\tfrac{1}{2}\ mol\right)\left(8.314 \times 10^{-3}\ \frac{kJ}{mol \cdot K}\right)(298\ K) = \mathbf{-348.4\ kJ}$$

17.40 *See Section 17.2, Example 17.3, and the Solution for Exercise 17.36.*

$2Cr(s) + \frac{3}{2}O_2(g) \rightarrow Cr_2O_3(s)$ involves a decrease of $\frac{3}{2}$ mol gas, so $\Delta n = -\frac{3}{2}$ mol.

$$\Delta H = \Delta H^o_f\ Cr_2O_3(s) = -1.136 \times 10^3\ kJ + \left(-\tfrac{3}{2}\ mol\right)\left(8.314 \times 10^{-3}\ \frac{kJ}{mol \cdot K}\right)(298\ K) = \mathbf{-1140\ kJ}$$

17.41 *See Section 17.3.*

(a) There is an increase in disorder as the test-tube shatters. Hence, the sign of the entropy change for the system is **positive**.

(b) There is an increase in disorder as a new deck of cards is shuffled. Hence, the sign of the entropy change for the system is **positive.**

(c) There is a decrease in disorder as steel is made from iron ore and carbon. Hence, the sign of the entropy change for the system is **negative.**

(d) There is an increase in disorder as a wooden fence rots. Hence, the sign of the entopy change for the system is **positive.**

17.42 *See Section 17.3.*

(a) There is an increase in disorder as glass is made from sand because sand is crystalline and glass is not. Hence, the sign of the entropy change for the system is **positive.**

(b) There is an decrease in disorder as hydrogen and oxygen combine to form water. Hence, the sign of the entropy change for the system is **negative.**

(c) There is a decrease in entropy as an artisan makes a patterned rug from yarn.. Hence, the sign of the entropy change for the system is **negative.**

(d) There is an increase in disorder when wood and oxygen are burned to provide heat.. Hence, the sign of the entopy change for the system is **positive.**

17.43 *See Section 17.3.*

(a) There is an increase in disorder as dry ice, CO_2 (s), melts. Hence, the sign of the entropy change for the system is **positive.**
Note: Dry ice sublimes under normal conditions. However, it does melt under high pressure conditions.

(b) There is a decrease in disorder when water freezes. Hence, the sign of the entropy change for the system is **negative.**

(c) There is an increase in disorder as gasoline evaporates. Hence, the sign of the entropy change for the system is **positive.**

17.44 *See Section 17.3.*

(a) There is an increase in disorder when water evaporates. Hence, the sign of the entropy change for the system is **positive.**

(b) There is an increase in disorder when butter melts.. Hence, the sign of the entropy change for the system is **positive.**

(c) There is a decrease in disorder when gold freezes.. Hence, the sign of the entropy change for the system is **negative.**

17.45 *See Section 17.3, Example 17.4, and Appendix G.*

(a) For $CO(g) + 2H_2(g) \rightarrow CH_3OH(\ell)$, there is a decrease in the number of moles of gas and formation of one mole of liquid and therefore an accompanying decrease in disorder. **The sign of the entropy change should be negative, and the magnitude of the entropy change should be rather large.**

$$\Delta S^\circ_{rxn} = \left[S^\circ CH_3OH(\ell)\right] - \left[S^\circ CO(g) + 2S^\circ H_2(g)\right]$$

$$\Delta S^\circ_{rxn} = \left[\left(1 \text{ mol} \times 126.8 \text{ J}\cdot\text{mol}^{-1}\cdot\text{K}^{-1}\right)\right] - \left[\left(1 \text{ mol} \times 197.56 \text{ J}\cdot\text{mol}^{-1}\cdot\text{K}^{-1}\right) + \left(2 \text{ mol} \times 130.57 \text{ J}\cdot\text{mol}^{-1}\cdot\text{K}^{-1}\right)\right]$$
$$= \mathbf{-331.9 \ J\cdot K^{-1}}$$

(b) For $N_2(g) + 3H_2(g) \rightarrow 2NH_3(g)$, there is a decrease in the number of moles of gas and therefore an accompanying decrease in disorder. **The sign of the entropy change should be negative, and the magnitude of the entropy change should be moderately large.**

$$\Delta S^\circ_{rxn} = \left[2S^\circ NH_3(g)\right] - \left[2S^\circ N_2(g) + 3S^\circ H_2(g)\right]$$
$$\Delta S^\circ_{rxn} = \left[\left(2 \text{ mol} \times 192.34 \text{ J}\cdot\text{mol}^{-1}\cdot\text{K}^{-1}\right)\right] - \left[\left(1 \text{ mol} \times 191.50 \text{ J}\cdot\text{mol}^{-1}\cdot\text{K}^{-1}\right) + \left(3 \text{ mol} \times 130.57 \text{ J}\cdot\text{mol}^{-1}\cdot\text{K}^{-1}\right)\right]$$
$$= \mathbf{-198.53 \ J\cdot K^{-1}}$$

(c) For $CH_4(g) + 2O_2(g) \rightarrow CO_2(g) + 2H_2O(\ell)$, there is a decrease in the number of moles of gas and formation of two moles of liquid and therefore an accompanying decrease in disorder. **The sign of the entropy change should be negative, and the magnitude of the entropy change should be rather large.**

$$\Delta S^\circ_{rxn} = \left[S^\circ CO_2(g) + 2S^\circ H_2O(\ell)\right] - \left[S^\circ CH_4(g) + 2S^\circ O_2(g)\right]$$
$$\Delta S^\circ_{rxn} = \left[\left(1 \text{ mol} \times 213.63 \text{ J}\cdot\text{mol}^{-1}\cdot\text{K}^{-1}\right) + \left(2 \text{ mol} \times 69.91 \text{ J}\cdot\text{mol}^{-1}\cdot\text{K}^{-1}\right)\right]$$
$$-\left[\left(1 \text{ mol} \times 186.15 \text{ J}\cdot\text{mol}^{-1}\cdot\text{K}^{-1}\right) + \left(2 \text{ mol} \times 205.03 \text{ J}\cdot\text{mol}^{-1}\cdot\text{K}^{-1}\right)\right] = \mathbf{-242.76 \ J\cdot K^{-1}}$$

(d) For $CO(g) + H_2O(g) \rightarrow CO_2(g) + H_2(g)$, there is no change in the number of moles of gas or even in the complexity of the reactants and products. Hence, **the entropy change should be relatively close to zero.**

$$\Delta S^\circ_{rxn} = \left[S^\circ CO_2(g) + S^\circ H_2(g)\right] - \left[S^\circ CO(g) + S^\circ H_2O(g)\right]$$
$$\Delta S^\circ_{rxn} = \left[\left(1 \text{ mol} \times 213.63 \text{ J}\cdot\text{mol}^{-1}\cdot\text{K}^{-1}\right) + \left(1 \text{ mol} \times 130.57 \text{ J}\cdot\text{mol}^{-1}\cdot\text{K}^{-1}\right)\right]$$
$$-\left[\left(1 \text{ mol} \times 197.56 \text{ J}\cdot\text{mol}^{-1}\cdot\text{K}^{-1}\right) + \left(1 \text{ mol} \times 188.72 \text{ J}\cdot\text{mol}^{-1}\cdot\text{K}^{-1}\right)\right] = \mathbf{-42.08 \ J\cdot K^{-1}}$$

17.46 *See Section 17.3, Example 17.4, and Appendix G.*

(a) For $C(s) + H_2O(g) \rightarrow CO(g) + H_2(g)$, there is an increase of one mole of gas and a decrease of one mole of solid and therefore an accompanying increase in disorder. **The sign of the entropy change should be positive, and the magnitude of the entropy change should be rather large.**

$\Delta S^\circ_{rxn} = [S^\circ CO(g) + S^\circ H_2(g)] - [S^\circ C(s) + S^\circ H_2O(g)]$
$S^\circ_{rxn} = [(1 \text{ mol x } 197.56 \text{ J}\cdot\text{mol}^{-1}\cdot\text{K}^{-1}) + (1 \text{ mol x } 130.57 \text{ J}\cdot\text{mol}^{-1}\cdot\text{K}^{-1})]$
$- [(1 \text{ mol x } 5.74 \text{ J}\cdot\text{mol}^{-1}\cdot\text{K}^{-1}) + (1 \text{ mol x } 188.72 \text{ J}\cdot\text{mol}^{-1}\cdot\text{K}^{-1})] = \mathbf{133.67 \ J\cdot K^{-1}}$

(b) For $2NO(g) + O_2 \rightarrow 2NO_2(g)$, there is a decrease of one mole of gas and therefore an accompanying decrease in disorder. **The sign of the entropy change should be negative, and the magnitude of the entropy change should be rather small.**

$\Delta S^\circ_{rxn} = [2S^\circ NO_2(g)] - [2S^\circ NO(g) + S^\circ O_2(g)]$
$= [(2 \text{ mol x } 239.95 \text{ J}\cdot\text{mol}^{-1}\cdot\text{K}^{-1})] - [(2 \text{ mol x } 210.65 \text{ J}\cdot\text{mol}^{-1}\cdot\text{K}^{-1}) + (1 \text{ mol x } 205.03 \text{ J}\cdot\text{mol}^{-1}\cdot\text{K}^{-1})]$
$= \mathbf{-146.43 \ J\cdot K^{-1}}$

(c) For $NaCl(s) \rightarrow Na^+(aq) + Cl^-(aq)$, there is an increase in disorder and therefore an accompanying increase in entropy. **The sign of the entropy change should be positive, and the magnitude of the entropy change should be small.**

$\Delta S^\circ_{rxn} = [S^\circ Na^+(aq) + S^\circ Cl^-(aq)] - [S^\circ NaCl(s)]$
$\Delta S^\circ_{rxn} = [(1 \text{ mol x } 59.0 \text{ J}\cdot\text{mol}^{-1}\cdot\text{K}^{-1}) + (1 \text{ mol x } 56.5 \text{ J}\cdot\text{mol}^{-1}\cdot\text{K}^{-1})] - [(1 \text{ mol x } 72.13 \text{ J}\cdot\text{mol}^{-1}\cdot\text{K}^{-1})]$
$= \mathbf{43.4 \ J\cdot K^{-1}}$

(d) For $C_5H_{12}(g) + 8O_2(g) \rightarrow 5CO_2(g) + 6H_2O(\ell)$, there is a decrease of four moles of gas and an increase of six moles of liquid and therefore an accompanying decrease in disorder. **The sign of the entropy change should be negative, and the magnitude of the entropy change should be large.**

$$\Delta S^\circ_{rxn} = \left[5S^\circ CO_2(g) + 6S^\circ H_2O(\ell)\right] - \left[S^\circ C_5H_{12}(g) - 8S^\circ O_2(g)\right]$$

$$\Delta S^\circ_{rxn} = \left[\left(5\,\text{mol} \times 213.63\ \text{J}\cdot\text{mol}^{-1}\cdot\text{K}^{-1}\right) + \left(6\,\text{mol} \times 69.91\ \text{J}\cdot\text{mol}^{-1}\cdot\text{K}^{-1}\right)\right]$$
$$- \left[\left(1\,\text{mol} \times 348.40\ \text{J}\cdot\text{mol}^{-1}\cdot\text{K}^{-1}\right) + \left(8\,\text{mol} \times 205.03\ \text{J}\cdot\text{mol}^{-1}\cdot\text{K}^{-1}\right)\right] = \mathbf{-501.03\ J\cdot K^{-1}}$$

17.47 *See Section 17.3, Example 17.4, and Appendix G.*

For $H_2(g) + CuO(s) \rightarrow H_2O(\ell) + Cu(s)$, there is a decrease of one mole of gas and an increase of one mole of liquid and therefore an accompanying decrease in disorder. **The sign of the entropy change should be negative, and the magnitude of the entropy change should be relatively large.**

$$\Delta S^\circ_{rxn} = \left[S^\circ H_2O(l) + S^\circ Cu(s)\right] - \left[S^\circ H_2(g) + S^\circ CuO(s)\right]$$

$$\Delta S^\circ_{rxn} = \left[\left(1\ \text{mol} \times 69.91\ \text{J}\cdot\text{mol}^{-1}\cdot\text{K}^{-1}\right) + \left(1\ \text{mol} \times 33.15\ \text{J}\cdot\text{mol}^{-1}\cdot\text{K}^{-1}\right)\right]$$
$$-\left[\left(1\ \text{mol} \times 130.57\ \text{J}\cdot\text{mol}^{-1}\cdot\text{K}^{-1}\right) + \left(1\ \text{mol} \times 42.63\ \text{J}\cdot\text{mol}^{-1}\cdot\text{K}^{-1}\right)\right] = \mathbf{-70.14\ J\cdot K^{-1}}$$

17.48 *See Section 17.3, Example 17.4, and Appendix G.*

For $2SO_2(g) + O_2(g) \rightarrow 2SO_3(g)$, there is a decrease of one mole of gas and an accompanying decrease in disorder. **The sign of the entropy change should be negative, and the magnitude of the entropy change should be rather large.**

$\Delta S^\circ_{rxn} = [2S^\circ\ SO_3(g)] - [2S^\circ\ SO_2(g) + S^\circ\ O_2(g)]$

$= [(2\ \text{mol} \times 256.65\ \text{J}\cdot\text{mol}^{-1}\cdot\text{K}^{-1})] - [(2\ \text{mol} \times 248.11\ \text{J}\cdot\text{mol}^{-1}\cdot\text{K}^{-1}) + (1\ \text{mol} \times 205.03\ \text{J}\cdot\text{mol}^{-1}\cdot\text{K}^{-1})]$

$= \mathbf{-187.95\ J\cdot K^{-1}}$

17.49 *See Section 17.4, Example 17.5, and Appendix G.*

(a) For $Fe_2O_3(s) + 2Al(s) \rightarrow Al_2O_3(s) + 2Fe(s)$,

$$\Delta G^\circ_{rxn} = \left[\Delta G^\circ_f Al_2O_3(s) + 2\Delta G^\circ_f Fe(s)\right] - \left[\Delta G^\circ_f Fe_2O_3(s) + 2\Delta G^\circ_f Al(s)\right]$$

$$\Delta G^\circ_{rxn} = \left[\left(1\ \text{mol} \times -1582.3\ \text{kJ}\cdot\text{mol}^{-1}\right) + \left(2\,\text{mol} \times 0.\ \text{kJ}\cdot\text{mol}^{-1}\right)\right]$$
$$-\left[\left(1\ \text{mol} \times -742.2\ \text{kJ}\cdot\text{mol}^{-1}\right) + \left(2\ \text{mol} \times 0.\ \text{kJ}\cdot\text{mol}^{-1}\right)\right] = \mathbf{-840.1\ kJ}$$

ΔG°_{rxn} is negative, and the reaction is spontaneous under standard conditions.

(b) For $CO(g) + 2H_2(g) \rightarrow CH_3OH(\ell)$,

$$\Delta G^\circ_{rxn} = \left[\Delta G^\circ_f CH_3OH(l)\right] - \left[\Delta G^\circ_f CO(g) + 2\Delta G^\circ_f H_2(g)\right]$$

$$\Delta G^\circ_{rxn} = \left[\left(1\ \text{mol} \times -166.35\ \text{kJ}\cdot\text{mol}^{-1}\right)\right] - \left[\left(1\ \text{mol} \times -137.15\ \text{kJ}\cdot\text{mol}^{-1}\right) + \left(2\ \text{mol} \times 0.\ \text{kJ}\cdot\text{mol}^{-1}\right)\right] = \mathbf{-29.2\ kJ}$$

ΔG°_{rxn} is negative, and the reaction is spontaneous under standard conditions.

17.50 *See Section 17.4, Example 17.5, and Appendix G.*

(a) For $4NH_3(g) + N_2(g) \rightarrow 3N_2H_4(g)$,

$$\Delta G^\circ_{rxn} = \left[\Delta G^\circ_f\ N_2H4(g)\right] - \left[4\Delta G^\circ_f NH_3(g) - \Delta G^\circ_f N_2(g)\right]$$

$\Delta G^\circ_{rxn} = \left[\left(3 \text{ mol} \times 149.24 \text{ kJ} \cdot \text{mol}^{-1}\right)\right] - \left[\left(4 \text{ mol} \times -16.48 \text{ kJ} \cdot \text{mol}^{-1}\right) + \left(1 \text{ mol} \times 0. \text{ kJ} \cdot \text{mol}^{-1}\right)\right] = 513.64 \text{ kJ}$

ΔG°_{rxn} is positive, and the reaction is nonspontaneous under standard conditions.

(b) For $2H_2O_2(\ell) \rightarrow 2H_2O(\ell) + O_2(g)$,

$\Delta G^\circ_{rxn} = \left[2\Delta G^\circ_f H_2O(\ell) + \Delta G^\circ_f O_2(g)\right] - \left[2\Delta G^\circ_f H_2O_2(\ell)\right]$

$\Delta G^\circ_{rxn} = \left[\left(2 \text{ mol} \times -237.18 \text{ kJ} \cdot \text{mol}^{-1}\right)\right] - \left[\left(1 \text{ mol} \times 0. \text{ kJ} \cdot \text{mol}^{-1}\right)\right] - \left[\left(2 \text{mol} \times -120.42 \text{ kJ} \cdot \text{mol}^{-1}\right)\right] = 233.52 \text{ kJ}$

ΔG°_{rxn} is negative, and the reaction is spontaneous under standard conditions.

17.51 *See Section 17.4, Example 17.5, and Appendix G.*

(a) For $Zn(s) + H_2SO_4(\ell) \rightarrow ZnSO_4(s) + H_2(g)$,

$\Delta G^\circ_{rxn} = \left[\Delta G^\circ_f ZnSO_4(s) + \Delta G^\circ_f H_2(g)\right] - \left[\Delta G^\circ_f Zn(s) + \Delta G^\circ_f H_2SO_4(\ell)\right]$

$\Delta G^\circ_{rxn} = \left[\left(1 \text{ mol} \times -871.5 \text{ kJ} \cdot \text{mol}^{-1}\right) + \left(1 \text{ mol} \times 0. \text{ kJ} \cdot \text{mol}^{-1}\right)\right]$

$-\left[\left(1 \text{ mol} \times 0. \text{ kJ} \cdot \text{mol}^{-1}\right) + \left(1 \text{ mol} \times -690.10 \text{ kJ} \cdot \text{mol}^{-1}\right)\right] = \mathbf{-181.4 \text{ kJ}}$

ΔG°_{rxn} is negative, and the reaction is spontaneous under standard conditions.

(b) For $Cu(s) + H_2SO_4(\ell) \rightarrow CuSO_4(s) + H_2(g)$,

$\Delta G^\circ_{rxn} = \left[\Delta G^\circ_f CuSO_4(s) + \Delta G^\circ_f H_2(g)\right] - \left[\Delta G^\circ_f Cu(s) + \Delta G^\circ_f H_2SO_4(\ell)\right]$

$\Delta G^\circ_{rxn} = \left[\left(1 \text{ mol} \times -661.9 \text{ kJ} \cdot \text{mol}^{-1}\right) + \left(1 \text{ mol} \times 0. \text{ kJ} \cdot \text{mol}^{-1}\right)\right]$

$-\left[\left(1 \text{ mol} \times 0. \text{ kJ} \cdot \text{mol}^{-1}\right) + \left(1 \text{ mol} \times -690.10 \text{ kJ} \cdot \text{mol}^{-1}\right)\right] = \mathbf{28.2 \text{ kJ}}$

ΔG°_{rxn} is positive, and the reaction is nonspontaneous under standard conditions.

17.52 *See Section 17.4, Example 17.5, and Appendix G.*

(a) For $2NO(g) + O_2(g) \rightarrow 2NO_2(g)$,

$\Delta G^\circ_{rxn} = \left[2\Delta G^\circ_f NO_2(g)\right] - \left[2\Delta G^\circ_f NO(g) + \Delta G^\circ_f O_2(g)\right]$

$\Delta G^\circ_{rxn} = \left[\left(2 \text{ mol} \times 51.29 \text{ kJ} \cdot \text{mol}^{-1}\right)\right] - \left[\left(2 \text{mol} \times 86.55 \text{kJ} \cdot \text{mol}^{-1}\right) + \left(1 \text{mol} \times 0. \text{kJ} \cdot \text{mol}^{-1}\right)\right] = \mathbf{-70.52 \text{ kJ}}$

ΔG°_{rxn} is negative, and the reaction is spontaneous under standard conditions.

(b) For $CO(g) + Cl_2(g) \rightarrow COCl_2(g)$,

$\Delta G^\circ_{rxn} = \left[\Delta G^\circ_f COCl_2(g)\right] - \left[\Delta G^\circ_f CO(g) + \Delta G^\circ_f Cl_2(g)\right]$

$\Delta G^\circ_{rxn} = \left[\left(1 \text{ mol} \times -204.6 \text{ kJ} \cdot \text{mol}^{-1}\right)\right] - \left[\left(1 \text{ mol} \times -137.15 \text{ kJ} \cdot \text{mol}^{-1}\right) + \left(1 \text{ mol} \times 0. \text{ kJ} \cdot \text{mol}^{-1}\right)\right] = \mathbf{-67.4 \text{ kJ}}$

ΔG°_{rxn} is negative, and the reaction is spontaneous under standard conditions.

17.53 *See Section 17.4 and Example 17.6.*

(a) For $H_2(g) + \frac{1}{2}O_2(g) \rightarrow H_2O(\ell)$,

$\Delta H^\circ_{rxn} = \left[\Delta H^\circ_f H_2O(\ell)\right] - \left[\Delta H^\circ_f H_2(g) + \frac{1}{2}\Delta H^\circ_f O_2(g)\right]$

$\Delta H^\circ_{rxn} = \left[\left(1 \text{ mol} \times -285.83 \text{ kJ} \cdot \text{mol}^{-1}\right)\right] - \left[\left(1 \text{ mol} \times 0. \text{ kJ} \cdot \text{mol}^{-1}\right) + \left(\frac{1}{2} \text{ mol} \times 0. \text{ kJ} \cdot \text{mol}^{-1}\right)\right] = -285.83 \text{ kJ}$

$\Delta S^\circ_{rxn} = \left[S^\circ H_2O(\ell)\right] - \left[S^\circ H_2(g) + \frac{1}{2}S^\circ O_2(g)\right]$

$\Delta S^\circ_{rxn} = \left[\left(1 \text{ mol} \times 69.91 \text{ J}\cdot\text{mol}^{-1}\cdot\text{K}^{-1}\right)\right] - \left[\left(1 \text{ mol} \times 130.57 \text{ J}\cdot\text{mol}^{-1}\cdot\text{K}^{-1}\right) + \left(\frac{1}{2} \text{ mol} \times 205.03 \text{ J}\cdot\text{mol}^{-1}\cdot\text{K}^{-1}\right)\right]$

$= -163.18 \text{ J}\cdot\text{K}^{-1}$

At 25°C, $\Delta G^\circ_{298} = \Delta H^\circ - (298 \text{ K})\Delta S^\circ$

$\Delta G^\circ_{298} = -285.83 \times 10^3 \text{ J} - (298 \text{ K})(-163.18 \text{ J}\cdot\text{K}^{-1}) = \mathbf{-2.372 \times 10^5 \text{ J}}$ or **-237.2 kJ**

The value of ΔG°_{298} is negative, and the reaction is **spontaneous** in the forward direction. **This is consistent with its favorable (ΔH° = -) enthalpy change, but not consistent with its unfavorable (ΔS° = -) entropy change.**

(b) For $CO(g) + 2H_2(g) \rightarrow CH_3OH(\ell)$,

$\Delta H^\circ_{rxn} = \left[\Delta H^\circ_f CH_3OH(\ell)\right] - \left[\Delta H^\circ_f CO(g) + 2\Delta H^\circ_f H_2(g)\right]$

$\Delta H^\circ_{rxn} = \left[\left(1 \text{ mol} \times -238.66 \text{ kJ}\cdot\text{mol}^{-1}\right)\right] - \left[\left(1 \text{ mol} \times -110.52 \text{ kJ}\cdot\text{mol}^{-1}\right) + \left(2 \text{ mol} \times 0. \text{ kJ}\cdot\text{mol}^{-1}\right)\right]$

$= -128.14 \text{ kJ}$

$\Delta S^\circ_{rxn} = \left[S^\circ CH_3OH(l)\right] - \left[S^\circ CO(g) + 2S^\circ H_2(g)\right]$

$\Delta S^\circ_{rxn} = \left[\left(1 \text{ mol} \times 126.8 \text{ J}\cdot\text{mol}^{-1}\cdot\text{K}^{-1}\right)\right] - \left[\left(1 \text{ mol} \times 197.56 \text{ J}\cdot\text{mol}^{-1}\cdot\text{K}^{-1}\right) + \left(2 \text{ mol} \times 130.57 \text{ J}\cdot\text{mol}^{-1}\cdot\text{K}^{-1}\right)\right]$

$= -331.9 \text{ J}\cdot\text{K}^{-1}$

At 25°C, $\Delta G^\circ_{298} = \Delta H^\circ - (298 \text{ K})\Delta S^\circ$

$\Delta G^\circ_{298} = -128.14 \times 10^3 \text{ J} - (298 \text{ K})(-331.9 \text{ J}\cdot\text{K}^{-1}) = \mathbf{-2.92 \times 10^4 \text{ J}}$ or **-29.2 kJ**

The value of ΔG°_{298} is negative, and the reaction is **spontaneous** in the forward direction. **This is consistent with its favorable (ΔH° = -) enthalpy change, but not consistent with its unfavorable (ΔS° = -) entropy change.**

17.54 ***See Section 17.4 and Example 17.6.***

(a) For $2H_2O(g) + 2Cl_2(g) \rightarrow 4HCl(g) + O_2(g)$,

$\Delta H^\circ_{rxn} - \left[4\Delta H^\circ_f HCl(g) + \Delta H^\circ_f O_2(g)\right] - \left[2\Delta H^\circ_f H_2O(g) + 2\Delta H^\circ_f Cl_2(g)\right]$

$\Delta H^\circ_{rxn} = \left[\left(4 \text{mol} \times -92.31 \text{ kJ}\cdot\text{mol}^{-1}\right) + \left(1\text{mol} \times 0.\text{kJ}\cdot\text{mol}^{-1}\right)\right]$

$- \left[\left(2\text{mol} \times -241.82 \text{ kJ}\cdot\text{mol}^{-1}\right) + \left(2\text{mol} \times 0. \text{ kJ}\cdot\text{mol}^{-1}\right)\right] = 114.40 \text{ kJ}$

$\Delta S^\circ_{rxn} = \left[4S^\circ HCl(g) + S^\circ O_2(g)\right] - \left[2S^\circ H_2O(g) + 2S^\circ Cl_2(g)\right]$

$\Delta S^\circ_{rxn} = \left[\left(4\text{mol} \times 186.80 \text{ J}\cdot\text{mol}^{-1}\cdot\text{K}^{-1}\right) + \left(1\text{mol} \times 205.03 \text{ J}\cdot\text{mol}^{-1}\cdot\text{K}^{-1}\right)\right]$

$- \left[\left(2\text{mol} \times 188.72 \text{ J}\cdot\text{mol}^{-1}\cdot\text{K}^{-1}\right) + \left(1\text{mol} \times 222.96 \text{ J}\cdot\text{mol}^{-1}\cdot\text{K}^{-1}\right)\right] = 128.77 \text{ J}\cdot\text{K}^{-1}$

At 25°C, $\Delta G^\circ_{298} = 114.40 \text{ kJ} - (298 \text{ K})(128.87 \times 10^{-3} \text{ kJ}\cdot\text{K}^{-1}) = 76.00 \text{ kJ}$

The value of ΔG°_{298} is positive, and the reaction is **nonspontaneous** in the forward direction. **This is consistent with its unfavorable (ΔH° = +) enthalpy change, but not consistent with its favorable (ΔS° = +) entropy change.**

(b) For $2CO_2(g) + 4H_2O(\ell) \rightarrow 2CH_3OH(\ell) + 3O_2(g)$,

$\Delta H^\circ_{rxn} = \left[2\Delta H^\circ_f CH_3OH(g) + 3\Delta H^\circ_f O_2(g)\right] - \left[2\Delta H^\circ_f CO_2(g) + 4\Delta H^\circ_f H_2O(\ell)\right]$

$\Delta H^\circ_{rxn} = \left[\left(2\text{mol} \times -238.66 \text{ kJ}\cdot\text{mol}^{-1}\right) + \left(3\text{mol} \times 0.\text{kJ}\cdot\text{mol}^{-1}\right)\right]$

$- \left[\left(2\text{mol} \times -393.51 \text{ kJ}\cdot\text{mol}^{-1}\right) + \left(4\text{mol} \times -285.83 \text{ kJ}\cdot\text{mol}^{-1}\right)\right] = 1453.02 \text{ kJ}$

$\Delta S^\circ_{rxn} = \left[2S^\circ CH_3OH(\ell) + 3S^\circ O_2(g)\right] - \left[2S^\circ CO_2(g) + 4S^\circ H_2O(\ell)\right]$

$\Delta S^\circ_{rxn} = \left[\left(2\text{mol} \times 126.8 \text{ J}\cdot\text{mol}^{-1}\cdot\text{K}^{-1}\right) + \left(3\text{mol} \times 205.03 \text{ J}\cdot\text{mol}^{-1}\cdot\text{K}^{-1}\right)\right]$

$$\Delta S^\circ_{rxn} = \left[(2\text{mol} \times 126.8\ \text{J}\cdot\text{mol}^{-1}\cdot\text{K}^{-1}) + (3\text{mol} \times 205.03\ \text{J}\cdot\text{mol}^{-1}\cdot\text{K}^{-1})\right]$$
$$- \left[(2\text{mol} \times 213.63\ \text{J}\cdot\text{mol}^{-1}\cdot\text{K}^{-1}) + (4\text{mol} \times 69.91\ \text{J}\cdot\text{mol}^{-1}\cdot\text{K}^{-1})\right] = 161.79\ \text{J}\cdot\text{K}^{-1}$$

At 25°C, $\Delta G^\circ_{298} = \Delta H^\circ - (298\ \text{K})\ \Delta S^\circ$

$\Delta G^\circ_{298} = 1453.02\ \text{kJ} - (298\ \text{K})(161.79 \times 10^{-3}\ \text{kJ}\cdot\text{K}^{-1}) = 1404.81\ \text{kJ}$

The value of ΔG°_{298} is positive, and the reaction is **nonspontaneous** in the forward direction. **This is consistent with its unfavorable (ΔH° = +) enthalpy change, but inconsistent with its favorable (ΔS° = +) entropy change.**

17.55 ***See Section 17.4 and Example 17.6.***

(a) For $CH_3COOH(\ell) + NaOH(s) \rightarrow Na^+(aq) + CH_3COO^-(aq) + H_2O(\ell)$,

$$\Delta H^\circ_{rxn} = \left[\Delta H^\circ_f Na^+(aq) + \Delta H^\circ_f CH_3COO^-(aq) + \Delta H^\circ_f H_2O(\ell)\right] - \left[\Delta H^\circ_f CH_3COOH(\ell) + \Delta H^\circ_f NaOH(s)\right]$$

$$\Delta H^\circ_{rxn} = \left[(1\ \text{mol} \times -240.12\ \text{kJ}\cdot\text{mol}^{-1}) + (1\ \text{mol} \times -486.01\ \text{kJ}\cdot\text{mol}^{-1}) + (1\ \text{mol} \times -285.83\ \text{kJ}\cdot\text{mol}^{-1})\right]$$
$$-\left[(1\ \text{mol} \times -484.5\ \text{kJ}\cdot\text{mol}^{-1}) + (1\ \text{mol} \times -425.61\ \text{kJ}\cdot\text{mol}^{-1})\right] = -101.8\ \text{kJ}$$

$$\Delta S^\circ_{rxn} = \left[S^\circ Na^+(aq) + S^\circ CH_3COO^-(aq) + S^\circ H_2O(\ell)\right] - \left[S^\circ CH_3COOH(\ell) + S^\circ NaOH(s)\right]$$

$$\Delta S^\circ_{rxn} = \left[(1\ \text{mol} \times 59.0\ \text{J}\cdot\text{mol}^{-1}\cdot\text{K}^{-1}) + (1\ \text{mol} \times 86.6\ \text{J}\cdot\text{mol}^{-1}\cdot\text{K}^{-1}) + (1\ \text{mol} \times 69.91\ \text{J}\cdot\text{mol}^{-1}\cdot\text{K}^{-1})\right]$$
$$-\left[(1\ \text{mol} \times 159.8\ \text{J}\cdot\text{mol}^{-1}\cdot\text{K}^{-1}) + (1\ \text{mol} \times 64.46\ \text{J}\cdot\text{mol}^{-1}\cdot\text{K}^{-1})\right] = -8.8\ \text{J}\cdot\text{K}^{-1}$$

At 25°C, $\Delta G^\circ_{298} = \Delta H^\circ - (298\ \text{K})\Delta S^\circ$

$\Delta G^\circ_{298} = -101.8 \times 10^3\ \text{J} - (298\ \text{K})(-8.8\ \text{J}\cdot\text{K}^{-1}) = \mathbf{-9.92 \times 10^4\ J}$ or **-99.2 kJ**

The reaction has a highly favorable enthalpy change and a slightly unfavorable entropy change. At low temperatures, the highly favorable ΔH° term dominates the slightly unfavorable $T\Delta S^\circ$. Hence, the value of ΔG°_{298} is negative, and the reaction is spontaneous in the forward direction.

(b) For $AgNO_3(s) + Cl^-(aq) \rightarrow AgCl(s) + NO_3^-(aq)$,

$$\Delta H^\circ_{rxn} = \left[\Delta H^\circ_f AgCl(s) + \Delta H^\circ_f NO_3^-(aq)\right] - \left[\Delta H^\circ_f AgNO_3(s) + \Delta H^\circ_f Cl^-(aq)\right]$$

$$\Delta H^\circ_{rxn} = \left[(1\ \text{mol} \times -127.07\ \text{kJ}\cdot\text{mol}^{-1}) + (1\ \text{mol} \times -205.0\ \text{kJ}\cdot\text{mol}^{-1})\right]$$
$$-\left[(1\ \text{mol} \times -124.39\ \text{kJ}\cdot\text{mol}^{-1}) + (1\ \text{mol} \times -167.16\ \text{kJ}\cdot\text{mol}^{-1})\right] = -40.5\ \text{kJ}$$

$$\Delta S^\circ_{rxn} = \left[S^\circ AgCl(s) + S^\circ NO_3^-(aq)\right] - \left[S^\circ AgNO_3(s) + S^\circ Cl^-(aq)\right]$$

$$\Delta S^\circ_{rxn} = \left[(1\ \text{mol} \times 96.2\ \text{J}\cdot\text{mol}^{-1}\cdot\text{K}^{-1}) + (1\ \text{mol} \times 146.4\ \text{J}\cdot\text{mol}^{-1}\cdot\text{K}^{-1})\right]$$
$$-\left[(1\ \text{mol} \times 140.92\ \text{J}\cdot\text{mol}^{-1}\cdot\text{K}^{-1}) + (1\ \text{mol} \times 56.5\ \text{J}\cdot\text{mol}^{-1}\cdot\text{K}^{-1})\right] = 45.2\ \text{J}\cdot\text{K}^{-1}$$

At 25°C, $\Delta G^\circ_{298} = \Delta H^\circ - (298\ \text{K})\Delta S^\circ$

$\Delta G^\circ_{298} = -40.5 \times 10^3\ \text{J} - (298\ \text{K})(45.2\ \text{J}\cdot\text{K}^{-1}) = \mathbf{-5.40 \times 10^4\ J}$ or **-54.0 kJ**

The reaction has a favorable enthalpy change and a favorable entropy change. Hence, the reaction is spontaneous in the forward direction at all temperatures.

17.56 ***See Section 17.4 and Example 17.6.***

(a) For $Fe_2O_3(s) + 3Cu(s) \rightarrow 3CuO(s) + 2Fe(s)$,

$$\Delta H^\circ_{rxn} = \left[3\Delta H^\circ_f CuO(s) + 2\Delta H^\circ_f Fe(s)\right] - \left[\Delta H^\circ_f Fe_2O_3(s) + 3\Delta H^\circ_f Cu(s)\right]$$

$$\Delta H^\circ_{rxn} = \left[(3 \text{ mol} \times -157.3 \text{ kJ}\cdot\text{mol}^{-1}) + (2 \text{ mol} \times 0. \text{ kJ}\cdot\text{mol}^{-1})\right]$$
$$-\left[(1 \text{ mol} \times -824.2 \text{ kJ}\cdot\text{mol}^{-1}) + (3 \text{ mol} \times 0. \text{ kJ}\cdot\text{mol}^{-1})\right] = 352.3 \text{ kJ}$$

$$\Delta S^\circ_{rxn} = \left[3S^\circ CuO(s) + 2S^\circ Fe(s)\right] - \left[S^\circ Fe_2O_3(s) + 3S^\circ Cu(s)\right]$$

$$\Delta S^\circ_{rxn} = \left[(3 \text{ mol} \times 42.63 \text{ J}\cdot\text{mol}^{-1}\cdot\text{K}^{-1}) + (2 \text{ mol} \times 27.28 \text{ J}\cdot\text{mol}^{-1}\cdot\text{K}^{-1})\right]$$
$$-\left[(1 \text{ mol} \times 87.40 \text{ J}\cdot\text{mol}^{-1}\cdot\text{K}^{-1}) + (3 \text{ mol} \times 33.15 \text{ J}\cdot\text{mol}^{-1}\cdot\text{K}^{-1})\right] = -4.40 \text{J}\cdot\text{K}^{-1}$$

At 25°C, $\Delta G^\circ_{298} = \Delta H^\circ - (298 \text{ K})\Delta S^\circ$

$\Delta G^\circ_{298} = 352.3 \text{ kJ} - (298 \text{ K})(-4.40 \times 10^{-3} \text{ kJ}\cdot\text{K}^{-1}) = \mathbf{353.6\ kJ}$

The value of ΔG°_{298} is positive, and the reaction is **nonspontaneous** in the forward direction. **This is consistent with its unfavorable ($\Delta H^\circ = +$) enthalpy change and with its unfavorable ($\Delta S^\circ = -$) entropy change.**

(b) For $SO_3(g) + NaOH(s) \rightarrow NaHSO_4(s)$,

$$\Delta H^\circ_{rxn} = \left[\Delta H^\circ_f NaHSO_4(s)\right] - \left[\Delta H^\circ_f SO_3(g) + \Delta H^\circ_f NaOH(s)\right]$$

$$\Delta H^\circ_{rxn} = \left[(1 \text{ mol} \times -1125.5 \text{ kJ}\cdot\text{mol}^{-1})\right] - \left[(1\text{mol} \times -395.72 \text{ kJ}\cdot\text{mol}^{-1}) + (1\text{mol} \times -425.61 \text{ kJ}\cdot\text{mol}^{-1})\right]$$
$$= -304.17 \text{ kJ}$$

$$\Delta S^\circ_{rxn} = \left[S^\circ NaHSO_4(s)\right] - \left[S^\circ SO_3(g) + S^\circ NaOH(s)\right]$$

$$\Delta S^\circ_{rxn} = \left[(1 \text{ mol} \times 113.0 \text{ J}\cdot\text{mol}^{-1}\cdot\text{K}^{-1})\right] - \left[(1\text{mol} \times 256.65 \text{ J}\cdot\text{mol}^{-1}\cdot\text{K}^{-1}) + (1\text{mol} \times 64.46 \text{ J}\cdot\text{mol}^{-1}\cdot\text{K}^{-1})\right]$$
$$= -208.1 \text{ J}\cdot\text{K}^{-1}$$

At 25°C, $\Delta G^\circ_{298} = \Delta H^\circ - (298 \text{ K})\Delta S^\circ$

$\Delta G^\circ_{298} = -304.17 \text{ kJ} - (298 \text{ K})(-208.1 \times 10^{-3} \text{ J}\cdot\text{K}^{-1}) = \mathbf{-242.15\ kJ}$

The value of ΔG°_{298} is negative, and the reaction is **spontaneous** in the forward direction under standard conditions at 298 K. **This is consistent with its favorable ($\Delta H^\circ = -$) enthalpy change but inconsistent with its unfavorable ($\Delta S^\circ = -$) entropy change.**

17.57 ***See Section 17.4 and Example 17.6.***

For $CH_3CHO(\ell) + \frac{5}{2}O_2(g) \rightarrow 2CO_2(g) + 2H_2O(\ell)$,

$$\Delta H^\circ_{rxn} = \left[2\Delta H^\circ_f CO_2(g) + 2\Delta H^\circ_f H_2O(\ell)\right] - \left[\Delta H^\circ_f CH_3CHO(\ell) + \tfrac{5}{2}\Delta H^\circ_f O_2(g)\right]$$

$$\Delta H^\circ_{rxn} = \left[(2 \text{ mol} \times -393.51 \text{ kJ}\cdot\text{mol}^{-1}) + (2 \text{ mol} \times -285.83 \text{ kJ}\cdot\text{mol}^{-1})\right]$$
$$-\left[(1 \text{ mol} \times -192.30 \text{ kJ}\cdot\text{mol}^{-1}) + (\tfrac{5}{2} \text{ mol} \times 0. \text{ kJ}\cdot\text{mol}^{-1})\right] = -1166.38 \text{ kJ}$$

$$\Delta S^\circ_{rxn} = \left[2S^\circ CO_2(g) + 2S^\circ H_2O(\ell)\right] - \left[S^\circ CH_3CHO(\ell) + \tfrac{5}{2}S^\circ O_2(g)\right]$$

$$\Delta S^\circ_{rxn} = \left[(2 \text{ mol} \times 213.63 \text{ J}\cdot\text{mol}^{-1}\cdot\text{K}^{-1}) + (2 \text{ mol} \times 69.91 \text{ J}\cdot\text{mol}^{-1}\cdot\text{K}^{-1})\right]$$
$$-\left[(1 \text{ mol} \times 160.2 \text{ J}\cdot\text{mol}^{-1}\cdot\text{K}^{-1}) + (\tfrac{5}{2} \text{ mol} \times 205.03 \text{ J}\cdot\text{mol}^{-1}\cdot\text{K}^{-1})\right] = -105.7 \text{ J}\cdot\text{K}^{-1}$$

The reaction has a highly favorable enthalpy change and an unfavorable entropy change. At **low temperatures**, the favorable ΔH° term will dominate the unfavorable $T\Delta S^\circ$ term, and the reaction will be **spontaneous. As temperature increases**, the importance of the unfavorable $T\Delta S^\circ$ term will increase, and eventually **the reaction will become nonspontaneous.**

17.58 *See Section 17.4 and Example 17.6.*

For $H_2(g) + \frac{1}{8}S_8(s) \rightarrow H_2S(g)$,

$$\Delta H^\circ_{rxn} = \left[\Delta H^\circ_f H_2S(g)\right] - \left[\Delta H^\circ_f H_2(g) + \tfrac{1}{8}\Delta H^\circ_f S_8(s)\right]$$

$$\Delta H^\circ_{rxn} = \left[\left(1\text{ mol} \times -20.63\text{ kJ}\cdot\text{mol}^{-1}\right)\right] - \left[\left(1\text{ mol} \times 0.0\text{ kJ}\cdot\text{mol}^{-1}\right) + \left(\tfrac{1}{8}\text{ mol} \times 0.0\text{ kJ}\cdot\text{mol}^{-1}\right)\right] = -20.63\text{ kJ}$$

$$\Delta S^\circ_{rxn} = \left[S^\circ H_2S(g)\right] - \left[\Delta S^\circ H_2(g) + \tfrac{1}{8}S^\circ S_8(s)\right]$$

$$= \left[\left(1\text{ mol} \times 205.68\text{ J}\cdot\text{mol}^{-1}\cdot\text{K}^{-1}\right)\right] - \left[\left(1\text{ mol} \times 130.57\text{ J}\cdot\text{mol}^{-1}\cdot\text{K}^{-1}\right) + \left(\tfrac{1}{8}\text{mol} \times 254.4\text{ J}\cdot\text{mol}^{-1}\cdot\text{K}^{-1}\right)\right]$$

$$= 43.31\text{ J}\cdot\text{K}^{-1}$$

The reaction has a favorable enthalpy change and a favorable entropy change. It will be **spontaneous at all temperatures.**

17.59 *See Section 17.4 and Example 17.6.*

For $N_2(g) + 3H_2(g) \rightarrow 2NH_3(g)$,

$$\Delta H^\circ_{rxn} = \left[2\Delta H^\circ_f NH_3(g)\right] - \left[\Delta H^\circ_f N_2(g) + 3\Delta H^\circ_f H_2(g)\right]$$

$$\Delta H^\circ_{rxn} = \left[\left(2\text{ mol} \times -46.11\text{ kJ}\cdot\text{mol}^{-1}\right)\right] - \left[\left(1\text{ mol} \times 0.\text{ kJ}\cdot\text{mol}^{-1}\right) + \left(3\text{ mol} \times 0.\text{ kJ}\cdot\text{mol}^{-1}\right)\right] = -92.22\text{ kJ}$$

$$\Delta S^\circ_{rxn} = \left[2S^\circ NH_3(g)\right] - \left[S^\circ N_2(g) + 3S^\circ H_2(g)\right]$$

$$\Delta S^\circ_{rxn} = \left[\left(2\text{ mol} \times 192.34\text{ J}\cdot\text{mol}^{-1}\cdot\text{K}^{-1}\right)\right] - \left[\left(1\text{ mol} \times 191.50\text{ J}\cdot\text{mol}^{-1}\cdot\text{K}^{-1}\right) + \left(3\text{ mol} \times 130.57\text{ J}\cdot\text{mol}^{-1}\cdot\text{K}^{-1}\right)\right]$$

$$= -198.53\text{ J}\cdot\text{K}^{-1}$$

The reaction has a favorable enthalpy change and an unfavorable entropy change. **At low temperatures**, the favorable ΔH° term will dominate the unfavorable $T\Delta S^\circ$ term, and the reaction will be **spontaneous. As temperature increases**, the importance of the unfavorable $T\Delta S^\circ$ term will increase, and eventually the reaction will become **nonspontaneous**.

17.60 *See Section 17.4 and Example 17.6.*

For $COCl_2(g) \rightarrow CO(g) + Cl_2(g)$,

$$\Delta H^\circ_{rxn} = \left[\Delta H^\circ_f CO(g) + \Delta H^\circ_f Cl_2(g)\right] - \left[\Delta H^\circ_f COCl_2(g)\right]$$

$$\Delta H^\circ_{rxn} = \left[\left(1\text{mol} \times -110.52\text{ kJ}\cdot\text{mol}^{-1}\right) + \left(1\text{mol} \times 0.\text{ kJ}\cdot\text{mol}^{-1}\right)\right] - \left[\left(1\text{mol} \times -218.8\text{ kJ}\cdot\text{mol}^{-1}\right)\right] = 108.3\text{ kJ}$$

$$\Delta S^\circ_{rxn} = \left[S^\circ CO(g) + S^\circ Cl_2(g)\right] - \left[S^\circ COCl_2(g)\right]$$

$$\Delta S^\circ_{rxn} = \left[\left(1\text{mol} \times 197.56\text{ J}\cdot\text{mol}^{-1}\cdot\text{K}^{-1}\right) + \left(1\text{mol} \times 222.96\text{ J}\cdot\text{mol}^{-1}\cdot\text{K}^{-1}\right)\right] - \left[\left(1\text{mol} \times 283.53\text{ J}\cdot\text{mol}^{-1}\cdot\text{K}^{-1}\right)\right]$$

$$= 136.99\text{ J}\cdot\text{K}^{-1}$$

The reaction has an unfavorable enthalpy change an a favorable entropy change. **At low temperatures** the unfavorable ΔH° term will dominate the favorable entropy term, and the reaction will be **nonspontaneous. As temperature increases** the importance of the favorable $T\Delta S^\circ$ term will increase and eventually the reaction will become **spontaneous.**

17.61 *See Section 17.4 and Example 17.6.*

(a) For $CO(g) + Cl_2(g) \rightarrow COCl_2(g)$,

$$\Delta H^\circ_{rxn} = \left[\Delta H^\circ_f COCl_2(g)\right] - \left[\Delta H^\circ_f CO(g) + \Delta H^\circ_f Cl_2(g)\right]$$

$$\Delta H^\circ_{rxn} = \left[\left(1\text{ mol} \times -218.8\text{ kJ}\cdot\text{mol}^{-1}\right)\right] - \left[\left(1\text{ mol} \times -110.52\text{ kJ}\cdot\text{mol}^{-1}\right) + \left(1\text{ mol} \times 0.\text{ kJ}\cdot\text{mol}^{-1}\right)\right] = -108.28\text{ kJ}$$

$$\Delta S^{\circ}_{rxn} = \left[S^{\circ}COCl_2(g)\right] - \left[S^{\circ}CO(g) + S^{\circ}Cl_2(g)\right]$$

$$\Delta S^{\circ}_{rxn} = \left[\left(1 \text{ mol} \times 283.53 \text{ J}\cdot\text{mol}^{-1}\cdot\text{K}^{-1}\right)\right] - \left[\left(1 \text{ mol} \times 197.56 \text{ J}\cdot\text{mol}^{-1}\cdot\text{K}^{-1}\right) + \left(1 \text{ mol} \times 222.96 \text{ J}\cdot\text{mol}^{-1}\cdot\text{K}^{-1}\right)\right]$$

$$= -136.99 \text{ J}\cdot\text{K}^{-1}$$

At 25°C, $\Delta G^{\circ}_{298} = \Delta H^{\circ} - (298 \text{ K})\Delta S^{\circ}$

$\Delta G^{\circ}_{298} = -108.28 \times 10^3 \text{ J} - (298 \text{ K})\left(-136.99 \text{ J}\cdot\text{K}^{-1}\right) = \mathbf{-6.746 \times 10^4 \text{ J}}$ or **-67.46 kJ**

At the temperature at which there is no net driving force in either direction, $\Delta G^{\circ} = 0$.

At this temperature, $0 = \Delta H^{\circ} - T\Delta S^{\circ}$. Hence,

$$T = \frac{\Delta H^{\circ}}{\Delta S^{\circ}} \qquad T = \frac{-108.28 \times 10^3 \text{ J}}{-136.99 \text{ J}\cdot\text{K}^{-1}} = 790 \text{ K}$$

The reaction is spontaneous in the direction written at 298 K ($\Delta G^{\circ}_{298} = -$) and experiences no net driving force in either direction at 790 K ($\Delta G^{\circ}_{790} = 0$). Above 790 K the reaction becomes nonspontaneous in the direction written. Hence, the reaction is spontaneous **up to 790 K**.

(b) For $NO(g) + \frac{1}{2}O_2(g) \rightarrow NO_2(g)$,

$$\Delta H^{\circ}_{rxn} = \left[\Delta H^{\circ}_f NO_2(g)\right] - \left[\Delta H^{\circ}_f NO(g) + \tfrac{1}{2}\Delta H^{\circ}_f O_2(g)\right]$$

$$\Delta H^{\circ}_{rxn} = \left[\left(1 \text{ mol} \times 33.18 \text{ kJ}\cdot\text{mol}^{-1}\right)\right] - \left[\left(1 \text{ mol} \times 90.25 \text{ kJ}\cdot\text{mol}^{-1}\right) + \left(\tfrac{1}{2} \text{ mol} \times 0. \text{ kJ}\cdot\text{mol}^{-1}\right)\right] = -57.07 \text{ kJ}$$

$$\Delta S^{\circ}_{rxn} = \left[S^{\circ}NO_2(g)\right] - \left[S^{\circ}NO(g) + \tfrac{1}{2}S^{\circ}O_2(g)\right]$$

$$\Delta S^{\circ}_{rxn} = \left[\left(1 \text{ mol} \times 239.95 \text{ J}\cdot\text{mol}^{-1}\cdot\text{K}^{-1}\right)\right] - \left[\left(1 \text{ mol} \times 210.65 \text{ J}\cdot\text{mol}^{-1}\cdot\text{K}^{-1}\right) + \left(\tfrac{1}{2} \text{ mol} \times 205.03 \text{ J}\cdot\text{mol}^{-1}\cdot\text{K}^{-1}\right)\right]$$

$$= -73.22 \text{ J}\cdot\text{K}^{-1}$$

At 25°C, $\Delta G^{\circ}_{298} = \Delta H^{\circ} - (298 \text{ K})\Delta S^{\circ}$

$\Delta G^{\circ}_{298} = -57.07 \times 10^3 \text{ J} - (298 \text{ K})\left(-73.22 \text{ J}\cdot\text{K}^{-1}\right) = \mathbf{-3.52 \times 10^3 \text{ J}}$ or **-3.52 kJ**

At the temperature at which there is no net driving force in either direction, $\Delta G^{\circ} = 0$.

At this temperature, $0 = \Delta H^{\circ} - T\Delta S^{\circ}$. Hence,

$$T = \frac{\Delta H^{\circ}}{\Delta S^{\circ}} \qquad T = \frac{-57.07 \times 10^3 \text{ J}}{-73.22 \text{ J}\cdot\text{K}^{-1}} = 779 \text{ K}$$

The reaction is spontaneous in the direction written at 298 K ($\Delta G^{\circ}_{298} = -$) and experiences no net driving force in either direction at 779 K ($\Delta G^{\circ}_{779} = 0$). Above 779 K the reaction becomes nonspontaneous in the direction written. Hence, the reaction is **spontaneous up to 779 K**.

17.62 ***See Section 17.4 and Example 17.6.***

(a) For $2POCl_3(g) \rightarrow 2PCl_3(g) + O_2(g)$,

$$\Delta H^{\circ}_{rxn} = \left[2\Delta H^{\circ}_f PCl_3(g) + \Delta H^{\circ}_f O_2(g)\right] - \left[2\Delta H^{\circ}_f POCl_3(g)\right]$$

$$\Delta H^{\circ}_{rxn} = \left[\left(2 \text{mol} \times -287.0 \text{ kJ}\cdot\text{mol}^{-1}\right) + \left(1 \text{mol} \times 0. \text{ kJ}\cdot\text{mol}^{-1}\right)\right] - \left[\left(2 \text{mol} \times -592.7 \text{ kJ}\cdot\text{mol}^{-1}\right)\right] = 611.4 \text{ kJ}$$

$$\Delta S^{\circ}_{rxn} = \left[2S^{\circ}PCl_3(g) + S^{\circ}O_2(g)\right] - \left[2S^{\circ}POCl_3(g)\right]$$

$$\Delta S^{\circ}_{rxn} = \left[\left(2 \text{mol} \times 311.67 \text{ J}\cdot\text{mol}^{-1}\cdot\text{K}^{-1}\right) + \left(1 \text{mol} \times 205.03 \text{ J}\cdot\text{mol}^{-1}\cdot\text{K}^{-1}\right)\right] - \left[\left(2 \text{mol} \times 324.6 \text{ J}\cdot\text{mol}^{-1}\cdot\text{K}^{-1}\right)\right]$$

$$= 179.2 \text{ J}\cdot\text{K}^{-1}$$

At 25°C $\Delta G^\circ_{298} = \Delta H^\circ - (298\ K)\Delta S^\circ$.

ΔG°_{298} = 611.4 kJ - (298K)(179.2 x 10^{-3} kJ·K^{-1}) = 558.0 kJ

At the temperature at which there is no net driving force in either direction, $\Delta G = 0$. At this temperature, $O = \Delta H^\circ - T\Delta S^\circ$. Hence,

$$T = \frac{\Delta H^\circ}{\Delta S^\circ} \qquad T = \frac{611.4\ kJ}{179.2 \times 10^{-3}\ kJ \cdot K^{-1}} = 3412\ K$$

The reaction is nonspontaneous in the direction written at 298 K(ΔG°_{298} = +) and experiences no net driving force in either direction at 3412 K(ΔG°_{3412} = 0). Above 3412 K, the reaction becomes spontaneous in the direction written. **Hence, the reaction is spontaneous above 3412 K.**

(b) For $PbO(s) + CO_2(g) \rightarrow PbCO_3(s)$,

$$\Delta H^\circ_{rxn} = \left[\Delta H^\circ_f PbCO_3(s)\right] - \left[\Delta H^\circ_f PbO(s) + \Delta H^\circ_f CO_2(g)\right]$$

$$\Delta H^\circ_{rxn} = \left[\left(1 mol \times -699.1\ kJ \cdot mol^{-1}\right)\right] - \left[\left(1 mol \times -218.99\ kJ \cdot mol^{-1}\right) + \left(1 mol \times 393.51\ kJ \cdot mol^{-1}\right)\right] = -86.6\ kJ$$

$$\Delta S^\circ_{rxn} = \left[\left(S^\circ PbCO_3(s)\right)\right] - \left[S^\circ PbO(s) + S^\circ CO_2(g)\right]$$

$$\Delta S^\circ_{rxn} = \left[\left(1 mol \times 130.96\ kJ \cdot mol^{-1}\right)\right] - \left[\left(1 mol \times 66.5\ kJ \cdot mol^{-1}\right) + \left(1 mol \times 213.63\ kJ \cdot mol^{-1}\right)\right] = -149.2\ J{\cdot}K^{-1}$$

At 25°C $\Delta G^\circ_{298} = \Delta H^\circ - (298\ K)\Delta S^\circ$.

ΔG°_{298} = -86.6 kJ - (298 K)(-149.2 x 10^{-3} kJ·K^{-1}) = -131.1 kJ

At the temperature at which there is no net driving force in either direction, $\Delta G^\circ = 0$. At this temperature, $O = \Delta H^\circ - T\Delta S^\circ$. Hence,

$$T = \frac{\Delta H^\circ}{\Delta S^\circ} \qquad T = \frac{-86.6\ kJ}{149.2 \times 10^{-3}\ kJ \cdot K^{-1}} = 580\ K$$

The reaction is spontaneous in the direction written at 298 K(ΔG°_{298} = -) and experiences no net driving force in either direction at 580 K(ΔG°_{580} = 0). Above 580 K the reaction becomes nonspontaneous. Hence, it is **spontaneous up to 580 K.**

17.63 ***See Section 17.4 and Example 17.6.***

(a) For ΔH° = +53.4 kJ and ΔS° = +112.4 J·K^{-1},

$$\Delta G^\circ_{298} = \Delta H^\circ - (298\ K)\Delta S^\circ \qquad \Delta G^\circ_{298} = 53.4 \times 10^3\ J - (298\ K)\left(112.4\ J \cdot K^{-1}\right) = 1.99 \times 10^4\ J \text{ or } 19.9\ kJ$$

At the temperature at which there is no net driving force in either direction, $\Delta G^\circ = 0$.
At this temperature, $0 = \Delta H^\circ - T\Delta S^\circ$. Hence,

$$T = \frac{\Delta H^\circ}{\Delta S^\circ} \qquad T = \frac{53.4 \times 10^3\ J}{112.4\ J \cdot K^{-1}} = 475\ K$$

The reaction is nonspontaneous in the direction written at 298 K (ΔG°_{298} = +) and experiences no net driving force in either direction at 475 K ($\Delta G^\circ_{475} = 0$). Above 475 K the reaction becomes spontaneous in the direction written. Hence, the reaction is **spontaneous above 475 K.**

(b) For ΔH° = -29.4 kJ and ΔS° = -91.2 J·K^{-1},

$$\Delta G^\circ_{298} = \Delta H^\circ - (298\ K)\Delta S^\circ \qquad \Delta G^\circ_{298} = -29.4 \times 10^3\ J - (298\ K)\left(-91.2\ J \cdot K^{-1}\right) = -2.2 \times 10^3\ J \text{ or } -22\ kJ$$

At the temperature at which there is no net driving force in either direction, $\Delta G^\circ = 0$.
At this temperature, $0 = \Delta H^\circ - T\Delta S^\circ$. Hence,

$$T = \frac{\Delta H^\circ}{\Delta S^\circ} \qquad T = \frac{-29.4 \times 10^3\ J}{-91.2\ J \cdot K^{-1}} = 322\ K$$

The reaction is spontaneous in the direction written at 298 K (ΔG°_{298} = –) and experiences no net driving force in either direction at 322 K ($\Delta G^\circ_{322} = 0$). Above 322 K the reaction becomes nonspontaneous in the direction written. Hence, the reaction is **spontaneous up to 322 K**.

17.64 ***See Section 17.4 and Example 17.6.***

(a) At 25°C $\Delta G^\circ_{298} = \Delta H^\circ - (298\text{ K})\Delta S^\circ$.

ΔG°_{298} = -53.4 x 10^3 J - (298 K)(-112.4 J·K^{-1}) = -1.99 x 10^4 J or -19.9 kJ

At the temperature at which there is no net driving force in either direction, $\Delta G^\circ = 0$. At this temperature $0 = \Delta H^\circ - T\Delta S^\circ$. Hence,

$$T = \frac{\Delta H^\circ}{\Delta S^\circ} \qquad T = \frac{-53.4 \text{ x } 10^3 \text{ J}}{-112.4 \text{ J}\cdot\text{K}^{-1}} = 475 \text{ K}$$

The reaction is spontaneous in the direction written at 298 K(ΔG°_{298} = -) and experiences no net driving force in either direction at 475 K($\Delta G^\circ_{475} = 0$). Above 475 K the reaction becomes nonspontaneous in the direction written. Hence, the reaction is **spontaneous up to 475 K.**

(b) At 25°C $\Delta G^\circ_{298} = \Delta H^\circ - (298\text{ K})\Delta S^\circ$.

ΔG°_{298} = -29.4 x 10^3 J - (298 K)(91.2 J·K^{-1}) = -2.22 x 10^3 J or 22.2 kJ

At the temperature at which there is no net driving force in either direction, $\Delta G^\circ = 0$. At this temperature $0 = \Delta H^\circ - T\Delta S^\circ$. Hence,

$$T = \frac{\Delta H^\circ}{\Delta S^\circ} \qquad T = \frac{29.4 \text{ x } 10^3 \text{ J}}{91.2 \text{ J}\cdot\text{K}^{-1}} = 322 \text{ K}$$

The reaction is nonspontaneous in the direction written at 298 K(ΔG°_{298} = +) and experiences no net driving force in either direction at 322 K($\Delta G^\circ_{322} = 0$). Above 322 K the reaction becomes spontaneous in the direction written. Hence, the reaction is **spontaneous up to 322 K.**

17.65 ***See Section 17.4 and Example 17.6.***

(a) For $NO_2(g) + N_2O(g) \rightarrow 3NO(g)$,

$$\Delta H^\circ_{rxn} = \left[3\Delta H^\circ_f NO(g)\right] - \left[\Delta H^\circ_f NO_2(g) + \Delta H^\circ_f N_2O(g)\right]$$

$$\Delta H^\circ_{rxn} = \left[\left(3 \text{ mol} \times 90.25 \text{ kJ}\cdot\text{mol}^{-1}\right)\right] - \left[\left(1 \text{ mol} \times 33.18 \text{ kJ}\cdot\text{mol}^{-1}\right) + \left(1 \text{ mol} \times 82.05 \text{ kJ}\cdot\text{mol}^{-1}\right)\right] = 155.52 \text{ kJ}$$

$$\Delta S^\circ_{rxn} = \left[3S^\circ NO(g)\right] - \left[S^\circ NO_2(g) + S^\circ N_2O(g)\right]$$

$$\Delta S^\circ_{rxn} = \left[\left(3 \text{ mol} \times 210.65 \text{ J}\cdot\text{mol}^{-1}\cdot\text{K}^{-1}\right)\right] - \left[\left(1 \text{ mol} \times 239.95 \text{ J}\cdot\text{mol}^{-1}\cdot\text{K}^{-1}\right) + \left(1 \text{ mol} \times 219.74 \text{ J}\cdot\text{mol}^{-1}\cdot\text{K}^{-1}\right)\right]$$

$$= 172.26 \text{ J}\cdot\text{K}^{-1}$$

$\Delta G^\circ_{400} = \Delta H^\circ - (400\text{ K})\Delta S^\circ \qquad \Delta G^\circ_{400} = 155.52 \text{ x } 10^3 \text{ J} - (400\text{ K})(172.26 \text{ J}\cdot\text{K}^{-1}) = \mathbf{8.66 \text{ x } 10^4 \text{ J}}$ **or 86.6 kJ**

$\Delta G^\circ_{600} = \Delta H^\circ - (600\text{ K})\Delta S^\circ \qquad \Delta G^\circ_{600} = 155.52 \text{ x } 10^3 \text{ J} - (600\text{ K})(172.26 \text{ J}\cdot\text{K}^{-1}) = \mathbf{5.2 \text{ x } 10^4 \text{ J}}$ **or 52 kJ**

Note: The reaction has an unfavorable enthalpy change and a favorable entropy change; the latter being due to the increase in moles of gas. At low temperatures, the unfavorable ΔH° term dominates the favorable $T\Delta S^\circ$ term. As temperature increases, the importance of the $T\Delta S^\circ$ increases. Eventually, the reaction will become spontaneous.

(b) For $2NH_3(g) \rightarrow N_2H_4(\ell) + H_2(g)$,

$$\Delta H^\circ_{rxn} = \left[\Delta H^\circ_f N_2H_4(l) + \Delta H^\circ_f H_2(g)\right] - \left[2\Delta H^\circ_f NH_3(g)\right]$$

$$\Delta H^\circ_{rxn} = \left[\left(1 \text{ mol} \times 50.63 \text{ kJ}\cdot\text{mol}^{-1}\right) + \left(1 \text{ mol} \times 0. \text{ kJ}\cdot\text{mol}^{-1}\right)\right] - \left[\left(2 \text{ mol} \times -46.11 \text{ kJ}\cdot\text{mol}^{-1}\right)\right] = 142.85 \text{ kJ}$$

$$\Delta S^\circ_{rxn} = \left[S^\circ N_2H_4\,(l) + S^\circ H_2\,(g)\right] - \left[2S^\circ NH_3\,(g)\right]$$

$$\Delta S^\circ_{rxn} = \left[\left(1\text{ mol} \times 121.21\text{ J}\cdot\text{mol}^{-1}\cdot\text{K}^{-1}\right) + \left(1\text{ mol} \times 130.57\text{ J}\cdot\text{mol}^{-1}\cdot\text{K}^{-1}\right)\right] - \left[\left(2\text{ mol} \times 192.34\text{ J}\cdot\text{mol}^{-1}\cdot\text{K}^{-1}\right)\right]$$

$$= -132.90\text{ J}\cdot\text{K}^{-1}$$

$\Delta G^\circ_{400} = \Delta H^\circ - (400\text{ K})\Delta S^\circ$ $\quad$ $\Delta G^\circ_{400} = 142.85 \times 10^3\text{ J} - (400\text{ K})\left(-132.90\text{ J}\cdot\text{K}^{-1}\right) = \mathbf{1.960 \times 10^5\ J}$

$\Delta G^\circ_{600} = \Delta H^\circ - (600\text{ K})\Delta S^\circ$ $\quad$ $\Delta G^\circ_{600} = 142.85 \times 10^3\text{ J} - (600\text{ K})\left(-132.90\text{ J}\cdot\text{K}^{-1}\right) = \mathbf{2.226 \times 10^5\ J}$

Note: The reaction has an unfavorable enthalpy change and an unfavorable entropy change; the latter being due to formation of one mole of liquid and one mole of gas from two moles of gas. The reaction is therefore nonspontaneous at all temperatures.

17.66 ***See Section 17.4 and Example 17.6.***

(a) For $BaO(s) + CO_2(g) \rightarrow BaCO_3(s)$,

$$\Delta H^\circ_{rxn} = \left[\Delta H^\circ_f BaCO_3\,(s)\right] - \left[\Delta H^\circ_f BaO(s) + \Delta H^\circ_f CO_2\,(g)\right]$$

$$\Delta H^\circ_{rxn} = \left[\left(1\text{mol} \times -1216.3\text{ kJ}\cdot\text{mol}^{-1}\right)\right] - \left[\left(1\text{mol} \times -582.0\text{ kJ}\cdot\text{mol}^{-1}\right) + \left(1\text{mol} \times -393.51\text{ kJ}\cdot\text{mol}^{-1}\right)\right]$$

$$= -240.8\text{ kJ}$$

$$\Delta S^\circ_{rxn} = \left[S^\circ BaCO_3\right] - \left[S^\circ BaO(s) + S^\circ CO_2\,(g)\right]$$

$$\Delta S^\circ_{rxn} = \left[\left(1\text{mol} \times 112.1\text{ J}\cdot\text{mol}^{-1}\cdot\text{K}^{-1}\right)\right] - \left[\left(1\text{mol} \times 70.3\text{ J}\cdot\text{mol}^{-1}\cdot\text{K}^{-1}\right) + \left(1\text{mol} \times 213.63\text{ J}\cdot\text{mol}^{-1}\cdot\text{K}^{-1}\right)\right]$$

$$= -171.8\text{ J}\cdot\text{K}^{-1}$$

$\Delta G^\circ_{300} = \Delta H^\circ - (300\text{ K})\Delta S^\circ$ $\quad$ $\Delta G^\circ_{300} = -240.8\text{ kJ} - (300\text{ K})(-171.8 \times 10^{-3}\text{ J}\cdot\text{K}^{-1}) = \mathbf{-189.3\ kJ}$

$\Delta G^\circ_{390} = \Delta H^\circ - (390\text{ K})\Delta S^\circ$ $\quad$ $\Delta G^\circ_{390} = -240.8\text{ kJ} - (390\text{ K})(-171.8 \times 10^{-3}\text{ J}\cdot\text{K}^{-1}) = \mathbf{-173.8\ kJ}$

Note: The reaction has a favorable enthalpy change and an unfavorable entropy change, the latter being due to the decrease in the moles of gas. At low temperatures the favorable ΔH° term dominates the unfavorable $T\Delta S^\circ$ term. As temperature increases the importance of $T\Delta S^\circ$ increases. Eventually the reaction will become nonspontaneous.

(b) For $CH_3COOH(\ell) \rightarrow CH_4(g) + CO_2(g)$,

$$\Delta H^\circ_{rxn} = \left[\Delta H^\circ_f CH_4\,(g) + \Delta H^\circ_f CO_2\,(g)\right] - \left[\Delta H^\circ_f CH_3COOH(\ell)\right]$$

$$\Delta H^\circ_{rxn} = \left[\left(1\text{mol} \times -74.81\text{ kJ}\cdot\text{mol}^{-1}\right) + \left(1\text{mol} \times -393.51\text{ kJ}\cdot\text{mol}^{-1}\right)\right] - \left[\left(1\text{mol} \times -484.5\text{ kJ}\cdot\text{mol}^{-1}\right)\right] = 16.2\text{ kJ}$$

$$\Delta S^\circ_{rxn} = \left[S^\circ CH_4\,(g) + S^\circ CO_2\,(g)\right] - \left[S^\circ CH_3COOH(\ell)\right]$$

$$\Delta S^\circ_{rxn} = \left[\left(1\text{mol} \times 186.15\text{ J}\cdot\text{mol}^{-1}\cdot\text{K}^{-1}\right) + \left(1\text{mol} \times 213.63\text{ J}\cdot\text{mol}^{-1}\cdot\text{K}^{-1}\right)\right] - \left[\left(1\text{mol} \times 159.8\text{ J}\cdot\text{mol}^{-1}\cdot\text{K}^{-1}\right)\right]$$

$$= 240.0\text{ J}\cdot\text{K}^{-1}$$

$\Delta G^\circ_{300} = \Delta H^\circ - (300\text{ K})\Delta S^\circ$ $\quad$ $\Delta G^\circ_{300} = 16.2\text{ kJ} - (300\text{ K})(240.0 \times 10^{-3}\text{ J}\cdot\text{K}^{-1}) = \mathbf{-55.8\ kJ}$

$\Delta G^\circ_{390} = \Delta H^\circ - (390\text{ K})\Delta S^\circ$ $\quad$ $\Delta G^\circ_{390} = 16.2\text{ kJ} - (390\text{ K})(240.0 \times 10^{-3}\text{ J}\cdot\text{K}^{-1}) = \mathbf{-77.4\ kJ}$

Note: The reaction has a slightly unfavorable enthalpy change and a highly favorable entropy change, the latter being due to the conversion of one mole of liquid to two moles of gas. At 300 K the $T\Delta S^\circ$ term already dominates the ΔH° term. As temperature increase, $T\Delta S^\circ$ increases causing ΔG° to become more negative and the reaction to become more product-favored.

17.67 ***See Section 17.4.***

At low temperatures the ΔH° term dominates the $T\Delta S^\circ$ term in $\Delta G^\circ = \Delta H^\circ - T\Delta S^\circ$. As temperature increases, the importance of the $T\Delta S^\circ$ term increases and eventually dominates the ΔH° term. Since the reaction proceeds spontaneously at low temperatures, it must have a favorable free energy change and favorable enthalpy change. The signs of ΔG° and ΔH° must therefore be negative. Similarly, since the reaction becomes nonspontaneous in the forward direction at elevated temperatures, it must have an unfavorable entropy change. The sign of ΔS° must therefore be negative.

Note: Only reactions having like signs for both ΔH° and ΔS° exhibit temperature dependent spontaneity. This is due to the fact one of these is unfavorable when both have like signs.

17.68 *See Section 17.4.*

At low temperatures the ΔH° term dominates the $T\Delta S^\circ$ term in $\Delta G^\circ = \Delta H^\circ - T\Delta S^\circ$. As temperature increases, the importance of the$T\Delta S^\circ$ term increases and eventually dominates the ΔH° term. Since the reaction is nonspontaneous at low temperatures it must have an unfavorable free energy change and an unfavorable enthalpy change. The sings of ΔG° and ΔH° must therefore be positive. Similarly since the reaction becomes spontaneous in the forward directive at elevated temperatures, it must have a favorable entropy change. The sign of ΔS° must therefore be positive.

Note: Only reactions having like signs for both ΔH° and ΔS° exhibit temperature dependent spontaneity. This is due to the fact one of these is unfavorable when both have like signs.

17.69 *See Sections 17.3, 17. 4, 17.5, Example 17.6.*

For $CO(g) + 2H_2(g) \rightarrow CH_3OH(g)$, $\Delta H^\circ_{rxn} = -90.14$ kJ and $\Delta S^\circ_{rxn} = -219.00\ J{\cdot}K^{-1}$.

$\Delta G^\circ_{298} = \Delta H^\circ - (298\ K)\Delta S^\circ \quad \Delta G^\circ_{298} = -90.14 \times 10^3\ J - (298\ K)(-219.00\ J \cdot K^{-1}) = -2.488 \times 10^4\ J$

or -24.88 kJ.

(a) False: The spontaneous direction of a reaction is determined by the sign of ΔG°_{rxn} , not ΔH°_{rxn}.

(b) True: The sign of ΔG°_{298} is negative.

(c) False: The reaction has a favorable enthalpy change and an unfavorable entropy change. As temperature increases, the importance of the unfavorable $T\Delta S^\circ$ term in $\Delta G^\circ = \Delta H^\circ - T\Delta S^\circ$ will increase and the reaction will become nonspontaneous in the direction written.

(d) True: The reaction is spontaneous at 298 K and becomes nonspontaneous at a temperature greater than 298 K.

(e) False: Since ΔH°_{rxn} is negative, $\ln K_{eq}$ will decrease at higher temperatures.

17.70 *See Sections 17.3, 17. 4, 17.5, Example 17.6.*

For $CO(g) + H_2O(g) \rightarrow HCOOH(g)$, $\Delta H^\circ_{rxn} = -26.27$ kJ and $\Delta S^\circ_{rxn} = -137.54\ J{\cdot}K^{-1}$.

$\Delta G^\circ_{298} = \Delta H^\circ - (298\ K)\Delta S^\circ \quad \Delta G^\circ_{298} = -26.27\ kJ - (298\ K)(-137.54 \times 10^{-3}\ J \cdot K^{-1}) = 14.72\ kJ$.

(a) False: The spontaneous direction of a reaction is determined by the sign of ΔG°_{rxn} , not ΔH°_{rxn}.

(b) False: The sign of ΔG°_{298} is positive. Hence, the reaction is nonspontaneous at 298 K.

(c) False: The reaction has a favorable enthalpy change and an unfavorable entropy change. As temperature increases, the importance of the unfavorable $T\Delta S^\circ$ term in $\Delta G^\circ = \Delta H^\circ - T\Delta S^\circ$ will increase and the reaction will remain nonspontaneous in the direction written.

(d) False: The reaction becomes nonspontaneous before 298 K is reached.

(e) False: Since ΔH°_{rxn} is negative, $\ln K_{eq}$ will decrease at higher temperatures.

17.71 *See Section 17.4.*

For $CH_3OH(\ell) \rightarrow CH_3OH(g)$,

$$\Delta H^\circ_{rxn} = \left[\Delta H^\circ_f CH_3OH(g)\right] - \left[\Delta H^\circ_f CH_3OH(\ell)\right]$$

$$\Delta H^\circ_{rxn} = \left[\left(1\ mol \times -200.66\ kJ \cdot mol^{-1}\right)\right] - \left[\left(1\ mol \times -238.66\ kJ \cdot mol^{-1}\right)\right] = 38.00\ kJ$$

$$\Delta S^\circ_{rxn} = \left[S^\circ CH_3OH(g)\right] - \left[S^\circ CH_3OH(\ell)\right]$$

$\Delta S^\circ_{rxn} = \left[\left(1 \text{ mol} \times 239.70 \text{ J}\cdot\text{mol}^{-1}\cdot\text{K}^{-1}\right)\right] - \left[\left(1 \text{ mol} \times 126.8 \text{ J}\cdot\text{mol}^{-1}\cdot\text{K}^{-1}\right)\right] = 112.9 \text{ J}\cdot\text{K}^{-1}$

$\Delta G^\circ_{353} = \Delta H^\circ - (353 \text{ K})\Delta S^\circ \qquad \Delta G^\circ_{353} = 38.00 \times 10^3 \text{ J} - (353 \text{ K})\left(112.9 \text{ J}\cdot\text{K}^{-1}\right) = \mathbf{-1.85 \times 10^3 \text{ J}}$ or $\mathbf{-1.85 \text{ kJ}}$

The negative value of ΔG°_{353} indicates the vaporization of methanol is **spontaneous** at 80°C and 1 atm. Assumption: ΔH° and ΔS° for $CH_3OH(\ell) \rightarrow CH_3OH(g)$ do not change appreciably with temperature. (Note: The solution for 17.77 below indicates this is a reasonably good assumption..

17.72 ***See Section 17.4.***

For $CH_3NO_2(g) \rightarrow CH_3NO_2(\ell$,

$\Delta H^\circ_{rxn} = \left[\Delta H^\circ_f CH_3NO_2(\ell)\right] - \left[\Delta H^\circ_f CH_3NO_2(g)\right]$

$\Delta H^\circ_{rxn} = \left[\left(1\text{mol} \times -113.1 \text{ kJ}\cdot\text{mol}^{-1}\right)\right] - \left[\left(1\text{mol} \times -74.73 \text{ kJ}\cdot\text{mol}^{-1}\right)\right] = -38.4 \text{ kJ}$

$\Delta S^\circ_{rxn} = \left[S^\circ CH_3NO_2(\ell)\right] - \left[\left[S^\circ CH_3NO_2(g)\right]\right]$

$\Delta S^\circ_{rxn} = \left[\left(1\text{mol} \times 171.76 \text{ J}\cdot\text{mol}^{-1}\cdot\text{K}^{-1}\right)\right] - \left[\left(1\text{mol} \times 274.42 \text{ J}\cdot\text{mol}^{-1}\cdot\text{K}^{-1}\right)\right] = -102.66 \text{ J}\cdot\text{K}^{-1}$

$\Delta G^\circ_{313} = \Delta H^\circ - (313 \text{ K})\Delta S^\circ \qquad \Delta G^\circ_{313} = -38.4 \text{ kJ} - (313 \text{ K})(-102.66 \times 10^{-3} \text{ J}\cdot\text{K}^{-1}) = -6.3 \text{ kJ}$

The negative value of ΔG°_{313} indicates the condensation of nitromethane is **spontaneous** at 40°C and 1 atm. Assumption: ΔH° and ΔS° for $CH_3NO_2(g) \rightarrow CH_3NO_2(\ell)$ do not change appreciably with temperature. (Note: The solution for 17.78 below indicates this is a resonably good assumption.)

17.73 ***See Section 17.5 and Example 17.7.***

(a) For $2NO(g) + O_2(g) \rightleftarrows 2NO_2(g)$, $\quad Q = \dfrac{\left(P_{NO_2}\right)^2}{\left(P_{NO}\right)^2\left(P_{O_2}\right)} = \dfrac{(3.0)^2}{\left(1.0 \times 10^{-3}\right)^2\left(2.0 \times 10^{-3}\right)} = 4.5 \times 10^9$

$\Delta G = \Delta G^\circ + RT\ln Q \qquad \Delta G = -70.52 \text{ kJ} + (8.314 \times 10^{-3} \text{ kJ}\cdot\text{mol}^{-1}\cdot\text{K}^{-1})(298 \text{ K})(\ln 4.5 \times 10^9) = \mathbf{-15.45 \text{ kJ}}$

(b) The sign of ΔG is negative, so the reaction is spontaneous in the **forward** direction.

17.74 ***See Section 17.5 and Example 17.7.***

(a) For $2N_2O(g) + 3O_2(g) \rightleftarrows 2N_2O_4(g)$, $\quad Q = \dfrac{\left(P_{N_2O_4}\right)^2}{\left(P_{N_2O}\right)^2\left(P_{O_2}\right)^2} = \dfrac{(0.10)^2}{\left(1.0 \times 10^{-2}\right)^2\left(4.0 \times 10^{-3}\right)} = 2.5 \times 10^3$

$\Delta G = \Delta G^\circ + RT \ln Q \qquad \Delta G = -6.36 \text{ kJ} + (8.314 \times 10^{-3} \text{ kJ}\cdot\text{mol}^{-1}\cdot\text{K}^{-1})(298)(\ln 2.5 \times 10^3) = \mathbf{18.73 \text{ kJ}}$

(b) The sign of ΔG is positive, so the reaction is not spontaneous in the forward direction. Rather it is spontaneous in the **reverse** direction.

17.75 ***See Section 17.5 and Example 17.7.***

(a) For $PbCl_2(s) \rightleftarrows Pb^{2+}(aq) + 2\ Cl^-(aq)$, $\quad Q = [Pb^{2+}][Cl^-]^2 = (1.0 \times 10^{-2})(2.0 \times 10^{-1})^2 = 4.0 \times 10^{-4}$

$\Delta G = \Delta G^\circ + RT\ln Q \qquad \Delta G = 27.4 \text{ kJ} + (8.314 \times 10^{-3} \text{ J}\cdot\text{mol}^{-1}\cdot\text{K}^{-1})(298 \text{ K})(\ln 4.0 \times 10^{-4}) = \mathbf{8.0 \text{ kJ}}$

(b) The sign of ΔG is positive, so the reaction is not spontaneous in the forward direction. Rather, it is spontaneous in the **reverse** direction.

17.76 *See Section 17.5 and Example 17.7.*

(a) For $ZnF_2(s) \rightleftarrows Zn^{2+}(aq) + 2F^-(aq)$, $Q = [Zn^{2+}][F^-]^2 = (3.5 \times 10^{-2})(2.3 \times 10^{-3})^2 = 1.9 \times 10^{-7}$

$\Delta G = \Delta G^\circ + RT \ln Q$ $\Delta G = 8.68\text{ kJ} + (8.314 \times 10^{-3}\text{ J}\cdot\text{mol}^{-1}\cdot\text{K}^{-1})(298\text{ K}) \ln(1.9 \times 10^{-7}) = \mathbf{-29.66\text{ kJ}}$

(b) The sign of ΔG is negative, so the reaction is spontaneous in the **forward** direction.

17.77 *See Section 17.4 and the Solution for 17.71.*

At the normal boiling point, the liquid is in equilibrium with the vapor at one atmosphere, so ΔG° is zero. Hence, $0 = \Delta H^\circ - T\Delta S^\circ$ and

$$T = \frac{\Delta H^\circ}{\Delta S^\circ} \qquad T = \frac{38.00 \times 10^3\text{ J}}{112.9\text{ J}\cdot\text{K}^{-1}} = 336.6\text{ K or }63.4^\circ\text{C}$$

According to the Handbook of Chemistry and Physics, the normal boiling point of methanol is 64.96°C. Hence, the calculated and experimental values are in excellent agreement.

17.78 *See Section 17.4 and the Solution for 17.72.*

At the normal boiling point, the liquid is in equilibrium with the vapor at one atmosphere, so ΔG° is zero. Hence, $0 = \Delta H^\circ - T\Delta S^\circ$ and

$$T = \frac{\Delta H^\circ}{\Delta S^\circ} \qquad T = \frac{-38.4 \times 10^3\text{ J}}{-102.66\text{ J}\cdot\text{K}^{-1}} = 374\text{ K or }101^\circ\text{ C}$$

According to the Handbook of Chemistry and Physics, the normal boiling point of nitromethane is 100.8°C. Hence, the calculated and experimental values are in excellent agreement.

17.79 *See Section 17.5 and Example 17.8.*

For $PCl_5(g) \rightleftarrows PCl_3(g) + Cl_2(g)$,

$$\Delta H^\circ_{rxn} = \left[\Delta H^\circ_f PCl_3(g) + \Delta H^\circ_f Cl_2(g)\right] - \left[\Delta H^\circ_f PCl_5(g)\right]$$

$$\Delta H^\circ_{rxn} = \left[\left(1\text{ mol} \times -287.0\text{ kJ}\cdot\text{mol}^{-1}\right) + \left(1\text{ mol} \times 0.\text{ kJ}\cdot\text{mol}^{-1}\right)\right] - \left[\left(1\text{ mol} \times -374.9\text{ kJ}\cdot\text{mol}^{-1}\right)\right] = 87.9\text{ kJ}$$

$$\Delta S^\circ_{rxn} = \left[S^\circ PCl_3(g) + S^\circ Cl_2(g)\right] - \left[S^\circ PCl_5(g)\right]$$

$$\Delta S^\circ_{rxn} = \left[\left(1\text{ mol} \times 311.67\text{ J}\cdot\text{mol}^{-1}\cdot\text{K}^{-1}\right) + \left(1\text{ mol} \times 222.96\text{ J}\cdot\text{mol}^{-1}\cdot\text{K}^{-1}\right)\right] - \left[\left(1\text{ mol} \times 364.47\text{ J}\cdot\text{mol}^{-1}\cdot\text{K}^{-1}\right)\right]$$

$$= 170.16\text{ J}\cdot\text{K}^{-1}$$

(a) At 25°C, $\Delta G^\circ_{298} = \Delta H^\circ - (298\text{ K})\Delta S^\circ$

$$\Delta G^\circ_{298} = 87.9 \times 10^3\text{ J} - (298\text{ K})\left(170.16\text{ J}\cdot\text{K}^{-1}\right) = 3.719 \times 10^4\text{ J or }37.19\text{ kJ}$$

$$K_{eq} = e^{\frac{-\Delta G^\circ}{RT}} \text{ and } \frac{-\Delta G^\circ}{RT} = \frac{-3.719 \times 10^4\text{ J}\cdot\text{mol}^{-1}}{\left(8.314\text{ J}\cdot\text{mol}^{-1}\cdot\text{K}^{-1}\right)(298\text{ K})} = -15.0. \text{ Therefore, } K_{eq} = e^{-15.0} = \mathbf{3.06 \times 10^{-7}}.$$

(b) At 250°C, $\Delta G^\circ_{523} = \Delta H^\circ - (523\text{ K})\Delta S^\circ$

$$\Delta G^\circ_{523} = 87.9 \times 10^3\text{ J} - (523\text{ K})\left(170.16\text{ J}\cdot\text{K}^{-1}\right) = -1.1 \times 10^3\text{ J or }-1.1\text{ kJ}$$

$$K_{eq} = e^{\frac{-\Delta G^\circ}{RT}} \text{ and } \frac{-\Delta G^\circ}{RT} = \frac{-\left(-1.1 \times 10^3\text{ J}\cdot\text{mol}^{-1}\right)}{\left(8.314\text{ J}\cdot\text{mol}^{-1}\cdot\text{K}^{-1}\right)(523\text{ K})} = 0.25. \text{ Therefore, } K_{eq} = e^{0.25} = \mathbf{1.3}.$$

Note: The reaction is endothermic, so increasing the temperature from 25°C to 298°C increases the value of K_{eq} from 3.06×10^{-7} to 1.3.

17.80 ***See Section 17.5 and Example 17.8.***

For $2SO_2(g) + O_2(g) \rightleftarrows 2SO_3(g)$,

$$\Delta H^\circ_{rxn} = \left[2\Delta H^\circ_f SO_3(g)\right] - \left[2\Delta H^\circ_f SO_2(g) + \Delta H^\circ_f O_2(g)\right]$$

$$\Delta H^\circ_{rxn} = \left[\left(2\,\text{mol} \times -395.72\ \text{kJ}\cdot\text{mol}^{-1}\right)\right] - \left[\left(2\,\text{mol} \times -296.83\ \text{kJ}\cdot\text{mol}^{-1}\right) + \left(1\,\text{mol} \times 0.\ \text{kJ}\cdot\text{mol}^{-1}\right)\right] = -197.78\ \text{kJ}$$

$$\Delta S^\circ_{rxn} = \left[2S^\circ SO_3(g)\right] - \left[2S^\circ SO_2(g) + S^\circ O_2(g)\right]$$

$$\Delta S^\circ_{rxn} = \left[\left(2\,\text{mol} \times 256.65\ \text{J}\cdot\text{mol}^{-1}\cdot\text{K}^{-1}\right)\right] - \left[\left(2\,\text{mol} \times 248.11\ \text{J}\cdot\text{mol}^{-1}\cdot\text{K}^{-1}\right) + \left(1\,\text{mol} \times 205.03\ \text{J}\cdot\text{mol}^{-1}\cdot\text{K}^{-1}\right)\right]$$

$$= -187.95\ \text{J}\cdot\text{K}^{-1}$$

(a) At 25°C, $\Delta G^\circ_{298} = \Delta H^\circ - (298\ \text{K})\Delta S^\circ$

$$\Delta G^\circ_{298} = -197.78 \times 10^3\ \text{J} - (298\ \text{K})\left(-187.95\ \text{J}\cdot\text{K}^{-1}\right) = -1.418 \times 10^5\ \text{J}$$

$$K_{eq} = e^{\frac{-\Delta G^\circ}{RT}} \text{ and } \frac{-\Delta G^\circ}{RT} = \frac{-1.418 \times 10^5\ \text{J}\cdot\text{mol}^{-1}}{\left(8.314\ \text{J}\cdot\text{mol}^{-1}\cdot\text{K}^{-1}\right)(298\ \text{K})} = 57.23. \text{ Therefore, } K_{eq} = e^{57.23} = \mathbf{7.2 \times 10^{24}}.$$

(b) At 250°C, $\Delta G^\circ_{523} = \Delta H^\circ - (523\ \text{K})\Delta S^\circ$

$$\Delta G^\circ_{523} = -197.78 \times 10^3\ \text{J} - (523\ \text{K})\left(-187.95\ \text{J}\cdot\text{K}^{-1}\right) = -9.948 \times 10^4\ \text{J}$$

$$K_{eq} = e^{\frac{-\Delta G^\circ}{RT}} \text{ and } \frac{-\Delta G^\circ}{RT} = \frac{-\left(-9.948 \times 10^4\ \text{J}\cdot\text{mol}^{-1}\right)}{\left(8.314\ \text{J}\cdot\text{mol}^{-1}\cdot\text{K}^{-1}\right)(523\ \text{K})} = 22.87. \text{ Therefore, } K_{eq} = e^{22.87} = \mathbf{8.6 \times 10^9}.$$

Note: The reaction is endothermic, so increasing the temperature from 25°C to 250°C decreases the value of K_{eq} from 7.2×10^{24} to 8.6×10^9.

17.81 ***See Section 17.5 and Example 17.8.***

At 10.0 K, $\Delta G^\circ_{10} = \Delta H^\circ - (10.0\ \text{K})\Delta S^\circ$

$$\Delta G^\circ_{10} = 10 \times 10^3\ \text{J} - (10.0\ \text{K})\left(100\ \text{J}\cdot\text{K}^{-1}\right) = 9 \times 10^3\ \text{J or } 9\ \text{kJ}$$

$$K_{eq} = e^{\frac{-\Delta G^\circ}{RT}} \text{ and } \frac{-\Delta G^\circ}{RT} = \frac{-\left(9 \times 10^3\ \text{J}\cdot\text{mol}^{-1}\right)}{\left(8.314\ \text{J}\cdot\text{mol}^{-1}\cdot\text{K}^{-1}\right)(10.0\ \text{K})} = -1 \times 10^2.$$

Therefore, $K_{eq} = e^{-1 \times 10^2} = \mathbf{4 \times 10^{-44}}$.

At 100 K, $\Delta G^\circ_{100} = \Delta H^\circ - (100\ \text{K})\Delta S^\circ$

$$\Delta G^\circ_{100} = 10 \times 10^3\ \text{J} - (100\ \text{K})\left(100\ \text{J}\cdot\text{K}^{-1}\right) = 0\ \text{kJ}$$

$$K_{eq} = e^{\frac{-\Delta G^\circ}{RT}} \text{ and } \frac{-\Delta G^\circ}{RT} = \frac{-\left(0\ \text{J}\cdot\text{mol}^{-1}\right)}{\left(8.314\ \text{J}\cdot\text{mol}^{-1}\cdot\text{K}^{-1}\right)(100\ \text{K})} = 0. \text{ Therefore, } K_{eq} = e^{0.} = \mathbf{1}.$$

At 1,000 K, $\Delta G^\circ_{1,000} = \Delta H^\circ - (1{,}000\ \text{K})\Delta S^\circ$

$$\Delta G^\circ_{1,000} = 10 \times 10^3\ \text{J} - (1{,}000\ \text{K})\left(100\ \text{J}\cdot\text{K}^{-1}\right) = -9.0 \times 10^4\ \text{J or } -90\ \text{kJ}$$

$$K_{eq} = e^{\frac{-\Delta G^\circ}{RT}} \text{ and } \frac{-\Delta G^\circ}{RT} = \frac{-\left(-9.0 \times 10^4 \text{ J} \cdot \text{mol}^{-1}\right)}{\left(8.314 \text{ J} \cdot \text{mol}^{-1} \cdot \text{K}^{-1}\right)(1{,}000 \text{ K})} = 11. \text{ Therefore, } K_{eq} = e^{11} = \mathbf{6.0 \times 10^4}.$$

Note: The reaction is endothermic, so increasing the temperature from 10.0 K to 100 K to1,000 K increases the value of K_{eq} from 4×10^{-44} to 1 to 6.0×10^4.

17.82 ***See Section 17.5 and Example 17.8.***

At 10.0 K, $\Delta G^\circ_{10} = \Delta H^\circ - (10.0 \text{ K})\Delta S^\circ$

$\Delta G^\circ_{10} = 10.0 \times 10^3 \text{ J} - (10.0 \text{ K})\left(-100 \text{ J} \cdot \text{K}^{-1}\right) = 1.10 \times 10^4 \text{ J or } 11.0 \text{ kJ}$

$$K_{eq} = e^{\frac{-\Delta G^\circ}{RT}} \text{ and } \frac{-\Delta G^\circ}{RT} = \frac{-\left(1.10 \times 10^4 \text{ J} \cdot \text{mol}^{-1}\right)}{\left(8.314 \text{ J} \cdot \text{mol}^{-1} \cdot \text{K}^{-1}\right)(10.0 \text{ K})} = -1.32 \times 10^2.$$

Therefore, $K_{eq} = e^{-1.32 \times 10^2} = \mathbf{4.71 \times 10^{-58}}$.

At 100 K, $\Delta G^\circ_{100} = \Delta H^\circ - (100 \text{ K})\Delta S^\circ$

$\Delta G^\circ_{100} = 10.0 \times 10^3 \text{ J} - (100 \text{ K})\left(-100 \text{ J} \cdot \text{K}^{-1}\right) = 2.00 \times 10^4 \text{ J or } 20.0 \text{ kJ}$

$$K_{eq} = e^{\frac{-\Delta G^\circ}{RT}} \text{ and } \frac{-\Delta G^\circ}{RT} = \frac{-\left(2.00 \times 10^4 \text{ J} \cdot \text{mol}^{-1}\right)}{\left(8.314 \text{ J} \cdot \text{mol}^{-1} \cdot \text{K}^{-1}\right)(100 \text{ K})} = -24.1. \text{ Therefore, } K_{eq} = e^{-24.1} = \mathbf{3.57 \times 10^{-11}}.$$

At 1,000 K, $\Delta G^\circ_{1,000} = \Delta H^\circ - (1{,}000 \text{ K})\Delta S^\circ$

$\Delta G^\circ_{1,000} = 10.0 \times 10^3 \text{ J} - (1{,}000 \text{ K})\left(-100 \text{ J} \cdot \text{K}^{-1}\right) = 1.10 \times 10^5 \text{ J or } 110 \text{ kJ}$

$$K_{eq} = e^{\frac{-\Delta G^\circ}{RT}} \text{ and } \frac{-\Delta G^\circ}{RT} = \frac{-\left(1.10 \times 10^5 \text{ J} \cdot \text{mol}^{-1}\right)}{\left(8.314 \text{ J} \cdot \text{mol}^{-1} \cdot \text{K}^{-1}\right)(1{,}000 \text{ K})} = -13.2. \text{ Therefore, } K_{eq} = e^{-13.2} = \mathbf{1.85 \times 10^{-6}}.$$

Note: The reaction is endothermic, so increasing the temperature from 10.0 K to 100 K to1,000 K increases the value of K_{eq} from 4.71×10^{-48} to 3.57×10^{-11} to 1.85×10^{-6}.

17.83 ***See Section 17.5 and Example 17.8.***

At 10.0 K, $\Delta G^\circ_{10} = \Delta H^\circ - (10.0 \text{ K})\Delta S^\circ$

$\Delta G^\circ_{10} = -10.0 \times 10^3 \text{ J} - (10.0 \text{ K})\left(100 \text{ J} \cdot \text{K}^{-1}\right) = \mathbf{-1.1 \times 10^4 \text{ J or } -11 \text{ kJ}}$

$$K_{eq} = e^{\frac{-\Delta G^\circ}{RT}} \text{ and } \frac{-\Delta G^\circ}{RT} = \frac{-\left(-1.1 \times 10^4 \text{ J} \cdot \text{mol}^{-1}\right)}{\left(8.314 \text{ J} \cdot \text{mol}^{-1} \cdot \text{K}^{-1}\right)(10.0 \text{ K})} = 1.3 \times 10^2. \text{ Therefore, } K_{eq} = e^{1.3 \times 10^2} = \mathbf{2.9 \times 10^{57}}.$$

At 100 K, $\Delta G^\circ_{100} = \Delta H^\circ - (100 \text{ K})\Delta S^\circ$

$\Delta G^\circ_{100} = -10.0 \times 10^3 \text{ J} - (100 \text{ K})\left(100 \text{ J} \cdot \text{K}^{-1}\right) = \mathbf{-2.00 \times 10^4 \text{ J or } -20.0 \text{ kJ}}$

$$K_{eq} = e^{\frac{-\Delta G^\circ}{RT}} \text{ and } \frac{-\Delta G^\circ}{RT} = \frac{-\left(-2.00 \times 10^4 \text{ J} \cdot \text{mol}^{-1}\right)}{\left(8.314 \text{ J} \cdot \text{mol}^{-1} \cdot \text{K}^{-1}\right)(100 \text{ K})} = 24.1. \text{ Therefore, } K_{eq} = e^{24.1} = \mathbf{2.93 \times 10^{10}}.$$

At 1,000 K, $\Delta G^\circ_{1,000} = \Delta H^\circ - (1{,}000 \text{ K})\Delta S^\circ$

$\Delta G^\circ_{1,000} = -10.0 \times 10^3 \text{ J} - (1{,}000 \text{ K})\left(100 \text{ J} \cdot \text{K}^{-1}\right) = \mathbf{-1.10 \times 10^5 \text{ J or } -110 \text{ kJ}}$

$$K_{eq} = e^{\frac{-\Delta G^\circ}{RT}} \text{ and } \frac{-\Delta G^\circ}{RT} = \frac{-\left(-1.10 \times 10^5 \text{ J} \cdot \text{mol}^{-1}\right)}{\left(8.314 \text{ J} \cdot \text{mol}^{-1} \cdot \text{K}^{-1}\right)(1{,}000 \text{ K})} = 13.2. \text{ Therefore, } K_{eq} = e^{13.2} = \mathbf{5.40 \times 10^5}.$$

Note: The reaction is exothermic, so increasing the temperature from 10.0 K to 100 K to 1,000 K decreases the value of K_{eq} from 2.9×10^{56} to 28.0×10^{10} to 5.40×10^{5}.

17.84 ***See Section 17.5 and Example 17.8.***

At 10.0 K, $\Delta G^\circ_{10} = \Delta H^\circ - (10.0\ K)\Delta S^\circ$

$\Delta G^\circ_{10} = -10.0 \times 10^3\ J - (10.0\ K)(-100\ J\cdot K^{-1}) = \mathbf{-9.0 \times 10^3\ J\ or\ -9.0\ kJ}$

$$K_{eq} = e^{\frac{-\Delta G^\circ}{RT}} \text{ and } \frac{-\Delta G^\circ}{RT} = \frac{-(-9.0 \times 10^3\ J\cdot mol^{-1})}{(8.314\ J\cdot mol^{-1}\cdot K^{-1})(10.0\ K)} = 1.1 \times 10^2.$$

Therefore, $K_{eq} = e^{1.1 \times 10^2} = \mathbf{5.9 \times 10^{47}}$.

At 100 K, $\Delta G^\circ_{100} = \Delta H^\circ - (100\ K)\Delta S^\circ$

$\Delta G^\circ_{100} = -10.0 \times 10^3\ J - (100\ K)(-100\ J\cdot K^{-1}) = \mathbf{0\ J\ or\ 0\ kJ}$

$$K_{eq} = e^{\frac{-\Delta G^\circ}{RT}} \text{ and } \frac{-\Delta G^\circ}{RT} = \frac{-(0\ J\cdot mol^{-1})}{(8.314\ J\cdot mol^{-1}\cdot K^{-1})(100\ K)} = 0. \text{ Therefore, } K_{eq} = e^{0} = \mathbf{1}.$$

At 1,000 K, $\Delta G^\circ_{1,000} = \Delta H^\circ - (1,000\ K)\Delta S^\circ$

$\Delta G^\circ_{1,000} = -10.0 \times 10^3\ J - (1,000\ K)(-100\ J\cdot K^{-1}) = \mathbf{9.00 \times 10^4\ J\ or\ 90\ kJ}$

$$K_{eq} = e^{\frac{-\Delta G^\circ}{RT}} \text{ and } \frac{-\Delta G^\circ}{RT} = \frac{-(9.00 \times 10^4\ J\cdot mol^{-1})}{(8.314\ J\cdot mol^{-1}\cdot K^{-1})(1,000\ K)} = -10.8. \text{ Therefore, } K_{eq} = e^{-10.8} = \mathbf{2.04 \times 10^{-5}}.$$

Note: The reaction is exothermic, so increasing the temperature from 10.0 K to 100 K to 1,000 K decreases the value of K_{eq} from 5.9×10^{47} to 0 to 2.00×10^{-5}.

17.85 ***See Section 17.5 and Example 17.7.***

For $CO(g) + Cl_2(g) \rightleftarrows COCl_2(g)$,

$\Delta H^\circ_{rxn} = [\Delta H^\circ_f COCl_2(g)] - [\Delta H^\circ_f CO(g) + \Delta H^\circ_f Cl_2(g)]$

$\Delta H^\circ_{rxn} = [(1\ mol \times -218.8\ kJ\cdot mol^{-1})] - [(1\ mol \times -110.52\ kJ\cdot mol^{-1}) + (1\ mol \times 0.\ kJ\cdot mol^{-1})] = -108.3\ kJ$

$\Delta S^\circ_{rxn} = [S^\circ COCl_2(g)] - [S^\circ CO(g) + S^\circ Cl_2(g)]$

$\Delta S^\circ_{rxn} = [(1 mol \times 283.53\ J\cdot mol^{-1}\cdot K^{-1})] - [(1 mol \times 197.56\ J\cdot mol^{-1}\cdot K^{-1}) + (1 mol \times 222.96\ J\cdot mol^{-1}\cdot K^{-1})]$

$= -136.99\ J\cdot K^{-1}$

For 25°C, $\Delta G^\circ_{298} = \Delta H^\circ - (298\ K)\Delta S^\circ$

$= -108.3 \times 10^3\ J - (298\ k)(-136.99\ J\cdot K^{-1}) = -6.75 \times 10^4$ J or **-67.5 kJ**

For 303°C, $\Delta G^\circ_{576} = \Delta H^\circ - (576\ K)\Delta S^\circ$

$= -108.3 \times 10^3\ J - (576\ K)(-136.99\ J\cdot K^{-1}) = -2.94 \times 10^4$ or -29.4 kJ

For CO(g, 2.0 atm) + Cl_2(g, 1.0 atm) $\rightleftarrows$ $COCl_2$(g, 0.10 atm) at 576 K,

$$\Delta G = \Delta G^\circ_{576} + RT \ln Q \qquad \Delta G = \Delta G^\circ + (8.314\ J\cdot mol^{-1}\cdot K^{-1})(576\ K)\ln\left(\frac{P_{COCl_2}}{P_{CO}P_{Cl_2}}\right)$$

$$\Delta G = -2.94 \times 10^4\ J + (8.314\ J\cdot mol^{-1}\cdot K^{-1})(576\ K)\ln\left(\frac{0.10}{(2.0)(1.0)}\right) = \mathbf{-4.37 \times 10^4\ J\ or\ -43.7\ kJ}.$$

$$\Delta G = -2.937 \times 10^4\ J + \left(8.314\ J \cdot mol^{-1} \cdot K^{-1}\right)(576\ K)\ln\left(\frac{0.10}{(2.0)(1.0)}\right) = \mathbf{-4.372 \times 10^4\ J\ or\ -43.72\ kJ.}$$

17.86 ***See Section 17.5 and Example 17.7.***

For $N_2O(g) + H_2(g) \rightleftarrows N_2(g) + H_2O(\ell)$,

$$\Delta H^o_{rxn} = \left[\Delta H^o_f N_2(g) + \Delta H^o_f H_2O(\ell)\right] - \left[\Delta H^o_f N_2O(g) + \Delta H^o_f H_2(g)\right]$$

$$\Delta H^o_{rxn} = \left[\left(1\ mol \times 0.\ kJ \cdot mol^{-1}\right) + \left(1\ mol \times -285.83\ kJ \cdot mol^{-1}\right)\right]$$

$$-\left[\left(1\ mol \times 82.05\ kJ \cdot mol^{-1}\right) + \left(1\ mol \times 0.\ kJ \cdot mol^{-1}\right)\right] = \mathbf{-367.88\ kJ}$$

$$\Delta S^o_{rxn} = \left[S^o N_2(g) + S^o H_2O(\ell)\right] - \left[S^o N_2O(g) + S^o H_2(g)\right]$$

$$\Delta S^o_{rxn} = \left[\left(1 mol \times 191.50\ J \cdot mol^{-1} \cdot K^{-1}\right) + \left(1 mol \times 69.91\ J \cdot mol^{-1} \cdot K^{-1}\right)\right]$$

$$-\left[\left(1 mol \times 219.74\ J \cdot mol^{-1} \cdot K^{-1}\right) + \left(1 mol \times 130.57\ J \cdot mol^{-1} \cdot K^{-1}\right)\right] = -88.9\ J \cdot K^{-1}$$

For 25°C, $\Delta G^o_{298} = \Delta H - (298\ K)\Delta S$

$= -367.88 \times 10^3\ J - (298\ K)(-88.9\ J \cdot K^{-1}) = -3.414 \times 10^5\ J$ or **-341.4 kJ**

For 37°C, $\Delta G^o_{310} = \Delta H - (310\ K)\Delta S$

$= -367.88 \times 10^3\ J - (310\ K)(-88.9\ J \cdot K^{-1}) = -3.403 \times 10^5\ J$ or -340.3 kJ

For N_2O (1 atm) + H_2 (0.4 atm) $\rightleftarrows$ N_2 (1 atm) + $H_2O(\ell)$ at 310 K,

$$\Delta G = \Delta G^o_{310} + RT \ln Q \qquad \Delta G = \Delta G^o_{310} + (8.314\ J \cdot mol^{-1} \cdot K^{-1})(310\ K) \ln\left(\frac{P_{N_2}}{P_{N_2O} P_{H_2}}\right)$$

$$\Delta G = -340.3\ kJ + (8.314 \times 10^{-3}\ kJ \cdot mol^{-1} \cdot K^{-1})(310\ K) \ln\left(\frac{1}{(1)(0.4)}\right) = \mathbf{-337.94\ kJ}$$

17.87 ***See Section 17.5.***

(a) For $N_2O(g) + H_2(g) \rightleftarrows N_2(g) + H_2O(\ell)$,

$$\Delta H^o_{rxn} = \left[\Delta H^o_f N_2(g) + \Delta H^o_f H_2O(\ell)\right] - \left[\Delta H^o_f N_2O(g) + \Delta H^o_f H_2(g)\right]$$

$$\Delta H^o_{rxn} = \left[\left(1\ mol \times 0.\ kJ \cdot mol^{-1}\right) + \left(1\ mol \times -285.83\ kJ \cdot mol^{-1}\right)\right]$$

$$-\left[\left(1\ mol \times 82.05\ kJ \cdot mol^{-1}\right) + \left(1\ mol \times 0.\ kJ \cdot mol^{-1}\right)\right] = -367.88\ kJ$$

Since the reaction is exothermic, we should expect the value of the equilibrium constant to **decrease** with increasing temperature.

(b) For $CO(g) + Cl_2(g) \rightleftarrows COCl_2(g)$,

$$\Delta H^o_{rxn} = \left[\Delta H^o_f COCl_2(g)\right] - \left[\Delta H^o_f CO(g) + \Delta H^o_f Cl_2(g)\right]$$

$$\Delta H^o_{rxn} = \left[\left(1\ mol \times -218.8\ kJ \cdot mol^{-1}\right)\right] - \left[\left(1\ mol \times -110.52\ kJ \cdot mol^{-1}\right) + \left(1\ mol \times 0.\ kJ \cdot mol^{-1}\right)\right] = -108.3\ kJ$$

Since the reaction is exothermic, we should expect the value of the equilibrium constant to **decrease** with increasing temperature.

(c) For $CO(g) + H_2O(g) \rightleftarrows CO_2(g) + H_2(g)$,

$$\Delta H^o_{rxn} = \left[\Delta H^o_f CO_2(g) + \Delta H^o_f H_2(g)\right] - \left[\Delta H^o_f CO(g) + \Delta H^o_f H_2O(g)\right]$$

$$\Delta H^o_{rxn} = \left[\left(1\ mol \times -393.51\ kJ \cdot mol^{-1}\right) + \left(1\ mol \times 0.\ kJ \cdot mol^{-1}\right)\right]$$

$$-\left[\left(1\ mol \times -110.52\ kJ \cdot mol^{-1}\right) + \left(1\ mol \times -241.82\ kJ \cdot mol^{-1}\right)\right] = -41.17\ kJ$$

Since the reaction is exothermic, we should expect the value of the equilibrium constant to **decrease** with increasing temperature.

(d) For $PCl_5(g) \rightleftarrows PCl_3(g) + Cl_2(g)$,

$$\Delta H^\circ_{rxn} = \left[\Delta H^\circ_f PCl_3(g) + \Delta H^\circ_f Cl_2(g)\right] - \left[\Delta H^\circ_f PCl_5(g)\right]$$

$$\Delta H^\circ_{rxn} = \left[\left(1\text{ mol} \times -287.0\text{ kJ}\cdot\text{mol}^{-1}\right) + \left(1\text{ mol} \times 0.\text{ kJ}\cdot\text{mol}^{-1}\right)\right] - \left[\left(1\text{ mol} \times -374.9\text{ kJ}\cdot\text{mol}^{-1}\right)\right] = 87.9\text{ kJ}$$

Since the reaction is endothermic, we should expect the value of the equilibrium constant to **increase** with increasing temperature.

(e) For $2SO_2(g) + O_2(g) \rightleftarrows 2SO_3(g)$,

$$\Delta H^\circ_{rxn} = \left[2\Delta H^\circ_f SO_3(g)\right] - \left[2\Delta H^\circ_f SO_2(g) + \Delta H^\circ_f O_2(g)\right]$$

$$\Delta H^\circ_{rxn} = \left[\left(2\text{ mol} \times -395.72\text{ kJ}\cdot\text{mol}^{-1}\right)\right] - \left[\left(2\text{ mol} \times -296.83\text{ kJ}\cdot\text{mol}^{-1}\right) + \left(1\text{ mol} \times 0.0\text{ kJ}\cdot\text{mol}^{-1}\right)\right] = -197.78\text{ kJ}$$

Since the reaction is exothermic, we should expect the value of the equilibrium constant to **decrease** with increasing temperature.

17.88 ***See Section 17.5.***

(a) For $N_2O_4(g) \rightleftarrows N_2(g) + 2NO(g)$,

$$\Delta H^\circ_{rxn} = \left[\Delta H^\circ_f N_2(g) + 2\Delta H^\circ_f NO(g)\right] - \left[\Delta H^\circ_f N_2O_4(g)\right]$$

$$\Delta H^\circ_{rxn} = \left[\left(1\text{ mol} \times 0.\text{ kJ}\cdot\text{mol}^{-1}\right) + \left(2\text{ mol} \times 33.18\text{ kJ}\cdot\text{mol}^{-1}\right)\right] - \left[\left(1\text{ mol} \times 9.16\text{ kJ}\cdot\text{mol}^{-1}\right)\right] = 57.2\text{ kJ}$$

Since the reaction is endothermic, we should expect the value of the equilibrium constant to **increase** with increasing temperature.

(b) For $2CO(g) + O_2(g) \rightleftarrows 2CO_2(g)$,

$$\Delta H^\circ_{rxn} = \left[\Delta H^\circ_f CO_2(g)\right] - \left[2\Delta H^\circ_f CO(g) + \Delta H^\circ_f O_2(g)\right]$$

$$\Delta H^\circ_{rxn} = \left[\left(2\text{ mol} \times -393.51\text{ kJ}\cdot\text{mol}^{-1}\right)\right] - \left[\left(2\text{ mol} \times -110.52\text{ kJ}\cdot\text{mol}^{-1}\right) + \left(1\text{ mol} \times 0.\text{ kJ}\cdot\text{mol}^{-1}\right)\right] = -565.98\text{ kJ}$$

Since the reaction is exothermic, we should expect the value of the equilibrium constant to **decrease** with increasing temperature.

(c) For $2NO(g) + Br_2(g) \rightleftarrows 2NOBr(g)$,

$$\Delta H^\circ_{rxn} = \left[2\Delta H^\circ_f NOBr(g)\right] - \left[2\Delta H^\circ_f NO(g) + \Delta H^\circ_f Br_2(g)\right]$$

$$\Delta H^\circ_{rxn} = \left[\left(2\text{ mol} \times 82.1\text{ kJ}\cdot\text{mol}^{-1}\right)\right] - \left[\left(2\text{ mol} \times 90.25\text{ kJ}\cdot\text{mol}^{-1}\right) + \left(1\text{ mol} \times 30.91\text{ kJ}\cdot\text{mol}^{-1}\right)\right] = -47.21\text{ kJ}$$

Since the reaction is exothermic, we should expect the value of the equilibrium constant to **decrease** with increasing temperature.

(d) For $2HI(g) \rightleftarrows H_2(g) + I_2(g)$,

$$\Delta H^\circ_{rxn} = \left[\Delta H^\circ_f H_2(g) + \Delta H^\circ_f I_2(g)\right] - \left[2\Delta H^\circ_f HI(g)\right]$$

$$\Delta H^\circ_{rxn} = \left[\left(1\text{ mol} \times 0.\text{ kJ}\cdot\text{mol}^{-1}\right) + \left(1\text{ mol} \times 62.44\text{ kJ}\cdot\text{mol}^{-1}\right)\right] - \left[\left(2\text{ mol} \times 26.48\text{ kJ}\cdot\text{mol}^{-1}\right)\right] = 9.48\text{ kJ}$$

Since the reaction is endothermic, we should expect the value of the equilibrium constant to **increase** with increasing temperature.

(e) For $C(s) + H_2O(g) \rightleftarrows CO(g) + H_2(g)$,

$$\Delta H^\circ_{rxn} = \left[\Delta H^\circ_f CO(g) + \Delta H^\circ_f H_2(g)\right] - \left[\Delta H^\circ_f C(s) + \Delta H^\circ_f H_2O(g)\right]$$
$$= \left[\left(1\ \text{mol} \times -110.52\ \text{kJ}\cdot\text{mol}^{-1}\right) + \left(1\ \text{mol} \times 0.0\ \text{kJ}\cdot\text{mol}^{-1}\right)\right]$$
$$- \left[\left(1\text{mol} \times 0.\ \text{kJ}\cdot\text{mol}^{-1}\right) + \left(1\text{mol} \times -241.82\ \text{kJ}\cdot\text{mol}^{-1}\right)\right] = 131.3\ \text{kJ}$$

Since the reaction is endothermic, we should expect the value of the equilibrium constant to **increase** with increasing temperature.

17.89 ***See Section 17.5 and Example 17.9.***

(a) For $CS_2(\ell) \rightarrow CS_2(g)$, $K_{eq} = P_{CS_2(g)}$ and

$$\Delta H^\circ_{rxn} = \left[\Delta H^\circ_f CS_2(g)\right] - \left[\Delta H^\circ_f CS_2(\ell)\right]$$
$$\Delta H^\circ_{rxn} = \left[\left(1\ \text{mol} \times 117.36\ \text{kJ}\cdot\text{mol}^{-1}\right)\right] - \left[\left(1\ \text{mol} \times 89.70\ \text{kJ}\cdot\text{mol}^{-1}\right)\right] = 27.66\ \text{kJ}$$
$$\Delta S^\circ_{rxn} = \left[S^\circ CS_2(g)\right] - \left[S^\circ CS_2(l)\right]$$
$$\Delta S^\circ_{rxn} = \left[\left(1\ \text{mol} \times 237.73\ \text{J}\cdot\text{mol}^{-1}\cdot\text{K}^{-1}\right)\right] - \left[\left(1\ \text{mol} \times 151.34\ \text{J}\cdot\text{mol}^{-1}\cdot\text{K}^{-1}\right)\right] = 86.39\ \text{J}\cdot\text{K}^{-1}$$

At 5°C, $\Delta G^\circ_{278} = \Delta H^\circ - (278\ \text{K})\Delta S^\circ$

$$\Delta G^\circ_{278} = 27.66 \times 10^3\ \text{J} - (278\ \text{K})\left(86.39\ \text{J}\cdot\text{K}^{-1}\right) = 3.64 \times 10^3\ \text{J}$$

$$K_{eq} = e^{\frac{-\Delta G^\circ}{RT}} \text{ and } \frac{-\Delta G^\circ}{RT} = \frac{-3.64 \times 10^3\ \text{J}\cdot\text{mol}^{-1}}{\left(8.314\ \text{J}\cdot\text{mol}^{-1}\cdot\text{K}^{-1}\right)(278\ \text{K})} = -1.57.$$

Therefore, $K_{eq} = P_{CS_2(g)} = e^{-1.57} = \mathbf{0.208\ atm}$.

(b) For $CCl_4(\ell) \rightarrow CCl_4(g)$, $K_{eq} = P_{CCl_4(g)}$ and

$$\Delta H^\circ_{rxn} = \left[\Delta H^\circ_f CCl_4(g)\right] - \left[\Delta H^\circ_f CCl_4(\ell)\right]$$
$$\Delta H^\circ_{rxn} = \left[\left(1\ \text{mol} \times -102.9\ \text{kJ}\cdot\text{mol}^{-1}\right)\right] - \left[\left(1\ \text{mol} \times -135.44\ \text{kJ}\cdot\text{mol}^{-1}\right)\right] = 32.5\ \text{kJ}$$
$$\Delta S^\circ_{rxn} = \left[S^\circ CCl_4(g)\right] - \left[S^\circ CCl_4(\ell)\right]$$
$$\Delta S^\circ_{rxn} = \left[\left(1\ \text{mol} \times 309.74\ \text{J}\cdot\text{mol}^{-1}\cdot\text{K}^{-1}\right)\right] - \left[\left(1\ \text{mol} \times 216.40\ \text{J}\cdot\text{mol}^{-1}\cdot\text{K}^{-1}\right)\right] = 93.34\ \text{J}\cdot\text{K}^{-1}$$

At 29°C, $\Delta G^\circ_{302} = \Delta H^\circ - (302\ \text{K})\Delta S^\circ$

$$\Delta G^\circ_{302} = 32.5 \times 10^3\ \text{J} - (302\ \text{K})\left(93.34\ \text{J}\cdot\text{K}^{-1}\right) = 4.31 \times 10^3\ \text{J}$$

$$K_{eq} = e^{\frac{-\Delta G^\circ}{RT}} \text{ and } \frac{-\Delta G^\circ}{RT} = \frac{-4.31 \times 10^3\ \text{J}\cdot\text{mol}^{-1}}{\left(8.314\ \text{J}\cdot\text{mol}^{-1}\cdot\text{K}^{-1}\right)(302\ \text{K})} = -1.72.$$

Therefore, $K_{eq} = P_{CCl_4(g)} = e^{-1.72} = \mathbf{0.179\ atm}$.

(c) For $C_6H_6(\ell) \rightarrow C_6H_6(g)$, $K_{eq} = P_{C_6H_6(g)}$ and

$$\Delta H^\circ_{rxn} = \left[\Delta H^\circ_f C_6H_6(g)\right] - \left[\Delta H^\circ_f C_6H_6(\ell)\right]$$
$$\Delta H^\circ_{rxn} = \left[\left(1\ \text{mol} \times 82.93\ \text{kJ}\cdot\text{mol}^{-1}\right)\right] - \left[\left(1\ \text{mol} \times 49.03\ \text{kJ}\cdot\text{mol}^{-1}\right)\right] = 33.90\ \text{kJ}$$
$$\Delta S^\circ_{rxn} = \left[S^\circ C_6H_6(g)\right] - \left[S^\circ C_6H_6(\ell)\right]$$

$$\Delta S^\circ_{rxn} = \left[\left(1\ mol \times 269.2\ J\cdot mol^{-1}\cdot K^{-1}\right)\right] - \left[\left(1\ mol \times 172.8\ J\cdot mol^{-1}\cdot K^{-1}\right)\right] = 96.4\ J\cdot K^{-1}$$

At 45°C, $\Delta G^\circ_{318} = \Delta H^\circ - (318\ K)\Delta S^\circ$

$$\Delta G^\circ_{318} = 33.9 \times 10^3\ J - (318\ K)\left(96.4\ J\cdot K^{-1}\right) = 3.24 \times 10^3\ J$$

$$K_{eq} = e^{\frac{-\Delta G^\circ}{RT}} \text{ and } \frac{-\Delta G^\circ}{RT} = \frac{-3.24 \times 10^3\ J\cdot mol^{-1}}{\left(8.314\ J\cdot mol^{-1}\cdot K^{-1}\right)(318\ K)} = -1.23.$$

Therefore, $K_{eq} = P_{C_6H_6(g)} = e^{-1.23} = \mathbf{0.292\ atm}.$

17.90 ***See Section 17.5 and Example 17.9.***

(a) For $CH_3OH(\ell) \rightarrow CH_3OH(g)$, $K_{eq} = P_{CH_3OH(g)}$ and

$$\Delta H^\circ_{rxn} = \left[\Delta H^\circ_f CH_3OH(g)\right] - \left[\Delta H^\circ_f CH_3OH(\ell)\right]$$

$$\Delta H^\circ_{rxn} = \left[\left(1\ mol \times -200.66\ kJ\cdot mol^{-1}\right)\right] - \left[\left(1\ mol \times -238.66\ kJ\cdot mol^{-1}\right)\right] = 38.00\ kJ$$

$$\Delta S^\circ_{rxn} = \left[S^\circ CH_3COH(g)\right] - \left[S^\circ CH_3OH(\ell)\right]$$

$$\Delta S^\circ_{rxn} = \left[\left(1\ mol \times 239.70\ J\cdot mol^{-1}\cdot K^{-1}\right)\right] - \left[\left(1\ mol \times 126.8\ J\cdot mol^{-1}\cdot K^{-1}\right)\right] = 112.9\ J\cdot K^{-1}$$

At 58°C, $\Delta G^\circ_{331} = \Delta H^\circ - (331\ K)\Delta S^\circ$

$$\Delta G^\circ_{331} = 38.00 \times 10^3\ J - (331\ K)\left(112.9\ J\cdot K^{-1}\right) = 630\ J$$

$$K_{eq} = e^{\frac{-\Delta G^\circ}{RT}} \text{ and } \frac{-\Delta G^\circ}{RT} = \frac{-630\ J\cdot mol^{-1}}{\left(8.314\ J\cdot mol^{-1}\cdot K^{-1}\right)(331\ K)} = -0.229.$$

Therefore, $K_{eq} = P_{CH_3OH(g)} = e^{-0.229} = \mathbf{0.795\ atm}.$

(b) For $CH_3CH_2OH(\ell) \rightarrow CH_3CH_2OH(g)$, $K_{eq} = P_{CH_3CH_2OH(g)}$ and

$$\Delta H^\circ_{rxn} = \left[\Delta H^\circ_f CH_3CH_2OH(g)\right] - \left[\Delta H^\circ_f CH_3CH_2OH(\ell)\right]$$

$$\Delta H^\circ_{rxn} = \left[\left(1\ mol \times -235.10\ kJ\cdot mol^{-1}\right)\right] - \left[\left(1\ mol \times -277.69\ kJ\cdot mol^{-1}\right)\right] = 42.59\ kJ$$

$$\Delta S^\circ_{rxn} = \left[S^\circ CH_3CH_2OH(g)\right] - \left[S^\circ CH_3CH_2OH(\ell)\right]$$

$$\Delta S^\circ_{rxn} = \left[\left(1\ mol \times 282.59\ J\cdot mol^{-1}\cdot K^{-1}\right)\right] - \left[\left(1\ mol \times 160.7\ J\cdot mol^{-1}\cdot K^{-1}\right)\right] = 121.89\ J\cdot K^{-1}$$

At 29°C, $\Delta G^\circ_{302} = \Delta H^\circ - (302\ K)\Delta S^\circ$

$$\Delta G^\circ_{302} = 42.59 \times 10^3\ J - (302\ K)\left(121.89\ J\cdot K^{-1}\right) = 5.78 \times 10^3\ J$$

$$K_{eq} = e^{\frac{-\Delta G^\circ}{RT}} \text{ and } \frac{-\Delta G^\circ}{RT} = \frac{-5.78 \times 10^3\ J}{\left(8.314\ J\cdot mol^{-1}\cdot K^{-1}\right)(302\ K)} = -2.30.$$

Therefore, $K_{eq} = P_{CH_3CH_2OH(g)} = e^{-2.30} = \mathbf{0.100\ atm}$.

(c) For $Hg(\ell) \rightarrow Hg(g)$, $K_{eq} = P_{Hg(g)}$ and

$$\Delta H^\circ_{rxn} = \left[\Delta H^\circ_f Hg(g)\right] - \left[\Delta H^\circ_f Hg(\ell)\right]$$

$$\Delta H^\circ_{rxn} = \left[\left(1\ mol \times 61.32\ kJ\cdot mol^{-1}\right)\right] - \left[\left(1\ mol \times 0.\ kJ\cdot mol^{-1}\right)\right] = 61.32\ kJ$$

$$\Delta S^\circ_{rxn} = \left[S^\circ Hg(g)\right] - \left[S^\circ H(\ell)\right]$$

$$\Delta S^\circ_{rxn} = \left[\left(1\ mol \times 174.85\ J\cdot mol^{-1}\cdot K^{-1}\right)\right] - \left[\left(1\ mol \times 76.02 J\cdot mol^{-1}\cdot K^{-1}\right)\right] = 98.83\ J\cdot K^{-1}$$

At 45°C, $\Delta G^\circ_{318} = \Delta H^\circ - (318\text{ K})\Delta S^\circ$

$\Delta G^\circ_{318} = 61.32 \times 10^3\text{ J} - (318\text{ K})(98.83\text{ J}\cdot\text{K}^{-1}) = 2.989 \times 10^4\text{ J}$

$K_{eq} = e^{\frac{-\Delta G^\circ}{RT}}$ and $\dfrac{-\Delta G^\circ}{RT} = \dfrac{-2.989 \times 10^4\text{ J}}{(8.314\text{ J}\cdot\text{mol}^{-1}\cdot\text{K}^{-1})(318\text{ K})} = -11.31.$

Therefore, $K_{eq} = P_{Hg(g)} = e^{-1131} = \mathbf{1.2 \times 10^{-5}\text{ atm}}$.

17.91 ***See Section 17.1 and Example 17.1.***

Initial pressure = $2000\text{ psi} \times \dfrac{1\text{ atm}}{14.7\text{ psi}} = 136\text{ atm}$

Final volume = $220\text{ ft}^3 \times \left(\dfrac{12\text{ in}}{1\text{ ft}}\right)^3 \times \left(\dfrac{2.54\text{ cm}}{1\text{ in}}\right)^3 \times \dfrac{1\text{ L}}{10^3\text{ cm}^3} = 6.23 \times 10^3\text{ L}$

Initial conditions: V_1 = ?, unknown $P_1 = 136\text{ atm}$

Final conditions: $V_2 = 6.23 \times 10^3\text{ L}$ $P_2 = P_{ext} = 1.0\text{ atm}$

Solving $P_1V_1 = P_2V_2$ for $V_1 = V_2 \times \dfrac{P_2}{P_1}$ $V_1 = 6.23 \times 10^3\text{ L} \times \dfrac{1.0\text{ atm}}{136\text{ atm}} = 45.8\text{ L}$

Hence, $w = -P\Delta V = -(1.0\text{ atm})(6.23 \times 10^3\text{ L} - 45.8\text{ L}) = -6.18 \times 10^3\text{ L}\cdot\text{atm}$

Converting L·atm to J gives $w = -6.18 \times 10^3\text{ L}\cdot\text{atm} \times 101.3\dfrac{\text{J}}{\text{L}\cdot\text{atm}} = \mathbf{-6.26 \times 10^5\text{ J}}$

17.92 ***See Section 17.4 and Example 17.6.***

The decomposition of TNT is accompanied by the release of a tremendous amount of energy, as is usually the case with explosions. Hence, the reation $2C_7H_5N_3O_6(s) \rightarrow 3N_2(g) + 5H_2O(\ell) + 7C(s) + 7CO(g)$ is predicted to be exothermic (i.e., have a negative value for ΔH°_{rxn}). The large increase in moles of gas during the reaction further indicates there is an increase in disorder and the value of ΔS°_{rxn} is positive. Hence, **the reaction has favorable enthalpy and entropy changes and is expected to be spontaneous at both low and high temperatures.**

17.93 ***See Section 17.5.***

Applying the equation $\ln K_{eq} = \dfrac{\Delta S^\circ}{R} - \dfrac{\Delta H^\circ}{RT}$ at temperatures T_2 and T_1 yields

$\ln K_{eq_2} = \dfrac{\Delta S^\circ}{R} - \dfrac{\Delta H^\circ}{RT_2}$ and $\ln K_{eq_1} = \dfrac{\Delta S^\circ}{R} - \dfrac{\Delta H^\circ}{RT_1}$.

Subtracting the second equation from the first gives $\ln K_{eq_2} - \ln K_{eq_1} = \left(\dfrac{\Delta S^\circ}{R} - \dfrac{\Delta H^\circ}{RT_2}\right) - \left(\dfrac{\Delta S^\circ}{R} - \dfrac{\Delta H^\circ}{RT_1}\right)$

and rearranging yields $\ln\left(\dfrac{K_{eq_2}}{K_{eq_1}}\right) = \dfrac{\Delta H^\circ}{R}\left(\dfrac{1}{T_1} - \dfrac{1}{T_2}\right)$ and $\ln\left(\dfrac{K_{eq_2}}{K_{eq_1}}\right) = \dfrac{\Delta H^\circ}{R}\left(\dfrac{T_2 - T_1}{T_1T_2}\right)$

In this case, $K_{eq_1} = 1.3$ at $T_1 = 779\text{ K}$ and $K_{eq_2} = 0.78$ at $T_2 = 803\text{ K}$.

Solving for ΔH° and substituting gives

$\Delta H^\circ = \left(\dfrac{RT_1T_2}{T_2 - T_1}\right)\ln\left(\dfrac{K_{eq_2}}{K_{eq_1}}\right)$ $\Delta H^\circ = \dfrac{(8.314\text{ J}\cdot\text{mol}^{-1}\cdot\text{K}^{-1})(779\text{ K})(803\text{ K})}{(803\text{ K} - 779\text{ K})}\ln\left(\dfrac{0.78}{1.3}\right) = \mathbf{-1.1 \times 10^5\text{ J}}$

Solving $\ln K_{eq} = \dfrac{\Delta S^\circ}{R} - \dfrac{\Delta H^\circ}{RT}$ for ΔS° yields $\Delta S^\circ = \dfrac{\Delta H^\circ}{T} + R\ln K_{eq}$.

Substituting data for 779 K gives $\Delta S^\circ = \frac{-1.1 \times 10^5 \text{ J}}{779 \text{ K}} + (8.314 \text{ J}\cdot\text{mol}^{-1}\cdot\text{K}^{-1})(\ln 1.3) = \mathbf{-1.4 \times 10^2 \text{ J}\cdot\text{K}^{-1}}$

At 298 K, $\Delta G^\circ_{298} = \Delta H^\circ - (298 \text{ K})\Delta S^\circ$

$\Delta G^\circ_{298} = -1.1 \times 10^5 \text{ J} - (298 \text{ K})(-1.4 \times 10^2 \text{ J}\cdot\text{K}^{-1}) = -7 \times 10^4 \text{ J}$

At the temperature at which there is no net driving force in either direction under standard conditions $\Delta G^\circ = 0$. At this temperature, $0 = \Delta H^\circ - T\Delta S^\circ$. Hence, $T = \frac{\Delta H^\circ}{\Delta S^\circ}$ $\quad T = \frac{-1.1 \times 10^5 \text{ J}}{-1.4 \times 10^2 \text{ J}\cdot\text{K}^{-1}} = 7.9 \times 10^2 \text{ K}$

The reaction is spontaneous in the direction written at 298 K ($\Delta G^\circ_{298} = -$) and experiences no net driving force in either direction at 7.9×10^2 K ($\Delta G^\circ_{7.9 \times 10^2} = 0$). Above 7.9×10^2 K, the reaction becomes nonspontaneous in the direction written. Hence, the reaction is **spontaneous up to 7.9×10^2 K**.

Note: This temperature agrees well with the temperature of 790 K that was obtained from calculations involving the more precise data in 17.61.

Note: The decrease in K_{eq} from 1.3 to 0.78 with the increase in temperature from 506°C to 530°C suggests the reaction is exothermic. This is confirmed by $\Delta H^\circ = -1.1 \times 10^5$ J.

17.94 *See Section 17.4 and Example 17.5.*

For $SO_2(g) + \frac{1}{2}O_2(g) \rightarrow SO_3(g)$,

$\Delta G^\circ_{rxn} = \left[\Delta G^\circ_f SO_3(g)\right] - \left[\Delta G^\circ_f SO_2(g) + \frac{1}{2}\Delta G^\circ_f SO_2(g)\right]$

$\Delta G^\circ_{rxn} = \left[(1\text{mol} \times -371.08 \text{ kJ}\cdot\text{mol}^{-1})\right] - \left[(1\text{mol} \times -300.19 \text{ kJ}\cdot\text{mol}^{-1}) + (\frac{1}{2}\text{mol} \times 0. \text{ kJ}\cdot\text{mol}^{-1})\right]$

$= -70.89 \text{kJ}\cdot\text{mol}^{-1}$

SO_3 is important because it reacts with water to produce acid rain.

17.95 *See Section 17.5.*

$\Delta G = \Delta H^\circ - T\Delta S^\circ = -RT \ln K_{eq}$ leads to $\ln K_{eq} = \frac{-\Delta H^\circ}{R}\left(\frac{1}{T}\right) + \frac{\Delta S^\circ}{R}$. Hence, a plot of $\ln K_{eq}$ vs. $\frac{1}{T}$ has a slope of $\frac{-\Delta H^\circ}{R}$ and an intercept of $\frac{\Delta S^\circ}{R}$. Determine the value of the slope of the line, and use $\Delta H^\circ = -(\text{slope})(R)$ to obtain ΔH°. Determine the value of the intercept of the line, and use $\Delta S^\circ = (\text{intercept})(R)$ to obtain ΔS°.

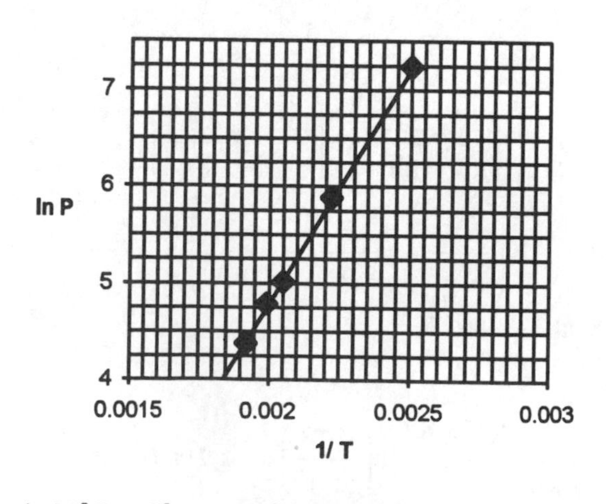

K_{eq}	$\ln K_{eq}$	T, K	$\frac{1}{T}$, K^{-1}
1.4×10^3	7.24	400	2.50×10^{-3}
3.6×10^2	5.89	450	2.22×10^{-3}
1.5×10^2	5.02	489	2.05×10^{-3}
1.2×10^2	4.79	502	1.99×10^{-3}
8.0×10^1	4.38	522	1.92×10^{-3}

A linear regression analysis of the data yields a slope of 4.9×10^3 and an intercept of -5.00 with a correlation coefficient of 0.999695 indicating excellent agreement between the experimental data and the equation.

$\Delta H^\circ = -(4.90 \times 10^3 \text{ K})(8.314 \text{ J}\cdot\text{mol}^{-1}\cdot\text{K}^{-1}) = -4.07 \times 10^4 \text{ J}\cdot\text{mol}^{-1}$

$\Delta S^\circ = (-5.00)(8.314 \text{ J}\cdot\text{mol}^{-1}\cdot\text{K}^{-1}) = -41.6 \text{ J}\cdot\text{mol}^{-1}\cdot\text{K}^{-1}$

At 25°C, $\Delta G^\circ_{298} = \Delta H^\circ - (298 \text{ K})\Delta S^\circ$

$\Delta G^\circ_{298} = -4.07 \times 10^4 \text{ J}\cdot\text{mol}^{-1} - (298 \text{ K})(-41.6 \text{ J}\cdot\text{mol}^{-1}\cdot\text{K}^{-1}) = -2.83 \times 10^4 \text{ J}\cdot\text{mol}^{-1}$ or $\mathbf{-28.3 \text{ kJ}\cdot\text{mol}^{-1}}$

At the temperature at which there is no net driving force in either direction, $\Delta G° = 0$. At this temperature,

$O = \Delta H° - T\Delta S°$. Hence, $T = \dfrac{\Delta H°}{\Delta S°}$ $\quad T = \dfrac{-4.07 \times 10^4 \ J \cdot mol^{-1}}{-41.6 \ J \cdot mol^{-1} \cdot K^{-1}} = 978 \ K$

The reaction is spontaneous in the direction written at 298 K($\Delta G°_{298}$ = -) and experiences no net driving force in either direction at 978 K($\Delta G°_{978} = 0$). Above 978 K the reaction becomes nonspontaneous in the direction written. Hence, the reaction is **spontaneous up to 978 K.**

Note: If your calculator does not have the ability to perform linear regression analyses or you do not have access to a computer graphing program, you can calculate the slope of the line using $\left(\dfrac{\ln K_2 - \ln K_1}{\frac{1}{T_2} - \frac{1}{T_1}}\right)$ with two points from the line which are far apart. The slope of the line can then be used to calculate the value of $\Delta H°$, and this value can be used with the given data to solve the equation for $\Delta S°$. It isn't feasible to obtain $\Delta S°$ from the graph, since the y-intercept occurs when x = 0 and this would require a graph of unreasonable size.

17.96 *See Section 17.4 and Example 17.6.*

For $2H_2O(g) \rightarrow 2H_2(g) + O_2(g)$,

$$\Delta H°_{rxn} = \left[2\Delta H°_f H_2(g) + \Delta H°_f O_2(g)\right] - \left[2\Delta H°_f H_2O(g)\right]$$

$$\Delta H°_{rxn} = \left[\left(2 \ mol \times 0. \ kJ \cdot mol^{-1}\right) + \left(1 \ mol \times 0. \ kJ \cdot mol^{-1}\right)\right] - \left[\left(2 \ mol \times -241.82 \ kJ \cdot mol^{-1}\right)\right] = 483.64 \ kJ$$

$$\Delta S°_{rxn} = \left[2S°H_2(g) + S°O_2(g)\right] - \left[2S°H_2O(g)\right]$$

$$\Delta S°_{rxn} = \left[\left(2 \ mol \times 130.57 \ J \cdot mol^{-1} \cdot K^{-1}\right) + \left(1 \ mol \times 205.03 \ J \cdot mol^{-1} \cdot K^{-1}\right)\right] - \left[\left(2 \ mol \times 188.72 \ J \cdot mol^{-1} \cdot K^{-1}\right)\right]$$

$$= 88.73 \ J \cdot K^{-1}$$

$$\Delta G°_{298} = \Delta H° - (298 \ K)\Delta S° \qquad \Delta G°_{298} = 483.64 \times 10^3 \ J - (298 \ K)\left(88.73 \ J \cdot K^{-1}\right) = 4.57 \times 10^5 \ J$$

At the temperature at which there is no net driving force in either direction under standard conditions $\Delta G° = 0$. At this temperature, $0 = \Delta H° - T\Delta S°$. Hence,

$$T = \frac{\Delta H°}{\Delta S°} \qquad T = \frac{483.64 \times 10^3 \ J}{88.73 \ J \cdot K^{-1}} = 5.45 \times 10^3 \ K$$

The reaction is nonspontaneous in the direction written at 298 K ($\Delta G°_{298}$ = +) and experiences no net driving force in either direction at 5.45×10^3 K ($\Delta G°_{5.45 \times 10^3} = 0$). Above 5.45×10^3 K the reaction becomes spontaneous in the direction written. Hence, the reaction is spontaneous **above 5.45×10^3 K.**

17.97 *See Section 17.4.*

The explosive combination of hydrogen and oxygen is nonspontaneous when the decomposition of water to give hydrogen and oxygen is spontaneous, since the combination is just the reverse of the decomposition. However, if the temperature is suddenly lowered to the point where the combination of hydrogen and oxygen is spontaneous, the possibility of an explosion could become a problem.

17.98 *See Section 17.5 and Example 17.8.*

$K_{eq} = e^{\frac{-\Delta G°}{RT}}$ and $\dfrac{-\Delta G°}{RT} = \dfrac{-840.1 \times 10^3 \ J \cdot mol^{-1}}{\left(8.314 \ J \cdot mol^{-1} \cdot K^{-1}\right)(298 \ K)} = 3.39 \times 10^2$. Therefore, $K_{eq} = e^{3.39 \times 10^2}$.

The value of K_{eq} is greater than 10^{99} and cannot be calculated using $K_{eq} = e^{3.39 \times 10^2}$ on most calculators.

However, $K_{eq} = 10^{\frac{-\Delta G°}{2.303RT}}$ derived from $\Delta G° = -2.303 \ RT \log K_{eq}$ can be used to obtain the value of K_{eq}: $K_{eq} = 10^{147.23} = 10^{0.23} \times 10^{147} = 1.7 \times 10^{147}$.

The equilibrium constant tells us $Al^{3+} - O^{2-}$ interactions are stonger than $Fe^{3+} - O^{2-}$ interactions. This is to be expected, since Al^{3+} is smaller than Fe^{3+}.

17.99 ***See Section 17.4 and Example 17.6.***

(a) For $2ZnO(s) + C(s,graphite) \rightarrow 2Zn(s) + CO_2(g)$,

$$\Delta H^o_{rxn} = \left[2\Delta H^o_f\ Zn(s) + \Delta H^o_f\ CO_2(g)\right] - \left[2\Delta H^o_f\ ZnO(s) + 2\Delta H^o_f\ C(s,gr)\right]$$
$$= \left[\left(2\ mol \times 0.\ kJ \cdot mol^{-1}\right) + \left(1\ mol \times -393.51\ kJ \cdot mol^{-1}\right)\right]$$
$$- \left[\left(2\ mol \times -348.28\ kJ \cdot mol^{-1}\right) + \left(1\ mol \times 0.\ kJ \cdot mol^{-1}\right)\right] = 303.05\ kJ$$
$$\Delta S^o_{rxn} = \left[2S^o\ Zn(s) + S^o\ CO_2(g)\right] - \left[2S^o\ ZnO(s) + S^o\ C(s,gr)\right]$$
$$= \left[\left(2\ mol \times 41.63\ J \cdot mol^{-1} \cdot K^{-1}\right) + \left(1\ mol \times 213.63\ J \cdot mol^{-1} \cdot K^{-1}\right)\right]$$
$$- \left[\left(2\ mol \times 43.64\ J \cdot mol^{-1} \cdot K^{-1}\right) + \left(1\ mol \times 5.74\ J \cdot mol^{-1} \cdot K^{-1}\right)\right] = 203.87\ J \cdot K^{-1}$$

$\Delta G^o_{298} = \Delta H^o - (298\ K)\Delta S^o \qquad \Delta G^o_{298} = 303.05\ kJ - (298\ K)(203.87 \times 10^{-3}\ kJ \cdot K^{-1}) = 242.30\ kJ$

ΔG^o_{298} is positive. Hence, the reaction is **nonspontaneous** at 298 K under standard conditions.

(b) At the temperature at which there is no net driving force in either direction under standard conditions ΔG^o = 0. At this temperature, $0 = \Delta H^o - T\Delta S^o$. Hence,

$$T = \frac{\Delta H^o}{\Delta S^o} \qquad T = \frac{303.05\ kJ}{203.87 \times 10^{-3}\ kJ \cdot K^{-1}} = 1486\ K \text{ or } 1213^o\ C$$

The reaction is nonspontaneous in the direction written at 298 K(ΔG^o_{298} = +) and experiences no net driving in either direction at 1486 K($\Delta G^o_{1486} = 0$). The reaction is predicted to be spontaneous in the direction written at temperatures greater than 1486 K, **greater than 1213° C**.

(c) For $2ZnO(s) + C(s,graphite) \rightarrow 2Zn(g) + CO_2(g)$,

$$\Delta H^o_{rxn} = \left[2\Delta H^o_f\ Zn(g) + \Delta H^o_f\ CO_2(g)\right] - \left[2\Delta H^o_f\ ZnO(s) + 2\Delta H^o_f\ C(s,gr)\right]$$
$$= \left[\left(2\ mol \times 130.73\ kJ \cdot mol^{-1}\right) + \left(1\ mol \times -393.51\ kJ \cdot mol^{-1}\right)\right]$$
$$- \left[\left(2\ mol \times -348.28\ kJ \cdot mol^{-1}\right) + \left(1\ mol \times 0.\ kJ \cdot mol^{-1}\right)\right] = 564.51\ kJ$$
$$\Delta S^o_{rxn} = \left[2S^o\ Zn(g) + S^o\ CO_2(g)\right] - \left[2S^o\ ZnO(s) + S^o\ C(s,gr)\right]$$
$$= \left[\left(2\ mol \times 160.87\ J \cdot mol^{-1} \cdot K^{-1}\right) + \left(1\ mol \times 213.63\ J \cdot mol^{-1} \cdot K^{-1}\right)\right]$$
$$- \left[\left(2\ mol \times 43.64\ J \cdot mol^{-1} \cdot K^{-1}\right) + \left(1\ mol \times 5.74\ J \cdot mol^{-1} \cdot K^{-1}\right)\right] = 442.35\ J \cdot K^{-1}$$

$$T = \frac{\Delta H^o}{\Delta S^o} \qquad T = \frac{564.51\ kJ}{442.35 \times 10^{-3}\ kJ \cdot K^{-1}} = 1276\ K \text{ or } 1003^o\ C$$

Hence, the reaction is predicted to be spontaneous in the direction written at temperatures **greater than 1003°C**.

(d) The Principle of LeChatelier tells us we can cause the reaction to proceed to completion at any temperature greater than the normal boiling point of zinc, at any temperature greater than 907° C, by removing the gaseous zinc product. This is how zinc is produced commercially.

17.100 *See Sections 9.8, 22.1, Table 9.4, Figure 22.4, and Example 9.16.*

(a) The SN for each carbon atom in each molecule is 4 indicating all C-C-C bond angles should be 109.5°.

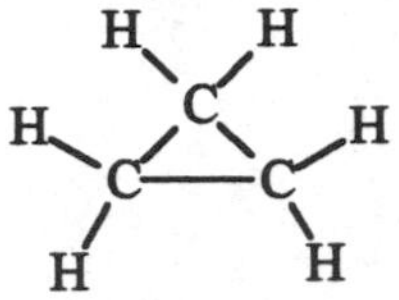

cyclopropane 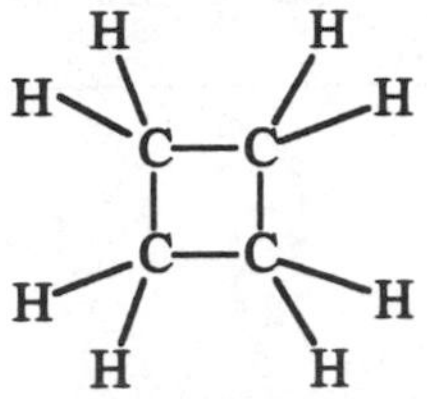   cyclobutane 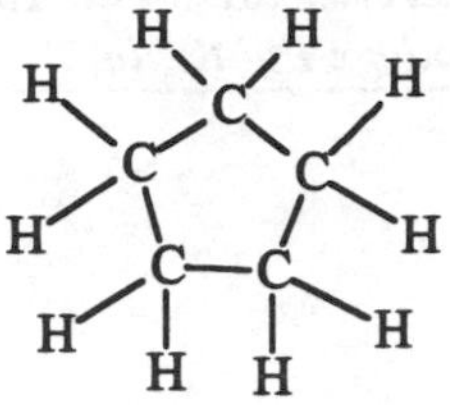   cyclopentane 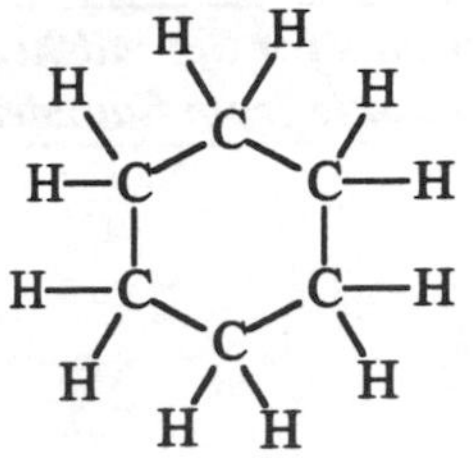   cyclcohexane

(b) For $C_3H_6(g) + \frac{9}{2}O_2(g) \rightarrow 3CO_2(g) + 3H_2O(g)$,

$$\Delta H_{rxn} = -1957.7 \text{ kJ} = \left[3D_{C-C} + 6D_{C-H} + \tfrac{9}{2}D_{O=O}\right] - \left[6D_{C=O} + 6D_{O-H}\right]$$

$$-1957.7 \text{ kJ} = \left[(3 \text{ mol})(D_{C-C}) + (6 \text{ mol})\left(414\frac{\text{kJ}}{\text{mol}}\right) + \left(\tfrac{9}{2}\text{ mol}\right)\left(498\frac{\text{kJ}}{\text{mol}}\right)\right] - \left[(6 \text{ mol})\left(799\frac{\text{kJ}}{\text{mol}}\right) + (6 \text{ mol})\left(463\frac{\text{kJ}}{\text{mol}}\right)\right]$$

$(3 \text{ mol})(D_{C-C}) = 889$ kJ $D_{C-C} = 296$ kJ/mol

For $C_4H_8(g) + 6O_2(g) \rightarrow 4CO_2(g) + 4H_2O(g)$,

$$\Delta H_{rxn} = -2567.6 \text{ kJ} = \left[4D_{C-C} + 8D_{C-H} + 6D_{O=O}\right] - \left[8D_{C=O} + 8D_{O-H}\right]$$

$$-2567.6 \text{ kJ} = \left[(4 \text{ mol})(D_{C-C}) + (8 \text{ mol})\left(414\frac{\text{kJ}}{\text{mol}}\right) + (6 \text{ mol})\left(498\frac{\text{kJ}}{\text{mol}}\right)\right] - \left[(8 \text{ mol})\left(799\frac{\text{kJ}}{\text{mol}}\right) + (8 \text{ mol})\left(463\frac{\text{kJ}}{\text{mol}}\right)\right]$$

$(4 \text{ mol})(D_{C-C}) = 1228$ kJ $D_{C-C} = 307$ kJ/mol

For $C_5H_{10}(g) + \frac{15}{2}O_2(g) \rightarrow 5CO_2(g) + 5H_2O(g)$,

$$\Delta H_{rxn} = -3097.6 \text{ kJ} = \left[5D_{C-C} + 10D_{C-H} + \tfrac{15}{2}D_{O=O}\right] - \left[10D_{C=O} + 10D_{O-H}\right]$$

$$-3097.6 \text{ kJ} = \left[(5 \text{ mol})(D_{C-C}) + (10 \text{ mol})\left(414\frac{\text{kJ}}{\text{mol}}\right) + \left(\tfrac{15}{2}\text{ mol}\right)\left(498\frac{\text{kJ}}{\text{mol}}\right)\right] - \left[(10 \text{ mol})\left(799\frac{\text{kJ}}{\text{mol}}\right) + (10 \text{ mol})\left(463\frac{\text{kJ}}{\text{mol}}\right)\right]$$

$(5 \text{ mol})(D_{C-C}) = 1647$ kJ $D_{C-C} = 329$ kJ/mol

For $C_6H_{12}(g) + 9O_2(g) \rightarrow 6CO_2(g) + 6H_2O(g)$,

$$\Delta H_{rxn} = -3685.5 \text{ kJ} = \left[6D_{C-C} + 12D_{C-H} + 9D_{O=O}\right] - \left[12D_{C=O} + 12D_{O-H}\right]$$

$$-3685.5 \text{ kJ} = \left[(6 \text{ mol})(D_{C-C}) + (12 \text{ mol})\left(414\frac{\text{kJ}}{\text{mol}}\right) + (9 \text{ mol})\left(498\frac{\text{kJ}}{\text{mol}}\right)\right] - \left[(12 \text{ mol})\left(799\frac{\text{kJ}}{\text{mol}}\right) + (12 \text{ mol})\left(463\frac{\text{kJ}}{\text{mol}}\right)\right]$$

$(6 \text{ mol})(D_{C-C}) = 2008$ kJ $D_{C-C} = 335$ kJ/mol

(c) As the C-C-C bond angles approach the predicted angle of 109.5° there is less ring strain and better overlap of the sp^3 orbitals used for bonding by the carbon atoms. This leads to stronger bonding and the average carbon-carbon bond energy increases accordingly.

17.101, 17.102 ***Solutions for Spreadsheet Exercises are on the Reger, Goode and Mercer Spreadsheet Solutions Diskette available from Saunders College Publishing.***

Chapter 18: Electrochemistry

18.1 ***See Sections 15.9, 18.1.***

Oxidation involves the loss of electrons by any chemical species. Reduction inolves the gain of electrons by any chemical species. In an oxidation-reduction (redox) reaction there is either a partial or complete transfer of electrons from one species to another. In a Lewis acid-base reaction, the acid provides an empty orbital and the Lewis base provides a filled orbital that overlap to form a coordinate covalent bond. A partial transfer of electrons occurs as the coordinate covalent bond is formed between atoms of differing electronegativity, as in the formation of the boron-nitrogen bond in $BF_3 + NH_3 \rightarrow F_3BNH_3$. This partial transfer is like the partial transfer of electrons that occurs in redox reactions like $C + O_2 \rightarrow CO_2$. However, the main differences between redox and Lewis acid-base reactions is that there is no requirement for one atom to be furnishing both of the electrons for the covalent bonds that are formed during redox reactions and redox reactions can also involve the complete transfer of electrons. Complete electron transfer occurs in reactions like $2Li + F_2 \rightarrow 2LiF$ and $2Ag^+ + Cu \rightarrow 2Ag + Cu^{2+}$.

18.2 ***See Section 18.5 and Table 18.1.***

The oxidizing power of the halogens increases in the order **$I_2(s) < Br_2(\ell) < Cl_2(g) < F_2(g)$**, the order in which the standard reduction potentials become more positive indicating greater ease of reduction.

18.3 ***See Section 18.5 and Appendix H.***

Zinc has a more positive standard oxidation potential than mercury and is therefore more easily oxidized and is a better reducing agent than mercury.

18.4 ***See Section 18.5, Table 18.1, and Example 18.9.***

Oxidizing agents are listed on the left-hand side of Table 18.1 and are listed in order of decreasing oxidizing power. Hence, **any four chemical species that are listed above Fe^{3+}**, such as Ag^+, $Cl_2(g)$, MnO_4^- and Ce^{4+}, **can oxidize Fe^{2+} to Fe^{3+}**.

18.5 ***See Section 18.5, Table 18.1, and Example 18.9.***

Reducing agents are listed on the right-hand side of Table 18.1 and are listed in order of increasing reducing power. Hence, **any four chemical species that are listed below Al**, such as Li(s), Ba(s), Na(s) and Mg(s), **can reduce Al^{3+} to Al**.

18.6 ***See Section 18.6 and Figure 18.9.***

An indicator electrode has a potential that depends on the concentration of a particular species in solution. A reference electrode has an accurately known potential that is dependent on its internal composition and is independent of the composition of the test solution.

18.7 ***See Section 18.9.***

Overvoltage is the voltage difference between the potential that is calculated to cause an electrolysis reaction to occur and the potential that is actually needed to cause the electrolysis reaction to occur. It is probably caused by slow transfer of electrons at the electrode surfaces and is similar to an activation energy. The magnitude of the overvoltage varies with the electrode material and particular electrode reaction and must be measured experimentally.

18.8 *See Section 18.12.*

Aluminum forms an oxide coating that adheres strongly to the surface of the metal and is insoluble in acid. There are a number of ways this hypothesis could be tested. Perhaps the simplest would be to scratch the oxide coating off from area to determine whether the exposed aluminum metal dissolves in the hydrochloric acid.

18.9 *See Sections 9.3, 18.1, and Example 18.1.*

(a) ClO_3^-

Total valence electrons = $[1 \times 7(Cl) + 3 \times 6(O) + 1(charge)] = 26$.

Twenty electrons remain after assigning three single bonds, and twenty unshared electrons are needed to give each atom a noble gas configuration (2 for Cl and 6 for each O). Hence, no multiple bonds are needed to give each atom a noble gas configuration. However, chlorine is capable of using an expanded valence shell, and double bonds to each oxygen atom are assigned to reduce the formal charges. Twenty six electrons are used in writing the Lewis structure.

A total of eight valence shell electrons are assigned to each oxygen atom, two more than are present in a nonbonded oxygen atom, giving each oxygen atom an oxidation state of -2. The chlorine atom having six of its seven valence shell electrons assigned to the more electronegative oxygen atoms and having gained one electron for the overall net charge of 1- has an oxidation state of +5. Hence, oxidation state of **Cl = +5, O = -2.**

(b) PF_3

Total valence electrons = $[1 \times 5(P) + 3 \times 7(F)] = 26$.

Twenty electrons remain after assigning three single bonds, and twenty unshared electrons are needed to give each atom a noble gas configuratin (2 for P and 6 for each F). Twenty six electrons are used in writing the Lewis structure.

A total of eight valence shell electrons are assigned to each fluorine atom, one more than are present in a nonbonded fluorine atom, giving each fluorine atom an oxidation state of -1. The phosphorus atom having three of its five valence shell electrons assigned to the more electronegative fluorine atoms has an oxidation state of +3. Hence, oxidation state of **P = +3, F = 1.**

(c) CO

Total valence electrons = $[1 \times 4(C) + 1 \times 6(O)] = 10$.

Eight valence electrons remain after assigning a single bond, and twelve unshared electrons are needed to give each atom a noble gas configuration (6 for C and 6 for O). Hence, four electrons (12-8) must be used to form two additional bonds. Ten electrons are used in writing the Lewis structure.

A total of eight valence shell electrons are assigned to the oxygen atoms, two more than are present in a nonbonded oxygen atom, giving the oxygen atom an oxidation state of -2. The carbon atom having two of its four valence shell electrons assigned to the more electronegative oxygen atom has an oxidation state of +2. Hence, oxidation state of **C = +2, O = -2.**

18.10 ***See Sections 9.3, 18.1, and Example 18.1.***

(a) N_2

N—N

:N≡N:

Total valence electrons = $[2 \times 5(N)] = 10$.

Eight valence electrons remain after assigning a single bond, and twelve unshared electrons are needed to give each atom a noble gas configuration (6 for each N). Hence, four electrons (12-8) must be used to form two additional bonds. Ten electrons are used in writing the Lewis structure.

The shared electrons are assigned equally to the nitrogen atoms giving the valence shell of each nitrogen atom five valence electrons, the same number as a nonbonded nitrogen atom. Therefore, each nitrogen atom has an oxidation state of zero. Hence, oxidation state of **N = 0.**

(b) $B(OH)_3$

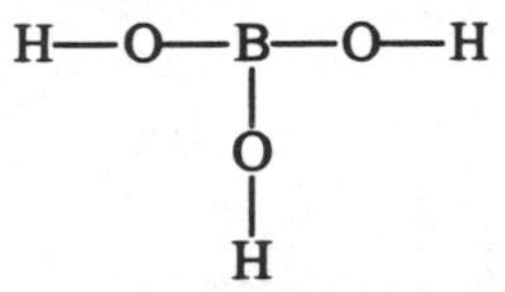

Total valence electrons = $[1 \times 3(B) + 3 \times 6(O) + 3 \times 1(H)] = 24$.

Twelve valence electrons reamin after assigning six single bonds, and twelve unshared electrons are needed to give each oxygen and hydrogen atom a noble gas configuration (4 for each O). Twenty four electrons are used in writing the electron deficient Lewis structrure.

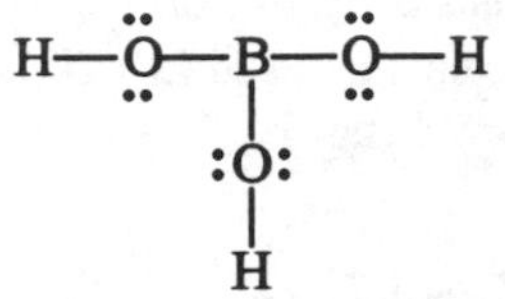

A total of eight valence shell electrons are assigned to the valence shell of each oxygen atom, two more than are present in a nonbonded oxygen atom, giving each oxygen atom an oxidation state of -2. The hydrogen atoms having their only valence electrons assigned to the more electronegative oxygen atoms have an oxidation state of +1. The boron atom having all three of its valence shell electrons assigned to the more electronegative oxygen atoms has an oxidation state of +3. Hence, oxidation state of **B = +3, H = +1, O = -2.**

(c) IF_4^-

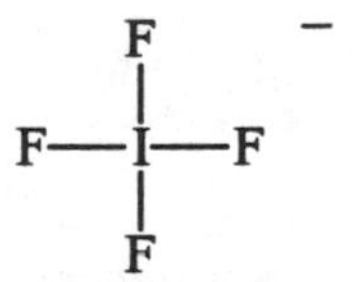

Total valence electrons = $[1 \times 7(I) + 4 \times 6(F) + 1\ (\text{charge})] = 36$.

Twenty eight valence electrons remain after assigning four single bonds, and twenty four unshared electrons are needed to give each atom a noble gas configuration (6 for each F). Hence, four electrons (28-24) must be assigned to iodine, the central atom. Thirty six electrons are used in writing the electron rich Lewis structrure.

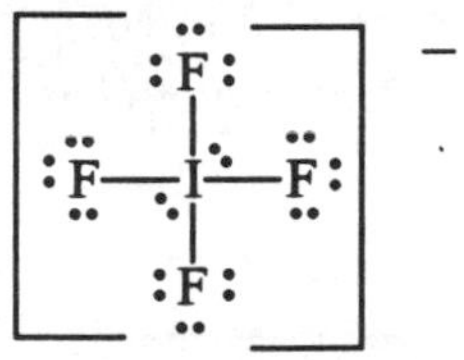

A total of eight valence shell electrons are assigned to the valence shell of each fluorine atom, one more than is present in a nonbonded fluorine atom, giving each fluorine atom an oxidation state of -1. The iodine atom having four of its valence shell electrons assigned to the more electronegative fluorine atoms and having gained one electron for the overall net charge of -1 has an oxidation state of +3. Hence, oxidation state of **I = +3, F = -1.**

18.11 ***See Sections 9.3, 18.1, and Example 18.1.***

(a) NO_3^-

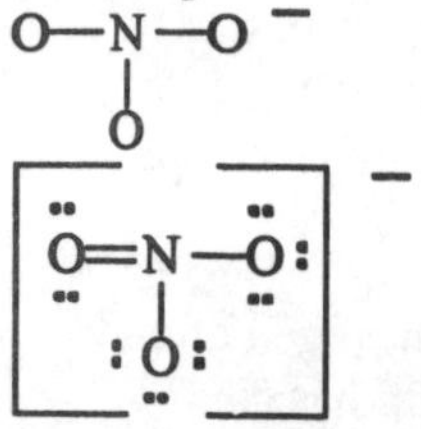

Total valence electrons $= [1 \times 5(N) + 3 \times 6(O) + 1(\text{charge})] = 24$.

Eighteen electrons remain after assigning three single bonds, and twenty unshared electrons are needed to give each atom a noble gas configuration (6 for each O and 2 for N). Hence, two electrons (20-18) must be used to form one additional bond. Twenty four electrons are used in writing the Lewis structure, which is one of three resonance forms.

A total of eight valence shell electrons are assigned to each oxygen atom, two more than are present in a nonbonded oxygen atom, giving each oxygen atom an oxidation state of -2. A nitrogen atom having its five valence shell electrons assigned to the more electronegative oxygen atoms has an oxidation state of +5. Hence, oxidation state of **N = +5, O = -2**.

(b) NO_2^-

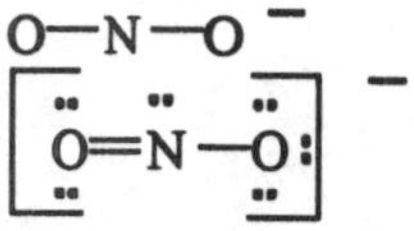

Total valence electrons $= [1 \times 5(N) + 2 \times 6(O) + 1(\text{charge})] = 18$.

Fourteen electrons remain after assigning two single bonds, and sixteen unshared electrons are needed to give each atom a noble gas configuration (6 for each O and 4 for N). Hence, two electrons (16-14) must be used to form one additional bond. Eighteen electrons are used in writing the Lewis structure, which is one of two resonance forms.

A total of eight valence shell electrons are assigned to each oxygen atom, two more than are present in a nonbonded oxygen atom, giving each oxygen atom an oxidation state of -2. A nitrogen atom having three of its five valence shell electrons assigned to the more electronegative oxygen atoms has an oxidation state of +3. Hence, oxidation state of **N = +3, O = -2**.

(c) NH_4^+

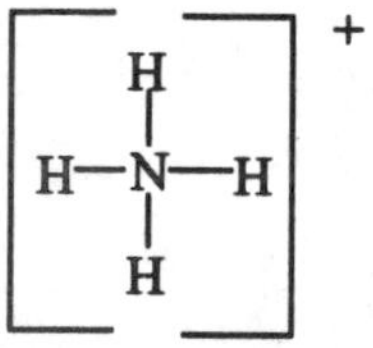

Total valence electrons $= [1 \times 5(N) + 4 \times 1(H) - 1(\text{charge})] = 8$.

All eight electrons are used in writing four single bonds.

All eight shared electrons are assigned to the valence shell of the more electronegative nitrogen atom. Since there are five electrons in the valence shell of a nonbonded nitrogen atom, nitrogen has an oxidation state of -3 in NH_4^+. Each hydrogen atom having its only valence shell electron assigned to the more electronegative nitrogen atom has an oxidation state of +1. Hence, oxidation state of **N = -3, H = +1**.

18.12 ***See Section 9.3, 18.1, and Example 18.1.***

(a) Br_2

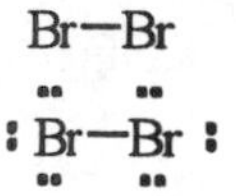

Total valence electrons $= [2 \times 7(Br)] = 14$.

Twelve electrons remain after assigning one single bond and twelve unshared electrons are needed to give each Br atom a noble gas configuration. Fourteen electrons are used in writing the Lewis structure.

The shared electrons are assigned equally to the bromine atoms giving the valence shell of each bromine atom seven valence electrons, the same number as a nonbonded bromine atom. Therefore, each bromine atom has an oxidation state of zero. Hence, oxidation state of **Br = 0**.

(b) CO_3^{2-}

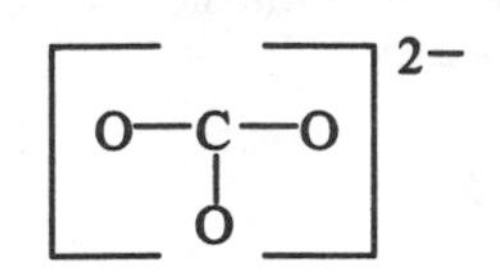

Total valence electrons $= [1 \times 4(C) + 3 \times 6(O) + 2(\text{charge})] = 24$.

Eighteen electrons remain after assigning three single bonds, and twenty unshared electrons are needed to give each atom a noble gas configuration (6 for each O and 2 for C). Hence, two electrons (20-18) must be used to form one additional bond. This gives the following possible resonance forms:

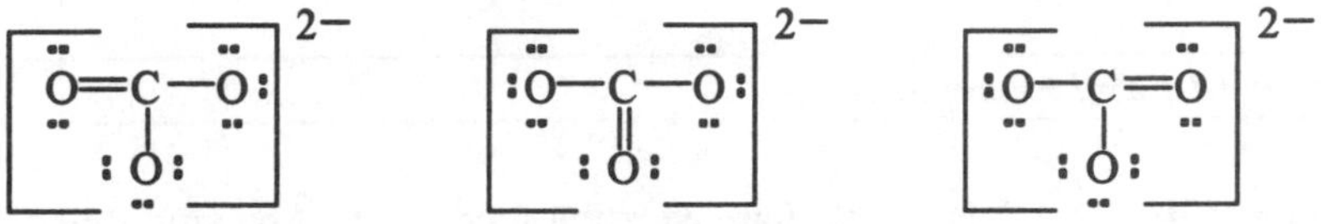

The shared electrons are assigned to the more electronegative oxygen atoms giving each oxygen atom eight electrons in its valence shell. This gives each oxygen atom two more valence electrons than a nonbonded oxygen atom and an oxidation state of -2. The carbon atom having its four valence shell electrons assigned to the more electronegative oxygen atoms has an oxidation state of +4. Hence, oxidation state of **C = +4, O = -2.**

(c) CO_2

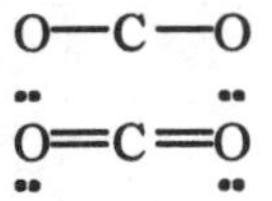

Total valence electrons$= [1 \times 4(C) + 2 \times 6(O)] = 16$.

Twelve electrons remain after assigning two single bonds, and sixteen electrons are needed to give each atom a noble gas configuration (6 for each O and 4 for C). Hence, four electrons (16-12) must be used to form additional bonds. A carbon-oxygen double bond is formed with each O atom, and sixteen electrons are used to write the Lewis structure.

The shared electrons are assigned to the more electronegative oxygen atoms giving each oxygen atom eight electrons in its valence shell. This gives each oxygen atom two more valence electrons than a nonbonded oxygen atom and an oxidation state of -2. The carbon atom having its four valence shell electrons assigned to the more electronegative oxygen atoms has an oxidation state of +4. Hence, oxidation state of **C = +4, O = -2**.

18.13 ***See Section 18.1 and Example 18.2.***

(a) ZrO_2: Applying Rule 5, O is assigned an oxidation state of -2. This gives y + 2(-2) = 0 and y = +4 for the oxidation state of the zirconium in ZrO_2. Hence, oxidation state of **Zr = +4, O = -2.**

(b) FeO: Applying Rule 5, O is assigned an oxidation state of -2. This gives y + 1(-2) = 0 and y = +2 for the oxidation state of iron in FeO. Hence, oxidation state of **Fe = +2, O = -2.**

(c) $Ca(NO_3)_2$: Applying Rule 2, Ca is assigned an oxidation state of +2. Applying Rule 5, each O is assigned an oxidation state of -2. This gives 2 + 2y + 6(-2) = 0 and y = +5 for the oxidation state of nitrogen in $Ca(NO_3)_2$. Hence, oxidation state of **Ca = +2, N = +5, O = -2.**

18.14 ***See Section 18.1 and Example 18.2.***

(a) PF_5: Applying Rule 3, F is assigned an oxidation state of -1. This gives y +5(-1) = 0 and y = +5 for the oxidation state of the phosphorus in PF_5. Hence, oxidation state of **P = +5, F = -1.**

(b) Na_2CrO_4: Applying Rule 2, Na is assigned an oxidation state of +1. Applying Rule 5, each O is assigned an oxidation state of - 2. This gives 2(+1) + y + 4(-2) = 0 and y = +6 for the oxidation state of the chromium in Na_2CrO_4. Hence, oxidation state of **Na = +1, Cr = +6, O = -2.**

(c) NO_2^- : Applying Rule 5, O is assigned an oxidation state of -2. This gives y + 2(-2) = -1 and y = +3 for the oxidation state of the nitrogen in NO_2^- . Hence, oxidation state of **N = +3, O = -2.**

18.15 *See Section 18.1 and Example 18.2.*

(a) BaO_2: Applying Rule 2, Ba is assigned an oxidation state of +2. This gives 2 + 2y = 0 and y = -1 for the oxidation state of the oxygen in BaO_2 (barium peroxide). Hence, oxidation state of **Ba = +2, O = -1**.

(b) F_2: Applying Rule 1, each F is assigned an oxidation state of 0. Hence, oxidation state of **F = 0**.

(c) Sn^{2+}: Applying Rule 2, Sn is assigned an oxidation state of +2. Hence oxidation state of **Sn = +2**.

18.16 *See Section 18.1 and Example 18.2.*

(a) $KMnO_4$: Applying Rule 2, K^+ is assigned an oxidation state of +1. Applying Rule 5, O is assigned an oxidation state of -2. This gives 1 + y + 4(-2) = 0 and y = +7 for the oxidation state of the manganese in $KMnO_4$. Hence, oxidation state of **K = +1, Mn = +7, O = -2**.

(b) H_2O: Applying Rule 4, hydrogen is assigned an oxidation state of +1. Applying Rule 5, oxygen is assigned an oxidation state of -2. This gives the expected sum of zero for H_2O. Hence, oxidation state of **H = +1, O = -2**.

(c) Cl_2: Applying Rule 1, each chlorine is assigned an oxidation state of zero. This gives the expected sum of zero for Cl_2. Hence, oxidation state of **Cl = 0**.

18.17 *See Section 18.1 and Example 18.2.*

(a) NO_2: Applying Rule 5, each oxygen atom is assigned an oxidation state of -2. This gives y + 2(-2) = 0 and y = +4 for the oxidation state of the nitrogen in NO_2. Hence, oxidation state of **N = +4, O = -2**.

(b) CrO_2^-: Applying Rule 5, each oxygen atom is assigned an oxidation state of -2. This gives y + 2(-2) = -1 and y = +3 for the oxidation state of chromium in CrO_2^-. Hence, oxidation state of **Cr = +3, O = -2**.

(c) $Co(NO_3)_3$: Applying Rule 2, cobalt is assigned an oxidation state of +3. Applying Rule 5, each oxygen atom is assigned an oxidation state of -2. This gives 3 + y + 3(-2) = 0 and y = +5 for the oxidation state of nitrogen in NO_3^- and $Co(NO_3)_3$. Hence, oxidation state of **Co = +3, N = +5, O = -2**.

18.18 *See Section 18.1 and Example 18.2.*

(a) $CaCO_3$: Applying Rule 2, calcium is assigned an oxidation state of +2. Applying Rule 5, oxygen is assigned an oxidation state of -2. This gives y + 3(-2) = -2 and y = +4 for the oxidation state of carbon in CO_3^{2-} and $CaCO_3$. Hence, oxidation state of **Ca = +2, C = +4, O = -2**.

(b) $HBrO_4$: Applying Rule 4, hydrogen is assigned an oxidation state of +1. Applying Rule 5, oxygen is assigned an oxidation state of -2. This gives 1 + y + 4(-2) = 0 and y = +7 for the oxidation state of bromine in $HBrO_4$. Hence, oxidation state of **H = +1, Br = +7, O = -2**.

(c) Fe^{3+}: Applying Rule 1, iron is assigned an oxidation state of +3. Hence, oxidation state of **Fe = +3**.

18.19 *See Section 18.1 and Example 18.2.*

(a) KHF_2: Applying Rule 1, potassium is assigned an oxidation state of +1. Applying Rule 3, fluorine is assigned an oxidation state of -1. This gives 1 + y + 2(-1) = 0 and y = +1 for the oxidation state of hydrogen in KHF_2. Hence, oxidation state of **K = +1, H = +1, F = -1**.

(b) H_2Se: Applying Rule 4, hydrogen is assigned an oxidation state of +1. This gives $2(+1) + y = 0$ and $y = -2$ for the oxidation state of the selemium in H_2Se. Hence, oxidation state of **H = +1, Se = -2.**

(c) NaO_2: Applying Rule 2, sodium is assigned an oxidation state of +1. This gives $1 + 2y = 0$ and $y = -\frac{1}{2}$ for the oxidation state of the oxygen in NaO_2 (sodium superperoxide). Hence, oxidation state of **Na = +1, O = $-\frac{1}{2}$.**

18.20 *See Section 18.1 and Example 18.2.*

(a) NO: Applying Rule 5, oxygen is assigned an oxidation state of -2. This gives $y + 1(-2) = 0$ and $y = +2$ for the oxidation state of the nitrogen in NO. Hence, oxidation state of **N = +2, O = -2.**

(b) BO_2^-: Applying Rule 5, oxygen is assigned an oxidation state of -2. This gives $y + 2(-2) = -1$ and $y = +3$ for the oxidation state of the boron in BO_2^-. Hence, oxidation state of **B = +3, O = -2.**

(c) $Cr(NO_3)_3$: Applying Rule 2, chromium is assigned an oxidation state of +3. Applying Rule 5, oxygen is assigned an oxidation state of -2. This gives $3 + 3y + 9(-2) = 0$ and $y = +5$ for the oxidation state of the N in $Cr(NO_3)_3$. Hence, oxidation state of **Cr = +3, N = +5, O = -2.**

18.21 *See Section 18.1 and Example 18.2.*

(a) $2H_2 + O_2 \rightarrow 2H_2O$
ONs 0 0 +1,-2
Hydrogen is oxidized and oxygen is reduced.

(b) $2Fe + 3O_2 \rightarrow 2Fe_2O_3$
ONs 0 0 +3,-2
Iron is oxidized and oxygen is reduced.

(c) $2Al_2O_3 + 3C \rightarrow 4Al + 3CO_2$
ONs +3,-2 0 0 +4,-2
Carbon is oxidized, aluminum is reduced, and oxygen is unchanged.

18.22 *See Section 18.1 and Example 18.2.*

(a) $Fe_2O_3 + 3H_2 \rightarrow 2Fe + 3H_2O$
ONs +3,-2 0 0 +1,-2
Hydrogen is oxidized, iron is reduced, and oxygen is unchanged.

(b) $CuCl_2 + 2Na \rightarrow 2NaCl + Cu$
ONs +2,-1 0 +1,-1 0
Sodium is oxidized, copper is reduced, and chlorine is unchanged.

(c). $C + O_2 \rightarrow CO_2$
ONs 0 0 +4,-2
Carbon is oxidized and oxygen is oxidized.

18.23 *See Section 18.1 and Example 18.2.*

(a) $2\,HOCl + H_2CO \rightarrow 2HCl + CO_2 + H_2O$ (unbalanced)
ONs +1,-2,+1 +1,0,-2 +1,-1 +4,-2 +1,-2
Carbon is oxidized. **HOCl is the oxidizing agent.**

(b) $H_2 + CO \rightarrow C + H_2O$
ONs 0 +2,-2 0 +1,-2
Hydrogen is oxidized. **CO is the oxidizing agent.**

(c) $O_3 + NO \rightarrow O_2 + NO_2$
ONs 0 +2,-2 0 +4,-2
Nitrogen is oxidized. **O_3 is the oxidizing agent.**

18.24 *See Section 18.1 and Example 18.2.*

(a) $Fe + Al_2O_3 \rightarrow Fe_2O_3 + Al$ (unbalanced)
ONs 0 +3,-2 +3,-2 0
Iron is oxidized. **Al_2O_3 is the oxidizing agent.**

(b) $H_2 + CO_2 \rightarrow CO + H_2O$
ONs 0 +4,-2 +2,-2 +1,-2
Hydrogen is oxidized. **CO_2 is the oxidizing agent.**

(c) $CO + O_2 \rightarrow CO_2$ (unbalanced)
ONs +2,-2 0 +4,-2
Carbon is oxidized. **O_2 is the oxidizing agent.**

18.25 *See Section 18.1 and Example 18.2.*

(a) $Na + FeCl_3 \rightarrow Fe + NaCl$ (unbalanced)
ONs 0 +3,-1 0 +1,-1
Iron is reduced. **Na is the reducing agent.**

(b) $SnCl_2 + FeCl_3 \rightarrow SnCl_4 + FeCl_2$ (unbalanced)
ONs +2,-1 +3,-1 +4,-1 +2,-1
Iron is reduced. **$SnCl_2$ is the reducing agent.**

(c) $CO + Cr_2O_3 \rightarrow Cr + CO_2$ (unbalanced)
ONs +2,-2 +3,-2 0 +4,-2
Chromium is reduced. **CO is the reducing agent.**

18.26 *See Section 18.1 and Example 18.2.*

(a) $Na + Hg_2Cl_2 \rightarrow NaCl + Hg$ (unbalanced)
ONs 0 +1,-1 +1,-1 0
Mercury is reduced. **Na is the reducing agent.**

(b) $HCl + Zn \rightarrow ZnCl_2 + H_2$ (unbalanced)
ONs +1,-1 0 +2,-1 0
Hydrogen is reduced. **Zn is the reducing agent.**

(c) $H_2 + CO_2 \rightarrow CO + H_2O$
ONs 0 +4,-2 +2,-2 +1,-2
Carbon is reduced. **H_2 is the reducing agent.**

18.27 ***See Section 18.2 and Example 18.3.***

(a) $Cr^{3+}(aq) + 3e^- \rightarrow Cr(s)$, gain of electrons and therefore reduction.
(b) $2I^-(aq) \rightarrow I_2(aq) + 2e^-$, loss of electrons and therfore oxidation.
(c) $NO_2^-(aq) + H_2O(\ell) \rightarrow NO_3^-(aq) + 2H^+(aq) + 2e^-$, loss of electrons and therefore oxidation.

18.28 ***See Section 18.2 and Example 18.3.***

(a) $Fe^{2+}(aq) \rightarrow Fe^{3+}(aq) + e^-$, loss of electrons and therefore oxidation.
(b) $Cr_2O_7^{2-}(aq) + 14H^+(aq) + 6e^- \rightarrow 2Cr^{3+}(aq) + 7H_2O(\ell)$, gain of electrons and therefore reduction.
(c) $VO_2^+(aq) + 4H^+(aq) + 2e^- \rightarrow V^{3+}(aq) + 2H_2O(\ell)$, gain of electrons and therefore reduction.

18.29 ***See Section 18.2 and Example 18.3.***

(a) $UO_2^{2+}(aq) + 4H^+(aq) + 2e^- \rightarrow U^{4+}(aq) + 2H_2O(\ell)$, gain of electrons and therefore reduction.
(b) $Zn(s) \rightarrow Zn^{2+}(aq) + 2e^-$, loss of electrons and therefore oxidation.
(c) $IO_3^-(aq) + 6H^+(aq) + 6e^- \rightarrow I^-(aq) + 3H_2O(\ell)$, gain of electrons and therefore reduction.

18.30 ***See Section 18.2 and Example 18.3.***

(a) $N_2O_4(g) + 2H_2O(\ell) \rightarrow 2NO_3^-(aq) + 4H^+(aq) + 2e^-$, loss of electrons and therefore oxidation.
(b) $Mn^{3+}(aq) + 4H_2O(\ell) \rightarrow MnO_4^-(aq) + 8H^+(aq) + 4e^-$, loss of electrons and therefore oxidation.
(c) $HOCl(aq) + 2H_2O(\ell) \rightarrow ClO_3^-(aq) + 5H^+(aq) + 4e^-$, loss of electrons and therefore oxidation.

18.31 ***See Section 18.2 and Example 18.3.***

(a) Oxidation half-reaction: $Sn(s) \rightarrow Sn^{2+}(aq) + 2e^-$
Reduction half-reaction: $2\left[Fe^{3+}(aq) + e^- \rightarrow Fe^{2+}(aq)\right]$

Balanced redox equation: $Sn(s) + 2Fe^{3+}(aq) \rightarrow Sn^{2+}(aq) + 2Fe^{2+}(aq)$

(b) Oxidation half-reaction: $HAsO_3^{2-}(aq) + H_2O(\ell) \rightarrow H_2AsO_4^-(aq) + H^+(aq) + 2e^-$
Reduction half-reaction: $I_2(aq) + 2e^- \rightarrow 2I^-(aq)$

Balanced redox equation: $HAsO_3^{2-}(aq) + H_2O(\ell) + I_2(aq) \rightarrow H_2AsO_4^-(aq) + H^+(aq) + 2I^-(aq)$

(c) Oxidation half-reaction: $Cu(s) \rightarrow Cu^{2+}(aq) + 2e^-$
Reduction half-reaction: $2\left[Ag^+(aq) + e^- \rightarrow Ag(s)\right]$

Balanced redox equation: $Cu(s) + 2Ag^+(aq) \rightarrow Cu^{2+}(aq) + 2Ag(s)$

18.32 *See Section 18.2 and Example 18.3.*

(a) Oxidation half-reaction: $5[H_2C_2O_4(aq) \rightarrow 2CO_2(g) + 2H^+(aq) + 2e^-]$

Reduction half-reaction: $2[MnO_4^-(aq) + 8H^+(aq) + 5e^- \rightarrow Mn^{2+}(aq) + 4H_2O(\ell)]$

Balanced redox equation:

$$2MnO_4^-(aq) + 6H^+(aq) + 5H_2C_2O_4(aq) \rightarrow 2Mn^{2+}(aq) + 8H_2O(\ell) + 10CO_2(g)]$$

(b) Oxidation half-reaction: $2Br^-(aq) \rightarrow Br_2(\ell) + 2e^-$

Reduction half-reaction: $Cl_2(g) + 2e^- \rightarrow 2Cl^-(aq)$

Balanced redox equation: $Cl_2(g) + 2Br^-(aq) \rightarrow 2Cl^-(aq) + Br_2(\ell)$

(c) Oxidation half-reaction: $3[Cu(s) \rightarrow Cu^{2+}(aq) + 2e^-]$

Reduction half-reaction: $2[NO_3^-(aq) + 4H^+(aq) + 3e^- \rightarrow NO(g) + 2H_2O(\ell)]$

Balanced redox equation: $3Cu(s) + 2NO_3^-(aq) + 8H^+(aq) \rightarrow 3Cu^{2+}(aq) + 2NO(g) + 4H_2O(\ell)$

18.33 *See Section 18.2 and Example 18.3.*

(a) Oxidation half-reaction: $Fe(s) \rightarrow Fe^{2+}(aq) + 2e^-$

Reduction half-reaction: $2[Ag^+ + e^- \rightarrow Ag(s)]$

Balanced redox equation: $Fe(s) + 2Ag^+(aq) \rightarrow Fe^{2+}(aq) + 2Ag(s)$

(b) Oxidation half-reaction: $2S_2O_3^{2-}(aq) \rightarrow S_4O_6^{2-}(aq) + 2e^-$

Reduction half-reaction: $I_2(aq) + 2e^- \rightarrow 2I^-(aq)$

Balanced redox equation: $2S_2O_3^{2-}(aq) + I_2(aq) \rightarrow S_4O_6^{2-}(aq) + 2I^-(aq)$

(c) Oxidation half-reaction: $5[Fe^{2+} \rightarrow Fe^{3+} + e^-]$

Reduction half-reaction: $MnO_4^-(aq) + 8H^+(aq) + 5e^- \rightarrow Mn^{2+}(aq) + 4H_2O(\ell)$

Balanced redox equation: $5Fe^{2+}(aq) + MnO_4^-(aq) + 8H^+(aq) \rightarrow 5Fe^{3+}(aq) + Mn^{2+}(aq) + 4H_2O(\ell)$

18.34 *See Section 18.2 and Example 18.3.*

(a) Oxidation half-reaction: $5[Zn(s) \rightarrow Zn^{2+}(aq) + 2e^-]$

Reduction half-reaction: $2NO_3^-(aq) + 12H^+(aq) + 10e^- \rightarrow N_2(g) + 6H_2O(\ell)$

Balanced redox equation: $5Zn(s) + 2NO_3^-(aq) + 12H^+(aq) \rightarrow 5Zn^{2+}(aq) + N_2(g) + 6H_2O(\ell)$

(b) Oxidation half-reaction: $5[2I^-(aq) \rightarrow I_2(aq) + 2e^-]$

Reduction half-reaction: $2IO_3^-(aq) + 12H^+(aq) + 10e^- \rightarrow I_2(aq) + 6H_2O(\ell)$

Balanced redox equation: $10I^-(aq) + 2IO_3^-(aq) + 12H^+(aq) \rightarrow 6I_2(aq) + 6H_2O(\ell)$

Smallest coefficients: $5I^-(aq) + IO_3^-(aq) + 6H^+(aq) \rightarrow 3I_2(aq) + 3H_2O(\ell)$

(c) Oxidation half-reaction: $2Cl^-(aq) \rightarrow Cl_2(aq) + 2e^-$

Reduction half-reaction: $2[Ce^{4+}(aq) + e^- \rightarrow Ce^{3+}]$

Balanced redox equation: $2Cl^-(aq) + 2Ce^{4+}(aq) \rightarrow Cl_2(aq) + 2Ce^{3+}(aq)$

18.35 ***See Section 18.2 and Example 18.4.***

(a) Oxidation half-reaction: $2\left[Al(s) + 4H_2O(\ell) \rightarrow Al(OH)_4^-(aq) + 4H^+(aq) + 3e^-\right]$

Reduction half-reaction: $3\left[ClO^-(aq) + 2H^+(aq) + 2e^- \rightarrow Cl^-(aq) + H_2O(\ell)\right]$

Balanced redox equation: $2Al(s) + 5H_2O(\ell) + 3ClO^-(aq) \rightarrow 2Al(OH)_4^-(aq) + 2H^+(aq) + 3Cl^-(aq)$

Base conversion equation: $2H^+(aq) + 2OH^-(aq) \rightarrow 2H_2O(\ell)$

Final redox equation:

$$2Al(s) + 3H_2O(\ell) + 3ClO^-(aq) + 2OH^-(aq) \rightarrow 2Al(OH)_4^-(aq) + 3Cl^-(aq)$$

(b) Oxidation half-reaction: $3\left[SO_3^{2-}(aq) + H_2O(\ell) \rightarrow SO_4^{2-}(aq) + 2H^+(aq) + 2e^-\right]$

Reduction reaction: $2\left[MnO_4^-(aq) + 4H^+(aq) + 3e^- \rightarrow MnO_2(s) + 2H_2O(\ell)\right]$

Balanced redox equation:

$$3SO_3^{2-}(aq) + 2MnO_4^-(aq) + 2H^+(aq) \rightarrow 3SO_4^{2-}(aq) + 2MnO_2(s) + H_2O(\ell)$$

Base conversion equation: $2H_2O(\ell) \rightarrow 2H^+(aq) + 2OH^-(aq)$

Final redox equation:

$$3SO_3^{2-}(aq) + 2MnO_4^-(aq) + H_2O(l) \rightarrow 3SO_4^{2-}(aq) + 2MnO_2(s) + 2OH^-(aq)$$

(c) Oxidation half-reaction: $4\left[Zn(s) + 4H_2O(l) \rightarrow Zn(OH)_4^{2-}(aq) + 4H^+(aq) + 2e^-\right]$

Reduction half-reaction: $NO_3^-(aq) + 9H^+(aq) + 8e^- \rightarrow NH_3(aq) + 3H_2O(l)$

Balanced redox equation: $4Zn(s) + 13H_2O(l) + NO_3^-(aq) \rightarrow 4Zn(OH)_4^{2-}(aq) + 7H^+(aq) + NH_3(aq)$

Base conversion equation: $7H^+(aq) + 7OH^-(aq) \rightarrow 7H_2O(\ell)$

Final redox equation:

$$4Zn(s) + 6H_2O(\ell) + NO_3^-(aq) + 7OH^-(aq) \rightarrow 4Zn(OH)_4^{2-}(aq) + NH_3(aq)$$

18.36 ***See Section 18.2 and Example 18.4.***

(a) Oxidation half-reaction: $2\left[CrO_2^-(aq) + 2H_2O(\ell) \rightarrow CrO_4^{2-}(aq) + 4H^+(aq) + 3e^-\right]$

Reduction half-reaction: $3\left[ClO^-(aq) + 2H^+(aq) + 2e^- \rightarrow Cl^-(aq) + H_2O(\ell)\right]$

Balanced redox equation:

$$2CrO_2^-(aq) + H_2O(\ell) + 3ClO^-(aq) \rightarrow 2CrO_4^{2-}(aq) + 2H^+(aq) + 3Cl^-(aq)$$

Base conversion equation: $2H^+(aq) + 2OH^-(aq) \rightarrow 2H_2O(\ell)$

Final redox equation:

$$2CrO_2^-(aq) + 3ClO^-(aq) + 2OH^-(aq) \rightarrow 2CrO_4^{2-}(aq) + 3Cl^-(aq) + H_2O(\ell)$$

(b) Oxidation half-reaction: $Br_2(aq) + 6H_2O(\ell) \rightarrow 2BrO_3^-(aq) + 12H^+(aq) + 10e^-$

Reduction half-reaction: $5\left[Br_2(aq) + 2e^- \rightarrow 2Br^-(aq)\right]$

Balanced redox equation: $6Br_2(aq) + 6H_2O(\ell) \rightarrow 2BrO_3^-(aq) + 12H^+(aq) + 10Br^-(aq)$

Smallest coefficients: $3Br_2(aq) + 3H_2O(\ell) \rightarrow BrO_3^-(aq) + 6H^+(aq) + 5Br^-(aq)$

Base conversion equation: $6H^+(aq) + 6OH^-(aq) \rightarrow 6H_2O(\ell)$

Final redox equation: $3Br_2(aq) + 6OH^-(aq) \rightarrow BrO_3^-(aq) + 5Br^-(aq) + 3H_2O(\ell)$

(c) Oxidation half-reaction: $N_2H_4(aq) \rightarrow N_2(g) + 4H^+(aq) + 4e^-$

Reduction half-reaction: $2\left[H_2O_2(aq) + 2H^+(aq) + 2e^- \rightarrow 2H_2O(\ell)\right]$

Balanced redox equation: $N_2H_4(aq) + 2H_2O_2(aq) \rightarrow N_2(g) + 4H_2O(\ell)$

18.37 ***See Section 18.2 and Example 18.4.***

(a) Oxidation half-reaction: $Cl_2(aq) + 6H_2O(\ell) \rightarrow 2ClO_3^-(aq) + 12H^+(aq) + 10e^-$

Reduction half-reaction: $5[Cl_2(aq) + 2e^- \rightarrow 2Cl^-(aq)]$

Balanced redox equation: $6Cl_2(aq) + 6H_2O(\ell) \rightarrow 2ClO_3^-(aq) + 12H^+(aq) + 10Cl^-(aq)$

Smallest coefficients: $3Cl_2(aq) + 3H_2O(\ell) \rightarrow ClO_3^-(aq) + 6H^+(aq) + 5Cl^-(aq)$

Base conversion equation: $6H^+(aq) + 6OH^-(aq) \rightarrow 6H_2O(\ell)$

Final redox equation: $3Cl_2(aq) + 6OH^-(aq) \rightarrow ClO_3^-(aq) + 5Cl^-(aq) + 3H_2O(\ell)$

(b) Oxidation half-reaction: $I^-(aq) + 3H_2O(\ell) \rightarrow IO_3^-(aq) + 6H^+(aq) + 6e^-$

Reduction half-reaction: $2[MnO_4^-(aq) + 4H^+(aq) + 3e^- \rightarrow MnO_2(s) + 2H_2O(\ell)]$

Balanced redox equation: $I^-(aq) + 2MnO_4^-(aq) + 2H^+(aq) \rightarrow IO_3^-(aq) + 2MnO_2(s) + H_2O(\ell)$

Base conversion equation: $2H_2O(\ell) \rightarrow 2H^+(aq) + 2OH^-(aq)$

Final redox equation: $I^-(aq) + 2MnO_4^-(aq) + H_2O(\ell) \rightarrow IO_3^-(aq) + 2MnO_2(s) + 2OH^-(aq)$

(c) Oxidation half-reaction: $3[CN^-(aq) + H_2O(\ell) \rightarrow CNO^-(aq) + 2H^+(aq) + 2e^-]$
Reduction half-reaction: $ClO_3^-(aq) + 6H^+(aq) + 6e^- \rightarrow Cl^-(aq) + 3H_2O(\ell)$

Balanced redox equation: $3CN^-(aq) + ClO_3^-(aq) \rightarrow 3CNO^-(aq) + Cl^-(aq)$

18.38 ***See Section 18.2 and Example 18.4.***

(a) Oxidation half-reaction: $4PH_3(g) \rightarrow P_4(s) + 12H^+(aq) + 12e^-$
Reduction half-reaction: $4[CrO_4^{2-}(aq) + 4H^+(aq) + 3e^- \rightarrow CrO_2^-(aq) + 2H_2O(\ell)]$

Balanced redox equation: $4PH_3(g) + 4CrO_4^{2-}(aq) + 4H^+(aq) \rightarrow P_4(s) + 4CrO_2^-(aq) + 8H_2O(\ell)$
Basic conversion equation: $4H_2O(\ell) \rightarrow 4H^+(aq) + 4OH^-(aq)$

Final redox equation: $4PH_3(g) + 4CrO_4^{2-}(aq) \rightarrow P_4(s) + 4CrO_2^-(aq) + 4H_2O(\ell) + 4OH^-(aq)$

(b) Oxidation half-reaction: $2H_2O(\ell) \rightarrow O_2(g) + 4H^+(aq) + 4e^-$
Reduction half-reaction: $2[F_2(g) + 2e^- \rightarrow 2F^-(aq)]$

Balanced redox equation: $2H_2O(\ell) + 2F_2(g) \rightarrow O_2(g) + 4H^+(aq) + 4F^-(aq)$
Base conversion equation: $4H^+(aq) + 4OH^-(aq) \rightarrow 4H_2O(\ell)$

Final redox equation: $2F_2(g) + 4OH^-(aq) \rightarrow O_2(g) + 4F^-(aq) + 2H_2O(\ell)$

(c) Oxidation half-reaction: $2[Cr(OH)_3(s) + H_2O(\ell) \rightarrow CrO_4^{2-}(aq) + 5H^+(aq) + 3e^-]$
Reduction half-reaction: $3[H_2O_2(aq) + 2H^+(aq) + 2e^- \rightarrow 2H_2O(\ell)]$

Balanced redox equation: $2Cr(OH)_3(s) + 3H_2O_2(aq) \rightarrow 2CrO_4^{2-}(aq) + 4H^+(aq) + 4H_2O(\ell)$
Base conversion equation: $4H^+(aq) + 4OH^-(aq) \rightarrow 4H_2O(\ell)$

Final redox equation: $2Cr(OH)_3(s) + 3H_2O_2(aq) + 4OH^-(aq) \rightarrow 2CrO_4^{2-}(aq) + 8H_2O(\ell)$

18.39 ***See Section 18.2 and Example 18.5.***

$$2MnO_4^-(aq) + 5C_2O_4^{2-}(aq) + 16H^+(aq) \rightarrow 2Mn^{2+}(aq) + 10CO_2(g) + 8H_2O(l)$$

Strategy: $g\ Na_2C_2O_4 \rightarrow mol\ Na_2C_2O_4 \rightarrow mol\ KMnO_4 \rightarrow M\ KMnO_4\ soln$

$$?\ \text{mol KMnO}_4 = 0.103\ \text{g Na}_2\text{C}_2\text{O}_4 \times \frac{1\ \text{mol Na}_2\text{C}_2\text{O}_4}{134.0\ \text{g Na}_2\text{C}_2\text{O}_4} \times \frac{2\ \text{mol KMnO}_4}{5\ \text{mol Na}_2\text{C}_2\text{O}_4} = 3.08 \times 10^{-4}\ \text{mol KMnO}_4$$

$$?\ M\ \text{KMnO}_4\ \text{soln} = \frac{3.08 \times 10^{-4}\ \text{mol KMnO}_4}{0.02430\ \text{L KMnO}_4\ \text{soln}} = \mathbf{1.27 \times 10^{-2}}\ \boldsymbol{M}\ \mathbf{KMnO_4}$$

18.40 ***See Section 18.2 and Example 18.5.***

$$H_3AsO_3(aq) + I_3^-(aq) + H_2O(\ell) \rightarrow H_3AsO_4(aq) + 3I^-(aq) + 2H^+(aq)$$

Strategy: $L\ I_3^- \rightarrow mol\ I_3^- \rightarrow mol\ H_3AsO_3 \rightarrow g\ H_3AsO_3$

$$?\ \text{g H}_3\text{AsO}_3 = 0.03105\ \text{L I}_3^-\ \text{soln} \times \frac{0.0501\ \text{mol I}_3^-}{1\ \text{L I}_3^-\ \text{soln}} \times \frac{1\ \text{mol H}_3\text{AsO}_3}{1\ \text{mol I}_3^-} \times \frac{126.0\ \text{g H}_3\text{AsO}_3}{1\ \text{mol H}_3\text{AsO}_3} = 0.196\ \text{g H}_3\text{AsO}_3$$

18.41	*See Section 18.3 and Figure 18.2.*

The cell would be similar to that shown in Figure 18.2.
At the negative electrode in the anode half-cell on the left: $Zn(s) \rightarrow Zn^{2+}(aq) + 2e^-$
At the positive electrode in the cathode half-cell on the right: $Ni^{2+}(aq) + 2e^- \rightarrow Ni(s)$
Hence, Ni is positive and Zn is negative.
Electrons will flow from left to right in the external circuit, will flow from the anode to the cathode. Anions will flow from right to left in the salt bridge and into the anode half-cell. Cations will flow from left to right in the salt bridge and into the cathode half-cell.

18.42	*See Section 18.3 and Figure 18.2.*

The cell would be similar to that shown in Figure 18.2.
At the negative electrode in the anode half-cell on the left: $Pb(s) \rightarrow Pb^{2+}(aq) + 2e^-$
At the positive electrode in the cathode half-cell on the right: $Ag^+ + e^- \rightarrow Ag(s)$
Hence, Ag is positive and Pb is negative.
Electrons will flow from left to right in the external circuit, will flow from the anode to the cathode. Anions will flow from right to left in the salt bridge and into the anode half-cell. Cations will flow from left to right in the salt bridge and into the cathode half-cell.

18.43	*See Section 18.3.*

(a) Oxidation: $2Hg(\ell) + 2Cl^-(aq) \rightarrow Hg_2Cl_2(s) + 2e^-$
Reduction: $Cu^{2+}(aq) + 2e^- \rightarrow Cu(s)$

(b) Cell: $2Hg(\ell) + 2Cl^-(aq) + Cu^{2+}(aq) \rightarrow Cu(s) + Hg_2Cl_2(s)$

(c) Electrons flow from the platinum wire to the copper metal electrode..

(d) As $Cu^{2+}(aq)$ ions come into contact with $Hg(\ell)$, a direct electron transfer will take place.

18.44	*See Section 18.3.*

(a) Oxidation: $Ag(s) + Br^-(aq) \rightarrow AgBr(s) + e^-$
Reduction: $Hg_2Br_2(s) + 2e^- \rightarrow 2Hg(\ell) + 2Br^-(aq)$

(b) Cell: $2Ag(s) + Hg_2Br_2(s) \rightarrow 2AgBr(s) + 2Hg(\ell)$

(c) Electrons flow from the silver wire to the platinum wire.

(d) A salt bridge is not necessary because the reactants are solids and the products are a solid and a liquid.

18.45	*See Section 18.4, Table 18.1, and Example 18.7.*

(a)

Anode:	$Zn(s) \rightarrow Zn^{2+}(aq) + 2e^-$	$E° = +0.76$ V
Cathode:	$Ni^{2+}(aq) + 2e^- \rightarrow Ni(s)$	$E° = -0.25$ V
Cell:	$Zn(s) + Ni^{2+}(aq) \rightarrow Zn^{2+}(aq) + Ni(s)$	$E° =$ **0.51 V**

(b) Anode: $Pb(s) \rightarrow Pb^{2+}(aq) + 2e^-$ $E^\circ = 0.126$ V

Cathode: $2Ag^+(aq) + 2e^- \rightarrow 2Ag(s)$ $E^\circ = 0.80$ V

Cell: $Pb(s) + 2Ag^+(aq) \rightarrow Pb^{2+}(aq) + 2Ag(s)$ **$E^\circ = 0.93$ V**

18.46 ***See Section 18.4, Table 18.1, and Example 18.7.***

(a) Anode: $Fe^{2+} \rightarrow Fe^{3+} + e^-$ $E^\circ = -0.77$V

Cathode: $Ag^+ + e^- \rightarrow Ag$ $E^\circ = 0.80$ V

Cell: $Fe^{2+} + Ag^+ \rightarrow Fe^{3+} + Ag$ $E^\circ = 0.03$ V

Since the calculated voltage is positive, the reaction is **spontaneous in the direction shown** under standard conditions.

(b) Anode: $Cu(s) \rightarrow Cu^{2+}(aq) + 2e^-$ $E^\circ = -0.34$ V

Cathode: $2AgCl(s) + 2e^- \rightarrow 2Ag(s) + 2Cl^-(aq)$ $E^\circ = 0.222$ V

Cell: $Cu(s) + 2AgCl(s) \rightarrow Cu^{2+}(aq) + 2Ag(s) + 2Cl^-(aq)$ $E^\circ = -0.12$ V

Since the calculated voltage is positive, the reaction is **spontaneous in the direction shown** under standard conditions.

(c) Anode: $Ni(s) \rightarrow Ni^{2+}(aq) + 2e^-$ $E^\circ = 0.25$ v

Cathode: $Br_2(\ell) + 2e^- \rightarrow 2Br^-(aq)$ $E^\circ = 1.06$ V

Cell: $Ni(s) + Br_2(\ell) \rightarrow Ni^{2+}(aq) + 2Br^-(aq)$ $E^\circ = 1.31$ V

Since the calculated voltage is positive, the reaction is **spontaneous in the direction shown** under standard conditions.

18.47 ***See Section 18.4, Table 18.1, Example 18.8, and Appendix H.***

(a) Anode: $H_2(g) \rightarrow 2H^+(aq) + 2e^-$ $E^\circ = 0.00$ V

Cathode: $Cl_2(g) + 2e^- \rightarrow 2Cl^-(aq)$ $E^\circ = 1.36$ V

Cell: $H_2(g) + Cl_2(g) \rightarrow 2H^+(aq) + 2Cl^-(aq)$ **$E^\circ = 1.36$ V**

Since the calculated voltage is positive, the reaction is **spontaneous in the direction shown** under standard conditions.

(b) Anode: $3Cr^{2+}(aq) \rightarrow 3Cr^{3+}(aq) + 3e^-$ $E^\circ = 0.41$ V

Cathode: $Al^{3+}(aq) + 3e^- \rightarrow Al(s)$ $E^\circ = -1.66$ V

Cell: $3Cr^{2+}(aq) + Al^{3+}(aq) \rightarrow 3Cr^{3+}(aq) + Al(s)$ **$E^\circ = -1.25$ V**

Since the calculated voltage is negative, the reaction is **spontaneous in the reverse direction** (nonspontaneous in the forward direction) under standard conditions.

18.48 ***See Section 18.4, Table 18.1, Example 18.8, and Appendix H.***

(a) Anode: $Cu^+(aq) \rightarrow Cu^{2+}(aq) + e^-$ $E^\circ = -0.153$ V
Cathode: $Ce^{4+}(aq) + e^- \rightarrow Ce^{3+}(aq)$ $E^\circ = 1.61$ V

Cell: $Cu^+(aq) + Ce^{4+}(aq) \rightarrow Cu^{2+}(aq) + Ce^{3+}(aq)$ $E^\circ = 1.46$ V
Since the calculated voltage is positive, the reaction is **spontaneous in the direction shown** under standard conditions.

(b) Anode: $Sn^{2+}(aq) \rightarrow Sn^{4+}(aq) + 2e^-$ $E^\circ = -0.15$ V
Cathode: $2Fe^{3+}(aq) + 2e^- \rightarrow 2Fe^{2+}(aq)$ $E^\circ = 0.771$ V

Cell: $Sn^{2+}(aq) + 2Fe^{3+}(aq) \rightarrow Sn^{4+}(aq) + 2Fe^{2+}(aq)$ $E^\circ =$ **0.62 V**
Since the calculated voltage is positive, the reaction is **spontaneous in the direction shown** under standard conditions.

18.49 ***See Section 18.4, Table 18.1, Example 18.8, and Appendix H.***

(a) Oxidation: $Cu(s) \rightarrow Cu^{2+}(aq) + 2e^-$ $E^\circ = -0.34$ V
Reduction: $2H^+(aq) + 2e^- \rightarrow H_2(g)$ $E^\circ \equiv 0.00$ V

Reaction: $Cu(s) + 2H^+(aq) \rightarrow Cu^{2+}(aq) + H_2(g)$ $E^\circ = -0.34$ V
Since the calculated voltage is negative, the reaction is **spontaneous in the reverse direction** (nonspontaneous in the forward direction) under standard conditions. This means copper metal will not dissolve in acid solutions under standard conditions, provided the acid is not HNO_3 which can use NO_3^- as a more powerful oxidizing agent than H^+.

(b) Oxidation: $Fe(s) \rightarrow Fe^{2+}(aq) + 2e^-$ $E^\circ = 0.44$ V
Reduction: $2H^+(aq) + 2e^- \rightarrow H_2(g)$ $E^\circ \equiv 0.00$ V

Reaction: $Fe(s) + 2H^+(aq) \rightarrow Fe^{2+}(aq) + H_2(g)$ $E^\circ = 0.44$ V
Since the calculated voltage is positive, the reaction is **spontaneous in the direction shown** under standard conditions. This means iron metal will dissolve in acid solutions under standard conditions.

(c) Oxidation: $Zn(s) \rightarrow Zn^{2+}(aq) + 2e^-$ $E^\circ = 0.76$ V
Reduction: $2H^+(aq) + 2e^- \rightarrow H_2(g)$ $E^\circ \equiv 0.00$ V

Reaction: $Zn(s) + 2H^+(aq) \rightarrow Zn^{2+}(aq) + H_2(g)$ $E^\circ = 0.76$ V
Since the calculated voltage is positive, the reaction is **spontaneous in the direction shown** under standard conditions. This means zinc metal will dissolve in acid solutions under standard conditions.

18.50 ***See Section 18.4, Table 18.1, Example 18.8, and Appendix H.***

(a) Oxidation $Ni(s) \rightarrow Ni^{2+}(aq) + 2e^-$ $E^\circ = 0.25$ V
Reduction: $2H^+(aq) + 2e^- \rightarrow H_2(g)$ $E^\circ \equiv 0.00$ V

Reaction: $Ni(s) + 2H^+(aq) \rightarrow Ni^{2+}(aq) + H_2(g)$ $E^\circ = 0.25$ V
Since the calculated voltage is positive, the reaction is **spontaneous in the direction shown** under standard conditions. This means nickel will dissolve in acid solutions under standard conditions.

(b) Oxidation: $Cr(s) \rightarrow Cr^{2+}(aq) + 2e^-$ $E° = 0.91$ V

Reduction: $2H^+(aq) + 2e^- \rightarrow H_2(g)$ $E° \equiv 0.00$ V

Reaction: $Cr(s) + 2H^+(aq) \rightarrow Cr^{2+}(aq) + H_2(g)$ $E° = 0.91$ V

Since the calculated voltage is positive, the reaction is **spontaneous in the direction shown** under standard conditions. This means chromium will dissolve in acid solutions under standard conditions.

(c) Oxidation: $Cd(s) \rightarrow Cd^{2+}(aq) + 2e^-$ $E° = 0.403$ V

Reduction: $2H^+(aq) + 2e^- \rightarrow H_2(g)$ $E° \equiv 0.00$ V

Reaction: $Cd(s) + 2H^+(aq) \rightarrow Cd^{2+}(aq) + H_2(g)$ $E° = 0.403$ V

Since the calculated voltage is positive, the reaction is **spontaneous in the direction shown** under standard conditions. This means cadmium will dissolve in acid solutions under standard conditions.

18.51 ***See Section 18.4, Table 18.1, Example 18.18, and Appendix H.***

(a) Oxidation: $Cu(s) + 2OH^-(aq) \rightarrow Cu(OH)_2(s) + 2e^-$ $E° = 0.36$ V

Reductions: $2H_2O(\ell) + 2e^- \rightarrow H_2(g) + 2OH^-(aq)$ $E° = -0.83$ V

Reaction: $Cu(s) + 2H_2O(\ell) \rightarrow Cu(OH)_2(s) + H_2(g)$ **$E° = -0.47$ V**

Since the calculated voltage is negative, the reaction of Cu(s) with $H_2O(\ell)$ is nonspontaneous under standard conditions. Hence, Cu(s) will not dissolve in water that contains 1 *M* OH^-.

(b) Oxidation: $2[Na(s) \rightarrow Na^+(aq) + e^-]$ $E° = 2.71$ V

Reduction: $2H_2O(\ell) + 2e^- \rightarrow H_2(g) + 2OH^-(aq)$ $E° = -0.83$ V

Reaction: $2Na(s) + 2H_2O(\ell) \rightarrow 2Na^+(aq) + H_2(g) + 2OH^-(aq)$ **$E° = 1.88$ V**

Since the calculated voltage is positive, the reaction of Na(s) with $H_2O(\ell)$ is spontaneous under standard conditions. Indeed, Na(s) dissolves in water in a very vigorous reaction.

18.52 ***See Section 18.4, Table 18.1, Example 18.18, and Appendix H.***

(a) Oxidation: $2[K(s) \rightarrow K^+(aq) + e^-]$ $E° = 2.93$ V

Reduction: $2H_2O(\ell) + 2e^- \rightarrow H_2(g) + 2OH^-(aq)$ $E° = -0.83$ V

Reaction: $2K(s) + 2H_2O(\ell) \rightarrow 2K^+(aq) + H_2(g) + 2OH^-(aq)$ $E° = 2.10$ V

Since the calculated voltage is positive, the reaction of K(s) with with $H_2O(\ell)$ is spontaneous under standard conditions. Indeed, K(s) dissolves in water in a very vigorous reaction.

(b) Oxidation: $2[Cr(s) + 3OH^-(aq) \rightarrow Cr(OH)_3(s) + 3e^-]$ $E° = 1.30$ V

Reduction: $3[2H_2O(\ell) + 2e^- \rightarrow H_2(g) + 2OH^-(aq)]$ $E° = -0.83$ V

Reaction: $2Cr(s) + 6H_2O(\ell) \rightarrow 2Cr(OH)_3(s) + 3H_2(g)$ $E° = 0.47$ V

Since the calculated voltage is positive even allowing for an overvoltage of 0.40 V for the production of H_2, the reaction of Cr(s) with $H_2O(\ell)$ is spontaneous under standard conditions. Hence, Cr(s) is predicted to dissolve in water that contains 1 *M* OH^-. However, the very small magnitude of the potential indicates the reaction isn't very favorable and won't proceed to any significant extent.

18.53 ***See Section 18.4, Table 18.1, and Example 18.7.***

(a)

$$Ni(s) \rightarrow Ni^{2+}(aq) + 2e^- \qquad E^\circ = 0.25\ V$$
$$Cu^{2+}(aq) + 2e^- \rightarrow Cu(s) \qquad E^\circ = 0.34\ V$$

$$Ni(s) + Cu^{2+}(aq) \rightarrow Ni^{2+}(aq) + Cu(s) \qquad E^\circ = \mathbf{0.59\ V}$$

Since the calculated voltage is positive, the reaction is **spontaneous in the direction shown** under standard conditions.

(b)

$$2Ag(s) + 2Cl^-(aq) \rightarrow 2AgCl(s) + 2e^- \qquad E^\circ = -0.222\ V$$
$$Cl_2(g) + 2e^- \rightarrow 2Cl^-(aq) \qquad E^\circ = 1.36\ V$$

$$2Ag(s) + Cl_2(g) \rightarrow 2AgCl(s) \qquad E^\circ = \mathbf{1.14\ V}$$

Since the calculated voltage is positive, the reaction is **spontaneous in the direction shown** under standard conditions.

(c)

$$2I^-(aq) \rightarrow I_2(s) + 2e^- \qquad E^\circ = -0.54\ V$$
$$Cl_2(g) + 2e^- \rightarrow 2Cl^-(aq) \qquad E^\circ = 1.36\ V$$

$$2I^-(aq) + Cl_2(g) \rightarrow I_2(s) + 2Cl^-(aq) \qquad E^\circ = \mathbf{0.82\ V}$$

Since the calculated voltage is positive, the reaction is **spontaneous in the direction shown** under standard conditions.

18.54 ***See Section 18.4, Table 18.1, and Example 18.7.***

(a)

$$Zn(s) \rightarrow Zn^{2+}(aq) + 2e^- \qquad E^\circ = 0.76\ V$$
$$Fe^{2+}(aq) + 2e^- \rightarrow Fe(s) \qquad E^\circ = -0.44\ V$$

$$Zn(s) + Fe^{2+}(aq) \rightarrow Zn^{2+}(aq) + Fe(s) \qquad E^\circ = 0.32\ V$$

Since the calculated voltage is positive, the reaction is **spontaneous in the direction shown** under standard conditions.

(b)

$$Fe^{2+}(aq) \rightarrow Fe^{3+}(aq) + e^- \qquad E^\circ = -0.77\ V$$
$$AgCl(s) + e^- \rightarrow Ag(s) + Cl^-(aq) \qquad E^\circ = 0.222\ V$$

$$Fe^{2+}(aq) + AgCl(s) \rightarrow Fe^{3+}(aq) + Ag(s) + Cl^-(aq) \qquad E^\circ = -0.55\ V$$

Since the calculated voltage is negative, the reaction is **spontaneous in the reverse direction** (nonspontaneous in the forward direction) under standard conditions.

(c)

$$2Cl^-(aq) \rightarrow Cl_2(g) + 2e^- \qquad E^\circ = -1.36\ V$$
$$Br_2(\ell) + 2e^- \rightarrow 2Br^-(aq) \qquad E^\circ = 1.06\ V$$

$$2Cl^-(aq) + Br_2(\ell) \rightarrow Cl_2(g) + 2Br^-(aq) \qquad E^\circ = -0.30\ V$$

Since the calculated voltage is negative, the reaction is **spontaneous in the reverse direction** (nonspontaneous in the forward direction) under standard conditions.

18.55 ***See Section 18.4, Table 18.1, and Example 18.7.***

$$Pb(s) \rightarrow Pb^{+}(aq) + 2e^{-} \qquad E^{\circ} = 0.126\ V$$

$$UO_2^{2+}(aq) + 4H^{+}(aq) + 2e^{-} \rightarrow U^{4+}(aq) + 2H_2O(\ell) \qquad E^{\circ} = y$$

$$Pb(s) + UO_2^{2+}(aq) + 4H^{+}(aq) \rightarrow Pb^{2+}(aq) + U^{4+}(aq) + 2H_2O(\ell) \qquad E^{\circ} = 0.460\ V$$

Solving 0.126 V + y = 0.460 V, gives y = 0.334 V. Hence, the UO_2^{2+}/U^{4+} standard reduction potential is **0.334 V.**

18.56 ***See Section 18.4, Table 18.1, and Example 18.7.***

$$Eu^{2+}(aq) \rightarrow Eu^{3+}(aq) + e^{-} \qquad E^{\circ} = y$$

$$Ag^{+}(aq) + e^{-} \rightarrow Ag(s) \qquad E^{\circ} = 0.80\ V$$

$$Eu^{2+}(aq) + Ag^{+}(aq) \rightarrow Eu^{3+}(aq) + Ag(s) \qquad E^{\circ} = 1.23\ V$$

Solving y + 0.80 V = 1.23 V, gives y = 0.43 V. Hence, the Eu^{3+} / Eu^{2+} standard reduction potential is **-0.43 V**.

18.57 ***See Section 18.4, Table 18.1, and Example 18.7.***

(a) Oxidation occurs at the negative electrode in a voltaic cell. Hence, Cu(s) must be oxidized and $Pu^{4+}(aq)$ must be reduced. This gives

Anode: $Cu(s) \rightarrow Cu^{2+}(aq) + 2e^{-}$ $\qquad E^{\circ} = -0.34\ V$

Cathode: $2Pu^{4+}(aq) + 2e^{-} \rightarrow 2Pu^{3+}(aq)$ $\qquad E^{\circ} = y$

Cell: $Cu(s) + 2Pu^{4+}(aq) \rightarrow Cu^{2+}(aq) + 2Pu^{3+}(aq)$ $\qquad E^{\circ} = 0.642$

(b) Solving -0.34 V + y = 0.642 V, gives y = 0.98 V. Hence, the Pu^{4+} / Pu^{3+} standard reduction potential is **0.98 V.**

18.58 ***See Section 18.4, Table 18.1, and Example 18.7.***

(a) Oxidation occurs at the negative electrode in a voltaic cell. Hence, Tl(s) must be oxidized and $Ag^{+}(aq)$ must be reduced. This gives

Anode: $Tl(s) \rightarrow Tl^{+}(aq) + e^{-}$ $\qquad E^{\circ} = y$

Cathode: $Ag^{+}(aq) + e^{-} \rightarrow Ag(s)$ $\qquad E^{\circ} = 0.80\ V$

Cell: $Tl(s) + Ag^{+}(aq) \rightarrow Tl^{+}(aq) + Ag(s)$ $\qquad E^{\circ} = 1.136\ V$

(b) Solving y + 0.80 V = 1.136 V, gives y = **0.34 V**.

18.59 ***See Section 18.4, Table 18.1, Example 18.9, and Appendix H.***

(a) $\mathbf{H_2O_2}$ has a more positive standard reduction potential than MnO_4^- (1.78 V vs. 1.51 V), is more easily reduced and is the better oxidant (oxidizing agent).

(b) **Cu** has a less negative standard oxidation potential than Ag (-0.153 V vs. -0.80 V), is more easily oxidized and is the better reducing agent.

18.60 *See Section 18.4, Table 18.1, Example 18.9, and Appendix H.*

(a) **Ce^{4+}** has a more positive standard reduction potential than $Cr_2O_7^{2-}$ (1.61 V vs. 1.33 V). is more easily reduced and is the better oxidant (oxidizing agent).

(b) **Sn^{2+}** has a less negative standard oxidation potential than Fe^{2+} (-0.15 V vs. -0.77 V), is more easily oxidized and is the better reducing agent.

18.61 *See Section 18.4, Table 18.1, and Example 18.9.*

(a) Reducing agents are listed on the right side of Table 18.1. Any chemical species below Ni has a sufficiently positive oxidation potential to reduce Ni^{2+} under standard conditions. The only metal ion shown below Ni is Cr^{2+}.

(b) Oxidizing agents are listed on the left side of Table 18.1. Any chemical species above Cu^{2+} has a sufficiently positive reduction potential to oxidize Cu under standard conditions. Examples include Fe^{3+}, Br_2, and MnO_4^-.

(c) **Ni^{2+} and Pb^{2+}** can be reduced by Cr^{2+} but not by H_2 under standard conditions.

18.62 *See Section 18.4, Table 18.1, and Example 18.9.*

(a) Reducing agents are listed on the right side of Table 18.1. Any chemical species listed between Cu and Pb has a sufficiently positive oxidation potential to reduce Cu^{2+} but not Pb^{2+} under standard conditions. **Ag(s) + Cl^-, Sn^{2+}, and H_2** meet this criterion.

(b) Oxidizing agents are listed on the left side of Table 18.1. Any chemical species listed between Cu^{2+} and Fe^{3+} has a sufficiently positive reduction potential to oxidize Cu but not Fe^{2+} under standard conditions. The species meeting this criterion are **O_2 and I_2**.

(c) Reducing agents are listed on the left side of Table 18.1. Any chemical species listed below Fe^{2+} has a sufficiently positive oxidation potential to reduce Fe^{3+} under standard conditions. Metal ions listed in Table 18.1 meeting this criterion are **Sn^{2+} and Cr^{2+}**.

18.63 *Section 18.5, Table 18.1, Examples 18.10, 18.11, Solution for Exercise 18.53, and Appendix H.*

(a) $\Delta G^\circ = -n\mathcal{F}E^\circ$ $\quad \Delta G^\circ = -(2 \text{ mol e}^-)\left(\frac{9.65 \times 10^4 \text{ C}}{1 \text{ mol e}^-}\right)\left(\frac{0.59 \text{ J}}{1 \text{ C}}\right) = \mathbf{-1.1 \times 10^5 \text{ J}}$

$\log K_{eq} = \frac{nE^\circ}{0.0591}$ $\quad \log K_{eq} = \frac{(2)(0.59)}{0.0591} = 20.0$ $\quad K_{eq} = 10^{20.0} = \text{inv}\log 20.0 = \mathbf{1 \times 10^{20}}$

Since the calculated voltage is positive, the reaction is **spontaneous** in the direction shown under standard conditions. This is consistent with a negative value for ΔG° and $K_{eq} > 1$.

(b) $\Delta G^\circ = -n\mathcal{F}E^\circ$ $\quad \Delta G^\circ = -(2 \text{ mol e}^-)\left(\frac{9.65 \times 10^4 \text{ C}}{1 \text{ mol e}^-}\right)\left(\frac{1.14 \text{ J}}{1 \text{ C}}\right) = \mathbf{-2.20 \times 10^5 \text{ J}}$

$\log K_{eq} = \frac{nE^\circ}{0.0591}$ $\quad \log K_{eq} = \frac{(2)(1.14)}{0.0591} = 38.6$ $\quad K_{eq} = 10^{38.6} = \text{inv}\log 38.6 = \mathbf{4 \times 10^{38}}$

Since the calculated voltage is positive, the reaction is **spontaneous** in the direction shown under standard conditions. This is consistent with a negative value for ΔG° and $K_{eq} > 1$.

(c) $\Delta G° = -nFE°$ $\quad \Delta G° = -(2 \text{ mol e}^-)\left(\frac{9.65 \times 10^4 \text{ C}}{1 \text{ mol e}^-}\right)\left(\frac{0.82 \text{ J}}{1 \text{ C}}\right) = \mathbf{-1.6 \times 10^5 \text{ J}}$

$\log K_{eq} = \frac{nE°}{0.0591}$ $\quad \log K_{eq} = \frac{(2)(0.82)}{0.0591} = 28$ $\quad K_{eq} = 10^{28} = \text{inv} \log 28 = \mathbf{1 \times 10^{28}}$

Since the calculated voltage is positive, the reaction is **spontaneous** in the direction shown under standard conditions. This is consistent with a negative value for $\Delta G°$ and $K_{eq} > 1$.

18.64 ***Section 18.5 , Table 18.1, Examples 18.10, 18.11, Solution for Exercise 18.54, and Appendix H.***

(a) $\Delta G° = -nFE°$ $\quad \Delta G° = -(2 \text{ mol e}^-)\left(\frac{9.65 \times 10^4 \text{ C}}{1 \text{ mol e}^-}\right)\left(\frac{0.32 \text{ J}}{1 \text{ C}}\right) = \mathbf{-6.2 \times 10^4 \text{ J}}$

$\log K_{eq} = \frac{nE°}{0.0591}$ $\quad \log K_{eq} = \frac{(2)(0.32)}{0.0591} = 10.83$ $\quad K_{eq} = 10^{10.83} = \text{inv} \log 10.83 = \mathbf{6.8 \times 10^{10}}$

Since the calculated voltage is positive, the reaction is **spontaneous** in the direction shown under standard conditions. This is consistent with a negative value for $\Delta G°$ and $K_{eq} > 1$.

(b) $\Delta G° = -nFE°$ $\quad \Delta G° = -(1 \text{ mol e}^-)\left(\frac{9.65 \times 10^4 \text{ C}}{1 \text{ mol e}^-}\right)\left(\frac{-0.55 \text{ J}}{1 \text{ C}}\right) = \mathbf{5.30 \times 10^4 \text{ J}}$

$\log K_{eq} = \frac{nE°}{0.0591}$ $\quad \log K_{eq} = \frac{(1)(-0.55)}{0.0591} = -9.31$ $\quad K_{eq} = 10^{-9.31} = \text{inv} \log -9.31 = \mathbf{4.9 \times 10^{-10}}$

Since the calculated voltage is negative, the reaction is **nonspontaneous** in the direction shown under standard conditions. This is consistent with a positive value for $\Delta G°$ and $K_{eq} < 1$.

(c) $\Delta G° = -nFE°$ $\quad \Delta G° = -(1 \text{ mol e}^-)\left(\frac{9.65 \times 10^4 \text{ C}}{1 \text{ mol e}^-}\right)\left(\frac{-0.30 \text{ J}}{1 \text{ C}}\right) = \mathbf{5.8 \times 10^4 \text{ J}}$

$\log K_{eq} = \frac{nE°}{0.0591}$ $\quad \log K_{eq} = \frac{(2)(-0.30)}{0.0591} = -10.15$ $\quad K_{eq} = 10^{-10.15} = \text{inv} \log -10.15 = \mathbf{7.8 \times 10^{-11}}$

Since the calculated voltage is negative, the reaction is **nonspontaneous** in the direction shown under standard conditions. This is consistent with a positive value for $\Delta G°$ and $K_{eq} < 1$.

18.65 ***See Section 18.6, Examples 18.12, 18.13, and Solution for Exercise 18.63.***

(a) For $Ni(s) + Cu^{2+}(aq) \rightarrow Ni^{2+}(aq) + Cu(s)$,

$$E = E° - \frac{0.0591}{n}\log\left(\frac{[Ni^{2+}]}{[Cu^{2+}]}\right) \qquad E = 0.59 \text{ V} - \frac{0.0591 \text{ V}\cdot\text{mol e}^-}{2 \text{ mol e}^-}\log\left(\frac{1.40}{0.050}\right) = \mathbf{0.55 \text{ V}}$$

(b) For $2Ag(s) + Cl_2(g) \rightarrow 2AgCl(s)$

$$E = E° - \frac{0.0591}{n}\log\left(\frac{1}{P_{Cl_2}}\right) \qquad E = 1.14 \text{ V} - \frac{0.0591 \text{ V}\cdot\text{mol e}^-}{2 \text{ mol e}^-}\log\left(\frac{1}{320 \text{ torr} \times \frac{1 \text{ atm}}{760 \text{ torr}}}\right) = \mathbf{1.13 \text{ V}}$$

(c) For $Cl_2(g) + 2I^-(aq) \rightarrow 2Cl^-(aq) + I_2(s)$,

$$E = E^\circ - \frac{0.0591}{n}\log\left(\frac{[Cl^-]^2}{P_{Cl_2}[I^-]^2}\right) \qquad E = 0.82\ V - \frac{0.0591\ V\cdot mol\ e^-}{2\ mol\ e^-}\log\left(\frac{(0.60)^2}{(0.300)(0.0010)^2}\right) = \mathbf{0.64\ V}$$

18.66 ***See Section 18.6, Examples 18.12, 18.13, and Solution for Exercise 18.64.***

(a) For $Zn(s) + Fe^{2+}(aq) \rightarrow Zn^{2+}(aq) + Fe(s)$,

$$E = E^\circ - \frac{0.0591}{n}\log\left(\frac{[Zn^{2+}]}{[Fe^{2+}]}\right) \qquad E = 0.32\ V - \frac{0.0591V\cdot mol\ e^-}{2\ mol\ e^-}\log\left(\frac{1.0\times 10^{-3}}{0.05}\right) = \mathbf{0.37\ V}$$

(b) For $AgCl(s) + Fe^{2+}(aq) \rightarrow Ag(s) + Fe^{3+}(aq) + Cl^-(aq)$,

$$E = E^\circ - \frac{0.0591}{n}\log\left(\frac{[Fe^{3+}][Cl^-]}{[Fe^{2+}]}\right) \qquad E = -0.55\ V - \frac{0.0591\ V\cdot mol\ e^-}{1\ mol\ e^-}\log\left(\frac{(0.010)(4.0\times 10^{-3})}{0.20}\right) = \mathbf{-0.33\ V}$$

(c) For $Br_2(\ell) + 2Cl^-(aq) \rightarrow Cl_2(g) + 2Br^-(aq)$,

$$E = E^\circ - \frac{0.0591}{n}\log\left(\frac{(P_{Cl_2})[Br^-]^2}{[Cl^-]^2}\right)$$

$$E = -0.30\ V - \frac{0.0591\ V\cdot mol\ e^-}{2\ mol\ e^-}\log\left(\frac{(0.50\ atm)(3.6\times 10^{-3})^2}{0.10}\right) = \mathbf{-0.18\ V}$$

18.67 ***See Section 18.6.***

The electrode at which reduction occurs is the positive electrode in a voltaic cell. Hence, $Pb^{2+}(aq)$ ions are reduced in this cell: $Pb^{2+}(aq) + 2e^- \rightarrow Pb(s)$. The Nernst equation expression for this cell is

$$E = k - \frac{0.0591\ V\cdot mol\ e^-}{n}\log\left(\frac{1}{[Pb^{2+}]}\right),$$

because a reference electrode is an electrode that has an accurately known potential that is independent of the composition of the test solution.

For $[Pb^{2+}] = 0.100\ M$, $0.053\ V = k - \frac{0.0591\ V\cdot mol\ e^-}{2\ mol\ e^-}\log\left(\frac{1}{0.100}\right)$

This gives, $0.053\ V = k - 0.030\ V$ and $k = 0.083\ V$.

For $[Pb^{2+}] = 1.6 \times 10^{-2}\ M$, $E = 0.083\ V - \frac{0.0591\ V\cdot mol\ e^-}{2\ mol\ e^-}\log\left(\frac{1}{1.6\times 10^{-2}}\right) = \mathbf{0.030\ V}$.

18.68 *See Section 18.6.*

For $AgCl(s) + e^- \rightarrow Ag(s) + Cl^-(aq)$ measured against a reference electrode,

$$E = k - \frac{0.0591}{n}\log\left[Cl^-\right]$$

because a reference electrode is an electrode that has an accurately known potential that is independent of the composition of the test solution.

For $[Cl^-]$ = 0.200 M, $0.345\ V = k - \frac{0.0591\ V \cdot mol\ e^-}{1\ mol\ e^-}\log(0.200)$

This gives 0.345 V = k + 0.0413 V and k = 0.304 V.

For $[Cl^-] = 3.00 \times 10^{-3}$ M, $E = 0.304\ V - \frac{0.0591\ V \cdot mol\ e^-}{1\ mol\ e^-}\log(3.00 \times 10^{-3}) = 0.453\ V =$ **453 mV**.

18.69 *See Section 18.7, Example 18.14, and Solution for Exercise 18.67.*

Substituting in $E = k - \frac{0.0591\ V \cdot mol\ e^-}{n}\log\left(\frac{1}{\left[Pb^{2+}\right]}\right)$ yields

$$0.010\ V = 0.083\ V - \frac{0.0591\ V \cdot mol\ e^-}{2\ mol\ e^-}\log\left(\frac{1}{\left[Pb^{2+}\right]}\right)$$

So, $\log\left(\frac{1}{\left[Pb^{2+}\right]}\right) = \frac{-2\ mol\ e^-(0.010\ V - 0.083\ V)}{0.0591\ V \cdot mol\ e^-} = 2.47$ and $\frac{1}{\left[Pb^{2+}\right]} = 10^{2.47} = \text{inv}\log 2.47 = 295$

Hence, $\left[Pb^{2+}\right] = \frac{1}{295} = \mathbf{3.4 \times 10^{-3}}$ ***M* $\mathbf{Pb^{2+}}$**

18.70 *See Section 18.7, Example 18.14, and Solution for Exercise 18.68.*

Substituting in $E = k - \frac{0.0591}{n}\log[Cl^-]$ yields $0.425\ V = 0.304\ V - \frac{0.0591\ V \cdot mol\ e^-}{1\ mol\ e^-}\log[Cl^-]$.

So, $\log[Cl^-] = \frac{1\ mol\ e^-(0.425\ V - 0.304\ V)}{0.0591\ V \cdot mol\ e^-} = -2.05$ and $[Cl^-] = 10^{-2.05}$ = inv log -2.05 = **8.9×10^{-3} *M***

18.71 *See Section 18.7.*

$Cd(s) \rightarrow Cd^{2+}(aq) + 2e^-$	$E^\circ = 0.403$ V
$Cd)(OH)_2(s) + 2e^- \rightarrow Cd(s) + 2OH^-(aq)$	$E^\circ = -0.83$ V
$Cd(OH)_2(s) \rightleftarrows Cd^{2+}(aq) + 2OH^-(aq)$	$E^\circ = -0.43$ V

$\Delta G^\circ = -nFE^\circ$ $\quad \Delta G^\circ = -(2\ mol\ e^-)\left(\frac{9.65 \times 10^4\ C}{1\ mol\ e^-}\right)\left(-0.43\frac{J}{C}\right) = 8.3 \times 10^4$ J

Solving $\Delta G^\circ = -RT\ln K_{sp}$ for K_{sp} gives $K_{sp} = e^{-\frac{\Delta G^\circ}{RT}}$.

Hence, $K_{sp} = e^{-\frac{(8.3 \times 10^4\ J)}{(8.314\ J \cdot mol^{-1} \cdot K^{-1})(298\ K)}} = \mathbf{2.8 \times 10^{-15}}$.

18.72 *See Section 18.7.*

$$PbSO_4(s) + 2e^- \rightarrow Pb(s) + SO_4^{2-}(aq) \qquad \Delta G_1^o$$
$$Pb(s) \rightarrow Pb^{2+}(aq) + 2e^- \qquad \Delta G_2^o$$

$$PbSO_4(s) \rightleftarrows Pb^{2+}(aq) + SO_4^{2-}(aq) \qquad \Delta G_3^o$$

$$\Delta G_1^o = -n\mathcal{F}E^o = -(2 \text{ mol e}^-)\left(\frac{9.65 \times 10^4 \text{ C}}{1 \text{ mol e}^-}\right)\left(-0.356\frac{\text{J}}{\text{C}}\right) = 6.87 \times 10^4 \text{ J}$$

$$\Delta G_2^o = -n\mathcal{F}E^o = -(2 \text{ mol e}^-)\left(\frac{9.65 \times 10^4 \text{ C}}{1 \text{ mol e}^-}\right)\left(0.126\frac{\text{J}}{\text{C}}\right) = -2.43 \times 10^4 \text{ J}$$

$$\Delta G_3^0 = -RT \ln K_{sp} = \Delta G_1^0 + \Delta G_2^0 = 4.44 \times 10^4 \text{ J}$$

Solving $\Delta G^o = -RT\ln K_{sp}$ for K_{sp} gives $K_{sp} = e^{-\frac{\Delta G^o}{RT}}$.

Hence, $e^{-\frac{(4.44\times10^4 \text{ J})}{(8.314 \text{ J}\cdot\text{mol}^{-1}\cdot\text{K}^{-1})(298 \text{ K})}} = \mathbf{1.6 \times 10^{-8}}$

18.73 *See Sections 18.2, 18.4, 18.6, and Appendix H.*

(a)

$$Zn(s) + 2OH^-(aq) \rightarrow Zn(OH)_2(s) + 2e^- \qquad E^\circ = 1.25 \text{ V}$$
$$Ag_2O(s) + H_2O(\ell) + 2e^- \rightarrow 2Ag(s) + 2OH^-(aq) \qquad E^\circ = 0.342 \text{ V}$$

$$Zn(s) + Ag_2O(s) + H_2O(\ell) \rightarrow Zn(OH)_2(s) + 2Ag(s) \qquad E^\circ = \mathbf{1.59 \text{ V}}$$

(b) Since the reactants and products are all solids and pure liquids, the value of the reaction quotient in the Nernst equation does not change during the course of the reaction, and the voltage therefore remains the same.

18.74 *See Sections 18.2, 18.4, 18.6, and Appendix H.*

(a)

$$Cd(s) + 2OH^-(aq) \rightarrow Cd(OH)_2(s) + 2e^- \qquad E^o = 0.83 \text{ V}$$
$$2[NiO(OH)(s) + H_2O(\ell) + e^- \rightarrow Ni(OH)_2(s) + OH^-(aq)] \qquad E^o = 0.52 \text{ V}$$

$$Cd(s) + 2NiO(OH)(s) + 2H_2O(\ell) \rightarrow Cd(OH)_2(s) + Ni(OH)_2(s) \qquad E^o = 1.35 \text{ V}$$

(b) Since the reactants and products are all solids and pure liquids, the value of the reaction quotient in the Nernst equation does not change during the course of the reaction, and the voltage therefore remains the same.

18.75 *See Sections 17.4, 18.5, 18.8, and Examples 17.4, 18.10.*

(a) Oxidation half-reaction: $C_3H_8(g) + 20OH^-(aq) \rightarrow 3CO_2(g) + 14H_2O(\ell) + 20e^-$

Reduction half-reaction: $O_2(g) + 2H_2O(\ell) + 4e^- \rightarrow 4OH^-(aq)$

(b) For $C_3H_8(g) + 5O_2(g) \rightarrow 3CO_2(g) + 4H_2O(\ell)$,

$$\Delta G^o_{rxn} = \left[3\Delta G_f^o CO_2(g) + 4\Delta G_f^o H_2O(l)\right] - \left[\Delta G_f^o C_3H_8(g) + 5\Delta G_f^o O_2(g)\right]$$

$$\Delta G^o_{rxn} = \left[\left(3 \text{ mol} \times -394.36 \text{ kJ}\cdot\text{mol}^{-1}\right) + \left(4 \text{ mol} \times -237.18 \text{ kJ}\cdot\text{mol}^{-1}\right)\right]$$
$$-\left[\left(1 \text{ mol} \times -23.49 \text{ kJ}\cdot\text{mol}^{-1}\right) + \left(5 \text{ mol} \times 0. \text{ kJ}\cdot\text{mol}^{-1}\right)\right] = \mathbf{-2108.31 \text{ kJ}}$$

Solving $\Delta G° = -nFE$ for $E°$ gives $E° = \frac{-\Delta G°}{nF}$ $\quad E° = \frac{-(-2108.31 \times 10^5 \text{ J})}{(20 \text{ mol e}^-)\left(\frac{9.65 \times 10^4 \text{ C}}{1 \text{ mol e}^-}\right)} = \mathbf{1.09 \text{ V}}$

(c) Recognizing that $\Delta G°$ represents the maximum amount of work that can be obtained from a system gives

$$? \text{ kJ for consumption of 1 g } C_3H_8 = 1.00 \text{ g } C_3H_8 \times \frac{1 \text{ mol } C_3H_8}{44.0 \text{ g } C_3H_8} \times \frac{2108.31 \text{ kJ}}{1 \text{ mol } C_3H_8} = \mathbf{5 \times 10^1 \text{ kJ}}$$

18.76 ***See Sections 17.4, 18.5, 18.8, and Examples 17.4, 18.10.***

(a) Oxidation half-reaction: $CH_4(g) + 8OH^-(aq) \rightarrow CO_2(g) + 6H_2O(\ell) + 8e^-$

Reduction half-reaction: $O_2(g) + 2H_2O(\ell) + 4e^- \rightarrow 4OH^-(aq)$

(b) For $CH_4(g) + 2O_2(g) \rightarrow CO_2(g) + 2H_2O(\ell)$,

$\Delta G°_{rxm} = [\Delta G°_f\ CO_2(g) + 2\Delta G°_f\ H_2O(\ell)] - [\Delta G°_f\ CH_4(g) + 2\Delta G°_f O_2(g)]$

$= [(1 \text{ mol} \times -394.4 \text{ kJ·mol}^{-1}) + (2 \text{ mol} \times -237.2 \text{ kJ·mol}^{-1})]$

$- [(1 \text{ mol} \times -50.8 \text{ kJ·mol}^{-1}) + (2 \text{ mol} \times 0. \text{ kJ·mol}^{-1})] = \mathbf{-818.0 \text{ kJ}}$

Solving $\Delta G° = -nFE$ for $E°$ gives $E° = \frac{-\Delta G°}{nF}$ $\quad E° = \frac{-(-818.0 \times 10^3 \text{ J})}{(8 \text{ mol e}^-)\left(\frac{9.65 \times 10^4 \text{ C}}{1 \text{ mol e}^-}\right)} = \mathbf{1.06 \text{ V}}$

(c) Recognizing that $\Delta G°$ represents the maximum amount of work that can be obtained from a system gives

$$? \text{ kJ for consumption of 1.00 g } CH_4 = 1.00 \text{ g } CH_4 \times \frac{1 \text{ mol } CH_4}{16.0 \text{ g } CH_4} \times \frac{818.0 \text{ kJ}}{1 \text{ mol } CH_4} = \mathbf{51.1 \text{ kJ}}$$

18.77 ***See Section 18.9, Table 18.1, Example 18.15, and Appendix H.***

The most easily oxidized species in an electrolysis is the one with the highest (most positive) oxidation potential. The most easily reduced species in an electrolysis is the one with the highest (most positive) reduction potential.

(a) At the anode the possible reaction are:

$2Cl^-(aq) \rightarrow Cl_2(g) + 2e^-$ $\quad E° = 1.36$ V

$2H_2O(\ell) \rightarrow O_2(g) + 4H^+(aq) + 4e^-$ $\quad E° = -1.23$ V

The oxidation of H_2O has a more positive oxidation potential than the oxidation of Cl^-. However, the overvoltage of approximately 0.4 V associated with producing $O_2(g)$ means it is actually easier to oxidize Cl^-. Hence, Cl^- will be oxidized at the anode.

At the cathode the possible reactions are:

$Zn^{2+}(aq) + 2e^- \rightarrow Zn(g)$ $\quad E° = -0.76$ V

$2H_2O(\ell) + 2e^- \rightarrow H_2(g) + 2OH^-(aq)$ $\quad E° = -0.83$ V

Since Zn^{2+} has a reduction potential that is more positive than that for H_2O, Zn^{2+} will be reduced at the cathode. The net reaction will be:

$Zn^{2+}(aq) + 2Cl^-(aq) \rightarrow Zn(s) + Cl_2(g)$

(b) At the anode the possible reactions are:

$2I^-(aq) \rightarrow I_2(aq) + 2e^-$ $\quad E° = -0.54$ V

$2H_2O(\ell) \rightarrow 4H^+(aq) + O_2(g) + 4e^-$ $\quad E° = -1.23$ V

Since I^- has a much more positive potential for oxidation, it will be oxidized at the anode.

At the cathode the possible reactions are:

$Na^+(aq) + e^- \rightarrow Na(s)$ $\qquad E^\circ = -2.71$ V

$2H_2O(\ell) + 2e^- \rightarrow H_2(g) + 2OH^-(aq)$ $\qquad E^\circ = -0.83$ V

Since H_2O has a potential for reduction that is far more positive than that for Na^+, the hydrogen overvoltage of 0.4 V is not a factor. Hence, H_2O will be reduced at the cathode.

The net reaction will be:

$$2I^-(aq) + 2H_2O(\ell) \rightarrow I_2(s) + H_2(g) + 2OH^-(aq)$$

(c) At the anode the possible reactions are:

$2Br^-(aq) \rightarrow Br_2(\ell) + 2e^-$ $\qquad E^\circ = -1.06$ V

$2H_2O(\ell) \rightarrow 4H^+(aq) + O_2g) + 4e^-$ $\qquad E^\circ = -1.23$ V

Since $Br^-(aq)$ has a more positive potential for oxidation and there is an additional oxygen overvoltage of 0.4 V, $Br^-(aq)$ will be oxidized at the anode.

At the cathode the possible reactions are:

$Ca^{2+}(aq) + 2e^- \rightarrow Ca(s)$ $\qquad E^\circ = -2.87$ V

$2H_2O(\ell) + 2e^- \rightarrow H_2(g) + 2OH^-(aq)$ $\qquad E^\circ = -0.83$ V

Since H_2O has a potential for reduction that is far more positive than that for Ca^{2+}, the hydrogen overvoltage of 0.4 V is not a factor. Hence H_2O will be reduced at the cathode.

The net reaction will be:

$$2Br^-(aq) + 2H_2O(\ell) \rightarrow Br_2(\ell) + H_2(g) + 2OH^-(aq)$$

18.78 ***See Section 18.9, Table 18.1, Example 18.15, and Appendix H.***

(a) Anode: $2Cl^- \rightarrow Cl_2(g) + 2e^-$

Cathode: $Ca^{2+} + 2e^- \rightarrow Ca(\ell)$

Cell: $\mathbf{2Cl^- + Ca^{2+} \rightarrow Cl_2(g) + Ca(\ell)}$

(b) At the anode the possible reactions are:

$2SO_4^{2-}(aq) \rightarrow S_2O_8^{2-}(aq) + 2e^-$ $\qquad E^\circ = -2.05$ V

$2H_2O(\ell) \rightarrow O_2(g) + 4H^+(aq) + 4e^-$ $\qquad E^\circ = -1.23$ V

Since H_2O still has a more positive potential for oxidation after allowing for the additional oxygen overvoltage of 0.4 V, it will be oxidized at the anode.

At the cathode the possible reactions are:

$Mg^{2+}(aq) + 2e^- \rightarrow Mg(s)$ $\qquad E^\circ = -2.37$ V

$2H_2O(\ell) + 2e^- \rightarrow H_2(g) + 2OH^-(aq)$ $\qquad E^\circ = -0.83$ V

Since H_2O still has a more positive potential for reduction after allowing for the additional hydrogen overvoltage of 0.4, it will reduced at the cathode.

The net reaction for the cell will be:

$$\mathbf{2H_2O(\ell) \rightarrow O_2(g) + 2H_2(g).}$$

(c) At the anode the possible reactions are:

$Ni(s) \rightarrow Ni^{2+}(aq) + 2e^-$ $\qquad E^\circ = 0.25$ V

$2Br^- \rightarrow Br_2(\ell) + 2e^-$ $\qquad E^\circ = -1.06$ V

$2H_2O(\ell) \rightarrow O_2(g) + 4H^+(aq) + 4e^-$ $\qquad E^\circ = -1.23$ V

Since Ni has a much more positive oxidation potential, the nickel anode will be oxidized.

At the cathode the possible reductions are:

$Ni^{2+}(aq) + 2e^- \rightarrow Ni(s)$ $E° = -0.25$ V

$2H_2O(\ell) + 2e^- \rightarrow H_2(g) + 2OH^-(aq)$ $E° = -0.83$ V

Since Ni^{2+} has a much more positive reduction potential, it will be reduced at the cathode.

Ni will be oxidized at the anode, and Ni^{2+} will be reduced at the cathode. Hence, no net reaction will occur.

18.79 ***See Section 18.9 and Table 18.1.***

The most easily oxidized species in an electrolysis is the one with the highest (most positive) oxidation potential. The most easily reduced species in an electrolysis is the one with the highest (most positive) reduction potential. The pertinent reduction potentials are:

$Ba^{2+}(aq) + 2e^- \rightarrow Ba(s)$ $E° = -2.90$ V

$Pb^{2+}(aq) + 2e^- \rightarrow Pb(s)$ $E° = -0.126$ V

$2H^+(aq) + 2e^- \rightarrow H_2(g)$ $E° = 0.00$ V

$Ag^+(aq) + e^- \rightarrow Ag(s)$ $E° = 0.80$ V

However, the overvoltage for producting hydrogen gas makes the potential for the H^+/H_2 half-reaction approximately 0.4 V less favorable. Hence,

(a) $Ag^+(aq)$ will be reduced first.

(b) $Pb^{2+}(aq)$ will be reduced second.

(c) $Ba^{2+}(aq)$ cannot be reduced by electrolysis of an aqueous solution, since H_2O is more easily reduced than Ba^{2+} (aq). $\left[2H_2O(\ell) + 2e^- \rightarrow H_2(g) + 2OH^-(aq) \text{ has } E° = -0.83 \text{ V}\right]$

18.80 ***See Section 18.9 and Table 18.1.***

The most easily oxidized species in an electrolysis is the one with the highest (most positive) oxidation potential. The most easily reduced species in an electrolysis is the one with the highest (most positive) reduction potential. The pertinent reduction potentials are:

$Ca^{2+}(aq) + 2e^- \rightarrow Ca(s)$ $E° = -2.87$ V

$Ni^{2+}(aq) + 2e^- \rightarrow Ni(s)$ $E° = -0.25$ V

$2H^+(aq) + 2e^- \rightarrow H_2(g)$ $E° = 0.00$ V

$Cu^{2+}(aq) + 2e^- \rightarrow Cu(s)$ $E° = 0.34$ V

However, the overvoltage for producting hydrogen gas makes the potential for the H^+/H_2 half-reaction approximately 0.4 V less favorable. Hence,

(a) $Cu^{2+}(aq)$ will be reduced first.

(b) $Ni^{2+}(aq)$ will be reduced second.

(c) $Ca^{2+}(aq)$ cannot be reduced by electrolysis of an aqueous solution, since H_2O is more easily reduced than $Ca^{2+}(aq)$. $\left[2H_2O(\ell) + 2e^- \rightarrow H_2(g) + 2OH^-(aq) \text{ has } E° = -0.83 \text{ V}\right]$

18.81 ***See Sections 18.9, 18.11.***

(a) HF is a molecular substance whereas KF is an ionic substance. Molten KF provides the ions necessary to conduct electricity through the electrolyte and thus completes the electrical circuit.

(b) Potassium would be produced by reduction of K^+ ions at the cathode, and fluorine would be produced by oxidation of F^- ions at the anode.

18.82 ***See Sections 18.9, 18.11.***

(a) Water is easier to reduce than $Mg^{2+}(aq)$. The pertinent reduction potentials are:

$$2H_2O(\ell) + 2e^- \rightarrow H_2(g) + 2OH^-(aq) \qquad E° = -0.83\ V$$

$$Mg^{2+}(aq) + 2e^- \rightarrow Mg(s) \qquad E° = -2.37\ V$$

The overvoltage of approximately 0.4 V for producing H_2 does not make H_2O more difficult to reduce than Mg^{2+}.

(b) Anode: $2Cl(\ell) \rightarrow Cl_2(g) + 2e^-$
Cathode: $Mg^{2+}(\ell) + 2e^- \rightarrow Mg(\ell)$

18.83 ***See Section 18.10.***

(a) Three moles of electrons are required to produce one mole of Al via $Al^{3+}(\ell) + 3e^- \rightarrow Al(\ell)$. Hence,

$$?\ C = 0.50\ \text{mol Al} \times \frac{3\ \text{mol e}^-}{1\ \text{mol Al}} \times \frac{9.65 \times 10^4\ C}{1\ \text{mol e}^-} = \mathbf{1.4 \times 10^5\ C}$$

(b) Two moles of electrons are required to produce one mole of Cu via $Cu^{2+}(aq) + 2e^- \rightarrow Cu(s)$. Hence,

$$?\ \text{mol } Cu^{2+}\ \text{present} = 0.100\ \text{L } Cu(NO_3)_2\ \text{soln} \times \frac{0.20\ \text{mol } Cu(NO_3)_2}{1\ \text{L } Cu(NO_3)_2\ \text{soln}} \times \frac{1\ \text{mol } Cu^{2+}}{1\ \text{mol } Cu(NO_3)_2}$$

$$= 0.020\ \text{mol } Cu^{2+}$$

$$?\ C = 0.020\ \text{mol } Cu^{2+} \times \frac{2\ \text{mol e}^-}{1\ \text{mol Cu}} \times \frac{9.65 \times 10^4\ C}{1\ \text{mol e}^-} = \mathbf{3.9 \times 10^3\ C}$$

(c) Two moles of electrons are involved in producting one mole of Cl_2 via $2Cl^-(aq) \rightarrow Cl_2(g) + 2e^-$. Hence,

$$?\ C = 10.0\ \text{g } Cl_2 \times \frac{1\ \text{mol } Cl_2}{70.9\ \text{g } Cl_2} \times \frac{2\ \text{mol e}^-}{1\ \text{mol } Cl_2} \times \frac{9.65 \times 10^4\ C}{1\ \text{mol e}^-} = \mathbf{2.72 \times 10^4\ C}$$

(d) One mole of electrons is required to produce one mole of Ag via $Ag^+(aq) + e^- \rightarrow Ag(s)$. Hence,

$$?\ C = 0.32\ \text{g Ag} \times \frac{1\ \text{mol Ag}}{107.9\ \text{g Ag}} \times \frac{1\ \text{mol e}^-}{1\ \text{mol Ag}} \times \frac{9.65 \times 10^4\ C}{1\ \text{mol e}^-} = \mathbf{2.9 \times 10^2\ C}$$

18.84 ***See Section 18.10.***

(a) Two moles of electrons are required to form one mole of Ca via $Ca^{2+}(\ell) + 2e^- \rightarrow Ca(s)$. Hence,

$$?\ c = 0.50\ \text{mol Ca} \times \frac{2\ \text{mol e}^-}{1\ \text{mol Ca}} \times \frac{9.65 \times 10^4\ C}{1\ \text{mol e}^-} = \mathbf{9.6 \times 10^4\ C}$$

(b) Three moles of electrons are required to produce one mole of Al via $Al^{3+}(\ell) + 3e^- \rightarrow Al(\ell)$. Hence, we can calculate the number of coulombs by calculating the number of moles of Al and moles of electrons involved.

Strategy: $g\ Al \rightarrow mol\ Al \rightarrow mol\ e^- \rightarrow C$

$$?\ C = 3.0\ \text{g Al} \times \frac{1\ \text{mol Al}}{26.98\ \text{g Al}} \times \frac{3\ \text{mol e}^-}{1\ \text{mol Al}} \times \frac{9.65 \times 10^4\ C}{1\ \text{mol e}^-} = \mathbf{3.2 \times 10^4\ C}$$

(c) Four moles of electrons are required to produce one mole of O_2 via $2H_2O(\ell) \rightarrow 4H^+(aq) + O_2(g) + 4e^-$ during the electrolysis of aqueous Na_2SO_4. Hence, we can calculate the number of coulombs by calculating the number of moles of O_2 and moles of electrons involved.

Strategy: g O_2 → mol O_2 → mol e^- → C

$$? \text{C} = 0.52 \text{ g } O_2 \times \frac{1 \text{ mol } O_2}{32.0 \text{ g } O_2} \times \frac{4 \text{ mol e}^-}{1 \text{ mol } O_2} \times \frac{9.65 \times 10^4 \text{ C}}{1 \text{ mol e}^-} = \mathbf{6.3 \times 10^3 \text{ C}}$$

(d) Two moles of electrons are required to form one mole of H_2 via $2H_2O(\ell) + 2e^- \rightarrow H_2(g) + 2OH^-(aq)$. Hence, we can calculate the number of coulombs by calculating the number of moles of H_2 and number of moles of elecgtrons.

Strategy: L H_2 → mol H_2 → mol e^- → C

$$? \text{C} = 1.0 \text{ L } H_2 \times \frac{1 \text{ mol } H_2}{22.4 \text{ L } H_2} \times \frac{2 \text{ mol e}^-}{1 \text{ mol } H_2} \times \frac{9.65 \times 10^4 \text{ C}}{1 \text{ mol e}^-} = \mathbf{8.6 \times 10^3 \text{ C}}$$

18.85 ***See Section 18.10 and Example 18.16.***

The equation for producing H_2 via electrolysis of an aqueous hydrochloric acid solution is: $2H^+(aq) + 2e^- \rightarrow H_2(g)$. Hence, we can calculate moles and of mass hydrogen produced by calculating moles of electrons involved.

Strategy: time and current → C → mole e^- → mol H_2 → g H_2

$$? \text{ g } H_2 = 59.0 \text{ min} \times \frac{60 \text{ s}}{1 \text{ min}} \times \frac{0.500 \text{ C}}{\text{s}} \times \frac{1 \text{ mol e}^-}{9.65 \times 10^4 \text{ C}} \times \frac{1 \text{ mol } H_2}{2 \text{ mol e}^-} \times \frac{2.02 \text{ g } H_2}{1 \text{ mol } H_2} = \mathbf{0.0185 \text{ g } H_2}$$

18.86 ***See Section 18.10 and Example 18.16.***

The equation for producing Cd via electrolysis of an aqueous $CdCl_2$ solution is $Cd^{2+}(aq) + 2e^- \rightarrow Cd(s)$. Hence, we can calculate the moles and mass of Cd by calculating moles of electrons involved.

Strategy time and current → C → mol e^- → mol Cd → g Cd

$$? \text{ g Cd} = 38.0 \text{ min} \times \frac{60 \text{ s}}{1 \text{ min}} \times 1.50 \frac{\text{coul}}{\text{s}} \times \frac{1 \text{ mol e}^-}{9.65 \times 10^4 \text{ C}} \times \frac{1 \text{ mol Cd}}{2 \text{ mol e}^-} \times \frac{112.4 \text{ g Cd}}{1 \text{ mol Cd}} = \mathbf{1.99 \text{ g Cd}}$$

18.87 ***See Section 18.10 and Example 18.16.***

The equation for producing Cu via electrolysis of an aqueous solution of Cu^{2+} is $Cu^{2+}(aq) + 2e^- \rightarrow Cu(s)$. Hence, we can calculate the amount of time required by calculating the moles of Cu and moles of electrons involved.

Strategy: g Cu → mol Cu → mol e^- → C → time

$$? \text{ min} = 15.0 \text{ g Cu} \times \frac{1 \text{ mol Cu}}{63.55 \text{ g Cu}} \times \frac{2 \text{ mol e}^-}{1 \text{ mol Cu}} \times \frac{9.65 \times 10^4 \text{ coul}}{1 \text{ mol e}^-} \times \frac{1 \text{ s}}{3.00 \text{ coul}} \times \frac{1 \text{ min}}{60 \text{ s}} = \mathbf{253 \text{ min}}$$

$$\text{Note: } 250 \text{ mA} \times \frac{1\text{A}}{10^3 \text{ mA}} = 0.250 \text{ A} = 0.250 \frac{\text{C}}{\text{s}}$$

18.88 ***See Section 18.10 and Example 18.16.***

The equation for producing Ag via electrolysis of an aqueous solution of Ag^+ is $Ag^+(aq) + 2e^- \rightarrow Ag(s)$. Hence, we can calculate the amount of time required by calculating the moles of Ag and moles of electrons involved.

Strategy: g Ag → mol Ag → mol e⁻ → C → time

$$? \text{ s} = 0.100 \text{ g Ag} \times \frac{1 \text{ mol Ag}}{107.9 \text{ gAgu}} \times \frac{1 \text{ mol e}^-}{1 \text{ mol Ag}} \times \frac{9.65 \times 10^4 \text{ C}}{1 \text{ mol e}^-} \times \frac{1 \text{ s}}{0.250 \text{ C}} = \mathbf{358 \text{ s}}$$

Note: $250 \text{ mA} \times \frac{1\text{A}}{10^3 \text{ mA}} = 0.250 \text{ A} = 0.250 \frac{\text{C}}{\text{s}}$

18.89 ***See Section 18.10 and Examples 18.17, 18.18.***

(a) The equation for producing pure Cu^{2+} by the electrorefining process is $Cu^{2+}(aq) + 2e^- \rightarrow Cu(s)$.

Strategy: g Cu/hr → mol Cu/hr → mole e⁻/hr → C/hr → C/s

$$? \text{ current in C/s} = \frac{500 \text{ g Cu}}{1 \text{ hr}} \times \frac{1 \text{ mol Cu}}{63.5 \text{ g Cu}} \times \frac{2 \text{ mol e}^-}{1 \text{ mol Cu}} \times \frac{9.65 \times 10^4 \text{ C}}{1 \text{ mol e}^-} \times \frac{1 \text{ hr}}{60 \text{ min}} \times \frac{1 \text{ min}}{60 \text{ s}}$$
$$= 4.22 \times 10^2 \text{ C/s} = \mathbf{4.22 \times 10^2 \text{ A}}$$

(b) The total charge required to produce 500 g of refined Cu is:

$$Q = 500 \text{ g Cu} \times \frac{1 \text{ mol Cu}}{63.5 \text{ g Cu}} \times \frac{2 \text{ mol e}^-}{1 \text{ mol Cu}} \times \frac{9.65 \times 10^4 \text{ C}}{1 \text{ mol e}^-} = 1.52 \times 10^6 \text{ C}$$

The electrical energy consumed is :

$$E = Q \times V \times \frac{1 \text{ kw} \cdot \text{hr}}{3.60 \times 10^6 \text{ J}} \qquad E = 1.52 \times 10^6 \text{ C} \times \frac{0.100 \text{ J}}{1 \text{ C}} \times \frac{1 \text{ kw} \cdot \text{hr}}{3.60 \times 10^6 \text{ J}} = \mathbf{4.22 \times 10^{-2} \text{ kw} \cdot \text{hr}}$$

18.90 ***See Section 18.10 and Examples 18.17, 18.18.***

(a) The equation for producing Al by electrolysis of molten Al_2O_3 is $Al^{3+}(\ell) + 3e^- \rightarrow Al(\ell)$.

Strategy: $\frac{C}{s} \rightarrow \frac{C}{hr} \rightarrow \frac{mol\ e^-}{hr} \rightarrow \frac{mol\ Al}{hr} \rightarrow \frac{g\ Al}{hr} \rightarrow \frac{kg\ Al}{hr}$

$$? \text{ kg} \frac{\text{Al}}{\text{hr}} = \frac{980 \text{ C}}{\text{s}} \times \frac{60 \text{ s}}{1 \text{ min}} \times \frac{60 \text{ min}}{1 \text{ hr}} \times \frac{1 \text{ mol e}^-}{9.65 \times 10^4 \text{ C}} \times \frac{1 \text{ mol Al}}{3 \text{ mol e}^-} \times \frac{26.98 \text{ g Al}}{1 \text{ mol Al}} \times \frac{1 \text{ kg Al}}{10^3 \text{ g Al}} = \mathbf{0.329 \text{ kg} \frac{\text{Al}}{\text{hr}}}$$

(b) The total charge required to produce 1 kg Al is:

$$Q = 1 \text{ kg Al} \times \frac{10^3 \text{ g Al}}{1 \text{ kg Al}} \times \frac{1 \text{ mol Al}}{26.98 \text{ g Al}} \times \frac{3 \text{ mol e}^-}{1 \text{ mol Al}} \times \frac{9.65 \times 10^4 \text{ C}}{1 \text{ mol e}^-} = 1.07 \times 10^7 \text{ C}$$

The electrical energy consumed is:

$$E = Q \times V \times \frac{1 \text{ kw} \cdot \text{hr}}{3.60 \times 10^6 \text{ J}} \qquad E = 1.07 \times 10^7 \text{ C} \times \frac{4.85 \text{ J}}{1 \text{ C}} \times \frac{1 \text{ kw} \cdot \text{hr}}{3.60 \times 10^6 \text{ J}} = \mathbf{14.4 \text{ kw} \cdot \text{hr}}$$

(c) Water is easier to reduce than $Al^{3+}(aq)$. The pertinent reduction potentials are:

$2H_2O(\ell) + 2e^- \rightarrow H_2(g) + 2OH^-(aq)$ $\quad E° = -0.83$ V

$Al^{3+}(aq) + 3e^- \rightarrow Al(s)$ $\quad E° = -1.66$ V

The overvoltage of approximately 0.4 V for producing H_2 does not make H_2O more difficult to reduce than Al^{3+}.

18.91 ***See Section 18.4, 18.12, and Table 18.1.***

Metallic sodium cannot be used for cathodic protection of the iron hull of an ocean vessel because it reacts vigorously with water to form sodium hydroxide and hydrogen gas: $2Na(s) + 2H_2O(\ell) \rightarrow 2NaOH(aq) + H_2(g)$.

18.92 See Sections 18.4, 18.6, 18.12, Table 18.1, and Examples 18.7, 18.12, 18.13.

(a) Oxidation half-reaction: $2[Fe(s) \rightarrow Fe^{2+}(aq) + 2e^-]$ $E° = +0.44$ V
Reduction half-reaction: $O_2(g) + 4H^+(aq) + 4e^- \rightarrow 2H_2O(\ell)$ $E° = +1.23$ V

Balanced redox equation: $2Fe(s) + O_2(g) + 4H^+(aq) \rightarrow 2Fe^{2+}(aq) + 2H_2O(\ell)$ $E° = 1.67$ V

(b) $E° = 0.44\text{ V} + 1.23\text{ V} = \mathbf{1.67\ V}$

(c) $$E = E° - \frac{0.0591}{n}\log\left(\frac{[Fe^{2+}]^4}{P_{O_2}[H^+]^4}\right)$$

At pH = 5.9 $[H^+]$ = inv log (-5.9) = 1.3×10^{-6} M.

$$\text{Hence, } E = 1.67\text{ V} - \frac{0.0591}{4}\log\left(\frac{(5\times10^{-5})^4}{(0.21\text{ atm})(1.3\times10^{-6})^4}\right) = \mathbf{1.44\ V}$$

18.93 See Sections 18.4, 18.5, and Examples 18.7, 18.8, 18.9.

Iron was oxidized by air to Fe^{2+}, which in turn reacted with the aluminum.

$3[Fe^{2+}(aq) + 2e^- \rightarrow Fe(s)]$ $E° = -0.44$ V
$2[Al(s) \rightarrow Al^{3+}(aq) + 3e^-]$ $E° = 1.66$ V

$3Fe^{2+}(aq) + 2Al(s) \rightarrow 3Fe(s) + 2Al^{3+}(aq)$ $E° = 1.22$ V

Aluminum acted as a sacrificial anode.

18.94 See Section 18.10 and Example 18.16.

The equation for the oxidation of the Mg is $Mg(s) \rightarrow Mg^{2+}(aq) + 2e^-$. Hence, we can calculate the moles and mass of Mg by calculating the moles of electrons involved.

Strategy: time and current → C → mol e⁻ → mol Mg → g Mg → lb Mg

$$?\text{ lb Mg} = 100\text{ yrs} \times \frac{365\text{ days}}{1\text{ yr}} \times \frac{24\text{ hr}}{1\text{ day}} \times \frac{60\text{ min}}{1\text{ hr}} \times \frac{60\text{ s}}{1\text{ min}} \times \frac{0.0020\text{ C}}{1\text{ s}}$$
$$\times \frac{1\text{ mol e}^-}{9.65\times10^4\text{ C}} \times \frac{1\text{ mol Mg}}{2\text{ mol e}^-} \times \frac{24.3\text{ g Mg}}{1\text{ mol Mg}} \times \frac{1\text{ lb Mg}}{453.6\text{ g Mg}} = \mathbf{1.8\ lb\ Mg}$$

Note: $2.0\text{ mA} \times \frac{1\text{A}}{10^3\text{ mA}} = 0.0020\text{ A} = 0.0020\frac{\text{C}}{\text{s}}$.

18.95 See Section 18.5 and Examples 18.10, 18.11.

The number of moles of electrons for the reaction $2VO^{2+}(aq) + Br_2(\ell) + 2H_2O(\ell) \rightarrow 2VO_2(aq) + 2Br^-(aq) + 4H^+(aq)$ is two; the oxidation state of each V changes from +4 to +5 and the oxidation state of each Br changes from 0 to -1.

$\Delta G° = -RT\ln K_{eq}$ $\Delta G° = -8.314\text{ J}\cdot\text{mol}^{-1}\cdot\text{K}^{-1} \times 298\text{ K} \times \ln(1.58 \times 10^2) = \mathbf{-1.25 \times 10^4\ J}$

$E° = \frac{0.0591\text{ V}\cdot\text{mol e}^-}{n}\log K_{eq}$ $E° = \frac{0.0591\text{ V}\cdot\text{mol e}^-}{2\text{ mol e}^-}\log(1.58 \times 10^2) = \mathbf{0.0650\ V}$

18.96 ***See Section 18.5 and Example 18.10.***

The number of moles of electrons for the reaction $2Fe(CN)_6^{3-}(aq) + 2I^-(aq) \rightarrow 2Fe(CN)_6^{4-}(aq) + I_2(s)$ is two; the oxidation state of each Fe changes from +3 to +2 and the oxidation state of each I changes from -1 to 0.

Solving $\Delta G° = -nFE°$ for E° gives $E° = \frac{-\Delta G°}{nF} = \frac{-(-34.3 \times 10^3\ J)}{(2\ mol\ e^-)\left(\frac{9.65 \times 10^4\ C}{1\ mol\ e^-}\right)} = 0.178\frac{J}{C} = \mathbf{0.178\ V}$.

18.97 ***See Section 18.6 and Example 18.13.***

For $2H^+(aq) + 2e^- \rightarrow H_2(g)$ with pH = 6.00 and $P_{H_2} = 2.5$ atm,

$$E = E° - \frac{0.0591}{n}\log\left(\frac{P_{H_2}}{[H^+]^2}\right) \qquad E = 0.00\ V - \frac{0.0591\ V \cdot mol\ e^-}{2\ mol\ e^-}\log\left(\frac{2.5}{(1.00 \times 10^{-6})^2}\right) = \mathbf{-0.37\ V}$$

18.98 ***See Section 18.6 and Example 18.13.***

For $Fe^{3+} + e^- \rightarrow Fe^{2+}$ with $[Fe^{3+}] = 0.033$ M and $[Fe^{2+}] = 0.0025$ M,

$$E = E° - \frac{0.0591}{n}\log\left(\frac{[Fe^{2+}]}{[Fe^{3+}]}\right) \qquad E = 0.77\ V - \frac{0.0591\ V \cdot mol\ e^-}{1\ mol\ e^-}\log\left(\frac{0.0025}{0.033}\right) = \mathbf{0.84\ V}$$

18.99 ***See Sections 18.5, 18.10, and Example 18.10.***

(a) The number of moles of electrons for the reaction $Zn(s) + HgO(s) \rightarrow Hg(\ell) + ZnO(s)$ is two; the oxidation state of Zn goes from 0 to +2 and the oxidation state of Hg goes from +2 to 0.

$$\Delta G° = -nFE° \qquad \Delta G° = -(2\ mol\ e^-)\left(\frac{9.65 \times 10^4\ C}{1\ mol\ e^-}\right)\left(\frac{1.35\ J}{1\ C}\right) = \mathbf{-2.61 \times 10^5\ J}$$

(b) $?\ \Delta G°$ for 1.00 g HgO $= 1.00\ g\ HgO \times \frac{1\ mol\ HgO}{216.6\ g\ HgO} \times \frac{-2.61 \times 10^5\ J}{1\ mol\ HgO} = \mathbf{-1.20 \times 10^3\ J}$

(c) *Strategy: g HgO → mol HgO → mol e⁻ → C → s → hr*

$$?\ \text{time in hr} = 3.50\ g\ HgO \times \frac{1\ mol\ HgO}{216.6\ g\ HgO} \times \frac{2\ mol\ e^-}{1\ mol\ HgO} \times \frac{9.65 \times 10^4\ C}{1\ mol\ e^-} \times \frac{1\ s}{10^{-2}\ C} \times \frac{1\ min}{60\ s} \times \frac{1\ hr}{60\ min} = \mathbf{86.6\ hr}$$

Note: $10\,mA \times \frac{1\ A}{10^3\ mA} = 10^{-2}\ A = 10^{-2}\ \frac{C}{s}$

18.100 ***See Section 18.1 and Example 18.2.***

(a) $2ZnS(s) + 3O_2(g) \rightarrow 2ZnO(s) + 2SO_2(g)$

Ons: +2,-2   0   +2,-2   +4,-2

$ZnO(s) + C(s, graphite) \rightarrow Zn(s) + CO(g)$

ONs: +2,-2   0   0   +2,-2

(b) In the first reaction S is oxidized and O is reduced. In the second reaction, C is oxidized and Zn is reduced.

18.101 *See Section 18.4, Example 18.7, and Appendix H.*

(a)

$$2\left[Fe^{3+}(aq) + e^- \rightarrow Fe^{2+}(aq)\right] \qquad E^\circ = 0.771\text{ V}$$
$$H_2SO_3(aq) + H_2O(\ell) \rightarrow HSO_4^-(aq) + 3H^+(aq) + 2e^- \qquad E^\circ = -0.17\text{ V}$$

$$2Fe^{3+}(aq) + H_2SO_3(aq) + H_2O(\ell) \rightarrow 2Fe^{2+}(aq) + HSO_4^-(aq) + 3H^+(aq) \qquad \mathbf{E^\circ = 0.60\text{ V}}$$

(b)

$$6\left[Fe^{2+}(aq) \rightarrow Fe^{3+}(aq) + e^-\right] \qquad E^\circ = -0.771\text{ V}$$
$$Cr_2O_7^{2-}(aq) + 14H^+(aq) + 6e^- \rightarrow 2Cr^{3+}(aq) + 7H_2O(\ell) \qquad E^\circ = 1.33\text{ V}$$

$$6Fe^{2+}(aq) + Cr_2O_7^{2-}(aq) + 14H^+(aq) \rightarrow 6Fe^{3+}(aq) + 2Cr^{3+}(aq) + 7H_2O(\ell) \qquad \mathbf{E^\circ = 0.56\text{ V}}$$

(c)

$$Fe^{2+}(aq) \rightarrow Fe^{3+}(aq) + e^- \qquad E^\circ = -0.771\text{ V}$$
$$HNO_2(aq) + H^+(aq) + e^- \rightarrow NO(g) + H_2O(\ell) \qquad E^\circ = 0.983\text{ V}$$

$$Fe^{2+}(aq) + HNO_2(aq) + H^+(aq) \rightarrow Fe^{3+}(aq) + NO(g) + H_2O(\ell) \qquad \mathbf{E^\circ = .212\text{ V}}$$

18.102 *See Section 18.4, Table 18.1, Examples 18.7, 18.9, and Appendix H.*

(a) $ClO_3^-(aq)$ and $Cl^-(aq)$ are predicted to form $\mathbf{Cl_2(g)}$ in acid solution.

$$5[2Cl^-(aq) \rightarrow Cl_2(g) + 2e^-] \qquad E^\circ = -1.36\text{ V}$$
$$2[ClO_3^-(aq) + 6H^+(aq) + 5e^- \rightarrow \tfrac{1}{2}Cl_2(g) + 3H_2O(\ell)] \qquad E^\circ = 1.47\text{ V}$$

$$10Cl^-(aq) + 2ClO_3^-(aq) + 12H^+(aq) \rightarrow 6Cl_2(g) + 6H_2O(\ell) \qquad E^\circ = 0.11\text{ V}$$
$$5Cl^-(aq) + ClO_3^-(aq) + 6H^+(aq) \rightarrow 3Cl_2(g) + 3H_2O(\ell)$$

(b) $\mathbf{Fe^{3+}}$ has a more positive reduction potential than Cr^{3+} (0.77 V vs. -0.41 V), is more easily reduced and is the better oxidizing agent.

(c) $Fe(CN)_6^{4-}$ has a more positive oxidation potential than Fe^{2+} (-0.358V vs. -0.771 V), is more easily oxidized and is the better reducing agent.

18.103 *See Section 18.1 and Example 18.2.*

(a) CaC_2O_4: Applying Rule 2 calcium is assigned an oxidation state of +2. Applying Rule 5 oxygen is assigned an oxidation state of -2. This gives 2 + 2y + 4(-2) = 0 and y = +3 for the oxidation state of carbon in CaC_2O_4. Hence, oxidation state of **Ca = +2, C= +3, O = -2.**

(b) $Ba(ClO_4)_2$: Applying Rule 2 barium is assigned an oxidation state of +2. Applying Rule 5 oxygen is assigned an oxidation state of -2. This gives 2 + 2y + 8(-2) = 0 and y = +7 for the oxidation state of chlorine in $Ba(ClO_4)_2$. Hence, oxidation state of **Ba = +2, Cl = +7, O = -2.**

(c) Tl^{3+}: Applying Rule 2 oxidation state of Tl is +3.

18.104 *See Section 18.9, Table 18.1, and Example 18.15.*

(a) Cu^{2+} has a more positive reduction potential than Zn^{2+} (0.34 V vs. -0.76 V) and is therefore reduced first via $Cu^{2+} + 2e^- \rightarrow Cu(aq)$. Since H_2O is more easily oxidized than SO_4^{2-}, the cell reaction is:

$$2Cu^{2+}(aq) + 2H_2O(\ell) \rightarrow 2Cu(s) + O_2(g) + 4H^+(aq)$$

(b) $\% Zn = \dfrac{g\ Zn}{g\ sample} \times 100\%$ $\quad \% Zn = \dfrac{(0.122\ g\ sample\ -\ 0.073\ g\ Cu)}{0.122\ g\ sample} \times 100\% = \mathbf{40\%\ Zn}$

(c) After the copper is reduced, the H^+ is reduced because H^+ has a more positive reduction potential than Zn^{2+}. The equation for the cell reaction becomes $\mathbf{2H_2O(\ell) \rightarrow 2H_2(g) + O_2(g)}$. The products of this reaction are gases which will not adhere to the electrodes and hence **will not interfere** with the analysis.

18.105 *See Sections 18.5, 18.6, 18.7.*

(a)

$$PbC_2O_4(s) \rightleftarrows Pb^{2+}(aq) + C_2O_4^{2-}(aq) \qquad \Delta G_1$$

$$Pb^{2+}(aq) + 2e^- \rightarrow Pb(s) \qquad \Delta G_2^\circ$$

$$PbC_2O_4(s) + 2e^- \rightarrow Pb(s) + C_2O_4^{2-}(aq) \qquad \Delta G_3^\circ$$

$\Delta G_1^\circ = -RT\ln K_{sp}$ $\quad \Delta G_1^\circ = -8.314\ J\cdot mol^{-1}\cdot K^{-1} \times 298\ K \times \ln(8.5 \times 10^{-10}) = 5.17 \times 10^4\ J$

$\Delta G_2^\circ = -nFE^\circ$ $\quad \Delta G_2^\circ = -(2\ mol\ e^-)\left(\dfrac{9.65 \times 10^4\ C}{1\ mol\ e^-}\right)\left(\dfrac{-0.126\ J}{1\ C}\right) = 2.43 \times 10^4\ J$

$\Delta G_3^\circ = \Delta G_1^\circ + \Delta G_2^\circ$ $\quad \Delta G_3^\circ = 5.17 \times 10^4\ J + 2.43 \times 10^4\ J = 7.60 \times 10^4\ J$

Solving $\Delta G^\circ = -nFE$ for E° gives $E^\circ = \dfrac{-\Delta G^\circ}{nF}$ $\quad E^\circ = \dfrac{-7.60 \times 10^4\ J}{(2\ mol\ e^-)\left(\dfrac{9.65 \times 10^4\ C}{1\ mol\ e^-}\right)} = -0.394\ J\cdot C^{-1} = \mathbf{-0.394\ V}$

(b) For $PbC_2O_4(s) + 2e^- \rightarrow Pb(s) + C_2O_4^{2-}(aq)$,

$E = E^\circ - \dfrac{0.0591\ V\cdot mol\ e^-}{n}\log[C_2O_4^{2-}]$ $\quad E = -0.394\ V - \dfrac{0.0591\ V\cdot mol\ e^-}{2\ mol\ e^-}\log(.025) = \mathbf{-0.347\ V}$

18.106 *See Sections 18.5, 18.6, 18.7.*

(a)

$$Pb(IO_3)_2(s) \rightleftarrows Pb^{2+}(aq) + 2IO_3^-(aq) \qquad \Delta G_1^\circ$$

$$Pb^{2+}(aq) + 2e^- \rightarrow Pb(s) \qquad \Delta G_2^\circ$$

$$Pb(IO_3)_2(s) + 2e^- \rightarrow Pb(s) + 2IO_3^-(aq) \qquad \Delta G_3^\circ$$

$\Delta G_1^\circ = -RT\ln K_{sp}$ $\quad \Delta G_1^\circ = -8.314\ J\cdot mol^{-1}\cdot K^{-1} \times 298\ K \times \ln(2.6 \times 10^{-13}) = 7.18 \times 10^3\ J$

$\Delta G_2^\circ = -nFE^\circ$ $\quad \Delta G_2^\circ = -(2\ mol\ e^-)\left(\dfrac{9.65 \times 10^4\ C}{1\ mol\ e^-}\right)\left(\dfrac{-0.126\ J}{1\ C}\right) = 2.43 \times 10^4\ J$

$\Delta G_3^\circ = \Delta G_1^\circ + \Delta G_2^\circ$ $\quad \Delta G_3^\circ = 7.18 \times 10^3\ J + 2.43 \times 10^4\ J = 9.61 \times 10^4\ J$

Solving $\Delta G° = -nFE$ for $E°$ gives $E° = \frac{-\Delta G°}{nF}$

$$E° = \frac{-9.61 \times 10^4 \text{ J}}{(2 \text{ mol e}^-)\left(\frac{9.65 \times 10^4 \text{ C}}{1 \text{ mol e}^-}\right)} = -0.498 \text{ J} \cdot \text{C}^{-1} = \mathbf{-0.498 \text{ V}}$$

(b) For $Pb(IO_3)_2(s) + 2e^- \rightarrow Pb(s) + 2IO_3^-(aq)$,

$$E = E° - \frac{0.0591 \text{ V} \cdot \text{mol e}^-}{n} \log\left[IO_3^-\right]^2$$

$$E = -0.498 \text{ V} - \frac{0.0591 \text{ V} \cdot \text{mol e}^-}{2 \text{ mol e}^-} \log\left(3.5 \times 10^{-3}\right)^2 = \mathbf{-0.353 \text{ V}}$$

18.107 ***See Sections 15.1, 18.7.***

$\left[H_3O^+\right] = C_{HA}$ for a monoprotic strong acid. Hence, $\left[H_3O^+\right] = 0.032$ *M*, and pH = -log(0.032) = 1.49.
Using this pH to solve for k gives
$k = E_{cell} + 0.059\text{pH}$ $\quad k = 0.135 \text{ V} + 0.059(1.49) = 0.223 \text{ V}$
and $E_{cell} = 0.223 \text{ V} - 0.059\text{pH}$ for this particular pH meter.
The pH at the equivalence point for titrating a strong acid with a strong base is 7.00. Hence, at the equivalence point,
$E_{cell} = 0.223 \text{ V} - 0.059\text{pH}$ $\qquad E_{cell} = 0.223 \text{ V} - 0.059(7.00) = \mathbf{-0.19 \text{ V}}$

18.108 ***See Section 18.10 and Examples 18.17, 18.18.***

The equation for producing O_2 by electrolysis of water is $2H_2O(\ell) \rightarrow O_2(g) + 4H^+(G) + 4e^-$.

Strategy: $\frac{C}{s} \rightarrow \frac{C}{min} \rightarrow \frac{mol\ e^-}{min} \rightarrow \frac{mol\ O_2}{min} \rightarrow \frac{L\ O_2}{min} \rightarrow \frac{mL\ O_2}{min}$

$$?\frac{\text{mL } O_2 \text{ at STP}}{\text{min}} = 0.250\frac{\text{C}}{\text{s}} \times \frac{60 \text{ s}}{1 \text{ min}} \times \frac{1 \text{ mol e}^-}{9.65 \times 10^4 \text{ C}} \times \frac{1 \text{ mol } O_2}{4 \text{ mol e}^-} \times \frac{22.4 \text{ L } O_2}{1 \text{ mol } O_2} \times \frac{10^3 \text{ mL } O_2}{1 \text{ L } O_2} = \mathbf{0.870} \ \frac{\textbf{mL } \mathbf{O_2}}{\textbf{min}}$$

18.109 ***See Section 18.10.***

The key to solving this problem involves recognizing that the same number of coulombs is passing through both cells. In the one cell, Ag is being produced via $Ag^+(aq) + e^- \rightarrow Ag(s)$, and in the other, Sn^{2+} is being produced via $Sn^{4+}(aq) + 2e^- \rightarrow Sn^{2+}(aq)$. Hence, calculating the number of coulombs that are being used to produce 0.158 g Ag will tell us the number of coulombs that are being used to reduce $Sn^{4+}(aq)$ to $Sn^{2+}(aq)$.

Strategy: $g\ Ag \rightarrow mol\ Ag \rightarrow mol\ e^- \rightarrow C$

$$? \text{ C to produce } 0.158 \text{ g Ag} = 0.158 \text{ g Ag} \times \frac{1 \text{ mol Ag}}{107.9 \text{ g Ag}} \times \frac{1 \text{ mol e}^-}{1 \text{ mol Ag}} \times \frac{9.65 \times 10^4 \text{ C}}{1 \text{ mol e}^-} = 1.41 \times 10^2 \text{ C}$$

Strategy: $C \rightarrow mol\ e^- \rightarrow mol\ Sn^{4+}(aq)\ reduced$

$$? \text{ mol } Sn^{4+}(aq) \text{ reduced} = 1.41 \times 10^2 \text{ C} \times \frac{1 \text{ mol e}^-}{9.65 \times 10^4 \text{ C}} \times \frac{1 \text{ mol } Sn^{4+}(aq) \text{ reduced}}{2 \text{ mol e}^-}$$
$$= \mathbf{7.32 \times 10^{-4} \text{ mol } Sn^{4+}(aq)}$$

18.110 ***Solutions for Spreadsheet Exercises are on the Reger, Goode and Mercer Spreadsheet Solutions Diskette available from Saunders College Publishing.***

18.111 ***See Section 18.10.***

The equation for producing Au via electrolysis of $Au(CN)_4^-(aq)$ is: $Au(CN)_4^-(aq) + 3e^- \rightarrow Au(s) + 4CN^-(aq)$. The key to solving this problem involves determining the number of moles of Au to be plated.

Strategy: area and thickness → volume Au → g Au → mol Au

$$? \text{ mol Au plated} = 100 \text{ cm}^2 \times \left(0.0020 \text{ mm} \times \frac{1 \text{ cm}}{10 \text{ mm}}\right) \times \frac{19.3 \text{ g Au}}{\text{cm}^3} \times \frac{1 \text{ mol Au}}{197.0 \text{ g Au}} = 2.0 \times 10^{-3} \text{ mol Au}$$

Strategy: mol Au → mol e⁻ → C → time

$$? \text{ min for plating} = 2.0 \times 10^{-3} \text{ mol Au} \times \frac{3 \text{ mol e}^-}{1 \text{ mol Au}} \times \frac{9.65 \times 10^4 \text{ C}}{1 \text{ mol e}^-} \times \frac{1 \text{ s}}{0.500 \text{ C}} \times \frac{1 \text{ min}}{60 \text{ s}} = \mathbf{19 \text{ min}}$$

18.112 ***See Sections 18.5, 18.6, 18.7.***

$$\begin{array}{lr} Ba(IO_3)_2(s) \rightleftarrows Ba^{2+}(aq) + 2IO_3^-(aq) & \Delta G_1 \\ Ba^{2+}(aq) + 2e^- \rightarrow Ba(s) & \Delta G_2^\circ \\ \hline Ba(IO_3)_2(s) + 2e^- \rightarrow Ba(s) + 2IO_3^-(aq) & \Delta G_3^\circ \end{array}$$

$\Delta G_1^\circ = -RT\ln K_{sp}$ $\quad \Delta G_1^\circ = -8.314 \text{ J}\cdot\text{mol}^{-1}\cdot\text{K}^{-1} \times 298 \text{ K} \times \ln(1.5 \times 10^{-9}) = 5.0 \times 10^4 \text{ J}$

$\Delta G_2^\circ = -nFE^\circ$ $\quad \Delta G_2^\circ = -(2 \text{ mol e}^-)\left(\frac{9.65 \times 10^4 \text{ C}}{1 \text{ mol e}^-}\right)\left(\frac{-2.90 \text{ J}}{1 \text{ C}}\right) = 5.60 \times 10^5 \text{ J}$

$\Delta G_3^\circ = \Delta G_1^\circ + \Delta G_2^\circ$ $\quad 5.0 \times 10^4 \text{ J} + 5.60 \times 10^5 \text{ J} = 6.10 \times 10^5 \text{ J}$

Solving $\Delta G^\circ = -nFE$ for E° gives $E^\circ = \frac{-\Delta G^\circ}{nF}$

$$E^\circ = \frac{-6.10 \times 10^5 \text{ J}}{(2 \text{ mol e}^-)\left(\frac{9.65 \times 10^4 \text{ C}}{1 \text{ mol e}^-}\right)} = -3.16 \text{ J}\cdot\text{C}^{-1} = \mathbf{-3.16 \text{ V}}$$

Chapter 19: Metallurgy, Transition Elements

19.1 *See Section 19.1.*

The goals of pretreatments of ores are to 1) concentrate the ore and 2) convert it into a chemical form that is more suitable for the reduction step. For example, sulfide ores are concentrated by flotation and then roasted to form oxides that can be reduced by using carbon.

19.2 *See Section 19.1.*

Metals and nonmetals can be purified by fractional distillation, electrolysis or zone refining.

1) Fractional distillation is used with Hg, Zn and Mg. The basic concepts of fractional distillation are described in Section 12.6 in the text.

2) The use of electrolysis to refine copper is described in Section 18.11 in the text. Impurities more active than copper are oxidized with copper at the impure copper anode but are not reduced with copper ions at the pure copper catinode. These ions remain in solution, whereas the less active metal impurities form an anode mud. Cobalt, lead and plutonium are also refined by electrolysis.

3) Zone refining is used to obtain very high purity silicon for use in semiconductors. This technique is described in Section 20.4 in the text.

19.3 *See Section 19.1.*

The principal reactions occurring during the reduction of Fe_2O_3 by CO in a blast furnace are:

$$3Fe_2O_3 + CO \rightarrow 2Fe_3O_4 + CO_2$$
$$Fe_3O_4 + CO \rightarrow 3FeO + CO_2$$
$$FeO + CO \rightarrow Fe + CO_2$$

19.4 *See Section 18.2.*

Transition elements have a partially filled *d* subshell in the metal atom or one of its oxidation states.

19.5 *See Section 18.2.*

Transition elements are able to use electrons from an inner shell as valence electrons, whereas representative elements cannot. Transition elements have n*s* and (n-1)*d* electrons that are of similar energy and are able to use both types of electrons when reacting, whereas representative elements are able to use only outer shell n*s* or n*s* and n*p* electrons when reacting.

19.6 *See Section 19.2.*

The atomic radii of the transition elements within a period decrease more rapidly from Group IIIB through Group VIB than through the rest of the transition elements in that period because increases in effective nuclear charge are outweighed by repulsions among d electrons toward the end of the d-block. This is partially due to pairing in the d orbitals.

19.7	*See Section 19.2.*

The atomic radii of transition elements in the sixth period are almost idential to the radii of transition elements in the same groups in the fifth period as a result of the lanthanide contraction. Hence, the volume of atoms of Ta is almost idential to the volume of atoms of Nb, and the ratios of mass to bulk volume for samples of Ta and Nb, the densities of Ta and Nb, should be almost identical to the ratio of their atomic weights.

19.8	*See Sections 8.1, 19.2, and Table 8.1.*

The atoms of elements in Group IB have just one electron in the outermost *s* subshell whereas the atoms of most other transition elements have two electrons in their outermost *s* subshell.

19.9	*See Sections 19.2 and Figure 19.6.*

Thre is a significant increase in the third ionization energy of transition elements near the end of a transition series. This increase in energy required to form a M^{3+} cation causes the +3 oxidation to become less common for elements near the end of a transition series.

19.10	*See Section 19.4, Table 19.5 and Figure 19.12.*

The most common geometric arrangements of ligands in transition metal complexes are:
1) linear 2) tetrahedral 3) square planar 4) octahedral

L—M—L

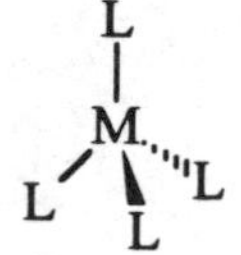

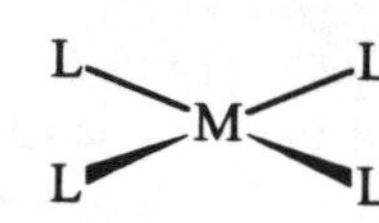

19.11	*See Section 19.1 and Table 19.2.*

(a) Electrolysis of molten LiCl must be used to obtain Li by reduction.

(b) Chemical reduction of Fe_2O_3 by CO is used to obtain Fe by reduction.

19.12	*See Section 19.1 and Table 19.2.*

(a) Electrolysis of molten Al_2O_3 must be used to obtain Al by reduction.

(b) Roasting HgS(s) yields Hg by reduction.

19.13	*See Section 8.1, Table 8.1, and Section 19.2.*

Sc has $[Ar]4s^2 3d^1$ as its electron configuration and forms Sc^{3+} with [Ar] as its electron configuration in its only common oxidation state.

19.14	*See Section 8.1, Table 8.1, and Section 19.2.*

Cr has the anomalous electron configuration $[Ar]4s^1 3d^5$ and therefore has six unpaired electrons.

19.15 ***See Section 8.1, Table 8.1, and Section 19.2.***

The elements in Group IIB(12), Zn, Cd, and Hg, are found in the "*d*-block" of the periodic table but do not have partially filled *d* orbitals in either the metal atoms or in one of their oxidation states. Hence, these elements do not meet the definition of a transition element.

19.16 ***See Section 8.1, Table 8.1, and Section 19.2.***

Ac has $[Rn]7s^2 6d^1$ as its electron configuration and therefore meets the definition of a transition element. (See discussion of La and Lu in text.)

19.17 ***See Section 19.2.***

Transition elements have a partially filled *d* subshell in the metal atom or one of its oxidation states. Hence, only (a) **Fe** is a transition element.

19.18 ***See Section 19.2.***

Transition elements have a partially filled *d* subshell in the metal atom or one of its oxidation states. Hence, (a) Mo, (b) La, and (c) Pd are all transition elements.

19.19 ***See Section 19.2 and Table 19.3.***

(a) **Cr** has a higher melting point than Co. The melting points of the transition elements generally reach a maximum in Group VB, VIB or VIIB.

(b) **Hf** has a higher melting point than Ti. The melting points of the transition elements generally increase going down a group in the periodic table.

(c) **Nb** has a higher melting point than V. The melting points of the transition elements generally increase going down a group in the periodic table.

(d) **W** has a higher melting point than Y. The melting points of the transition elements generally reach a maximum in Groups VB, VIB or VIIB and generally increase going down a group in the periodic table.

19.20 ***See Section 19.2 and Table 19.3.***

(a) **Cr** has a higher melting point than Cu. The melting points of the transition elements generally reach a maximum in Group VB, VIB or VIIB.

(b) **Os** has a higher melting point than Fe. The melting points of the transition elements generally increase going down a group in the periodic table.

(c) **V** has a higher melting point than Cr. The melting points of the transition elements generally reach a maximum in Group VB, VIB or VIIB.

(d) **W** has a higher melting point than La. The melting points of the transition elements generally reach a maximum in Groups VB, VIB or VIIB.

19.21 ***See Section 19.2 and Table 19.3.***

The transition metal in the fourth period having the highest melting point is likely to have the strongest bonding and highest heat of fusion. It is **V**.

19.22 ***See Section 19.2 and Table 19.3.***

The transition metal in the fifth period having the highest melting point is likely to have the strongest bonding and be the hardest transition metal in the fifth period. It is **Mo**.

19.23 ***See Section 19.2 and Table 19.4.***

The order of decreasing atomic radii is: **Nb > W > V > Co**. The atomic radii of the transition elements generally decrease going across a period and increase going down groups. However, thc atomic radii of transition clements in the sixth period are almost identical to the radii of transtion elements in the same groups in the fifth period as a result of the lanthanide contraction. This accounts for the relative sizes of Nb and W. The radius of W would be approximately equal to the radius of Nb if it were directly below Nb but is less because it is to the right of that position.

19.24 ***See Section 19.2 and Table 19.4.***

The order of decreasing atomic radii is: **Hf > Ta > Mo > Fe**. The atomic radii of the transition elements generally decrease going across a period and increase going down groups.

19.25 ***See Section 19.2 and Figure 19.5.***

The maximum positive oxidation state is equal to the group number for elements in Groups IIIB through VIIB.
This gives: (a) **Ti**, IVB, **+4** (b) **W**, VIB, **+6** (c) **Ta**, VB, **+5** (d) **Re**, VIIB, +7.

19.26 ***See Section 19.2 and Figure 19.5.***

The maximum positive oxidation state is equal to the group number for elements in Groups IIIB through VIIB.
This gives: (a) **Cr**, VIB, **+6** (b) **Zr**, IVB, **+4** (c) **Y**, IIIB, +3 (d) Tc, VIIB +7.

19.27 ***See Section 19.2 and Figure 19.6.***

(a) **Mn** has a higher first ionization energy than Ti. Ionization energies generally increase with increasing effective nuclear charge and decreasing size proceeding from left to right across a period in the periodic table.

(b) **Ta** has a higher first ionization energy than V. The increase in effective nuclear charge that occurs during filling the poor shielding 4f orbitals outweighs the increase in size from V to Ta causing the ionization energy of Ta to be higher than that of V.

(c) **Rh** has a higher first ionization energy than Ru; see explanation given in answer for part a.

(d) **Os** has a higher first ionization energy than Mo. Transition metals of the sixth period generally have higher first ionization energies than transition metals in the fifth period. This is due to the increase in effective nuclear charge that occurs during filling the poor shielding 4f orbitals. In addition, Os is further to the right in the periodic table. This leads to an additional increase in effective nuclear charge and a decrease in size, causing Os to be smaller than Mo. This also contributes to the higher ionization energy of Os.

19.28 ***See Section 19.2 and Figure 19.6.***

(a) **Tc** has a higher first ionization energy than Zr. Ionization energies generally increase with increasing effective nuclear charge and decreasing size proceeding from left to right across a period in the periodic table.

(b) **W** has a higher first ionization energy than Mo. Transition metals of the sixth period generally have higher first ionization energies than transition metals in the fifth period. This is due to the increase in effective nuclear charge that occurs during filling the poor shielding 4f orbitals.

(c) **Pt** has a higher first ionization energy than Fe. Transition metals of the sixth period generally have higher first ionization energies than transition metals in the fifth period. This is due to the increase in effective nuclear charge that occurs during filling the poor shielding 4f orbitals. In addition, Pt is further to the right in the periodic table. This leads to an additional increase in effective nuclear charge and contributes to the higher ionization energy of Pt.

(d) **Co** has a higher first ionization energy than Mn. Ionization energies generally increase with increasing effective nuclear charge and decreasing size proceeding from left to right across a period in the periodic table.

19.29 ***See Section 19.3 and Appendix H.***

(a) $FeCr_2O_4 + 4C \rightarrow Fe + 2Cr + 4CO$

(b) $Cr_2O_3 + 2Al \rightarrow 2Cr + Al_2O_3$

(c) $4Cr^{2+}(aq) + O_2(g) + 4H^+(aq) \rightarrow 4Cr^{3+}(aq) + 2H_2O(\ell)$

(d) $2Cr(OH)_4^-(aq) + 3ClO^-(aq) + 2OH^-(aq) \rightarrow 2CrO_4^{2-}(aq) + 3Cl^-(aq) + 5H_2O(\ell)$

(e) $6Fe^{2+}(aq) + Cr_2O_7^{2-}(aq) + 14H^+(aq) \rightarrow 6Fe^{3+}(aq) + 2Cr^{3+}(aq) + 7H_2O(\ell)$

19.30 ***See Section 19.3 and Appendix H.***

(a) $K_2Cr_2O_7 + 2H_2SO_4 \rightarrow 2CrO_3 + H_2O + 2KHSO_4$

(b) $Cr_2O_3 + 6H^+ \rightarrow 2Cr^{3+} + 3H_2O$
$Cr_2O_3 + 2OH^- + 3H_2O \rightarrow 2Cr(OH)_4^-$

(c) $6Cr^{2+} + Cr_2O_7^{2-} + 14H^+ \rightarrow 8Cr^{3+} + 7H_2O$

(d) $Cr^{3+} + 4OH^- \rightarrow Cr(OH)_4^-$
$2Cr(OH)_4^- + 3HO_2^- \rightarrow 2CrO_4^{2-} + OH^- + 5H_2O$

19.31 ***See Section 18.12 and Figure 18.21.***

$4Fe^{2+}(aq) + O_2(g) + (4 + 2x)H_2O(\ell) \rightarrow 2Fe_2O_3{\bullet}xH_2O(s) + 8H^+(aq)$

19.32 ***See Section 19.3.***

$Cu(s) + 2H_2SO_4(aq) \rightarrow CuSO_4(aq) + SO_2(g) + 2H_2O(\ell)$

The oxidizing agent is H_2SO_4; the S is reduced from +6 oxidation state in H_2SO_4 to +4 oxidation state in SO_2.

19.33 ***See Section 19.4, Tables 19.6, 19.7, and Example 19.1.***

(a) $[Cr(H_2O)_5Cl]Cl_2$ (b) $[Cr(NH_3)_4Cl_2]Cl$ (c) $K_3[Fe(CN)_6]$

19.34 *See Section 19.4, Tables 19.6, 19.7, and Example 19.1.*

(a) $\left[Co(en)_3\right]\left[Co(CN)_6\right]$ (b) $\left[Pt(NH_3)_4\right](NO_3)_2$ (c) $Na_2\left[RhCl_5(H_2O)\right]$

19.35 *See Section 19.4, Tables 19.7, 19.8, and Example 19.2.*

(a) $[Pt(NH_3)_2Cl_2]$, diamminedichloroplatinum(II)

(b) $[Co(en)_2(NO_2)_2]NO_3$, bis(ethylenediamine)dinitrocobalt(III) nitrate

(c) $K_3[RhCl_6]$, potassium hexachlororhodate(III)

(d) $[PtNH_3)_4][PtCl_4)]$, tetraammineplatinum(II) tetrachloroplatinate(II)

(e) $[CrCO)_6]$, hexacarbonylchromium(0)

19.36 *See Section 19.4, Tables 19.7, 19.8, and Example 19.2.*

(a) $\left[Fe(CO)_5\right]$, pentacarbonyliron(0)

(b) $K_2\left[Cr(CN)_5NO\right]$, potassium pentacyanonitrosylchromate(III)

(c) $\left[Ru(NH_3)_5Cl\right]Cl_2$, pentaamminechlororuthenium(III) chloride

(d) $\left[Co(dien)Br_3\right]$, tribromo(diethylenetriamine)cobalt(III)

(e) $\left[Cr(NH_3)_6\right]\left[Cr(CN)_6\right]$, hexaamminechromium(III) hexacyanochromate(III)

19.37 *See Section 19.4, Tables 19.7, 19.8, and Example 19.3.*

(a) pentaaquachlorochromium(III) chloride, $[Cr(H_2O)_5Cl]Cl_2$

(b) tetraamminedinitrorhodium(III) bromide, $[Rh(NH_3)_4(NO_2)_2]Br$

(c) dichlorobis(ethylenediamine)ruthenium(III), $[Ru(en)_2Cl_2]^+$

(d) diaquatetrachlororhodate(III), $[Rh(H_2O)_2Cl_4]^-$

(e) triamminetribromoplatinum(IV), $[Pt(NH_3)_3Br_3]^+$

19.38 *See Section 19.4, Tables 19.7, 19.8, and Example 19.3.*

(a) hexaaquachromium(III) hexacyanoferrate(III), $\left[Cr(H_2O)_6\right]\left[Fe(CN)_6\right]$

(b) bromochlorobis(ethylenediamine)cobalt(III), $\left[Co(en)_2BrCl\right]^+$

(c) carbonylpentacyanocolbaltate(III), $\left[CoCO(CN)_5\right]^{2-}$

(d) trinitro(diethylenetriamine)chromium(III), $\left[Cr(dien)(NO_2)_3\right]$

(e) pentaaquathiocyanotoiron(III), $\left[Fe(H_2O)_5 SCN\right]^{2+}$

19.39 *See Section 19.5 and Example 19.4.*

cis-tetraaquadibromochromium(III)

trans- tetraaquadibromochromium(III)

19.40 *See Section 19.5 and Figure 19.17.*

fac-triamminetrichlorocobalt(III)

mer-triamminetrichlorocobalt(III)

19.41 *See Section 19.5.*

(a) $[Co(en)_2Br2]^+$ can exist as cis or trans isomers and enantiomers of the cis isomer.

(b) $[Pt(NH_3)_3SCN]^+$ could have linkage isomers with SCN^- bound to the Pt^{2+} through either the S or N atoms of SCN^-.

(c) $[Co(NH_3)_3Cl_3]$ and $[Pt(NH_3)_3SCN]^+$ are octahedral and square-planar complexes having planes of symmetry are not chiral and therefore cannot form optically active isomers.

(d) $[Cr(H_2O)_2Cl_2Br_2]^-$ has the largest number of possible isomers, namely:

all trans

cis-bromo, cis-chloro

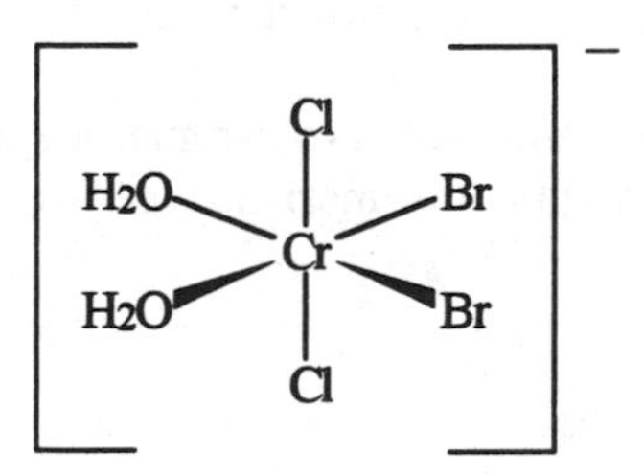

cis-aqua, cis-bromo

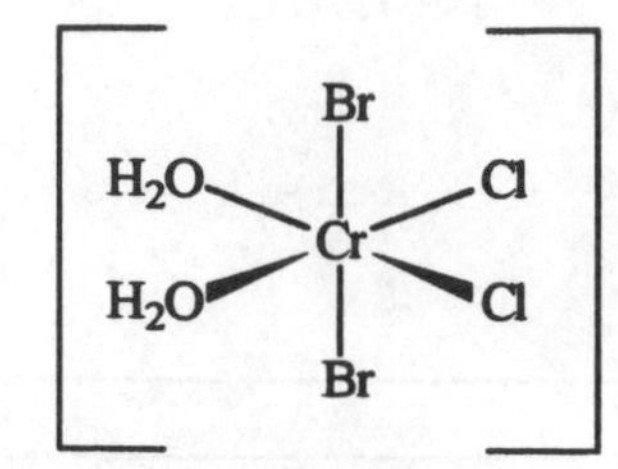

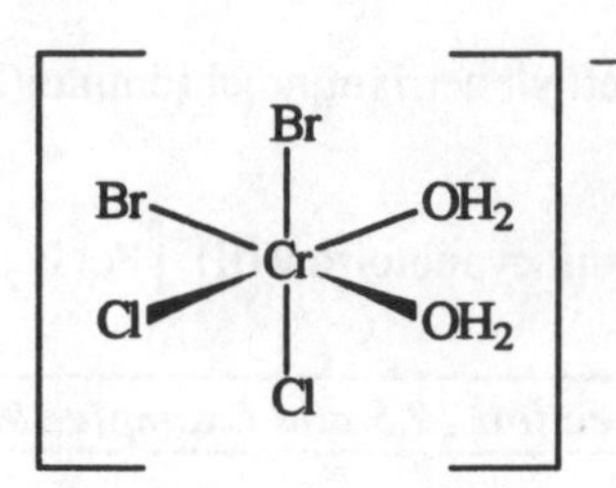

cis-aqua, cis-chloro

all cis-enantiomers

19.42 *See Section 19.5.*

(a) Linkage isomerism of NCS^-; see Figure 19.15.

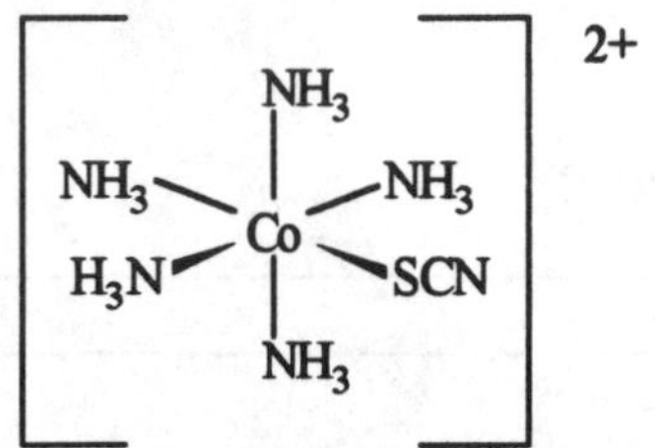

pentaammine-S-thiocyanotocobalt(III)

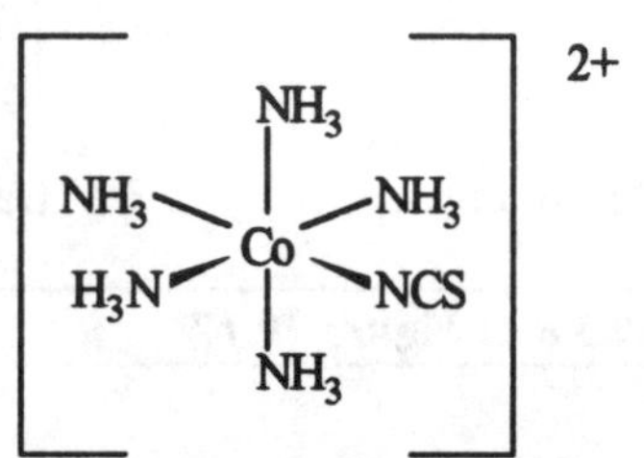

pentaammine-N-thiocyanatocobalt(III)

(b) Geometric isomerism; see Figure 19.16.

Br—Pd—NH_3, Br, NH_3

cis-diamminedibromopalladium(II)

Br—Pd—NH_3, H_3N, Br

trans-diamminedibromopalladium(II)

(c) The oxalate anion, $C_2O_4^{2-}$, is a symmetrical bidentate ligand (Table 19.6) which can be abbreviated as O—O. Optical isomersism; see Figure 19.21.

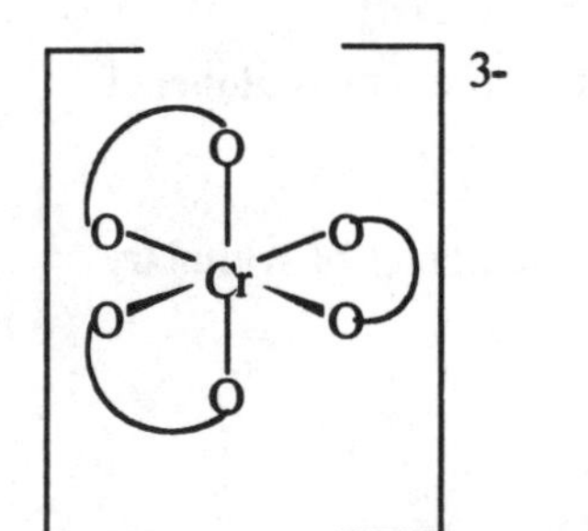

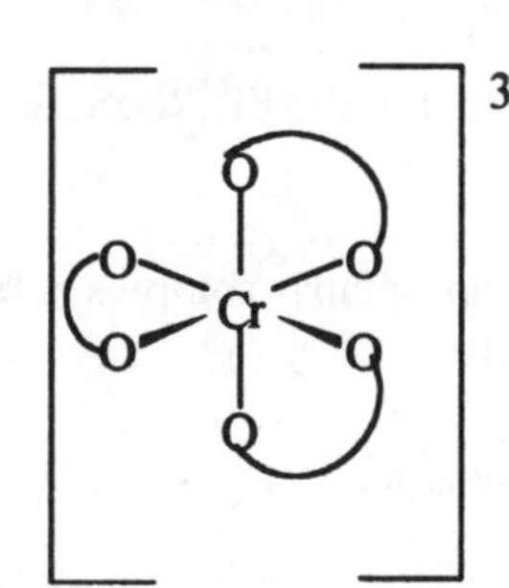

tris(oxalato)chromate(III)

(d) Ethylenediamine (en) is a symmetrical bidenate ligand (Table 19.6) which can be abbreviated as N-N. Geometric and optical isomerism; see Example 19.5.

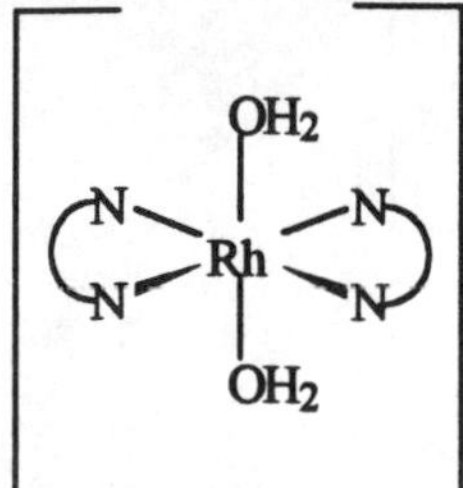

trans-diaquabis(ethylenediamine)rhodium(III)

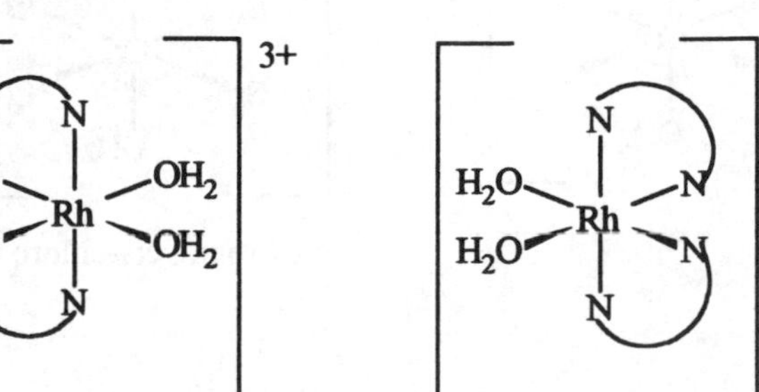

cis-diaquabis(ethylenediamine)rhodium(III)

(e) Geometric isomerism; see Figure 19.17.

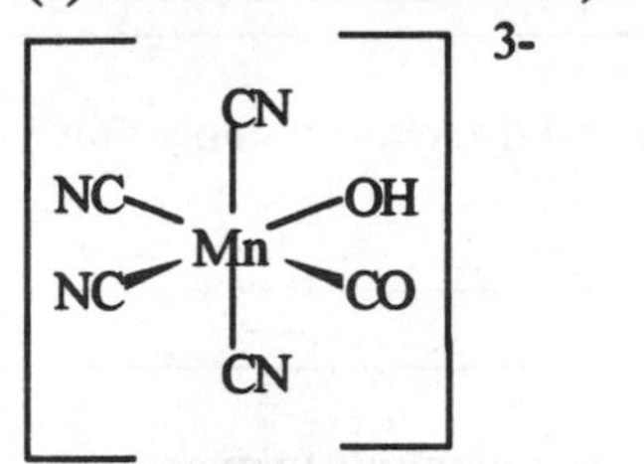

cis-carbonyltetracyanohydroxomanganate(II)

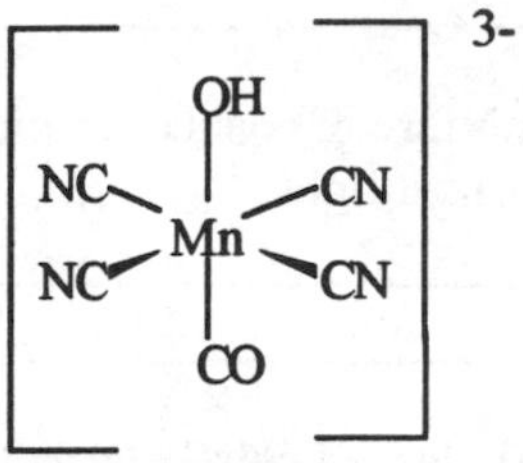

trans-carbonyltetracyanohydroxomanganate(II)

19.43 ***See Section 19.5 and Figure 19.17.***

(a)

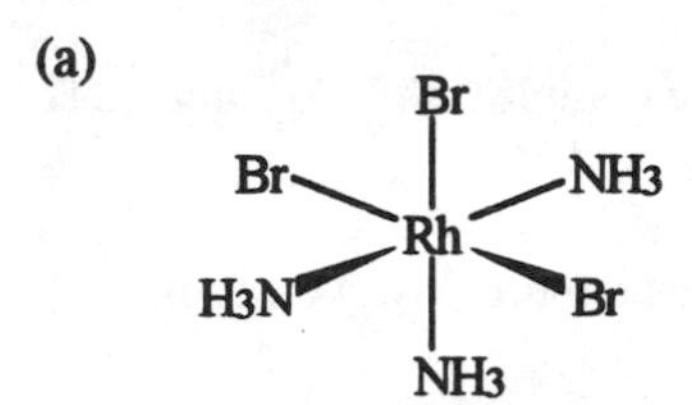

mer-triamminetribromorhodium(III)

(b)

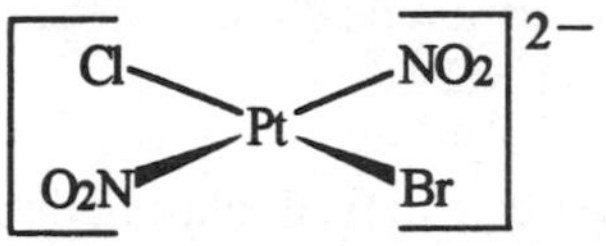

trans-dinitrobromochloroplatinate(II)

19.44 ***See Section 19.5 and Figure 19.17.***

(a)

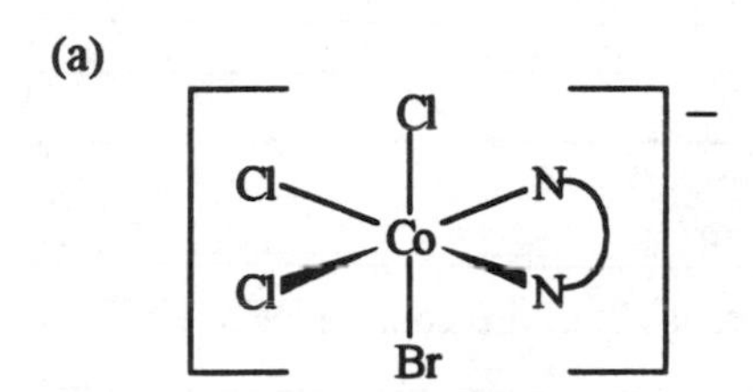

fac-bromotrichloroethylenediaminecobaltate(III)

(b)

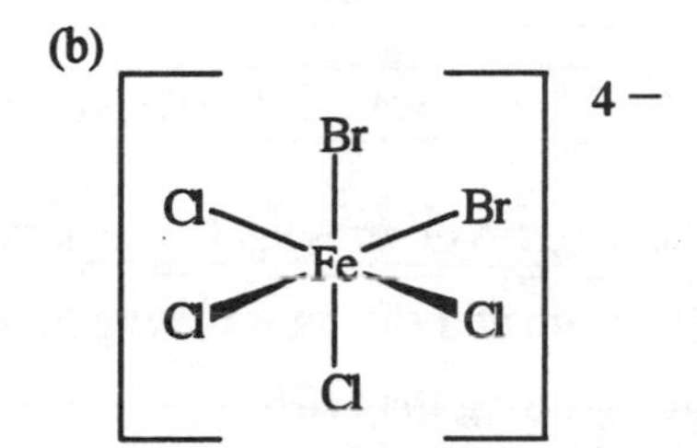

cis-dibromotetrachloroferrate(II)

19.45 ***See Section 19.5.***

Linkage isomerism is possible when lone pairs of electrons on two or more atoms of different elements in a ligand can be used to form monodentate linkages to metals. Hence, linkage isomerism is possible with :

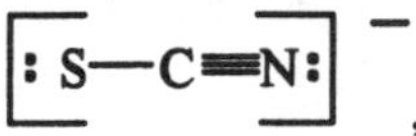

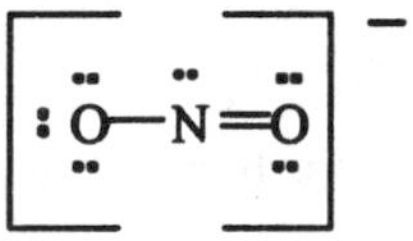

[:S—C≡N:]⁻ , [:O—N=O]⁻ and [:N≡C—O:]⁻

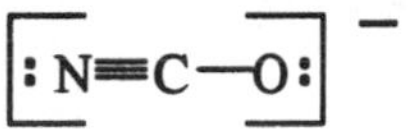

but not with

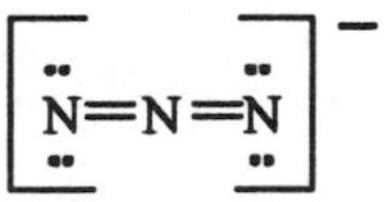

[N=N=N]⁻ or $H_2N—CH_2—CH_2—NH_2$

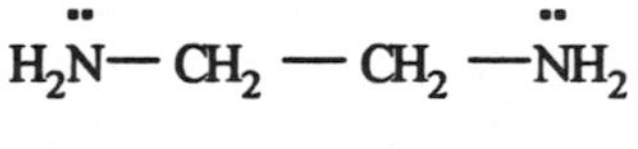

19.46 ***See Section 19.5.***

Molecules and ions that have mirror image structures that cannot be superimposed are optically active and are called chiral.

19.47 *See Section 19.5.*

A racemic mixture is a mixture of equal quantities of isomers that exist as enantiomers and therefore produces no net rotation of plane polarized light.

19.48 *See Section 19.5.*

Enantiomers are a special class of chiral molecules that rotate plane polarized light in equal amounts in exactly opposite directions, presuming they are present in equal concentrations.

19.49 *See Section 19.6, Table 19.10, and Example 19.6.*

(a) $[Co(CN)_6]^{3-}$ has a larger crystal field splitting than $[Co(NH_3)_6]^{3+}$ because both are complexes of Co^{3+} and CN^- causes a greater splitting than NH_3.

(b) $[Cr(H_2O)_6]^{3+}$ has a larger crystal field splitting than $[Cr(H_2O)_6]^{2+}$ because Cr^{3+} has a more positive cation charge than Cr^{2+}.

(c) $[Ru(H_2O)_6]^{2+}$ has a larger crystal field splitting than $[Fe(H_2O)_6]^{2+}$ because Δ is larger for a metal with larger $4d$ valence orbitals compared to $3d$.

(d) $[Cr(H_2O)_6]^{3+}$ has a larger crystal field splitting than $[CrF_6]^{3-}$ because both are complexes of Cr^{3+} and H_2O causes a greater splitting than F^-.

19.50 *See Section 19.6, Table 19.6, and Example 19.6.*

The complex having the larger Δ absorbs light at a shorter wavelength. Hence,

(a) $\left[Rh(CN)_6\right]^{3-}$ will absorb at a shorter wavelength than $\left[Rh(NH_3)_6\right]^{3+}$ because both are complexes of Rh^{3+} and CN^- causes a greater splitting than NH_3.

(b) $\left[Fe(H_2O)_6\right]^{3+}$ will absorb at a shorter wavelength than $\left[Fe(H_2O)_6\right]^{2+}$ because Fe^{3+} has a more positive cation charge than Fe^{2+}.

(c) $\left[Rh(H_2O)_6\right]^{3+}$ will absorb at a shorter wavelength than $\left[Co(H_2O)_6\right]^{3+}$ because Δ is larger for a metal with larger $4d$ valence orbitals compared to $3d$.

(d) $\left[Ti(H_2O)_6\right]^{3+}$ will absorb at a shorter wavelength than $\left[TiF_6\right]^{3-}$ because.both are complexes of Ti^{3+} and H_2O causes a larger crystal field splitting than F^-.

19.51 *See Section 19.6 and Figure 19.26.*

(a)

$\_\_ \quad \_\_$

$d_{x^2-y^2} \quad d_{z^2}$ $\qquad\qquad \left[V(H_2O)_6\right]^{3+}$

$\uparrow \quad \uparrow \quad \_\_$

$d_{xy} \quad d_{yz} \quad d_{xz}$

(b)

$\uparrow$ $d_{x^2-y^2}$ — d_{z^2}

$\uparrow$ d_{xy} $\uparrow$ d_{yz} $\uparrow$ d_{xz}

$[Cr(H_2O)_6]^{2+}$

or

— $d_{x^2-y^2}$ — d_{z^2}

$\uparrow\downarrow$ d_{xy} $\uparrow$ d_{yz} $\uparrow$ d_{xz}

$[Cr(CN)_6]^{4-}$

(c)

$\uparrow$ $d_{x^2-y^2}$ $\uparrow$ d_{z^2}

$\uparrow\downarrow$ d_{xy} $\uparrow$ d_{yz} $\uparrow$ d_{xz}

$[Fe(H_2O)_6]^{2+}$

or

— $d_{x^2-y^2}$ — d_{z^2}

$\uparrow\downarrow$ d_{xy} $\uparrow\downarrow$ d_{yz} $\uparrow\downarrow$ d_{xz}

$[Co(NH_3)_6]^{3+}$

(d)

$\uparrow$ $d_{x^2-y^2}$ $\uparrow$ d_{z^2}

$\uparrow\downarrow$ d_{xy} $\uparrow\downarrow$ d_{yz} $\uparrow\downarrow$ d_{xz}

$[Ni(H_2O)_6]^{2+}$

19.52 ***See Section 19.6 and Figure 19.26.***

(a)

— $d_{x^2-y^2}$ — d_{z^2}

$\uparrow$ d_{xy} $\uparrow$ d_{yz} $\uparrow$ d_{xz}

$[Cr(H_2O)_6]^{3+}$

(b)

$\uparrow$ $d_{x^2-y^2}$ $\uparrow$ d_{z^2}

$\uparrow$ d_{xy} $\uparrow$ d_{yz} $\uparrow$ d_{xz}

$[Fe(H_2O)_6]^{3+}$

or

— $d_{x^2-y^2}$ — d_{z^2}

$\uparrow\downarrow$ d_{xy} $\uparrow\downarrow$ d_{yz} $\uparrow$ d_{xz}

$[Fe(CN)_6]^{3-}$

(c)

$\uparrow$ $d_{x^2-y^2}$ $\uparrow$ d_{z^2}

$\uparrow\downarrow$ d_{xy} $\uparrow\downarrow$ d_{yz} $\uparrow$ d_{xz}

$[CoF_6]^{4-}$

or

$\uparrow$ $d_{x^2-y^2}$ — d_{z^2}

$\uparrow\downarrow$ d_{xy} $\uparrow\downarrow$ d_{yz} $\uparrow\downarrow$ d_{xz}

$[Co(CN)_6]^{4-}$

(d) ↑↓ ↑
$d_{x^2-y^2}$ d_{z^2} $[Cu(H_2O)_6]^{2+}$

↑↓ ↑↓ ↑↓
d_{xy} d_{yz} d_{xz}

19.53 ***See Section 19.6, Figures 19.30, 19.31, and Example 19.7.***

Complex	Composition	Classification	Number of unpaired electrons
(a) $[CrCl_6]^{3-}$	Cr^{3+} and Cl^-	d^3	3 unpaired electrons
(b) $[Co(CN)_5(H_2O)]^{2-}$	Co^{3+}, CN^-, H_2O	low spin d^6	0 unpaired electrons
(c) $[Mn(H_2O)_6]^{2+}$	Mn^{2+} and H_2O	high spin d^5	5 unpaired electrons
(d) $[Rh(H_2O)_6]^{3+}$	Rh^{3+} and H_2O	low spin d^6	0 unpaired electrons
(e) $[V(H_2O)_6]^{3+}$	V^{3+} and H_2O	d^2	2 unpaired electrons

19.54 ***See Section 19.6, Figures 19.30, 19.31, and Example 19.7.***

Complex	Composition	Classification	Number of unpaired electrons
(a) $[MnF_6]^{3-}$	Mn^{3+} and F^-	high spin d^4	4 unpaired electrons
(b) $[Fe(CN)_6]^{3-}$	Fe^{3+} and CN^-	low spin d^5	1 unpaired electron
(c) $[Re(H_2O)_6]^{2+}$	Re^{2+} and H_2O	low spin d^5	1 unpaired electron
(d) $[Fe(H_2O)_6]^{2+}$	Fe^{2+} and H_2O	high spin d^6	4 unpaired electrons
(e) $[Ni(H_2O)_6]^{2+}$	Ni^{2+} and H_2O	d^8	2 unpaired electrons

19.55 ***See Section 19.6, Figures 19.32, 19.33 and Example 19.8.***

Ni^{2+} has a d^8 electron configuration and is known to form both square-planar and tetrahedral four-coordinate complexes. Comparing the d orbital energy level diagrams for both square-planar and tetrahedral complexes of Ni^{2+} with the experimental observation of two unpaired electrons indicates $NiBr_4^{2-}$ is **tetrahedral**. This could be expected given the large size of the Br^- ligands and the larger bond angles for tetrahedral complexes compared to square-planar complexes.

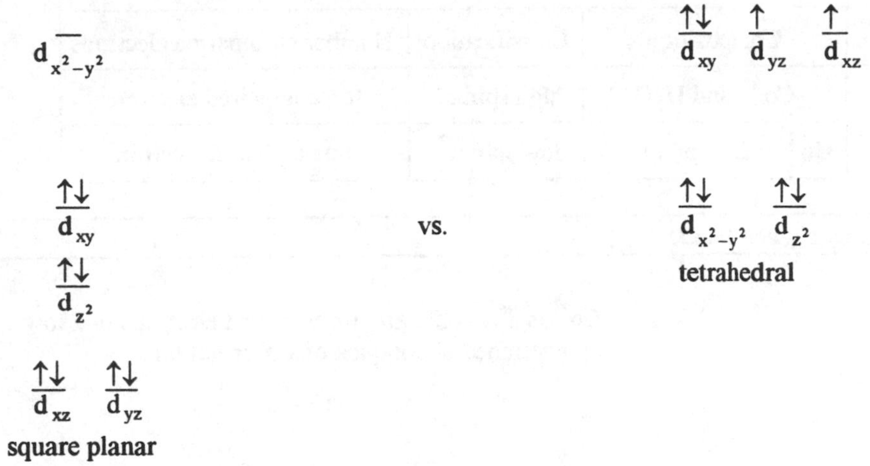

$\overline{}$
$d_{x^2-y^2}$

$\uparrow\downarrow$ $\uparrow$ $\uparrow$
d_{xy} d_{yz} d_{xz}

$\uparrow\downarrow$
d_{xy}

vs.

$\uparrow\downarrow$ $\uparrow\downarrow$
$d_{x^2-y^2}$ d_{z^2}
tetrahedral

$\uparrow\downarrow$
d_{z^2}

$\uparrow\downarrow$ $\uparrow\downarrow$
d_{xz} d_{yz}
square planar

19.56 ***See Section 19.6, Figures 19.32, 19.33, and Example 19.8.***

Pt^{2+} has a d^8 electron configuration. Comparing the d orbital energy level diagrams for both square-planar and tetrahedral complexes of Pt^{2+} with the experimental observation of no unpaired electrons indicates $PtCl_4^{2-}$ is square-planar.

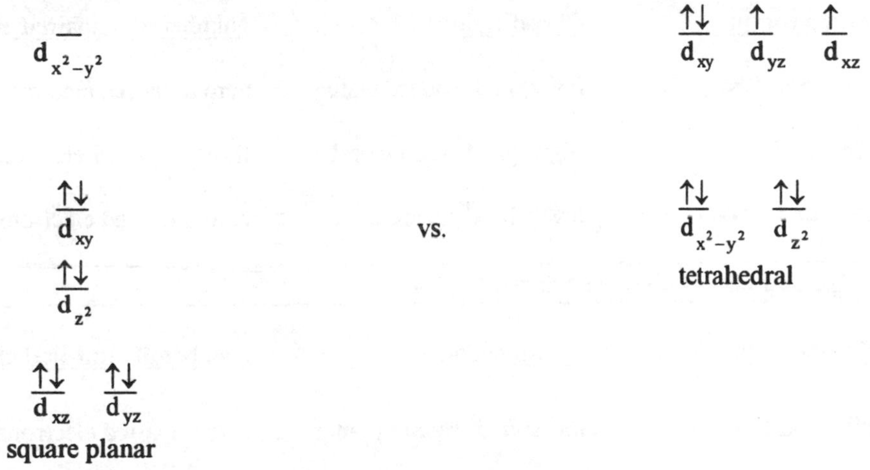

$\overline{}$
$d_{x^2-y^2}$

$\uparrow\downarrow$ $\uparrow$ $\uparrow$
d_{xy} d_{yz} d_{xz}

$\uparrow\downarrow$
d_{xy}

vs.

$\uparrow\downarrow$ $\uparrow\downarrow$
$d_{x^2-y^2}$ d_{z^2}
tetrahedral

$\uparrow\downarrow$
d_{z^2}

$\uparrow\downarrow$ $\uparrow\downarrow$
d_{xz} d_{yz}
square planar

19.57 ***See Section 19.6, Figures 19.30, 19.31, and Example 19.7.***

	Complex	Composition	Classification	Number of unpaired electrons
(a)	$[Cr(H_2O)_6]^{2+}$	Cr^{2+} and H_2O	high spin d^4	four unpaired electrons
	$[Mn(CN)_6]^{3-}$	Mn^{3+} and CN^-	low spin d^4	two unpaired electrons

	Complex	Composition	Classification	Number of unpaired electrons
(b)	$[Fe(H_2O)_6]^{2+}$	Fe^{2+} and H_2O	high spin d^6	four unpaired electrons
	$[Ru(H_2O)_6]^{2+}$	Ru^{2+} and H_2O	low spin d^6	zero unpaired electrons

	Complex	Composition	Classification	Number of unpaired electrons
(c)	$[Co(H_2O)_6]^{2+}$	Co^{2+} and H_2O	high spin d^7	three unpaired electrons
	$[Co(CN)_5(H_2O)]^{3-}$	Co^{2+}, CN^- and H_2O	low spin d^7	one unpaired electron

19.58 ***See Section 19.6 and Figure 19.33.***

↑ ↑ —
d_{xy} d_{yz} d_{xz}

↑↓ ↑↓
$d_{x^2-y^2}$ d_{z^2}

Co^{3+} is d^6. There are two unpaired electrons in a low spin tetrahedral complex of a d^6 metal ion.

19.59 ***See Section 19.6, Figures 19.32, 19.33, and Example 19.8.***

	Complex	Composition	Classification	Number of unpaired electrons
(a)	$[Ar(CN)_4]^-$	Au^{3+} and CN^-	low spin d^8 square planar	zero unpaired electrons
(b)	$[CoCl_4]^{2-}$	Co^{2+} and Cl^-	high spin d^7 tetrahedral	three unpaired electrons
(c)	$[Pd(NH_3)_4]^{2+}$	Pd^{2+} and NH_3	low spin d^8 square planar	zero unpaired electrons

19.60 ***See Section 19.6, Figures 19.32, 19.33, and Example 19.8.***

	Complex	Composition	Classification	Number of unpaired electrons
(a)	$[Ni(CN)_4]^{2-}$	Ni^{2+} and CN^-	low spin d^8 square planar	zero unpaired electrons
(b)	$[FeCl_4]^{2-}$	Fe^{2+} and Cl^-	high spin d^6 tetrahedral	four unpaired electrons
(c)	$[Pd(NH_3)_4]^{2+}$	Pd^{2+} and NH_3	low spin d^8 square planar	zero unpaired electrons

19.61 ***See Section 19.6 and Figure 19.26.***

The first three electrons of the d^5 electron configuration are placed in the lower energy set of orbitals of the octahedral complex the d_{xy}, d_{xz} and d_{yz} set of orbitals. Then there is a choice between placing the next two electrons in the lower energy set of orbitals or the set that is Δ higher in energy, the $d_{x^2-y^2}$ and d_{z^2} of orbitals. The cost of placing the fourth and fifth electrons in the lower energy set of orbitals is 2P because two electron pairs would be formed. On the other hand, the cost of placing these same electrons in each of the higher energy set oribitals is 2Δ. Since P> Δ, 2Δ is < 2P, and the high spin complex with the electrons in the upper set of orbitals in favored energywise.

19.62 ***See Section 19.5 and Figure 19.17.***

For an octahedral arrangement,

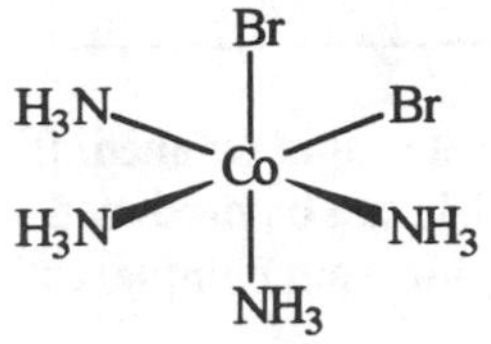

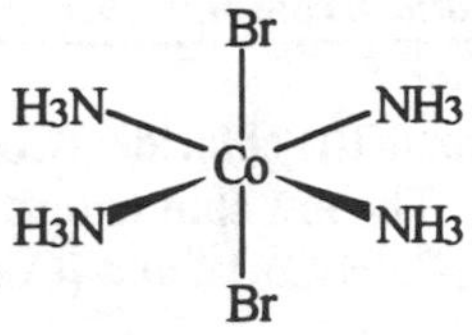

cis isomer

trans isomer

For a trigonal prism arrangement labeled as,

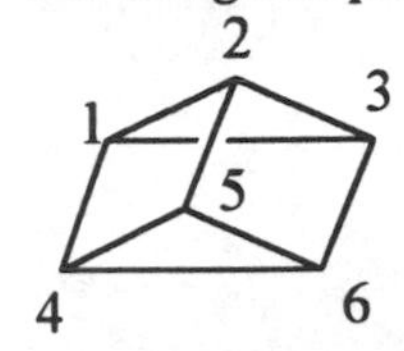

a total of three isomers would be possible with the bromide ions located in the 1,2 or 3; 1,4 and 1,5 or 1,6 positions. Hence, the existence of only two isomers rules out the trigonal prism shape.

19.63 ***See Sections 19.4, 19.5, 19.6, Figures 19.17, 19.31, 19.32, and Examples 19.2, 19.5, 19.7, 19.8.***

$[FeCl_4(H_2O)_2]^{2-}$, diaquatetrachloroferrate(II), is a weak field octahedral complex that is composed of $Fe^{2+}(d^6)$, Cl^- and H_2O and has four unpaired electrons. It can exist as a cis or trans isomer.

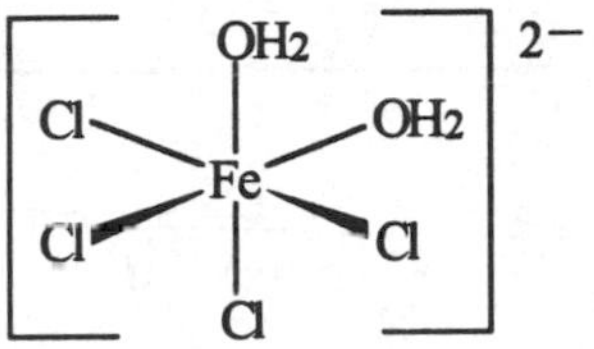

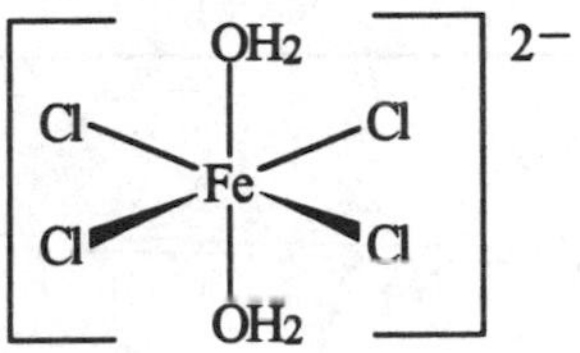

cis isomer

trans isomer

$[FeCl_4]^{2-}$, tetrachloroferrate(II), is a tetrahedral complex that is composed of $Fe^{2+}(d^6)$ and Cl^- and has four unpaired electrons. It cannot exist as isomers because all the ligand positions in a tetrahedral complex are adjacent to one another and all the ligands are Cl^- thereby precluding the possibility of forming optically active isomers.

19.64 ***See Section 19.6 and Figure 19.31.***

$[Fe(H_2O)_6]^{3+} + H_2O \rightleftarrows [Fe(H_2O)_5OH]^{2+} + H_3O^+$

$[Fe(H_2O)_6]^{3+}$ and $[Fe(H_2O)_5OH]^{2+}$ are both weak field complexes of $Fe^{3+}(d^5)$ having five unpaired electrons.

It is possible to predict the number of unpaired electrons on the basis of experimental data, such as that used to determine the spectrochemical series, but it is not possible to determine the actual number of unpaired electrons without conducting an experiment. For example, the placements of H_2O and NH_3 in the spectrochemical series suggest $Co(H_2O)_6^{3+}$ and $Co(NH_3)_6^{3+}$ might be weak field complexes having four unpaired electrons, whereas experiments indicate each is a strong field complex having no unpaired electrons.

19.65 ***See Section 19.6, Figures 19.32, 19.33, and Example 19.8.***

Pd^{2+} has a d^8 electron configuration. Comparing the d orbital energy level diagrams for both square-planar and tetrahedral complexes of Pt^{2+} with the experimental observation of no unpaired electrons indicates $PtCl_4^{2-}$ is

square-planar. Hence, the experimental magnetic information concerning $PdCl_4^{2-}$ is sufficient to determine that the geometry of $PdCl_4^{2-}$ is square planar.

19.66 ***See Sections 19.4, 19.6, Figures 19.12, 19.31 and Examples 19.1, 19.7.***

The compound is $[Co(H_2O)_4Cl_2]Cl$, tetraaquadichlorocobalt(III) chloride. $[Co(H_2O)_4Cl_2]^+$ is a low spin octahedral complex of $Co^{3+}(d^6)$ containing zero unpaired electrons. The low spin nature of $[Co(H_2O)_4Cl_2]^+$ can be predicted on the basis of water being the predominant ligand and $[Co(H_2O)_6]^{3+}$ and $[Co(NH_3)_6]^{3+}$ being low spin complexes.

19.67 ***See Section 19. , Table 19.6, and Example 19.5.***

Ethylenediamine (en) is a symmetrical bidendate liqand which can be abbreviated as N-N.

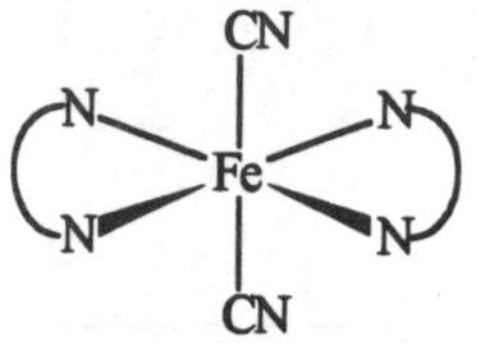

CN N N Fe N N CN

N N CN Fe N CN N   N NC N Fe NC N N

trans isomer 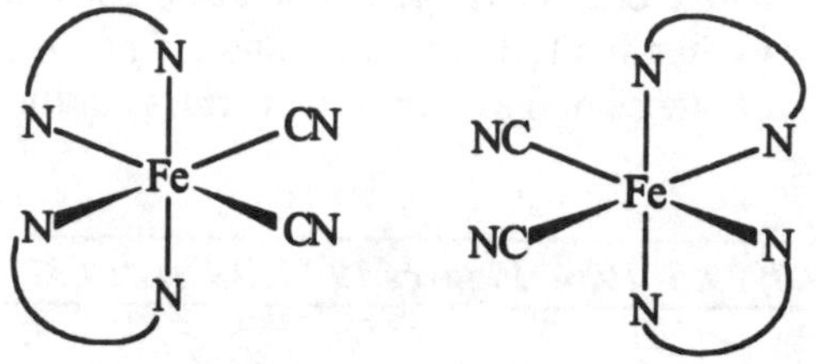   cis enantiomers

19.68 ***See Section 19.1.***

$2ZnS(s) + 3O_2(g) \rightarrow 2ZnO(s) + 2SO_2(g)$

Strategy: gZnS → mol ZnS → mol SO₂ → L SO₂

$$? \text{ L SO}_2 \text{ at STP} = 1000 \text{ g ZnS} \times \frac{1 \text{ mol ZnS}}{97.5 \text{ g ZnS}} \times \frac{2 \text{ mol SO}_2}{2 \text{ mol ZnS}} \times \frac{22.4 \text{ L SO}_2}{1 \text{ mole SO}_2} = \mathbf{230 \text{ L SO}_2}$$

19.69 ***See Section 19.3.***

The standard potential for metallic copper displacing hydrogen ions from acids is negative indicating the dissolving process is nonspontaneous under standard conditions:

$Cu(s) + 2H^+(aq) \rightarrow Cu^{2+}(aq) + H_2(g)$ $E° = -0.34$ V

However, copper dissolves in hot, concentrated sulfuric acid as sulfur is reduced instead of hydrogen. The equation for the reaction is:

$Cu(s) + 2H_2SO_4(aq) \rightarrow CuSO_4(aq) + SO_2(g) + 2H_2O(\ell)$

Strategy: L H_2SO_4 → mol H_2SO_4 → mol Cu → g Cu

$$? \text{ g Cu} = 0.125 \text{ L H}_2\text{SO}_4 \text{ soln} \times \frac{10.4 \text{ mol H}_2\text{SO}_4}{1 \text{ L H}_2\text{SO}_4 \text{ soln}} \times \frac{1 \text{ mol Cu}}{2 \text{ mol H}_2\text{SO}_4} \times \frac{63.5 \text{ g Cu}}{1 \text{ mol Cu}} = \mathbf{41.3 \text{ g Cu}}$$

19.70 ***See Section 19.5.***

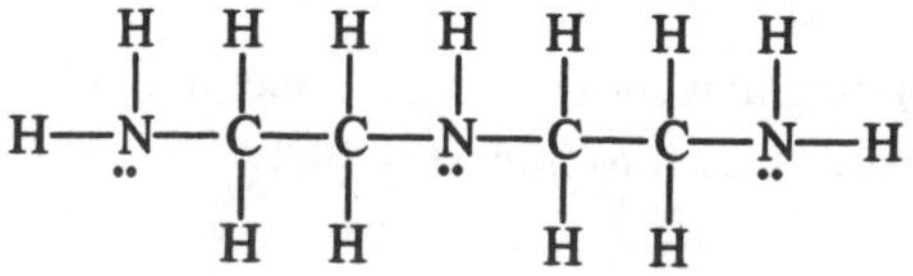

H H H H H H H
H—N—C—C—N—C—C—N—H
H H H H

Using N-N-N to abbreviate dien gives

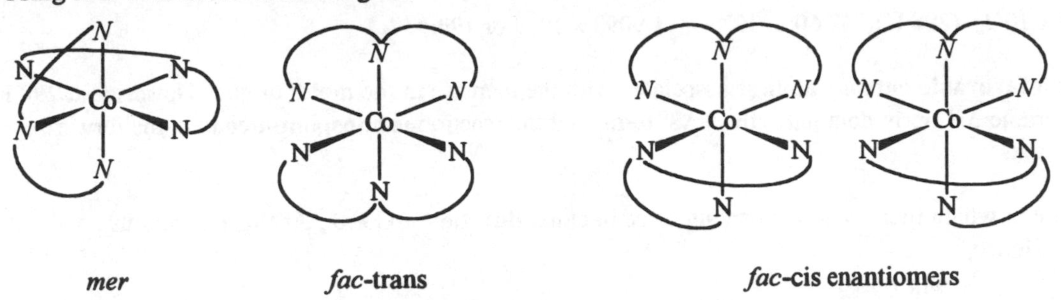

Note: cis and trans refer to the positions of the middle N atoms in dien.

19.71 ***See Sections 3.3, 18.2, and Example 3.11.***

Assume the sample has a mass of 100.0 g and therefore contains 49.75 g V, 15.62 g O and 34.63 g Cl.

$$? \text{ mol V} = 49.75 \text{ g V} \times \frac{1 \text{ mol V}}{50.942 \text{ g V}} = 0.9766 \text{ mol V} \qquad \text{relative mol V} = \frac{0.9766 \text{ mol V}}{0.9763} = 1.000 \text{ mol V}$$

$$? \text{ mol O} = 15.62 \text{ g O} \times \frac{1 \text{ mol O}}{15.999 \text{ g O}} = 0.9763 \text{ mol O} \qquad \text{relative mol O} = \frac{0.9763 \text{ mol O}}{0.9763} = 1.000 \text{ mol O}$$

$$? \text{ mol Cl} = 34.63 \text{ g Cl} \times \frac{1 \text{ mol Cl}}{35.453 \text{ g Cl}} = 0.9768 \text{ mol Cl} \qquad \text{relative mol Cl} = \frac{0.9768 \text{ mol Cl}}{0.9763} = 1.000 \text{ mol Cl}$$

The empirical formula is VOCl, and the low boiling point of the compound suggests it is covalent. The oxidation state of V in VOCl is +3 indicating there are two electrons in the 3*d* orbitals of the V.

19.72 ***See Sections 6.4, 17.4, 18.1, and Examples 6.6, 17.6.***

$MnO_2(s) + 2C(s) \rightarrow Mn(s) + 2CO(g)$

Strategy: L CO → mol CO → mol C → g C

Known Quantities: $V_{CO} = 22.8$ L P = 1.00 atm T = 298 K

Solving PV = nRT for n gives $n = \frac{PV}{RT}$

$$n = \frac{(1.00 \text{ atm})(22.8\text{L})}{\left(0.0821 \frac{\text{L} \cdot \text{atm}}{\text{mol} \cdot \text{K}}\right)(298 \text{ K})} = 0.932 \text{ mol}$$

$$? \text{ g C} = 0.932 \text{ mol CO} \times \frac{2 \text{ mol C}}{2 \text{ mol CO}} \times \frac{12.0 \text{ g C}}{1 \text{ mol C}} = \mathbf{11.2 \text{ g C}}.$$

$$\Delta H^\circ_{rxn} = \left[\Delta H^\circ_f Mn(s) + 2\Delta H^\circ_f CO(g)\right] - \left[\Delta H^\circ_f MnO_2(s) + 2\Delta H^\circ_f C(s, \text{graphite})\right]$$
$$= \left[(1 \text{ mol} \times 0.00 \text{ kJ} \cdot \text{mol}^{-1}) + (2 \text{ mol} \times -110.52 \text{ kJ} \cdot \text{mol}^{-1})\right]$$
$$- \left[\left(1 \text{ mol} \times -520.03 \text{ kJ} \cdot \text{mol}^{-1}\right) + (2 \text{ mol} \times 0.00 \text{ kJ} \cdot \text{mol}^{-1})\right] = \mathbf{298.99 \text{ kJ}}$$

$$\Delta S^\circ_{rxn} = \left[S^\circ Mn(s) + 2S^\circ CO(g)\right] - \left[S^\circ MnO_2(s) + 2S^\circ C(s, \text{graphite})\right]$$
$$= \left[(1 \text{ mol} \times 32.01 \text{ J} \cdot \text{mol}^{-1} \cdot \text{K}^{-1}) + (2 \text{ mol} \times -197.56 \text{ J} \cdot \text{mol}^{-1} \cdot \text{K}^{-1})\right]$$
$$- \left[\left(1 \text{ mol} \times -53.05 \text{ J} \cdot \text{mol}^{-1} \cdot \text{K}^{-1}\right) + (2 \text{ mol} \times 5.74 \text{ J} \cdot \text{mol}^{-1} \cdot \text{K}^{-1})\right] = \mathbf{362.60 \text{ J} \cdot \text{K}^{-1}}$$

At 298 K, $\Delta G^\circ_{rxn} = \Delta H^\circ_{rxn} - (298\text{ K})\,\Delta S^\circ_{rxn}$.

$\Delta G^\circ_{298} = 298.99 \times 10^3\text{J} - (298\text{ K})(362.60\text{ J}\cdot\text{K}^{-1}) = 1.9093 \times 10^5\text{ J}$ or **190.93 kJ**

The reaction has a **favorable entropy change** associated with the increase in the moles of gas. However, at 298 K the highly unfavorable ΔH° term dominates the $T\Delta S^\circ$ term, and the reaction is **nonspontaneous** in the forward direction.

At the temperature at which there is no net driving force in either direction, $\Delta G = 0$. At this temperature, $0 = \Delta H^\circ - T\Delta S^\circ$. Hence,

$$T = \frac{\Delta H^\circ}{\Delta S^\circ} \qquad T = \frac{298.99\text{ kJ}}{362.60 \times 10^{-3}\text{ kJ}\cdot\text{K}^{-1}} = 825\text{ K}$$

The reaction is nonspontaneous in the direction written at 298 K ($\Delta G^\circ_{298} = +$) and experiences no net driving force in either direction at 825 K ($\Delta G^\circ_{825} = 0$). Above **825 K** the reaction becomes spontaneous in the direction written under standard conditions.

Chapter 20: Hydrogen, Groups IIIA-VIA, and the Noble Gases

20.1 ***Section 20.1.***

Atoms of elements in the second period tend to be much smaller than atoms of elements in the same groups in later periods. Hence, second period elements form sigma bonds which are short enough to also enable formation of strong pi bonds by sideways overlap of *p* orbitals and therefore readily form multiple bonds. However, the larger elements in the same groups form longer sigma bonds and weaker pi bonds. Hence, these elements tend to form just sigma single bonds to acquire a noble gas configuration.

On the other hand, elements in the second period typically do not form compounds in which the Lewis structures would place more than eight electrons around these atoms. This is a consequence of the observation that elements of the second period only have four valence orbitals, the 2*s* and 2*p* orbitals. However, elements from the third period have access to empty d orbitals having the same principal quantum number as the *s* and *p* valence orbitals (e.g. 3*s*, 3*p*, and 3*d* for S and Cl) and can use these *d* orbitals to form electon rich molecules and ions.

20.2 ***See Section 20.1 and Figures 11.17, 20.26, 20.32.***

Atoms of elements in the second period tend to be much smaller than atoms of elements in the third period. Hence, second period elements form sigma bonds that are short enough to enable formation of strong pi bonds by sideways overlap of *p* orbitals, whereas sigma bonds formed by third period elements are too long to enable forming strong pi bonds. This is why carbon can exist in a graphite form (sp^2 with *p-p* pi bonds) and a diamond form (sp^3 and no pi bonds) but silicon only exists in a diamond form; oxygen exits as diatamic molecule with *p-p* pi bonding (Section 10.5) but sulfur exists as S_8 rings in which each S atom forms two sigma bonds; and why nitrogen exists as a diatomic molecule having a triple bond (one sigma and two pi bonds) but phosphorus exists as tetrameric P_4 molecules in which each P forms three sigma bonds.

20.3 ***See Sections 8.3, 8.4, 20.1, and Figures 8.13, 8.14, 9.8.***

Metals generally have lower ionization energies and lower electronegativities than nonmetals.

20.4 ***See Section 20.1.***

Oxygen only has four valence orbitals, the 2*s* and 2*p* orbitals, whereas sulfur has access to empty *d* orbitals that have the same principal quantum as its *s* and *p* valence orbitals, the 3*d* orbitals. Hence sulfur is able form electron-rich compounds but oxygen is not able to do so.

20.5 ***See Section 20.2.***

Hydrogen forms ionic bonds, nonpolar covalent bonds and polar covalent bonds. Hydrogen forms ionic bonds in combination with metals as in KH in which it exists as the hydride ion, H^-. Hydrogen forms nonpolar covalent bonds in combination with itself and phosphorus (EN H = 2.1 and EN P = 2.1) as in H_2 and PH_3. Hydrogen forms polar covalent bonds in combination with nonmetals having electronegativities greater than 2.1 as in NH_3 and HBr.

20.6 *See Section 20.2.*

(a) KH is an ionic compound containing K^+ and H^- ions held together by electrostatic forces of attraction.

(b) HCl is a polar covalent compound having a polar covalent bond between H and Cl.

(c) H_2 is a nonpolar covalent compound having a nonpolar covalent bond between H atoms.

20.7 *See Sections 11.4, 20.2.*

H_2 exists as small nonpolar molecules having weak dispersion forces between molecules and therefore boils at the very low temperature of -253°C.

20.8 *See Section 20.2.*

The three isotopes of hydrogen are ^1H (protium, 99.985% abundant), ^2H (deuterium, 0.015% abundant), and ^3H (tritium, <0.001% abundant).

20.9 *See Section 20.2.*

Hydrogen is combustible, wheras helium is not. In fact the combustion of hydrogen is violently implosive when hydrogen and oxygen are present in stoichiometric or nearly stoichiometric amounts. This is why helium is used today as the gas in blimps.

20.10 *See Section 20.2.*

Unsaturated vegetable oils have more carbon-carbon double bonds and therefore less hydrogen atoms per carbon atom than saturated vegetable oils. Hence, unsaturated vegetable oils exist as oils, and saturated vegetable oils exists as solids.

20.11 *See Section 20.3.*

Elements in Groups IIIA, IVA and VA have ns^2np^1; ns^2np^2, and ns^2np^3 valence electron configurations. The tendency for the lower members of these groups to use only the valence electrons in the *p* orbitals and not the pair of electrons in the *s* orbital for bonding is known as the inert pair effect.

On descending the elements in Group IIIA, there is an increasing tendency to form the +1 oxidation state rather than +3 oxidation state.

20.12 *See Section 20.3.*

Thallium is larger than aluminum and therefore forms weaker bonds than aluminum. This reduces the tendency of thallium to use all of its valence electrons and causes the +1 oxidation state to be more stable for thallium than aluminum.

20.13 *See Sections 1.3, 20.3, and Figure 1.8.*

B is a metalloid, and A1, Ga, In and Tl are metals.

20.14 *See Section 20.3 and Figure 20.7.*

Boron exists in a number of allotropic forms that all contain an isosahedron of boron atoms, which is a regular polyhedron having 20 faces and 12 vertices. The icosahedra are connected differently in the various allotropic forms, but all have extended bonding arrangements between the icosahedra.

20.15 *See Section 20.*

Aluminum forms a thin, strongly adhering coating of Al_2O_3 which protects it from further corrosion.

20.16 *See Section 20.4.*

The elements in Group IVA have four valence electrons and four valence orbitals and can form four covalent bonds without producing formal charges.

20.17 *See Section 20.4.*

Graphite is used to make molds for casting metals, is used as electrodes in many industrial processes, and is the "lead" in lead pencils. It also is used as a lubricant.

20.18 *See Section 20.4.*

Adsorption refers to the process of particles being adsorbed on the surface of a solid. Activated charcoal has high adsorption qualities because it has a large surface area per unit volume.

20.19 *See Section 20.4 and Figure 20.4.*

Silicon is produced by reducing SiO_2 with C in an electric arc furnace. The silicon is then treated with chlorine gas to convert it to volatile silicon tetrachloride. The silicon tetrachloride is then purified by repeated distillations and reduced with magnesium or hydrogen to form purified silicon. Rods of silicon produced in this manner are then purified further by zone refining.

20.20 *See Section 20.4.*

Phosphorus has one more valence shell electron than does silicon. The extra electron enters the conduction band of silicon producing an n-type semiconductor.

20.21 *See Section 20.4.*

The mineral galena is lead sulfide, PbS, and is used for the production of lead. The ore is obtained by first roasted to convert the sulfide to the oxide in reaction with oxygen. This is followed by reduction of the oxide to the metal with carbon and carbon monoxide.

20.22 *See Section 20.5.*

The Haber process involves reacting nitrogen gas and hydrogen gas at high temperatures and pressures to form ammonia, NH_3. High temperatures are needed to cause the reaction to proceed at an acceptable rate because N_2 and H_2 have strong bonds that cause the activation energy of the reaction to be high. However, the reaction is also exothermic, so high temperatures shift the reaction toward the reactant side in accord with the Principle of LeChatelier. High pressures of N_2 and H_2 are needed to restore the yield of NH_3. The principle of LeChatelier predicts that high pressures will favor the formation of NH_3 because four volumes of gasous reactants are converted to two volumes of gaseous product in the Haber process reaction: $N_2(g) + 3H_2(g) \rightarrow 2NH_3(g)$.

20.23 *See Section 20.5.*

The phosphorus minerals found in nature are insoluble in water and must be converted to more soluble compounds by treatment with sulfuric acid or phosphoric acid for use as fertilizers.

20.24 *See Sections 9.6, 10.5, 20.6.*

	Oxygen, O_2	Ozone, O_3
Physical:	light blue as liquid	
	boils at -183°C	boils at -112°C
	pale blue as gas	
	oderless gas	pungent gas
Chemicals:	stable solid and liquid phases	explosively unstable solid and liquid phases
	relatively unreactive gas at room temperature; can however react with most elements, forming basic oxides with metals on left side of periodic table and acidic oxides with nonmetals	very reactive gas at room temperature
		absorbs ultraviolet light readily causing decomposition reaction

20.25 *See Section 20.6.*

Metals on the left side of the periodic table form oxides which are basic. Other metals and metalloids form oxides that can be amphoteric, and nonmetals form oxides that are acidic.

20.26 *See Section 20.7.*

Elements in Group VIIIA have completed valence orbitals and little or no tendency to form chemical bonds. Hence, these elements exist as monatomic gases and are relatively nonreactive.

20.27 *See Section 20.7.*

Neil Bartlett reacted O_2 with PtF_6 to form $[O_2^+][PtF_6^-]$ and noted that the ionization energy of xenon was similar to that of O_2. Hence, he reacted Xe with PtF_6 and formed the compound that was first thought to be $[Xe^+][PtF_6^-]$, the first compound of an "inert gas" element.

20.28 *See Section 20.7.*

The main source of radon in homes is uranium in the soil near the homes. Radon is formed by the radioactive decay of an isotope of uranium and can escape from the soil because it is gaseous.

20.29 *See Section 20.1, 20.4.*

	Element	Classification
(a)	carbon	nonmetal
(b)	tin	metal
(c)	chlorine	nonmetal
(d)	silicon	metalloid

20.30 *See Sections 20.1, 20.3, 20.5.*

	Element	Classification
(a)	gallium	metal
(b)	nitrogen	nonmetal
(c)	arsenic	metalloid
(d)	indium	metal

20.31 *See Sections 20.1, 20.3, 20.4, 20.5.*

Metallic character in these elements increases in the order nitrogen <silicon< gallium. Hence, gallium is most metallic and nitrogen is least metallic. Metallic character increases going down a group and going to the left in the periodic table.

20.32 *See Sections 20.1, 20.4.*

Metallic character in these elements increases in the order silicon <germainium< tin. Hence, tin is most metallic and silicon is least metallic. Metallic character increases going down a group.

20.33 *See Section 20.2.*

$NaH(s) + H_2O(l) \rightarrow NaOH(aq) + H_2(g)$

The Ideal Gas Equation, PV = nRT, can be used to calculate the number of moles of H_2.

Known Quantities: $P = 1.00\text{ atm} \quad V = 1.00\text{ L} \quad T = 25 + 273 = 298\text{ K}$

Solving PV = nRT for n gives $n = \dfrac{PV}{RT}$ $\quad n = \dfrac{(1.00\text{ atm})(1.00\text{ L})}{\left(0.0821\dfrac{\text{L}\cdot\text{atm}}{\text{mol}\cdot\text{K}}\right)(298\text{ K})} = 0.0409\text{ mol H}_2$

Strategy: mol H_2 → mol NaH → g NaH

$$?\text{ g NaH} = 0.0409\text{ mol H}_2 \times \frac{1\text{ mol NaH}}{1\text{ mol H}_2} \times \frac{24.0\text{ g NaH}}{1\text{ mol NaH}} = \mathbf{0.981\text{ g NaH}}$$

20.34 *See Section 20.2.*

$Zn(s) + 2HCl(aq) \rightarrow ZnCl_2(aq) + H_2(g)$

The Ideal Gas Equation, PV = nRT, can be used to calculate the number of moles of H_2 .

Known Quantities: $P = 1.00\text{ atm} \quad V = 1.00\text{ L} \quad T = 25 + 273 = 298\text{ K}$

Solving PV = nRT for n gives $n = \dfrac{PV}{RT}$ $\quad n = \dfrac{(1.00\text{ atm})(1.00\text{ L})}{\left(0.0821\dfrac{\text{L}\cdot\text{atm}}{\text{mol}\cdot\text{K}}\right)(298\text{ K})} = 0.0409\text{ mol H}_2$

Stategy: mol H_2 → mol Zn → gZn

$$?\text{ g Zn} = 0.0409\text{ mol H}_2 \times \frac{1\text{ mol Zn}}{1\text{ mol H}_2} \times \frac{65.4\text{ g Zn}}{1\text{ mol Zn}} = \mathbf{2.67\text{ g Zn}}$$

20.35 See Section 20.2.

Two important industrial preparations of H_2 involve 1) the high temperature reaction of methane and steam, and 2) the reaction of red-hot carbon (from coal) with steam. The equations for these reactions are:
$CH_4(g) + H_2O(g) \rightarrow 3H_2(g) + CO(g)$ and $C(s) + H_2O(g) \rightarrow H_2(g) + CO(g)$.

20.36 See Section 20.2.

The most important industrial use of hydrogen gas is in the synthesis of ammonia by the Haber process. The equation for this reaction is: $N_2(g) + 3H_2(g) \rightarrow 2NH_3(g)$.

20.37 See Section 20.2.

The equation for the water gas shift reaction is: $CO(g) + H_2O(g) \rightarrow H_2(g) + CO_2(g)$.

20.38 See Section 20.2.

The series of equations for converting coal into methanol is:

$$3C(s) + 3H_2O(g) \rightarrow 3CO(g) + 3H_2(g)$$
$$CO(g) + H_2O(g) \rightarrow H_2(g) + CO_2(g)$$
$$2CO(g) + 4H_2(g) \rightarrow 2CH_3OH(\ell)$$

$$3C(s) + 4H_2O(g) \rightarrow 2CH_3OH(\ell) + CO_2(g)$$

20.39 See Section 20.3 and Figure 20.8.

```
H   H   H
 \ / \ /
  B   B
 / \ / \
H   H   H
```

The boron atoms use sp^3 hybrid orbitals. The B-H bonds involving the outside H atoms are formed by the overlap of sp^3 orbitals from B and 1*s* orbitals from H, each containing one electron. The B-H-B bonds involving the center H atoms are formed by sp^3-1*s*-sp^3 overlaps and are three-center two-electron bonds having one electron furnished by one of the B atoms and one electron furnished by the H atom.

20.40 See Section 20.3.

BCl_3

```
Cl—B—Cl
   |
   Cl
```

Total valence electrons = [1x3(B) + 3x7(Cl)] = 24.
Eighteen electrons remain after assigning three single bonds, and eighteen unshared electrons are needed to give each chlorine atom a noble gas configuration (6 for each Cl). Twenty four electrons are used to write the electron deficient Lewis structure.

```
 ..      ..
:Cl—B—Cl:
 ..  |   ..
    :Cl:
     ..
```

The steric number for B in BCl_3 is 3. Hence, the electron-pair arrangement is predicted to be trigonal planar and B is predicted to use sp^2 hybrid orbitals. The B-Cl bonds are formed by overlap of sp^2 hybrids from B and 3*p* orbitals from Cl, each containing one electron. The boron atom is considered to be electron deficient though there is some evidence that there is small extent of interaction between the empty *p* orbital that remains for B and the filled 3*p* orbitals of Cl having the same orientation in space.

20.41 *See Section 20.3.*

A three-center, two-electron bond is formed by the overlap of three orbitals from three atoms (one from each atom) and involves two electrons. It is considered to be a delocalized bond because the resulting electron probability pattern is spread over three atoms.

20.42 *See Section 20.3.*

A normal two-center, two-electron bond is formed by the overlap of two orbitals from two atoms (one from each atom) and involves two electrons. A three-center, two-electron bond is formed by the overlap of three orbitals from three atoms (one from each atom) and also involves two electrons.

20.43 *See Section 20.3.*

$2B_4H_{10}(s) + 22O_2(g) \rightarrow 10B_2O_3(s) + 14H_2O(g)$

20.44 *See Section 20.3.*

$BCl_3 + :NH_3 \rightarrow Cl_3B \leftarrow NH_3$

The BCl_3 acts as a Lewis acid and therefore provides and empty orbital to accept the unshared pair of electrons on the nitrogen atom in NH_3. The geometry about the B atom in BCl_3 is trigonal planar and that about the B atom in Cl_3BNH_3 is tetrahedral. The hybridization of the B atom changes from sp^2 to sp^3 and thereby provides an empty orbital that can overlap with the sp^3 orbital of N containing the unshared pair of electrons in NH_3.

20.45 *See Sections 18.11, 20.3.*

The Al_2O_3 ore containing Fe_2O_3 and SiO_2 impurities is treated with strong base, and Al_2O_3 and SiO_2 dissolve leaving Fe_2O_3 to be separated by filtration. The solution containing dissolved Al_2O_3 and SiO_2 is acidified to cause Al_2O_3 to precipitate out and the silicon remains in solution as silicates. The Al_2O_3 is obtained by filtration and subjected to electrolysis as a molten salt containing Na_3AlF_6 to lower the melting point. This use of electical energy to produce aluminum is costly and much less energy is needed to produce aluminum by melting and recycling existing aluminum objects.

20.46 *See Section 20.3 and Figure 20.11.*

The thermite reaction involves the reaction of powdered aluminum with iron(III) oxide and is so exothermic that it produces molten iron: $2Al(s) + Fe_2O_3(s) \rightarrow Al_2O_3(s) + 2Fe(\ell)$. Hence, it can be used to weld iron and steel and was previously used to weld cracked railroad tracks back together.

20.47 *See Section 20.3 and Figure 20.12.*

A ruby is composed of α-alumina and Cr^{3+} impurities that are responsible for the red color.

20.48 *See Section 20.3.*

Impurities in water are adsorbed on the surface of γ-alumina when it is used to purify water by filtration.

20.49 See Section 20.3, Figure 20.13, and Solution for Exercise 20.39.

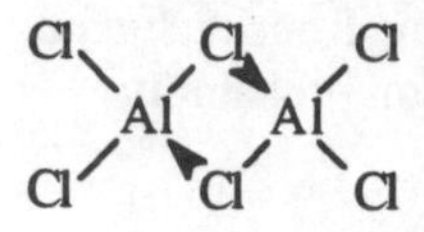

The aluminum atoms use sp^3 hybrids. The Al-Cl bonds involving the outside Cl atoms and the Al-Cl bonds shown on the top left and bottom right in the Al-Cl-Al bridges involve the overlap of sp^3 orbitals from Al and $3p$ orbitals from Al, each containing one electron. The Al-Cl bonds shown as Al ← Cl in the Al-Cl-Al bridges are coordinate covalent bonds involving the overlap of empty sp^3 orbitals from Al with filled $3p$ orbitals from Cl. Hence, each Al-Cl-Al bridge involves one bond in which each atom furnishes one electron and one bond in which Cl furnishes both electrons. The designation of the coordinate covalent bond is arbititrary.

Al_2Cl_6 and B_2H_6 have the same structure. However, the bonding in Al_2Cl_6 involves normal two electron bonds, whereas the bonding in B_2H_6 involves normal two electron bonds between the B and the outside H atoms and three-centered two-electron bonds in the bridges. The difference in bonding is due to chlorine having filled $3p$ orbitals to form coordinate covalent bonds, whereas H does not.

20.50 See Section 20.3.

Using the reactions of Al_2O_3 with acid and base as examples, we can write:

$Ga_2O_3(s) + 6H^+(aq) \rightarrow 2Ga^{3+}(aq) + 3H_2O(\ell)$

$Ga_2O_3(s) + 2OH^-(aq) + 3H_2O(\ell) \rightarrow 2[Ga(OH)_4^-](aq)$

20.51 See Sections 20.1, 20.4.

C uses sp hybridization in CO_2 and forms *p-p* pi bonds with each oxygen atom giving two carbon-oxygen double bonds. Si does not form simple molecules analogous to CO_2 because it is in the third period and is larger. Its sigma bond lengths are therefore longer, and it does not form strong *p-p* pi bonds. Instead it forms four sigma bonds to four different oxygen atoms as shown in Figure 20.1. In this arrangement, Si uses sp^3 hybridization.

20.52 See Sections 20.1, 20.4, and Figure 20.1.

CO_2 exists as linear molecules having carbon-oxygen double bonds, the second bond in each case being a pi bond formed by the sideways overlaps of $2p$ orbitals. However, silicon is from the third period and is much larger than carbon. It's sigma bond lengths are too long to enable effective *p-p* pi overlaps. This causes silicon to use sp^3 hybrids to acquire a noble gas configuration by forming four sigma bonds in SiO_2 as shown in Figure 20.1.

20.53 See Section 20.4.

When silicon is doped with a Group IIIA element, such as boron, it creates "holes" in the lower-energy band. This enables electrons to move by falling into the holes and leaving behind "holes" at their previous locations. Because the entity that appears to move is a positive "hole" (the absence of an electron), this kind of doped silicon is called a *p*-type semiconductor.

20.54 See Section 20.4.

When silicon is doped with a Group V element, such as phosphorus or arsenic, the extra electrons enter the conduction band. Because the doping adds negative charge carriers (electrons), this kind of doped silicon is called a *n*-type semiconductor.

20.55 *See Sections 9.3, 10.1, 10.3, 20.4.*

$SiCl_4$

Cl
Cl—Si—Cl
Cl

:Cl:
:Cl—Si—Cl:
:Cl:

Total valence electrons $= [1 \times 4(\text{Si}) + 4 \times 7(\text{Cl})] = 32$.

Twenty four electrons remain after assigning four single bonds, and twenty four unshared electrons are needed to give each atom a noble gas configuration (6 for each Cl). Twenty four electrons are used in writing the Lewis structure.

The Lewis structure predicts a tetrahedral electron-pair geometry and a tetrahedral molecular geometry. The hybridization of silicon in $SiCl_4$ is therefore predicted to be sp^3. Since $SiCl_4$ is totally symmetrical, it is nonpolar.

20.56 *See Section 20.4 and Figure 20.18.*

Silicon forms four sigma bonds in silicate anions using sp^3 hybrids.

20.57 *See Section 20.5 and Figure 20.26.*

The most common allotropic forms of nitrogen and phosphorus are N_2 and P_4. Nitrogen acquires a noble gas configuration by forming three covalent bonds, one sigma bond and two pi bonds, in forming N_2 molecules. However, phosphorus is much larger and its sigma bond lengths are too long to permit the effective sideways overlap of *p* orbitals to form pi bonds. Hence, each phosphorus atom acquires a noble gas configuration by forming three sigma gonds resulting in P_4 molecules.

20.58 *See Sections 9.3, 20.5, and Figure 20.26.*

P_4

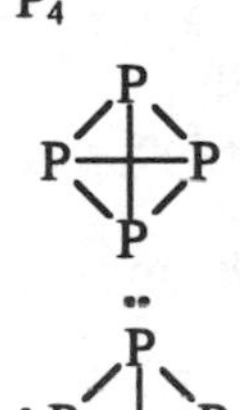

Total valence electrons $= [4 \times 5(\text{P})] = 20$.

Eight electrons remain after assigning six single bonds, and eight unshared electrons are needed to give each atom a noble gas configuration (2 for each P). Twenty electrons are used in writing the Lewis structure.

20.59 *See Sections 20.5.*

Nitrogen comprises 78% by volume of the earth's atmosphere and is isolated by fractional distillation of air. Its boiling point is -196°C compared to -183°C for oxygen, the other main component of the atmosphere, so N_2 boils at a lower temperature than O_2 and can be isolated by boiling off the O_2 and other components of the atmosphere.

20.60 *See Section 20.5.*

The term *fixation* implies that elemental nitrogen is being converted to a useful form, is being *fixed* for useful applications. The *fixation* of nitrogen with hydrogen refers to the combination of nitrogen with hydrogen in the Haber process reaction: $N_2(g) + 3H_2(g) \rightarrow 2NH_3(g)$.

20.61 *See Sections 9.3, 10.1, 10.3, 20.5, and Table 20.1.*

NO_2

O—N—O

Ö=Ṅ—Ö:⊖ (⊕ on N)

Total valence electrons $= [1 \times 5(N) + 2 \times 6(O)] = 17$.
Thirteen electrons remain after assigning two single bonds, and sixteen unshared electrons are needed to give each atom a noble gas configuration (6 for each O and 4 for N). Two electrons are used to form a nitrogen-oxygen double bond, and one unpaired electron is assigned to nitrogen to account for the difference of 3 electrons (16-13). Seventeen electrons are used in writing the Lewis structure.

The Lewis structure supports a trigonal planar electron-pair arrangement for NO_2 and sp^2 hybridization for N in NO_2.

N_2O_3

O—N—N—O (O bonded to right N)

Ö=N̈—N(⊕)=Ö, with :Ö:⊖ bonded to right N

Total valence electrons $= [2 \times 5(N) + 3 \times 6(O)] = 28$.
Twenty electrons remain after assigning four single bonds, and twenty four unshared electrons are needed to give each atom a noble gas configuration (6 for each O, 4 for left N and 2 for right N). Hence, a double bond is formed between each nitrogen atom and an oxygen atom. Twenty eight electrons are used in writing the Lewis structure.

The Lewis structure suggests a trigonal planar electron-pair arrangement about each nitrogen atom in N_2O_3 and sp^2 hybridization for each N atom in N_2O_3.

20.62 *See Sections 9.3, 10.1, 10.3, 20.5, and Table 20.1.*

N_2O

N—N—O

:N≡N(⊕)—Ö:⊖

Total valence electrons = [2x5(N) + 1x6(O)] = 16.
Twelve electrons remain after assigning two single bonds, and sixteen unshared electrons are needed to give each atom a noble gas configuration (6 for left N, 4 for right N, and 6 for O). Hence, four electrons (16-12) must be used to form two additional bonds. The lowest formal charges occur when these additional bonds are formed by the nitrogen atoms. Sixteen electrons are used in writing the Lewis structure.

The Lewis structure suggests a linear electron-pair arrangement about the central nitrogen atom in N_2O and sp hybridization for the central nitrogen atom.

N_2O_4

O—N—N—O (each N bonded to an O)

Ö=N(⊕)—N(⊕)=Ö, with :Ö:⊖ bonded to each N

Total valence electrons = [2x5(N) + 4x6(O)[= 34.
Twenty four electrons remain after forming five single bonds, and twenty eight unshared electrons are needed to give each atom a noble gas configuration (2 for each N and 6 for each O). Hence, four electrons (28-24) must be used to form two additional bonds. Thirty four electrons are used in writing the Lewis structure.

The Lewis structure suggests a trigonal planar electron-pair arrangement about each nitrogen in N_2O_4 and sp^2 hybridization for each nitrogen atom.

20.63 *See Section 20.5.*

(a) $3\,Mg(s) + N_2(g) \rightarrow Mg_3N_2(s)$
(b) $P_4(s) + 5O_2(g) \rightarrow P_4O_{10}(s)$
(c) $2NO_2(g) + H_2O(\ell) \rightarrow HNO_2(aq) + HNO_3(aq)$

20.64 *See Section 20.5.*

(a) $NH_4NO_3(s) \rightarrow N_2O(g) + 2H_2O(g)$
(b) $N_2H_4(aq) + O_2(aq) \rightarrow N_2(g) + 2H_2O(\ell)$
(c) $P_4O_{10}(s) + 6H_2O(\ell) \rightarrow 4H_3PO_4(aq)$

20.65 *See Section 20.5 and Solutions for Exercises 20.61, 20.62.*

In the gas phase NO_2 and N_2O_4 co-exist in a state of equilibrium. However, in the solid phase only N_2O_4 exists. The structures for these molecules are:

$O{=}\overset{\oplus}{N}{-}O^{\ominus}$ and $O{=}\overset{\oplus}{N}(O^{\ominus}){-}\overset{\oplus}{N}(O^{\ominus}){=}O$

20.66 *See Section 20.5.*

$NO(g) + O_3(g) \rightarrow NO_2(g) + O_2(g)$
$NO_2(g) + O_3(g) \rightarrow NO(g) + 2O_2(g)$

20.67 *See Section 20.5.*

The chemical equations for the production of nitric acid via the Ostwald process are:

$4NH_3(g) + 5O_2(g) \xrightarrow{Pt} 4NO(g) + 6H_2O(g)$
$2NO(g) + O_2(g) \rightarrow 2NO_2(g)$
$3NO_2(g) + H_2O(\ell) \rightarrow 2HNO_3(aq) + NO(g)$

20.68 *See Section 20.5.*

$NH_3(g) + HNO_3(aq) \rightarrow NH_4NO_3(aq)$

20.69 *See Section 20.6.*

All elements except helium, neon, argon, and possibly krypton form compounds with oxygen.

20.70 *See Section 20.6 and Figures 20.32, 20.33.*

Orthorhombic sulfur consists of eight-membered rings of sulfur atoms. Plastic sulfur consists of long chains of sulfur atoms.

20.71 ***See Section 20.6.***

Sulfuric acid is produced commercially via

(1) burning sulfur to form SO_2,
(2) reacting SO_2 with O_2 in the presence of V_2O_5 to form SO_3,
(3) trapping the SO_3 in concentrated H_2SO_4 to form $H_2S_2O_7$, and
(4) adding water to the $H_2S_2O_7$ to convert it to a solution that is 96 to 98% H_2SO_4 by weight.

This procedure is used because the direct reaction of SO_3 with H_2O to form H_2SO_4 generates so much heat that it forms a mist of H_2SO_4.

20.72 ***See Section 20.6.***

Sulfuric acid reacts as a strong acid, a dehydrating agent and a mild oxidizing agent.

20.73 ***See Sections 9.3, 10.1, 20.7.***

XeF_2

F—Xe—F

:F̤̈—Ẍ̤e—F̤̈:

Total valence electrons $= [1 \times 8(\mathrm{Xe}) + 2 \times 7(\mathrm{F})] = 22$.

Eighteen electrons remain after assigning two single bonds, and sixteen unshared electrons are needed to give each atom a noble gas configuration (6 for each F and 4 for Xe). The remaining two electrons (18-16) are assigned to valence shell of Xe giving Xe an expanded valence shell. Twenty two electrons are used in writing the Lewis structure.

The Lewis structure suggests a trigonal bipyramidal electron-pair arrangement and a symmetrical linear molecular geometry for XeF_2. Hence, XeF_2 is nonpolar.

XeO_3

O—Xe—O
|
O

Ö=Ẍe=Ö
‖
:O:

Total valence electrons $= [1 \times 8(\mathrm{Xe}) + 3 \times 6(\mathrm{O})] = 26$.

Twenty electrons remain after assigning three single bonds, and twenty unshared electrons are needed to give each atom a noble gas configuration (6 for each O and 2 for Xe). However, twenty six electrons also can be used to draw a Lewis structure which shows an expanded valence shell for Xe and has formal charges of zero for all atoms. Twenty six electrons are used in writing this Lewis structure.

The Lewis structure predicts a tetrahedral electron-pair geometry and an trigonal pyramidal molecular geometry for XeO_3. Hence, XeO_3 is polar.

20.74 *See Sections 9.3, 10.1, 20.7.*

XeF_4

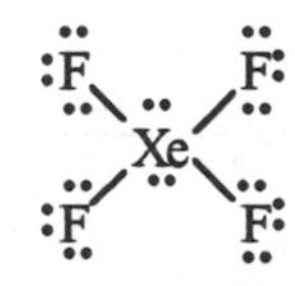

Total valence electrons = [1x8(Xe) + 4x7(F)] = 36.

Twenty eight electrons remain after assigning four single bonds and twenty four electrons are needed to give each atom a noble gas configuration (6 for each F). The remaining four electrons (28-24) are assigned to the valence shell of Xe giving Xe an expanded valence shell. Thirty six electrons are used in writing the Lewis structure.

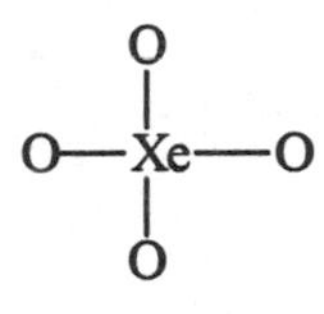

The Lewis structure suggests an octahedral electron-pair arrangement and a symmetrical square planar molecular geometry for XeF_4. Hence, XeF_4 is nonpolar.

XeO_4

Total valence electrons = [1x8(Xe) + 4x6(O)] = 32.

Twenty four electrons remain after assigning four single bands, and twenty four electrons are needed to give each atom a noble gas configuration (6 for each O). However, thirty two electrons can also be used to draw a Lewis structure which shows an expanded valence shell for Xe and has formal charges of zero for all atoms. Thirty two electrons are used in writing this Lewis structure.

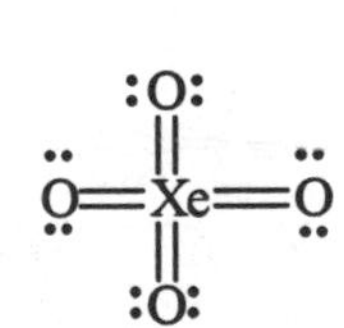

The Lewis structure suggests a tetrahedral electron-pair arrangement and a symmetrical tetrahedral molecular geometry for XeO_4. Hence, XeO_4 is nonpolar.

20.75 *See Section 20.5.*

$$4NH_3(g) + 5O_2(g) \xrightarrow{Pt} 4NO(g) + 6H_2O(g)$$

Strategy: $g\ NO \rightarrow mol\ NO \rightarrow mol\ NH_3 \rightarrow g\ NH_3$

$$?\ g\ NH_3 = 25 \times 10^3\ g\ NO \times \frac{1\ mol\ NO}{30.0\ g\ NO} \times \frac{4\ mol\ NH_3}{4\ mol\ NO} \times \frac{17.0\ g\ NH_3}{1\ mol\ NH_3} = \mathbf{1.4 \times 10^4\ g\ NH_3} \text{ or } \mathbf{14\ kg\ NH_3}$$

20.76 *See Section 20.5.*

$NH_3(g) + HNO_3(aq) \rightarrow NH_4NO_3(aq)$

Strategy: $g\ NH_4NO_3 \rightarrow mol\ NH_4NO_3 \rightarrow mol\ NH_3 \rightarrow g\ NH_3$

$$?\ g\ NH_3 = 5.22 \times 10^3\ g\ NH_4NO_3 \times \frac{1\ mol\ NH_4NO_3}{80.0\ g\ NH_4NO_3} \times \frac{1\ mol\ NH_3}{1\ mol\ NH_4NO_3} \times \frac{17.0\ g\ NH_3}{1\ mol\ NH_3}$$

$$= \mathbf{1.11 \times 10^3\ g\ NH_3} \text{ or } \mathbf{1.11\ kg\ NH_3}$$

20.77 See Sections 11.4, 11.5, and Figure 9.9.

The boiling point of NH_3 is -33°C and that of PH_3 is -88°C. The boiling point of NH_3 is 55°C higher than that of PH_3 because there is hydrogen bonding between NH_3 molecules and no hydrogen bonding between PH_3 molecules. This is due to the greater electronegativity and much smaller size of N compared to P.

20.78 See Section 20.5.

According to valence bond theory, the nearly 90° H-P-H bond angles suggest the P-H bonds are formed from almost pure 3*p* orbitals from phosphorus, leaving the 3*s* orbital to accommodate the lone pair of electrons.

20.79 See Section 20.6.

$$2KClO_3(s) \xrightarrow[150°]{MnO_2} 2KCl(s) + 3O_2(g)$$

Strategy: $PV = nRT \rightarrow mol\ O_2 \rightarrow mol\ KClO_3 \rightarrow g\ KClO_3$

Known Quantities: $P = 755\ \text{torr} \times \frac{1\ \text{atm}}{760\ \text{torr}} = 0.993\ \text{atm}$ $V = 0.50\ \text{L}$ $T = 27 + 273 = 300\ \text{K}$

Solving PV = nRT for n gives $n = \frac{PV}{RT}$ $n = \frac{(0.993\ \text{atm})(0.50\ \text{L})}{\left(0.0821 \frac{\text{L}\cdot\text{atm}}{\text{mol}\cdot\text{K}}\right)(300\ \text{K})} = 0.020\ \text{mol}\ O_2$

$$?\ g\ KClO_3 = 0.020\ \text{mol}\ O_2 \times \frac{2\ \text{mol}\ KClO_3}{3\ \text{mol}\ O_2} \times \frac{122.6\ g\ KClO_3}{1\ \text{mol}\ KClO_3} = \mathbf{1.6\ g\ KClO_3}$$

20.80 See Section 20.6.

$2PbS(s) + 3O_2(g) \rightarrow 2PbO(s) + 2SO_2(g)$

Strategy: $g\ PbS \rightarrow mol\ PbS \rightarrow mol\ SO_2 \rightarrow L\ SO_2$ at STP

$$?\ L\ SO_2 \text{ at STP} = 1.0 \times 10^2\ g\ PbS \times \frac{1\ \text{mol PbS}}{239.3\ g\ PbS} \times \frac{1\ \text{mol}\ SO_2}{1\ \text{mol PbS}} \times \frac{22.4\ L\ SO_2}{1\ \text{mol}\ SO_2} = \mathbf{9.4\ L\ SO_2}$$

20.81 See Section 20.5

$N_2H_4(aq) + O_2(aq) \rightarrow N_2(g) + 2H_2O(\ell)$

Strategy: tons $H_2O \rightarrow gO_2 \rightarrow mol\ O_2 \rightarrow mol\ N_2H_4 \rightarrow g\ N_2H_4$

$$?\ g\ N_2H_4 = 10\ \text{ton}\ H_2O \times \frac{2{,}000\ \text{lb}\ H_2O}{1\ \text{ton}\ H_2O} \times \frac{453.6\ g\ H_2O}{1\ \text{lb}\ H_2O} \times \frac{2.0 \times 10^{-8}\ g\ O_2}{1\ g\ H_2O}$$
$$\times \frac{1\ \text{mol}\ O_2}{32.00\ g\ O_2} \times \frac{1\ \text{mol}\ N_2H_4}{1\ \text{mol}\ O_2} \times \frac{32.06\ g\ N_2H_4}{1\ \text{mol}\ N_2H_4} = \mathbf{0.18\ g\ N_2H_4}$$

20.82 ***See Sections 11.4, 11.6, 20.4, 20.5.***

(a) Pb, metallic bonds

(b) SiO_2, covalent bonds

(c) P_4, London dispersion forces

(d) buckyballs, London dispersion forces

20.83 ***See Sections 9.7, 20.7, and Examples 9.13, 9.14.***

XeO_2F_4

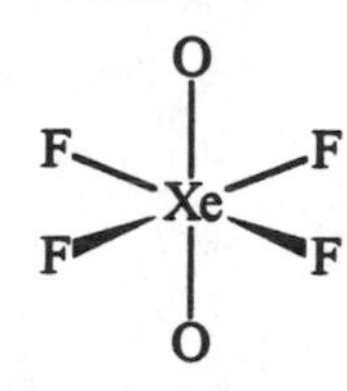

Total valence electrons = $[1 \times 8(\text{Xe}) + 2 \times 6(\text{O}) + 4 \times 7(\text{F})] = 48.$
Thirty six electrons remain after assigning six single bonds, and thirty six unshared electrons are needed to give each atom a noble gas configuration (6 for each O and 6 for each F). Forty eight electrons are used in writing the **electron rich** Lewis structure that leads to the lowest formal charges..

SN = 6 suggests an octahedral electron-pair arrangement. Since there are no lone paires of electrons about the Xe atom, the molecular geometry is also **octahedral**.

20.84 ***See Sections 9.6, 9.7, 20.5, 20.6.***

The possible resonance structures for HNO_3 are:

A

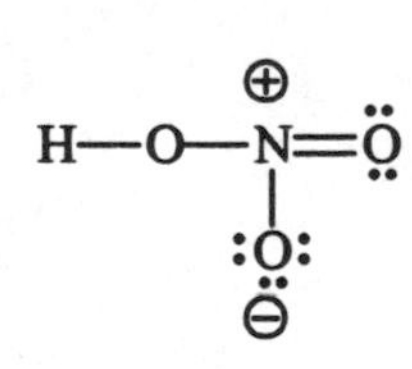

B

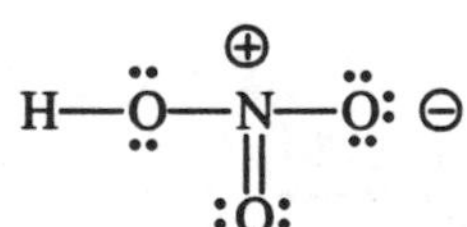

C

However, only A and B are considered to be important and an average of A and B describes the bonding in HNO_3.

The possible resonance forms for H_2SO_4 are:

A

B

However, resonance structure B is considered to more important because formal charges should be minimized.

H_2SO_4 is able to exist as an electron-rich species with reduced formal charges because sulfur is able to use its 3d orbitals with its 3s and 3p orbitals to accomodate more than eight electrons in its valence shell. HNO_3 is not able to do likewise because there are no valence shell d orbitals for elements of the second period (ℓ cannot equal 2 when $n = 2$; ℓ can only equal 0 and 1).

20.85 *See Sections 9.7, 10.5, 20.5, and Figure 10.48.*

According to the Lewis structure, NO is an odd-electron molecule having a bond order of 2.

$\cdot\ddot{N}{=}\ddot{\underset{\cdot\cdot}{O}}$

According to molecular orbital theory, NO is an odd-electron olecule having a bond order of 2.5.

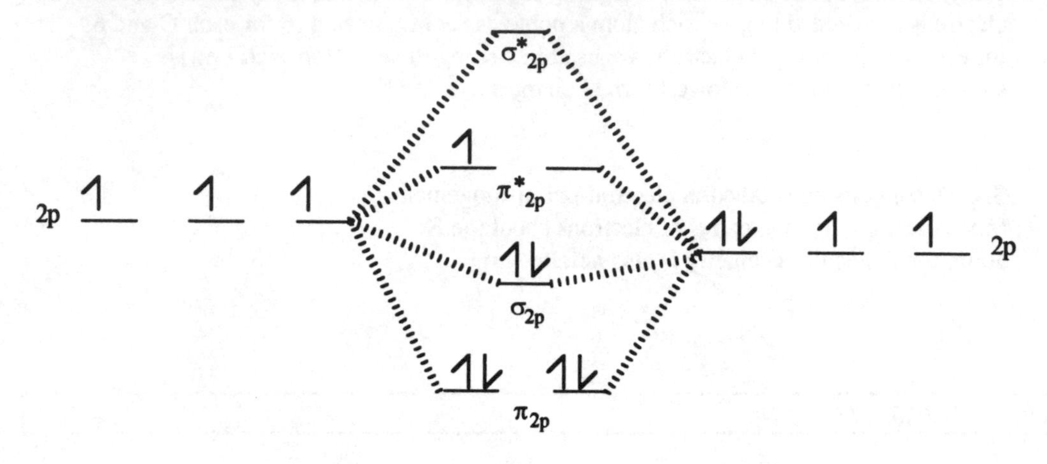

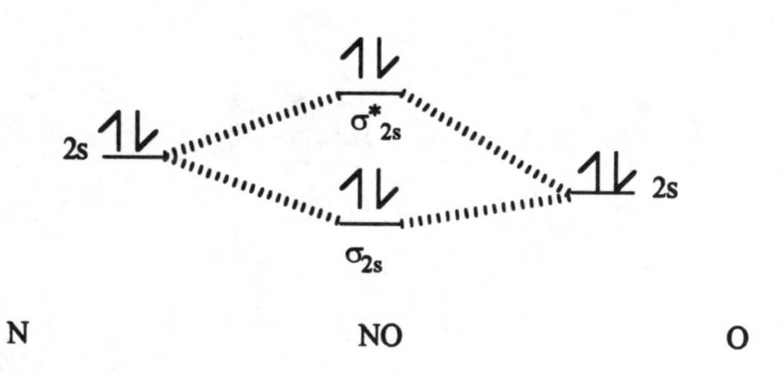

N NO O

The bond order for NO = $\frac{1}{2}[8 - 3] = 2.5$.

Molecular orbital theory gives a more accurate description of the bonding in NO because experiment indicates it has a bond strength that is more compatible with a bond order of 2.5 than 2.0.

Chapter 21: Nuclear Chemistry

21.1 *See Section 21.1.*

Beta emission has the net effect of converting a neutron in an original nucleus into a proton in the product nucleus via ${}^{1}_{0}n \rightarrow {}^{1}_{1}p + {}^{0}_{-1}\beta$ and thereby lowers the neutron-to-proton ratio in the nucleus. This produces a nuclide that is closer to the band of stability.

21.2 *See Section 21.1.*

(a) Beta emission has the net effect of converting a neutron in an original nucleus into a proton in the product nucleus via ${}^{1}_{0}n \rightarrow {}^{1}_{1}p + {}^{0}_{-1}\beta$ and thereby **decreases** the neutron-to-proton ratio in the nucleus.

(b) Positron emission has the net effect of converting a proton in an original nucleus into a neutron in a product nucleus via ${}^{1}_{1}p \rightarrow {}^{1}_{0}n + {}^{0}_{1}\beta$ and thereby **increases** the neutron-to-proton ratio in the nucleus.

(c) Gamma rays merely carry away energy released as nucleons return to the ground-state nuclear configuration. Hence, gamma emission **does not change** the neutron-to-proton ratio in the nucleus.

(d) Alpha emission decreases the number of neutrons in the nucleus by two and the number of protons in the nucleus by two. Since alpha emission usually occurs with atoms having more neutrons than protons, alpha emission **increases** the neutron-to-proton ratio in the nucleus.

21.3 *See Section 21.1.*

Nuclei which have too small a neutron-to-proton ratio to be stable can undergo **elecron capture or positron emission** to move toward the band of stability. The net effect of each of these processes is the conversion of a proton to a neutron.

21.4 *See Section 21.1.*

The electomagnetic radiation that is released with electron capture is called x-rays because it arises from electronic transitions that occur *outside* the nucleus. Gamma rays carry away energy that is released as nuclear particles (nucleons) return to the ground state *within* the nucleus.

21.5 *See Section 21.1.*

When a radioactive nuclide decomposes by alpha or beta particle emission, some of the nucleons are left in an excited nuclear state. Gamma rays carry away the energy that is released when the nucleons return to the ground-state nuclear configuration.

21.6 *See Section 21.1.*

All of the methods of detecting radioactive decay depend on the ability of the high-energy particles and radiation produced to cause ionization of atoms and molecules.

21.7 *See Section 21.2.*

The half-life for carbon-14 is 5,730 years, so the amount of ^{14}C remaining after 50 years would not be significantly different from the amount of ^{14}C remaining after 20 years and neither would have an amount of ^{14}C differing significantly from the original amount of ^{14}C. Hence, carbon-14 dating would not be a useful technique to use to determine whether a sample of paper was 20 or 50 years old.

21.8 *See Section 21.3.*

A positively charged bombarding particle must be accelerated to a very high energy to overcome the electrostatic repulsion of the protons in the nucleus of the atom that will be hit.

21.9 *See Section 21.3.*

Neutrons are electrically neutral particles. When neutrons are used as bombarding particles there is no electrostatic repulsion barrier, and nuclear transmutations can occur at much lower energies. Hence, virtually all nuclides react with low-energy neutrons.

21.10 *See Sections 21.1, 21.5.*

The neutron-to-proton ratios of fission products are too high, so nearly all of them are radioactive, decaying by a series of beta emissions to approach the band of stability shown in Figure 21.1.

21.11 *See Section 21.5.*

A moderator, such as heavy water, is placed between the fuel rods in a nuclear reactor to slow the high-energy neutrons that are produced and enable them to be captured more efficiently by fissionable nuclei in the fuel rods. On the other hand, control rods contain a neutron absorbing material like boron or cadmium and can be inserted into the core to absorb some of the neutrons to keep the reaction from becoming supercritical.

21.12 *See Section 21.6.*

We could make the acetic acid using ^{18}O as the oxygen atom that is located between the carbon atom and the H atom in the C-O-H connectivity and determine whether it remains and is the oxygen atom in the C-O-CH_3 connectivity in the ester. We also could make the methyl alcohol using ^{18}O and determine whether it becomes the oxygen atom in the C-O-CH_3 connectivity in the ester. Hence, either approach could be used to gather conclusive evidence for the mechanism of the reaction.

21.13 *See Section 21.7.*

Alpha particles are relatively heavy particles that have very low penetrating power and are generally stopped by the skin. Beta particles have greater penetrating power than alpha particles and are generally stopped within 1 cm below the skin. Gamma radiation has the greatest penetrating power and can penetrate 30 cm or more below the skin. However, the quality factor Q that measures biological effects is about 10 times greater for alpha particles.

21.14 *See Section 21.7.*

People working in nuclear medicine and nuclear power plants near the proximity of nuclear processes and those mining and processing uranium are likely to receive a greater exposure to radiation than the average exposure of the U.S. population. Pilots and flight attendants on commercial airplanes also are likely to receive greater than average exposure to radiation because the presence of cosmic radiation is greater at higher elevations than it is at sea level. Indeed, anyone working in cities located at higher elevations, such as Denver, Colorado at 5,260 ft. and Flagstaff, Arizona at 7,000 ft., are likely to receive greater than average exposures to radiation for the same reason.

21.15 *See Introduction to Chapter 21 and Section 21.1.*

Symbol	Z	A	number of protons	number of neutrons
$^{23}_{11}Na$	11	23	11	12
$^{103}_{45}Rh$	45	103	45	58
$^{70}_{32}Ge$	32	70	32	38
$^{234}_{90}Th$	90	234	90	144

21.16 *See Introduction to Chapter 21 and Section 21.1.*

Symbol	Z	A	number of protons	number of neutrons
$^{40}_{20}Ca$	20	40	20	20
$^{31}_{15}P$	15	31	15	16
$^{118}_{50}Sn$	50	118	50	68
$^{239}_{93}Np$	93	239	93	146

21.17 *See Introduction to Chapter 21 and Section 21.1.*

	Symbol	number of protons	number of neutrons	n/p ratio
(a)	$^{12}_{6}C$	6	6	1.00
(b)	$^{40}_{20}Ca$	20	20	1.00
(c)	$^{90}_{40}Zr$	40	50	1.25
(d)	$^{138}_{56}Ba$	56	82	1.46
(e)	$^{208}_{82}Pb$	82	126	1.54

21.18 *See Introduction to Chapter 21 and Section 21.1.*

	Symbol	number of protons	number of neutrons	n/p ratio
(a)	$^{17}_{8}O$	8	9	1.12
(b)	$^{40}_{19}K$	19	18	1.05
(c)	$^{75}_{33}As$	33	42	1.27
(d)	$^{118}_{50}Sn$	50	68	1.36
(e)	$^{196}_{78}Pt$	78	118	1.51

21.19 *See Section 21.1 and Figure 21.1.*

	Symbol	number of protons	number of neutrons	n/p ratio	Comment
(a)	$^{17}_{8}O$	8	9	1.13	Within band of stability.
(b)	$^{93}_{38}Sr$	38	55	1.45	Above band of stability.
(c)	$^{67}_{30}Zn$	30	37	1.23	Within band of stability.
(d)	$^{233}_{92}U$	92	141	1.53	Beyond band of stability.
(e)	$^{28}_{12}Mg$	12	16	1.33	Above band of stability.

21.20 See Section 21.1 and Figure 21.1.

	Symbol	number of protons	number of neutrons	n/p ratio	Comment
(a)	$^{16}_{8}O$	8	8	1.00	Within band of stability.
(b)	$^{37}_{17}Cl$	17	20	1.18	Within band of stability.
(c)	$^{68}_{29}Cu$	29	39	1.34	Above band of stability.
(d)	$^{239}_{94}Pu$	94	145	1.54	Beyond band of stability.
(e)	$^{88}_{41}Nb$	41	47	1.15	Below band of stability.

21.21 See Section 21.1 and Example 21.1.

(a) $^{201}_{83}Bi \rightarrow ^{197}_{81}Tl + ^{4}_{2}He$ (b) $^{184}_{77}Ir \rightarrow ^{184}_{76}Os + ^{0}_{1}\beta$ (c) $^{135}_{57}La + ^{0}_{-1}e \rightarrow ^{135}_{56}Ba$
(d) $^{80}_{35}Br \rightarrow ^{80}_{36}Kr + ^{0}_{-1}\beta$

21.22 See Section 21.1 and Example 21.1.

(a) $^{14}_{6}C \rightarrow ^{14}_{7}N + ^{0}_{-1}\beta$ (b) $^{234}_{90}Th \rightarrow ^{230}_{88}Ra + ^{4}_{2}He$ (c) $^{40}_{19}K \rightarrow ^{40}_{18}Ar + ^{0}_{+1}\beta$
(d) $^{198}_{82}Pb + ^{0}_{-1}e \rightarrow ^{198}_{81}Tl$

21.23 See Section 21.1 and Example 21.1.

(a) $^{227}_{90}Th \rightarrow ^{223}_{88}Ra + ^{4}_{2}He$ (b) $^{22}_{11}Na + ^{0}_{-1}e \rightarrow ^{22}_{10}Ne$ (c) $^{223}_{90}Th \rightarrow ^{219}_{88}Ra + ^{4}_{2}He$
(d) $^{72}_{31}Ga \rightarrow ^{72}_{32}Ge + ^{0}_{-1}\beta$ (e) $^{60}_{29}Cu \rightarrow ^{60}_{28}Ni + ^{0}_{1}\beta$

21.24 See Section 21.1 and Example 21.1.

(a) $^{67}_{31}Ga + ^{0}_{-1}e \rightarrow ^{67}_{30}Zn$ (b) $^{215}_{87}Fr \rightarrow ^{211}_{85}At + ^{4}_{2}He$ (c) $^{67}_{29}Cu \rightarrow ^{67}_{30}Zn + ^{0}_{-1}\beta$
(d) $^{124}_{54}Xe \rightarrow ^{124}_{53}I + ^{0}_{1}\beta$ (e) $^{227}_{90}Th \rightarrow ^{223}_{88}Ra + ^{4}_{2}He$

21.25 See Section 21.1.

(a) Beta emission is characteristic of nuclei which are above the band of stability.
Hence, $^{32}_{14}Si$ is above the band of stability.

(b) Electron capture is characteristic of nuclei which are below the band of stability.
Hence, $^{44}_{22}Ti$ is below the band of stability.

(c) Positron emission is characteristic of nuclei which are below the band of stability.
Hence, $^{52}_{25}Mn$ is below the band of stability.

21.26 See Section 21.1.

(a) Beta emission is characteristic of nuclei which are above the band of stability.
Hence, $^{116}_{47}Ag$ is above the band of stability.

(b) Electron capture is characteristic of nuclei which are below the band of stability.
Hence, $^{86}_{40}Zr$ is below the band of stability.

(c) Positron emission is characteristic of nuclei which are below the band of stability. Hence, $^{70}_{34}Se$ is below the band of stability.

21.27 ***See Section 21.1 and Examples 21.1, 21.2.***

(a) The mass number of ^{76}Br is less than the atomic mass of Br (79.9), so (A - atomic mass) is negative and positron emission or electron capture is predicted for ^{76}Br. $^{76}_{35}Br \rightarrow ^{76}_{34}Se + ^{0}_{1}\beta$ $^{76}_{35}Br + ^{0}_{-1}e \rightarrow ^{76}_{34}Se$

(b) The mass number of ^{84}Br is greater than the atomic mass of Br(79.9), so (A - atomic mass) is positive and beta emission is predicted for ^{84}Br. $^{84}_{35}Br \rightarrow ^{84}_{36}Kr + ^{0}_{-1}\beta$

(c) The mass number of ^{109}Pd is greater than the atomic mass of Pd(106.4), so (A - atomic mass) is positive and beta emission is predicted for ^{109}Pd. $^{109}_{46}Pd \rightarrow ^{109}_{47}Ag + ^{0}_{-1}\beta$

(d) The atomic number of Am is greater than 92, so alpha particle emission is predicted for $^{241}_{95}Am$. $^{241}_{95}Am \rightarrow ^{237}_{93}Np + ^{4}_{2}\alpha$

21.28 ***See Section 21.1 and Examples 21.1, 21.2.***

(a) The mass number of ^{117}Sb is less than the atomic mass of Sb(121.75), so (A - atomic mass) is negative and positron emission or electron capture is preducted for ^{117}Sb. $^{117}_{51}Sb \rightarrow ^{117}_{50}Sn + ^{0}_{1}\beta$ $^{117}_{51}Sb + ^{0}_{-1}e \rightarrow ^{117}_{50}Sn$

(b) The mass number of ^{83}Se is greater than the atomic mass of Se(78.96), so (A - atomic mass) is positive and beta emission is predicted for ^{83}Se. $^{83}_{34}Se \rightarrow ^{83}_{35}Br + ^{0}_{-1}\beta$

(c) The mass number of ^{221}Ac is less than the atomic mass of Ac(227.03), so (A - atomic mass) is negative and positron emission or electron capture is predicted for ^{221}Ac. $^{221}_{89}Ac \rightarrow ^{221}_{88}Ra + ^{0}_{1}\beta$ $^{221}_{89}Ac + ^{0}_{-1}e \rightarrow ^{221}_{88}Ra$

(d) The mass number of ^{42}Ar is greater than the atomic mass of Ar(39.9), so (A - atomic mass) is positive and beta emission is predicted for ^{42}Ar. $^{42}_{18}Ar \rightarrow ^{42}_{19}K + ^{0}_{-1}\beta$

21.29 ***See Section 21.2 and Example 21.4.***

Solving $\ln\left(\frac{N}{N_0}\right) = \frac{-0.693t}{t_{1/2}}$ for $t_{1/2}$ gives $t_{1/2} = \frac{-0.693t}{\ln\left(\frac{N}{N_0}\right)}$.

Since $\left(\frac{N}{N_0}\right)$ can be replaced by $\left(\frac{R}{R_0}\right)$, $t_{1/2} = \frac{-0.693t}{\ln\left(\frac{R}{R_0}\right)}$ $t_{1/2} = \frac{-0.693(22.5\ hr)}{\ln\left(\frac{350}{1245}\right)} = \mathbf{12.3\ hr}$

21.30 ***See Section 21.2 and Example 21.4.***

Solving $\ln\left(\frac{N}{N_0}\right) = \frac{-0.693t}{t_{1/2}}$ for $t_{1/2}$ gives $t_{1/2} = \frac{-0.693t}{\ln\left(\frac{N}{N_0}\right)}$.

Since $\left(\frac{N}{N_0}\right)$ can be replaced by $\left(\frac{R}{R_0}\right)$, $t_{1/2} = \frac{-0.693t}{\ln\left(\frac{R}{R_0}\right)}$ $\quad t_{1/2} = \frac{-0.693(7.00 \text{ days})}{\ln\left(\frac{650}{865}\right)} = \mathbf{17.0 \text{ days}}$

21.31 ***See Section 21.2 and Example 21.5.***

Solving rate = kN for k gives $k = \frac{\text{rate}}{N}$ $\quad k = \frac{383 \text{ atoms} \cdot \text{min}^{-1}}{3.75 \times 10^{13} \text{ atoms}} = 1.02 \times 10^{-11} \text{ min}^{-1}$

Substituting for k in $t_{1/2} = \frac{0.693}{k}$ gives $\quad t_{1/2} = \frac{0.693}{1.02 \times 10^{-11} \text{ min}^{-1}} = 6.79 \times 10^{10} \text{ min}$

or $\quad 6.79 \times 10^{10} \text{ min} \times \frac{1 \text{ hr}}{60 \text{ min}} \times \frac{1 \text{ day}}{24 \text{ hours}} \times \frac{1 \text{ yr}}{365 \text{ days}} = \mathbf{1.29 \times 10^5 \text{ yr}}$

21.32 ***See Section 21.2 and Example 21.5.***

(a) Solving rate = kN for k gives $k = \frac{\text{rate}}{N}$ $\quad k = \frac{1238 \text{ atoms} \cdot \text{min}^{-1}}{2.64 \times 10^{10} \text{ atoms}} = 4.69 \times 10^{-8} \text{ min}^{-1}$

Substituting for k in $t_{1/2} = \frac{0.693}{k}$ gives $\quad t_{1/2} = \frac{0.693}{4.69 \times 10^{-8} \text{ min}^{-1}} = 1.48 \times 10^{7} \text{ min}$

or $\quad 1.48 \times 10^{7} \text{ min} \times \frac{1 \text{ hr}}{60 \text{ min}} \times \frac{1 \text{ day}}{24 \text{ hours}} \times \frac{1 \text{ yr}}{365 \text{ days}} = \mathbf{28.2 \text{ yrs}}$

(b) Solving $\ln\left(\frac{N}{N_0}\right) = \frac{-0.693t}{t_{1/2}}$ for t and substituting $\left(\frac{R}{R_o}\right)$ for $\left(\frac{N}{N_o}\right)$ gives

$t = \frac{-t_{1/2}}{0.693} \ln\left(\frac{R}{R_0}\right)$ $\quad t = \frac{-(28.2 \text{ yrs})}{0.693} \ln\left(\frac{1000}{1238}\right) = \mathbf{8.69 \text{ yrs}}$

21.33 ***See Section 21.1 and Example 21.3.***

For each alpha ($^4_2\alpha$) particle emitted, the mass number decreases by 4, while a beta decay does not change the mass number. In the $^{239}_{94}Pu$ to $^{207}_{82}Pb$ series, the change in mass number is from 239 to 208 = 32. Hence, a total of eight (32/4) alpha particles must be emitted to balance the change in mass number in this series.

The change in atomic number in this series is from 94 to 82 = 12. This change can be represented as $[(\text{\# alpha particles})(2) + (\text{\# beta particles})(-1)] = 12$ because an alpha particle is $^4_2\alpha$ and a beta particle is $^{0}_{-1}\beta$. Numerically, this yields $[(8)(2) + (x)(-1)] = 12$, where x represents the number of beta particles emitted. Hence, the number of beta particles emitted is 4.

21.34 ***See Section 21.1 and Example 21.3.***

For each alpha ($^4_2\alpha$) particle emitted, the mass number decreases by 4, while a beta decay does not change the mass number. In the $^{237}_{93}Np$ to $^{209}_{83}Bi$ series, the change in mass number is from 237 to 209 = 28. Hence, a total of seven (28/4) alpha particles must be emitted to balance the change in mass number in this series.

The change in atomic number in this series is from 93 to 83 = 10. This change can be represented as $\left[(\text{\# alpha particles})(2)+(\text{\# beta particles})(-1)\right]=12$ because an alpha particle is ${}^{4}_{2}\alpha$ and a beta particle is ${}^{0}_{-1}\beta$. Numerically, this yields $\left[(7)(2)+(x)(-1)\right]$ = 10, where x represents the number of beta particles emitted. Hence, the number of beta particles emitted is 4.

21.35 ***See Section 21.2 and Example 21.7.***

$\left(\dfrac{\text{Current mass } {}^{238}\text{U}}{\text{Original mass } {}^{238}\text{U}}\right)$ can be used in place of $\left(\dfrac{N}{N_0}\right)$ in $\ln\left(\dfrac{N}{N_0}\right)=\dfrac{-0.693t}{t_{1/2}}$ giving

$$t=\frac{-t_{1/2}}{0.693}\ln\left(\frac{\text{Current mass } {}^{238}\text{U}}{\text{Original mass } {}^{238}\text{U}}\right)$$

$$\text{Original mass } {}^{238}\text{U}=6.73 \text{ mg } {}^{238}\text{U}+\left(3.22 \text{ mg } {}^{206}\text{Pb}\right)\left(\frac{238 \text{ mg U}}{206 \text{ mg Pb}}\right)=10.45 \text{ mg } {}^{238}\text{U}$$

Hence, $t=\dfrac{-4.51 \times 10^{9} \text{ yr}}{0.693}\ln\left(\dfrac{6.73}{10.45}\right)=\mathbf{2.86 \times 10^{9}\ yr}$

21.36 ***See Section 21.1 and Example 21.6.***

Solving $\ln\left(\dfrac{N}{N_o}\right)=\dfrac{-0.693t}{t_{1/2}}$ for $\dfrac{N}{N_o}$ gives $\dfrac{N}{N_o}=e^{-0.693t/t_{1/2}}=e^{-(0.693)(2300 \text{ yrs})/(5730 \text{ yrs})}=\mathbf{0.757}$.

This means 75.7% of the ^{14}C remains after 2300 years.

21.37 ***See Section 21.2 and Example 21.7.***

One can assume the mass of ^{87}Sr formed is equal to the mass of ^{87}Rb reacted, since these species both conatain 87 nuclear particles. Hence, original mass of ^{87}Rb is given as

$$\begin{aligned}\text{original mass } {}^{87}\text{Rb} &= \left[\text{current mass } {}^{87}\text{Rb} + \text{current mass } {}^{87}\text{Sr}\right]\\ &= \left[\text{current mass } {}^{87}\text{Rb} + (0.51)\text{current mass } {}^{87}\text{Rb}\right]\\ &= 1.051\left(\text{current mass } {}^{87}\text{Rb}\right).\end{aligned}$$

Solving $\ln\left(\dfrac{N}{N_0}\right)=\dfrac{-0.693t}{t_{1/2}}$ for t and substituting gives

$$t=\frac{-t_{1/2}}{0.693}\ln\left(\frac{N}{N_0}\right) \qquad t=\frac{-5.0 \times 10^{10} \text{ yr}}{0.693}\ln\left(\frac{1.000}{1.051}\right)=\mathbf{3.6 \times 10^{9}\ yr}$$

21.38 ***See Section 21.2 and Example 21.5.***

$$\text{percentage } {}^{14}_{6}\text{C atoms}=\frac{\text{number of } {}^{14}_{6}\text{C atoms in } 1.00 \text{ g C}}{\text{total number of carbon atoms in } 1.00 \text{ g C}}\times 100\%$$

The number of ${}^{14}_{6}C$ atoms in 1.00 g C can be determined from the rate of decay.

Solving rate = kN for N gives $N=\dfrac{\text{rate}}{k}$, and the value of k can be determined from the half-life for ${}^{14}_{6}C$.

Solving $t_{1/2} = \frac{0.693}{k}$ for k gives $k = \frac{0.693}{t_{1/2}}$ $k = \frac{0.693}{\left(5730\text{ yr} \times \frac{365\text{ days}}{\text{yr}} \times \frac{24\text{ hr}}{\text{day}} \times \frac{60\text{ min}}{\text{hr}}\right)} = 2.30 \times 10^{-10}\text{ min}^{-1}$

and $N = \frac{15.3\ ^{14}_{6}\text{C atoms disintegrating/ min}}{2.30 \times 10^{-10}\text{ min}^{-1}} = 6.65 \times 10^{10}\ ^{14}_{6}\text{C atoms}$

$? \text{ total C atoms in } 1.00\text{ g C} = 1.00\text{ g C} \times \frac{1\text{ mol C}}{12.01\text{ g C}} \times \frac{6.02 \times 10^{23}\text{ C atoms}}{1\text{ mol C}} = 5.01 \times 10^{22}\text{ C atom}$

Hence, percentage $^{14}_{6}$C atoms $= \frac{6.65 \times 10^{10}\ ^{14}_{6}\text{C atoms}}{5.01 \times 10^{22}\text{ C atoms}} \times 100\% = \mathbf{1.33 \times 10^{-10}\%}$

21.39 ***See Section 21.3 and Example 21.8.***

(a) $^{54}_{26}\text{Fe} + ^{4}_{2}\text{He} \rightarrow 2^{1}_{1}\text{H} + \mathbf{^{56}_{26}Fe}$

(b) $\mathbf{^{27}_{13}Al} + ^{1}_{0}\text{n} \rightarrow ^{24}_{11}\text{Na} + ^{4}_{2}\text{He}$

(c) $^{238}_{92}\text{U} + ^{16}_{8}\text{O} \rightarrow \mathbf{^{249}_{100}Fm} + 5^{1}_{0}\text{n}$

(d) $^{96}_{42}\text{Mo} + \mathbf{^{2}_{1}H} \rightarrow ^{97}_{43}\text{Tc} + ^{1}_{0}\text{n}$

(e) $^{250}_{98}\text{Cf} + ^{11}_{5}\text{B} \rightarrow 5^{1}_{0}\text{n} + \mathbf{^{256}_{103}Lr}$

21.40 ***See Section 21.3 and Example 21.8.***

(a) $^{249}_{98}\text{Cf} + ^{10}_{5}\text{B} \rightarrow ^{257}_{103}\text{Lr} + \mathbf{2^{1}_{0}n}$

(b) $^{14}_{7}\text{N} + ^{1}_{1}\text{P} \rightarrow ^{4}_{2}\text{He} + \mathbf{^{11}_{6}C}$

(c) $^{238}_{92}\text{U} + \mathbf{^{1}_{0}n} \rightarrow \mathbf{^{239}_{93}Np} + ^{0}_{-1}\beta$

(d) $^{6}_{3}\text{Li} + \mathbf{^{1}_{0}n} \rightarrow ^{4}_{2}\text{He} + \mathbf{^{3}_{1}H}$

(e) $\mathbf{^{9}_{4}Be} + ^{4}_{2}\text{He} \rightarrow ^{12}_{6}\text{C} + \mathbf{^{1}_{0}n}$

21.41 ***See Section 21.4, Table 21.3, and Example 21.9.***

(a) Mass defect of one atom of $^{19}_{9}$F:

mass of 9 $^{1}_{1}$H:	$= 9 \times 1.007825$ u	= 9.070425 u
mass of 10 $^{1}_{0}$n:	$= 10 \times 1.008665$ u	= 10.008665 u
total mass of nucleons of $^{19}_{9}$F		= 26.214370 u
mass of one atom of $^{19}_{9}$F		= 18.9984 u
mass defect of one atom of $^{19}_{9}$F		= 0.1587 u
mass defect for one mole of $^{19}_{9}$F atoms		= 0.1587 g

(b) ? total binding energy for 1 atom of $^{19}_{9}$F in MeV $= \frac{0.1587\text{ u}}{1\text{ atom }^{19}_{9}\text{F}} \times \frac{931.5\text{ MeV}}{1\text{ u}} = \mathbf{147.8\ \frac{MeV}{atom\ ^{19}_{9}F}}$

(c) ? binding energy per nucleon $= \frac{147.8\text{ MeV}}{1\text{ atom }^{19}_{9}\text{F}} \times \frac{1\text{ atom }^{19}_{9}\text{F}}{19\text{ nucleons}} = \mathbf{7.779\ \frac{MeV}{nucleon}}$

21.42 ***See Section 21.4, Table 21.3, and Example 21.9.***

(a) Mass defect of one atom of $^{31}_{15}$P:

mass of 15 $^{1}_{1}$H = 15 x 1.007825 μ	= 15.117375 μ
mass of 16 $^{1}_{0}$n = 16 x 1.008665 μ	= 16.138640 μ
total mass of nucleons of $^{31}_{15}$P	= 31.256015 μ
mass of one atom of $^{31}_{15}$P	= 30.9738 μ

$$\text{mass defect of one atom of } {}^{31}_{15}\text{P} = 0.2822 \quad \mu$$
$$\text{mass defect for one mole of } {}^{31}_{15}\text{P} = 0.2822 \quad \text{g}$$

(b) $? \text{ total binding energy for 1 atom of } {}^{31}_{15}\text{P in MeV} = \frac{0.2822\mu}{1 \text{ atom } {}^{31}_{15}\text{P}} \times \frac{931.5 \text{ MeV}}{1\mu} = \frac{\mathbf{262.9 \text{ meV}}}{\mathbf{atom\ {}^{31}_{15}P}}$

(c) $? \text{ binding energy per nucleon} = \frac{262.9 \text{ MeV}}{1 \text{ atom } {}^{31}_{15}\text{P}} \times \frac{1 \text{ atom } {}^{31}_{15}\text{P}}{31 \text{ nucleons}} = \frac{\mathbf{8.481 \text{ MeV}}}{\mathbf{nucleon}}$

21.43 ***See Section 21.4, Table 21.3, and Example 21.9.***

(a) Mass defect of ^{26}Al:

$$\text{mass of 13 } {}^{1}_{1}\text{H}: = 13 \times 1.007825 \text{ u} = 13.101725 \text{ u}$$
$$\text{mass of 13 } {}^{1}_{0}\text{n}: = 13 \times 1.008665 \text{ u} = \underline{13.112645 \text{ u}}$$
$$\text{total mass of nucleons of } {}^{26}\text{Al} = 26.214370 \text{ u}$$
$$\text{mass of } {}^{26}\text{Al} = \underline{25.9869 \quad \text{u}}$$
$$\text{mass defect of } {}^{26}\text{Al} = 0.2275 \quad \text{u}$$

Binding energy per nucleon for ^{26}Al:

$$? \text{ binding energy per nucleon} = \frac{0.2775 \text{ u}}{26 \text{ nucleons}} \times \frac{931.5 \text{ MeV}}{1 \text{ u}} = 8.151 \frac{\text{MeV}}{\text{nucleon}}$$

Mass defect for ^{27}Al:

$$\text{mass of 13 } {}^{1}_{1}\text{H}: = 13 \times 1.007825 \text{ u} = 13.101725 \text{ u}$$
$$\text{mass of 14 } {}^{1}_{0}\text{n}: = 14 \times 1.008665 \text{ u} = \underline{14.12131 \quad \text{u}}$$
$$\text{total mass of nucleons of } {}^{27}\text{Al} = 27.223035 \text{ u}$$
$$\text{mass of } {}^{27}\text{Al} = \underline{26.9815 \quad \text{u}}$$
$$\text{mass defect of } {}^{27}\text{Al} = 0.2415 \quad \text{u}$$

Binding energy per nucleon for ^{27}Al:

$$? \text{ binding energy per nucleon} = \frac{0.2415 \text{ u}}{27 \text{ nucleons}} \times \frac{931.5 \text{ MeV}}{1 \text{ u}} = 8.332 \frac{\text{MeV}}{\text{nucleon}}$$

Mass defect for ^{28}Al:

$$\text{mass of 13 } {}^{1}_{1}\text{H}: = 13 \times 1.007825 \text{ u} = 13.101725 \text{ u}$$
$$\text{mass of 15 } {}^{1}_{0}\text{n}: = 15 \times 1.008665 \text{ u} = \underline{15.129975 \text{ u}}$$
$$\text{total mass of nucleons of } {}^{28}\text{Al} = 28.231700 \text{ u}$$
$$\text{mass of } {}^{28}\text{Al} = \underline{27.9819 \quad \text{u}}$$
$$\text{mass defect of } {}^{28}\text{Al} = 0.2498 \quad \text{u}$$

Binding energy per nucleon for ^{28}Al:

$$? \text{ binding energy per nucleon} = \frac{0.2498 \text{ u}}{28 \text{ nucleons}} \times \frac{931.5 \text{ MeV}}{1 \text{ u}} = 8.310 \frac{\text{MeV}}{\text{nucleon}}$$

(b) ^{27}Al has the highest binding energy per nucleon and is therefore likely to be the stable isotope of Al. ^{26}Al and ^{28}Al have lower binding energies per nucleon than ^{27}Al and are therefore likely to be the radioactive nuclides of Al.

21.44 *See Section 21.4, Table 21.3, and Example 21.9.*

(a) Mass defect of ^{30}P:

$$\begin{aligned} \text{mass of } 15\ {}^{1}_{1}H: &= 15 \times 1.007825\ u = 15.117375\ u \\ \text{mass of } 15\ {}^{1}_{0}n: &= 15 \times 1.008665\ u = \underline{15.129975\ u} \\ \text{total mass of nucleons of } {}^{30}P &= 30.247350\ u \\ \text{mass of } {}^{30}P &= \underline{29.9783\ \ u} \\ \text{mass defect of } {}^{30}P &= 0.2690\ \ u \end{aligned}$$

Binding energy per nucleon for ^{30}P:

$$? \text{ binding energy per nucleon} = \frac{0.2690\ u}{30\ \text{nucleons}} \times \frac{931.5\ \text{MeV}}{1\ u} = 8.352\frac{\text{MeV}}{\text{nucleon}}$$

Mass defect for ^{31}P:

$$\begin{aligned} \text{mass of } 15\ {}^{1}_{1}H: &= 15 \times 1.007825\ u = 15.117375\ u \\ \text{mass of } 16\ {}^{1}_{0}n: &= 16 \times 1.008665\ u = \underline{16.138640\ u} \\ \text{total mass of nucleons of } {}^{31}P &= 31.256015\ u \\ \text{mass of } {}^{31}P &= \underline{30.9738\ \ u} \\ \text{mass defect of } {}^{31}P &= 0.2822\ \ u \end{aligned}$$

Binding energy per nucleon for ^{31}P:

$$? \text{ binding energy per nucleon} = \frac{0.2822\ u}{31\ \text{nucleons}} \times \frac{931.5\ \text{MeV}}{1\ u} = 8.480\frac{\text{MeV}}{\text{nucleon}}$$

Mass defect for ^{32}P:

$$\begin{aligned} \text{mass of } 15\ {}^{1}_{1}H: &= 15 \times 1.007825\ u = 15.117375\ u \\ \text{mass of } 17\ {}^{1}_{0}n: &= 17 \times 1.008665\ u = \underline{17.147305\ u} \\ \text{total mass of nucleons of } {}^{32}P &= 32.264680\ u \\ \text{mass of } {}^{32}P &= \underline{31.9739\ \ u} \\ \text{mass defect of } {}^{32}P &= 0.2908\ \ u \end{aligned}$$

Binding energy per nucleon for ^{32}P:

$$? \text{ binding energy per nucleon} = \frac{0.2908\ u}{32\ \text{nucleons}} \times \frac{931.5\ \text{MeV}}{1\ u} = 8.465\frac{\text{MeV}}{\text{nucleon}}$$

(b) ^{31}P has the highest binding energy per nucleon and is therefore likely to be the stable isotope of P. ^{30}P and ^{32}P have lower binding energies per nucleon than ^{31}P and are therefore likely to be the radioactive nuclides of P.

21.45 *See Sections 21.4, 21.5, Table 21.3, and Example 21.9.*

${}^{1}_{0}n$	+	${}^{239}_{94}Pu$	$\rightarrow$	${}^{98}_{40}Zr$	+	${}^{139}_{54}Xe$	+	$3{}^{1}_{0}n$
1.008665 u		239.052 u		97.913 u		138.919 u		3(1.008665 u)

(a) Δm = mass defect for the reaction = mass of reactants - mass of products

$$\Delta m = [(1\times 1.008665\ u)+(1\times 239.052\ u)]-[(1\times 97.913\ u)+(1\times 138.919\ u)+(3\times 1.008665\ u)] = \mathbf{0.203\ u}$$

(b) $$? \text{ J for one fission} = \frac{0.203\ u}{1\ \text{fission}} \times \frac{931.5\ \text{MeV}}{1\ u} \times \frac{1.602 \times 10^{-13}\ J}{1\ \text{MeV}} = \mathbf{3.03 \times 10^{-11}\ \frac{J}{fission}}$$

(c) ? J for 1.00 g Pu = $1.00 \text{ g Pu} \times \frac{1 \text{ mol Pu}}{239.05 \text{ g Pu}} \times \frac{6.02 \times 10^{23} \text{ atoms Pu}}{1 \text{ mol Pu}} \times \frac{3.03 \times 10^{-11} \text{ J}}{1 \text{ Pu atom}} = \mathbf{7.63 \times 10^{10} \text{ J}}$

21.46 ***See Sections 21.4, 21.5, Table 21.3, and Example 21.9.***

$$^{1}_{0}n \quad + \quad ^{239}_{94}Pu \quad \rightarrow \quad ^{96}_{39}Y \quad + \quad ^{140}_{55}Cs \quad + \quad 4^{1}_{0}n$$

1.008665 u   239.052 u   95.916 u   139.917 u   4(1.008665 u)

(a) Δm = mass defect for the reaction = mass of reactants - mass of products
$\Delta m = [(1 \times 1.008665 \text{ u}) + (1 \times 239.052 \text{ u})] - [(1 \times 95.916 \text{ u}) + (1 \times 139.917 \text{ u}) + (4 \times 1.008665 \text{ u})] = \mathbf{0.193 \text{ u}}$

(b) ? J for one fission = $\frac{0.193 \text{ u}}{1 \text{ fission}} \times \frac{931.5 \text{ MeV}}{1 \text{ u}} \times \frac{1.602 \times 10^{-13} \text{ J}}{1 \text{ MeV}} = \mathbf{2.88 \times 10^{-11} \frac{J}{fission}}$

(c) ? J for 1.00 g Pu = $1.00 \text{ g Pu} \times \frac{1 \text{ mol Pu}}{239.05 \text{ g Pu}} \times \frac{6.02 \times 10^{23} \text{ atoms Pu}}{1 \text{ mol Pu}} \times \frac{2.88 \times 10^{-11} \text{ J}}{1 \text{ Pu atom}} = \mathbf{7.26 \times 10^{10} \text{ J}}$

21.47 ***See Section 21.1 and Example 21.6.***

Solving $\ln\left(\frac{N}{N_o}\right) = \frac{-0.693t}{t_{½}}$ for $\frac{N}{N_o}$ gives $\frac{N}{N_o} = e^{-0.693t/t_{½}} = e^{-(0.693)(51 \text{ yrs})/(30.17 \text{ yrs})} = \mathbf{0.31.}$
This means 31% of the ^{137}Cs still remained 51 years after the detonation of the bomb.

21.48 ***See Section 21.1 and Example 21.6.***

Solving $\ln\left(\frac{N}{N_o}\right) = \frac{-0.693t}{t_{½}}$ for $\frac{N}{N_o}$ gives $\frac{N}{N_o} = e^{-0.693t/t_{½}} = e^{-(0.693)(51 \text{ yrs})/(28.1 \text{ yrs})} = \mathbf{0.28.}$
This means 28% of the ^{90}Sr still remained 51 years after the detonation of the bomb.

21.49 ***See Sections 6.1, 21.4.***

$$^{1}_{0}n \quad + \quad ^{113}_{48}Cd \quad \rightarrow \quad ^{114}_{48}Cd \quad + \quad \gamma$$

1.008665 u   112.9044 u   113.9034 u

(a) Δm = mass defect for the reaction = mass of reactants - mass of products
$\Delta m = [(1 \times 1.008665 \text{ u}) + (1 \times 112.9044 \text{ u})] - [(1 \times 113.9034 \text{ u})] = \mathbf{9.7 \times 10^{-3} \text{ u}}$

? energy of γ ray in MeV = $9.7 \times 10^{-3} \text{ u} \times \frac{931.5 \text{ MeV}}{1 \text{ u}} = \mathbf{9.0 \text{ MeV}}$

(b) Solving $\Delta E = \frac{hc}{\lambda}$ for λ gives $\lambda = \frac{hc}{\Delta E}$. Substituting in the energy of the gamma ray yields

$$\lambda = \frac{(6.63 \times 10^{-34} \text{ J}\cdot\text{s})(3.00 \times 10^{8} \text{ m}\cdot\text{s}^{-1})}{\left(9.0 \text{ MeV} \times \frac{1.602 \times 10^{-13} \text{ J}}{1 \text{ MeV}}\right)} = \mathbf{1.4 \times 10^{-13} \text{ m}}$$

21.50 ***See Section 21.1 and Example 21.1.***

$$^{128}_{53}I \rightarrow {}^{128}_{54}Xe + {}^{0}_{-1}\beta$$

$$^{128}_{53}I \rightarrow {}^{128}_{52}Te + {}^{0}_{+1}\beta$$

21.51 ***See Sections 21.4, 21.5, Table 21.3, and Example 21.9..***

$$^{2}_{1}H + {}^{3}_{1}H \rightarrow {}^{4}_{2}He + {}^{1}_{0}n$$

2.0140 u 3.016 u 4.0026 u 1.008665 u

Δm = mass defect for the reaction = mass of reactants - mass of products

$$\Delta m = [(1\times 2.0140\ u)+(1\times 3.016\ u)]-[(1\times 4.0026u\ )+(1\times 1.008665\ u)] = \mathbf{0.0188\ u}$$

$$?\ J\ \text{per gram}\ ^4He\ \text{formed} = 1.00\ g\ ^4He \times \frac{1\ mol\ ^4He}{4.0026\ g\ ^4He} \times \frac{6.02\ x\ 10^{23}\ ^4He\ atoms}{1\ mol\ ^4He} \times \frac{0.0188\ u}{1\ atom\ ^4He}$$

$$\times \frac{931.5\ MeV}{1\ u} \times \frac{1.602\ x\ 10^{-13}\ J}{1\ MeV} = \mathbf{4.22\ x\ 10^{11}\ J}$$

The energy released with the formation of one gram of ^4He via fusion of ^2_1H and ^3_1H is 5.3 times greater than the energy released by fission of one gram of ^{235}U (4.22 x 10^8 kJ vs. 8 x 10^7 kJ).

21.52 ***See Section 21.5.***

(a) $$?\ J\ \text{for}\ 1\ kg\ ^{235}U = 1.0\ kg\ ^{235}U \times \frac{10^3\ g\ ^{235}U}{kg\ ^{235}U} \times \frac{1\ mol\ ^{235}U}{235.0493\ g\ ^{235}U} \times \frac{1.9\ x\ 10^{10}\ kJ}{1\ mol\ ^{235}U} = 8.1\ x\ 10^{10}\ kJ$$

$$?\ \text{metric tons coal supplying equal energy} = 8.1\ x\ 10^{10}\ kJ \times \frac{1.0\ kg\ coal}{2.8\ x\ 10^4\ kJ} \times \frac{10^3\ g\ coal}{1\ kg\ coal} \times \frac{1\ metric\ ton\ coal}{10^6\ g\ coal}$$

$$= \mathbf{2.9\ x\ 10^3\ metric\ tons\ co}$$

(b) $$?\ \text{metric tons}\ SO_2 = 2.9\ x\ 10^3\ \text{metric tons coal} \times \frac{0.90\ \text{metric ton S}}{100.00\ \text{metric tons coal}} \times \frac{64.0\ \text{metric tons}\ SO_2}{32.0\ \text{metric tons S}}$$

$$= \mathbf{52\ metric\ tons\ SO_2}$$

21.53 ***See Sections 6.8, 21.5.***

$$\frac{\text{rate of effusion of}\ ^{235}UF_6}{\text{rate of effusion of}\ ^{238}UF_6} = \sqrt{\frac{\mathcal{M}\ ^{238}UF_6}{\mathcal{M}\ ^{235}UF_6}} = \sqrt{\frac{238.0508+6(18.998)}{235.0493+6(18.998)}} = \mathbf{1.0043}$$

Chapter 22: Organic and Biochemistry

22.1 *See Chapter Introduction.*

Organic chemistry is the study of carbon-hydrogen compounds and substituted carbon-hydrogen compounds in which carbon forms bonds to other nonmetals such as nitrogen, oxygen, sulfur and the halogens. **Biochemistry** is the chemistry of systems in living organisms.

22.2 *See Chapter Introduction.*

Carbon has four valence orbitals and four valence electrons and therefore tends to form four covalent bonds. It is is also a small element of the second period and therefore forms strong single bonds with itself and other elements and strong multiple bonds with itself, nitrogen, oxygen and sulfur. Its unique ability to form bonds to itself giving chains and rings is called catenation.

22.3 *See Section 22.1.*

Alkanes are known as saturated hydrocarbons because each carbon atom forms four bonds to four other atoms and is therefore forming its maximum number of bonds possible. In forming four bonds to four other atoms in alkanes, each carbon atom uses **sp^3** hybrid orbitals.

22.4 *See Section 22.1 and Figure 22.5.*

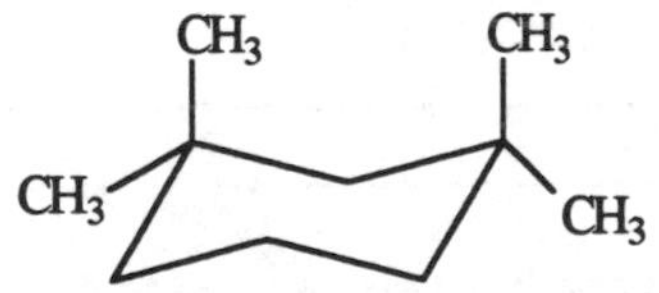

chair form
of 1,1,3,3-dimethylcyclohexane

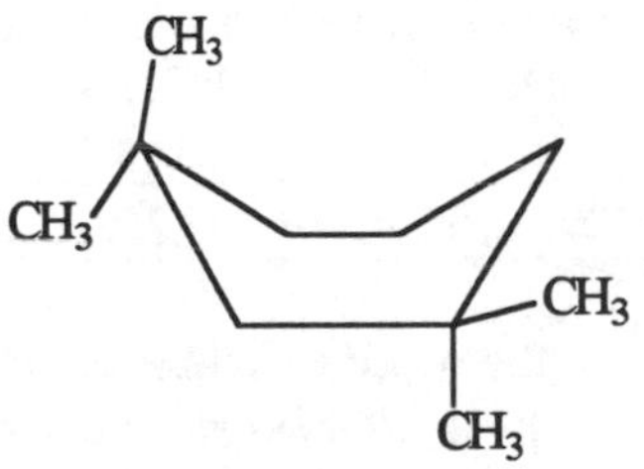

boat form
of 1,1,3,3-dimethylcyclohexane

The chair form of 1,1,3,3-tetramethylcyclohexane is favored over the boat form because repulsive interactions between the methyl groups are reduced in the chair arrangement.

22.5 *See Sections 10.4, 22.2 and Figures 10.3, 10.30, 22.7.*

The steric numbers around the carbon atoms in ethylene, $H_2C{=}CH_2$, and acetylene, $HC{\equiv}CH$, are three and two, respectively. Hence, the carbon atoms in these molecules use **sp^2** and **sp** hybrids to form the sigma bonds in these respective compounds.

22.6 *See Section 22.2 and Example 22.4.*

In 1,1-dibromoethene there are two bromine atoms on one of the carbon atoms and two hydrogen atoms on the other carbon atom. Hence, there is no possibility for the bromine atoms to be adjacent to each other on the separate carbon atoms (cis) or opposite each other on the separate carbon atoms (trans). However, this possibility exists with 1,2-dibromoethene because there is one bromine atom and one hydrogen atom on each carbon atom.

22.7 *See Section 22.3 and Table 22.3..*

In a carboxylic acid there is a highly electronegative oxygen atom attached to the same carbon atom as the -OH group. This oxygen atom withdraws electron density from the carbon atom and in turn causes a shifit in the electron density in the C-O-H bonds towards the carbon atom. The electron density shift causes the -OH hydrogen atom to become more positively charged , more protonic, through an inductive effect that is not operative in an alcohol because there is no oxygen atom attached to the carbon atom of the C-O-H linkage of an alcohol. This is why the -OH functional group of a carboxylic acid is acidic in water and that of an alcohol is not acidic.

22.8 *See Sections 11.4, 22.3.*

Ethyl alcohol, CH_3CH_2OH, molecules have O-H bonds and therefore have hydrogen bonding forces operating as forces of attraction between molecules, whereas acetone, $CH_3C(O)CH_3$, is only slightly polar. Hence, ethyl alcohol has the stronger intermolecular forces of attraction and the higher boiling point.

22.9 *See Section 22.4 and Figure 22.23.*

Cross-linking involves using a substance to connect the individual chains of a polymer together. This prevents the individual chains from sliding past each other and leads to a harder and more elastic polymer.

22.10 *See Section 22.5.*

The **primary structure** of a protein is the sequence of amino acids in the polypeptide chain. The **secondary structure** describes the shape of the polypeptide chain, which is determined by the types of hydrogen bonds made by the amide portion of the chains. The **tertiary structure** is the overall three-dimensional arrangement of the protein. Lastly, some proteins have **quaternary structure**, which describes the orientation of different polypeptide chains with respect to each other.

22.11 *See Section 22.5 and Figures 22.25, 22.26.*

In an **α-helix structure**, the peptide chain is held together by hydrogen bonding between amide groups in the same region of the chain. In a **β-pleated sheet structure**, the peptide chain is held together by hydrogen bonding between amide groups in different sections of the same chain or amide groups in different polypeptides. These arrangements refer to the **secondary structure** of the protein.

22.12 *See Section 22.6 and Figure 22.30.*

The glucose units of cellulose are joined through oxygen bridges at the β (equatorial) positions, whereas the glucose units of starch are joined through oxygen bridges at the α (axial) positions. The structure of starch is not as suitable for hydrogen bonding interactions and does not form strong fibers like cellulose. Hence, the human digestive system is able to process starch but not cellulose. Cellulose passes through the digestive systems of humans essentially unchanged and is known as "fiber" in food for that reason.

22.13 *See Section 22.7.*

DNA units are located inside chromosones and are used to store genetic information. RNA units are located outside chromosones and are used to transfer genetic information from DNA and direct the synthesis of proteins.

22.14 *See Section 22.7.*

A necleotide is composed of a five-carbon sugar, a base and a phosphate group. The five-carbon sugar in DNA is deoxyribose and that in RNA is ribose.

22.15 *See Section 22.7 and Figure 22.33.*

In 1953, Watson and Crick proposed that the secondary structure of DNA is a double helix of two entwined nucleic acid strands in which the adenine bases on one strand hydrogen bond with only the thymine bases on the other strand and the guanine bases on one strand hydrogen bond with only the cytosine bases on the other strand. Their model was based on the knowledge that any sample of DNA contained the same molar amount of adenine base as there is thymine base and the same molar amount of guanine base as there is cytosine base, as well as x-ray crystallographic studies. The unique base pairing interactions in DNA explain why the A/T and G/C ratios are 1.

22.16 *See Section 22.7.*

Protein synthesis occurs in the cytoplasm of the cell, not the nucleus of the cell where the DNA is located. A molecule of RNA called messenger RNA is synthesized by the DNA to transfer the genetic information that is stored in the DNA. The messenger RNA travels to the cytoplasm where, with the help of two other types of RNA, *transfer* RNA and *ribosomal* RNA, it directs the order in which amino acids are incorporated into newly synthesized proteins. The order is determined by the order of the bases in the RNA, which was originally determined by the order of bases in the DNA.

22.17 *See Section 22.1.*

Noncyclic alkanes have the general formula C_nH_{2n+2}. Hence, only (b) $\mathbf{C_6H_{14}}$ and (d) $\mathbf{C_9H_{20}}$ are noncylcic alkanes.

22.18 *See Section 22.1.*

Noncyclic alkanes have the general formula C_nH_{2n+2}. Hence, only (a) $\mathbf{C_5H_{12}}$ is a noncyclic alkane.

22.19 *See Section 22.1.*

Noncyclic alkanes have the general formula C_nH_{2n+2}. Hence, the formula of the noncyclic alkane having eight carbon atoms is $\mathbf{C_8H_{18}}$.

22.20 *See Section 22.1.*

Noncyclic alkanes have the general formula C_nH_{2n+2}. Hence, the formula of the noncylcic alkane having twelve carbon atoms is $\mathbf{C_{12}H_{26}}$..

22.21 *See Section 22.1, Example 22.1 and Table 22.1.*

The structural formula for the straight chain isomer of C_5H_{12} is:

```
   H  H  H  H  H
   |  |  |  |  |
H—C—C—C—C—C—H
   |  |  |  |  |
   H  H  H  H  H
```

The name of the alkyl group that is formed by removing a hydrogen atom from one of the terminal carbon atoms is **n-pentyl**.

22.22 *See Section 22.1, Example 22.1 and Table 22.1.*

The structural formula for C_3H_8 is:

```
    H  H  H
    |  |  |
 H—C—C—C—H
    |  |  |
    H  H  H
```

The name of the alkyl group that is formed by removing a hydrogen atom from the central carbon atom is isopropyl.

22.23 *See Section 22.1 and Examples 22.1, 22.2.*

```
    H  H  H  H  H  H
    |  |  |  |  |  |
 H—C—C—C—C—C—C—H
    |  |  |  |  |  |
    H  H  H  H  H  H
```

n-hexane

```
             H
             |
          H—C—H
    H  H  H  |  H
    |  |  |  |  |
 H—C—C—C—C—C—H
    |  |  |  |  |
    H  H  H  H  H
```

2-methylpentane

```
          H
          |
       H—C—H
    H  H  |  H  H
    |  |  |  |  |
 H—C—C—C—C—C—H
    |  |  |  |  |
    H  H  H  H  H
```

3-methylpentane

```
          H
          |
       H—C—H
    H  H  |  H
    |  |  |  |
 H—C—C—C—C—H
    |  |  |  |
    H  |  H  H
    H—C—H
       |
       H
```

2,3-dimethylbutane

```
          H
          |
       H—C—H
    H  H  |  H
    |  |  |  |
 H—C—C—C—C—H
    |  |  |  |
    H  H  |  H
       H—C—H
          |
          H
```

2,2-dimethylbutane

22.24 *See Section 22.1 and Example 22.3.*

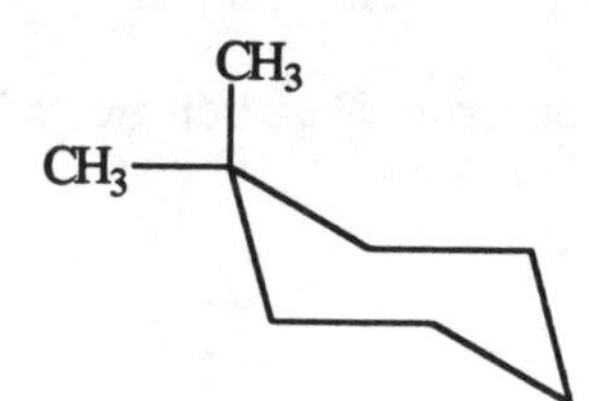

1,1-dimethylcyclohexane

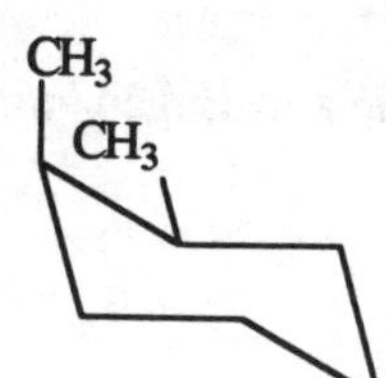

cis-1,2-dimethylcyclohexane

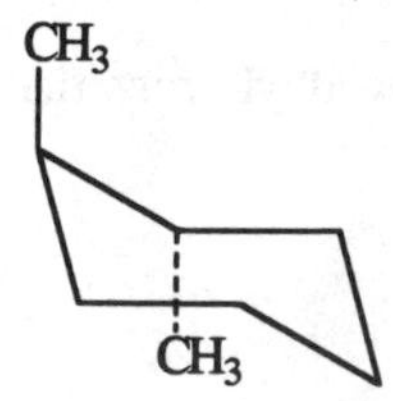

trans-1,2-dimethylcyclohexane

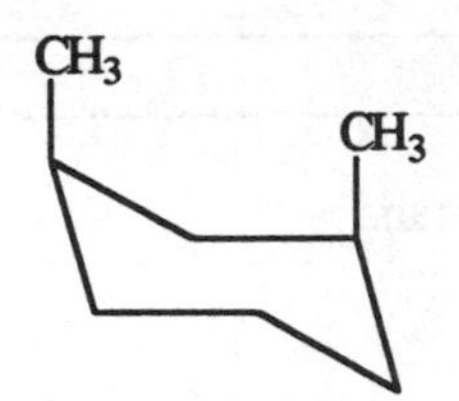

cis-1,3-dimethylcyclohexane

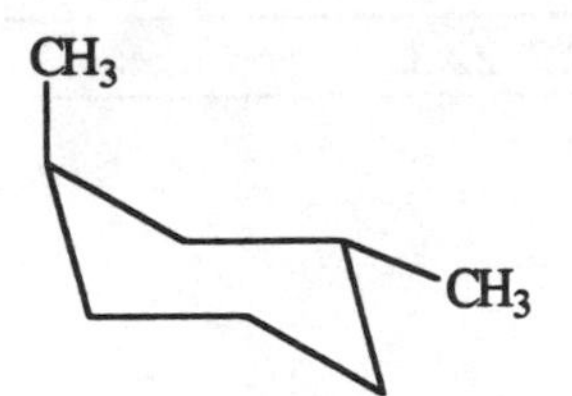

trans-1,3-dimethylcyclohexane

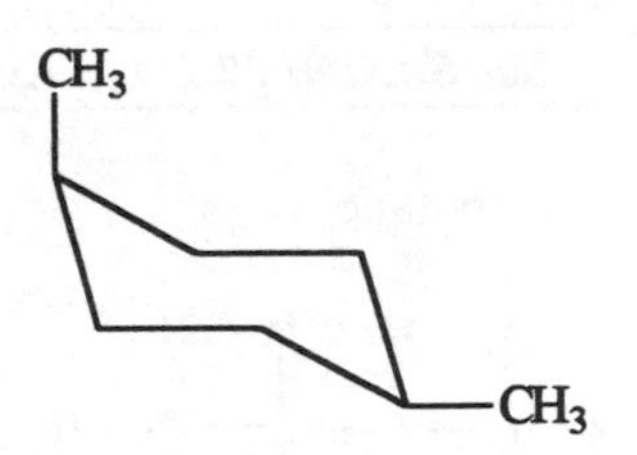

cis-1,4-dimethylcyclohexane

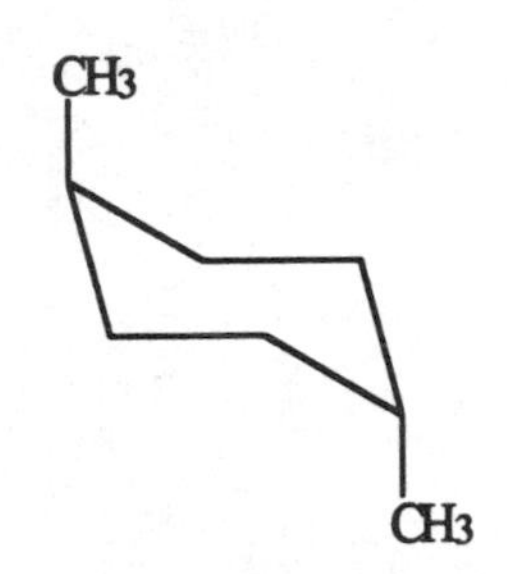

trans-1,4-dimethylcyclohexane

22.25 ***See Section 22.1 and Examples 22.2, 22.3.***

(a) 2-methylhexane

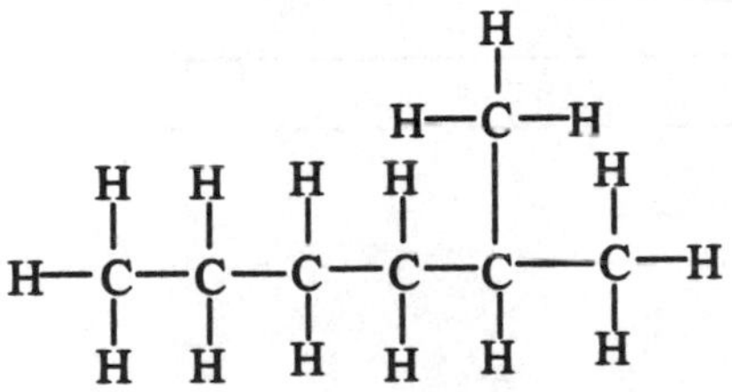

(b) 3,3-dichloroheptane

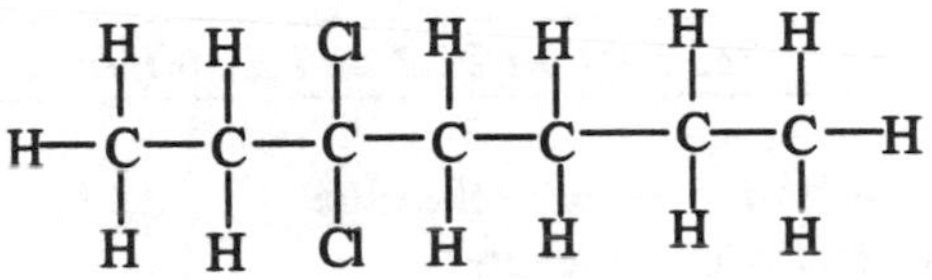

(c) 2-methyl-3-phenyloctane

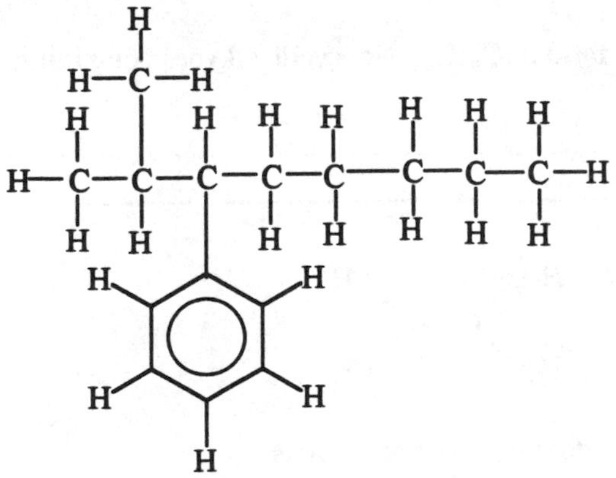

(d) 1,1-diethylcyclohexane

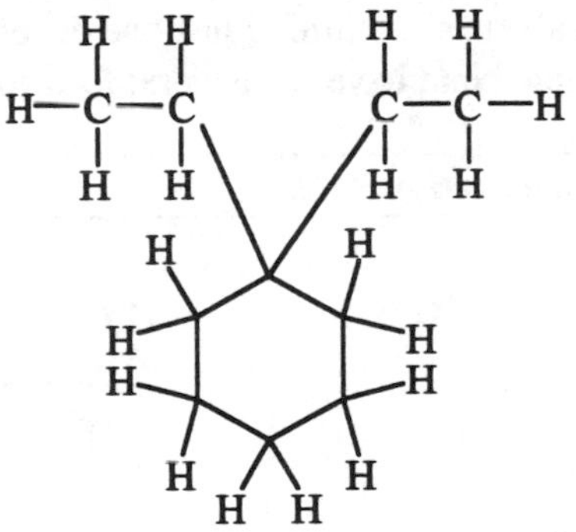

22.26 ***See Section 22.1 and Examples 22.2, 22.3.***

(a) 1-bromobutane

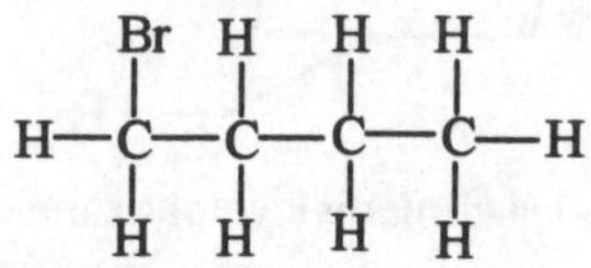

```
    Br  H   H   H
    |   |   |   |
H — C — C — C — C — H
    |   |   |   |
    H   H   H   H
```

(b) 2-methyl-3-nitropentane

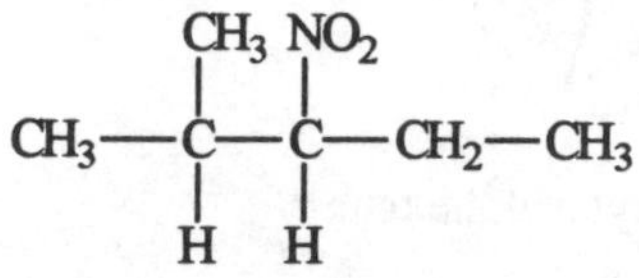

```
      CH3 NO2
       |   |
CH3 — C — C — CH2 — CH3
       |   |
       H   H
```

(c) 2-2-dimethylhexane

```
       CH3
        |
CH3 — C — CH2 — CH2 — CH2 — CH3
        |
       CH3
```

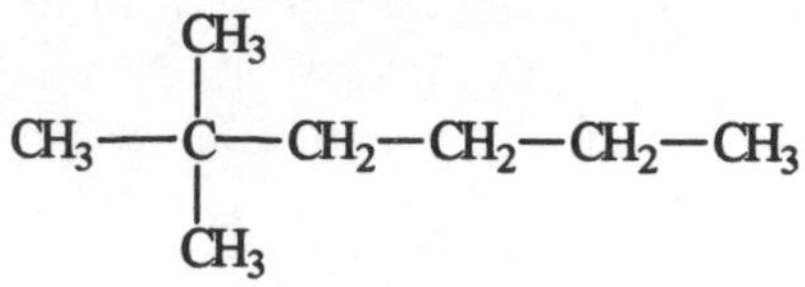

(d) 1-chloro-1-methylcyclopentane

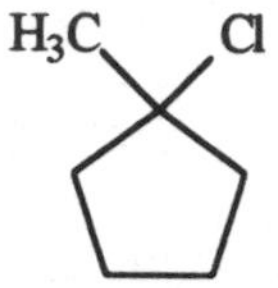

H3C Cl (on the same ring carbon of a cyclopentane ring)

22.27 ***See Section 22.1 and Examples 22.2, 22.3.***

(a) 1-fluoro-2-methylpentane
(b) 3-methylhexane
(c) 2,2-dimethylbutane
(d) 1-chloro-2-ethylcyclobutane

22.28 ***See Section 22.1 and Examples 22.2, 22.3.***

(a) 4-bromo-2-methylhexane
(b) methylcyclohexane
(c) 2-amino-3-methyloctane
(d) 3-ethylhexane

22.29 ***See Section 22.2.***

Noncyclic alkenes containing just one double bond have the general formula**C_nH_{2n}**. Noncyclic alkynes containing just one triple bond have the general formaula **C_nH_{2n-2}**.

22.30 ***See Section 22.2.***

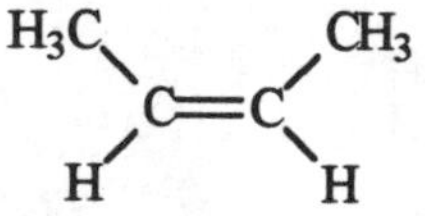

```
H3C       CH3        H3C       H           H3C       CH3
   \     /              \     /               \     /
    C = C                C = C                 C = C
   /     \              /     \               /     \
 H         H          H        CH3         H3C       CH3
```

cis isomer 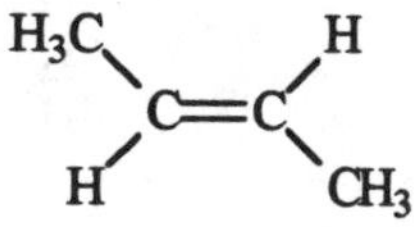 | trans isomer 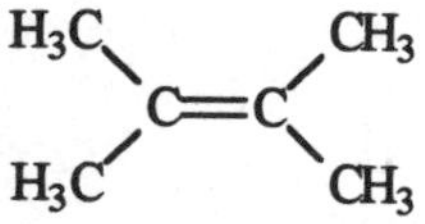 | does not exist as isomers

22.31 ***See Section 22.2 and Example 22.4.***

(a) trans-1-bromopropene
(b) cis-3-heptene
(c) 5-fluoro-2-pentyne
(d) cis-1-chloro-3-hexene

22.32 See Section 22.2 and Example 22.4.

(a) 2-methylpropene
(b) trans-2-heptene
(c) 1,1-difluoro-1-pentene
(d) 1-bromo-2-butyne

22.33 See Section 22.2 and Example 22.4.

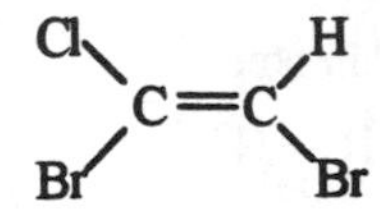

cis isomer

Cl Br
C=C
Br H

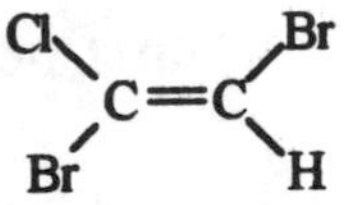

trans isomer

22.34 See Section 22.2 and Example 22.4.

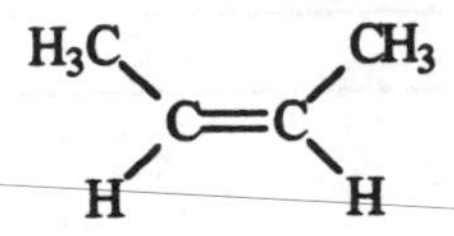

cis isomer

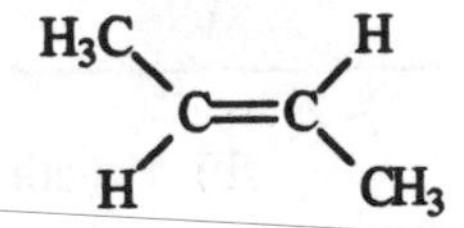

trans isomer

22.35 See Section 22.2 and Example 22.4.

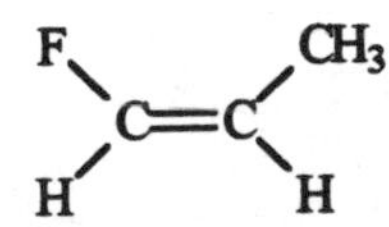

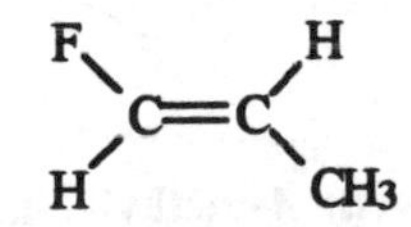

H F
C=C
H CH_3

H H
C=C
H CH_2F

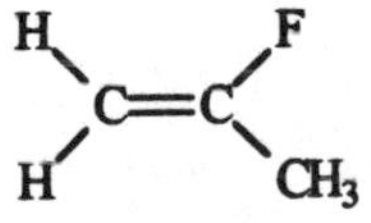

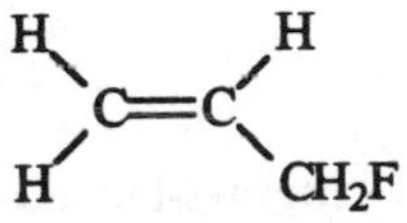

22.36 See Section 22.2 and Example 22.4.

The structural isomers for the substituted alkene C_4H_7Cl with the chlorine atom attached to a double-bonded carbon atom are:

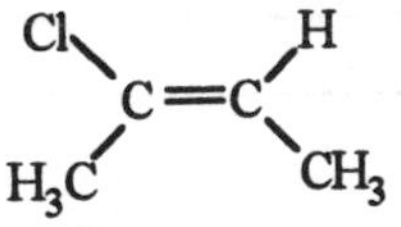

Cl H
C=C
H_3C CH_3

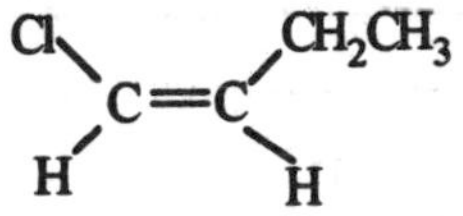

Cl CH_2CH_3
C=C
H H

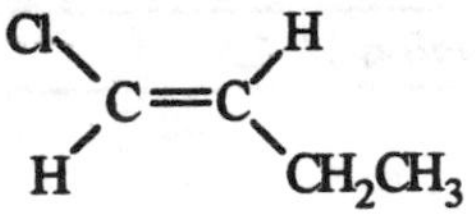

Cl H
C=C
H CH_2CH_3

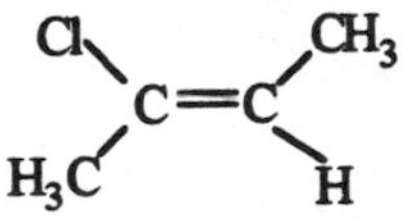

Cl CH_3
C=C
H_3C H

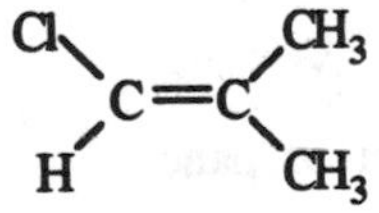

Cl CH_3
C=C
H CH_3

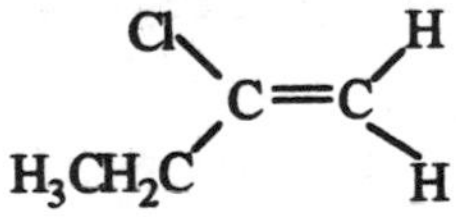

Cl H
C=C
H_3CH_2C H

22.37 *See Section 22.2 and Example 22.4.*

(a) 2-bromo-1-hexene

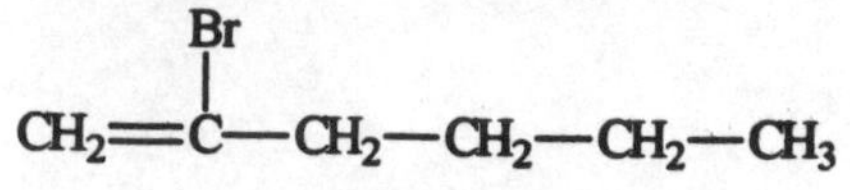

$$CH_2{=}C(Br){-}CH_2{-}CH_2{-}CH_2{-}CH_3$$

(b) cis-4-nitro-2-pentene

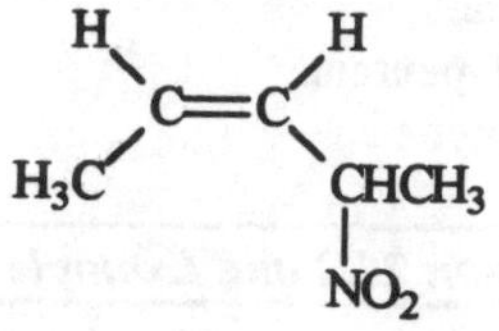

$$\text{H}(H_3C)C{=}C(\text{H})CH(NO_2)CH_3$$

(c) 1,2-dichloro-3-hexyne

$$H{-}CHCl{-}CHCl{-}C{\equiv}C{-}CH_2{-}CH_2{-}H$$

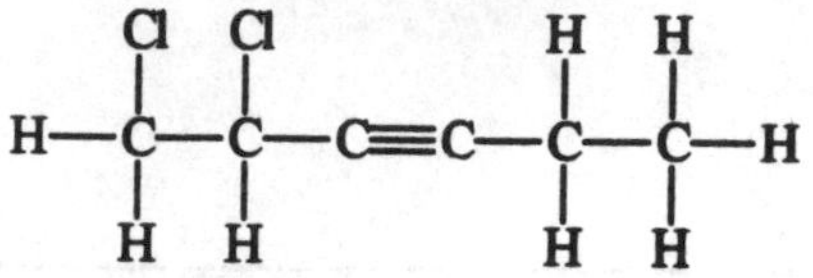

(d) 1,1-difluoro-3-chloro-1-heptene

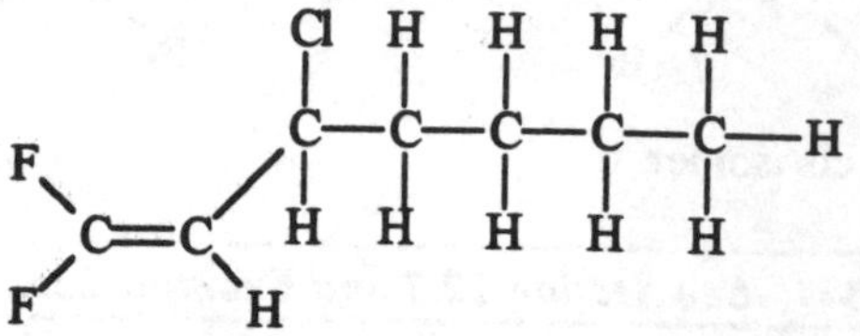

$$F_2C{=}CH{-}CHCl{-}CH_2{-}CH_2{-}CH_2{-}CH_2{-}H$$

22.38 *See Section 22.2.*

(a) 2,3-dimethyl-1-pentene

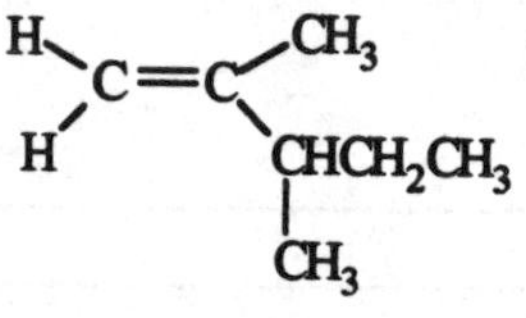

$$H_2C{=}C(CH_3){-}CH(CH_3)CH_2CH_3$$

(b) 1-methylcyclohexene

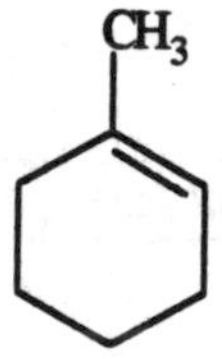

CH_3 on a cyclohexene ring (at the double-bond carbon)

(c) *cis*-1-chloro-2-butene

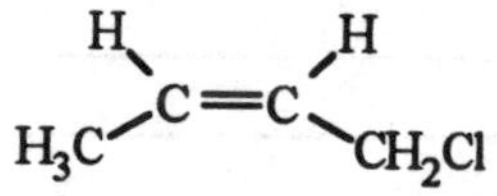

$$\text{H}(H_3C)C{=}C(\text{H})CH_2Cl$$

(d) 4-methyl-1-hexene

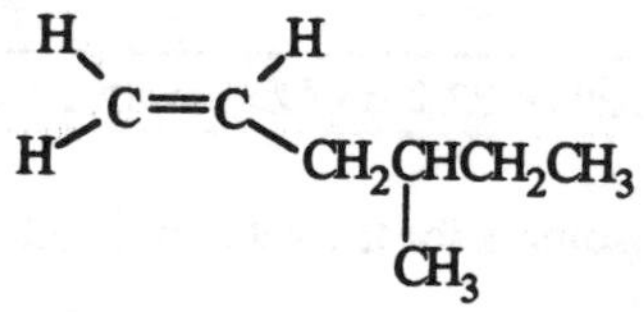

$$H_2C{=}CH{-}CH_2CH(CH_3)CH_2CH_3$$

22.39 *See Section 22.2.*

$$H{-}C{\equiv}C{-}CH_2{-}H \;\; + \; 2\,Cl_2 \; \rightarrow \; H{-}CCl_2{-}CCl_2{-}CH_2{-}H$$

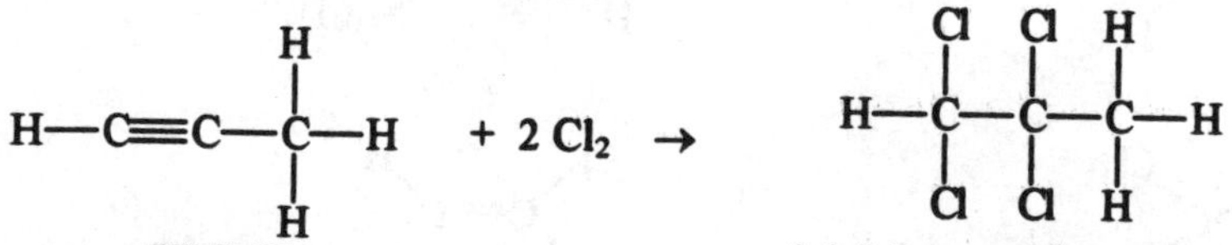

propyne → 1,1,2,2-tetrachloropropane

22.40 *See Section 22.2.*

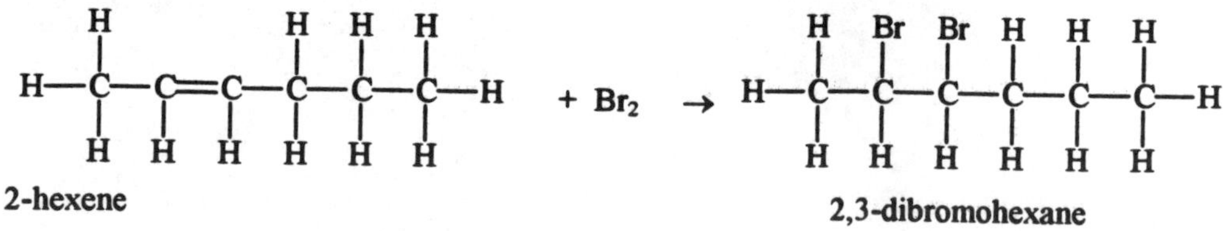

$$H{-}CH_2{-}CH{=}CH{-}CH_2{-}CH_2{-}CH_2{-}H \;\; + \; Br_2 \; \rightarrow \; H{-}CH_2{-}CHBr{-}CHBr{-}CH_2{-}CH_2{-}CH_2{-}H$$

2-hexene → 2,3-dibromohexane

22.41 *See Section 22.2.*

(a) $CH_2{=}CHCH_3 + H_2 \rightarrow CH_3CH_2CH_3$

(b) $C_6H_6 + Cl_2 \xrightarrow{FeCl_3} C_6H_5Cl + HCl$

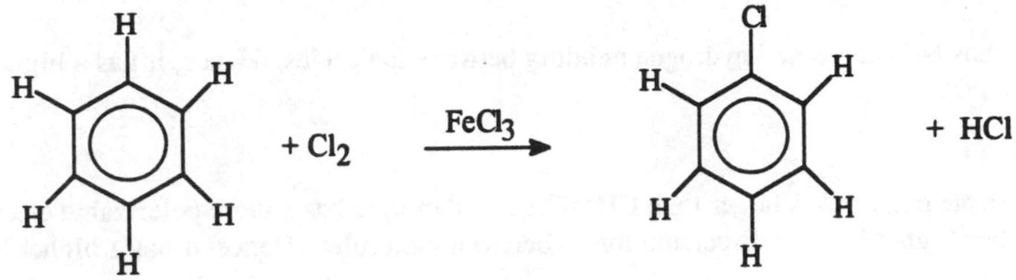

22.42 *See Section 22.2.*

(a) $CH_2{=}CHCH_3 + Br_2 \rightarrow CH_2BrCHBrCH_3$

(b) $CH{\equiv}CCH_2CH_3 + 2Cl_2 \rightarrow CHCl_2CCl_2CH_2CH_3$

22.43 *See Section 22.2.*

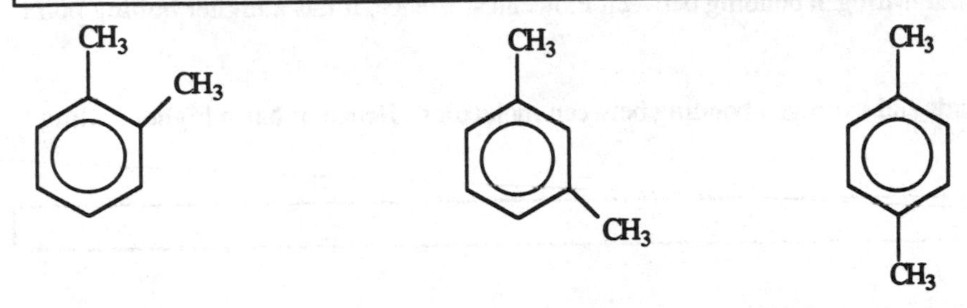

22.44 *See Section 22.2.*

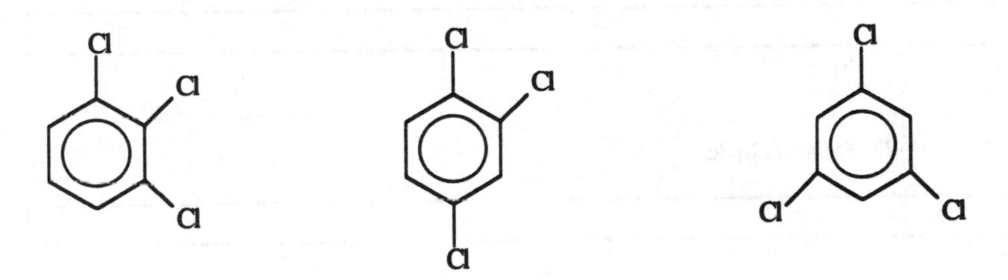

22.45 *See Section 22.3 and Table 22.3.*

(a) carboxylic acid —C(=O)—OH

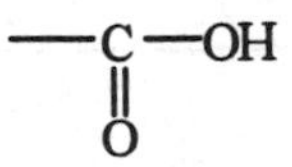

(b) aldehyde —C(=O)—H

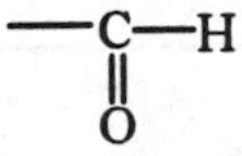

(c) ether C—O—C

22.46 *See Section 22.3 and Table 22.3.*

(a) ketone —C(=O)—

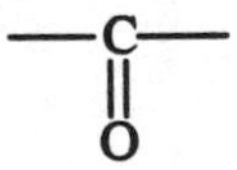

(b) ester —C(=O)—O—R

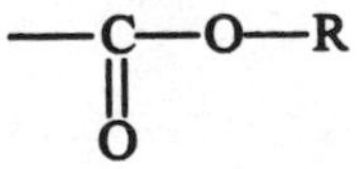

(c) amide —C(=O)—NR_2

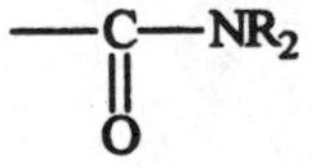

22.47 *See Section 11.4, 22.3.*

(a) 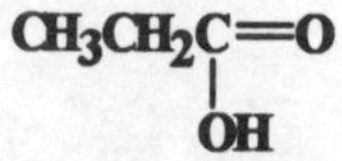 $CH_3CH_2C(=O)OH$ (structure: $CH_3CH_2C{=}O$ with OH on C) has O-H bonds and hydrogen bonding betweeen molecules. Hence, it has a higher boiling point than $CH_3CH_2OCH_2CH_3$.

(b) **$CH_3CH_2CH_2CH_2NH_2$** has N-H bonds and hydrogen bonding between molecules. Hence, it has a higher boiling point than $CH_3CH_2C(=O){-}CH_3$.

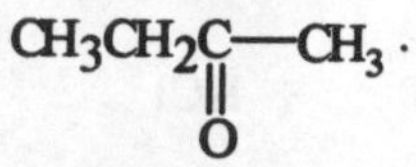

(c) **$CH_3CH_2C{\equiv}CF$** has a triple bond and is larger than $CH_2{=}CH_2$. It therefore has a more polarizable electron cloud than $CH_2{=}CH_2$ and has higher London dispersion forces between molecules. Hence, it has a higher boiling point than $CH_2{=}CH_2$.

22.48 *See Sections 11.4, 22.3.*

(a) $CH_3CH_2CH_2CH{=}CH_2$ has a double bond and is larger than CH_3OCH_3. It therefore has a more polarizable electron cloud than CH_3OCH_3 and has higher London dispersion forces between molecules. Hence, it has a higher boiling point than CH_3OCH_3.

(b) **$CH_3C(O)NH_2$** has N-H bonds and hydrogen bonding between molecules. Hence, it has a higher boiling point than $CH_3C{\equiv}CCH_3$.

(c) $CH_3CH_2CH_2CH_2OH$ has O-H bonds and hydrogen bonding between molecules. Hence, it has a higher boiling point than $CH_3C(O)OCH_3$.

22.49 *See Section 22.3.*

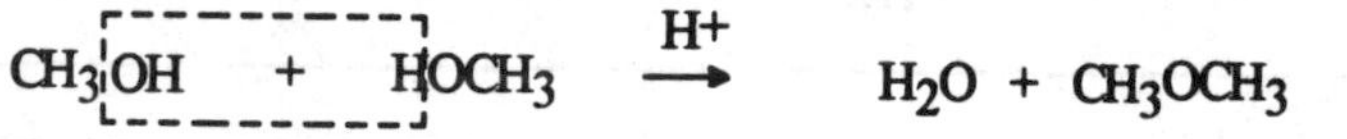

$$CH_3OH + HOCH_3 \xrightarrow{H^+} H_2O + CH_3OCH_3$$ dimethylether

22.50 *See Section 22.3.*

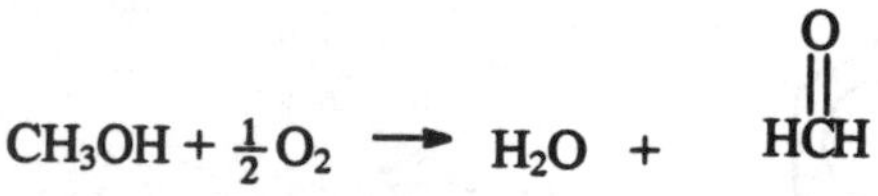

$$CH_3OH + \tfrac{1}{2}O_2 \longrightarrow H_2O + HC(=O)H$$ formaldehyde

22.51 *See Section 22.3.*

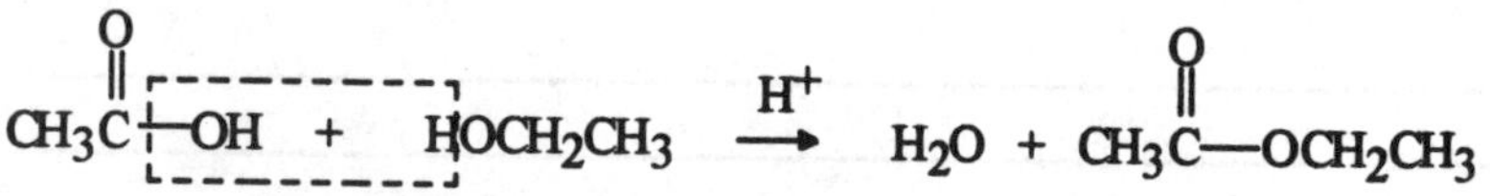

$$CH_3C(=O){-}OH + HOCH_2CH_3 \xrightarrow{H^+} H_2O + CH_3C(=O){-}OCH_2CH_3$$ ethyl acetate (ethyl ethanoate)

22.52 *See Section 22.3.*

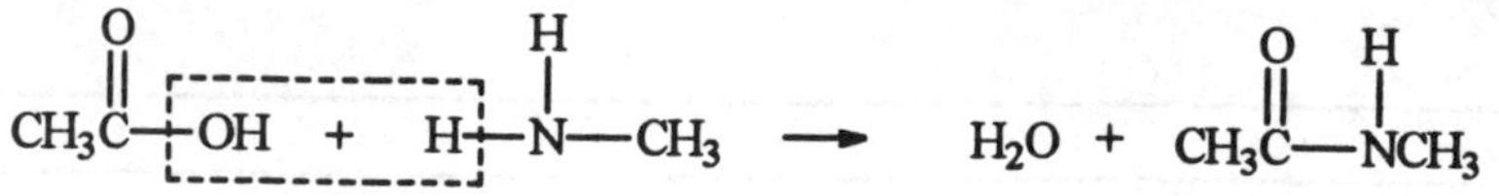

$$CH_3C(=O){-}OH + H{-}N(H){-}CH_3 \longrightarrow H_2O + CH_3C(=O){-}N(H)CH_3$$ methylacetamide

22.53 ***See Section 22.3 and Example 22.7.***

(a) butanol

$CH_3CH_2CH_2CH_2OH$

(b) 3-methyl-2-pentanone

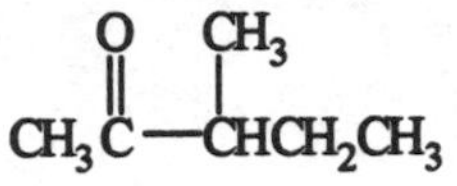

$CH_3C(=O)-CH(CH_3)CH_2CH_3$

(c) methyl acetate

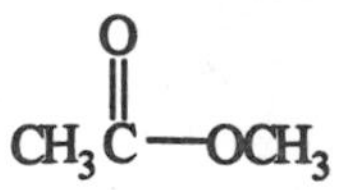

$CH_3C(=O)-OCH_3$

(d) ethyl phenyl amine

$C_6H_5-NH-CH_2CH_3$

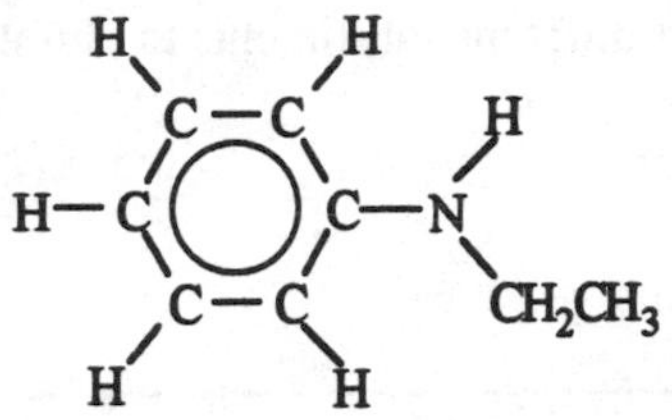

22.54 ***See Section 22.3 and Example 22.7.***

(a) ethylvinyl ether

$CH_3-CH_2-O-CH=CH_2$

(b) 2-bromopropanal

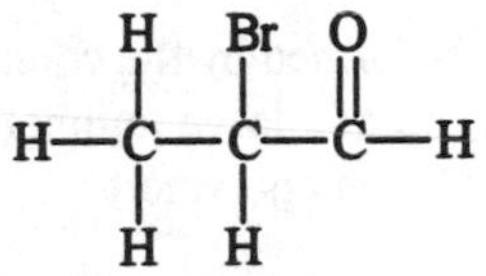

$H-CH_2-CHBr-C(=O)-H$

(c) pentanoic acid

$CH_3-CH_2-CH_2-CH_2-C(=O)-O-H$

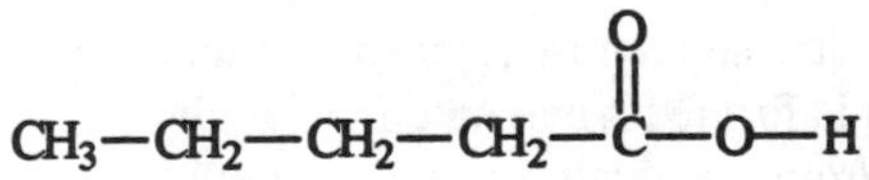

(d) 3-fluorophenol

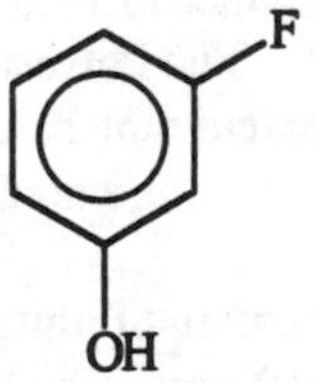

22.55 ***See Section 22.3 and Example 22.7.***

(a) 3-fluoro-1-propanol
(b) propanoic acid
(c) isopropyl acetate
(d) n-propyl amine

22.56 ***See Section 22.3 and Example 22.7.***

(a) 1-butene
(b) ethyl butyrate or ethyl butanoate
(c) ethyl isopropyl ether
(d) pentanamide

22.57 *See Section 22.3.*

A carbon atom that is bonded to four different substituents is chiral.

(a) not chiral
(b) not chiral
(c) chiral at central carbon atom

22.58 *See Section 22.3.*

A carbon atom that is bonded to four different substituents in chiral.

(a) chiral at F substituted carbon atom
(b) not chiral
(c) chiral at F substituted carbon atom

22.59 *See Section 22.4.*

(a) A **chain-growth** or addition polymer is a polymer chain formed from monomeric units with no loss of atoms. Polyethylene formed by the polymerization of a large number of ethylene molecules in the presence of a catalyst is a good example of a chain-growth polymer.

(b) **Homopolymers** are chain growth polymers formed by the combination of many units of a single monomer compound. Teflon formed by the polymerization of a large number of tetrafluorefhylene molecules in the presence of a catalyst is a chain-growth polymer that is a homopolymer.

22.60 *See Section 22.4.*

(a) A **step-growth** or condensation polymer is a polymer formed by a reaction that eleminates a small molecule each time a monomer is linked to the polymer chain. The Polyester Dacron is formed when an ethylene glycol dialcohol molecule is linked to a terephthalic acid diacid molecule. The H from the alcohol combines with the OH of the carboxylic acid to form the H_2O molecule that is elminiated during the linking process.

(b) **Copolymers** are polymers formed by the combination of units of more than one type of monomer. Saran wrap is formed by condensing chloroethene (vinyl chloride) and 1,1-dichloroethene.

22.61 *See Section 22.4.*

The repeating unit of polyvinyl chloride is:

$$\left(\begin{array}{cc} H & H \\ | & | \\ -C- & -C- \\ | & | \\ H & Cl \end{array} \right)_n$$

22.62 *See Section 22.4.*

Phenylethylene (styrene) is $C(C_6H_5)H{=}CH_2$. The repeating unit of polystyrene is:

$$\left(\begin{array}{cc} C_6H_5 & H \\ | & | \\ -C- & -C- \\ | & | \\ H & H \end{array} \right)_n$$

22.63 *See Section 22.4.*

1,1-difluorethylene is $CF_2{=}CH_2$, and hexafluoropropene is $CF_2{=}CFCF_3$. The repeating unit of viton is:

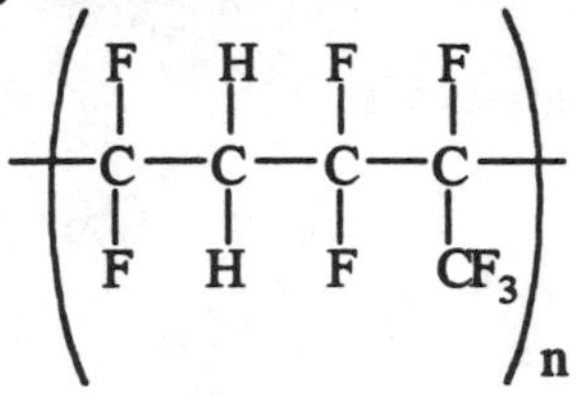

$$\left(\begin{array}{cccc} F & H & F & F \\ | & | & | & | \\ -C- & C- & C- & C- \\ | & | & | & | \\ F & H & F & CF_3 \end{array}\right)_n$$

22.64 *See Section 22.4.*

The repeating unit of SBR is:

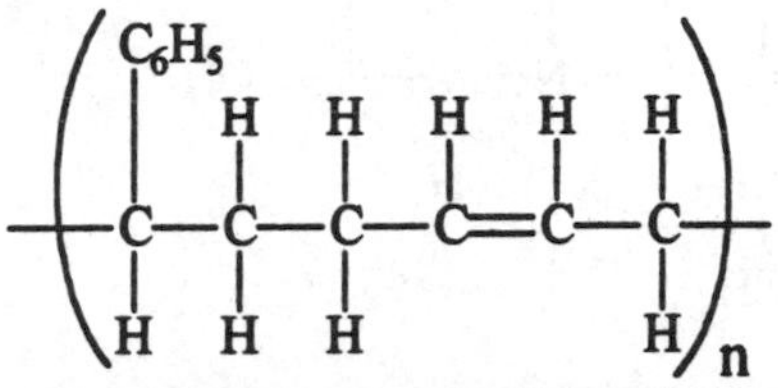

$$\left(\begin{array}{cccccc} C_6H_5 & & & & & \\ | & H & H & H & H & H \\ | & | & | & | & | & | \\ -C- & C- & C- & C= & C- & C- \\ | & | & | & & & | \\ H & H & H & & & H \end{array}\right)_n$$

22.65 *See Section 22.4.*

The monomer that is used to form polyacrylonitrile is: $CH_2{=}CHCN$.

22.66 *See Section 22.4.*

The monomer that is used to form Kevlar is: $H_2N(C_6H_4)COOH$

22.67 *See Section 22.5.*

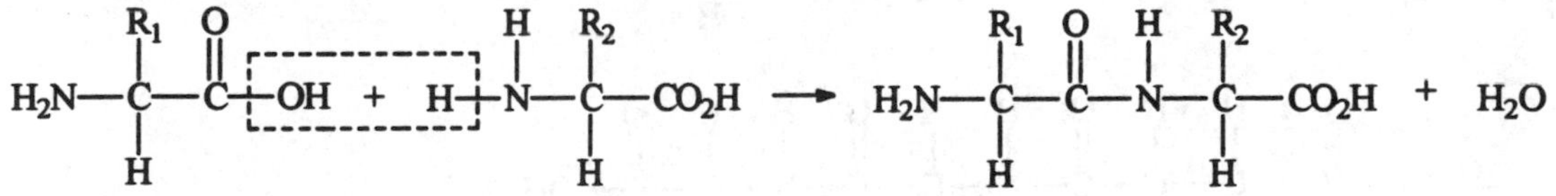

$$H_2N-\underset{H}{\overset{R_1}{C}}-\overset{O}{\overset{\|}{C}}-OH + H-\overset{H}{N}-\underset{H}{\overset{R_2}{C}}-CO_2H \rightarrow H_2N-\underset{H}{\overset{R_1}{C}}-\overset{O}{\overset{\|}{C}}-\overset{H}{N}-\underset{H}{\overset{R_2}{C}}-CO_2H + H_2O$$

22.68 *See Section 22.5.*

The two possible dipeptides are ser-val and val-ser.

ser-val is:

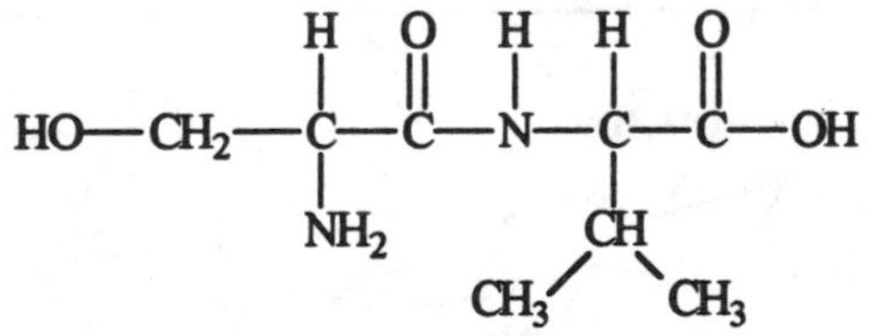

$$HO-CH_2-\underset{NH_2}{\overset{H}{C}}-\overset{O}{\overset{\|}{C}}-\overset{H}{N}-\underset{CH(CH_3)_2}{\overset{H}{C}}-\overset{O}{\overset{\|}{C}}-OH$$

val-ser is:

$$H_3C-\underset{CH_3}{\overset{H}{C}}-\underset{NH_2}{\overset{H}{C}}-\overset{O}{\overset{\|}{C}}-\overset{H}{N}-\underset{CH_2OH}{\overset{H}{C}}-\overset{O}{\overset{\|}{C}}-OH$$

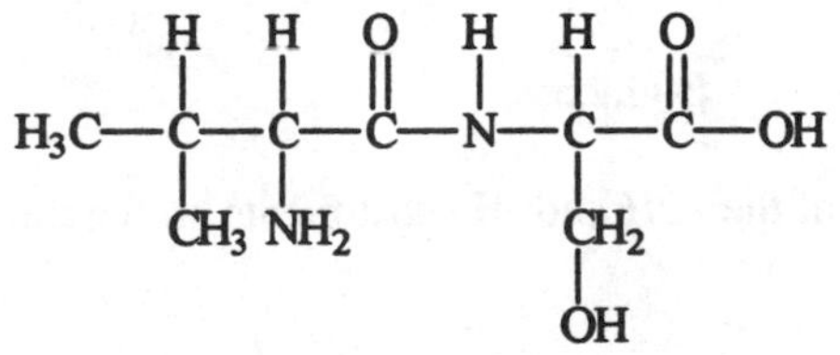

22.69 ***See Section 22.5.***

The tripeptides that can be formed from alanine, glycine and cysteine are ala-gly-cys, gly-ala-cys, and ala-cys-gly.

ala-gly-cys is:

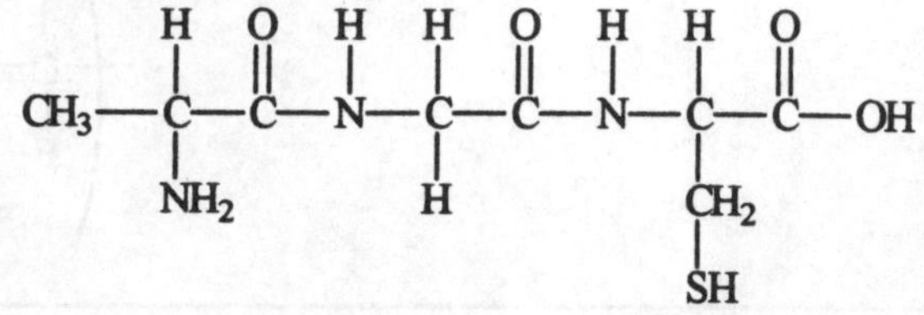

ala-cys-gly is:

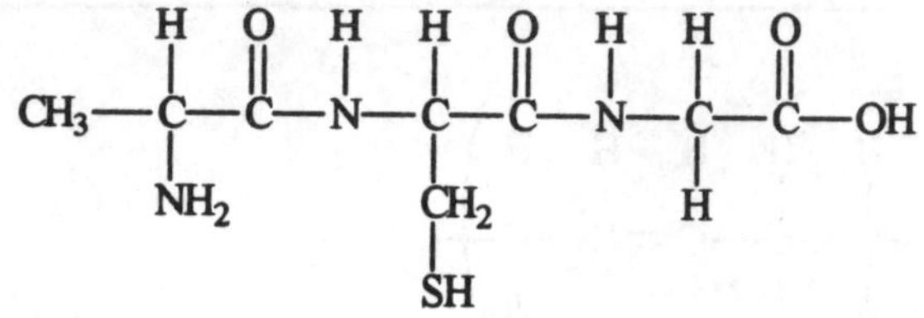

22.70 ***See Section 22.5.***

The tripeptides that can be formed from aline, alanine and serine are val-ala-ser, val-ser-ala, and ser-ala-val.

val-ala-ser is:

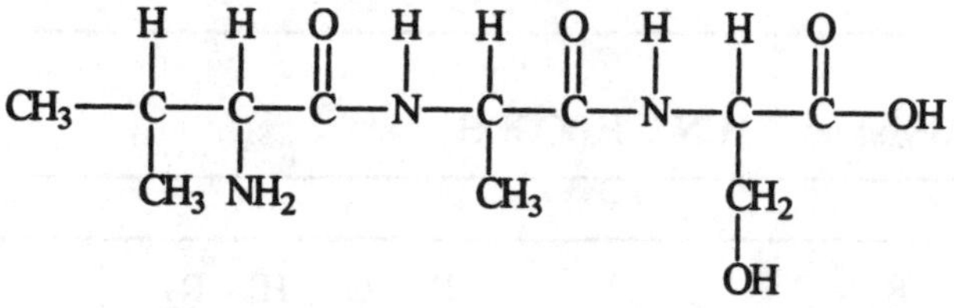

val-ser-ala is:

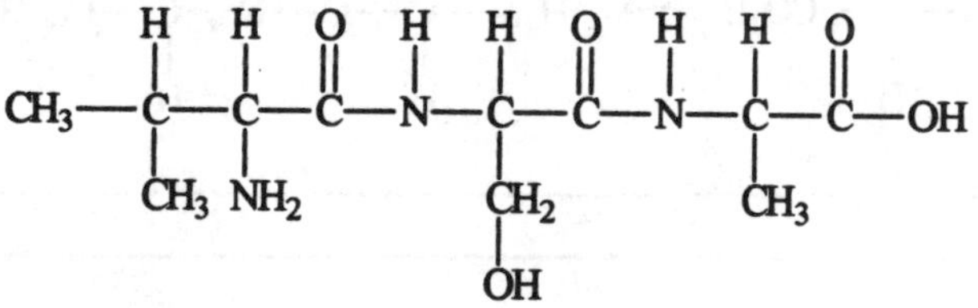

22.71 ***See Section 22.6 and Figure 22.29.***

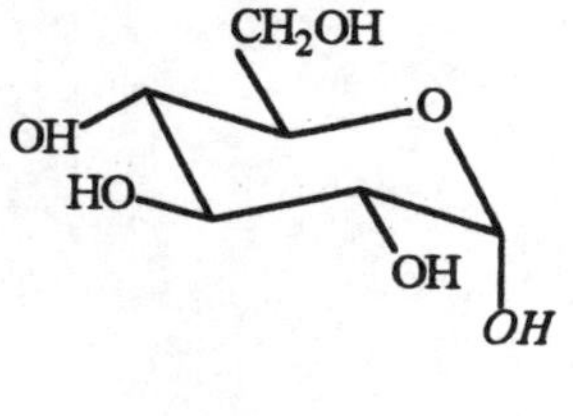

α-glucose

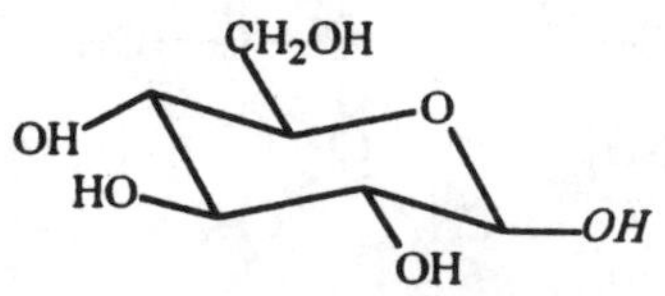

β-glucose

The two isomers of glucose differ in the arrangement of the -OH and -H substituents on the number 1 carbon atoms.

22.72 ***See Section 22.6 and Margin Figure of β-galactose.***

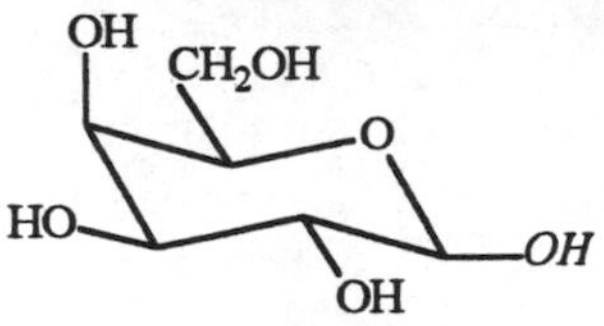

β-galactose has the OH group at (upper left) C4 in an axial position, whereas β-glucose has the OH group at C4 in an equatorial position.

22.73 ***See Section 22.6.***

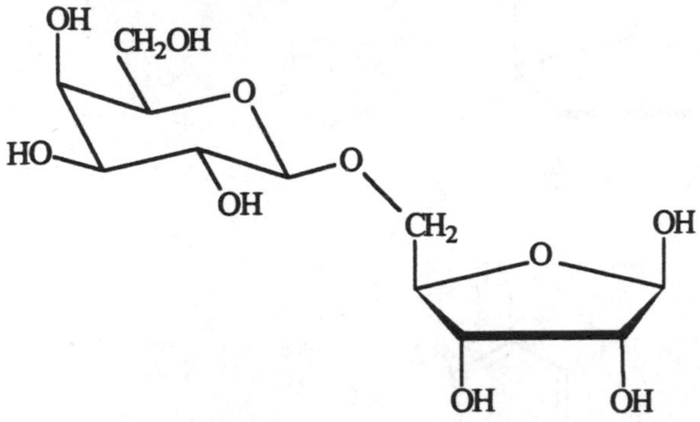

22.74 ***See Section 22.6.***

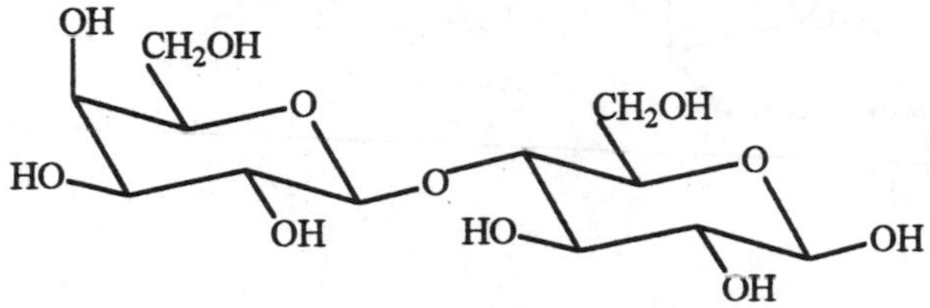

Lactose

22.75 ***See Section 22.7 and Figure 22.3.***

Found only in DNA

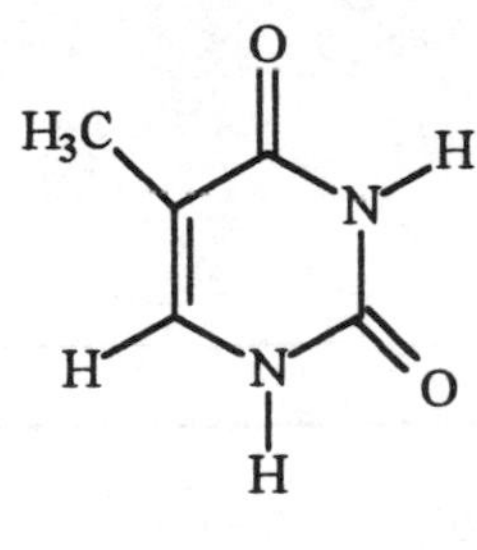

thymine

Found only in RNA

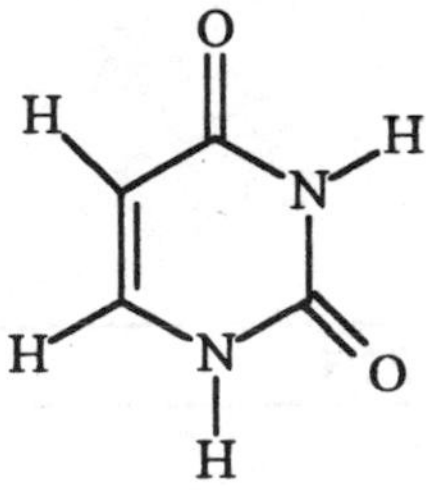

uracil

Found in both DNA and RNA

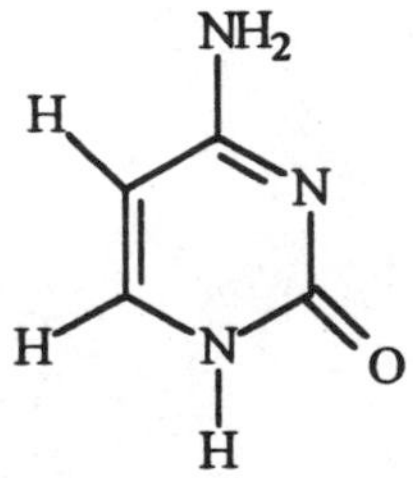

cytosine

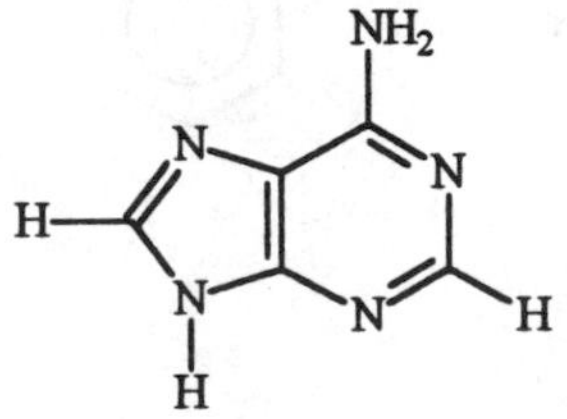

adenine

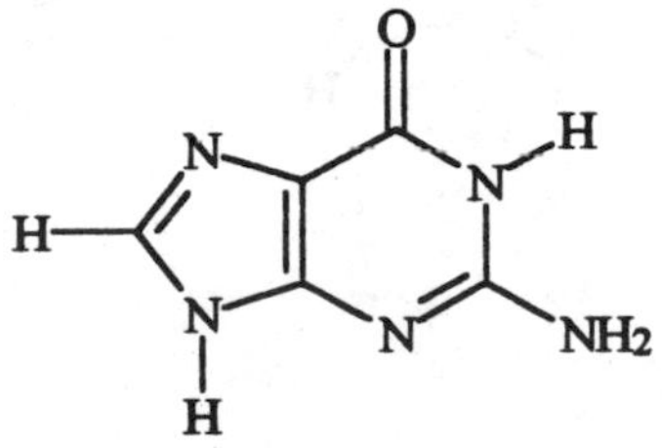

guanine

22.76 *See Section 22.7.*

A nucleotide is composed of a five-carbon sugar, a base, and a phosphate group.

(a) From ribose and adenine,

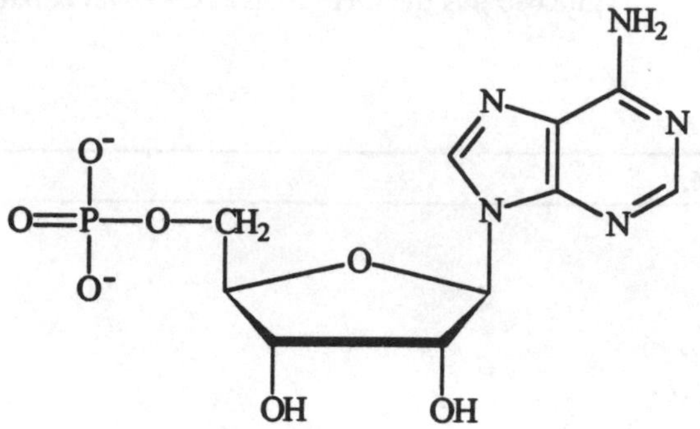

(b) From deoxyribose and cytosine,

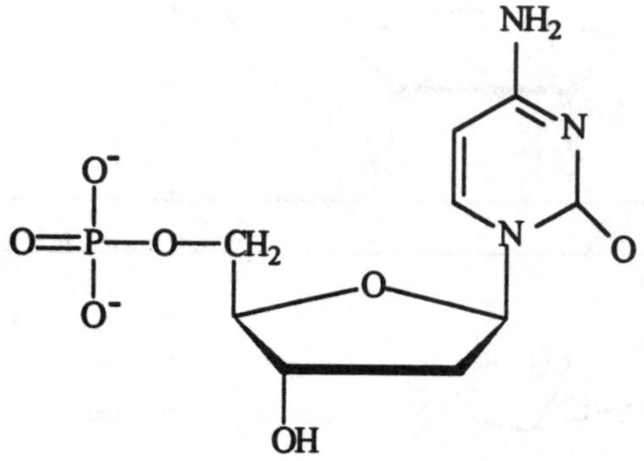

(c) From ribose and uracil,

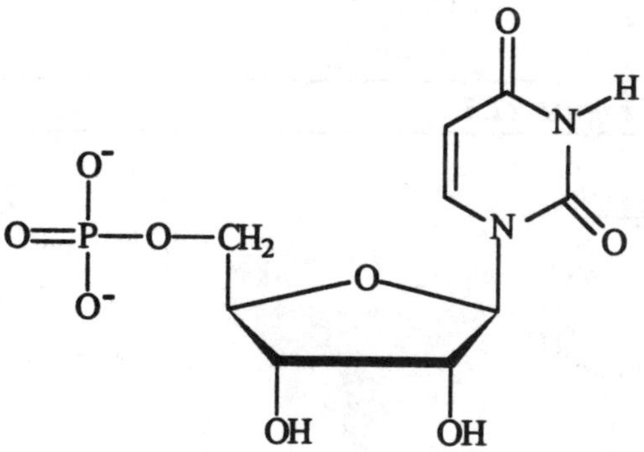

22.77 *See Section 22.3.*

(a) cis-2-bromo-3-hexene

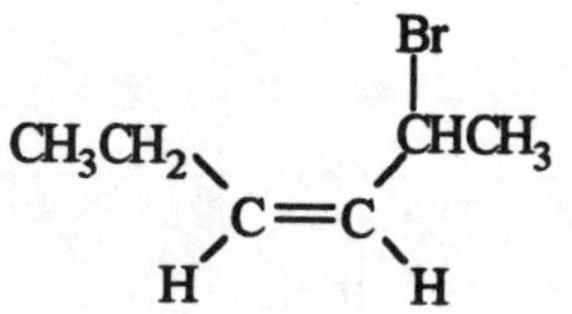

(b) 2-nitrophenol

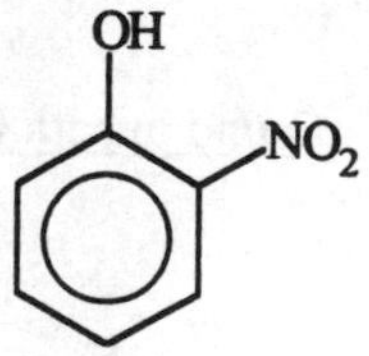

22.78 *See Section 22.3 and Table 22.3.*

(a) aryl, ester, carboxylic acid
(b) phenol, ketone
(c) aryl, ether, amide

22.79 *See Sections 11.4, 22.3.*

Dimethyl ether, CH_3OCH_3, is only slightly polar, whereas ethanol, CH_3CH_2OH, has O-H bonds and can readily form hydrogen bonds with water molecules. Hence, dimethyl ether is only slightly soluble in water, whereas ethanol is very soluble in water.

22.80 *See Sections 22.3, 22.5.*

A carbon atom that is bonded to four different substituents is chiral. An example of an α-amino acid with a chiral center is:

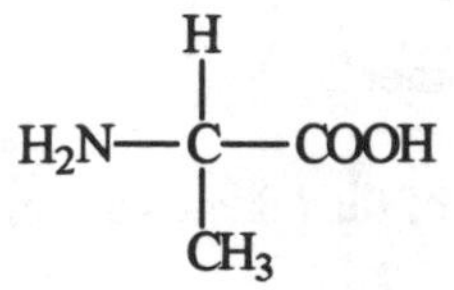

H
|
H_2N—C—COOH
|
CH_3

22.81 *See Section 22.4.*

—(O—⟨◯⟩—C(CH_3)(CH_3)—⟨◯⟩—O—C(=O))$_n$—

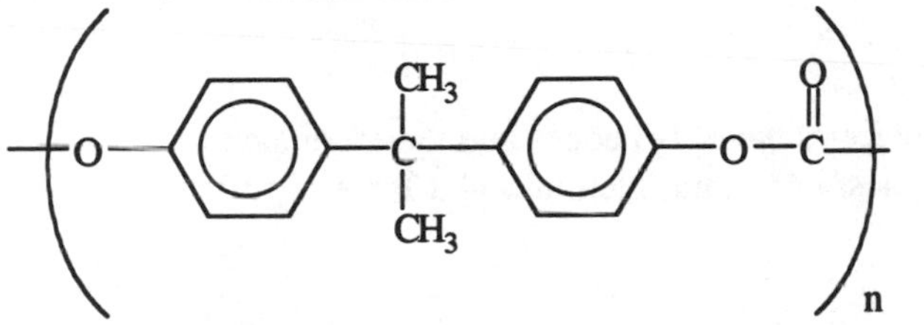

22.82 *See Section 22.4.*

CH_2=CH_2 and CF_2=CF_2 could form —(CH₂—CH₂—CF₂—CF₂)$_n$—

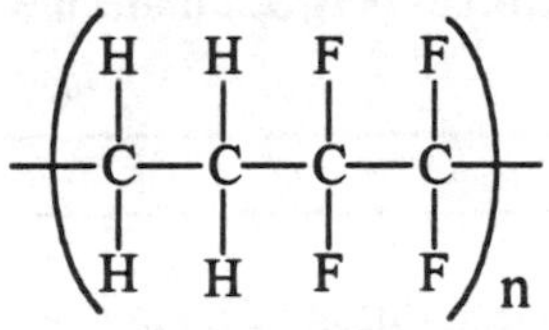

22.83 *See Section 22.7.*

A particular codon matches a specific amino acid. A sequence of codons will match a particular sequence of amino acids in the proteins.

22.84 *See Section 22.5.*

The tripeptides that can be formed from one glycine, one valine and one alanine are gly-val-ala, gly-ala-val, val-gly-ala, val-ala-gly, ala-val-gly and ala-gly-val . Hence, **six** tripeptides can be formed.

22.85 *See Section 22.2.*

$CH_2=CHCH_3 + Br_2 \rightarrow CH_2BrCHBrCH_3$

Strategy: $g\ CH_2 = CHCH_3 \rightarrow mol\ CH_2 = CHCH_3 \rightarrow mol\ Br_2 \rightarrow g\ Br_2$

$$?\ g\ Br_2 = 22.1\ g\ CH_2 = CHCH_3 \times \frac{1\ mol\ CH_2 = CHCH_3}{42.09\ g\ CH_2 = CHCH_3} \times \frac{1\ mol\ Br_2}{1\ mol\ CH_2 = CHCH_3} \times \frac{159.80\ g\ Br_2}{1\ mol\ Br_2} = \mathbf{83.9\ g\ Br_2}$$

22.86 *See Section 22.2.*

$H\text{-}C\equiv C\text{-}CH_2CH_3 + Cl_2 \rightarrow CHCl{=}CClCH_2CH_3$

The possible isomers for the product are:

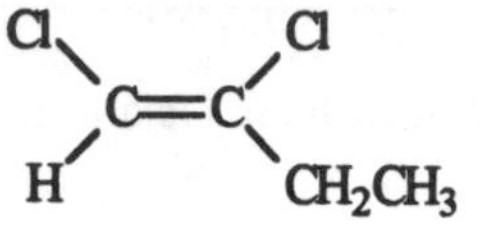

cis isomer: Cl, H on one carbon; Cl, CH_2CH_3 on the other (Cl atoms on the same side)

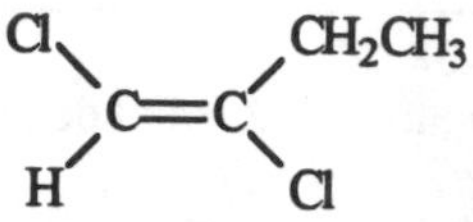

trans isomer: Cl, H on one carbon; CH_2CH_3, Cl on the other (Cl atoms on opposite sides)

The steric number for the alkyne carbon atoms changes from 2 to 3. Hence, the hydridization of the alkyne carbon atoms changes from **sp to sp^2**.

22.87 *See Sections 11.4, 22.3.*

(a) $(CH_3)_3C\text{-}OH$

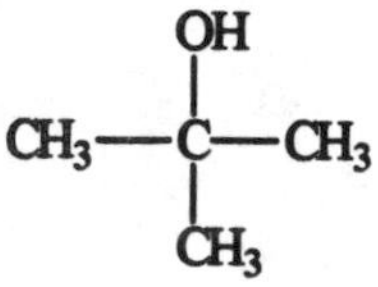

2-methyl-2-propanol molecules have O-H bonds and therefore have **hydrogen bonds** as the most important intermolecular forces between molecules.

(b) $CH_3\text{-}C(=O)\text{-}CH_2\text{-}CH_3$

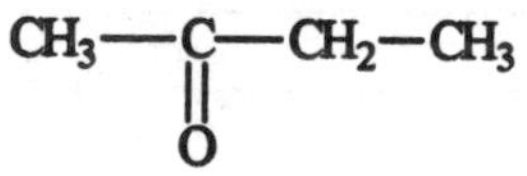

2-butanone molecules are only slightly polar and therefore have **London dispersion forces** as the most important intermolecular forces between molecules.

22.88 *See Section 22.3.*

The correct structure is the cyclic structure because addition of H_2 to it gives just one compound, whereas addition of an equimolar amount of H_2 to $CH_2=CHCH=CHCH_3$ having two double bonds gives a mixture of compounds resulting from addition of H_2 at either double bond.